17th Asia Pacific Bioinformatics Conference (APBC 2019)

Wuhan, China
14-16 January 2019

Editors:

David Sankoff
Louxin Zhang
Shaoliang Peng
Guoliang Li
Yi-Ping Phoebe Chen
Hong-Yu Zhang

ISBN: 978-1-5108-8508-0

Printed from e-media with permission by:

Curran Associates, Inc.
57 Morehouse Lane
Red Hook, NY 12571

Some format issues inherent in the e-media version may also appear in this print version.

Printed by Curran Associates, Inc. (2019)

For further information, please contact BioMed Central Ltd
at the address below. The collection is available online at
https://www.biomedcentral.com/search?query=APBC+2019&searchType=publisherSearch

BioMed Central Ltd
236 Gray's Inn Road
London WC1X 8HB
UK

Phone: +44 (0)20 3192 2000
Fax: +44 (0)20 3192 2010

Email: info@biomedcentral.com
Web: www.biomedcentral.com

Additional copies of this publication are available from:

Curran Associates, Inc.
57 Morehouse Lane
Red Hook, NY 12571 USA
Phone: 845-758-0400
Fax: 845-758-2633
Email: curran@proceedings.com
Web: www.proceedings.com

TABLE OF CONTENTS

SYSTEMS BIOLOGY

GENOMICS

BIOINFOMATICS

Ye and Ho *BMC Systems Biology* 2019, **13**(Suppl 2):35
https://doi.org/10.1186/s12918-019-0690-2

BMC Systems Biology

METHODS

Open Access

Ultrafast clustering of single-cell flow cytometry data using FlowGrid

Xiaoxin Ye[1,2] and Joshua W. K. Ho[1,2,3*]

From The 17th Asia Pacific Bioinformatics Conference (APBC 2019)
Wuhan, China. 14–16 January 2019

Abstract

Background: Flow cytometry is a popular technology for quantitative single-cell profiling of cell surface markers. It enables expression measurement of tens of cell surface protein markers in millions of single cells. It is a powerful tool for discovering cell sub-populations and quantifying cell population heterogeneity. Traditionally, scientists use manual gating to identify cell types, but the process is subjective and is not effective for large multidimensional data. Many clustering algorithms have been developed to analyse these data but most of them are not scalable to very large data sets with more than ten million cells.

Results: Here, we present a new clustering algorithm that combines the advantages of density-based clustering algorithm DBSCAN with the scalability of grid-based clustering. This new clustering algorithm is implemented in python as an open source package, FlowGrid. FlowGrid is memory efficient and scales linearly with respect to the number of cells. We have evaluated the performance of FlowGrid against other state-of-the-art clustering programs and found that FlowGrid produces similar clustering results but with substantially less time. For example, FlowGrid is able to complete a clustering task on a data set of 23.6 million cells in less than 12 seconds, while other algorithms take more than 500 seconds or get into error.

Conclusions: FlowGrid is an ultrafast clustering algorithm for large single-cell flow cytometry data. The source code is available at https://github.com/VCCRI/FlowGrid.

Keywords: Clustering, Flow cytometry, Single cell, DBSCAN

Background

Recent technological advancement has made it possible to quantitatively measure the expression of a handful of protein markers in millions of cells in a flow cytometry experiment [1]. The ability to profile such a large number of cells allows us to gain insights into cellular heterogeneity at an unprecedented resolution. Traditionally, cell types are identified based on manual gating of several markers in flow cytometry data. Manual gating relies on visual inspection of a series of two dimensional scatter plots, which makes it difficult to discover structure in high dimensions. It also suffers subjectivity, in terms of the order in which pairs of protein markers are explored, and the inherent uncertainty of manually drawing the cluster boundaries [2]. An emerging solution is to use unsupervised clustering algorithms to automatically identify clusters in potentially multidimensional flow cytometry data.

The Flow Cytometry Critical Assessment of Population Identification Methods (Flow-CAP) challenge has compared the performance of many flow cytometry clustering algorithms [3]. In the challenge, ADIcyt has the highest accuracy but has a long runtime, which makes it impractical for routine usage. Flock [4] maintains a high accuracy and reasonable runtime. After the challenge, several algorithms have been built for flow cytometry data analysis such as FlowPeaks [5], FlowSOM [6] and BayesFlow [7].

*Correspondence: jwkho@hku.hk
[1]Victor Chang Cardiac Research Institute, Sydney, Australia
[2]University of New South Wales, Sydney, Australia
Full list of author information is available at the end of the article

FlowPeaks and Flock are largely based on k-means clustering. k-means clustering requires the number of clusters (k) to be defined prior to the analysis. It is hard to determine a suitable k in practice. FlowPeaks performs k-means clustering with a large initial k, and iteratively merges nearby clusters that are not separated by low density regions into one cluster. Flock utilises grids to identify high density regions, which the algorithm then uses to identify initial cluster centres for k-means clustering. This grid-based method of identifying high density region allows k-means clustering to converge much quicker compared to using random initialisation of cluster centres, and also directly identifies a suitable value for k. FlowSOM starts with training Self-Organising Map (SOM), followed by consensus hierarchical clustering of the cells for meta-clustering. In the algorithm, the number of clusters (k) is required for meta-clustering.

BayesFlow uses a Bayesian hierarchical model to identify different cell populations in one or many samples. The key benefit of this method is its ability to incorporate prior knowledge, and captures the variability in shapes and locations of populations between the samples [7]. However, BayesFlow tends to be computational expensive as Markov Chain Monte Carlo sampling requires a large number of iterations. Therefore, BayesFlow is often impractical for flow cytometry data sets of realistic size.

These algorithms perform well on the Flow-CAP data sets, but they may not be scalable to larger data sets that we are dealing with nowadays – those with tens of millions of cells. Aiming to quantify cell population heterogeneity in huge data sets, we have to develop an ultrafast and scalable clustering algorithm.

In this paper, we present a new clustering algorithm that combines the benefit of DBSCAN [8] (a widely-based density-based clustering algorithm) and a grid-based approach to achieve scalability. DBSCAN is fast and can detect clusters with complex shapes in the presence of outliers [8]. DBSCAN starts with identifying core points that have a large number of neighbours within a user-defined region. Once the core points are found, nearby core points and closely located non-core points are grouped together to form clusters. This algorithm will identify clusters that are defined as high-density regions that are separated by the low-density regions. However, DBSCAN is memory inefficient if the data set is very large, or has large highly connected components.

To reduce the computational search space and memory requirement, our algorithm extends the idea of DBSCAN by using equal-spaced grids like Flock. We implemented our algorithm in an open source python package called FlowGrid. Using a range of real data sets, we demonstrate that FlowGrid is much faster than other state-of-the-art flow cytometry clustering algorithms, and produce similar clustering results. The detail of the algorithm is presented in the Methods section.

Methods

The key idea of our algorithm is to replace the calculation of density from individual points to discrete bins as defined by a uniform grid. This way, the clustering step of the algorithm will scale with the number of non-empty bins, which is significantly smaller than the number of points in lower dimensional data sets. Therefore the overall time complexity of our algorithm is dominated by the binning step, which is in the order of $O(N)$. This is significantly better than the time complexity of DBSCAN, which is in the order of $O(NlogN)$. The definition and algorithm are presented in the following subsections.

Definition

The key terms involved in the algorithm are defined in this subsection. A graphical example can be found in Fig. 1.

- N_{bin} is the number of equally sized bins in each dimension. In theory, there are $(N_{bin})^d$ bins in the data space, where d is the number of dimensions. However, in practice, we only consider the non-empty bins. The number of non-empty bins (N) is less than $(N_{bin})^d$, especially for high dimensional data. Each non-empty bin is assigned an integer index $i = 1 \ldots N$.
- Bin_i is labelled by a tuple with d positive integers $C_i = (C_{i1}, C_{i2}, C_{i3}, \ldots, C_{id})$ where C_{i1} is the coordinate (the bin index) at dimension 1. For

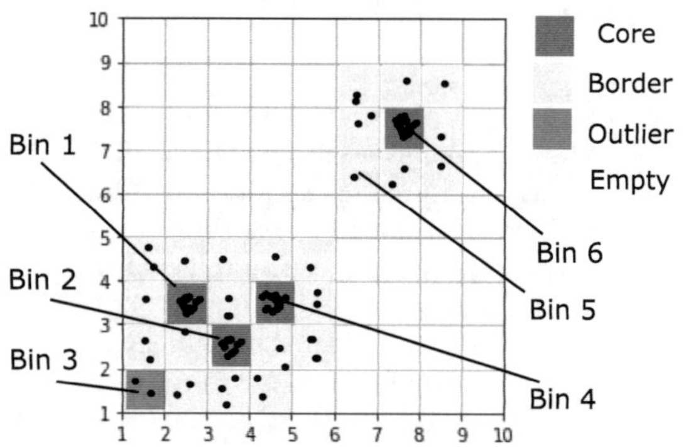

Fig. 1 An illustrative example of the FlowGrid clustering algorithm. In this example, Bin 1, Bin 2, Bin 3 and Bin 6 are core bins as their Den_b are larger than $MinDen_b$ (5 in this example), their Den_c are larger than $MinDen_c$ (20 in thi s example), and their Den_b are larger than ρ% (75% in this example) of its directly connected bins. $Dist(C_1, C_2) = \sqrt{1^2 + 1^2} = \sqrt{2} \leq \sqrt{\epsilon}$ ($\epsilon = 2$ in this example), so Bin 1 and Bin 2 are directly connected. $Dist(C_2, C_4) = \sqrt{1^2 + 1^2} = \sqrt{2} \leq \sqrt{\epsilon}$, so Bin 2 and Bin 4 are directly connected. Therefore, Bin 1, Bin 2 and Bin 4 are mutually connected, and they are assigned into the same cluster. Bin 5 is not a core bin but is a border bin, as it is directly connected to Bin 6, which is a core bin. Bin 3 is a outlier bin, as it is not a core bin nor a border bin. In practice, $MinDen_b$ is set to be 3, $MinDen_c$ is set to 40 and ρ is set to be 85

example, if Bin_i has coordinate $C_i = (2, 3, 5)$, this bin is located in second bin in dimension 1, third bin in dimension 2 and the fifth bin in dimension 3.

- The distance between Bin_i and Bin_j is defined as

$$Dist(C_i, C_j) = \sqrt{\sum_{k=1}^{d} (C_{ik} - C_{jk})^2} \quad (1)$$

- Bin_i and Bin_j are defined to be directly connected if $Dist(C_i, C_j) \leqslant \sqrt{\epsilon}$, where ϵ is a user-specified parameter.
- $Den_b(C_i)$ is the density of Bin_i, which is defined as the number of points in Bin_i.
- $Den_c(C_i)$ is the collective density of Bin_i, calculated by

$$Den_c(C_i) = \sum_{\{j|Bin_j \text{ and} Bin_i \text{ are directly connected}\}} Den_b(C_j) \quad (2)$$

- Bin_i is a core bin if
  1. $Den_b(C_i)$ is larger than $MinDen_b$, a user-specified parameter.
  2. $Den_b(C_i)$ is larger than ρ% of its directly connected bins, where ρ is a user-specified parameter.
  3. $Den_c(C_i)$ is larger than $MinDen_c$, a user-specified parameter.
- Bin_i is a border bin if it is not a core bin but it is directly connected to a core bin.
- Bin_i is an outlier bin, if it is not a core bin nor a border bin.
- Bin_a and Bin_b are in the same cluster, if they satisfy one of the following conditions:
  1. they are directly connected and at least one of them is core bin;
  2. they are not directly connected but are connected by a sequence of directly connected core bins.
- Two points are in the same cluster, if they belong to the same bin or their corresponding bins belong to the same cluster.

Algorithm 1: FlowGrid

input : $X, N_{bin}, \epsilon, \rho, MinDen_b, MinDen_c$
output: DataLabel
1. Normalise the data X ranging from 1 to $(N_{bin} + 1)$
2. Assign data into corresponding bins based on the integer of normalised value
3. Identify S_{bin} as the set of non-empty bins
4. Search the core bins and their directly connected bins by SearchCore
5. Group connected bins into a cluster by Breadth First Search(BFS)
6. Label cells by the label of their corresponding bins

Algorithm 2: SearchCore

```
input  : S_bin, ε, ρ, MinDen_b, MinDen_c
output: S_core, L
Initial an empty adjacency list L.
S_core = {}
forall the Bin_i in S_bin do
    if Den_b(i) > MinDen_b then
        nnBin=radiusNeighbors(S_bin, Bin_i, ε)
        nnCount counting the number of points for
        each bin in nnBin
        if Den_b(i) is greater than ρ% of nnCount then
            Den_c(i)= the sum of nnCount
            if Den_c(i) > MinDen_c then
                S_core = S_core ∪ {i}
                mapping bin_i with nnBin in L
            end
        end
    end
end
```

The input of radiusNeighbors is all non-empty bins, the query bin and the maximum query distance $\sqrt{\epsilon}$. The output is the bins whose distance with the query bin are less than $\sqrt{\epsilon}$ (including the query bin).

Algorithm
Algorithm 1 describes the key steps of FlowGrid, starting with normalising the values in each dimension to range between 1 and $(N_{bin} + 1)$. Then, we use the integer part of the normalised value as the coordinate of its corresponding bin. Then, the SearchCore algorithm is applied to discover the core bins and their directly connected bins. Once the core bins and connections are found, Breadth First Search(BFS) is used to group the connected bins into a cluster. The cells are labelled by the label of their corresponding bins.

Evaluation
Procedure
FlowGrid aims to be an ultrafast and accurate clustering algorithm for very large flow cytometry data. Therefore, both the accuracy and scalability performance need to be evaluated. The benchmark data sets from Flow-CAP [3], the multi-centre CyTOF data from Li et al. [9] and the SeaFlow project [10] are selected to compare the performance of FlowGrid against other state-of-the-art algorithms, FlowSOM, FlowPeaks, and FLOCK. These three algorithms are chosen because they are widely used,

Algorithm 3: Breadth First Search(BFS)

```
input  : S_core, S_bin, adjacency list L
output: Bin Label
Label every bin as -1
Index=1
for Bin_i  in  S_core do
    if the laebl of Bin_i is -1 then
        Queue={}
        Label Bin_i as Index
        Queue.push(Bin_i)
        while Queue is not empty do
            Bin_1= Queue.pop()
            forall the directed connected Bin_2 of Bin_1
            do
                if the laebl of Bin_2 is -1 then
                    Label Bin_2 as Index
                    if Bin_2 is core bin then
                        Queue.push(Bin_2)
                    end
                end
            end
        end
        index=index +1
    end
end
```

are generally considered to be quite fast, and have good accuracy.

Three benchmark data sets from Flow-CAP [3] are selected for evaluation, including the Diffuse Large B-cell Lymphoma (DLBL), Hematopoietic Stem Cell Transplant (HSCT), and Graft versus Host Disease(GvHD) data set. Each data set contains 10-30 samples with 3-4 markers, and each sample includes 2,000-35,000 cells.

The multi-centre CyTOF data set from Li et al. [9] provides a labelled data set with 16 samples. Each samples contains 40,000-70,000 cells and 26 markers. Since only 8 out fo 26 markers are determined to be relevant markers in the original paper [9], only these 8 markers are used for clustering.

We also use three data sets from the SeaFlow project [10] and they contain many samples. Instead of analysing the independent samples, we analyse the concatenated data sets as the original paper [10] and these concatenated data sets contain 12.7, 22.7 and 23.6 millions of cells respectively. Each data sets include 15 features but the original study only uses four features for clustering analysis. The four features are forward scatter (small and perpendicular), phycoerythrin, and chlorophyll (small) [10].

In the evaluation, we treat the manual gating label as the gold standard for measuring the quality of clustering. In the pre-precessing step, we apply the inverse hyperbolic function with the factor of 5 to transform the multi-centre data and the SeaFlow data. As the Flow-CAP and multi-centre CyTOF data contain many samples and we treat each sample as a data set, we run all algorithms on each sample. The performances are measured by the ARI and runtime, which are reported by the arithmetic means ($\bar{x}$) and standard deviation (*sd*). For the Seaflow data sets, we treat each concatenated data set as a data set. In the evaluation, all algorithms are applied on these concatenated data sets.

To evaluate the scalability of each algorithm, we down-sample the largest concatenated data set from the SeaFlow project, generating 10 sub-sampled data sets in which the numbers of cells range from 20 thousand to 20 million.

Performance measure

The efficiency performance is measured by the runtime while the clustering performance is measured by Adjusted Rand Index (ARI). ARI is used to measure the clustering performance. ARI is the corrected-for-chance version of the Rand index [11]. Although it may result in negative values if the index is less than expected, it tends to be more robust than many other measures like F-measure and Rand index.

ARI is calculated as follow. Given a set S of n elements, and two groups of cluster labels (one group of ground truth label and one group of predicted labels) of these elements, namely $X = \{X_1, X_2, \ldots, X_r\}$ and $Y = \{Y_1, Y_2, \ldots, Y_s\}$, the overlap between X and Y can be summarized by n_{ij} where n_{ij} denotes the number of objects in common between X_i and Y_j: $n_{ij} = |X_i \cap Y_j|$.

$$ARI = \frac{\sum_{ij}\binom{n_{ij}}{2} - \left[\sum_i\binom{a_i}{2}\sum_j\binom{b_j}{2}\right]/\binom{n}{2}}{\frac{1}{2}\left[\sum_i\binom{a_i}{2} + \sum_j\binom{b_j}{2}\right] - \left[\sum_i\binom{a_i}{2}\sum_j\binom{b_j}{2}\right]/\binom{n}{2}}$$

where $a_i = \sum_j n_{ij}$ and $b_j = \sum_i n_{ij}$

Experimentation

FlowGrid is publicly available as an open source program on GitHub. FlowSOM and FlowPeaks are available as R packages from Bioconductor. The source code of Flock is downloaded from its Sourceforge repository. To reproduce all the comparisons presented in this paper, the source code and data can be downloaded from the GitHub repository *FlowGrid_compare*. We run all the experiments on six 2.60 GHz cores CPU with 32 G RAM.

FlowPeaks and Flock provide automated version without any user-input parameter. FlowSOM requires one user-supplied parameter (k, the number of clusters in meta-clustering step). FlowGrid requires two user-supplied parameters (bin_n and ϵ). To optimise the result,

Table 1 Comparison of runtime (in seconds) of FlowGrid against other clustering algorithms

Data set	Samples	Markers	Cells	Time in second ($\bar{x} \pm sd$)			
				FlowGrid	FlowSOM	FlowPeaks	Flock
Multi-center	16	8	29-77 $\times 10^3$	0.23± 0.09	4.01± 1.08	2.27± 0.61	10.3± 3.45
Flow-CAP-GvHD	12	4	12-33$\times 10^3$	0.07± 0.04	2.16± 0.54	0.28± 0.16	0.58± 0.28
Flow-CAP-DLBL	30	3	2-25$\times 10^3$	0.04± 0.01	1.25± 0.32	0.10± 0.09	0.22± 0.16
Flow-CAP-HSCT	30	4	6-9$\times 10^3$	0.04± 0.02	1.35± 0.28	0.11± 0.02	0.28± 0.06
Seaflow0	-	4	23.6$\times 10^6$	11.51	572.65	NA	6628.30
Seaflow1	-	4	12.7$\times 10^6$	3.09	312.95	258.13	NA
Seaflow11	-	4	22.7$\times 10^6$	6.37	544.79	NA	NA

NA represents that the algorithm got error in the data set

we try many k for FlowSOM and many combinations of bin_n and ϵ for our algorithm.

Results

Performance comparison

Table 1 summaries the performance of our algorithm and three other algorithms – FlowSOM, FlowPeaks, and Flock in terms of runtime. Our algorithm is substantially faster than other clustering algorithms in all the data sets. This improvement of runtime is especially substantial in the Seaflow data sets. FLOCK and FlowPeaks sometimes fail to complete in some of the data sets. In a data set of 23.6 million cells, FlowSOM completes the execution in 572 s, whereas FlowGrid completes the execution in only 12 s. This is a 50× speed up. Table 2 summaries the clustering accuracy performance. In Flow-CAP and the multi-centre data sets, FlowGrid shares the similar clustering accuracy (in terms of ARI) with other clustering algorithms but in Seaflow data sets, FlowGrid gives higher accuracy than other clustering algorithms.

Figure 2 shows that the clustering results of our algorithm and three other algorithms in a HSCT sample. FlowGrid, FlowSOM and FlowPeaks recover similar number of clusters, and the clustering results are largely similar.

Table 2 Comparison of accuracy (in ARI) of FlowGrid against other clustering algorithms

Data set	ARI ($\bar{x} \pm sd$)			
	FlowGrid	FlowSOM	FlowPeaks	Flock
Multi-center	0.66± 0.20	0.75± 0.17	0.68± 0.20	0.66± 0.16
Flow-CAP-GvHD	0.79± 0.15	0.85± 0.11	0.72± 0.16	0.47± 0.20
Flow-CAP-DLBL	0.85± 0.10	0.84± 0.10	0.82± 0.15	0.84± 0.09
Flow-CAP-HSCT	0.90± 0.08	0.87± 0.14	0.83± 0.24	0.57± 0.27
Seaflow0	0.94	0.81	NA	0.27
Seaflow1	0.59	0.54	0.34	NA
Seaflow11	0.77	0.33	NA	NA

NA represents that the algorithm got error in the data set

Flock generates too many clusters in this case. It is important to note that FlowGrid also identifies cells that do not belong to a main cluster (i.e., a high density region). These cells can be viewed as 'outliers', and are labelled as '-1' in Fig. 2. This is a feature that is not present in other clustering algorithms.

Scalability analysis

To further evaluate the scalability of the algorithms, we sub-sample one Seaflow data set and the sampled data sets range from 20 thousand to 20 million cells. Figure 3 shows the scalability of our algorithm and three other algorithms. Flock has a low runtime when processing a small data set, but its runtime dramatically increases to 6640 s for a 20 million-cell data set. FlowPeaks and FlowSOM share similar scalability but FlowPeaks is not able to execute 20 million data set. Our algorithm have the best performance in the evaluation as FlowGrid is faster than other algorithm in all the sampled data by an order of magnitude.

Parameter robustness analysis

Like other density-based clustering algorithm, parameter setting is important. In our experience, Bin_n and ϵ are data-set-dependent. We recommend trying out different combinations of Bin_n between 4 and 15, and ϵ between 1 and 5. To pick the best parameter combinations, some prior knowledge is helpful such as the expected number of clusters and the proportion of outliers which should be less than 10% in our experience.

We found that other parameters, namely $MinDen_b$, $MinDen_c$ and ρ are mostly robust across a wide range of values.

To demonstrate this robustness, we used the benchmark data sets from Flow-CAP for a parameter sensitivity analysis. For these experiments, we first set 3, 40, 85, 4 and 1 as the default value for $MinDen_b$, $MinDen_c$, ρ, Bin_n and ϵ, respectively. In each experiment, we only change

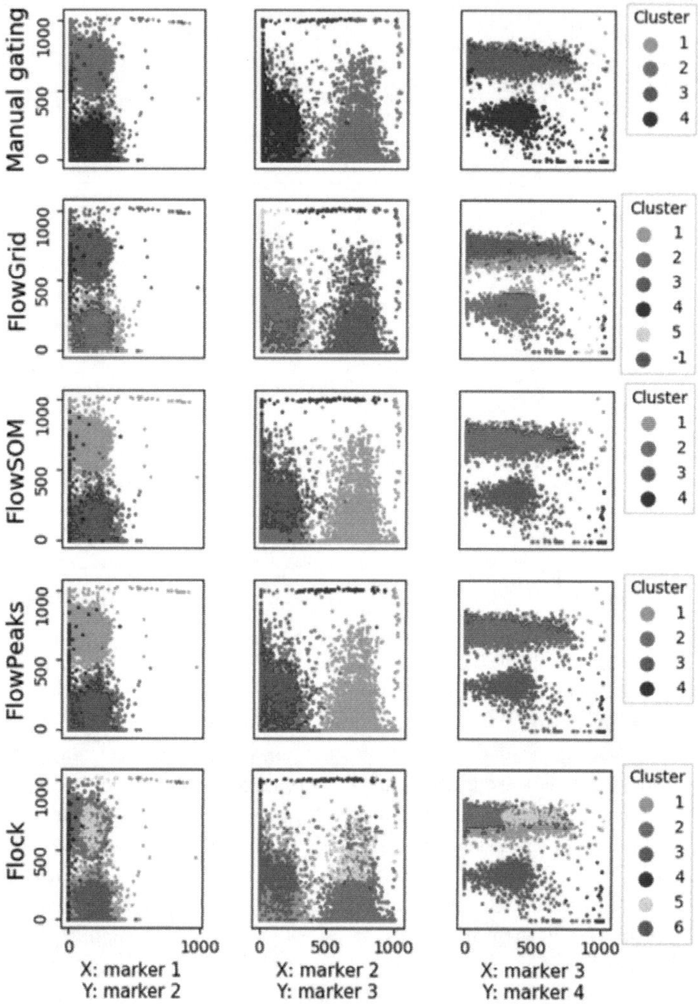

Fig. 2 Visual comparison of the clustering performance of FlowGrid, FlowPeaks, FlowSOM, and Flock using manual gating (top row) as the gold standard

one parameter to test its sensitivity to the overall classification result. The performance is measured by ARI and runtime. In the first experiment, we varied $MinDen_b$ from 1 to 50 while fixing other parameters. In the second experiment, we varied $MinDen_c$ ranging from 10 to 300 while fixing other parameters. In the third experiment, we varied ρ ranging from 70 to 95 while fixing other parameters.

Figure 4 demonstrates that the clustering accuracy and runtime are largely insensitive to $MinDen_b$, $MinDen_c$ and ρ across a large range of parameter values. The experiments are applied to all the benchmark data sets from Flow-CAP and similar results are observed in all the benchmark data sets. In our experiments, when $MinDen_b$, $MinDen_c$ and ρ are set to be 3, 40 and 85 respectively, FlowGrid maintains good clustering performance and excellent runtime. They are therefore set as the default parameters for FlowGrid.

Discussion

In this paper, we have developed an ultrafast clustering algorithm, FlowGrid, for single-cell flow cytometry analysis, and compared it against other state-of-the-art algorithms such as Flock, FlowSOM and FlowPeaks. FlowGrid borrows ideas from DBSCAN for detection of high density regions and outliers. It does not only perform well in the presence of outliers, but also have great scalability without getting into memory issues. It is both time efficient and memory efficient. FlowGrid

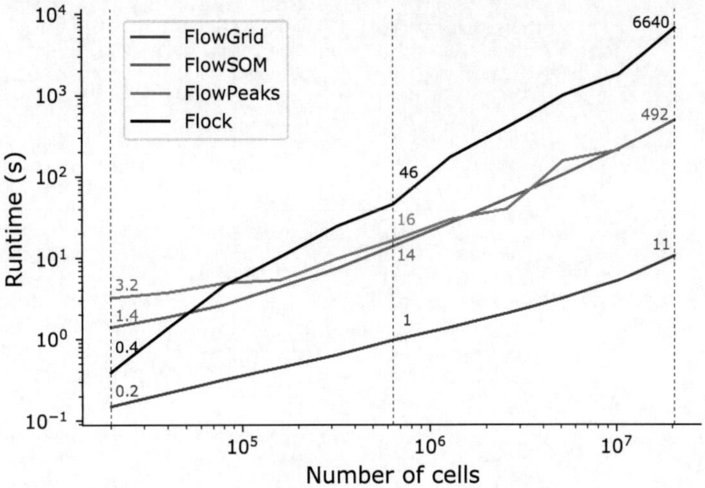

Fig. 3 Comparison of the runtime of FlowGrid, FlowPeaks, FlowSOM, and Flock using data sets with different number of cells

shares similar clustering accuracy with state-of-the-art flow cytometry clustering algorithms, but it is substantially faster than them. With any given number of markers, the runtime of FlowGrid scales linearly with the number of cells, which is a useful property for extremely large data sets.

$MinDen_b$ and $MinDen_c$ are density threshold parameters to reduce the search space of high density bins. If the parameters are set very low, the runtime may fractionally increase but the accuracy is not likely to be affected. However, if the parameters are set very high, the runtime will also fractionally decreases but it may lead to separation of real clusters and create spurious outliers. In any case, we showed that the performance of FlowGrid is generally robust against changes in $MinDen_b$, $MinDen_c$ and ρ.

The current implementation of FlowGrid is already very fast for most practical purposes. In the future, if the data size grows even larger, it is possible to further speed up FlowGrid by parallelising the binning step of the algorithm, which is currently the most computationally intensive step of the algorithm.

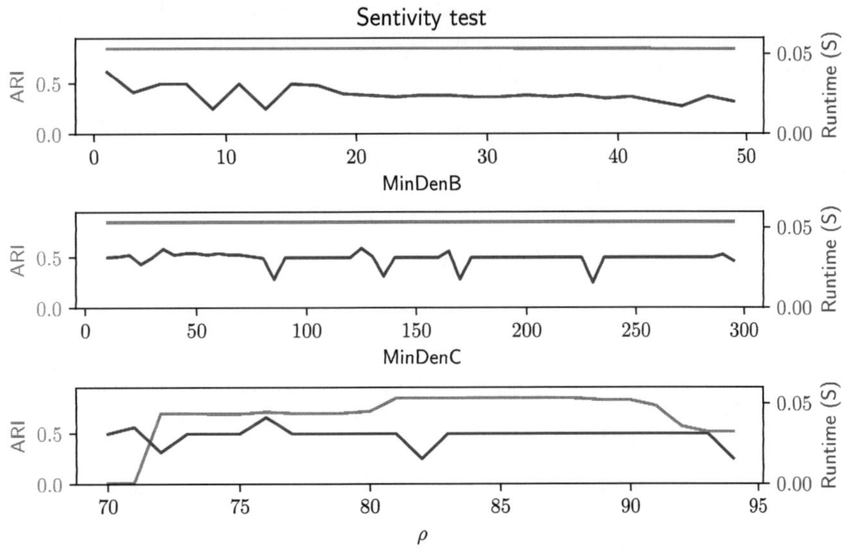

Fig. 4 Sensitivity analysis of three different parameters on clustering accuracy (as measured by adjusted rand index; ARI) and runtime (seconds)

Abbreviations
ARI: Adjusted rand index; BFS: Breadth first search; CyTOF: Mass cytometry; DBSCAN: Density-based spatial clustering of applications with noise; Flow-CAP: Flow cytometry critical assessment of population identification methods; SOM: Self-organising map

Acknowledgments
We thank members of the Ho Laboratory for their valuable comments.

Funding
This work was supported in part by funds from the New South Wales Ministry of Health, a National Health and Medical Research Council Career Development Fellowship (1105271 to JWKH), and a National Heart Foundation Future Leader Fellowship (100848 to JWKH). Publication charge is supported by the Victor Chang Cardiac Research Institute.

Availability of data and materials
- Project Name: FlowGrid
- Project Home Page: https://github.com/VCCRI/FlowGrid
- Operating Systems: Unix, Mac, Windows
- Programming Languages: Python
- Other Requirements: sklearn, numpy
- License: MIT Public License
- Any Restrictions to Use By Non-Academics: None

About this supplement
This article has been published as part of *BMC Systems Biology Volume 13 Supplement 2, 2019: Selected articles from the 17th Asia Pacific Bioinformatics Conference (APBC 2019): systems biology*. The full contents of the supplement are available online at https://bmcsystbiol.biomedcentral.com/articles/supplements/volume-13-supplement-2.

Authors' contributions
XY and JWKH initiated and designed the project. XY implemented the algorithm, carried out all the experiments, and wrote the paper. JWKH revised the paper. Both authors approved the final version of the paper.

Ethics approval and consent to participate
Not applicable.

Consent for publication
Not applicable.

Competing interests
The authors declare that they have no competing interests.

Publisher's Note
Springer Nature remains neutral with regard to jurisdictional claims in published maps and institutional affiliations.

Author details
[1]Victor Chang Cardiac Research Institute, Sydney, Australia. [2]University of New South Wales, Sydney, Australia. [3]School of Biomedical Sciences, Li Ka Shing Faculty of Medicine, The University of Hong Kong, Pokfulam, Hong Kong.

Published: 5 April 2019

References
1. Weber LM, Robinson MD. Comparison of clustering methods for high-dimensional single-cell flow and mass cytometry data. Cytom Part A. 2016;89(12):1084–96.
2. Saeys Y, Van Gassen S, Lambrecht BN. Computational flow cytometry: helping to make sense of high-dimensional immunology data. Nat Rev Immunol. 2016;16(7):449.
3. Aghaeepour N, Finak G, Hoos H, Mosmann TR, Brinkman R, Gottardo R, Scheuermann RH, Consortium F, Consortium D, et al. Critical assessment of automated flow cytometry data analysis techniques. Nat Methods. 2013;10(3):228.
4. Qian Y, Wei C, Eun-Hyung Lee F, Campbell J, Halliley J, Lee JA, Cai J, Kong YM, Sadat E, Thomson E, et al. Elucidation of seventeen human peripheral blood B-cell subsets and quantification of the tetanus response using a density-based method for the automated identification of cell populations in multidimensional flow cytometry data. Cytom Part B Clin Cytom. 2010;78(S1):69–82.
5. Ge Y, Sealfon SC. flowPeaks: a fast unsupervised clustering for flow cytometry data via k-means and density peak finding. Bioinformatics. 2012;28(15):2052–8.
6. Van Gassen S, Callebaut B, Van Helden MJ, Lambrecht BN, Demeester P, Dhaene T, Saeys Y. FlowSOM: Using self-organizing maps for visualization and interpretation of cytometry data. Cytom Part A. 2015;87(7):636–45.
7. Johnsson K, Wallin J, Fontes M. BayesFlow: latent modeling of flow cytometry cell populations. BMC Bioinformatics. 2016;17(1):25.
8. Ester M, Kriegel H-P, Sander J, Xu X, et al. A density-based algorithm for discovering clusters in large spatial databases with noise. In: Proceedings of the Second International Conference on Knowledge Discovery and Data Mining (KDD'96). Portland: AAAI Press; 1996. p. 226–231.
9. Li H, Shaham U, Stanton KP, Yao Y, Montgomery RR, Kluger Y. Gating mass cytometry data by deep learning. Bioinformatics. 2017;33(21):3423–30.
10. Hyrkas J, Clayton S, Ribalet F, Halperin D, Virginia Armbrust E, Howe B. Scalable clustering algorithms for continuous environmental flow cytometry. Bioinformatics. 2015;32(3):417–23.
11. Hubert L, Arabie P. Comparing partitions. J Classif. 1985;2(1):193–218.

Tabei et al. BMC Systems Biology 2019, 13(Suppl 2):39
https://doi.org/10.1186/s12918-019-0691-1

BMC Systems Biology

RESEARCH Open Access

Network-based characterization of drug-protein interaction signatures with a space-efficient approach

Yasuo Tabei[1*], Masaaki Kotera[2], Ryusuke Sawada[3] and Yoshihiro Yamanishi[3,4]

From The 17th Asia Pacific Bioinformatics Conference (APBC 2019)
Wuhan, China. 14–16 January 2019

Abstract

Background: Characterization of drug-protein interaction networks with biological features has recently become challenging in recent pharmaceutical science toward a better understanding of polypharmacology.

Results: We present a novel method for systematic analyses of the underlying features characteristic of drug-protein interaction networks, which we call "drug-protein interaction signatures" from the integration of large-scale heterogeneous data of drugs and proteins. We develop a new efficient algorithm for extracting informative drug-protein interaction signatures from the integration of large-scale heterogeneous data of drugs and proteins, which is made possible by space-efficient representations for fingerprints of drug-protein pairs and sparsity-induced classifiers.

Conclusions: Our method infers a set of drug-protein interaction signatures consisting of the associations between drug chemical substructures, adverse drug reactions, protein domains, biological pathways, and pathway modules. We argue the these signatures are biologically meaningful and useful for predicting unknown drug-protein interactions and are expected to contribute to rational drug design.

Keywords: Drug-protein interaction prediction, Drug discovery, Large-scale prediction

Background

Target proteins of drug molecules are classified into a primary target and off-targets. The former is the desired target, whereas the latter could lead to adverse drug reactions [1] or unexpected beneficial effects in drug repositioning [2]. Therefore, comprehensive analysis throughout primary targets and off-targets on a genome-wide scale is crucial in drug discovery. The in silico approach is expected to improve the research productivity in this field.

Several computational methods have been presented for predicting drug-protein interactions (or compound-protein interactions) from chemogenomic and pharmacogenomic viewpoints on a large-scale. The basic idea behind the chemogenomic approach is that chemically similar drugs are expected to interact with similar proteins, with which the similarity of drugs and proteins are defined based on their side-effects and the amino acid sequences, respectively [3–8]. On the other hand, the key idea behind the pharmacogenomic approach is that phenotypically similar drugs are predicted to interact with similar proteins, on the basis of drug side effects and/or protein sequences [9–12]. However, previous predictive models are not easily interpretable, making it difficult to extract biological features characterizing drug-protein interactions and making it impossible to give insights into the theoretical basis of interactions.

The characterization of drug-protein interaction networks with biological characteristics has become a challenging problem in modern pharmaceutical science toward better understanding of poly-pharmacology. It is hypothesized that polypharmacology is involved in various features of drugs and target proteins (e.g., chemical substructures, pharmacophores, functional sites, and

*Correspondence: yasuo.tabei@riken.jp
[1]RIKEN Center for Advanced Intelligence Project, Nihonbashi 1-chome Mitsui Building, 15th floor, 1-4-1 Nihonbashi, Chuo-ku, 103-0027, Tokyo, Japan
Full list of author information is available at the end of the article

pathways) and complicated associations between the heterogeneous features.

A variety of feature extraction methods have recently been proposed for automatically characterizing drug-protein interactions. A data mining method was proposed for extracting molecular substructure pairs appearing frequently in interacting drug-target pairs [13]. Machine learning methods with sparse statistical models were presented to associate protein domains with drug chemical substructures [14, 15] or with drug side effects [16]. The inference of proteins eliciting drug side effects has been reported by several groups [17, 18]. However, the scalability of these methods is very limited, and these studies were conducted from the perspective of either protein functional sites, drug chemical substructures or drug phenotypic effects. There is a strong and growing need to develop efficient and scalable methods for characterizing overall drug-protein interactions with many types of features of drugs and proteins at once.

We present a novel method for systematic analyses of the underlying features characteristic of drug-protein interaction networks, which we call "drug-protein interaction signatures". We develop a new efficient algorithm for extracting informative drug-protein interaction signatures from the integration of large-scale heterogeneous data of drugs and proteins, which is made possible by space-efficient representations for fingerprints of drug-protein pairs and sparsity-induced classifiers. In the results, our method infers a set of drug-protein interaction signatures consisting of the associations between drug chemical substructures, adverse drug reactions, protein domains, biological pathways, and pathway modules. We argue that these signatures are biologically meaningful and useful for predicting unknown drug-protein interactions. To the best of our knowledge, this is the first report on characterizing a large-scale drug-protein interaction network with various biological features of drugs and proteins in an integrative framework. The drug-protein interaction signatures comprehensively inferred with our method are expected to contribute to rational drug design.

Results

Drug-protein interactions

We got the information on drug-protein interactions from five databases: ChEMBL [19], KEGG [20], DrugBank [21], PDSP Ki [22], and Matador [23]. The number of unique drug-protein interactions in the merged dataset is 78,692. These interactions involve 2302 drugs and 2334 target proteins, and the number of all possible drug-protein pairs is 5,372,868. We utilized this dataset in our experiments.

Drug profiles

We described drug chemical structures by 17,017 chemical substructures using the KEGG Chemical Function and Substructures (KCF-S) descriptor [24]. We represented each drug by a 17,017-dimension binary vector where the presence or absence of each of the KCF-S substructures is coded as 1 or 0. The resulting vector is referred to as a *chemical profile.*

We obtained the information about adverse drug reactions (ADRs) from the public release of the adverse event reporting system (AERS) of the US Food and Drug Administration (FDA) [25]. We derived 2,904,050 reports from 2004 to 2010 and mapped the drug names to KEGG following a previous study [12]. Based on the resulting 10,543 ADRs, we represented each drug by a 10,543-dimension binary vector where the presence or absence of each ADR is coded as 1 or 0. The resulting vector is referred to as an *ADR profile.*

Finally, we constructed an integrative feature vector of each drug by concatenating the chemical and the ADR profiles into a single one. The dimension of the resulting feature vector of each drug was 27,560.

Protein profiles

We obtained functional domains, biological pathways, and pathway modules (compactly clustered pathways) about proteins from the KEGG [20] and the PFAM [26] databases.

Based on 2678 PFAM domains, we represented each protein by a 2678-dimension binary vector where the presence or absence of a functional domain is coded as 1 or 0. The resulting vector is referred to as *domain profile.* Based on 270 KEGG pathway maps, we represented each protein by a 270 dimension binary vector where the presence or absence of the involvement in a biological pathway is coded as 1 or 0. The resulting vector is referred to as a *pathway profile.* Based on 107 KEGG pathway modules, we represented each protein by a 107-dimension binary vector where the presence or absence of the involvement in a pathway module is coded as 1 or 0. The resulting vector is referred to as *module profile.*

Finally, we constructed an integrative feature vector of each protein by concatenating the domain, pathway, and module profiles into a single profile. The dimension of the resulting feature vector of each protein was 3055.

We address the problem of extracting features characterizing drug-protein interaction networks in the framework of supervised classification.

Linear model for drug-protein pairs

Let C be a drug (or a drug candidate compound) and let P be a target protein (or a target candidate protein). We represent a drug-protein pair (C, P) as a high dimensional feature vector $\Phi(C, P)$ and present a linear function, $f(C, P) = \mathbf{w}^T \Phi(C, P)$, whose output is used to predict whether a (C, P) is an interacting pair or not. The weight vector $\mathbf{w}$ is estimated such that each drug-protein pair is correctly classified into the interaction class (positive class) or non-interaction class (negative class) based on the training set.

An advantage of the linear model is that one can interpret features effective for predictions from learned models. Since each element in $\Phi(C,P)$ corresponds to an element of $\mathbf{w}$, effective features can be selected by extracting highly weighted features. However, the performance of the linear model depends heavily on the feature vector design.

We represent each drug-protein pair as a high dimension feature vector by taking the tensor product of a drug profile and protein profile. The representation is similar to that in previous studies [15, 16]. The profile of a C is defined as a D-dimension binary vector:

$\Phi(C) = (c_1, c_2, ..., c_D)^T$,

where $c_i \in \{0,1\}$, $i = 1, ..., D$. The profile of a P is defined as a D'-dimension binary vector: $\Phi(P) = (p_1, p_2, ..., p_{D'})^T$, where $p_i \in \{0,1\}$, $i = 1, ..., D'$. We compute the tensor product between a drug profile $\Phi(C)$ and protein profile $\Phi(P)$, and define a feature vector $\Phi(C,P)$ as follows:

$$\Phi(C,P) = (c_1p_1, c_1p_2, ..., c_1p_{D'}, c_2p_1, ...c_Dp_1, ..., c_Dp_{D'})^T.$$

where $\Phi(C,P)$ is composed of all possible products between elements in $\Phi(C)$ and those in $\Phi(P)$. The resulting feature vector is a $D \times D'$-dimension binary vector, i.e., fingerprint, for encoding cross-integrated biological features. This is referred to as a *"tensor-product fingerprint"*.

In this study, $\Phi(C)$ was a 27,560-dimension binary vector, and $\Phi(P)$ was a 3055-dimension binary vector. Thus, the tensor-product fingerprint $\Phi(C,P)$ of each drug-protein pair is a 84,195,800-dimension binary vector.

A simpler way for representing each drug-protein pair is to concatenate $\Phi(C)$ and $\Phi(P)$ into a single feature vector as $\Phi(C,P) = \left(\Phi(C)^T, \Phi(P)^T\right)^T$ [7]. However, it cannot determine the correlation between drug and protein features. The feature vector is referred to as a *"concatenated fingerprint"*.

Logistic regression

We apply logistic regression to train the weight vector in the linear model and introduce L_1-regularizations to prevent over-fitting. The L_1-regularization induces sparsity in the weight vector and drives most of the weight elements corresponding to unimportant features to zeros, which makes it easier for us to interpret the model and extract features.

Minimizing the logistic loss with L_1-regularization for a large number of high dimensional data is difficult, but several efficient algorithms have recently been proposed. To the best of our knowledge, LIBLINEAR [27] is the most efficient and high-performance algorithm, but it requires a huge amount of memory for extremely high-dimensional data. In fact, it was not computationally feasible for our dataset in this study because of the memory problem (see the "Results" section). To overcome this difficulty, we introduce a *gradient-based method*.

Given a collection of drug-protein pairs and their labels $\left(\Phi(C_i,P_j), y_{ij}\right)$ where $y_{ij} \in \{+1,-1\}$ $(i = 1, ..., n, j = 1, ..., m)$, the logistic loss is defined as

$$LR(\mathbf{w}) = \sum_{i=1}^{n}\sum_{j=1}^{m} \log(1 + \exp\left(-y_{ij}\mathbf{w}^T\Phi(C_i,P_j)\right).$$

The logistic loss with L_1-regularization is defined as

$$L_1\text{-}LR(\mathbf{w}) = \sum_{i=1}^{n}\sum_{j=1}^{m} \log\left(1 + \exp\left(-y_{ij}\mathbf{w}^T\Phi(C_i,P_j)\right)\right) + C\|\mathbf{w}\|_1,$$

where $\|\mathbf{w}\|_1$ is L_1 norm (the sum of absolute value in the vector) and C is a regularization parameter.

Since L_1-$LR(\mathbf{w})$ is a convex function, the weight vector $\mathbf{w}$ minimizing L_1-$LR(\mathbf{w})$ can be found at zero of its gradient. However, it is impossible to compute the gradient of L_1-$LR(\mathbf{w})$, because L_1 norm contains non-differential points where $w_d = 0$. Instead, we compute the d-th dimensional gradient $\nabla_d LR(\mathbf{w})$ of $LR(\mathbf{w})$ as follows:

$$\nabla_d LR(\mathbf{w}) = \sum_{i=1}^{n}\sum_{j=1}^{m} \frac{-y_{ij}\Phi_d(C_i,P_j)\exp\left(-y_{ij}\mathbf{w}^T\Phi(C_i,P_j)\right)}{1 + \exp\left(-y_{ij}\mathbf{w}^T\Phi(C_i,P_j)\right)},$$

where $\Phi_d(C_i,P_j)$ is the d-th dimensional value of $\Phi(C_i,P_j)$. We then compute the $D \times D'$-dimensional gradient vector $\nabla LR(\mathbf{w}) \in \Re^{D\times D'}$ as

$$\nabla LR(\mathbf{w}) = \left(\nabla_1 LR(\mathbf{w}), \nabla_2 LR(\mathbf{w}), ..., \nabla_{D\times D'} LR(\mathbf{w})\right)^T.$$

The use of $\nabla LR(\mathbf{w})$ enables the global minimum for the optimal $\mathbf{w}$ in L_1-$LR(\mathbf{w})$ to be found using an efficient gradient-based optimization algorithm called orthant-wise limited-memory quasi-newton (OWL-QN) [28]. The L_1-regularized logistic regression methods, with the tensor product of the fingerprint proposed and with the concatenated fingerprint, is referred to as *L1LOG-tensor* and *L1LOG-concat*, respectively.

For comparison, we also trained models with L_2-regularized logistic regression using the gradient-based algorithm called the limited memory quasi-Newton (L-BFGS) [29]. The L_2-regularized logistic regression method, with the tensor-product fingerprint and the concatenated fingerprint, are referred to as *L2LOG-tensor* and *L2LOG-concat*, respectively.

Space-efficient representation of drug-protein pairs

Compact representation of drug-protein pairs is crucial for training linear models in memory, so we use the set representation with items corresponding to dimensions of one bit in the fingerprint. However, this still consumes a huge amount of memory for storing them, resulting

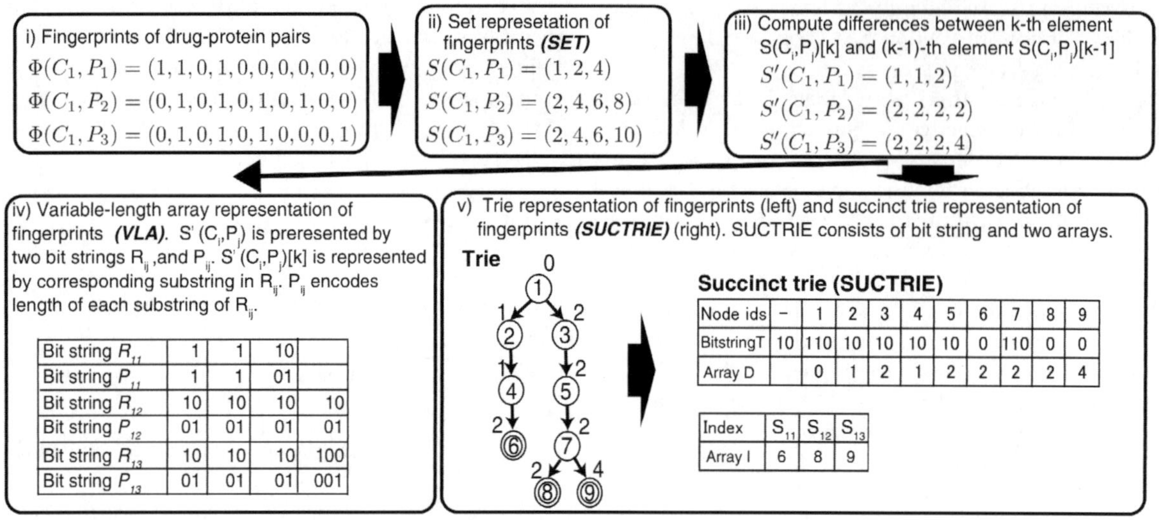

Fig. 1 Brief summary of constructing space-efficient representations of fingerprints for drug-protein pairs constructed with our proposed method: VLA and SUCTRIE

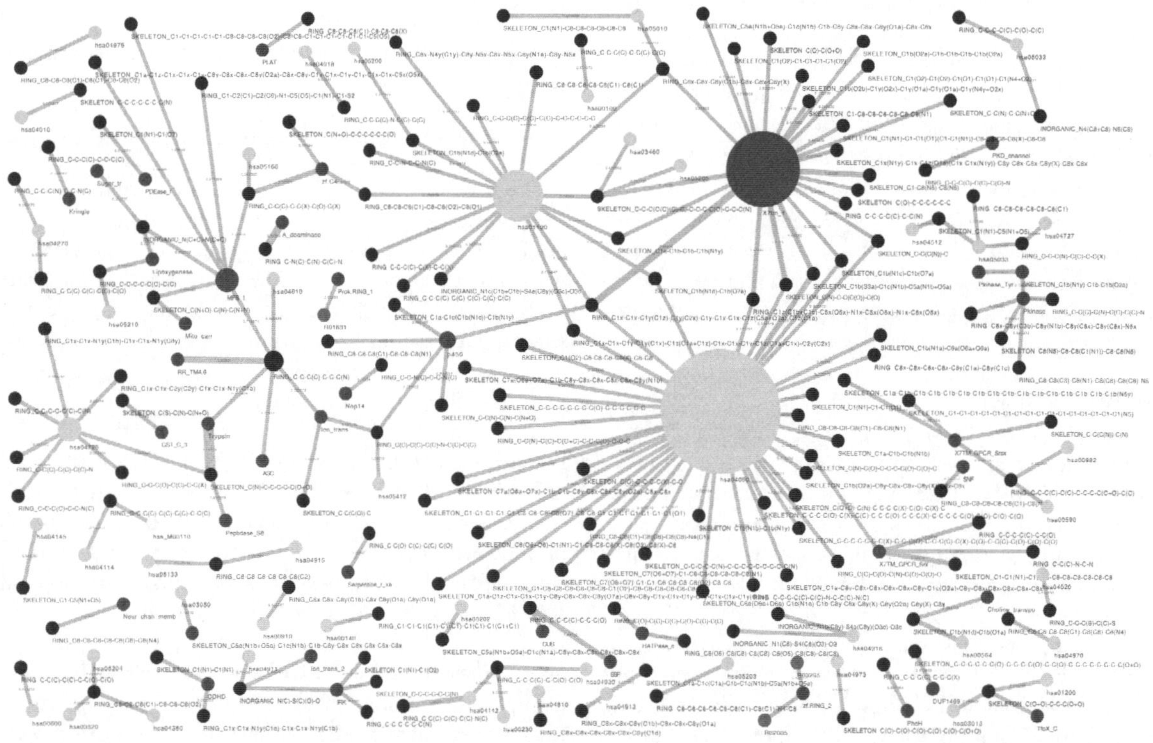

Fig. 2 Part of obtained drug-protein interaction signature network among five features, i.e., drug chemical substructures (blue), adverse drug reactions (red), protein domain (gray), biological pathway (green), and pathway module (yellow). Node size represents degree of each feature, and edge width represents corresponding weight in model

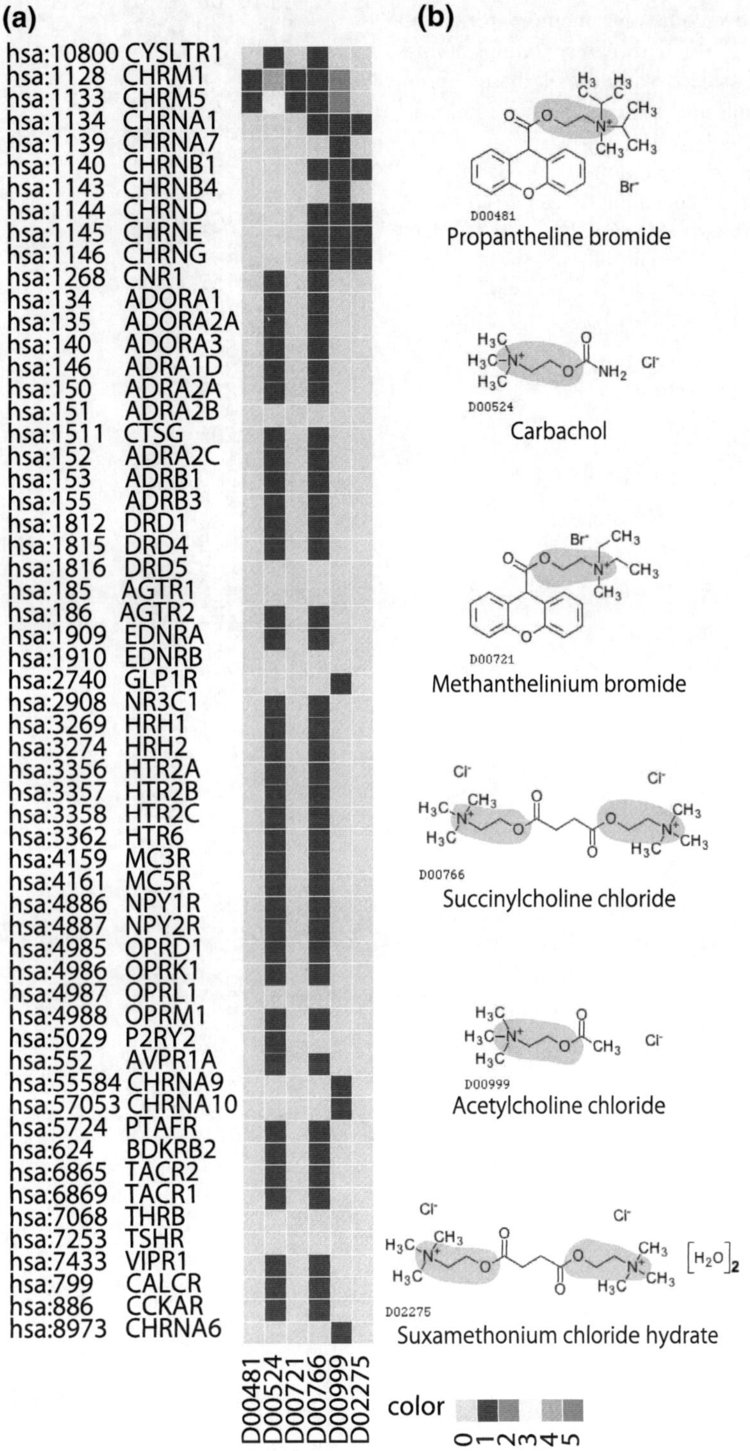

Fig. 3 Example of drug-protein interaction signature: association between drug-chemical substructure (SKELETON C1b(N1d)-C1b(O7a) in KCF-S format) and biological pathway (hsa04080 Neuroactive ligand-receptor interaction). **a** Horizontal axis shows drugs sharing chemical substructure, and vertical axis shows proteins sharing biological pathway. Color of each element corresponds to number of databases storing corresponding interaction. **b** Chemical structures of drugs sharing substructure are shown, and extracted substructure is highlighted in red (see Table 1 for further details)

in limited scalability in memory for extremely high-dimensional data. To overcome this memory problem, we constructed two space-efficient representations of fingerprints. We present a brief overview of these representations (further details are given in the supplemental material [30]).

Figure 1 illustrates the construction of our two representations. We first represent each fingerprint $\Phi(C_i, P_j)$ as a set $S(C_i, P_j) = \{d|\Phi_d(C_i, P_j) = 1, d = 1, ..., D \times D'\}$ that

Table 1 Association between KCF-S "RING C1x-C1x-C1y(C1z)-C1y(C2x)-C1y-C1x-C1x-C1z(C5a+O7a)-C1z(C1a)" and KEGG pathway hsa04080 Neuroactive "ligand-receptor interaction". See also Fig. 4

KCF-S	RING C1x-C1x-C1y(C1z)-C1y(C2x)-C1y-C1x-C1x-C1z(C5a+O7a)-C1z(C1a)
Pathway	hsa04080 Neuroactive ligand-receptor interaction
Drug	D00952 Megestrol acetate (antineoplastic)
	D01299 Chlormadinone acetate (progestin)
	D01368 Cyproterone acetate (anti-androgen)
Protein	hsa:10800 cysteinyl leukotriene receptor 1
	hsa:1128-hsa:1133 muscarinic acetylcholine receptor M1 - M5
	hsa:1134 nicotinic acetylcholine receptor alpha-1
	hsa:1268 cannabinoid receptor 1
	hsa:134,hsa:135,hsa:140 adenosine A1 receptor A1, A2a, A3
	hsa:146,hsa:150,hsa:151 adrenergic receptor alpha-1D,2A,2B
	hsa:1511 cathepsin G
	hsa:152 adrenergic receptor alpha-2C
	hsa:153-hsa:155 adrenergic receptor beta-1,2,3
	hsa:1812-hsa:1816 dopamine receptor D1-D5
	hsa:185-hsa:186 angiotensin II receptor type 1,2
	hsa:1909-hsa:1910 endothelin receptor type A, B
	hsa:2908 glucocorticoid receptor
	hsa:3269,hsa:3274 histamine receptor H1,H2
	hsa:3356,hsa:3357,hsa:3358 5-hydroxytryptamine receptor 2
	hsa:3362 5-hydroxytryptamine receptor 6
	hsa:4159-hsa:4161 melanocortin receptor 3,4,5
	hsa:4886,hsa:4887 neuropeptide Y receptor type 1/4/6,2
	hsa:4985 delta-type opioid receptor
	hsa:4986 kappa-type opioid receptor
	hsa:4988 mu-type opioid receptor
	hsa:552 arginine vasopressin receptor 1A
	hsa:5724 platelet-activating factor receptor
	hsa:624 bradykinin receptor B2
	hsa:6865,hsa:6869 tachykinin receptor 1,2
	hsa:7068 thyroid hormone receptor beta
	hsa:7253 thyroid stimulating hormone receptor
	hsa:7433 vasoactive intestinal peptide receptor 1
	hsa:886 cholecystokinin A receptor

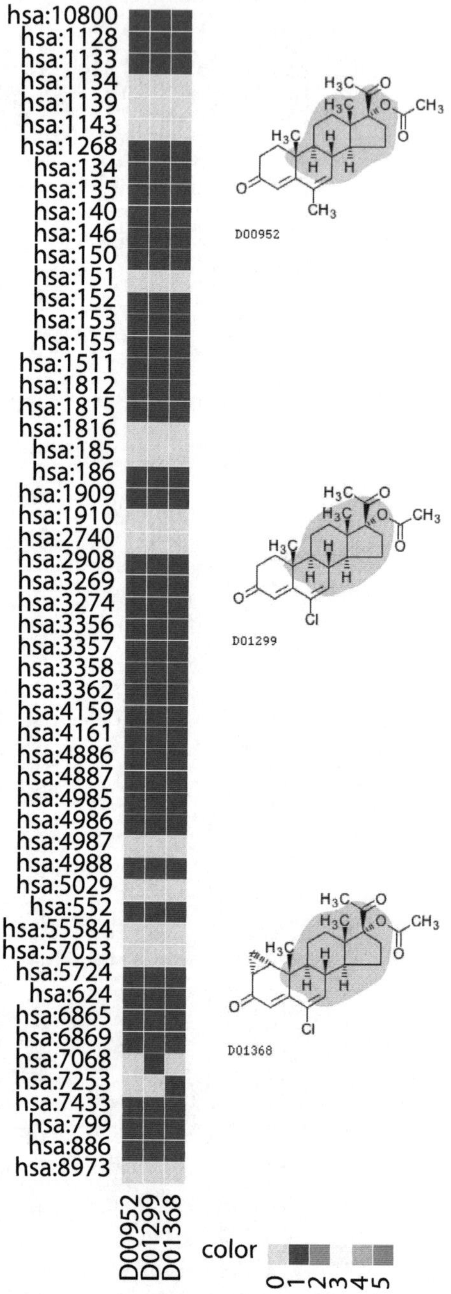

Fig. 4 The association between KCF-S "RING C1x-C1x-C1y(C1z)-C1y(C2x)-C1y-C1x-C1x-C1z(C5a+O7a)-C1z(C1a)" and KEGG pathway hsa04080 Neuroactive "ligand-receptor interaction" **a** The heat map shows the numbers of databases that register confirmed drug-protein interactions from KEGG, DrugBank, Matador, Chembl, PSD pi databases. Horizontal and vertical axes show drugs and proteins, respectively. Gray, blue, green, yellow, orange and red indicate that 0, 1, 2, 3, 4 and 5 databases contain the corresponding interaction. **b** Chemical structures of some drugs, where red areas (if any) show the extracted substructure indicated by KCF-S. See also Table 1

contains items corresponding to dimensions of one bit in $\Phi(C_i, P_j)$. We refer to a set representation of fingerprints as *SET*. To minimize each item, we then compute the difference between the k-th item $S(C_i, P_j)[k]$ and $(k-1)$-th item $S(C_i, P_j)[k-1]$ as $(S(C_i, P_j)[k] - S(C_i, P_j)[k-1])$ and keep the results in a new set $S'(C_i, P_j)$. We can recover $S(C_i, P_j)$ by cumulatively adding the items in $S'(C_i, P_j)$.

We constructed our two space-efficient representations of fingerprints by leveraging the idea behind *succinct data structures* that achieve space-efficient representations of data structures while preserving the property of fast operations. The first one is a variable-length array for compactly representing fingerprints. The $S'(C_i, P_j)$ is represented by two bit strings R_{ij} and P_{ij} which are indexed by *rank/select dictionary*, i.e., a succinct data structure for bit strings. We can randomly access any element in $S'(C_i, P_j)$ in $O(1)$ time by using fast operations in the rank/select dictionary [31]. We refer to this variable-length array representation of fingerprints as *VLA*.

The second one is a type of succinct trie for representing fingerprints. The trie is a data structure for strings, and it is also practical for representing fingerprints. A standard point-based implementation of trie consumes a huge amount of memory, resulting in limited scalability. Alternatively, we present a compact representation of trie by using a succinct data structure called LOUDS [32]. We can recover the original fingerprints by traversing a succinct trie in a depth-first manner. We refer to this succinct trie representation of fingerprints as *SUCTRIE*.

Extraction of drug-protein interaction signatures

We applied the proposed method (L1LOG-tensor) to extract drug-protein interaction signatures from drug profiles (chemical substructures and adverse drug reactions) and protein profiles (protein domains, biological pathways, and pathway modules), based on a large-scale drug-protein interaction network. Each signature is the association between a drug feature and protein feature, where two features in the same signature are thought of as being associated in terms of drug-protein interactions. The results of all extracted drug-protein interaction signatures are presented in the supplemental material [30].

L1LOG-tensor extracted 105,684 signatures, while L2LOG-tensor extracted 7,843,218 signatures. Note that the number of all possible combinations of drug features and protein features is 84,195,504. The number of signatures from our L1LOG-tensor method was much less than that of L2LOG-tensor, due to the sparsity induced by L1-regularization. This makes it easier to analyze the extracted drug-protein interaction signatures for biological interpretation, so we focused on analyzing the results from L1LOG-tensor below.

Figure 2 shows a network representation of some of the drug-protein signatures extracted with L1LOG-tensor, where highly weighted associations of five features of drugs or proteins, that is, drug-chemical substructures (blue), adverse drug reaction (red), protein pathway (green), pathway module (yellow) and protein domain (gray). Only selected results are shown due to space limitation. The inferred signature association network provides us with clues about the important features behind the drug-protein interaction network. There has been no study on the inference of these associations.

Biological interpretation of the extracted signatures

We constructed biological interpretations for the drug-protein interaction signatures extracted with L1LOG-tensor. We give only two examples due to space limitation. The result of all analyzed signatures and the figures/tables are presented in the supplemental material [30].

Table 2 Example of drug-protein interaction signature: association between adverse drug reaction (ADR) (R01631 Graft-versus-host disease) and protein domain (PF14446 Prokaryotic RING finger family 1)

Extracted ADR	R01631 Graft-versus-host disease
Extracted domain	PF14446 Prokaryotic RING finger family 1
Drugs sharing the extracted ADR	D00322 Fluconazole (antifungal)
	D00333 Ganciclovir (antiviral)
	D00399 Valproic acid (anticonvulsant)
	D00407 Methylprednisolone (glucocorticoid)
	D06272 Sorafenib tosilate (anticancer, antineoplastic)
	D06413 Nilotinib hydrochloride (antineoplastic)
	D08062 Idarubicin (antineoplastic, antibiotic)
	D08066 Imatinib (antineoplastic)
	D08524 Sorafenib (antineoplastic, anticancer)
	D08556 Tacrolimus (immunosuppressant)
Proteins sharing extracted domain	hsa:5587,hsa:23683,hsa:25865 protein kinase D
	hsa:51317 PHD finger protein 21A
	hsa:5580 protein kinase C
	hsa:64283 Rho guanine nucleotide exchange factor
	hsa:673 B-Raf proto-oncogene protein kinase
	hsa:80829 ZFP91 zinc finger protein

Further details are given in Fig. 5

Figure 3 shows an extracted signature representing the association between a drug-chemical substructure (SKELETON C1b(N1d)-C1b(O7a) in the KCF-S format) and biological pathway (hsa04080 Neuroactive ligand-receptor interaction), where the vertical axis on the heat map (a) shows all drugs sharing the extracted substructure, and the horizontal axis shows all proteins sharing the extracted pathway. The extracted drug-chemical substructures on the associated drug structures (b) are in pink. Drugs and proteins in known interacting pairs tend to have such extracted features in the same signature. For example, Propantheline bromide (D00481), Methanthelinium bromide (D00721), Acetylcholine chloride (D00999), Carbachol (D00524), Succinylcholine chloride (D00766), and Suxamethonium chloride (D02275) share a choline skeleton, and all known to act on acetylcholine receptors. However, there are many other drugs sharing the extracted drug feature and proteins sharing the extracted protein feature, and the drug-protein interactions are not known. Thus, it may be possible to predict previously unknown interactions between drugs and proteins through the extracted features in the signatures. See Table 1 and Fig. 4 for detail.

Table 2 shows an extracted signature representing the association between an ADR (R01631 Graft-versus-host disease) and protein domain (PF14446 Prokaryotic RING finger family 1), where all drugs sharing the extracted ADR and all proteins sharing the extracted protein domain are also shown. Interestingly, most drugs sharing the ADR (R01631 Graft-versus-host disease) were related to inflammation, immunosuppression, and cancer, which supports the recently expanded concept that inflammation is a critical component of cancer progression [33]. See Fig. 5 and Table 3 for detail.

Figure 4 shows an extracted signature representing the association between a drug-chemical substructure (RING C1x-C1x-C1y(C1z)-C1y(C2x)-C1y-C1x-C1x-C1z (C5a+O7a)-C1z(C1a) in the KCF-S format) and biological pathway (hsa04080 Neuroactive ligand-receptor interaction). It was observed that Megestrol acetate (D00952), Cyproterone acetate (D01368) and Chlormadinone acetate (D01299) share common ring structures. All these drugs are known to act on many neuroactive ligand-receptors. See Table 1 for detail.

Figure 6 show an extracted signature representing the association between a drug-chemical substructure (SKELETON C5a(N1b+O5a)-C1c(N1b)-C1b-C8y-C8x-C8x-C8x-C8x-C8x in the KCF-S format) and biological pathway (hsa03050 Proteasome). Proteasome inhibitors have been applied to the treatment of cancer, especially multiple myeloma. The substructure "SKELETON C5a (N1b+O5a)-C1c(N1b)-C1b-C8y-C8x-C8x-C8x-C8x-C8x" corresponds to a phenylalanine residue, which is captured as a characteristic substructure in known proteasome inhibitors Bortezomib (D03150) and Carfilzomib (D08880). See Table 4 for detail.

Performance evaluation on generalization property

If the extracted signatures are biologically meaningful in terms of drug-protein interactions, they need to have good generalization to predict drug-protein interactions.

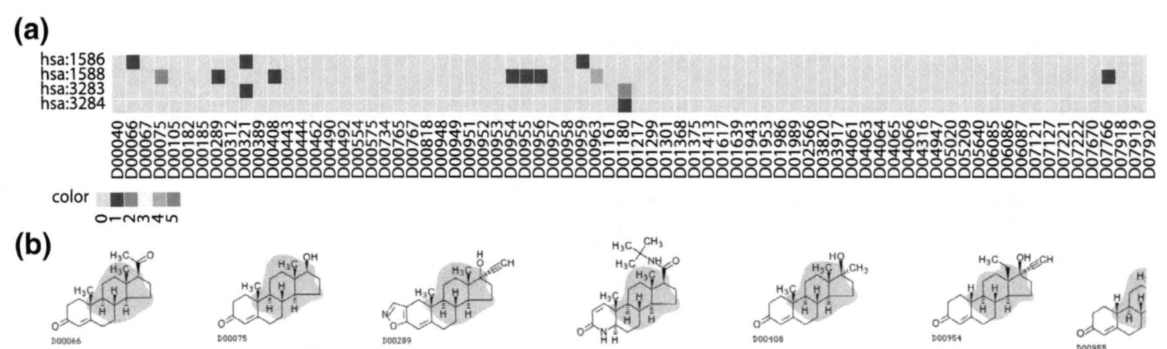

Fig. 5 The association between KCF-S "RING C1x-C1x-C1y(C1z)-C1y(C1x)-C1y(C1x)-C1z(C1a+C1y)" and KEGG pathway module "hsa_M00110 C19/C18-Steroid hormone biosynthesis". **a** The heat map shows the numbers of databases that register confirmed drug-protein interactions from KEGG, DrugBank, Matador, Chembl, PSD pi databases. Horizontal and vertical axes show drugs and proteins, respectively. Gray, blue, green, yellow, orange and red indicate that 0, 1, 2, 3, 4 and 5 databases contain the corresponding interaction. **b** Chemical structures of some drugs, where red areas (if any) show the extracted substructure indicated by KCF-S. See also Table 3

Table 3 The association between KCF-S "RING C1x-C1x-C1y(C1z)-C1y(C1x)-C1y(C1x)-C1z(C1a+C1y)" and KEGG pathway module "hsa_M00110 C19/C18-Steroid hormone biosynthesis"

KCF-S	RING C1x-C1x-C1y(C1z)-C1y(C1x)-C1y(C1x)-C1z(C1a+C1y)	
Module	hsa_M00110 C19/C18-Steroid hormone biosynthesis	
Drug	D00040 Cholesterol (pharmaceutic aid)	D00066 Progesterone (progestin)
	D00067 Estrone (estrogen)	D00075 Testosterone (androgen)
	D00105 Estradiol (estrogen)	D00182 Norethisterone (progestin)
	D00185 Estriol (estrogen)	D00289 Danazol (Anterior pituitary suppressant)
	D00312 Estrone sodium sulfate (estrogen)	D00321 Finasteride (alpha-reductase inhibitor)
	D00389 Metandienone (androgen)	D00408 Methyltestosterone (androgen)
	D00443 Spironolactone (diuretic)	D00444 Stanozolol (androgen)
	D00462 Oxandrolone (androgen)	D00490 Oxymetholone (androgen)
	D00492 Pancuronium (neuromuscular blocking agent)	D00554 Ethinylestradiol (estrogen)
	D00575 Mestranol (estrogen)	D00734 Ursodeoxycholic acid (anticholelithogenic)
	D00765 Rocuronium (neuromuscular blocking agent)	D00767 Vecuronium (neuromuscular blocking agent)
	D00818 Maprotiline hydrochloride (antidepressant)	D00948 Estropipate (estrogen)
	D00949 Hydroxyprogesterone caproate (progestin)	D00951 Medroxyprogesterone acetate (progestin)
	D00952 Megestrol acetate (antineoplastic)	D00953 Norethisterone acetate (progestin)
	D00954 Norgestrel (progestin)	D00955 Nandrolone decanoate (androgen)
	D00956 Nandrolone phenylpropionate (androgen)	D00957 Testosterone cypionate (androgen)
	D00958 Testosterone enanthate (androgen)	D00959 Testosterone propionate (androgen)
	D00963 Exemestane (antineoplastic)	D01161 Fulvestrant (antiestrogen)
	D01180 Trilostane (adrenocortical suppressant)	D01217 Dydrogesterone (progestin)
	D01299 Chlormadinone acetate (progestin)	D01301 Metenolone enanthate (anabolic)
	D01368 Cyproterone acetate (anti-androgen)	D01375 Metenolone acetate (anabolic)
	D01413 Estradiol valerate (estrogen)	D01617 Estradiol dipropionate (estrogen)
	D01639 Tibolone (Menopausal symptoms suppressant)	D01943 Potassium canrenoate (aldosterone antagonist)
	D01953 Estradiol benzoate (estrogen)	D01986 Estriol tripropionate (estrogen)
	D01989 Estriol diacetate benzoate (estrogen)	D02566 Maprotiline (antidepresant)
	D03820 Dutasteride (prostatic hyperplasia)	D03917 Drospirenone (aldosterone antagonist)
	D04061 Estradiol acetate (estrogen)	D04063 Estradiol cypionate (estrogen)
	D04064 Estradiol enanthate (estrogen)	D04065 Estradiol undecylate (estrogen)
	D04066 Estramustine (antineoplastic)	D04316 Gestodene (progestin)
	D04947 Mesterolone (androgen)	D05020 Mexrenoate potassium (aldosterone antagonist)
	D05209 Norgestimate (progestin)	D05640 Prorenoate potassium (aldosterone antagonist)
	D06085 Testosterone ketolaurate (androgen)	D06086 Testosterone phenylacetate (androgen)
	D06087 Testosterone undecanoate (testosterone)	D07121 Alfatradiol (five alfa-reductase inhibitor)
	D07127 Norethandrolone (anabolic)	D07221 Promestriene (estrogen)
	D07222 Nomegestrol (progestin)	D07670 Chlormadinone (progestin)
	D07766 Cyproterone (antiandrogen)	D07918 Estradiol hemihydrate (estrogen)
	D07919 Estradiol 17 beta-hemisuccinate (estrogen)	D07920 Estriol succinate (estrogen)
	D07921 Estriol sodium succinate (estrogen)	D08052 Hydroxyprogesterone (progestin)
	D08166 Medroxyprogesterone (progestin, antineoplastic)	D08167 Megestrol (progestin)
	D08250 Nandrolone (anabolic, ophthalmic)	D08281 Nomegestrol acetate (contraceptive)
	D08285 Norethisterone enantate (progestin)	D08409 Prasterone (androgen)
	D08573 Testosterone decanoate (androgen)	D08574 Testosterone phenylpropionate (androgen)
	D09701 Abiraterone acetate (anticancer)	

Table 3 The association between KCF-S "RING C1x-C1x-C1y(C1z)-C1y(C1x)-C1y(C1x)-C1z(C1a+C1y)" and KEGG pathway module "hsa_M00110 C19/C18-Steroid hormone biosynthesis" (*Continued*)

Protein	hsa:1586 cytochrome P450, family 17, subfamily A
	hsa:1588 cytochrome P450, family 19, subfamily A
	hsa:3283 steroid delta-isomerase
	hsa:3284 steroid delta-isomerase

See also Fig. 5

We tested five feature extraction methods: L1LOG-tensor, L2LOG-tensor, L1LOG-concat, L2LOG-concat, and L1LOG-LIBLINEAR-tensor on their abilities to reconstruct known drug-protein interactions. As mentioned above, L1LOG-tensor is our proposed method. The others are previous methods based on current algorithms or conventional fingerprints (see the Logistic regression section for further details). L1LOG-tensor and L2LOG-tensor use tensor-fingerprints represented by our space-efficient algorithm.

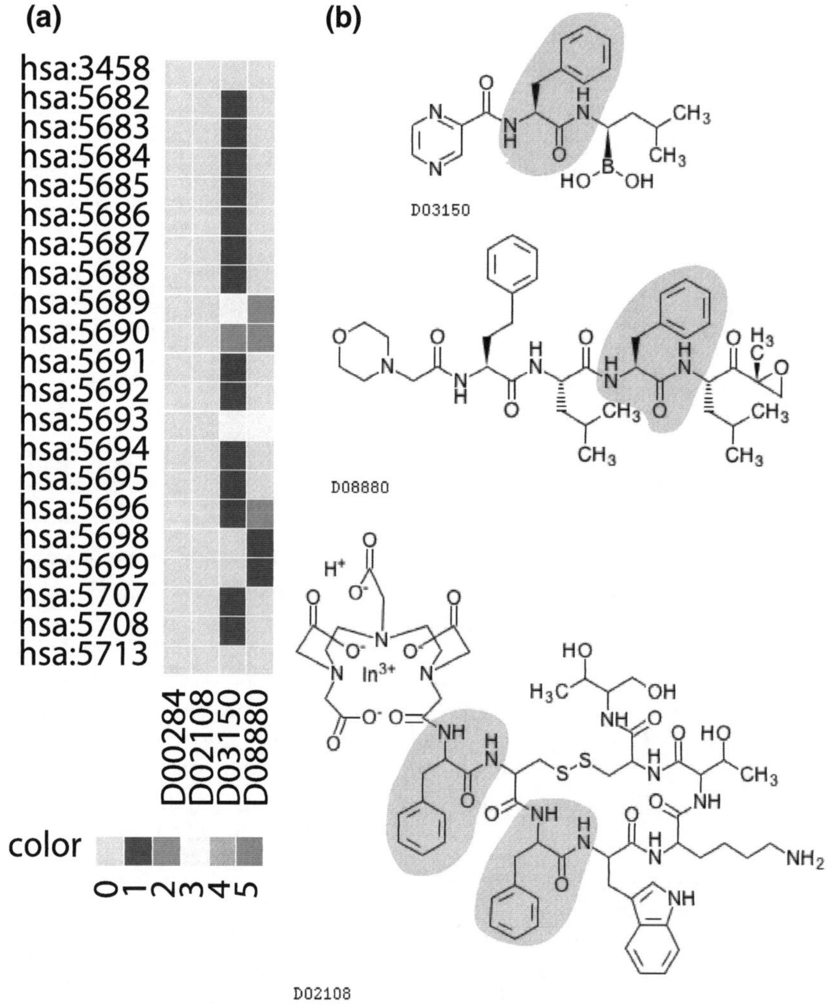

Fig. 6 The association between KCF-S "SKELETON C5a(N1b+O5a)-C1c(N1b)-C1b-C8y-C8x-C8x-C8x-C8x-C8x" and KEGG pathway "hsa03050 Proteasome" **a** The heat map shows the numbers of databases that register confirmed drug-protein interactions from KEGG, DrugBank, Matador, Chembl, PSD pi databases. Horizontal and vertical axes show drugs and proteins, respectively. Gray, blue, green, yellow, orange and red indicate that 0, 1, 2, 3, 4 and 5 databases contain the corresponding interaction. **b** Chemical structures of some drugs, where red areas (if any) show the extracted substructure indicated by KCF-S. See also Table 4 for detail

Table 4 The association between KCF-S "SKELETON C5a(N1b+O5a)-C1c(N1b)-C1b-C8y-C8x-C8x-C8x-C8x-C8x" and KEGG pathway "hsa03050 Proteasome"

KCF-S	SKELETON C5a(N1b+O5a)-C1c(N1b)-C1b-C8y-C8x-C8x-C8x-C8x-C8x
Pathway	hsa03050 Proteasome
Drug	D00284 Cosyntropin (hormone, adrenocorticotropic)
	D02108 Indium In 111 pentetreotide (radioactive agent)
	D03150 Bortezomib (anticancer, proteasome inhibitor)
	D08880 Carfilzomib (anticancer, proteasome inhibitor)
Protein	hsa:3458 interferon gamma
	hsa:5682 - hsa:5688 20S proteasome subunit alpha 1-7
	hsa:5689 - hsa:5699 20S proteasome subunit beta 1-10
	hsa:5707, hsa:5708, hsa:5713 26S proteasome regulatory subunit N1, N2, N8

See also Fig. 6

L1LOG-concat and L2LOG-concat use previous concatenated fingerprints [7] represented by the LIBLINEAR algorithm [27]. L1LOG-LIBLINEAR-tensor is a method [15, 16] which uses the tensor-product fingerprints represented by the LIBLINEAR algorithm [27].

We conducted the following fold cross-validation in a pair-wise manner. We first randomly divided all drug-protein pairs in the gold standard set into five subsets. Next, we considered four of the subsets as a training set and the remaining subset as a test set. We learned a predictive model on the drug-protein pairs in the training set. Finally, we applied the predictive model to the drug-protein pairs in the test set.

We used the receiver operating characteristic curve (ROC curve), which is defined as a plot of true positive rates against false positive rates based on various thresholds, and the precision-recall curve (PR curve), which is defined as a plot of precision (positive predictive value) against recall (sensitivity) based on various thresholds, as evaluation measures for prediction performance.

We computed the area under the ROC curve (AUC score) and the area under the PR curve (AUPR score). The parameters involved in each method (e.g., regularization parameter) were fit with AUC and AUPR as the objective functions.

Figure 7 shows the AUC and AUPR scores in the pair-wise cross-validation, where the number of negative pairs in the training set was changed from the same number of positive examples to that of all possible negative examples in the training set. We observed that the prediction accuracy of the models trained with all five methods improved as the number of negative examples in the training set increased. This suggests that using all possible negative examples for learning a predictive model will enhance prediction reliability. L1LOG-tensor performed the best.

L1LOG-LIBLINEAR-tensor did not perform well with an increasing number of negative examples in the training set because of the memory storage problem. The learning process with the LIBLINEAR algorithm consumed all the memory of our machine with 128GB-memory. In

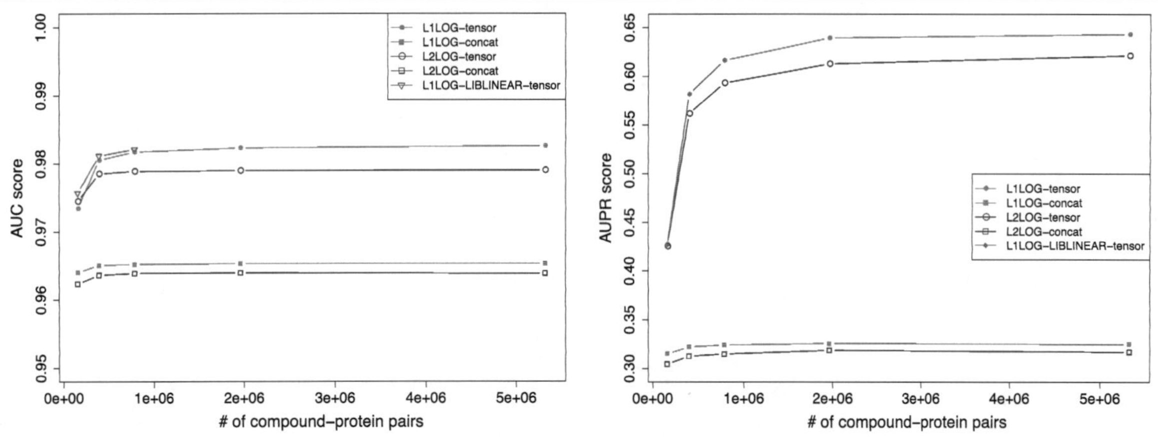

Fig. 7 AUC score (left) and AUPR score (right) in pair-wise cross validation

Table 5 AUC score, AUPR score, training time in seconds, and consumed memory in megabytes in the pair-wise cross validation experiments

Method	AUC score	AUPR score	Training time (sec)	Memory (MB)
L1LOG-tensor	0.982 ± 0.000	0.643 ± 0.001	85,211	24,079
L1LOG-concat	0.965 ± 0.000	0.324 ± 0.000	9323	177
L2LOG-tensor	0.979 ± 0.000	0.621 ± 0.000	82,853	24,079
L2LOG-concat	0.963 ± 0.000	0.317 ± 0.000	9129	177
L1LOG-LIBLINEAR-tensor	–	–	–	> 131,072

contrast, the other four methods with our space-efficient algorithm were able to finish the training process. This suggests that our space-efficient algorithm is more suitable and powerful for learning a predictive model on extremely high-dimensional data.

L1-LOG-tensor and L2LOG-tensor performed better than L1-LOG-concat and L2LOG-concat, which suggests that the tensor-product fingerprint can capture relevant information for drug-protein interaction prediction. On the other hand, the concatenated fingerprint cannot capture enough information, even though calculation is faster.

Table 5 shows the AUC score, AUPR score, training time, and consumed memory in the pair-wise cross-validation. L1LOG-tensor and L2LOG-tensor consumed 24 GB for learning predictive models on all possible drug-protein pairs, which suggests their applicability for large-scale drug-protein interaction prediction. They also took about 24 hours, which can be considered reasonable on a practical level, though they were slower than L1LOG-concat and L2LOG-concat.

In the pair-wise cross-validation, drugs and proteins in test pairs often overlap with those in the training set. We conducted a different 5-fold cross-validation to avoid the overlap of drugs and proteins in test pairs between those in the training set, which we call "block-wise cross-validation". The results of this block-wise cross-validation are shown in Fig. 8 and Table 6. The same tendency in the pair-wise cross-validation was also seen in the block-wise cross-validation. However, the AUC and AUPR scores in the block-wise cross-validation were much lower than those in the pair-wise cross validation. The results indicate that predictions of unknown interactions for new drug candidates (without known targets) and orphan proteins (without known ligands) are much more difficult than detecting missing interactions between drugs of known targets and proteins of known ligands in practice.

Finally, we tested SUCTRIE, VLA, and SET on their space-efficiencies of fingerprint representations. Note that SET is a standard representation, and SUCTRIE and VLA are those constructed with our proposed method. Figure 9 shows a plot of the consumed memory against the number of fingerprints. SET is known to use a large amount of memory for storing all possible fingerprints. In fact, it consumed 57GB for storing all possible drug-protein pairs in our dataset, which limits its practical

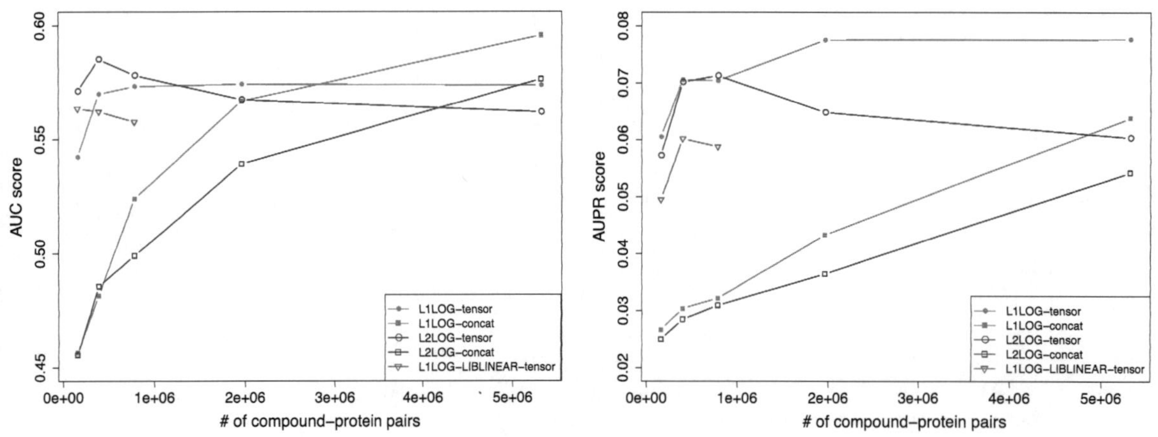

Fig. 8 AUC score (left) and AUPR score (right) in block-wise cross validation

Table 6 AUC score, AUPR score, training time in seconds, and consumed memory in megabytes in the block-wise cross validation experiments

Method	AUC score	AUPR score	Training time (sec)	Memory (MB)
L1LOG-tensor	0.574 ± 0.056	0.068 ± 0.002	71,353	19,482
L1LOG-concat	0.596 ± 0.058	0.064 ± 0.002	8323	175
L2LOG-tensor	0.562 ± 0.059	0.060 ± 0.019	70,253	19,482
L2LOG-concat	0.577 ± 0.069	0.054 ± 0.019	8010	175
L1LOG-LIBLINEAR-tensor	–	–	–	> 131,072

usage. In contrast, our proposed representations SUCTREE and VLA are more space-efficient than SET. The consumed memory of SUCTREE was slightly smaller than that of VLA. SUCTREE and VLA consumed 16 and 20 GB, respectively, for storing all possible drug-protein pairs, Suggesting the usefulness of our SUCTREE and VLA. In fact, we were not able to conduct all the analyses for this study without SUCTRIE.

Conclusions

We proposed a novel method of extracting the underlying features characterizing overall drug-protein interactions, which we call "drug-protein interaction signatures". We extracted a set of drug-protein interaction signatures consisting of the associations between drug chemical substructures, adverse drug reactions, protein domains, biological pathways, and pathway modules, and argued that the extracted drug-protein interaction signatures were biologically meaningful. Our proposed method is original in that the space-efficient representations for high-dimensional fingerprints of drug-protein pairs, in the characterization of a large-scale drug-protein interaction network with various features in an integrative framework, and in the interpretability for the extracted feature associations.

Our proposed method will be useful for various applications in drug discovery. A limitation of the method is that it cannot extract the associations between different attributes of drugs or proteins. For example, it cannot detect the associations between drug-chemical substructures and adverse drug reactions or the associations between protein domains and biological pathways. Extension of the method for analyzing such more complicated features is an important future work.

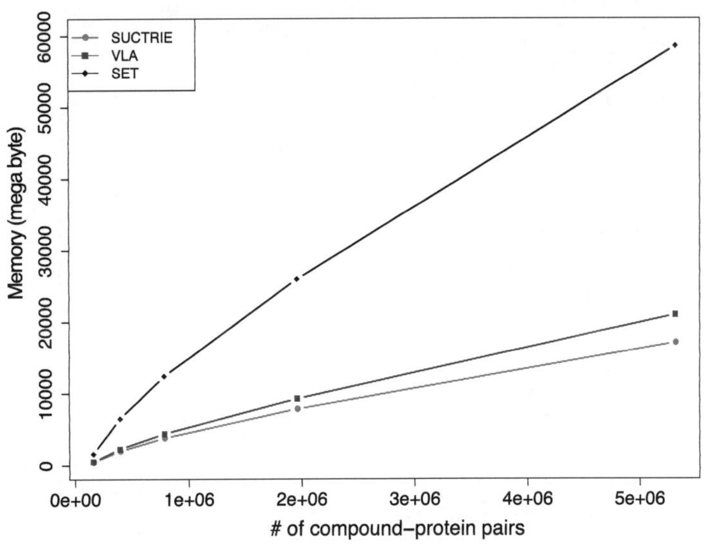

Fig. 9 Comparison of consumed memory between different fingerprint representations: SUCTRIE, VLA and SET

Abbreviations
ADRs: Adverse drug reactions; AERS: Adverse event reporting system; AUC score: area under the ROC curve; AUPR score: area under the PR curve FDA: Food and drug administration; KCF-S: KEGG chemical function and substructures; KEGG: Kyoto encyclopedia of genes and genomes; L-BFGS: Limited-memory quasi-newton; OWL-QN: Orthant-wise limited-memory quasi-newton; PR: Precision recall; ROC curve: Receiver operating characteristic curve

Acknowledgements
We thank anonymous reviewers for their thoughtful and constructive reviews.

Funding
Publication of this work was supported by JST PRESTO Grant Number JPMJPR15D8.

Availability of data and materials
All results and datasets are available at [30].

About this supplement
This article has been published as part of *BMC Systems Biology Volume 13 Supplement 2, 2019: Selected articles from the 17th Asia Pacific Bioinformatics Conference (APBC 2019): systems biology*. The full contents of the supplement are available online at https://bmcsystbiol.biomedcentral.com/articles/supplements/volume-13-supplement-2.

Authors' contributions
YT and YY designed research. YT designed the method. YT and MK performed the experiments. YT, MK and YY wrote the paper. All of the authors have read and approved the final manuscript.

Ethics approval and consent to participate
Not applicable.

Consent for publication
Not applicable.

Competing interests
The authors declare that they have no competing interests.

Author details
[1]RIKEN Center for Advanced Intelligence Project, Nihonbashi 1-chome Mitsui Building, 15th floor, 1-4-1 Nihonbashi, Chuo-ku, 103-0027, Tokyo, Japan. [2]School of Engineering, Department of Chemical System Engineering, The University of Tokyo, 7-3-1 Hongo, Bunkyo-ku 113-8656, Tokyo, Japan. [3]Department of Bioscience and Bioinformatics, Faculty of Computer Science and Systems Engineering, Kyushu Institute of Technology, 680-4 Kawazu, Lizuka, 820-8502, Fukuoka, Japan. [4]PRESTO, Japan Science and Technology Agency, 332-0012, Saitama, Japan.

Published: 5 April 2019

References

1. Whitebread S, Hamon J, Bojanic D, Urban L. Keynote review: In vitro safety pharmacology profiling: an essential tool for successful drug development. Drug Discov Today. 2005;10(21):1421–33.
2. Chong CR, Sullivan DJ. New uses for old drugs. Nature. 2007;448:645–6.
3. Faulon JL, Misra M, Martin S, Sale K, Sapra R. Genome scale enzyme-metabolite and drug-target interaction predictions using the signature molecular descriptor. Bioinformatics. 2008;24:225–33.
4. Yamanishi Y, Araki M, Gutteridge A, Honda W, Kanehisa M. Prediction of drug-target interaction networks from the integration of chemical and genomic spaces. Bioinformatics. 2008;24:232–40.
5. Jacob L, Vert JP. Protein-ligand interaction prediction: an improved chemogenomics approach. Bioinformatics. 2008;24:2149–56.
6. Keiser MJ, Setola V, Irwin JJ, Laggner C, Abbas AI, Hufeisen SJ, Jensen NH, Kuijer MB, Matos RC, Tran TB, Whaley R, Glennon RA, Hert J, Thomas KL, Edwards DD, Shoichet BK, Roth BL. Predicting new molecular targets for known drugs. Nature. 2009;462(7270):175–81.
7. Yabuuchi H, Niijima S, Takematsu H, Ida T, Hirokawa T, Hara T, Ogawa T, Minowa Y, Tsujimoto G, Okuno Y. Analysis of multiple compound-protein interactions reveals novel bioactive molecules. Mol Syst Biol. 2011;7:472.
8. Lounkine E, Keiser MJ, Whitebread S, Mikhailov D, Hamon J, Jenkins JL, Lavan P, Weber E, Doak AK, Côté S, et al. Large-scale prediction and testing of drug activity on side-effect targets. Nature. 2012;486(7403): 361–7.
9. Campillos M, Kuhn M, Gavin A-C, Jensen LJ, Bork P. Drug target identification using side-effect similarity. Science. 2008;321(5886):263–6.
10. Yamanishi Y, Kotera M, Kanehisa M, Goto S. Drug-target interaction prediction from chemical, genomic and pharmacological data in an integrated framework. Bioinformatics. 2010;26(12):246–54.
11. Atias N, Sharan R. An algorithmic framework for predicting side-effects of drugs. J Comput Biol. 2011;18(3):207–18.
12. Takarabe M, Kotera M, Nishimura Y, Goto S, Yamanishi Y. Drug target prediction using adverse event report systems: a pharmacogenomic approach. Bioinformatics. 2012;28:611–8.
13. Takigawa I, Tsuda K, Mamitsuka H. Mining Significant Substructure Pairs for Interpreting Polypharmacology in Drug-Target Network. PloS ONE. 2011;6:16999.
14. Yamanishi Y, Pauwels E, Saigo H, Stoven V. Extracting Sets of Chemical Substructures and Protein Domains Governing Drug-Target Interactions. J Chem Inf Model. 2011;51:1183–94.
15. Tabei Y, Pauwels E, Stoven V, Takemoto K, Yamanishi Y. Identification of chemogenomic features from drug-target interaction networks using interpretable classifiers. Bioinformatics. 2012;28(18):487–94. https://doi.org/10.1093/bioinformatics/bts412.
16. Iwata H, Mizutani S, Tabei Y, Kotera M, Goto S, Yamanishi Y. Inferring protein domains associated with drug side effects based on drug-target interaction network. BMC Syst Biol. 2013;7(Suppl 6):18. https://doi.org/10.1186/1752-0509-7-S6-S18.
17. Mizutani S, Pauwels E, Stoven V, Goto S, Yamanishi Y. Relating drug–protein interaction network with drug side effects. Bioinformatics. 2012;28(18):522–8.
18. Kuhn M, Al Banchaabouchi M, Campillos M, Jensen LJ, Gross C, Gavin A-C, Bork P. Systematic identification of proteins that elicit drug side effects. Mol Syst Biol. 2013;9(1):.
19. Gaulton A, Bellis L, Bento A, Chambers J, Davies M, Hersey A, Light Y, McGlinchey S, Michalovich D, Al-Lazikani B, Overington J. ChEMBL: a large-scale bioactivity database for drug discovery. Nucleic Acids Res. 2012;40:1100–7.
20. Kanehisa M, Goto S, Sato Y, Furumichi M, Tanabe M. KEGG for integration and interpretation of large-scale molecular data sets. Nucleic Acids Res. 2013;40:109–14.
21. Law V, Knox C, Djoumbou Y, Jewison T, Guo AC, Liu Y, Maciejewski A, Arndt D, Wilson M, Neveu V, Tang A, Gabriel G, Ly C, Adamjee S, Dame ZT, Han B, Zhou Y, Wishart DS. DrugBank 4.0: shedding new light on drug metabolism. Nucleic Acids Res. 2014;42:1091–7.
22. Roth BL, Lopez E, Patel S, Kroeze WK. The multiplicity of serotonin receptors: Uselessly diverse molecules or an embarrassment of riches?. Neuroscientist. 2000;6:252–62.
23. Gunther S, Kuhn M, Dunkel M, Campillos M, Senger C, Petsalaki E, Ahmed J, Urdiales E, Gewiess A, Jensen L, Schneider R, Skoblo R, Russell R, Bourne P, Bork P, Preissner R. SuperTarget and Matador: resources for exploring drug-target relationships. Nucleic Acids Res. 2008;36:919–22.
24. Kotera M, Tabei Y, Yamanishi Y, Moriya Y, Tokimatsu T, Kanehisa M, Goto S. KCF-S: KEGG Chemical Function and Substructure for improved interpretability and prediction in chemical bioinformatics. BMC Syst Biol. 2013;7(Suppl 6):2.
25. FDA. 2018. http://www.fda.gov/.
26. Finn R, Tate J, Mistry J, Coggill P, Sammut J, Hotz H, Ceric G, Forslund K, Eddy S, Sonnhammer E, Bateman A. The Pfam protein families database. Nucleic Acids Res. 2008;36:281–8.
27. Fan RE, Chang KW, Hsieh CJ, Wang XR, Lin CJ. LIBLINEAR:A library for large linear classification. J Mach Learn Res. 2008;9:1871–4.
28. Andrew G, Gao J. Scalable training of L_1-regularized log-linear models. In: Proceedings of the Twenty-Fourth International Conference on Machine Learning; 2007. p. 33–40.
29. Liu DC, Nocedal J, Liu DC, Nocedal J. On the limited memory bfgs method for large scale optimization. Math Program. 1989;45:503–28.

30. Supplementary information. 2018. http://labo.bio.kyutech.ac.jp/~yamani/drugprotein/.
31. Jacobson G. Succinct static data structures. PhD thesis, Carnegie Mellon University; 1989.
32. Jacobson G. Space-efficient Static Trees and Graphs. In: Proceedings of the 30th Annual Symposium of Foundations of Computer Science; 1989. p. 549–54.
33. Coussens LM, Werb Z. Inflammation and cancer. Nature. 2002;420:860–7.

Dutta *et al. BMC Systems Biology* 2019, **13**(Suppl 2):36
https://doi.org/10.1186/s12918-019-0692-0

BMC Systems Biology

RESEARCH

Open Access

Boolean network modeling of β-cell apoptosis and insulin resistance in type 2 diabetes mellitus

Pritha Dutta[1], Lichun Ma[2], Yusuf Ali[3], Peter M.A. Sloot[4] and Jie Zheng[5*]

From The 17th Asia Pacific Bioinformatics Conference (APBC 2019)
Wuhan, China. 14-16 January 2019

Abstract

Background: Major alteration in lifestyle of human population has promoted Type 2 diabetes mellitus (T2DM) to the level of an epidemic. This metabolic disorder is characterized by insulin resistance and pancreatic β-cell dysfunction and apoptosis, triggered by endoplasmic reticulum (ER) stress, oxidative stress and cytokines. Computational modeling is necessary to consolidate information from various sources in order to obtain a comprehensive understanding of the pathogenesis of T2DM and to investigate possible interventions by performing *in silico* simulations.

Results: In this paper, we propose a Boolean network model integrating the insulin resistance pathway with pancreatic β-cell apoptosis pathway which are responsible for T2DM. The model has five input signals, i.e. ER stress, oxidative stress, tumor necrosis factor α (TNFα), Fas ligand (FasL), and interleukin-6 (IL-6). We performed dynamical simulations using random order asynchronous update and with different combinations of the input signals. From the results, we observed that the proposed model made predictions that closely resemble the expression levels of genes in T2DM as reported in the literature.

Conclusion: The proposed model can make predictions about expression levels of genes in T2DM that are in concordance with literature. Although experimental validation of the model is beyond the scope of this study, the model can be useful for understanding the aetiology of T2DM and discovery of therapeutic intervention for this prevalent complex disease. The files of our model and results are available at https://github.com/JieZheng-ShanghaiTech/boolean-t2dm.

Keywords: Boolean model, Type 2 diabetes mellitus, Insulin resistance, β-cell apoptosis

Background

Type 2 diabetes mellitus (T2DM) is characterized by insulin resistance at its onset. Persistence of insulin resistance leads to pancreatic β-cell dysfunction and in extreme cases to β-cell apoptosis [1–3]. Insulin resistance increases the load on β-cells to produce more insulin in order to maintain blood glucose at normal levels. This homeostasis is maintained as long as β-cells can meet the increased insulin demand. However, persistence of excessive nutrients could lead to hyperglycemia, elevated free fatty acids (FFA), and inflammation, which severely impair β-cell functions, leading to insulin resistance and β-cell apoptosis.

The ER in the β-cells is responsible for the production and secretion of insulin. The increased demand for insulin synthesis in the presence of high glucose and FFA levels triggers the accumulation of misfolded proteins in the ER, causing ER stress and the consequent activation of the unfolded protein response (UPR). UPR initially attempts to mitigate ER stress by degrading misfolded proteins and preventing their further accumulation. However, when ER stress is not mitigated, UPR activates the apoptosis signals [4–6]. 78 kDa glucose regulated protein (GRP78) serves

*Correspondence: zhengjie@shanghaitech.edu.cn
[5]School of Information Science and Technology, ShanghaiTech University, Shanghai, China
Full list of author information is available at the end of the article

as a sensor of protein misfolding [7]. Under non-stressed conditions, GRP78 binds to three UPR initiator proteins, i.e. inositol requiring 1 (IRE1), PKR-like ER kinase (PERK), and activating transcription factor 6 (ATF6), and maintains them in the inactive state [8]. Under stressed conditions, GRP78 dissociates from these three proteins, causing their activation and initiation of UPR.

When ER stress can be resolved, the UPR assists β cells in their survival. However, when ER stress cannot be resolved the UPR activates the pro-apoptotic signals [9]. Hyperglycemia causes oxidative stress through the generation of reactive oxygen species (ROS) [10]. In the absence of an appropriate antioxidant response, the system experiences redox imbalance, leading to the activation of oxidative stress-sensitive signaling pathways. Cytokines, including FasL, TNFα, and IL-6, play important roles in the induction of β-cell apoptosis [11–15] as well as insulin resistance [16, 17]. Caspases serve as the final mediators of apoptosis. The upstream apoptosis initiator caspases 8 and 9 are activated on receiving death signal from the death-inducing signaling complex (DISC) and apoptosome respectively, which in turn activate the downstream apoptosis effector caspases 3, 6 and 7, which ultimately execute apoptosis [18].

Computational modeling is necessary to consolidate information from various sources, such as listed above, in order to obtain a comprehensive understanding of the pathogenesis of T2DM and investigate possible interventions by performing *in silico* simulations. A few dynamic models of insulin resistance in T2DM have been proposed recently. For instance, Brannmark et al. [19] proposed an ordinary differential equation (ODE) model of insulin signaling in T2DM. Rajan et al. proposed an ODE model to study the contribution of Forkhead box protein O1 (FOXO1) to insulin resistance in T2DM [20]. Another paper [21] presented an ODE model to simulate the development of insulin resistance by hyperglycemia, FFA, ROS, and inhibition of glucose transporter type 1 (GLUT-1) and glucose transporter type 4 (GLUT-4). However, there exists no model of β-cell apoptosis occurring in the T2DM condition. Also, there is no existing work that attempts to integrate the insulin resistance and β-cell apoptosis pathways in order to obtain a comprehensive understanding of the molecular mechanisms underlying T2DM. To discover potential therapeutic interventions for T2DM, it is essential to have a more comprehensive model for the mechanisms causing T2DM pathogenesis.

Therefore, we propose a Boolean network model integrating the insulin resistance pathway and β-cell apoptosis pathway for the purpose of obtaining deeper insights into the mechanisms of development and progression of T2DM. The aforementioned existing models are ODE models, whereas we constructed a Boolean network model. The reason behind this selection is that ODE models require detailed kinetic knowledge and time-series data for accurate parameter estimation. However, the size of our proposed network is relatively big (consisting of 72 nodes) and hence obtaining time-series expression data for all the genes would be expensive as well as time-consuming. Also, estimating the parameters of the ODE model with the time-series expression data of only a small subset of genes would result in erroneous parameter values. Furthermore, in a Boolean network

Table 1 The gene interactions incorporated into the model with reference to the existing literature

Gene interations	Reference
IRE1 ↑ → XBP1 ↑ → β-cell dysfunction	[26]
(IRE1 + TRAF2 + ASK1) ↑ → JNK ↑ → BCL2 (anti-apoptotic gene)↓	[28–30]
BCL2 ↓ → (BAX + BAK) (pro-apoptotic) ↑	[50, 51]
PERK ↑ → EIF2S1 ↓ → ATF4 ↑ → CHOP (pro-apoptotic) ↑	[27]
ATF6 ↑ → CHOP (pro-apoptotic) ↑ → BCL2 (anti-apoptotic gene)↓	[51, 52]
Oxidative stress ↑ → ASK1 ↑, JNK ↑, p38 ↑	[31–33]
p38 ↑ → CHOP (pro-apoptotic) ↑	[34]
FasL ↑ → (FasR + FADD + pro-caspase-8) ↑ → caspase-8 ↑ → caspase-3 ↑ → apoptosis	[53]
TNFα ↑ → (TNFR1 + TRADD) ↑ → RIPK1 ↑, FADD ↑, TRAF2 ↑	[54]
FADD ↑ → caspase-8 ↑	[54]
RIPK1 ↑ → RAIDD ↑ → caspase-8 ↑	[54]
TNFα ↑ → TNFR2 TNFα ↑ → TRAF2 ↑ → ... → JNK ↑, NF-kB ↑	[55–57]
(BAX + BAK) (pro-apoptotic) ↑ → Cytochrome c ↑ → (APAF1 + caspase-9) ↑ → caspase-3 ↑	[6, 58]
XIAP ↑ → caspase-3 ↓, caspase-7 ↓, caspase-9 ↓	[35, 36]
DIABLO ↑, HtrA2 ↑ → XIAP ↓	[37]
INSR ↑ → IRS ↑ → PI3K ↑ → ...→ AKT ↑ → FOXO1 ↓, GSK3β ↓, GLUT4 ↑	[59–61]
GSK3β ↑ → GS ↓ → glycogen synthesis ↓	[42, 43]
FOXO1 ↑ → PEPCK ↑, G6PC ↑ → glucose synthesis ↑	[47, 47–49]
(mTORC1 + S6K) ↑ → IRS ↓	[44–46]
IKKβ ↑ → TSC1/2 ↓ → mTORC1 ↑	[62]
ER stress ↑ → ...→ IRE1 ↑ → ...→ JNK ↑ → IRS ↓	[38, 39, 63]
ER stress ↑ → ...→ IRE1 ↑ → XBP1 ↑ → FOXO1 ↓	[64]
PERK ↑ → FOXO1 ↑	[65]
ER stress ↑ → ...→ ATF4 ↑ → CHOP ↑ → TRB3↑ → AKT ↓	[40, 41]
IL-6 ↑ → JAK ↑ → STAT3 ↑ → SOCS3 ↑ → IRS ↓	[66–68]

model, gene expression is represented by either TRUE (1) or FALSE (0). By simplifying the gene expression levels into binary states, Boolean networks are feasible for simulating the behaviour of large regulatory networks in a qualitative way.

In a Boolean network model the state of each gene is represented by either 1 (TRUE), indicating the gene is highly expressed, or 0 (FALSE) when the gene is lowly expressed. An edge in a Boolean network can be either activating or inhibiting [22]. In this paper, we have used random asynchronous Boolean simulation [23, 24], which updates genes in a random order in each iteration. This random asynchronous update method is inspired by the stochastic nature of gene regulatory networks, where gene expression alteration occurs in a random order rather than simultaneously [24].

Due to the lack of experimental gene expression data, we validate our simulation results by comparing predicted patterns of gene expression levels with experimental observations reported in the literature. We also analyze the dynamical behaviors of the model by visualizing the state transition graphs under different combinations of input signals. Our results show that the simple Boolean network model can capture some qualitative trends of the genetic circuits regulating the cell fate decision of β-cells, and shed light on the causes and processes of dysfunctional insulin metabolism and loss of β-cell homeostasis that occur in T2DM.

Methods

In this paper, we propose a Boolean network model of β-cell fate in T2DM. The model was constructed by extracting information from the KEGG pathways [25] and literature. The gene interactions incorporated into the model with reference to the existing literature are listed in Table 1. In this model, we integrated the β-cell apoptosis pathway with the insulin resistance pathway, as shown in Fig. 1. The apoptosis pathway consists of the signaling pathways triggered by ER stress (UPR pathway), oxidative stress, and 3 cytokines, i.e. FasL, TNFα, and IL-6. The insulin resistance pathways consist of phosphatidylinositide 3-kinase (PI3K)-protein kinase B (PKB or AKT) (KEGG ID: hsa04151), mammalian target of rapamycin (mTOR) (KEGG ID: hsa04150), janus kinase (JAK)- signal transducer and activator of transcription (STAT) (KEGG ID: hsa04630), and insulin (KEGG ID: hsa04910) signaling pathways. T2DM first causes insulin resistance, i.e. insulin fails to bind to insulin receptors in cells, thereby blocking the uptake of blood glucose by cells. Sustained insulin resistance finally leads to β-cell failure and apoptosis.

The Boolean update functions, listed in Table 2, for the target genes in the model are defined by combining activating input genes using OR functions and inhibiting input genes using AND functions. The reason behind this combination strategy is that a target gene will be expressed when at least one of its activating genes is expressed and all of its inhibiting genes are absent.

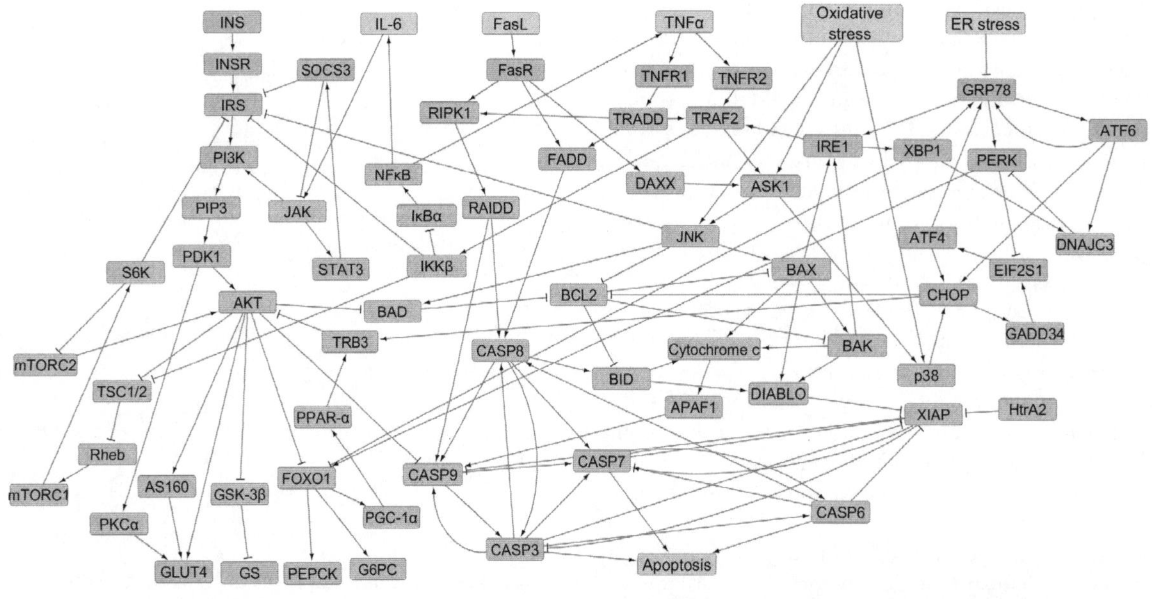

Fig. 1 Gene Regulatory Network. Insulin resistance and β-cell apoptosis pathways involved in the pathogenesis of Type 2 diabetes mellitus. The red nodes denote the five input signals and the purple node represents β-cell apoptosis. A → B indicates activation of gene B by gene A, and A —| B indicates inhibition of gene B by gene A

Table 2 Boolean functions for the Boolean model

Node	Boolean function	Node	Boolean function
ER	ER	OS	OS
FasL	FasL	TNFα	TNFα or NFKB
IL-6	IL-6 or NFKB		
GRP78	ATF6 or XBP1 or ATF4 and (not ER)	ATF6	GRP78
PERK	GRP78 and (not DNAJC3)	IRE1	BAX or BAK or GRP78
EIF2S1	GADD34 and (not PERK)	DNAJC3	ATF6 or XBP1
ATF4	EIF2S1	CHOP	ATF6 or ATF4
XBP1	IRE1	GADD34	CHOP
TNFR1	TNFα	TNFR2	TNFα
TRAF2	IRE1 or TNFR2 or TRADD	ASK1	OS or TRAF2 or DAXX
JNK	OS or ASK1 or GADD45	p38	OS or ASK1
BCL2	(not JNK) and (not CHOP) and (not P53) and (not BAD)	BID	CASP8 and (not BCL2)
BAX	JNK or P53 and (not BCL2)	BAK	BAX and (not BCL2)
DIABLO	BAX or BAK or BID	HtrA2	BAX or BAK or BID
FasR	FasL	TRADD	TNFR1
DAXX	FasR	RIPK1	FasR or TRADD
RAIDD	RIPK1	FADD	FasR or TRADD
CASP8	RAIDD or FADD or CASP3 or CASP6	CASP9	RAIDD or CASP8 or CASP3 or APAF1 or CASP12 and (not XIAP) and (not AKT)
CASP3	CASP9 or CASP8 and (not XIAP)	CASP7	CASP9 or CASP8 or CASP3 or CASP6 and (not XIAP)
CASP6	CASP7 or CASP3		
XIAP	(not DIABLO) and (not HtrA2) (not CASP3)	CytochromeC	BAX or BAK or BID
APAF1	CytochromeC or P53	Apoptosis	CASP3 or CASP6 or CASP7
INS	INS	INSR	INS
IRS	INSR and (not SOCS3) and (not JNK) (not IKKβ) and (not S6K)	PI3K	IRS or JAK
PIP3	PI3K	PDK1	PIP3
AKT	PDK1 or mTORC2 and (not TRB3)	AS160	AKT
PKCα	PDK1	GLUT4	AKT or AS160 or PKCα
GSK3β	not AKT	GS	not GSK3β
FOXO1	PERK and (not AKT) and (not XBP1)	PGC1α	FOXO1
PEPCK	FOXO1	G6PC	FOXO1
PPARα	PGC1α	TRB3	PPARα or CHOP
TSC1/2	(not AKT) and (not IKKβ)	Rheb	not TSC1/2
mTORC1	Rheb	S6K	mTORC1
mTORC2	not S6K	BAD	JNK and (not AKT)
JAK	IL-6 and (not SOCS3)	STAT3	JAK
SOCS3	STAT3	IKKβ	TRAF2
NFκB	not IKBα	IKBα	not IKKβ

The proposed Boolean network consists of 72 nodes, of which five are input signals, one node represents Apoptosis, and the remaining 66 nodes represent genes. We employ the random asynchronous Boolean update [23, 24] method to perform the simulations. The random asynchronous Boolean method first generates a random permutation of the nodes at each time step and updates the states of the nodes in the order specified by the permutation. This allows us to capture the stochastic changes in gene expressions that occur in real gene regulatory networks. The random asynchronous Boolean simulations were performed using the Python code provided in [23] which is available at https://gitlab.com/stemcellbioengineering/garuda-boolean.

For example, suppose a gene regulatory network consists of 3 genes, $\{g_1, g_2, g_3\}$. The Boolean update functions for the genes are as follows:

$$g_1 = g_3$$

$$g_2 = g_1 \vee g_3$$

$$g_3 = g_2$$

Suppose an iteration randomly generates a permutation of nodes as $\{3, 1, 2\}$. Then the asynchronous Boolean updates will be carried out as follows:

$$g_3(t+1) = g_2(t)$$

$$g_1(t+1) = g_3(t+1)$$

$$g_2(t+1) = g_1(t+1) \vee g_3(t+1)$$

From the above equations, we see that the nodes are updated in a randomly generated order as specified by the permutation, rather than simultaneously.

After performing the simulations for a fixed number of iterations, a directed graph of states is obtained, where each state is a vector representing the expression levels of all genes at a particular time step. The strategy of strongly connected components (SCCs) is employed on this directed graph to capture the dynamic nature of the states [23]. An SCC of a directed graph is a sub-graph that is strongly connected, i.e., each node is reachable from every other node in the sub-graph. An illustration of SCC is given in Fig. 2. Each node is a state with the expression levels of all the genes in the network (for the example we assume a network with five genes) and there is a path between each pair of nodes in both directions. Let us consider that an SCC consists of a set of N states $\{S_1, S_2, ..., S_N\}$. The probability of state S_i being one of the states of the SCC is given by:

$$P(S_i) = \frac{\text{number of occurrences of } S_i}{\sum_{j=1}^{N} \text{number of occurrences of } S_j}.$$

We calculate the gene expression level of each gene in a particular SCC as the sum of probabilities of states where the gene is in the ON state. Therefore, the expression level

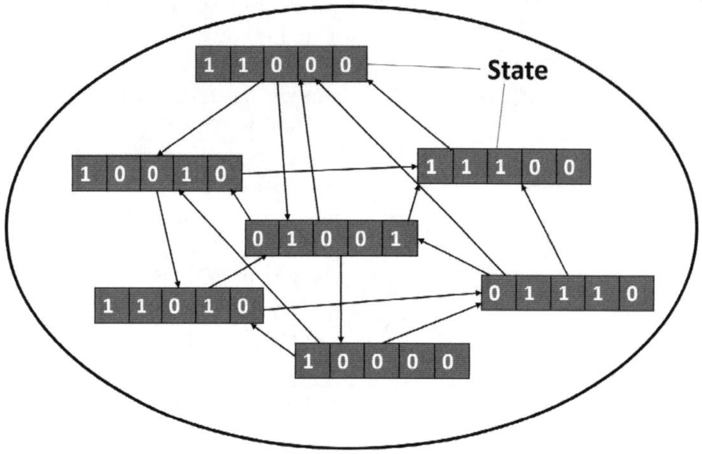

Fig. 2 Strongly Connected Component. An example of a strongly connected component (SCC). Suppose the network consists of five genes. Then each node is a state which contains the expression levels of the five genes. An arrow from state S_1 to state S_2 indicates an update step. In an SCC all states can be reached from every other state

of a gene, g_i, with respect to an SCC is determined as follows:

$$Exp(g_i) = \sum_{S_j \in OnSt(g_i)} P(S_j)$$

where

$$OnSt(g_i) = \{S_j \in SCC \mid g_i(S_j) = 1\}.$$

It is easy to see that

$$\sum_{j=1}^{N} P(S_j) = 1$$

We use ER stress, oxidative stress, TNFα, FasL, and IL-6 as input signals. Also, based on the literature, some of the nodes are assigned specific values (Table 3) and the rest are set to random values as initial conditions. We performed simulations using different combinations of the input signals, as shown in Table 4. We carried out 1000 simulation runs and 1000 Boolean update steps per simulation for each input signal. The results of the simulations are presented and discussed in the following section.

Table 3 Initial conditions

Node	Initial value	Reason
Apoptosis	False	We set apoptosis to False to see whether the input signals can cause apoptosis
Caspases 3, 6, 7, 8, 9	False	Caspases serve as the final mediators of apoptosis. So, we set them to False to see whether the input signals can activate them

Due to the lack of experimental data, we validate our proposed Boolean network model using relevant literature (see Table 1). For each gene g_i, we use the same symbol g_i to represent its binary expression level.

$$g_i = \begin{cases} 1 & \text{if } g_i \text{ is reported as expressed in the literature} \\ 0 & \text{if } g_i \text{ is reported as not expressed in the literature} \end{cases}$$

In our model, we determine the expression level of each gene with respect to a particular SCC. Thus the gene expression levels are in the range [0, 1]. We assume that if the expression value of a gene is greater than 0.50, then the gene is expressed, otherwise, it is not expressed.

For the purpose of validating our proposed model, we employ the performance metrics of precision, recall (sensitivity), specificity, and F1 score. The simulation result of our proposed model is verified against the literature as

Table 4 Different combinations for the input signal nodes

	ER stress	Oxidative stress	TNFα	FasL	IL-6
Case 1	True	False	False	False	False
Case 2	False	True	False	False	False
Case 3	True	True	False	False	False
Case 4	False	False	True	False	False
Case 5	False	False	False	True	False
Case 6	False	False	False	False	True
Case 7	False	False	True	True	True
Case 8	True	True	True	True	True

Table 5 Gene expressions of the significant genes in the model for input signal cases 1-5 and 7-8

Node	Case 1		Case 2		Case 3		Case 4		Case 5		Case 7		Case 8	
	A1	A2	A1	A2	A1	A2	A1	A2	A1	A2	A1	A2	A1	A2
Apoptosis	1	1	1	1	1	1	1	1	1	1	1	1	1	1
AKT	0.50	0.50	0.49	0.49	0.49	0.49	0.49	0.49	0.50	0.49	0.49	0.50	0.50	0.49
APAF-1	1	1	1	1	1	1	1	1	1	1	1	1	1	1
ASK1	1	1	1	1	1	1	1	1	1	1	1	1	1	1
ATF4	1	0	1	0	1	0	1	0	1	0	1	0	1	0
ATF6	0	0	0	0	0	0	0	0	0	0	0	0	0	0
BAK	1	1	1	1	1	1	1	1	1	1	1	1	1	1
BAX	1	1	1	1	1	1	1	1	1	1	1	1	1	1
BCL2	0	0	0	0	0	0	0	0	0	0	0	0	0	0
Caspase-3	1	1	1	1	1	1	1	1	1	1	1	1	1	1
Caspase-6	1	1	1	1	1	1	1	1	1	1	1	1	1	1
Caspase-7	1	1	1	1	1	1	1	1	1	1	1	1	1	1
Caspase-8	1	1	1	1	1	1	1	1	1	1	1	1	1	1
Caspase-9	1	1	1	1	1	1	1	1	1	1	1	1	1	1
CHOP	1	0	1	0	1	0	1	0	1	0	1	0	1	0
DIABLO	1	1	1	1	1	1	1	1	1	1	1	1	1	1
EIF2S1	1	0	1	0	1	0	1	0	1	0	1	0	1	0
FADD	1	1	1	1	1	1	1	1	1	1	1	1	1	1
FASR	0	0	0	0	0	0	0	0	0	0	0	0	0	0
FOXO1	0	0	0	0	0	0	0	0	0	0	0	0	0	0
G6PC	0	0	0	0	0	0	0	0	0	0	0	0	0	0
GADD34	1	0	1	0	1	0	1	0	1	0	1	0	1	0
GLUT4	0.65	0.65	0.65	0.65	0.65	0.65	0.65	0.65	0.65	0.65	0.65	0.65	0.65	0.65
GRP78	1	1	1	1	1	1	1	1	1	1	1	1	1	1
GS	0.50	0.50	0.50	0.49	0.49	0.49	0.49	0.49	0.50	0.50	0.50	0.50	0.50	0.49
GSK3β	0.49	0.49	0.50	0.50	0.50	0.50	0.50	0.50	0.49	0.49	0.49	0.49	0.49	0.50
HtrA2	1	1	1	1	1	1	1	1	1	1	1	1	1	1
IKBα	0	0	0	0	0	0	0	0	0	0	0	0	0	0
IKKβ	1	1	1	1	1	1	1	1	1	1	1	1	1	1
INS	1	1	1	1	1	1	1	1	1	1	1	1	1	1
INSR	1	1	1	1	1	1	1	1	1	1	1	1	1	1
IRE1	1	1	1	1	1	1	1	1	1	1	1	1	1	1
IRS	0	0	0	0	0	0	0	0	0	0	0	0	0	0
JAK	0.50	0.49	0.50	0.49	0.49	0.50	0.49	0.49	0.49	0.49	0.49	0.50	0.50	0.50
JNK	1	1	1	1	1	1	1	1	1	1	1	1	1	1
NFKB	1	1	1	1	1	1	1	1	1	1	1	1	1	1
PEPCK	0	0	0	0	0	0	0	0	0	0	0	0	0	0
PERK	0	0	0	0	0	0	0	0	0	0	0	0	0	0
PI3K	0.50	0.49	0.49	0.49	0.50	0.50	0.50	0.50	0.50	0.49	0.49	0.50	0.50	0.49
RAIDD	1	1	1	1	1	1	1	1	1	1	1	1	1	1
RIPK1	1	1	1	1	1	1	1	1	1	1	1	1	1	1
S6K	1	1	1	1	1	1	1	1	1	1	1	1	1	1
SOCS3	0.49	0.49	0.50	0.49	0.50	0.50	0.49	0.49	0.50	0.50	0.49	0.49	0.49	0.50
STAT3	0.50	0.49	0.49	0.49	0.49	0.50	0.49	0.49	0.50	0.50	0.49	0.49	0.50	0.50
TNFR1	1	1	1	1	1	1	1	1	1	1	1	1	1	1
TNFR2	1	1	1	1	1	1	1	1	1	1	1	1	1	1

Table 5 Gene expressions of the significant genes in the model for input signal cases 1-5 and 7-8 (*Continued*)

Node	Case 1		Case 2		Case 3		Case 4		Case 5		Case 7		Case 8	
	A1	A2	A1	A2	A1	A2	A1	A2	A1	A2	A1	A2	A1	A2
TRADD	1	1	1	1	1	1	1	1	1	1	1	1	1	1
TRAF2	1	1	1	1	1	1	1	1	1	1	1	1	1	1
TRB3	1	0	1	0	1	0	1	0	1	0	1	0	1	0
TSC2	0	0	0	0	0	0	0	0	0	0	0	0	0	0
XBP1	1	1	1	1	1	1	1	1	1	1	1	1	1	1
XIAP	0	0	0	0	0	0	0	0	0	0	0	0	0	0
mTORC1	1	1	1	1	1	1	1	1	1	1	1	1	1	1
p38	1	1	1	1	1	1	1	1	1	1	1	1	1	1

Here A1 and A2 denotes SCC1 and SCC2

follows. For each gene g_i,

$$g_i \in \begin{cases} \text{True positive,} & \text{if } g_i = 1 \text{ (simulation result) and } g_i = 1 \text{ (literature)} \\ \text{True negative,} & \text{if } g_i = 0 \text{ (simulation result) and } g_i = 0 \text{ (literature)} \\ \text{False positive,} & \text{if } g_i = 1 \text{ (simulation result) and } g_i = 0 \text{ (literature)} \\ \text{False negative,} & \text{if } g_i = 0 \text{ (simulation result) and } g_i = 1 \text{ (literature)} \end{cases}$$

The four evaluation metrics are calculated using the following formulae:

$$Precision = \frac{True\ positive}{True\ positive + False\ positive}$$

$$Recallorsensitivity = \frac{True\ positive}{True\ positive + False\ negative}$$

$$Specificity = \frac{True\ negative}{True\ negative + False\ positive}$$

$$F1\ \ score = \frac{2 \times precision \times recall}{precision + recall}$$

Results

Comparison with the literature

The expression levels of genes in the SCCs obtained by performing simulations with our proposed Boolean model are listed in Tables 5 and 6. Simulations performed using input signal cases 1, 2, 3, 4, 5, 7, and 8 (Table 4) result in two attractors (SCCs). Apoptosis is ON in both of the attractors. Simulations performed using input signal case 6 (Table 4) result in six attractors (SCCs). Apoptosis is ON in four attractors and OFF in the remaining two attractors. These observations are consistent with the literature where ER stress, oxidative stress, and cytokines have been shown to cause apoptosis of β-cells individually as well as together [4–6].

From our simulation results, we observe that Caspases 3, 6, 7, 8, and 9, which serve as the final mediators of apoptosis [18] are TRUE in the attractors, even though in the initial condition they were set to FALSE. The ER stress sensor IRE1 and its downstream gene X-box protein binding 1 (XBP1) are TRUE in some attractors, and FALSE in others [26]. Another ER stress sensor, PERK is observed to be FALSE in all the attractors. Also, eukaryotic translation initiation factor 2 subunit 1 (EIF2S1), activating transcription factor 4 (ATF4), and C/EBP homologous protein (CHOP) are TRUE in some attractors and FALSE in the others. PERK phosphorylates and inactivates EIF2S1, which inhibits protein synthesis. Phosphorylated EIF2S1 increases the translation of ATF4 [8], which in turn activates pro-apoptotic CHOP, causing β-cell dysfunction and death [27]. The attractors where IRE1, XBP1, EIF2S1, ATF4, and CHOP have expression levels of 0 may denote the transition states when these genes are not contributing to apoptosis.

While associating with TNF-receptor-associated factor 2 (TRAF2) and apoptosis signal-regulating kinase 1 (ASK1), IRE1 activates jun N-terminal kinase (JNK) [28, 29], which in turn inhibits the anti-apoptotic protein B-cell lymphoma 2 (BCL2) [30]. Oxidative stress activates ASK1 [31, 32], JNK and p38 [33]. Activated p38 phosphorylates and elevates the expression of pro-apoptotic CHOP [34]. From the simulation results, we observe that the pro-apoptotic genes, TRAF2, ASK1, JNK, p38, BAX, and BAK are TRUE and the anti-apoptotic gene BCL2 is FALSE in one attractor, while the reverse states are observed in the other. X-linked inhibitor of apoptosis protein (XIAP), which inhibits Caspases 3, 7, and 9 [35, 36], has an expression level of 0, whereas direct IAP-binding protein with low pI (DIABLO) and high temperature requirement protein A2 (HtrA2), which inhibit XIAP [37], have expression levels of 1.

JNK phosphorylates and inhibits insulin receptor substrate (IRS) [38, 39]. IRS gene is FALSE in both of the attractors. PI3K has an expression level of around 0.50 in all the attractors. Tribbles homolog 3 (TRB3) is induced by

Table 6 Gene expressions of the significant genes in the model for input signal case 6. Here A1-A6 denotes SCC1-SCC6

Node	Case 6					
	A1	A2	A3	A4	A5	A6
Apoptosis	1	1	1	1	0	0
AKT	0.49	0.49	0.63	0.55	0.65	0.56
APAF-1	1	1	0	1	0	0
ASK1	1	1	0	0	0	0
ATF4	1	0	0	1	0	1
ATF6	0	0	0	0	0	0
BAK	1	1	0	0	0	0
BAX	1	1	0	0	0	0
BCL2	0	0	1	0	1	0
Caspase-3	1	1	1	1	0	0
Caspase-6	1	1	1	1	0	0
Caspase-7	1	1	1	1	0	0
Caspase-8	1	1	1	1	0	0
Caspase-9	1	1	1	1	0	0
CHOP	1	0	0	1	0	1
DIABLO	1	1	0	1	0	0
EIF2S1	1	0	0	1	0	1
FADD	1	1	0	0	0	0
FASR	0	0	0	0	0	0
FOXO1	0	0	0	0	0	0
G6PC	0	0	0	0	0	0
GADD34	1	0	0	1	0	1
GLUT4	0.65	0.65	0.75	0.70	0.78	0.71
GRP78	1	1	0	1	0	1
GS	0.49	0.49	0.62	0.54	0.65	0.55
GSK3β	0.50	0.50	0.38	0.45	0.36	0.45
HtrA2	1	1	0	1	0	0
IKBα	0	0	1	1	1	1
IKKβ	1	1	0	0	0	0
INS	1	1	1	1	1	1
INSR	1	1	1	1	1	1
IRE1	1	1	0	0	0	0
IRS	0	0	0.19	0.22	0.18	0.22
JAK	0.49	0.49	0.49	0.49	0.49	0.50
JNK	1	1	0	0	0	0
NFKB	1	1	0	0	0	0
PEPCK	0	0	0	0	0	0
PERK	0	0	0	0	0	0
PI3K	0.49	0.49	0.54	0.55	0.54	0.56
RAIDD	1	1	0	0	0	0
RIPK1	1	1	0	0	0	0
S6K	1	1	0.62	0.55	0.62	0.56
SOCS3	0.49	0.50	0.50	0.49	0.49	0.49
STAT3	0.49	0.49	0.49	0.49	0.49	0.50
TNFR1	1	1	0	0	0	0
TNFR2	1	1	0	0	0	0
TRADD	1	1	0	0	0	0

Table 6 Gene expressions of the significant genes in the model for input signal case 6. Here A1-A6 denotes SCC1-SCC6 *(Continued)*

Node	Case 6					
	A1	A2	A3	A4	A5	A6
TRAF2	1	1	0	0	0	0
TRB3	1	0	0	1	0	1
TSC2	0	0	0.37	0.45	0.36	0.44
XBP1	1	1	0	0	0	0
XIAP	0	0	0	0	1	1
mTORC1	1	1	0.63	0.55	0.63	0.57
p38	1	1	0	0	0	0

ER stress through the ATF4-CHOP pathway [40]. Overexpression of TRB3 inhibits AKT and decreases glucose uptake [41]. TRB3 is TRUE in one attractor and FALSE in the other. AKT has an expression level of 0.50 in both of the attractors. Thus, from the results, we observe that ER stress inhibits the PI3K-AKT signaling pathway and promotes insulin resistance.

Insulin promotes conversion of glucose to glycogen by inhibiting glycogen synthase kinase-3β (GSK3β) through the PI3K-AKT signaling pathway, which leads to the activation of glycogen synthase (GS) [42]. From the simulation results, we observe that the expression level of GSK3β, which inhibits glycogen synthesis through inhibition of GS [42, 43] is approximately 0.49 and that of GS is approximately 0.50. From these simulation results, we can infer that glycogen synthesis is reduced which contributes to insulin resistance.

In T2DM, the mammalian target of rapamycin complex 1 (mTORC1)/ S6 kinase (S6K) signaling is activated [44] leading to the inhibition of IRS [45, 46]. We observe from the simulation results that mTORC1 and S6K have expression levels of 1 thus inhibiting IRS which has an expression of 0. These events cause PI3K and AKT to have low expression levels of approximately 0.50, which in turn reduces glucose uptake through GLUT4 whose expression level is around 0.65.

FOXO1 increases the expression of phosphoenolpyruvate carboxykinase (PEPCK) and glucose-6-phosphatase (G6PC) and thus promotes glucose synthesis [47]. Insulin inhibits the expression of FOXO1 through the activation of the PI3K/AKT signaling pathway, which in turn suppresses PEPCK and G6PC, and thereby reduces glucose synthesis [47–49]. From our simulation results, we observe that FOXO1, PEPCK, and G6PC are FALSE. This could be due to the fact that PI3K and AKT are not completely inactive, though they may have low expression levels, and hence is still able to inhibit the expressions of FOXO1, PEPCK, and G6PC.

In Case 6 where only signal IL6 is active, we observe six attractors (Table 6), of which four indicate apoptosis and two do not. For the attractors where apoptosis is observed, the expression levels of the genes are similar to those mentioned above for the other input signal cases. When apoptosis is not observed, i.e. in the two remaining attractors, the caspases, JNK, BAX, and BAK are FALSE. In one of these two attractors, BCL2 is FALSE and CHOP is TRUE. In the other attractor we observe the reverse expression pattern. Thus, in the presence of only IL-6, apoptosis may or may not be activated.

We further assessed the performance of our proposed Boolean network model by comparing model predictions of gene expressions against the literature. Considering the simulation results obtained using the 8 input signals listed in Table 4, the average precision, recall (sensitivity), specificity, and F1 score obtained for our model are 0.9524, 0.8, 0.875, and 0.8696, respectively. We observe that the validation scores for our model are not very high, maybe because our model is sensitive to some missing interactions.

State transition graphs

Figure 3 shows the state transition graph of the state space generated by simulations conducted using input signal combination given in case 8 (Table 4). The two dense red regions represent the two SCCs where apoptosis is ON. The blue nodes represent states where apoptosis is OFF. Thus from the state transition graph, we observe that, in the presence of all input signals, apoptosis is eventually activated, even though in the initial condition it is set to FALSE.

Figure 4 shows the state transition graph of the state space generated by simulations conducted using input signal combination given in case 6 (Table 4). The four dense red regions represent the four SCCs where apoptosis is ON. The two dense blue regions represent the two SCCs where apoptosis is OFF. Thus from the state transition graph, we observe that, in the presence of only IL-6, apoptosis may or may not be activated.

Comparison with random Boolean networks

We also compared our Boolean network model with random Boolean network models using the 8 input signal combinations given in Table 4. For cases 1, 2, 3, 4, 5, 7, and 8 we found that the number of attractors obtained by simulating the random Boolean networks ranges from 28 to 177, whereas for our Boolean network model the number of attractors is 2. Similarly, for case 6, the number of attractors obtained by simulating the random Boolean networks ranges from 25 to 180, whereas for our Boolean network model the number of attractors is 6. Thus, from the results we observe that the random Boolean networks typically have large numbers of attractors.

Conclusion

In this paper, we proposed a Boolean network model of the integrated insulin resistance and β-cell apoptosis pathways. Such a model, which explores the combined mechanism and interplay between insulin resistance and β-cell apoptosis in the pathogenesis of T2DM, has not been proposed before. We used the model to simulate the

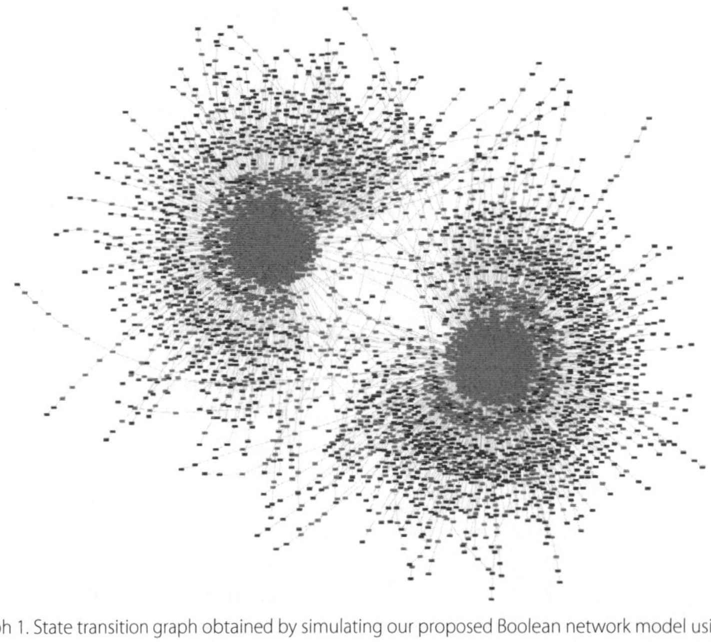

Fig. 3 State Transition Graph 1. State transition graph obtained by simulating our proposed Boolean network model using input signal condition given in Case 8 of Table 4. Simulations generate 2 attractors, both having the Apoptosis node activated. Apoptosis is ON in the red coloured states and OFF in the blue colored states

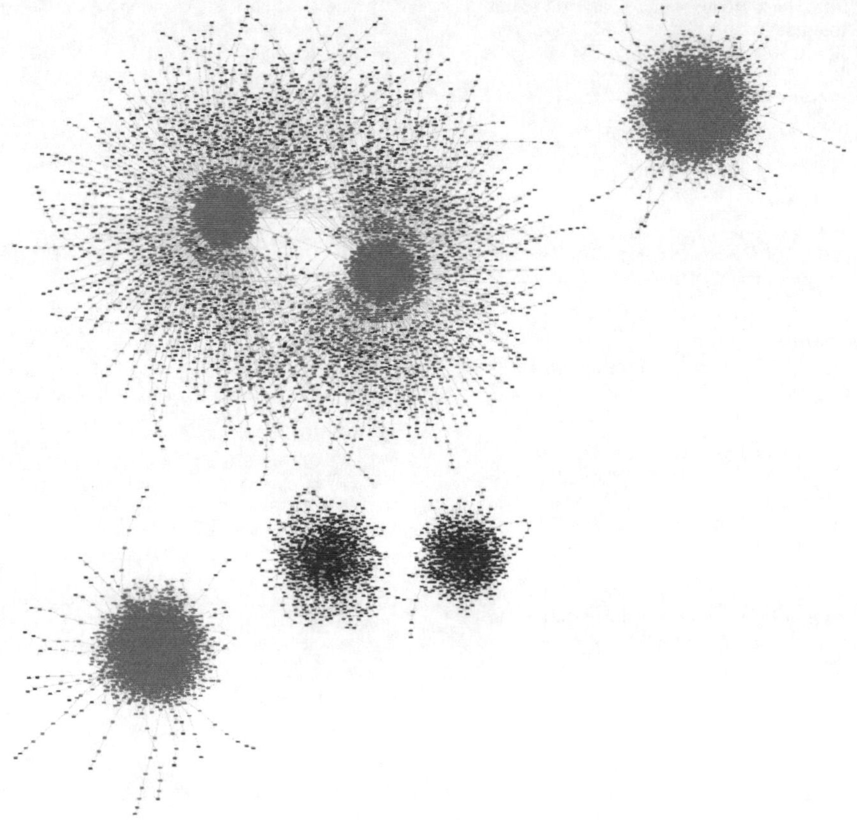

Fig. 4 State Transition Graph 2. State transition graph obtained by simulating our proposed Boolean network model using input signal condition given in Case 6 of Table 4. Simulations generate 6 attractors. In four of the attractors Apoptosis is ON, denoted by red colour, and in the remaining two attractors Apoptosis is OFF, denoted by blue colour

dynamics of gene expression induced by different combinations of the five input signals, i.e. ER stress, oxidativ e stress, and cytokines (TNFα, FasL, IL-6), which serve as triggers for insulin resistance and β-cell apoptosis.

The random order asynchronous update method was employed to perform the simulations, i.e. all nodes were updated in a random order at each update step. We assessed the performance of our model using the metrics of precision, recall (sensitivity), specificity, and F1 score, when validating our model against the literature. The precision score obtained is high, but sensitivity, specificity, and F1 scores are not so. One possible reason may be that some missing interactions affect the predictions of our model. We also compared our Boolean network model with random Boolean network models and observed that random Boolean networks typically have large numbers of attractors ranging from around 25 to 180, whereas our model shows small numbers of attractors ranging from 2 to 6.

As a future step, we can use this model to perform virtual gene knockout experiments to determine genes that play pivotal roles in insulin resistance and/or β-cell apoptosis, and these genes could be further investigated for possible disease interventions.

Abbreviations

AKT (PKB): Protein kinase B; APAF1: Apoptotic protease-activating factor 1 ; ASK1: Apoptosis signal-regulating kinase 1; ATF4: Activating transcription factor 4; ATF6: Activating Transcription Factor 6; BCL2: B-cell lymphoma 2; CHOP: C/EBP homologous protein; DIABLO: Direct IAP-binding protein with low pl; EIF2S1: Eukaryotic translation initiation factor 2 subunit 1; ER: Endoplasmic reticulum; FADD: Fas-associated death domain-containing protein; FasL: Fas ligand; FasR: Fas receptor; FFA: Free fatty acids; FOXO1: Forkhead box protein O1; GADD34: Growth arrest and DNA damage-inducible protein; G6PC: Glucose-6-phosphatase; GLUT-1: Glucose transporter type 1; GLUT-4: Glucose transporter type 4; GRP78: 78 kDa glucose regulated protein; GS: Glycogen synthase; GSK3β: Glycogen synthase kinase-3β; HtrA2: High temperature requirement protein A2; IL-6: Interleukin-6; INSR: Insulin receptor; IRE1: Inositol Requiring 1; IRS: Insulin receptor substrate; JAK: Janus kinase; JNK: Jun N-terminal kinase; mTORC1: Mammalian target of rapamycin complex 1; mTORC2: Mammalian target of rapamycin complex 2; ODE: Ordinary differential equation; PEPCK: Phosphoenolpyruvate carboxykinase; PERK: PKR-like ER kinase; PI3K: Phosphatidylinositide 3-kinase; Rheb: Ras homolog enriched in brain; RIPK1: Receptor-interacting serine/threonine-protein kinase 1; ROS: Reactive oxygen species; S6K: S6 kinase; SCC: Strongly connected component; SOCS3: Suppressor of cytokine signaling 3; STAT3: Signal transducer and activator of transcription 3; T2DM: Type 2 Diabetes Mellitus; TNFα: Tumor necrosis factor α; TNFR1: Tumor necrosis factor receptor superfamily member 1A; TNFR2: Tumor necrosis factor receptor superfamily member 1B; TRADD: TNFR1-associated death domain; TRAF2: TNF-receptor-associated factor 2; TRB3: Tribbles homolog 3; TSC: Tuberous sclerosis complex; UPR: Unfolded pro-tein respons; XBP1: X-box protein binding 1; XIAP: X-linked inhibitor of apoptosis protein

Acknowledgements
We would like to thank Dr. Ket Hing Chong for his valuable discussions.

Funding
This work was supported by MOE AcRF Tier 1 grant (2015-T1-002-094), Ministry of Education, Singapore and the Start-Up Grant of ShanghaiTech University, China. Publication of this article was sponsored by the Start-up Grant of ShanghaiTech University, China.

Availability of data and materials
Data sharing is not applicable to this article as no datasets were generated or analysed during the current study.

About this supplement
This article has been published as part of *BMC Systems Biology Volume 13 Supplement 2, 2019: Selected articles from the 17th Asia Pacific Bioinformatics Conference (APBC 2019): systems biology*. The full contents of the supplement are available online at https://bmcsystbiol.biomedcentral.com/articles/supplements/volume-13-supplement-2.

Authors' contributions
JZ initiated the project and idea, PD and LM constructed the model, and PD carried out the simulations and analysis; YA provided biological input, PS gave suggestions on modeling; PD drafted the manuscript with critical input from JZ, and other authors provided comments on the manuscript. All authors have read and approved the final manuscript.

Ethics approval and consent to participate
Not applicable.

Consent for publication
Not applicable.

Competing interests
The authors declare that they have no competing interests.

Publisher's Note

Springer Nature remains neutral with regard to jurisdictional claims in published maps and institutional affiliations.

Author details
[1]Interdisciplinary Graduate School, Nanyang Technological University, Singapore, Republic of Singapore. [2]Biomedical Informatics Lab, School of Computer Science and Engineering, Nanyang Technological University, Singapore, Republic of Singapore. [3]Lee Kong Chian School of Medicine, Nanyang Technological University, Singapore, Republic of Singapore. [4]Complexity Institute, Nanyang Technological University, Singapore, Republic of Singapore. [5]School of Information Science and Technology, ShanghaiTech University, Shanghai, China.

Published: 5 April 2019

References

1. Tomita T. Apoptosis in pancreatic β-islet cells in type 2 diabetes. Bosnian J Basic Med Sci. 2016;16(3):162.
2. Butler AE, Janson J, Bonner-Weir S, Ritzel R, Rizza RA, Butler PC. β-cell deficit and increased β-cell apoptosis in humans with type 2 diabetes. Diabetes. 2003;52(1):102–10.
3. Cnop M, Welsh N, Jonas J-C, Jörns A, Lenzen S, Eizirik DL. Mechanisms of pancreatic β-cell death in type 1 and type 2 diabetes: many differences, few similarities. Diabetes. 2005;54(suppl 2):97–107.
4. Harding HP, Ron D. Endoplasmic reticulum stress and the development of diabetes: a review. Diabetes. 2002;51(suppl 3):455–61.
5. Oyadomari S, Araki E, Mori M. Endoplasmic reticulum stress-mediated apoptosis in pancreatic β-cells. Apoptosis. 2002;7(4):335–45.
6. Schröder M, Kaufman RJ. The mammalian unfolded protein response. Annu Rev Biochem. 2005;74:739–89.
7. Bertolotti A, Zhang Y, Hendershot LM, Harding HP, Ron D. Dynamic interaction of bip and er stress transducers in the unfolded-protein response. Nat Cell Biol. 2000;2(6):326.
8. Ron D, Walter P. Signal integration in the endoplasmic reticulum unfolded protein response. Nat Rev Mol Cell Biol. 2007;8(7):519.
9. Erguler K, Pieri M, Deltas C. A mathematical model of the unfolded protein stress response reveals the decision mechanism for recovery, adaptation and apoptosis. BMC Syst Biol. 2013;7(1):16.
10. Ihara Y, Toyokuni S, Ichida K, Odaka H, et al. Hyperglycemia causes oxidative stress in pancreatic beta-cells of gk rats, a model of type 2 diabetes. Diabetes. 1999;48(4):927.
11. Donath MY, Shoelson SE. Type 2 diabetes as an inflammatory disease. Nat Rev Immunol. 2011;11(2):98.
12. Maedler K, Spinas GA, Lehmann R, Sergeev P, Weber M, Fontana A, Kaiser N, Donath MY. Glucose induces β-cell apoptosis via upregulation of the fas receptor in human islets. Diabetes. 2001;50(8):1683–90.
13. Spranger J, Kroke A, Möhlig M, Hoffmann K, Bergmann MM, Ristow M, Boeing H, Pfeiffer AF. Inflammatory cytokines and the risk to develop type 2 diabetes: results of the prospective population-based european prospective investigation into cancer and nutrition (epic)-potsdam study. Diabetes. 2003;52(3):812–7.
14. Eizirik DL, Mandrup-Poulsen T. A choice of death–the signal-transduction of immune-mediated beta-cell apoptosis. Diabetologia. 2001;44(12): 2115–33.
15. Donath MY. Targeting inflammation in the treatment of type 2 diabetes: time to start. Nat Rev Drug Discov. 2014;13(6):465–76.
16. Shoelson SE, Lee J, Goldfine AB. Inflammation and insulin resistance. J Clin Investig. 2006;116(7):1793–801.
17. Hameed I, Masoodi SR, Mir SA, Nabi M, Ghazanfar K, Ganai BA. Type 2 diabetes mellitus: from a metabolic disorder to an inflammatory condition. World J Diabetes. 2015;6(4):598.
18. Hui H, Dotta F, Mario UD, Perfetti R. Role of caspases in the regulation of apoptotic pancreatic islet beta-cells death. J Cell Physiol. 2004;200(2): 177–200.
19. Brannmark C, Nyman E, Fagerholm S, Bergenholm L, Ekstrand E-M, Cedersund G, Stralfors P. Insulin signaling in type 2 diabetes-experimental and modeling analyses reveal mechanisms of insulin resistance in human adipocytes. J Biol Chem. 2013;288(14):9867–80.
20. Rajan MR, Nyman E, Brännmark C, Olofsson CS, Strålfors P. Inhibition of foxo1 transcription factor in primary human adipocytes mimics the insulin resistant state of type 2 diabetes. Biochem J. 2018;475:1807–20.
21. Sarkar J, Dwivedi G, Chen Q, Sheu IE, Paich M, Chelini CM, D'Alessandro PM, Burns SP. A long-term mechanistic computational model of physiological factors driving the onset of type 2 diabetes in an individual. PLoS ONE. 2018;13(2):0192472.
22. Albert I, Thakar J, Li S, Zhang R, Albert R. Boolean network simulations for life scientists. Source Code Biol Med. 2008;3(1):16.
23. Yachie-Kinoshita A, Onishi K, Ostblom J, Langley MA, Posfai E, Rossant J, Zandstra PW. Modeling signaling-dependent pluripotency with boolean logic to predict cell fate transitions. Mol Syst Biol. 2018;14(1):7952.
24. Albert R, Thakar J. Boolean modeling: a logic-based dynamic approach for understanding signaling and regulatory networks and for making useful predictions. Wiley Interdiscip Rev Syst Biol Med. 2014;6(5):353–69.
25. Kanehisa M, Goto S. Kegg: kyoto encyclopedia of genes and genomes. Nucleic Acids Res. 2000;28(1):27–30.
26. Allagnat F, Christulia F, Ortis F, Pirot P, Lortz S, Lenzen S, Eizirik DL, Cardozo AK. Sustained production of spliced x-box binding protein 1 (xbp1) induces pancreatic beta cell dysfunction and apoptosis. Diabetologia. 2010;53(6):1120–30.
27. Oyadomari S, Mori M. Roles of chop/gadd153 in endoplasmic reticulum stress. Cell Death Differ. 2004;11(4):381.
28. Urano F, Wang X, Bertolotti A, Zhang Y, Chung P, Harding HP, Ron D. Coupling of stress in the er to activation of jnk protein kinases by transmembrane protein kinase ire1. Science. 2000;287(5453):664–6.
29. Lipson KL, Fonseca SG, Ishigaki S, Nguyen LX, Foss E, Bortell R, Rossini AA, Urano F. Regulation of insulin biosynthesis in pancreatic beta cells by an endoplasmic reticulum-resident protein kinase ire1. Cell Metab. 2006;4(3): 245–54.

30. Xu C, Bailly-Maitre B, Reed JC. Endoplasmic reticulum stress: cell life and death decisions. J Clin Investig. 2005;115(10):2656–64.
31. Ray PD, Huang B-W, Tsuji Y. Reactive oxygen species (ros) homeostasis and redox regulation in cellular signaling. Cell Signal. 2012;24(5):981–90.
32. Tobiume K, Matsuzawa A, Takahashi T, Nishitoh H, Morita K-i, Takeda K, Minowa O, Miyazono K, Noda T, Ichijo H. Ask1 is required for sustained activations of jnk/p38 map kinases and apoptosis. EMBO Rep. 2001;2(3): 222–8.
33. Evans JL, Goldfine ID, Maddux BA, Grodsky GM. Oxidative stress and stress-activated signaling pathways: a unifying hypothesis of type 2 diabetes. Endocr Rev. 2002;23(5):599–622.
34. Wang X, Ron D. Stress-induced phosphorylation and activation of the transcription factor chop (gadd153) by p38 map kinase. Science. 1996;272(5266):1347–9.
35. Tenev T, Zachariou A, Wilson R, Ditzel M, Meier P. Iaps are functionally non-equivalent and regulate effector caspases through distinct mechanisms. Nat Cell Biol. 2005;7(1):70.
36. Holcik M, Gibson H, Korneluk RG. Xiap: apoptotic brake and promising therapeutic target. Apoptosis. 2001;6(4):253–61.
37. Saelens X, Festjens N, Walle LV, Van Gurp M, van Loo G, Vandenabeele P. Toxic proteins released from mitochondria in cell death. Oncogene. 2004;23(16):2861.
38. Aguirre V, Uchida T, Yenush L, Davis R, White MF. The c-jun nh2-terminal kinase promotes insulin resistance during association with insulin receptor substrate-1 and phosphorylation of ser307. J Biol Chem. 2000;275(12):9047–54.
39. Hirosumi J, Tuncman G, Chang L, Görgün CZ, Uysal KT, Maeda K, Karin M, Hotamisligil GS. A central role for jnk in obesity and insulin resistance. Nature. 2002;420(6913):333.
40. Ohoka N, Yoshii S, Hattori T, Onozaki K, Hayashi H. Trb3, a novel er stress-inducible gene, is induced via atf4–chop pathway and is involved in cell death. EMBO J. 2005;24(6):1243–55.
41. Koh H-J, Toyoda T, Didesch MM, Lee M-Y, Sleeman MW, Kulkarni RN, Musi N, Hirshman MF, Goodyear LJ. Tribbles 3 mediates endoplasmic reticulum stress-induced insulin resistance in skeletal muscle. Nat Commun. 2013;4:1871.
42. Cohen P, Frame S. The renaissance of gsk3. Nat Rev Mol Cell Biol. 2001;2(10):769.
43. Lee J, Kim M-S. The role of gsk3 in glucose homeostasis and the development of insulin resistance. Diabetes Res Clin Pract. 2007;77(3): 49–57.
44. Khamzina L, Veilleux A, Bergeron S, Marette A. Increased activation of the mammalian target of rapamycin pathway in liver and skeletal muscle of obese rats: possible involvement in obesity-linked insulin resistance. Endocrinology. 2005;146(3):1473–81.
45. Harrington LS, Findlay GM, Gray A, Tolkacheva T, Wigfield S, Rebholz H, Barnett J, Leslie NR, Cheng S, Shepherd PR, et al. The tsc1-2 tumor suppressor controls insulin–pi3k signaling via regulation of irs proteins. J Cell Biol. 2004;166(2):213–23.
46. Shah OJ, Wang Z, Hunter T. Inappropriate activation of the tsc/rheb/mtor/s6k cassette induces irs1/2 depletion, insulin resistance, and cell survival deficiencies. Curr Biol. 2004;14(18):1650–6.
47. Cheng Z, White MF. Targeting forkhead box o1 from the concept to metabolic diseases: lessons from mouse models. Antioxid Redox Signal. 2011;14(4):649–61.
48. Kousteni S. Foxo1, the transcriptional chief of staff of energy metabolism. Bone. 2012;50(2):437–43.
49. Barthel A, Schmoll D. Novel concepts in insulin regulation of hepatic gluconeogenesis. Am J Physiol Endocrinol Metab. 2003;285(4):685–92.
50. Wei MC, Zong W-X, Cheng EH-Y, Lindsten T, Panoutsakopoulou V, Ross AJ, Roth KA, MacGregor GR, Thompson CB, Korsmeyer SJ. Proapoptotic bax and bak: a requisite gateway to mitochondrial dysfunction and death. Science. 2001;292(5517):727–30.
51. McCullough KD, Martindale JL, Klotz L-O, Aw T-Y, Holbrook NJ. Gadd153 sensitizes cells to endoplasmic reticulum stress by down-regulating bcl2 and perturbing the cellular redox state. Mol Cell Biol. 2001;21(4):1249–59.
52. Marciniak SJ, Yun CY, Oyadomari S, Novoa I, Zhang Y, Jungreis R, Nagata K, Harding HP, Ron D. Chop induces death by promoting protein synthesis and oxidation in the stressed endoplasmic reticulum. Gene Dev. 2004;18(24):3066–77.
53. Emamaullee JA, Shapiro AJ. Interventional strategies to prevent β-cell apoptosis in islet transplantation. Diabetes. 2006;55(7):1907–14.
54. Baud V, Karin M. Signal transduction by tumor necrosis factor and its relatives. Trends Cell Biol. 2001;11(9):372–7.
55. Akash H, Sajid M, Rehman K, Liaqat A. Tumor necrosis factor-alpha: Role in development of insulin resistance and pathogenesis of type 2 diabetes mellitus. J Cell Biochem. 2018;119:105–10.
56. Reinhard C, Shamoon B, Shyamala V, Williams LT. Tumor necrosis factor α-induced activation of c-jun n-terminal kinase is mediated by traf2. EMBO J. 1997;16(5):1080–92.
57. Rothe M, Sarma V, Dixit VM, Goeddel DV. Traf2-mediated activation of nf-kappa b by tnf receptor 2 and cd40. Science. 1995;269(5229):1424–7.
58. Li P, Nijhawan D, Budihardjo I, Srinivasula SM, Ahmad M, Alnemri ES, Wang X. Cytochrome c and datp-dependent formation of apaf-1/caspase-9 complex initiates an apoptotic protease cascade. Cell. 1997;91(4):479–89.
59. Perry RJ, Samuel VT, Petersen KF, Shulman GI. The role of hepatic lipids in hepatic insulin resistance and type 2 diabetes. Nature. 2014;510(7503):84.
60. Nikoulina SE, Ciaraldi TP, Mudaliar S, Mohideen P, Carter L, Henry RR. Potential role of glycogen synthase kinase-3 in skeletal muscle insulin resistance of type 2 diabetes. Diabetes. 2000;49(2):263–71.
61. Favaretto F, Milan G, Collin GB, Marshall JD, Stasi F, Maffei P, Vettor R, Naggert JK. Glut4 defects in adipose tissue are early signs of metabolic alterations in alms1gt/gt, a mouse model for obesity and insulin resistance. PloS ONE. 2014;9(10):109540.
62. Lee D-F, Kuo H-P, Chen C-T, Hsu J-M, Chou C-K, Wei Y, Sun H-L, Li L-Y, Ping B, Huang W-C, et al. Ikkβ suppression of tsc1 links inflammation and tumor angiogenesis via the mtor pathway. Cell. 2007;130(3):440–55.
63. Özcan U, Cao Q, Yilmaz E, Lee A-H, Iwakoshi NN, Özdelen E, Tuncman G, Görgün C, Glimcher LH, Hotamisligil GS. Endoplasmic reticulum stress links obesity, insulin action, and type 2 diabetes. Science. 2004;306(5695): 457–61.
64. Zhou Y, Lee J, Reno CM, Sun C, Park SW, Chung J, Lee J, Fisher SJ, White MF, Biddinger SB, et al. Regulation of glucose homeostasis through a xbp-1–foxo1 interaction. Nat Med. 2011;17(3):356.
65. Zhang W, Hietakangas V, Wee S, Lim SC, Gunaratne J, Cohen SM. Er stress potentiates insulin resistance through perk-mediated foxo phosphorylation. Genes Dev. 2013;27(4):441–9.
66. Klover PJ, Clementi AH, Mooney RA. Interleukin-6 depletion selectively improves hepatic insulin action in obesity. Endocrinology. 2005;146(8): 3417–27.
67. Klover PJ, Zimmers TA, Koniaris LG, Mooney RA. Chronic exposure to interleukin-6 causes hepatic insulin resistance in mice. Diabetes. 2003;52(11):2784–9.
68. Howard JK, Flier JS. Attenuation of leptin and insulin signaling by socs proteins. Trends Endocrinol Metab. 2006;17(9):365–71.

Hao *et al. BMC Systems Biology* 2019, **13**(Suppl 2):27
https://doi.org/10.1186/s12918-019-0693-z

RESEARCH

Open Access

Identification of Hürthle cell cancers: solving a clinical challenge with genomic sequencing and a trio of machine learning algorithms

Yangyang Hao[1], Quan-Yang Duh[2], Richard T. Kloos[3], Joshua Babiarz[1], R. Mack Harrell[4,7,8], S. Thomas Traweek[5], Su Yeon Kim[1], Grazyna Fedorowicz[1], P. Sean Walsh[1], Peter M. Sadow[6], Jing Huang[1] and Giulia C. Kennedy[1,3*]

From The 17th Asia Pacific Bioinformatics Conference (APBC 2019)
Wuhan, China. 14-16 January 2019

Abstract

Background: Identification of Hürthle cell cancers by non-operative fine-needle aspiration biopsy (FNAB) of thyroid nodules is challenging. Resultingly, non-cancerous Hürthle lesions were conventionally distinguished from Hürthle cell cancers by histopathological examination of tissue following surgical resection. Reliance on histopathological evaluation requires patients to undergo surgery to obtain a diagnosis despite most being non-cancerous. It is highly desirable to avoid surgery and to provide accurate classification of benignity versus malignancy from FNAB preoperatively. In our first-generation algorithm, Gene Expression Classifier (GEC), we achieved this goal by using machine learning (ML) on gene expression features. The classifier is sensitive, but not specific due in part to the presence of non-neoplastic benign Hürthle cells in many FNAB.

Results: We sought to overcome this low-specificity limitation by expanding the feature set for ML using next-generation whole transcriptome RNA sequencing and called the improved algorithm the Genomic Sequencing Classifier (GSC). The Hürthle identification leverages mitochondrial expression and we developed novel feature extraction mechanisms to measure chromosomal and genomic level loss-of-heterozygosity (LOH) for the algorithm. Additionally, we developed a multi-layered system of cascading classifiers to sequentially triage Hürthle cell-containing FNAB, including: 1. presence of Hürthle cells, 2. presence of neoplastic Hürthle cells, and 3. presence of benign Hürthle cells. The final Hürthle cell Index utilizes 1048 nuclear and mitochondrial genes; and Hürthle cell Neoplasm Index leverages LOH features as well as 2041 genes. Both indices are Support Vector Machine (SVM) based. The third classifier, the GSC Benign/Suspicious classifier, utilizes 1115 core genes and is an ensemble classifier incorporating 12 individual models.

Conclusions: The accurate algorithmic depiction of this complex biological system among Hürthle subtypes results in a dramatic improvement of classification performance; specificity among Hürthle cell neoplasms increases from 11.8% with the GEC to 58.8% with the GSC, while maintaining the same sensitivity of 89%.

Keywords: Machine learning, Algorithm, Genomic, RNA-seq, Thyroid cancer, Hürthle, Personalized healthcare

* Correspondence: giulia@veracyte.com
[1]Department of Research & Development, Veracyte, Inc, 6000 Shoreline Court, Suite 300, South San Francisco, CA 94080, USA
[3]Department of Medical Affairs, Veracyte, Inc, South San Francisco, USA
Full list of author information is available at the end of the article

Background

About one-third of adults have a thyroid nodule [1–3]. Physicians use thyroid ultrasonography to prioritize thyroid nodules whose size and ultrasonographic features warrant fine needle aspiration biopsy (FNAB) for the possibility of a clinically significant thyroid lesion [4–7]. An estimated 10 thyroid FNAB occur in the United States for each thyroid cancer diagnosed, suggesting that about 540,000 thyroid FNAB will occur in 2018 [8–10]. The small cytological FNAB sample is evaluated by light microscopy, and those with sufficient content are categorized by one of several nomenclature systems in order to estimate risk of malignancy [11, 12]. Nodules with benign FNAB results typically undergo clinical and thyroid ultrasound observation as the risk of cancer is < 5%, while nearly all of those with greater cancer risk historically were treated with surgery [13]. Those included the cytologically indeterminate Bethesda categories III and IV, which have an estimated risk of cancer of 11–29% [14]. Recent advances in molecular testing of cytologically indeterminate nodules have dramatically reduced diagnostic surgery among them [15, 16]. Still, cytologically indeterminate thyroid nodules with a significant Hürthle cell population occur in about 10% of all FNAB specimens [17], pose substantial cytological and molecular challenges [18], and are the focus of this research.

Hürthle cells, also known as oncocytes or oxyphil cells, are follicular-derived epithelial cells with acidophilic cytoplasm containing abundant granular (mitochondria-rich) cytoplasm [19]. Unless the cytopathologist is convinced that the Hürthle cells are part of a non-neoplastic process, such as Hashimoto thyroiditis [20], the specimen is typically categorized as cytologically indeterminate [Suspicious for Hürthle Cell Neoplasm (included within Bethesda IV), or when less Hürthle cell cellularity is present, Atypia of Undetermined Significance/Follicular Lesion of Undetermined Significance (included within Bethesda III)] [21]. While most cytologically indeterminate Hürthle cell FNAB are from benign thyroid nodules, these have historically been recommended for diagnostic surgical resection [20] because of a 9–39% risk of malignancy [18, 22, 23].

Insights into Hürthle cell carcinomas are growing, with recent investigations reporting alterations in nuclear and mitochondrial genomes and enriched genomic instability, which differ from the genomic profiles of non-Hürthle neoplasms [24, 25]. Still, mitochondrial and classic DNA mutations, and other cytological, radiological, and laboratory approaches have shown low sensitivity in detecting carcinomas among Hürthle cell FNAB, and imperfect specificity in differentiating benign from malignant nodules [20, 26, 27]. In 2011, the Afirma® gene expression classifier (GEC) was developed as a cancer "rule-out" test for cytologically indeterminate nodules with the intention that GEC benign samples with a low risk of malignancy could undergo clinical observation similar to cytologically benign nodules [15, 16, 28]. While many FNAB contain Hürthle cells from non-neoplastic nodules, it was also recognized that differentiating Hürthle adenomas (HCA) from Hürthle carcinomas (HCC) was a major challenge [29]. With an overall goal of an accurate benign GEC test result, a Hürthle cassette was inserted upstream of the main GEC classifier. The Hürthle cassette erred on the side of caution by identifying all samples with strong Hürthle cell neoplastic mRNA signatures as suspicious, while allowing the remainder to pass through to the main GEC classifier for final evaluation as molecularly benign or suspicious. The impact of this process is seen in the 2012 clinical validation study in which only 19% of HCA received an overall benign result, compared to 58% of non-Hürthle adenomas [15]. In practice, centers whose Hürthle FNAB included more non-neoplastic samples, or samples that passed through the Hürthle cassette, received an overall benign result among Hürthle FNAB in about one-third of their samples [30, 31], whereas others reported benign results less often [32, 33]. While the accuracy of the benign result among Hürthle FNAB remained high, the low benign call rate diminished the cost-effectiveness of the test amongst this sample type [33].

Recently, the GEC test was migrated from a microarray mRNA expression platform to a next-generation RNA sequencing platform, which provided access to RNA transcriptome expression and sequencing of nuclear and mitochondrial RNA; and detection of genomic copy number, including loss-of-heterozygosity (LOH). We coupled this increased genomic content with enhanced bioinformatics and machine-learning (ML) strategies to maintain high test sensitivity with improved overall test specificity. A key objective was to improve the specificity in classification of FNAB containing Hürthle cells. This enhanced test is the Afirma Genomic Sequencing Classifier (GSC) [34]. Here, we describe the development and validation of a Hürthle cell Index (HI) to detect FNA samples with Hürthle cell features in all FNA samples tested with an additional Hürthle cell Neoplasm index (NI) to further score only HI positive cases and separate them into neoplastic and non-neoplastic categories. These two indices enable the test to function without input from physician cytological interpretation and automatically interfaces with the core GSC classifier to render an overall GSC benign or suspicious result on every FNAB specimen. The result is a dramatic improvement of specificity among Hürthle cell neoplasms from 11.8% with the GEC to 58.8% with the GSC, while maintaining the same sensitivity of 89% [34]. This improvement in specificity substantially increases number of highly accurate benign result among Hürthle cell FNAB, safely saving patients with these challenging nodules from diagnostic thyroid surgery.

Methods

Study design

Three different types of thyroid samples were utilized for this study. The first group (Cyto-Hürthle) included FNAB for developing Hürthle cell and Hürthle cell Neoplasm indices, where detailed cytologic features were curated by expert thyroid cytopathologists using microscopic examination from FNAB. Hürthle cell-specific cytology class labels were assigned based on the presence or absence of Hürthle cells and potentially neoplastic features. The second group was comprised of FNAB for developing the Afirma Benign/Suspicious (B/S) classifier, where histopathology diagnoses were available. Two subsets were separately analyzed, as described in [34]: (1) a subset of samples used for training the Afirma GSC and (2) 191 samples used for the validation of Afirma GSC. The third group used fresh, frozen thyroid surgical tissues with histopathology diagnoses, including 12 Hürthle cell and 43 non-Hürthle cell cases, for examining copy number variation (CNV) and LOH using the Affymetrix CytoScan platform.

Identifying cytopathology samples for review

The Veracyte database was queried to identify 285 FNAB where Hürthle cell features were noted in the initial cytological reading of the case. An additional 272 FNAB where no Hürthle cell features were noted were also selected. These cases were subject to blinded re-review by a panel of 3 cytopathologists. Features examined were: cellularity, proportion of Hürthle cells, Hürthle cell morphology, Hürthle cell maturation spectrum, and presence or absence of colloid. Based on these features, four classes of samples were generated (See Additional file 1: Figure S1 for representative images): 1. Hürthle cell positive, Neoplasm positive (H + N+); 2. Hürthle cell positive, Neoplasm negative (H + N-); 3. Hürthle cell negative, Neoplasm positive (H-N+); 4. Hürthle cell negative, Neoplasm negative (H-N-). This analysis yielded 318 samples, including 119 Hürthle cell-negative and 199 Hürthle cell-positive samples. Of the 199 Hürthle cell-positive samples, 27 were identified as Bethesda II and were therefore labelled Neoplasm-negative, while 71 were Bethesda IV and were therefore labelled Neoplasm-positive. Samples were de-identified prior to cytopathology re-review prior to RNA Access library preparation.

Affymetrix CytoScan

Thyroid tissue DNA was extracted with the AllPrep Micro kit (Qiagen, Hilden, Germany) and quantitated with the Pico Green dsDNA kit (Thermo Fisher) on a Tecan Infinite Pro 200 plate reader (Tecan, Männedorf, Switzerland). DNA (125 ng) was used as input into the CytoScan HD array kit (Affymetrix, Santa Clara, CA) and the samples were processed according to the manufacturer's protocol. Cel files were input into the Affymetrix Chromosome Analysis Software (ChAS) and Copy Number and SNP outputs were analyzed for LOH and other CNVs.

RNA library preparation and next-generation sequencing

Samples were processed as described [34]. Briefly, 15 ng of total RNA was input into a Microlab STAR (Hamilton, Reno, NV) automated version of the TruSeq RNA Access Library Preparation Kit (Illumina, San Diego, CA). Libraries were sequenced on the NextSeq 500 (Illumina, San Diego, CA) using paired-end 2 × 76 cycle reads.

RNA sequencing pipeline, feature extraction, and quality control

RNA-seq data were processed as described [34]. Raw sequencing data was aligned to human reference genome assembly 37 using STAR aligner. Normalized expression levels were obtained using variance stabilizing transformation (VST) from the DESeq2 package [32]. The gene-wise dispersion parameter was estimated by the 'local fit' method. Genome-wide variants were identified using the GATK variant calling pipeline. Samples that did not satisfy the minimum in-house sequencing QC metrics were excluded from downstream analyses.

Feature engineering

Loss-of-heterozygosity (LOH) statistic

We developed a LOH statistic at the chromosome and genome level using genome-wide variants. The statistic quantifies the magnitude of LOH by calculating the proportion of variants that have a variant allele frequency (VAF; fraction of reads carrying the alternative allele) away from 0.5 (< 0.2 or > 0.8) after pre-filtering of variants with a VAF exactly at one. For the genome-level LOH statistic calculation, the mitochondrial genome and X and Y chromosomes were excluded. The details of the LOH statistic calculation are shown in the formulas below, where *"n_loss_het"* is the number of variants with a VAF far away from 0.5 (< 0.2 or > 0.8), and *"n_all_het"* is the total number of potentially heterozygous variants. LOH statistic was calculated both for each chromosome (referred as chromosome-level LOH) and for the entire genome (referred as the genome-level LOH).

$$LOH = \frac{n\_loss\_het}{n\_all\_het}$$

$$n\_loss\_het = \sum_{i=1}^{N} \begin{matrix} 1 \; if \; 0 < vaf < 0.2 \; or \; 0.8 < vaf < 1 \\ 0 \end{matrix}$$

$$n\_all\_het = \sum_{i=1}^{N} \begin{matrix} 1 \; if \; 0 < vaf < 1 \\ 0 \end{matrix}$$

Fifty-four tissue samples have LOH measured by both Affymetrix CytoScan and RNA-seq. Concordance between these two methods is shown in Additional file 1: Figure S2. These data correlated well on the samples with high level LOH (> 0.2 by Affymetrix CytoScan), all of which were Hürthle samples.

Mitochondrial features

Mitochondrial genes were captured during RNA Access library preparation and the same experimental procedures and bioinformatic sequencing pipelines were applied as described in previous sections. In total, there are 13 protein coding genes, and transcripts from all 13 were captured by the sequencing assay. Exploratory data analysis revealed all 13 genes showed differential expression levels between Hürthle negative and positive groups. Therefore, all mitochondrial genes were included in the gene feature set to undergo feature selection in downstream classifier development.

Hürthle cell index (HI) development

A total of 318 FNA samples (199 Hürthle cell + and 119 Hürthle cell -) were used to develop a Hürthle cell Index (HI), which is a binary classifier, determining if a sample is Hürthle cell + or Hürthle cell -. Ten-fold cross validation was performed to estimate the training performance, and the final model was built on all samples. Classifier development comprised three sequential steps: (1) differential expression analysis on 21,162 genes, using a statistical software package, edgeR [35], (2) selection of top-ranked genes with a FDR-adjusted p-value < 0.05 and expression fold-change (log2 scale) > 1.5, (3) optimizing parameter setting of multiple state-of-the-art machine learning algorithms with nested cross-validation. The algorithms we tested include support vector machine (SVM), elastic net, random forest, as well as SVM with asymmetrical cost to account for class imbalance. Hyperparameter tuning was performed in the inner layer; while the performance evaluation was performed in the outer layer holding out 10% of samples for each fold. SVM was selected due to its optimal cross-validated performance. The cost-parameter tuning for SVM was performed on a grid of (1e-04, 0.001, 0.01, 0.05, 0.1, 1, 5, 10). The best parameter selected for the final model was 0.001, and the associated number of support vectors was 106. Based on these parameters, the final SVM model was established using the 'svmLinear' method from the 'caret' R package [36] with all training samples and 1408 genes selected from the differential expression analysis.

Hürthle cell neoplasm index (NI) development

Among the 199 Hürthle cell samples used for HI development, 98 were further grouped into Neoplasm+ ($n = 71$) and Neoplasm- ($n = 27$) and used for Neoplasm Index (NI) training. NI is a binary classifier, determining if a Hürthle cell + sample is Neoplasm+ or Neoplasm-. Algorithm training for the NI was carried out similarly to the training for the HI but included novel LOH statistics as features. For the final model, 2041 genes were selected from the differential expression analysis. In addition, 15 chromosome-level LOH statistics (chromosomes 1, 2, 3, 4, 5, 6, 8, 9, 11, 13, 14, 15, 16, 18, and 19) and genome-level LOH were included as features for model training. The SVM was then built similarly to HI training on the same cost-parameter grid. The best parameter selected for the final model was 0.001, and the associated number of support vectors was 51.

Integrating Hürthle and Hürthle neoplasm indices into the Afirma GSC B/S classification workflow

The Afirma GSC classification workflow (Fig. 1a) begins with four upstream classifiers handling special thyroid FNA entities (Parathyroid Adenoma (PTA), Medullary Thyroid Carcinoma (MTC), BRAF V600E, and RET/PTC1 and RET/PTC3 fusions). It then uses the ensemble B/S model to classify a majority of samples as GSC benign or suspicious, as described previously [34]. The HI and NI are integrated with the ensemble model to increase overall classification performance (Fig. 1b).

There are three mechanisms for arriving at the GSC Benign versus Suspicious binary outcome for a given sample:

(1) The result is GSC Benign if the ensemble B/S score is lower than the nominal threshold; otherwise
(2) The result is initially GSC Suspicious but can be reassigned to a benign call by "Hürthle-adjustment" if the sample is predicted as Hürthle cell Index-positive (HI+), and Neoplasm Index-negative (NI-), and the ensemble B/S score is lower than the Hürthle-adjusted threshold; otherwise
(3) The result is GSC Suspicious.

The three types of outcomes are described with mathematical formulae as follows. For a given sample i, we denote the scores from HI and NI, and the ensemble B/S classifier as H_i, N_i, and BS_i, respectively. We denote the thresholds for HI and NI as t_H and t_N, respectively. For the ensemble B/S score, two thresholds exist; one is the nominal threshold, $t_{BS\_nominal}$, and the other is an increased threshold to handle Hürthle cell-positive, Neoplasm-negative cases. The latter is referred to as a "Hürthle-adjusted" threshold and denoted as $t_{BS\_Hürthle}$. The binary call outcome of the sample will be:

(1) GSC Benign, if $BS_i < t_{BS\_nominal}$; otherwise
(2) Initially GSC Suspicious, but reassign to GSC Benign if $H_i > t_H$, and $N_i < t_N$, and $t_{BS\_nominal} \leq BS_i < t_{BS\_Hürthle}$; otherwise
(3) GSC Suspicious

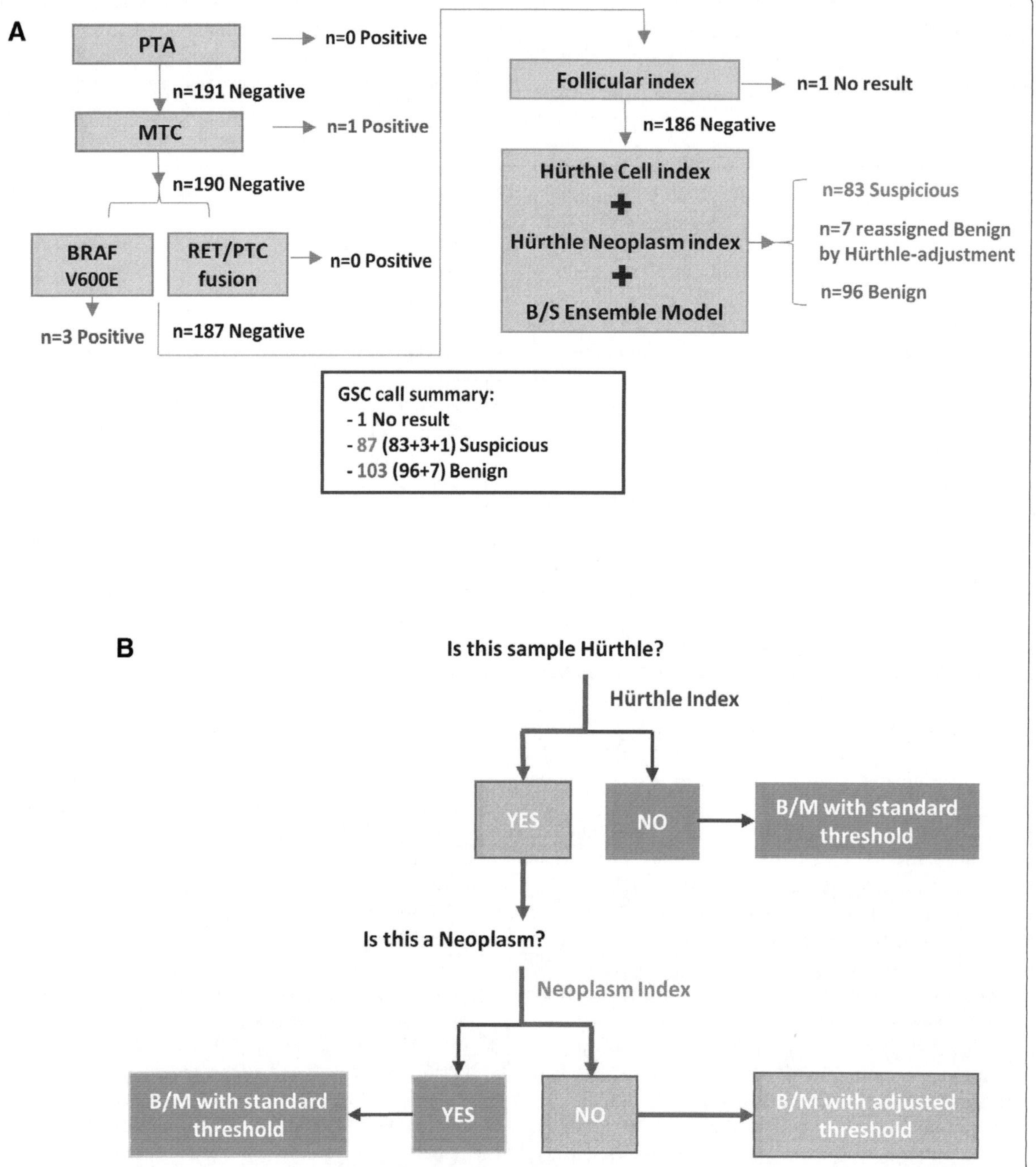

Fig. 1 The Afirma GSC Algorithm Workflow. **a** A diagram of the Afirma GSC workflow with the validation cohort outcomes listed. **b** Nested strategy for Hürthle classification. Samples are first examined by the HI classifier. HI+ samples are passed to the NI classifier. NI- samples are subject to an adjusted threshold for the main B/S classifier

The integration of HI and NI into Afirma GSC workflow aims to "rescue" truly benign Hürthle cell-containing samples by actively reassigning these samples to GSC benign that would otherwise have been called GSC Suspicious.

Determining thresholds during algorithm development to maximize overall performance

Multiple factors were considered in determining thresholds for individual HI and NI, as well as the Hürthle-adjusted B/S threshold. First, cross-validation training performance of HI and NI were evaluated against the cytology label. Second, we examined the concordance of predicted Hürthle cell and Neoplasm status with the histopathology diagnosis using samples from the Afirma GSC B/S training set. Finally, the potential gain in the overall Afirma GSC performance due to Hürthle-adjustment was assessed. Due to the complexity of including two types of datasets (one focused on cytologic features, and the other on histopathology labels) and integration of three separate classifiers, complex dynamic parameter optimization was engaged with extensive multi-dimensional grid search to examine the tradeoff in the sensitivity and specificity for each classifier alone, and in combination. The final thresholds were chosen by optimizing overall Afirma GSC B/S specificity, while maintaining a high sensitivity (> 90%), by enabling high performance in both HI and NI.

Results

Identifying mitochondrial features in Hürthle positive samples

One prominent microscopic feature observed in Hürthle cell-positive samples, visualized by both cytopathology and histopathology, is the inherent increased intracytoplasmic mitochondria in Hürthle cells [19]. We sought to determine if there was a genomic signature associated with the high mitochondrial content noted in Hürthle cells. Analyzing differential expression between Hürthle cell positive and Hürthle cell negative samples revealed that all 13 mitochondrial transcripts were observed with an FDR-adjusted p-value of < 0.05 (Fig. 2). Therefore, the elevated number of mitochondria observed microscopically can be detected genomically using RNA Sequencing.

Identifying LOH features in Hürthle positive, neoplasm positive samples

We next sought to identify genomic features associated with Hürthle cell positive, neoplasm positive samples. Previous reports have identified copy number changes in Hürthle subtypes [26, 37, 38]. We performed genome-wide DNA copy number analysis on thyroid tissues and observed extensive LOH in many Hürthle tissues, while relatively little LOH was observed in non-Hürthle tissues (See Fig. 3a). Interestingly, the LOH was primarily, but not exclusively, enriched in Hürthle cell carcinomas (See Additional file 1: Figure S3 for example of LOH in various tissues).

Based on these DNA findings, we sought to recapitulate the LOH signal in RNA-seq data, utilizing SNPs called from expressed genes. Because RNA-seq data is limited to the exome, only chromosome-wide and genome-wide LOH were examined. We compared the CytoScan DNA data to RNA-seq data for the tissues

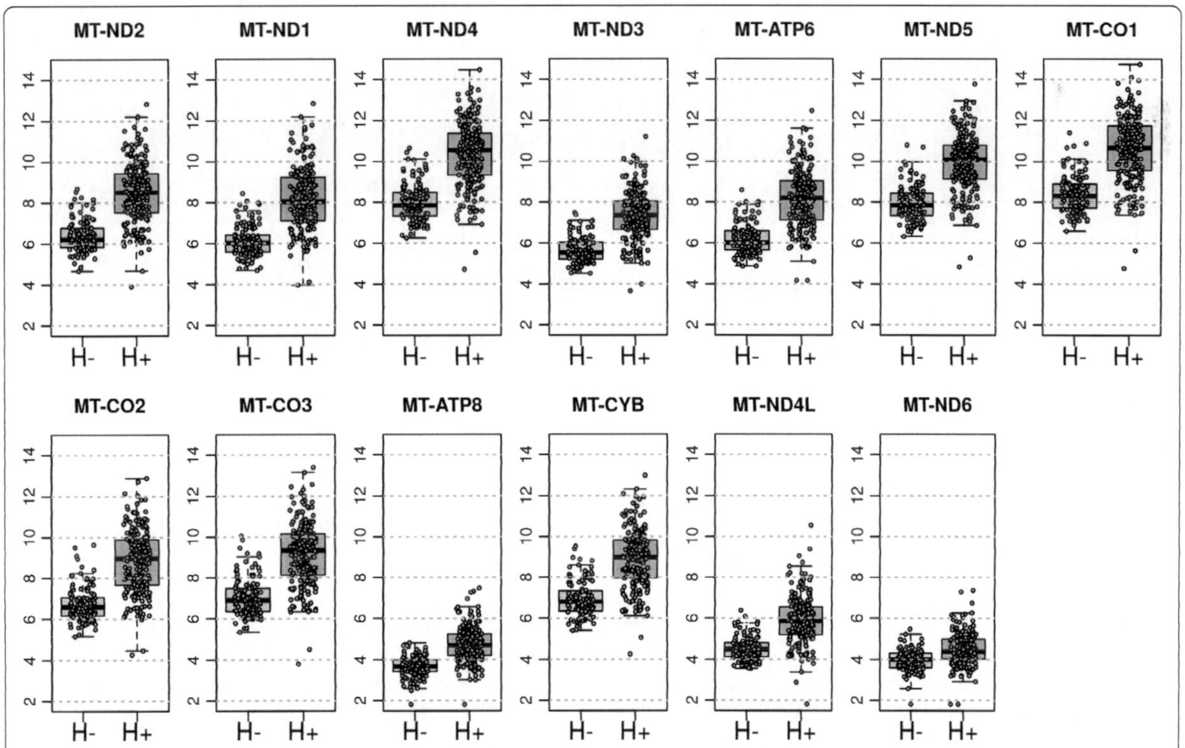

Fig. 2 Mitochondrial expression in cytopathology Hürthle positive (H+) and negative (H-) cohorts. Shown are the 13 mitochondrial transcripts present in RNA-seq data. Each transcript shows a boxplot of expression values for Hürthle negative (H-) and Hürthle positive (H+), respectively

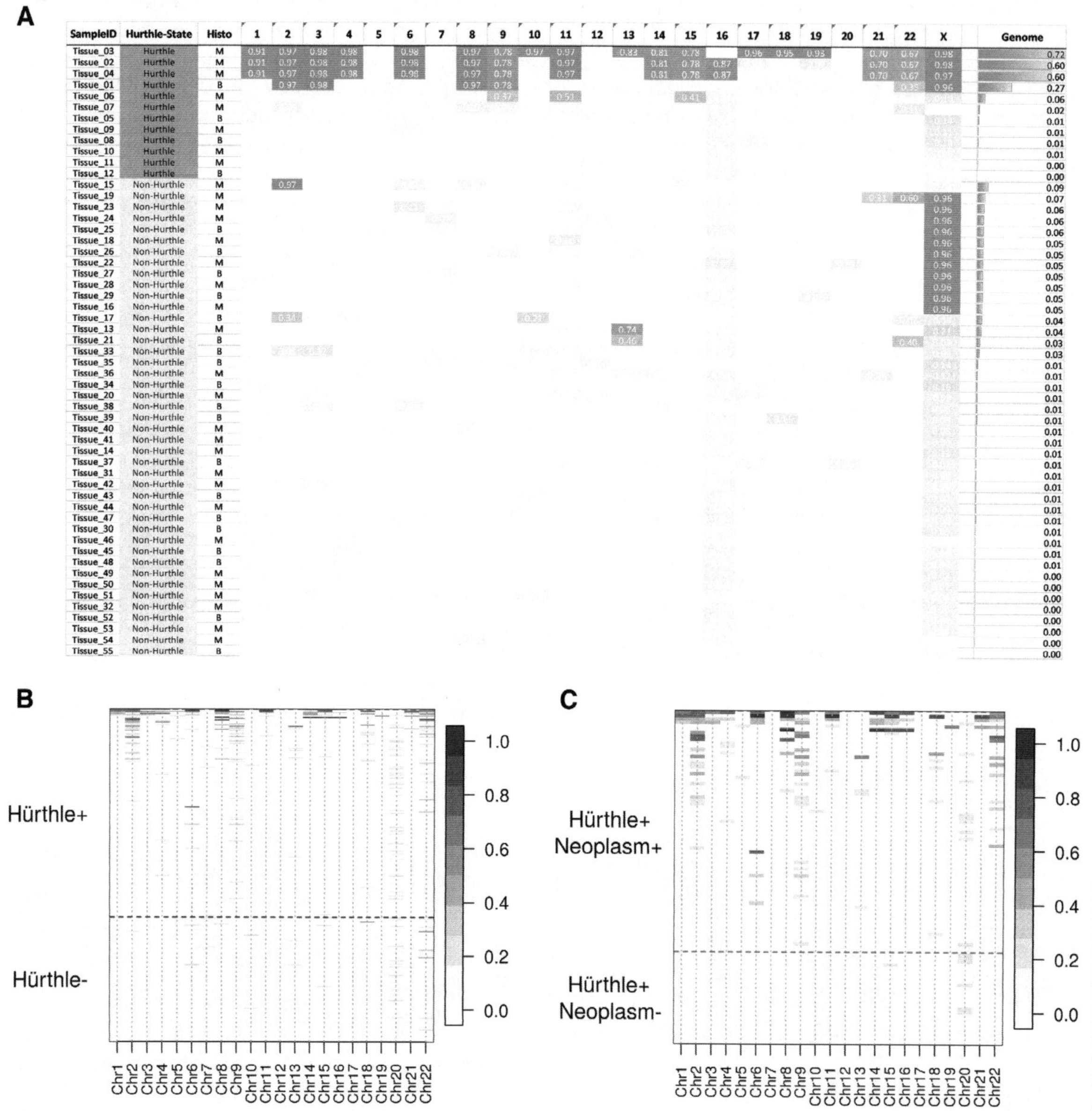

Fig. 3 Loss of Heterozygosity in Hürthle positive samples. **a** Affymetrix CytoScan array data on Hürthle tissues vs. non-Hürthle normal tissues. Each column represents one chromosome and each row represents one sample. The value in each cell is the proportion of the chromosome displaying LOH for a given sample. The samples are sorted in descending order for genome-wide LOH. **b** Chromosome-level LOH data from RNA-Seq Hürthle positive and Hürthle negative samples. Above the horizontal dashed line are Hürthle positive samples, below the dashed line are Hürthle negative samples. **c** Chromosome-level LOH data from RNA-Seq Hürthle positive, Neoplasm positive or Neoplasm negative samples. Above the horizontal dashed line are Hürthle positive, Neoplasm positive samples, below the dashed line are Hürthle positive, but Neoplasm negative samples. For both **(b)** and **(c)** the LOH scale is to the right, with red indicating more LOH

described above and found that the genome-wide LOH signal is similar between two platforms (Additional file 1: Figure S2). Examination of LOH data from RNA-seq at the chromosome level shows that LOH is predominantly observed in Hürthle cell positive samples (Fig. 3b), although not all Hürthle cell positive samples show elevated LOH. Next, the LOH signal was examined for neoplasm positive versus neoplasm negative in the context of Hürthle cell positive samples. Only Hürthle positive, neoplasm positive samples show extensive LOH across multiple chromosomes (Fig. 3c). These data show that the LOH signal can be detected in RNA-seq data, and that it is strongly correlated with Hürthle cell positive, neoplasm positive samples.

An expression-based classifier with mitochondrial features to separate Hürthle from non-Hürthle

We used expression levels, including mitochondrial features, to develop a classifier that separates Hürthle cell-positive FNAB from Hürthle cell-negative FNAB (Fig. 4a and b). Classifier score distribution was wider for the Hürthle cell positive case, with a much tighter distribution for Hürthle cell negative cases (Fig. 4c). Cross-validation performance showed a high AUC (0.966) with very high specificity (96.6%) and high (81.4%) sensitivity (Fig. 4d).

An expression-based classifier with expressed LOH features to separate Hürthle/neoplasm positive from negative

We developed a classifier from 27 neoplasm-negative and 71 neoplasm-positive samples, all Hürthle cell-positive (Fig. 5a and b). The neoplasm classifier score distribution is wider for the neoplasm positive samples than the neoplasm negative samples (Fig. 5c), with all but two LOH-positive samples classified as neoplasm positive (blue triangles, Fig. 5c). Cross-validation performance revealed a high AUC (0.946), with 96.3% specificity and 78.9% Sensitivity (Fig. 5d).

Integrating the Hürthle and neoplasm classifiers into the Afirma genomic sequencing classifier algorithm workflow

The Afirma GSC includes multiple classifiers to identify key factors related to thyroid nodule malignancies [34]. Figure 1a shows the overall Afirma GSC algorithm workflow Fig. 1b shows a description of the Hürthle cell and Neoplasm algorithm workflow.

We examined the HI and NI scores for the validation cohort, using histopathology as truth, revealing that most samples are correctly classified (Fig. 6a and b). All 9 HCC samples and 10/17 HCA samples are classified as Hürthle cell positive (Fig. 6b). One HCC sample was erroneously classified neoplasm negative (Fig. 6c). It is noteworthy that this sample was a false negative in both the GEC [15] and GSC [34] validation cohorts, and this sample was characterized by several rounds of discordant histopathological diagnoses. Five of ten HI positive HCA samples were classified neoplasm negative (Fig. 6c). Seven samples were rescued by the Hürthle-adjusted threshold (Fig. 6b). Six samples were benign, including 2 HCA, 2 BFN, 1 CLT, and 1 FA. One sample was malignant, thereby resulting in one false negative. Because the HCC false negative was called benign by the B/S classifier at the nominal threshold it was not subjected to the Hürthle-adjusted threshold. Combining the B/S classifier with the HI and NI resulted in a significant performance gain in Hürthle subtypes, with specificity increasing from 11.8% with the GEC to 58.8% with the GSC.

Discussion

Hürthle cells are follicular-derived epithelial cells with acidophilic cytoplasm. These oncocytes, or oxyphil cells, occur in multiple tissue types and are characterized by abundant, mitochondria-rich cytoplasm [19]. Known as Hürthle cells when found in the thyroid, this cellular transformation exists along a continuum, and in the thyroid is thought to result from high oxidative stress and reactive oxygen species [19]. How and if this process relates to tumorigenesis is unknown. Almost all benign and malignant thyroid neoplasms have a Hürthle cell counterpart, with the exception of anaplastic thyroid cancer [19]. Hürthle cells are associated with hyperplasia (multinodular adenomatous goiter), chronic inflammation (Hashimoto's thyroiditis), benign neoplasia (HCA) and malignant neoplasia (HCC) [19]. Hürthle cells have long challenged cytologists to accurately allocate such FNAB specimens into definitively benign or malignant categories [20]. To not miss cancer, most nodules not categorized as benign have historically undergone surgical resection. Thus, Hürthle cell cytology has contributed to the costs, morbidity, and occasional mortality of diagnostic thyroid surgery [16].

Here we present the detailed development of two classifiers to address the possibility that a cytologically indeterminate FNAB specimen would contain Hürthle cells. The two classifiers interface with a third classifier (the core GSC classifier), whose overall specificity was improved compared to its predecessor (the GEC), and together this trio of classifiers automatically assess the molecular signature of every FNAB specimen presented to them.

Key to understanding how the two Hürthle classifiers function is to recognize that Hürthle cells are found among both neoplastic and non-neoplastic processes, and that in the absence of neoplasia, the specimen should be benign. The HI classifier first determines if the specimen contains Hürthle cells. It does this using 1408 differentially expressed genes, including 13 mitochondrial transcripts. If this HI classifier is negative, then the core GSC classifier renders a benign or suspicious result (Fig. 1). If the HI classifier is positive for Hürthle cells, then the NI classifier determines if the specimen is neoplastic, using 2041 differentially expressed genes and a novel LOH statistic. If neoplastic, then the core GSC classifier renders a benign or suspicious result. However, if the NI suggests the absence of neoplasia, then the core GSC classifier uses an adjusted threshold that allows more of these samples to receive a GSC benign result. This adjusted threshold is justified based on the absence of neoplasia. While the NI sensitivity is high, it is not perfect, and a neoplastic specimen could be falsely deemed non-neoplastic. For this reason, the specimen is not automatically given a final benign result, but rather must still pass the core GSC classifier

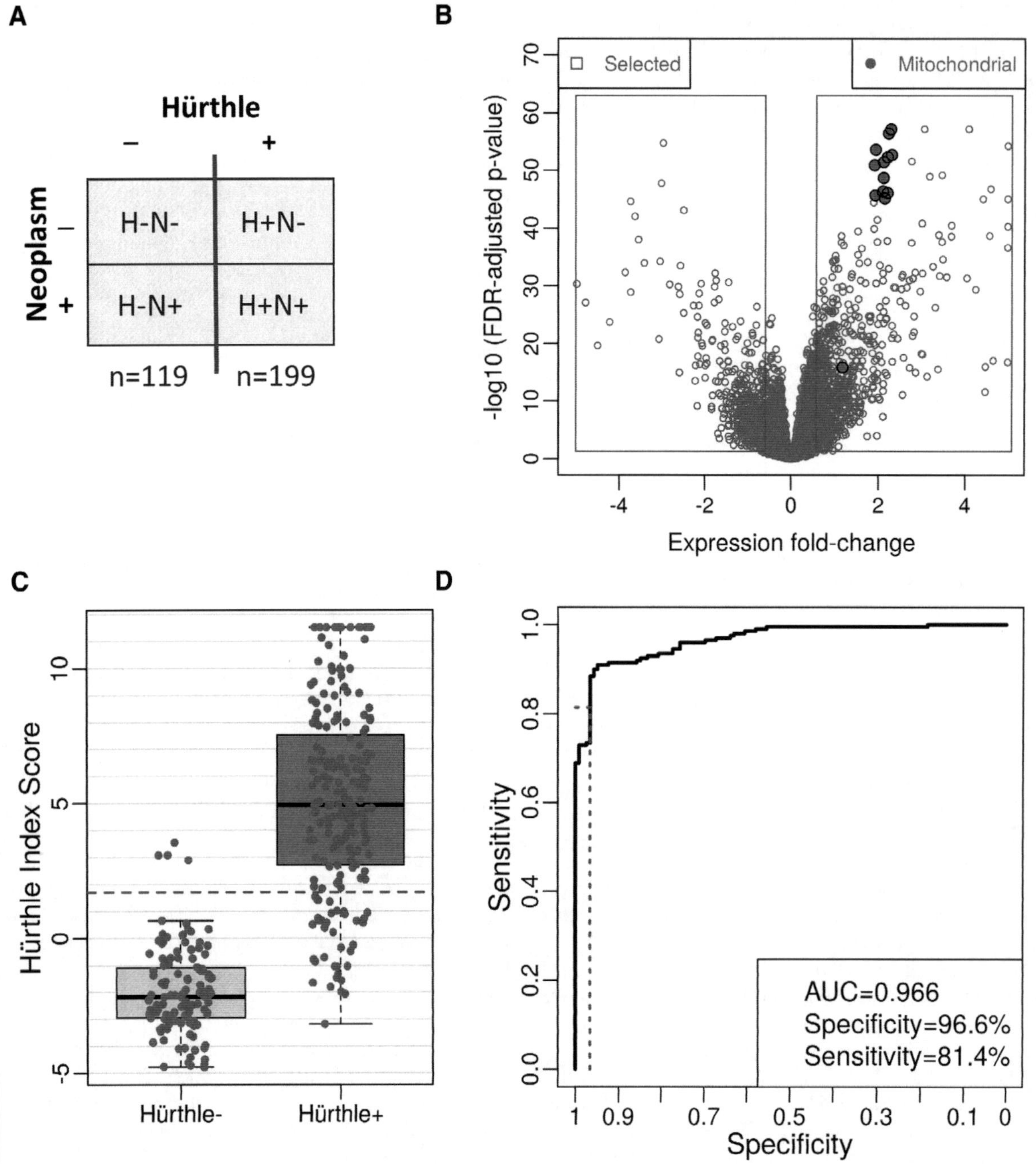

Fig. 4 Hürthle Classifier Cross Validation Performance. **a** Samples used in the Classifier development. Hürthle and Neoplasm labels are defined by cytopathology. **b** Volcano plots of differential expression. Fold-change (log2 scale) is plotted on the x-axis, and FDR-adjusted *p*-values are plotted on the y-axis. Mitochondrial genes are shown in purple. **c** Hürthle Index Score. Red dashed line indicates the cut-off for HI+ vs. HI-. The green boxplot represents the score for cytopathology Hürthle negative samples and the purple boxplot represents cytopathology Hürthle positive samples. **d** ROC curve showing classifier performance. The red-dashed lines indicate performance at the selected cutoff

with an adjusted (more tolerant) benign versus suspicious threshold, rather than having the threshold removed completely. Among the FNAB specimens derived from a prospective, multicenter, and blinded cohort of cytologically indeterminate thyroid nodules used to validate the GEC, and subsequently the GSC, an additional 4% of samples were given a final benign result due solely to the two Hürthle classifiers. These "rescued" samples represent

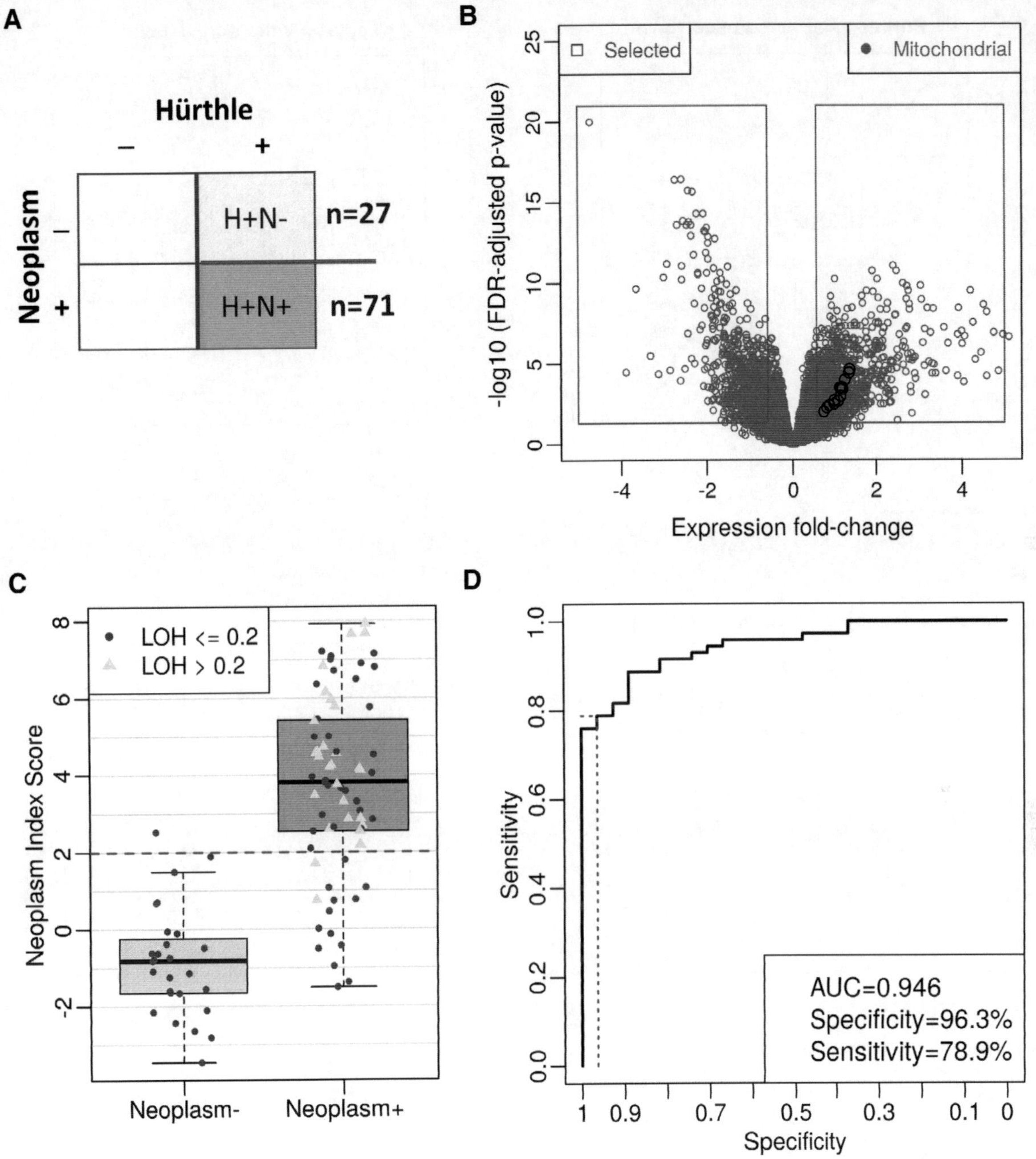

Fig. 5 Neoplasm Classifier Cross Validation Performance. **a** Samples used in the Classifier development. Hürthle and Neoplasm labels are defined by cytopathology. Note that all samples are Hürthle positive. **b** Volcano plots of differential expression. Fold-change (log2 scale) is plotted on the x-axis, and FDR-adjusted p-values are plotted on the y-axis. Mitochondrial genes are shown in purple. **c** Neoplasm Index Score. Red dashed line indicates the cut-off for NI+ vs. NI-. Cyan triangles indicate genome-wide LOH positive. **b** ROC curve showing classifier performance. The red-dashed lines indicate performance at the selected cutoff

19% of those deemed "Hürthle" by the HI classifier, and 44% of those deemed "not neoplastic" by the NI (Fig. 6b). These results are similar to what we have seen subsequently in the Veracyte CLIA laboratory. In its first 6203 cytologically indeterminate specimens, 5779 specimens passed all quality control requirements and received a final result, including 3% BRAFV600E classifier positive, 0.6% parathyroid classifier positive, 0.4% RET/PTC1 or

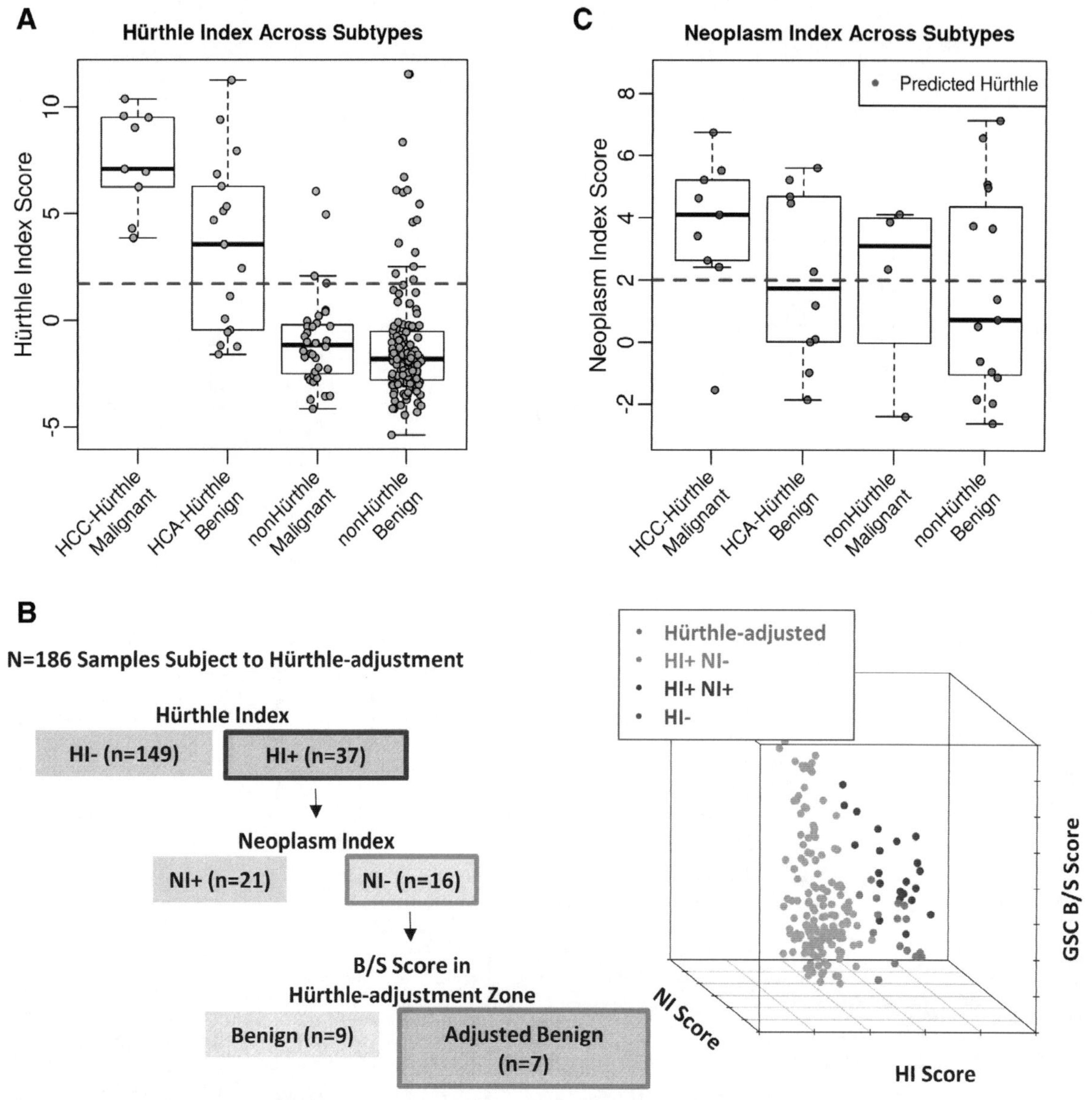

Fig. 6 Hürthle and Neoplasm Scores from the Afirma GSC Validation Cohort. **a** HI classifier scores for the validation cohort. Red dashed line indicates cutoff for HI+ vs. HI-. HI score distribution is plotted as boxplot with individual sample values for the four groups separately: Hürthle Cell Carcinoma "HCC", Hürthle Cell Adenoma "HCA", non-Hürthle histopathology malignant "nonHürthle Malignant", non-Hürthle histopathology benign "nonHürthle Benign". **b** The combination of the main B/S, Hürthle, and Neoplasm Indices. Gray points are HI- samples. Purple points are HI+, NI+. Green points are HI+, NI-. Blue dots are HI+, NI- samples that were subject to the adjusted cutoff. **c** NI classifier scores for HI+ samples from the validation cohort

RET/PTC3 fusion positive, and 0.3% medullary thyroid cancer classifier positive. Of the remaining 5552 specimens, 21% were positive by the HI classifier, 56% of them were deemed negative by the NI classifier which invoked the adjusted core GSC classifier threshold through which 14% of the HI positive but NI negative specimens received a final benign result due to the adjusted threshold (unpublished results). The impact of the coordinated trio of classifiers described here is now being reported among independent real-world clinical experiences. In one report, the authors state that 21% of their Bethesda III and IV cytologically indeterminate FNAB were Hürthle cell-dominant. Among them, only 18% of 107 cytologically Hürthle cell-containing FNAB received a benign result by GEC, compared to 67% of 18 Hürthle cell-containing FNAB with GSC ($p < 0.0001$) [39]. Similar, independent experiences support this substantial increase in the GSC benign call rate and accuracy of this classifier trio [40, 41].

The strength of this work includes the incorporation of expert cytopathology to establish "truth" labels for training the two Hürthle classifiers. The decision cut-off for the HI classifier results in very few non-Hürthle cell samples being falsely identified by the Hürthle cell index as "Hürthle", but a larger fraction of truly Hürthle cell samples being missed. These HI false negative samples lose the opportunity to be potentially rescued by the adjusted core GSC classifier threshold. We consider this a safer mode of failure. The high specificity of the NI classifier leads to high accuracy in samples called neoplasm positive; the lower sensitivity translates to approximately 30% of truly neoplastic (but not necessarily malignant) samples being falsely identified by as non-neoplastic and therefore invoke the more tolerant GSC classifier threshold. Despite this risk, we demonstrate in clinical validation a preserved high sensitivity among our Hürthle cell neoplasms when the entire system performance is considered. Perhaps the greatest limitation to differentiating benign from malignant Hürthle cell nodules is the imperfection of gold-standard surgical histology benign or malignant "truth" labels. Low concordance of truth labels among these specimens by expert surgical pathologists is well-known [29, 42], and even perfectly trained classifiers can only carry forward this imperfection, but they cannot correct it. Until improved "truth" labels are accepted, this barrier to improved classification will remain. It is noteworthy that we observed significant LOH among a fraction of both histologically benign and malignant Hürthle cell neoplasms. Whether or not the genomic instability of LOH should represent a pre-malignant, or carcinoma in-situ, is unknown and their natural history is unknown since these neoplasms were all surgically resected in our training and validation cohorts.

Conclusions

Three coordinate classifiers were developed to address cytologically indeterminate Hürthle cell thyroid FNAB using ML algorithms which harness the enriched genomic content from RNA-sequencing, including mRNA expression from differentially expressed nuclear and mitochondrial genes, and a novel LOH statistic. As an adjunct to clinical judgment, this trio of classifiers empowers physicians to reduce unnecessary diagnostic thyroid surgery among these most challenging cytological specimens: an action that directly improves patient safety, saves healthcare costs, and enhances quality of life.

Additional file

Additional file 1: Figure S1. Representative cytopathology images of Hürthle and Neoplasm samples. (a) Hürthle negative, Neoplasm negative sample. Bethesda II FNA. (b) Hürthle positive, Neoplasm negative sample. Bethesda II FNA. (c) Hürthle negative, Neoplasm positive sample. Bethesda IV FNA. (d) Hürthle positive, Neoplasm positive sample. Bethesda IV FNA. **Figure S2.** Comparing genome-wide LOH measured by CytoScan with LOH statistic from RNA-seq. **Figure S3.** Karyotype views of CytoScan data. Karyoview from Affymetrix's Chromosome Analysis Suite (ChAS) software visualizing LOH (magenta), Gain (dark blue), Gain-Mosaicism (light blue), Loss (red), Loss-Mosaicism (light pink) events across 24 chromosomes. Chromosomes are shown in order from 1-22, X, and Y. Events spanning > 100 Kb are shown, and are indicated as a small mark (four-sided star for LOH, upward and downward triangle for gain and loss, respectively). (a) Hürthle cell adenoma sample exhibiting all five types of large aberrations: LOH on chromosomes (chr) 2, 3, 8, 9, and X; Gain and Gain-Mosaicism on chr 5, 7, 18, 19, 20, and Loss and Loss-Mosaicism on chr 1, 2, 3, 8, 9, 11, 13, 21, 22. (b) Hürthle cell carcinoma sample exhibiting large LOH or Gain events: LOH and Gain events alternatively occur on almost all chromosomes, except chr 14 where both events are present. (c) Normal sample. No large aberrations observed; only small LOH events are scattered around across the genome with limited Gain (chr 4 and chr 14) and Gain-Mosaicism (chr 4). (PDF 1097 kb)

Abbreviations

AUC: Area under the curve; BFN: Benign follicular nodule; CLT: Chronic lymphocytic thyroiditis; CNV: Copy number variation; FA: Follicular adenoma; FNAB: Fine needle aspirate biopsy; GEC: Gene expression classifier; GSC: Genomic sequencing classifier; HCA: Hürthle cell adenoma; HCC: Hürthle cell carcinoma; HI: Hürthle cell index; LOH: Loss of heterozygosity; ML: Machine learning; MTC: Medullary thyroid carcinoma; NI: Hürthle cell Neoplasm Index; PTA: Parathyroid Adenoma; SNP: Single Nucleotide Polymorphism; SVM: Support vector machine; VAF: Variant Allele Frequency

Acknowledgments

We'd like to thank Hajime Matsuzaki, Duncan Whitney, Jing Lu and Ying Shen for their contributions to laboratory studies and algorithm development. We additionally thank Virginia LiVolsi and Ronald Ghossein for providing surgical histological diagnoses.

Funding

Funding for the publication of this article was provided by Veracyte, Inc.

Availability of data and materials

All data generated and supportive of the conclusions in this study are included in this published article.

About this supplement

This article has been published as part of *BMC Systems Biology Volume 13 Supplement 2, 2019: Selected articles from the 17th Asia Pacific Bioinformatics Conference (APBC 2019): systems biology*. The full contents of the supplement are available online at https://bmcsystbiol.biomedcentral.com/articles/supplements/volume-13-supplement-2.

Authors' contributions

YH, RTK, JB, PSW, SYK, GF, JH and GCK designed, performed, analyzed and interpreted the study. QYD, RMH, STT, and PMS performed clinical patient review and/or provided key study feedback. YH, RTK, JB, SYK, JH and GCK implemented the analysis, prepared figures and drafted the manuscript. YH, RTK, GF, JH, QYD, RMH, STT, PMS, and GCK edited and revised the manuscript. All authors have read and approved the final manuscript.

Ethics approval and consent to participate

Ethics approval was obtained prior to the initiation of the studies described in this manuscript. Study samples were collected according to protocols that were reviewed and approved by study site institutional review boards (IRB) or central IRBs (Liberty IRB, DeLand, Florida; now Chesapeake IRB, and Copernicus Group Independent Review Board, Cary, North Carolina).

Consent for publication
Not applicable.

Competing interests
YH, RTK, JB, SYK, GF, PSW, JH, and GCK are employed by Veracyte, Inc.

Publisher's Note

Author details
[1]Department of Research & Development, Veracyte, Inc, 6000 Shoreline Court, Suite 300, South San Francisco, CA 94080, USA. [2]Department of Surgery, Section of Endocrine Surgery, University of California San Francisco, San Francisco, CA, USA. [3]Department of Medical Affairs, Veracyte, Inc, South San Francisco, USA. [4]The Memorial Center for Integrative Endocrine Surgery, Hollywood, FL, USA. [5]Thyroid Cytopathology Partners, Austin, TX, USA. [6]Department of Pathology, Head and Neck Pathology Division, Massachusetts General Hospital and Harvard Medical School, Boston, MA, USA. [7]The Memorial Center for Integrative Endocrine Surgery, Weston, FL, USA. [8]The Memorial Center for Integrative Endocrine Surgery, Boca Raton, FL, USA.

Published: 5 April 2019

References
1. Guth S, Theune U, Aberle J, Galach A, Bamberger CM. Very high prevalence of thyroid nodules detected by high frequency (13 MHz) ultrasound examination. Eur J Clin Investig. 2009;39(8):699–706.
2. Mazzaferri EL. Management of a solitary thyroid nodule. N Engl J Med. 1993; 328(8):553–9.
3. Hegedus L. Clinical practice. The thyroid nodule. N Engl J Med. 2004;351(17): 1764–71.
4. Haugen BR, Alexander EK, Bible KC, Doherty GM, Mandel SJ, Nikiforov YE, et al. 2015 American Thyroid Association management guidelines for adult patients with thyroid nodules and differentiated thyroid Cancer: the American Thyroid Association guidelines task force on thyroid nodules and differentiated thyroid Cancer. Thyroid. 2016;26(1):1–133.
5. Thyroid Carcinoma. In: NCCN Clinical Practice Guidelines in Oncology. Version 1; 2018. http://www.nccn.org/professionals/physician_gls/pdf/thyroid.pdf. Accessed 2 Aug 2018.
6. Gharib H, Papini E, Garber JR, Duick DS, Harrell RM, Hegedus L, et al. American Association of Clinical Endocrinologists, American College of Endocrinology, and Associazione Medici Endocrinologi medical guidelines for clinical practice for the diagnosis and Management of Thyroid Nodules - 2016 update. Endocr Pract. 2016;22(5):622–39.
7. Ross DS. Diagnostic approach to and treatment of thyroid nodules. In: Cooper DS, Mulder JE, editors. UpToDate 2018. https://www.uptodate.com. Accessed 3 Aug 2018.
8. Sosa JA, Hanna JW, Robinson KA, Lanman RB. Increases in thyroid nodule fine-needle aspirations, operations, and diagnoses of thyroid cancer in the United States. Surgery. 2013;154(6):1420–7.
9. Sosa JA, Hanna HW, Lanman RB, Robinson KA, Ladenson PW. Increases in thyroid nodule fine needle aspirations, surgeries, and diagnoses of thyroid cancer in the United States. Chicago: American Association of Endocrine Surgeons Annual Scientific Meeting (Abstract); 2013.
10. Cancer Stat Facts: Thyroid Cancer. 2018. https://seer.cancer.gov/statfacts/html/thyro.html. Accessed 3 Aug 2018.
11. Cibas ES, Ali SZ. The 2017 Bethesda system for reporting thyroid cytopathology. Thyroid. 2017;27(11):1341–6.
12. Poller DN, Baloch ZW, Fadda G, Johnson SJ, Bongiovanni M, Pontecorvi A, et al. Thyroid FNA: new classifications and new interpretations. Cancer Cytopathol. 2016;124(7):457–66.
13. Cooper DS, Doherty GM, Haugen BR, Kloos RT, Lee SL, Mandel SJ, et al. Revised American Thyroid Association management guidelines for patients with thyroid nodules and differentiated thyroid cancer. Thyroid. 2009;19(11): 1167–214.
14. Krauss EA, Mahon M, Fede JM, Zhang L. Application of the Bethesda classification for thyroid fine-needle aspiration: institutional experience and meta-analysis. Arch Pathol Lab Med. 2016;140(10):1121–31.
15. Alexander EK, Kennedy GC, Baloch ZW, Cibas ES, Chudova D, Diggans J, et al. Preoperative diagnosis of benign thyroid nodules with indeterminate cytology. N Engl J Med. 2012;367(8):705–15.
16. Kloos RT. Molecular profiling of thyroid nodules: current role for the Afirma gene expression classifier on clinical decision making. Mol Imaging Radionucl Ther. 2017;26(Suppl 1):36–49.
17. Yazgan A, Balci S, Dincer N, Kiyak G, Tuzun D, Ersoy R, et al. Hurthle cell presence alters the distribution and outcome of categories in the Bethesda system for reporting thyroid cytopathology. Cytopathology. 2014;25(3):185–9.
18. Olson MT, Clark DP, Erozan YS, Ali SZ. Spectrum of risk of malignancy in subcategories of 'atypia of undetermined significance'. Acta Cytol. 2011; 55(6):518–25.
19. Maximo V, Lima J, Prazeres H, Soares P, Sobrinho-Simoes M. The biology and the genetics of Hurthle cell tumors of the thyroid. Endocr Relat Cancer. 2012;19(4):R131–47.
20. Cannon J. The significance of hurthle cells in thyroid disease. Oncologist. 2011;16(10):1380–7.
21. VandenBussche CJ, Adams C, Ali SZ, Olson MT. Cytotechnologist performance for screening Hurthle cell atypia in indeterminate thyroid fine-needle aspirates. Acta Cytol. 2015;59(5):377–83.
22. Straccia P, Santoro A, Rossi ED, Brunelli C, Mosseri C, Musarra T, et al. Incidence, malignancy rates of diagnoses and cyto-histological correlations in the new Italian reporting system for thyroid cytology: an institutional experience. Cytopathology. 2017;28(6):503–8.
23. Straccia P, Rossi ED, Bizzarro T, Brunelli C, Cianfrini F, Damiani D, et al. A meta-analytic review of the Bethesda system for reporting thyroid cytopathology: has the rate of malignancy in indeterminate lesions been underestimated? Cancer Cytopathol. 2015;123(12):713–22.
24. Ganly I, Makarov V, Deraje S, Dong Y, Reznik E, Seshan V, et al. Integrated genomic analysis of Hurthle cell Cancer reveals oncogenic drivers, recurrent mitochondrial mutations, and unique chromosomal landscapes. Cancer Cell. 2018;34(2):256–70 e5.
25. Gopal RK, Kubler K, Calvo SE, Polak P, Livitz D, Rosebrock D, et al. Widespread chromosomal losses and mitochondrial DNA alterations as genetic drivers in Hurthle cell Carcinoma. Cancer Cell. 2018;34(2):242–55 e5.
26. Ganly I, Ricarte Filho J, Eng S, Ghossein R, Morris LG, Liang Y, et al. Genomic dissection of Hurthle cell carcinoma reveals a unique class of thyroid malignancy. J Clin Endocrinol Metab. 2013;98(5):E962–72.
27. Conti L, Vatrano S, Bertero L, Masu L, Pacchioni D, Daniele L, et al. Mitochondrial DNA "common deletion" in post-fine needle aspiration infarcted oncocytic thyroid tumors. Hum Pathol. 2017;69:23–30.
28. Walsh PS, Wilde JI, Tom EY, Reynolds JD, Chen DC, Chudova DI, et al. Analytical performance verification of a molecular diagnostic for cytology-indeterminate thyroid nodules. J Clin Endocrinol Metab. 2012; 97(12):E2297–306.
29. Cibas ES, Baloch ZW, Fellegara G, LiVolsi VA, Raab SS, Rosai J, et al. A prospective assessment defining the limitations of thyroid nodule pathologic evaluation. Ann Intern Med. 2013;159(5):325–32.
30. Brauner E, Holmes BJ, Krane JF, Nishino M, Zurakowski D, Hennessey JV, et al. Performance of the Afirma gene expression classifier in Hurthle cell thyroid nodules differs from other indeterminate thyroid nodules. Thyroid. 2015;25(7):789–96.
31. Lastra RR, Pramick MR, Crammer CJ, LiVolsi VA, Baloch ZW. Implications of a suspicious afirma test result in thyroid fine-needle aspiration cytology: an institutional experience. Cancer Cytopathol. 2014;122(10):737–44.
32. Harrell RM, Bimston DN. Surgical utility of Afirma: effects of high Cancer prevalence and Oncocytic cell types in patients with indeterminate thyroid cytology. Endocr Pract. 2013. https://doi.org/10.4158/EP13330.OR:1-16.
33. Wu JX, Young S, Hung ML, Li N, Yang SE, Cheung DS, et al. Clinical factors influencing the performance of gene expression classifier testing in indeterminate thyroid nodules. Thyroid. 2016;26(7):916–22.
34. Patel KN, Angell TE, Babiarz J, Barth NM, Blevins T, Duh QY, et al. Performance of a genomic sequencing classifier for the preoperative diagnosis of Cytologically indeterminate thyroid nodules. JAMA Surg. 2018; 153(9):817–24.
35. Robinson MD, McCarthy DJ, Smyth GK. edgeR: a Bioconductor package for differential expression analysis of digital gene expression data. Bioinformatics. 2010;26(1):139–40.
36. Kuhn M, contributions from Wing J, Weston S, Williams A, Keefer C, Engelhardt A, et al. caret: Classification and Regression Training. R package version 6.0–68. 2016. https://CRAN.R-project.org/package=caret. Accessed 5 Aug 2018.

37. Kurelac I, de Biase D, Calabrese C, Ceccarelli C, Ng CK, Lim R, et al. High-resolution genomic profiling of thyroid lesions uncovers preferential copy number gains affecting mitochondrial biogenesis loci in the oncocytic variants. Am J Cancer Res. 2015;5(6):1954–71.
38. Wada N, Duh QY, Miura D, Brunaud L, Wong MG, Clark OH. Chromosomal aberrations by comparative genomic hybridization in hurthle cell thyroid carcinomas are associated with tumor recurrence. J Clin Endocrinol Metab. 2002;87(10):4595–601.
39. Harrell RM, Eyerly-Webb SA, Golding AC, Edwards CM, Bimston DN. Statistical Comparison of Afirma GSC and Afirma GEC Outcomes in a Community Endocrine Surgical Practice: Early Findings. Endocr Pract. 2018. https://doi.org/10.4158/EP-2018-0395.
40. Marqusee E, Heller HT, Cibas ES, Barletta JA, Kim MI, Krane JF, et al. An independent performance comparison of the Afirma genomic sequencing classifier (GSC) and gene expression classifier (GEC) for cytologically indeterminate thyroid nodules. Thyroid. 2018;28(Supp 1):A-189.
41. Nasr CE, Krishnamurthy VD. Automatic identification of Hurthle cells and Hurthle neoplasia by two classifiers that coordinate with the core Afirma GSC classifier. Thyroid. 2018;28(Supp 1):A-196.
42. Su HK, Wenig BM, Haser GC, Rowe ME, Asa SL, Baloch Z, et al. Inter-observer variation in the pathologic identification of minimal Extrathyroidal extension in papillary thyroid Carcinoma. Thyroid. 2016;26(4):512–7.

KC *et al. BMC Systems Biology* 2019, **13**(Suppl 2):38
https://doi.org/10.1186/s12918-019-0694-y

BMC Systems Biology

RESEARCH Open Access

GNE: a deep learning framework for gene network inference by aggregating biological information

Kishan KC[1*], Rui Li[1], Feng Cui[2], Qi Yu[1] and Anne R. Haake[1]

From The 17th Asia Pacific Bioinformatics Conference (APBC 2019)
Wuhan, China. 14–16 January 2019

Abstract

Background: The topological landscape of gene interaction networks provides a rich source of information for inferring functional patterns of genes or proteins. However, it is still a challenging task to aggregate heterogeneous biological information such as gene expression and gene interactions to achieve more accurate inference for prediction and discovery of new gene interactions. In particular, how to generate a unified vector representation to integrate diverse input data is a key challenge addressed here.

Results: We propose a scalable and robust deep learning framework to learn embedded representations to unify known gene interactions and gene expression for gene interaction predictions. These low- dimensional embeddings derive deeper insights into the structure of rapidly accumulating and diverse gene interaction networks and greatly simplify downstream modeling. We compare the predictive power of our deep embeddings to the strong baselines. The results suggest that our deep embeddings achieve significantly more accurate predictions. Moreover, a set of novel gene interaction predictions are validated by up-to-date literature-based database entries.

Conclusion: The proposed model demonstrates the importance of integrating heterogeneous information about genes for gene network inference. GNE is freely available under the GNU General Public License and can be downloaded from GitHub (https://github.com/kckishan/GNE).

Keywords: Gene interaction networks, Gene expression, Network embedding, Heterogeneous data integration, Deep learning

Background

A comprehensive study of gene interactions (GIs) provides means to identify the functional relationship between genes and their corresponding products, as well as insights into underlying biological phenomena that are critical to understanding phenotypes in health and disease conditions [1–3]. Since advancements in measurement technologies have led to numerous high-throughput datasets, there is a great value in developing efficient computational methods capable of automatically extracting and aggregating meaningful information from heterogeneous datasets to infer gene interactions.

Although a wide variety of machine learning models have been developed to analyze high-throughput datasets for GI prediction [4], there are still some major challenges, such as efficient analysis of large heterogeneous datasets, integration of biological information, and effective feature engineering. To address these challenges, we propose a novel deep learning framework to integrate diverse biological information for GI network inference.

Our proposed method frames GI network inference as a problem of network embedding. In particular, we represent gene interactions as a network of genes and their interactions and create a deep learning framework to automatically learn an informative representation which

*Correspondence: kk3671@rit.edu
[1] Golisano College of Computing and Information Sciences, Rochester Institute of Technology, 20 Lomb Memorial Drive, 14623 Rochester, New York, USA
Full list of author information is available at the end of the article

integrates both the topological property and the gene expression property. A key insight behind our gene network embedding method is the "guilt by association" assumption [5], that is, genes that are co-localized or have similar topological roles in the interaction network are likely to be functionally correlated. This insight not only allows us to discover similar genes and proteins but also to infer the properties of unknown ones. Our network embedding generates a lower-dimensional vector representation of the gene topological characteristics. The relationships between genes including higher-order topological properties are captured by the distances between genes in the embedding space. The new low-dimensional representation of a GI network can be used for various downstream tasks, such as gene function prediction, gene interaction prediction, and gene ontology reconstruction [6].

Furthermore, since the network embedding method can only preserve the topological properties of a GI network, and fails to generalize for genes with no interaction information, our scalable deep learning method also integrates heterogeneous gene information, such as expression data from high throughput technologies, into the GI network inference. Our method projects genes with similar attributes closer to each other in the embedding space, even if they may not have similar topological properties. The results show that by integrating additional gene information in the network embedding process, the prediction performance is improved significantly.

GI prediction is a long-standing problem. The proposed machine learning methods include statistical correlation, mutual information [7], dimensionality reduction [8], and network-based methods (e.g. common neighborhood, network embedding) [4, 9]. Among these methods, some methods such as statistical correlation and mutual information consider only gene expression whereas other methods use only topological properties to predict GIs.

Network-based methods have been proposed to leverage topological properties of GI networks [10]. Neighborhood-based methods quantify the proximity between genes, based on common neighbors in GI network [11]. The proximity scores assigned to a pair of genes rely on the number of neighbors that the pair has in common. Adjacency matrix, representing the interaction network, or proximity matrix, obtained from neighborhood-based methods, are processed with network embedding methods to learn embeddings that preserve the structural properties of the network. Structure-preserving network embedding methods such as Isomap [12] are proposed as a dimensionality reduction technique. Since the goal of these methods is solely for graph reconstruction, the embedding space may not be suitable for GI network inference. Besides, these methods construct the graphs from the data features where proximity between genes is well defined in the original feature space [9]. On the other hand, in GI networks, gene proximities are not explicitly defined, and they depend on the specific analytic tasks and application scenarios.

Our deep learning method allows incorporating gene expression data with GI network topological structure information to preserve both topological and attribute proximity in the low-dimensional representation for GI predictions. Moreover, the scalable architecture enables us to incorporate additional attributes. Topological properties of GI network and expression profiles are transformed into two separate embeddings: ID embedding (which preserves the topological structure proximity) and attribute embedding (which preserves the attribute proximity) respectively. With a multilayer neural network, we then aggregate the complex statistical relationships between topology and attribute information to improve GI predictions.

In summary, our contributions are as follows:

- We propose a novel deep learning framework to learn lower dimensional representations while preserving topological and attribute proximity of GI networks.
- We evaluate the prediction performance on the datasets of two organisms based on the embedded representation and achieve significantly better predictions than the strong baselines.
- Our method can predict new gene interactions which are validated on an up-to-date GI database.

Methods

Preliminaries

We formally define the problem of gene network inference as a network embedding problem using the concepts of topological and attribute proximity as demonstrated in Fig. 1.

Definition 1 (Gene network) *Gene network can be represented as a network structure, which represents the interactions between genes within an organism. The interaction between genes corresponds to either a physical interaction through their gene products, e.g., proteins, or one of the genes alters or affects the activity of other gene of interest. We denote gene network as* $G = (V, E, A)$ *where* $V = \{v_i\}$ *denotes genes or proteins,* $E = \{e_{ij}\}$ *denotes edges that correspond to interactions between genes* v_i *and* v_j*, and* $A = \{A_i\}$ *represents the attributes of gene* v_i*. Edge* e_{ij} *is associated with a weight* $w_{ij} \geq 0$ *indicating the strength of the connection between gene* v_i *and* v_j*. If gene* v_i *and* v_j *is not linked by an edge,* $w_{ij} = 0$*. We name interactions with* $w_{ij} > 0$ *as positive interactions and* $w_{ij} = 0$ *as negative interactions. In this paper, we consider weights* w_{ij} *to be binary, indicating whether genes* v_i *and* v_j *interact or not.*

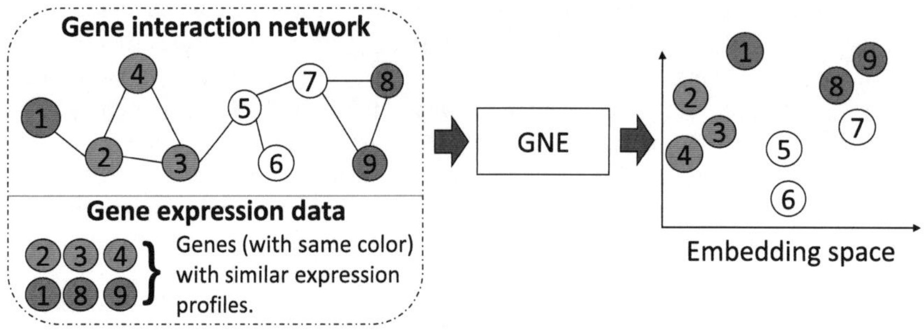

Fig. 1 An illustration of gene network embedding (GNE). GNE integrates gene interaction network and gene expression data to learn a lower-dimensional representation.The nodes represent genes, and the genes with the same color have similar expression profiles. GNE groups genes with similar network topology, which are connected or have a similar neighborhood in the graph, and attribute similarity (similar expression profiles) in the embedded space

Genes directly connected with a gene v_i in gene network denote the local network structure of gene v_i. We define local network structures as the first-order proximity of a gene.

Definition 2 (First-order proximity) *The first-order proximity in a gene network is the pairwise interactions between genes. Weight w_{ij} indicates the first-order proximity between gene v_i and v_j. If there is no interaction between gene v_i and v_j, their first-order proximity w_{ij} is 0.*

Genes are likely to be involved in the same cellular functions if they are connected in the gene network. On the other hand, even if two genes are not connected, they may be still related in some cellular functions. This indicates the need for an additional notion of proximity to preserve the network structure. Studies suggest that genes that share a similar neighborhood are also likely to be related [6]*. Thus, we introduce second-order proximity that characterizes the global network structure of the genes.*

Definition 3 (Second-order proximity) *Second order proximity denotes the similarity between the neighborhood of genes. Let $N_i = \{s_{i,1}, \ldots, s_{i,i-1}, s_{i,i+1}, \ldots, s_{i,M-1}\}$ denotes the first-order proximity of gene v_i, where $s_{i,j}$ is w_{ij} if there is direct connection between gene v_i and gene v_j, otherwise 0. Then, the second order proximity is the similarity between N_i and N_j. If there is no path to reach gene v_i from gene v_j, the second proximity between these genes is 0.*

Integrating first-order and second-order proximities simultaneously can help to preserve the topological properties of the gene network. To generate a more comprehensive representation of the genes, it is crucial to integrate gene expression data as gene attributes with their topological properties. Besides preserving topological properties, gene expression provides additional information to predict the network structure.

Definition 4 (Attribute proximity) *Attribute proximity denotes the similarity between the expression of genes.*

We thus investigate both topological and attribute proximity for gene network embedding, which is defined as follows:

Definition 5 (Gene network embedding) *Given a gene network denoted as $G = (V, E, A)$, gene network embedding aims to learn a function f that maps gene network structure and their attribute information to a d-dimensional space where a gene is represented by a vector $y_i \in \mathbb{R}^d$ where $d \ll M$. The low dimensional vectors y_i and y_j for genes v_i and v_j preserve their relationships in terms of the network topological structure and attribute proximity.*

Gene network embedding (GNE) model

Our deep learning framework as shown in Fig. 2 jointly utilizes gene network structure and gene expression data to learn a unified representation for the genes. Embedding of a gene network projects genes into a lower dimensional space, known as the embedding space, in which each gene is represented by a vector. The embeddings preserve both the gene network structure and statistical relationships of gene expression. We list the variables to specify our framework in Table 1.

Gene network structure modeling

GNE framework preserves first-order and second-order proximity of genes in the gene network. The key idea of network structure modeling is to estimate the pairwise proximity of genes in terms of the network structure. If two genes are connected or share similar neighborhood genes, they tend to be related and should be placed closer to each other in the embedding space. Inspired by the Skip-gram model [13], we use one hot encoded representation to represent topological information of a gene. Each

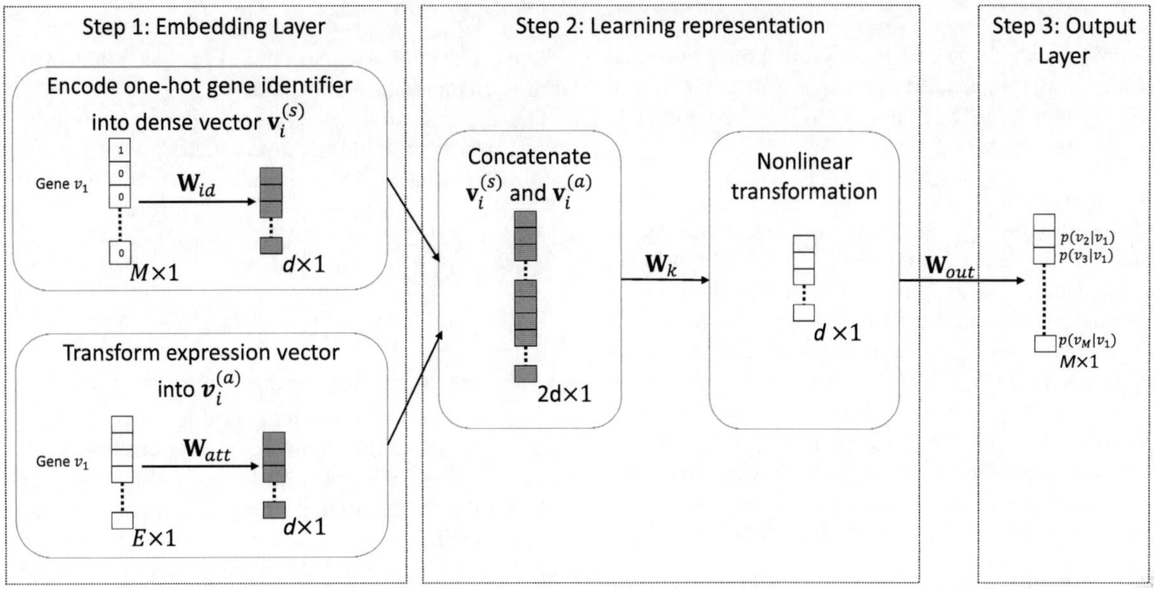

Fig. 2 Overview of Gene Network Embedding (GNE) Framework for gene interaction prediction. On the left,one-hot encoded representation of gene is encoded to dense vector $\mathbf{v}_i^{(s)}$ of dimension $d \times 1$ which captures topological properties and expression vector of gene is transformed to $\mathbf{v}_i^{(a)}$ of dimension $d \times 1$ which aggregates the attribute information (Step 1). Next, concatenation of two embedded vectors (creates vector with dimension $2d \times 1$) allows to combine strength of both network structure and attribute modeling. Then, nonlinear transformation of concatenated vector enables GNE to capture complex statistical relationships between network structure and attribute information and learn better representations (Step 2). Finally, these learned representation of dimension $d \times 1$ is transformed into a probability vector of length $M \times 1$ in output layer, which contains the predictive probability of gene v_i to all the genes in the network. Conditional probability $p(v_j|v_i)$ on output layer indicates the likelihood that gene v_j is connected with gene v_i (Step 3)

gene v_i in the network is represented as an M-dimensional vector where only the ith component of the vector is 1.

To model topological similarity, we define the conditional probability of gene v_j on gene v_i using a softmax function as:

$$p(v_j|v_i) = \frac{\exp(f(v_i, v_j))}{\sum_{j'=1}^{M} \exp(f(v_i, v_{j'}))} \tag{1}$$

Table 1 Terms and Notations

Symbol	Definitions
M	Total number of genes in gene network
E	Number of expression values for each gene
N_i	Set of the neighbor genes of gene v_i
$\mathbf{v}_i^{(s)}$	Topological representation of gene v_i
$\mathbf{v}_i^{(a)}$	Attribute representation of gene v_i
$\widetilde{\mathbf{v}}_i$	Neighborhood representation of gene v_i
$\mathbf{v}_i$	Concatenated representation of topological properties and expression data
k	Number of hidden layers to transform concatenated representation into embedding space
$\mathbf{h}^{(k)}$	Output of kth hidden layer
$\mathbf{W}_k$	Weight matrix for k^{th} hidden layer
$\mathbf{W}_{id}$	Weight matrix for topological structure embedding
$\mathbf{W}_{att}$	Weight matrix for attribute embedding
$\mathbf{W}_{out}$	Weight matrix for output layer

which measures the likelihood of gene v_i being connected with v_j. Let function f represents the mapping of two genes v_i and v_j to their estimated proximity score. Let $p(N|v)$ be the likelihood of observing a neighborhood N for a gene v. By assuming conditional independence, we can factorize the likelihood so that the likelihood of observing a neighborhood gene is independent of observing any other neighborhood gene, given a gene v_i:

$$p(N_i|v_i) = \prod_{v_j \in N_i} p(v_j|v_i) \tag{2}$$

where N_i represents the neighborhood genes of the gene v_i. Global structure proximity for a gene v_i can be preserved by maximizing the conditional probability over all genes in the neighborhood. Hence, we can define the likelihood function that preserve global structure proximity as:

$$L = \prod_{i=1}^{M} p(N_i|v_i) = \prod_{i=1}^{M} \prod_{v_j \in N_i} p(v_j|v_i) \tag{3}$$

Let $\mathbf{v}_i^{(s)}$ denotes the dense vector generated from one-hot gene ID vector, which represents topological information of that gene. GNE follows direct encoding methods [13, 14] to map genes to vector embeddings, simply known as embedding lookup:

$$\mathbf{v}_i^{(s)} = \mathbf{W}_{id} v_i \tag{4}$$

where $\mathbf{W}_{id} \in \mathbb{R}^{d \times M}$ is a matrix containing the embedding vectors for all genes and $v_i \in \mathbb{I}_M$ is a one-hot indicator vector indicating the column of $\mathbf{W}_{id}$ corresponding to gene v_i.

Gene expression modeling

GNE encodes the expression data from microarray experiments to the dense representation using a non-linear transformation. The amount of mRNA produced during transcription measured over a number of experiments helps to identify similarly expressed genes. Since expression data have inherent noise [15], transforming expression data using a non-linear transformation can be helpful to uncover the underlying representation. Let x_i be the vector of expression values of gene v_i measured over E experiments. Using non-linear transformation, we can capture the non-linearities of expression data of gene v_i as:

$$\mathbf{v}_i^{(a)} = \delta_a(\mathbf{W}_{att} \cdot x_i) \tag{5}$$

where $\mathbf{v}_i^{(a)}$ represents the lower dimensional attribute representation vector for gene v_i. $\mathbf{W}_{att}$, and δ_a represents the weight matrix, and activation function of attribute transformation layer respectively.

We use the deep model to approximate the attribute proximity by capturing complex statistical relationships between attributes and introducing non-linearities, similar to structural embedding.

GNE integration

GNE models the integration of network structure and attribute information to learn more comprehensive embeddings for gene networks. GNE takes two inputs: one for topological information of a gene as one hot gene ID vector and another for its expression as an attribute vector. Each input is encoded to its respective embeddings. One hot representation for a gene v_i is projected to the dense vector $\mathbf{v}_i^{(s)}$ which captures the topological properties. Non-linear transformation of attribute vector generates compact representation vector $\mathbf{v}_i^{(a)}$. Previous work [16] combines heterogeneous information using the late fusion approach. However, the late fusion approach is the approach of learning separate models for heterogeneous information and integrating the representations learned from separate models. On the other hand, the early fusion combines heterogeneous information and train the model on combined representations [17]. We thus propose to use the early fusion approach to combine them by concatenating. As a result, learning from topological and attribute information can complement each other, allowing the model to learn their complex statistical relationships as well. Embeddings from topological and attribute information are concatenated into a vector as:

$$\mathbf{v}_i = \left[\mathbf{v}_i^{(s)} \quad \lambda \mathbf{v}_i^{(a)}\right] \tag{6}$$

where λ is the importance of gene expression information relative to topological information.

The concatenated vectors are fed into a multilayer perceptron with k hidden layers. The hidden representations from each hidden layer in GNE are denoted as $\mathbf{h}_i^{(0)}, \mathbf{h}_i^{(1)}, \ldots\ldots, \mathbf{h}_i^{(k)}$, which can be defined as :

$$\begin{aligned} \mathbf{h}_i^{(0)} &= \delta\left(\mathbf{W}_0 \mathbf{v}_i + b^{(0)}\right), \\ \mathbf{h}_i^{(k)} &= \delta_k\left(\mathbf{W}_k \mathbf{h}_i^{(k-1)} + b^{(k)}\right) \end{aligned} \tag{7}$$

where δ_k represents the activation function of layer k. $\mathbf{h}_i^{(0)}$ represents initial representation and $\mathbf{h}_i^{(k)}$ represents final representation of the input gene v_i. Transformation of input data using multiple non-linear layers has shown to improve the representation of input data [18]. Moreover, stacking multiple layers of non-linear transformations can help to learn high-order statistical relationships between topological properties and attributes.

At last, final representation $\mathbf{h}_i^{(k)}$ of a gene v_i from the last hidden layer is transformed to probability vector, which contains the conditional probability of all other genes to v_i:

$$\mathbf{o}_i = \left[p(v_1|v_i), p(v_2|v_i), \ldots, p(v_M|v_i)\right] \tag{8}$$

where $p(v_j|v_i)$ represents the probability of gene v_i being related to gene v_j and $\mathbf{o}_i$ represents the output probability vector with the conditional probability of gene v_i being connected to all other genes.

Weight matrix $\mathbf{W}_{out}$ between the last hidden layer and the output layer corresponds to the abstractive representation of neighborhood of genes. A jth row from $\mathbf{W}_{out}$ refers to the compact representation of neighborhood of gene v_j, which can be denoted as $\widetilde{\mathbf{v}}_j$. The proximity score between gene v_i and v_j can be defined as:

$$f(v_i, v_j) = \widetilde{\mathbf{v}}_j \cdot \mathbf{h}_i^{(k)} \tag{9}$$

which can be replaced into Eq. 1 to calculate the conditional probability:

$$p(v_j|v_i) = \frac{\exp\left(\widetilde{\mathbf{v}}_j \cdot \mathbf{h}_i^{(k)}\right)}{\sum_{j'=1}^{M} \exp\left(\widetilde{\mathbf{v}}_{j'} \cdot \mathbf{h}_i^{(k)}\right)} \tag{10}$$

Our model learns two latent representations $\mathbf{h}_i^{(k)}$ and $\widetilde{\mathbf{v}}_i$ for a gene v_i where $\mathbf{h}_i^{(k)}$ is the representation of gene as a node and $\widetilde{\mathbf{v}}_i$ is the representation of the gene v_i as a neighbor. Neighborhood representation $\widetilde{\mathbf{v}}_i$ can be combined with node representation $\mathbf{h}_i^{(k)}$ by addition [19, 20] to get final representation for a gene as:

$$y_i = \mathbf{h}_i^{(k)} + \widetilde{\mathbf{v}}_i \tag{11}$$

which returns us better performance results.

For an edge connecting gene v_i and v_j, we create feature vector by combining embeddings of those genes using Hadamard product. Empirical evaluation shows features created with Hadamard product gives better performance over concatenation [14]. Then, we train a logistic classifier on these features to classify whether genes v_i and v_j interact or not.

Parameter optimization

To optimize GNE, the goal is to maximize objective function mentioned in Eq. 10 as a function of all parameters. Let Θ be the parameters of GNE which includes $\{\mathbf{W}_{id}, \mathbf{W}_{att}, \mathbf{W}_{out}, \Theta_h\}$ and Θ_h represents weight matrices $\mathbf{W}_k$ of hidden layers. We train our model to maximize the objective function with respect to all parameters Θ :

$$\underset{\Theta}{\operatorname{argmax}} \left[\sum_{i=1}^{M} \sum_{v_j \in N_i} \log \frac{\exp\left(\widetilde{\mathbf{v}}_j \cdot \mathbf{h}_i^{(k)}\right)}{\sum_{j'=1}^{M} \exp\left(\widetilde{\mathbf{v}}_{j'} \cdot \mathbf{h}_i^{(k)}\right)} \right] \tag{12}$$

Maximizing this objective function with respect to Θ is computationally expensive, which requires the calculation of partition function $\sum_{j'=1}^{M} \exp\left(\widetilde{\mathbf{v}}_{j'} \cdot \mathbf{h}_i^{(k)}\right)$ for each gene. To calculate a single probability, we need to aggregate all genes in the network. To address this problem, we adopt the approach of negative sampling [13] which samples the negative interactions, interactions with no evidence of their existence, according to some noise distribution for each edge e_{ij}. This approach allows us to sample a small subset of genes from the network as negative samples for a gene, considering that the genes on selected subset don't fall in the neighborhood N_i of the gene. Above objective function enhances the similarity of a gene $v_i$with its neighborhood genes $v_j \in N_i$ and weakens the similarity with genes not in its neighborhood genes $v_j \notin N_i$. It is inappropriate to assume that the two genes in the network are not related if they are not connected. It may be the case that there is not enough experimental evidence to support that they are related yet. Thus, forcing the dissimilarity of a gene with all other genes, not in its neighborhood N_i seems to be inappropriate.

We adopt Adaptive Moment Estimation (Adam) optimization [21], which is an extension to stochastic gradient descent, for optimizing Eq. 12. Adam computes the adaptive learning rate for each parameter by performing smaller updates for the frequent parameters and larger updates for the infrequent parameters. The Adam method provides the ability of AdaGrad [22] to deal with sparse gradients and also the ability of RMSProp [23] to deal with non-stationary objectives. In each step, Adam algorithm samples mini-batch of interactions and then updates GNE's parameters. To address the issue of overfitting, regularization like dropout [24] and batch normalization [25] is added to hidden layers. Proper optimization of GNE gives the final representation for each gene.

Experimental setup

We evaluate our model using two real organism datasets. We take gene interaction network data from the BioGRID database [26] and gene expression data from DREAM5 challenge [7]. We use two interaction datasets from BioGRID database (2017 released version 3.4.153 and 2018 released version 3.4.158) to evaluate the predictive performance of our model. Self-interactions and redundant interactions are removed from interaction datasets. The statistics of the datasets are shown in Table 2.

We evaluate the learned embeddings to infer gene network structure. We randomly hold out a fraction of interactions as the validation set for hyper-parameter tuning. Then, we divide the remaining interactions randomly into training and testing dataset with the equal number of interactions. Since the validation set and the test set contains only positive interactions, we randomly sample an equal number of gene pairs from the network, considering the missing edge between the gene pairs represents the absence of interactions. Given the gene network G with a fraction of missing interactions, the task is to predict these missing interactions.

We compare the GNE model with five competing methods. Correlation directly predicts the interactions between genes based on the correlation of expression profiles. Then, the following three baselines (Isomap, LINE, and node2vec) are network embedding methods. Specifically, node2vec is the strong baseline for structural network embedding. We evaluate the performance of GNE against the following methods:

- **Correlation** [27]
  It computes Pearson's correlation coefficient between all genes and the interactions are ranked via correlation scores, i.e., highly correlated gene pairs receive higher confidence.
- **Isomap** [10]
  It computes all-pairs shortest-path distances to create a distance matrix and performs singular-value decomposition of that matrix to learn a lower-dimensional representation. Genes separated

Table 2 Statistics of the interaction datasets from BioGRID and the gene expression data from DREAM5 challenge

		#(Interactions)		Expression data
Datasets	#(Genes)	2017 version	2018 version	#(Experiments)
Yeast	5950	544,652	557,487	536
E. coli	4511	148,340	159,523	805

by the distance less than threshold ϵ in embedding space are considered to have the connection with each other and the reliability index, a likelihood indicating the interaction between two genes, is computed using FSWeight [28].

- **LINE [16]**
  Two separate embeddings are learned by preserving first-order and second-order proximity of the network structure respectively. Then, these embeddings are concatenated to get final representations for each node.
- **node2vec [14]**
  It learns the embeddings of the node by applying Skip-gram model to node sequences generated by a biased random walk. We tuned two hyper-parameters p and q that control the random walk.

Note that the competing methods such as Isomap, LINE, and node2vec are designed to capture only the topological properties of the network. For the fair comparison with GNE that additionally integrates expression data, we concatenate attribute feature vector with learned gene representation to extend baselines by including the gene expression. We name these variants as Isomap+, LINE+, and node2vec+.

We have implemented GNE with TensorFlow framework [29]. The parameter settings for GNE are determined by its performance on the validation set. We randomly initialize GNE's parameters, optimizing with mini-batch Adam. We test the batch size of [8, 16, 32, 64, 128, 256] and learning rate of [0.1, 0.01, 0.005, 0.002, 0.001, 0.0001]. We test the number of negative samples to be [2, 5, 10, 15, 20] as suggested by [13]. We test the embedding dimension d of [32, 64, 128, 256] for all methods. Also, we evaluate model's performance with respect to different values of λ [0, 0.2, 0.4, 0.6, 0.8, 1], which is discussed in more detail later. The parameters are selected based on empirical evaluation and Table 3 summarizes the optimal parameters tuned on validation data sets.

To capture the non-linearity of gene expression data, we choose Exponential Linear Unit (ELU) [30] activation function, which corresponds to δ_a in Eq. 5. Also, ELU activation avoids vanishing gradient problem and provides improved learning characteristics in comparison to other methods. We use a single hidden layer ($k = 1$) with hyperbolic tangent activation (Tanh) to model complex statistical relationships between topological properties and attributes of the gene. The choice of ELU for attribute transformation and Tanh for hidden layer shows better performance upon empirical evaluation.

We use the area under the ROC curve (AUROC) and area under the precision-recall curve (AUPR) [31] to evaluate the rankings generated by the model for interactions in the test set. These metrics are widely used in evaluating the ranked list of predictions in gene interaction [4].

Results and discussion

We evaluate the ability of our GNE model to predict gene interaction of two real organisms. We present empirical results of our proposed method against other methods.

Analysis of gene embeddings

We visualize the embedding vectors of genes learned by GNE. We take the learned embeddings, which specifically model the interactions by preserving topological and attribute similarity. We embed these embeddings into a 2D space using t-SNE package [32] and visualize them (Fig. 3). For comparison, we also visualize the embeddings learned by structure-preserving deep learning methods, such as LINE, and node2vec.

In E. coli, a substantial fraction of functionally related genes are organized into operons, which are the group of genes that interact with each other and are co-regulated [33]. Since this concept fits well with the topological and attribute proximity implemented in GNE, we expect GNE to place genes within an operon close to each other in the embedding space. To evaluate this, we collect information about operons of E. coli from the DOOR database and visualize the embeddings of genes within these operons (Fig. 3).

Figure 3 reveals the clustering structure that corresponds to the operons on E. coli. For example, operon with operon id 3306 consists of seven genes: rsxA, rsxB, rsx, rsxD, rsxG, rsxE, and nth that are involved in electron transport. GNE infers similar representations for these genes, resulting in localized projection in the 2D space. Similarly, other operons also show similar patterns (Fig. 3).

To test if the pattern in Fig. 3 holds across all operons, we compute the average Euclidean distance between each gene's vector representation and vector representations of other genes within the same operon. Genes within the same operon have significantly similar vector representation y_i than expected by chance (p-value $= 1.75e - 127$, 2-sample KS test).

Table 3 Optimal parameter settings for GNE model

Dataset	Learning rate	Batch size	Embedding dimension (d)	Epoch	Negative samples
Yeast	0.005	256	128	20	10
E. coli	0.002	128	128	20	10

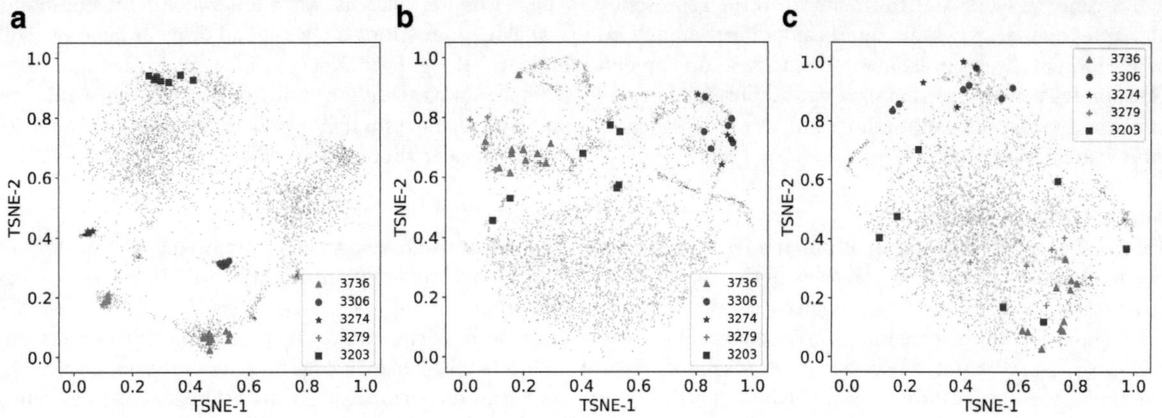

Fig. 3 Visualization of learned embeddings for genes on E. coli. Genes are mapped to the 2D space using the t-SNE package [32] with learned gene representations ($y_i, i = 1, 2, \ldots, M$) from different methods: **a** GNE, **b** LINE, and **c** node2vec as input. Operons 3203, 3274, 3279, 3306, and 3736 of E. coli are visualized and show clustering patterns. Best viewed on screen

Thus, the analysis here indicates that GNE can learn similar representations for genes with similar topological properties and expression.

Gene interaction prediction

We randomly remove 50% of interactions from the network and compare various methods to evaluate their predictions for 50% missing interactions. Table 4 shows the performance of GNE and other methods on gene interaction prediction across different datasets. As our method significantly outperforms other competing methods, it indicates the informativeness of gene expression in predicting missing interactions. Also, our model is capable of integrating attributes with topological properties to learn better representations.

Table 4 Area under ROC curve (AUROC) and Area under PR curve (AUPR) for gene Interaction Prediction

Methods	Yeast		E. coli	
	AUROC	AUPR	AUROC	AUPR
Correlation	0.582	0.579	0.537	0.557
Isomap	0.507	0.588	0.559	0.672
LINE	0.726	0.686	0.897	0.851
node2vec	0.739	0.708	0.912	0.862
Isomap+	0.653	0.652	0.644	0.649
LINE+	0.745	0.713	0.899	0.856
node2vec+	0.751	0.716	0.871	0.826
GNE (Topology)	0.787	0.784	0.930	0.931
GNE (our model)	**0.825***	**0.821***	**0.940***	**0.939***

+ indicates the concatenation of expression data with learned embeddings to create final representation. * denotes that GNE significantly outperforms node2vec at 0.01 level paired t-test. Note that method that achieves the best performance is bold faced

We compare our model with a correlation-based method, that takes only expression data into account. Our model shows significant improvement of 0.243 (AUROC), 0.242 (AUPR) on yeast and 0.403 (AUROC), 0.382 (AUPR) on E. coli over correlation-based methods. This improvement suggests the significance of the topological properties of the gene network.

The network embedding method, Isomap, performs poorly in comparison to correlation-based methods on yeast because of its limitation on network inference. Deep learning based network embedding methods such as LINE, and node2vec show the significant gain over Isomap and correlation-based methods. node2vec outperforms LINE across two datasets. Moreover, GNE trained only with topological properties outperforms these structured-based deep learning methods (Table 4). However, these methods don't consider the attributes of the gene that we suggest to contain useful information for gene interaction prediction. By adding expression data with topological properties, GNE outperforms structure-preserving deep embedding methods across both datasets.

Focusing on the results corresponding to the integration of expression data with topological properties, we find that the method of integrating the expression data plays an essential role in the performance. Performance of node2vec+ (LINE+, Isomap+) shows little improvement with the integration of expression data on yeast. However, node2vec+ (LINE+, Isomap+) has no improvement or decline in performance on E. coli. The decline in performance indicates that merely concatenating the expression vector with learned representations for the gene is insufficient to capture the rich information in expression data. The late fusion approach of combining the embedding vector corresponding to the topological properties of the gene network and the feature vector representing expression data has no significant improvement in the performance (except Isomap). In contrast, our model incorporates gene expression data with topological properties by the early fusion method and shows significant improvement over other methods.

Impact of network sparsity

We investigate the robustness of our model to network sparsity. We hold out 10% interactions as the test set and change the sparsity of the remaining network by randomly removing a portion of remaining interactions. Then, we train GNE to predict interactions in the test set and evaluate the change in performance to network sparsity. We evaluate two versions of our implementations: GNE with only topological properties and GNE with topological properties and expression data. The result is shown in Fig. 4.

Figure 4 shows that our method's performance improves with an increase in the number of training interactions across datasets. Also, our method's performance improves when expression data is integrated with the topological structure. Specifically, GNE trained on 10% of total interactions and attributes of yeast shows a significant gain of 0.172 AUROC (from 0.503 to 0.675) over GNE trained only with 10% of total interactions as shown in Fig. 4. Similarly, GNE improves the AUROC from 0.497 to 0.816 for E. coli with the same setup as shown in Fig. 4. The integration of gene expression data results in less improvement when we train GNE on a relatively large number of interactions.

Moreover, the performance of GNE trained with 50% of total interactions and expression data is comparable to be trained with 80% of total interactions without gene expression data as shown in Fig. 4. The integration of expression data with topological properties into GNE model has more improvement on E. coli than yeast when we train with 10% of total interactions for each dataset. The reason for this is likely the difference in the number of available interactions for yeast and E. coli (Table 2). This indicates the informativeness of gene expression when we have few interactions and supports the idea that the integration of expression data with topological properties improves gene interaction prediction.

Impact of λ

GNE involves the parameter λ that controls the importance of gene expression information relative to topological properties of gene network as shown in Eq. 6. We examine how the choice of the parameter λ affects our method's performance. Figure 5 shows the comparison of our method's performance with different values of λ when GNE is trained on varying percentage of total interactions.

We evaluate the impact of λ on range [0, 0.2, 0.4, 0.6, 0.8, 1]. When λ becomes 0, the learned representations model only topological properties. In contrast, setting the high value for λ makes GNE learn only from attributes and degrades its performance. Therefore, our model performs well when λ is within [0, 1].

Figure 5 shows that the integration of expression data improves the performance of GNE to predict gene interactions. Impact of λ depends on the number of interactions used to train GNE. If GNE is trained with few interactions, integration of expression data with topological properties plays a vital role in predicting missing interactions. As the number of training interactions increases, integration of expression data has less impact but still improves the performance over only topological properties.

Figures 4 and 5 demonstrate that the expression data contributes the increase in AUROC by nearly 0.14 when interactions are less than 40% for yeast and about 0.32 when interactions are less than 10% for E. coli. More topological properties and attributes are required for yeast than E. coli. It may be related to the fact that yeast is a more complex species than E. coli. Moreover, we can speculate that more topological properties and attributes are

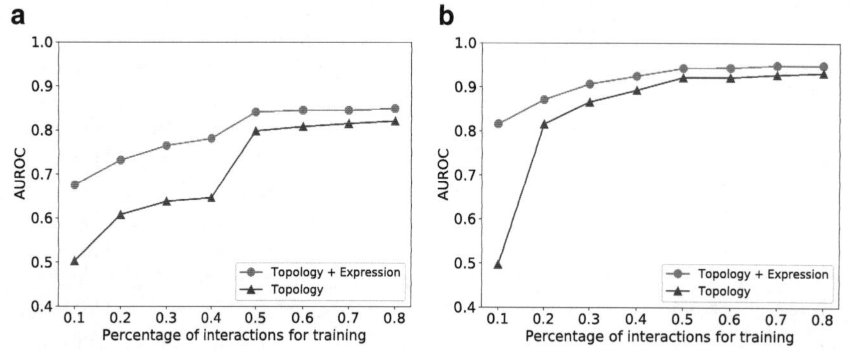

Fig. 4 AUROC comparison of GNE's performance with respect to network sparsity. **a** yeast **b** E. coli. Integration of expression data with topological properties of the gene network improves the performance for both datasets

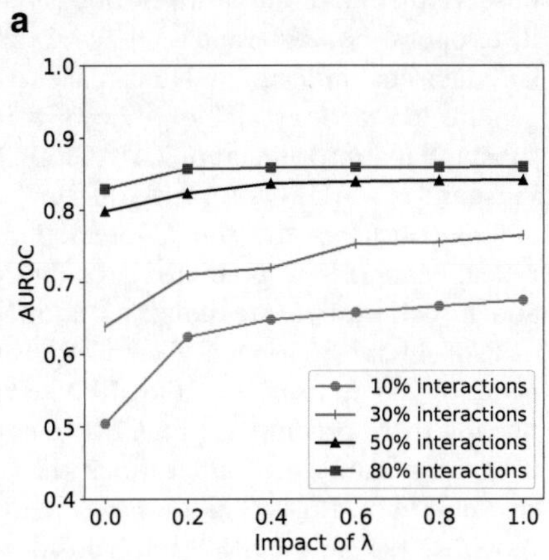

Fig. 5 Impact of λ on GNE's performance trained with different percentages of interactions. **a** yeast **b** E. coli. Different lines indicate performance of GNE trained with different percentages of interactions

required for higher eukaryotes like humans. In humans, GNE that integrates topological properties with attributes may be more successful than the methods that only use either topological properties or attributes.

This demonstrates the sensitivity of GNE to parameter λ. This parameter λ has a considerable impact on our method's performance and should be appropriately selected.

Investigation of GNE's predictions

We investigate the predictive ability of our model in identifying new gene interactions. For this aim, we consider

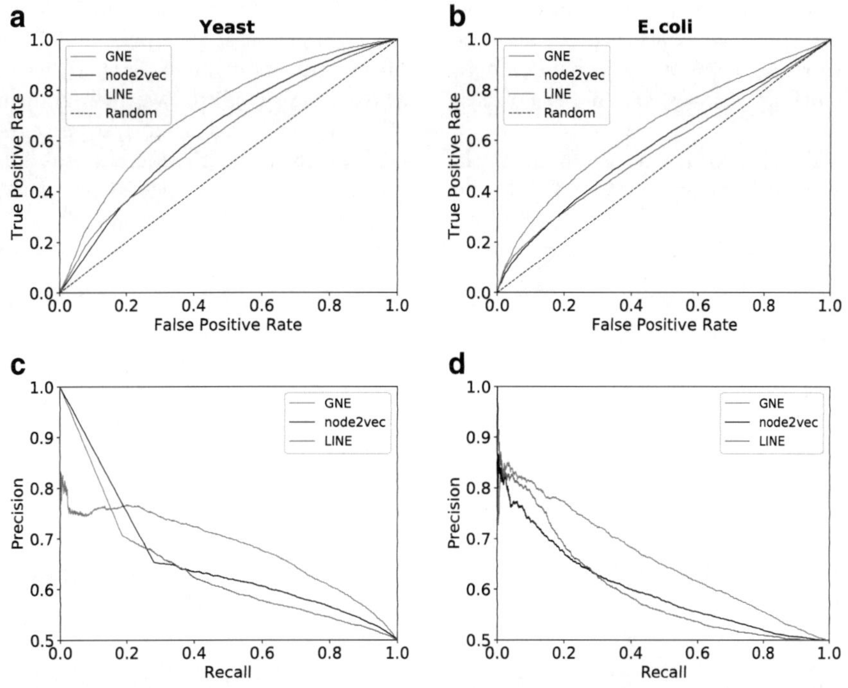

Fig. 6 Temporal holdout validation in predicting new interactions. Performance is measured by the area under the ROC curve and the area under the precision-recall curve. Shown are the performance of each method based on the AUROC (**a, b**) and AUPR (**c, d**) for yeast and E. coli. The limit of the y-axis is adjusted to [0.5, 1.0] for the precision-recall curve to make the difference in performance more visible. GNE outperforms LINE and node2vec

Table 5 AUROC and AUPR comparision for temporal holdout validation

Methods	Yeast		E. coli	
	AUROC	AUPR	AUROC	AUPR
LINE	0.620	0.611	0.569	0.598
node2vec	0.640	0.609	0.587	0.599
GNE (our model)	**0.710**	**0.683**	**0.653**	**0.658**

Note that method that achieves the best performance is bold faced

two versions of BioGRID interaction datasets at two different time points (2017 and 2018 version), where the older version is used for training and the newer one is used for testing the model (temporal holdout validation). The 2018 version contains 12,835 new interactions for yeast and 11,185 new interactions for E. coli than the 2017 version. GNE's performance trained with 50% and 80% of total interactions are comparable for both yeast and E. coli (Figs. 4 and 5). We thus train our model with 50% of total interactions from the 2017 version to learn the embeddings for genes and demonstrate the impact of integrating expression data with topological properties. We create the test set with new interactions from the 2018 version of BioGRID as positive interactions and the equal number of negative interactions randomly sampled. We make predictions for these interactions using learned embeddings and create a list of (Gene v_i, Gene v_j, probability), ranked by the predicted probability. We consider predicted gene pairs with the probabilities of 0.5 or higher but are missing from BioGRID for further investigation as we discuss later in this section.

The temporal holdout performance of our model in comparison to other methods is shown in Fig. 6. We observe that GNE outperforms both node2vec and LINE in temporal holdout validation across both yeast and E. coli datasets, indicating GNE can accurately predict new genetic interactions. Table 5 shows that GNE achieves substantial improvement of 7.0 (AUROC), 7.4 (AUPR) on yeast and 6.6 (AUROC), 5.9 (AUPR) on E. coli datasets.

Table 6 shows the top 5 interactions with the significant increase in predicted probability for both yeast and E. coli after expression data is integrated. We also provide literature evidence with experimental evidence code obtained from the BioGRID database [26] supporting these predictions. BioGRID compiles interaction data from numerous publications through comprehensive curation efforts. Taking new interactions added to BioGRID (version 3.4.158) into consideration, we evaluate the probability of these interactions predicted by GNE trained with and without expression data. Specifically, integration of expression data increases the probability of 8331 (out of 11,185) interactions for E. coli (improving AUROC from 0.606 to 0.662) and 6,010 (out of 12,835) interactions for yeast (improving AUROC from 0.685 to 0.707). Integration of topology and expression data significantly increases the probabilities of true interactions between genes (Table 6).

To further evaluate GNE's predictions, we consider the new version of BioGRID (version 3.4.162) and evaluate 2609 yeast gene pairs (Additional file 1: Table S1) and 871 E. coli gene pairs (Additional file 2: Table S2) predicted by GNE with the probabilities of 0.5 or higher. We find that 128 (5%) yeast gene pairs and 78 (9%) E. coli gene pairs are true interactions that have been added to the latest release of BioGRID. We then evaluate the predictive ability of GNE by calculating the percentage of true interactions with regard to different probability bins (Fig. 7). Sixteen percent of predicted yeast gene pairs and 17.5%

Table 6 New gene interactions that are assigned high probability by GNE

Organism	Probability		Gene *i*	Gene *j*	Experimental evidence code
	Topology	Topology + Expression			
Yeast	0.287	0.677	TFC8	DHH1	Affinity Capture-RNA [35]
	0.394	0.730	SYH1	DHH1	Affinity Capture-RNA [35]
	0.413	0.746	CPR7	DHH1	Affinity Capture-RNA [35]
	0.253	0.551	MRP10	DHH1	Affinity Capture-RNA [35]
	0.542	0.835	RPS13	ULP2	Affinity Capture-MS [36]
E. coli	0.014	0.944	ATPB	RFBC	Affinity Capture-MS [37]
	0.012	0.941	NARQ	CYDB	Affinity Capture-MS [37]
	0.013	0.937	PCNB	PAND	Affinity Capture-MS [37]
	0.015	0.939	FLIF	CHEY	Affinity Capture-MS [37]
	0.017	0.938	YCHM	PROB	Affinity Capture-MS [37]

New gene interactions on 2018 version that are assigned high probability by GNE after integration of expression data. We provide probability predicted by GNE (with/without expression data) for new interactions in the 2018 version and evidence supporting the existence of predicted interactions

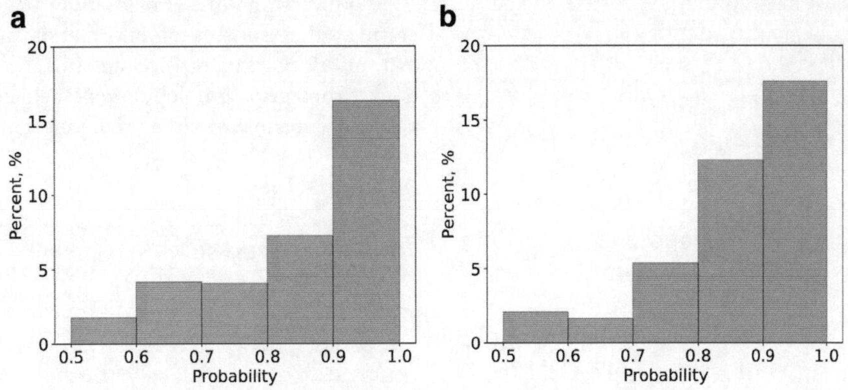

Fig. 7 The percentage of true interactions from GNE's predictions with different probability bins. **a** yeast **b** E. coli. We divide the gene pairs based on their predicted probabilities to different probability ranges (as shown in the x-axis) and identify the number of predicted true interactions in each range. Each bar indicates the percentage of true interactions out of predicted gene pairs in that probability range

of predicted E. coli gene pairs with the probability higher than 0.9 are true interactions. This suggests that gene pairs with high probability predicted by GNE are more likely to be true interactions.

To support our finding that GNE predicted gene pairs have high value, we manually check gene pairs that have high predicted probability but are missing from the latest BioGRID release. We find that these gene pairs interact with the same set of other genes. For example, GNE predicts the interaction between YDR311W and YGL122C with the probability of 0.968. Mining BioGRID database, we find that these genes interact with the same set of 374 genes. Similarly, E. coli genes DAMX and FLIL with the predicted probability of 0.998 share 320 interacting genes. In this way, we identify all interacting genes shared by each of the predicted gene pairs in yeast and E. coli (Additional file 1: Table S1 and Additional file 2: Table S2). Figure 8 shows the average number of interacting genes shared by a gene pair. In general, gene pairs with a high GNE probability tend to have a large number of interacting genes. For example, gene pairs with the probability greater than 0.9 have, on average, 82 common interacting genes for yeast and 58 for E. coli. Two sample t-test analysis has shown that there is a significant difference in the number of shared interacting genes with respect to different probability bins (Table 7).

Moreover, we search the literature to see if we can find supporting evidence for predicted interactions. We find literature evidence for an interaction between YCL032W (STE50) and YDL035C (GPR1), which has the probability of 0.98 predicted by GNE. STE50 is an adaptor that links G-protein complex in cell signalling, and GPR1 is a G-protein coupled receptor. Both STE50 and GPR1 share a

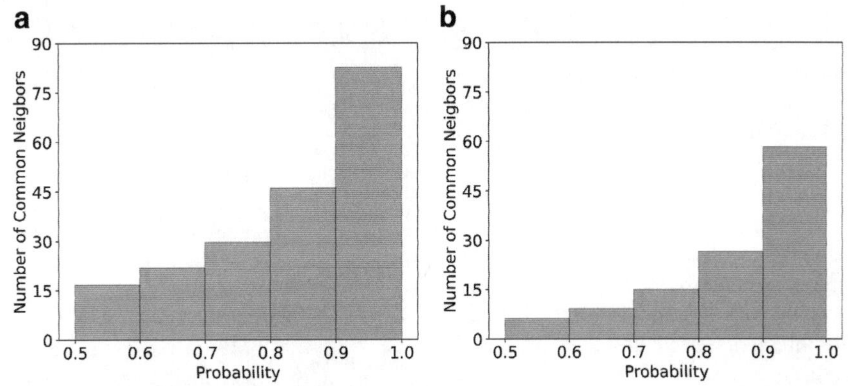

Fig. 8 The average number of common interacting genes between the gene pairs predicted by GNE. **a** yeast **b** E. coli. We divide gene pairs into different probability groups based on predicted probabilities by GNE and compute the number of common interacting genes shared by these gene pairs. We categorize these gene pairs into different probability ranges (as shown in the x-axis). Each bar represents the average number of common interacting genes shared by gene pairs in each probability range

Table 7 Results of two-sample t-test

Probability bin for Sample A	Probability bin for Sample B	*p*-value for yeast	*p*-value for E. coli
0.5 - 0.6	0.6 - 0.7	$9.2e-14$	$6.9e-02$
0.5 - 0.6	0.7 - 0.8	$3.02e-51$	$1.23e-05$
0.5 - 0.6	0.8 - 0.9	$6.1e-117$	$7.4e-14$
0.5 - 0.6	0.9 - 1.0	$2.1e-177$	$3.7e-39$
0.6 - 0.7	0.7 - 0.8	$8.2e-17$	$1.1e-02$
0.6 - 0.7	0.8 - 0.9	$3.5e-69$	$9.2e-09$
0.6 - 0.7	0.9 - 1.0	$2.1e-128$	$1.9e-30$
0.7 - 0.8	0.8 - 0.9	$4.7e-28$	$4.8e-04$
0.7 - 0.8	0.9 - 1.0	$6.2e-87$	$7.4e-23$
0.8 - 0.9	0.9 - 1.0	$4.3e-35$	$5.1e-13$

We divide gene pairs into different probability groups based on predicted probabilities by GNE and compute the number of common interacting genes shared by these gene pairs. Significance test shows there is the significant difference between average number of shared genes in different probability bins

common function of cell signalling via G-protein. Besides, STE50p interacts with STE11p in the two-hybrid system, which is a cell-based system examining protein-protein interactions [34]. Also, BioGRID has evidence of 30 physical and 4 genetic associations between STE50 and STE11. Thus, STE50 is highly likely to interact with STE11, which in turn interacts with GPR1.

This analysis demonstrates the potential of our method in the discovery of gene interactions. Also, GNE can help the curator to identify interactions with strong potential that need to be looked at with experimental validation or within the literature.

Conclusion

We developed a novel deep learning framework, namely GNE to perform gene network embedding. Specifically, we design deep neural network architecture to model the complex statistical relationships between gene interaction network and expression data. GNE is flexible to the addition of different types and number of attributes. The features learned by GNE allow us to use out-of-the-box machine learning classifiers like Logistic Regression to predict gene interactions accurately.

GNE relies on a deep learning technique that can learn the underlying patterns of gene interactions by integrating heterogeneous data and extracts features that are more informative for interaction prediction. Experimental results show that GNE achieve better performance in gene interaction prediction over other baseline approaches in both yeast and E. coli organisms. Also, GNE can help the curator to identify the interactions that need to be looked at.

As future work, we aim to study the impact of integrating other sources of information about gene such as transcription factor binding sites, functional annotations (from gene ontology), gene sequences, metabolic pathways, etc. into GNE in predicting gene interaction.

Additional files

Additional file 1: Table S1. Includes yeast gene pairs predicted by GNE with probabilities of 0.5 or higher. (XLSX 109 kb)

Additional file 2: Table S2. Includes E. coli gene pairs predicted by GNE with probabilities of 0.5 or higher. Rows marked with yellow color indicate predicted interaction is true based on latest version 3.4.162 of BioGRID interaction dataset released on June 2018. (XLSX 43.7 kb)

Abbreviations

AUPR: Area under the PR curve; AUROC: Area under the ROC curve; DOOR: Database of Prokaryotic Operons; DREAM: Dialogue on reverse Engineering assessment and methods; ELU: Exponential linear unit; GI: Genetic interaction; GNE: Gene network embedding; ID: Identifier; KS: Kolmogrov-Smirnov; LINE: Large-scale information network embedding; PR: Precision recall; ROC: Receiver operator characterstic; t-SNE: t-Distributed stochastic neighbor embedding

Acknowledgements

Not applicable.

Funding

This work was supported by the National Science Foundation [1062422 to A.H.] and the National Institutes of Health [R15GM116102 to F.C.]. Publication of this article was sponsored by NSF grant (1062422).

Availability of data and materials

The datasets analysed during the current study are publicly available. The gene expression data was downloaded from DREAM5 Network Challenge https://www.synapse.org/#!Synapse:syn2787209/wiki/70351. The interaction datasets for yeast and E. coli were downloaded from BioGRID https://thebiogrid.org.

About this supplement

This article has been published as part of *BMC Systems Biology Volume 13 Supplement 2, 2019: Selected articles from the 17th Asia Pacific Bioinformatics Conference (APBC 2019): systems biology*. The full contents of the supplement are available online at https://bmcsystbiol.biomedcentral.com/articles/supplements/volume-13-supplement-2.

Authors' contributions

KK, RL, FC, and AH designed the research. KK performed the research and wrote the manuscript. RL, FC, QY, and AH improved the draft. All authors read and approved the final manuscript.

Ethics approval and consent to participate

Not applicable.

Consent for publication

Not applicable.

Competing interests

The authors declare that they have no competing interests.

Publisher's Note

Springer Nature remains neutral with regard to jurisdictional claims in published maps and institutional affiliations.

Author details

[1]Golisano College of Computing and Information Sciences, Rochester Institute of Technology, 20 Lomb Memorial Drive, 14623 Rochester, New York, USA. [2]Thomas H. Gosnell School of Life Sciences, Rochester Institute of Technology, 84 Lomb Memorial Drive, 14623 Rochester, New York, USA.

Published: 5 April 2019

References

1. Mani R, Onge RPS, Hartman JL, Giaever G, Roth FP. Defining genetic interaction. Proc Natl Acad Sci. 2008;105(9):3461–6.
2. Boucher B, Jenna S. Genetic interaction networks: better understand to better predict. Front Genet. 2013;4:290.
3. Lage K. Protein–protein interactions and genetic diseases: the interactome. Biochim Biophys Acta (BBA) - Mol Basis Dis. 2014;1842(10): 1971–80.
4. Madhukar NS, Elemento O, Pandey G. Prediction of genetic interactions using machine learning and network properties. Front Bioeng Biotechnol. 2015;3:172.
5. Oliver S. Proteomics: guilt-by-association goes global. Nature. 2000;403(6770):601.
6. Cho H, Berger B, Peng J. Compact integration of multi-network topology for functional analysis of genes. Cell Syst. 2016;3(6):540–8.
7. Marbach D, Costello JC, Küffner R, Vega NM, Prill RJ, Camacho DM, Allison KR, Aderhold A, Bonneau R, Chen Y, et al. Wisdom of crowds for robust gene network inference. Nat Methods. 2012;9(8):796.
8. Li R, KC K, Cui F, Haake AR. Sparse covariance modeling in high dimensions with gaussian processes. In: Proceedings of The 32nd Conference on Neural Information Processing Systems (NIPS). 2018.
9. Cui P, Wang X, Pei J, Zhu W. A survey on network embedding. IEEE Trans Knowl Data Eng. 2018. arXiv preprint arXiv:1711.08752. IEEE.
10. Lei Y-K, You Z-H, Ji Z, Zhu L, Huang D-S. Assessing and predicting protein interactions by combining manifold embedding with multiple information integration. In: BMC Bioinformatics, vol. 13. BioMed Central; 2012. p. 3.
11. Alanis-Lobato G, Cannistraci CV, Ravasi T. Exploitation of genetic interaction network topology for the prediction of epistatic behavior. Genomics. 2013;102(4):202–8.
12. Tenenbaum JB, De Silva V, Langford JC. A global geometric framework for nonlinear dimensionality reduction. science. 2000;290(5500):2319–23.
13. Mikolov T, Sutskever I, Chen K, Corrado GS, Dean J. Distributed representations of words and phrases and their compositionality. In: Advances in Neural Information Processing Systems. 2013. p. 3111–3119.
14. Grover A, Leskovec J. node2vec: Scalable feature learning for networks. In: Proceedings of the 22nd ACM SIGKDD International Conference on Knowledge Discovery and Data Mining. ACM; 2016. p. 855–864.
15. Tu Y, Stolovitzky G, Klein U. Quantitative noise analysis for gene expression microarray experiments. Proc Natl Acad Sci. 2002;99(22): 14031–6.
16. Tang J, Qu M, Wang M, Zhang M, Yan J, Mei Q. Line: Large-scale information network embedding. In: Proceedings of the 24th International Conference on World Wide Web. International World Wide Web Conferences Steering Committee; 2015. p. 1067–1077.
17. Snoek CG, Worring M, Smeulders AW. Early versus late fusion in semantic video analysis. In: Proceedings of the 13th Annual ACM International Conference on Multimedia. 2005. p. 399–402. ACM.
18. He K, Zhang X, Ren S, Sun J. Deep residual learning for image recognition. In: Proceedings of the IEEE Conference on Computer Vision and Pattern Recognition. 2016. p. 770–778.
19. Pennington J, Socher R, Manning C. Glove: Global vectors for word representation. In: Proceedings of the 2014 Conference on Empirical Methods in Natural Language Processing (EMNLP). 2014. p. 1532–1543.
20. Levy O, Goldberg Y, Dagan I. Improving distributional similarity with lessons learned from word embeddings. Trans Assoc Comput Linguist. 2015;3:211–25.
21. Kingma DP, Ba J. Adam: A method for stochastic optimization. In: International Conference on Learning Representations; 2015.
22. Duchi J, Hazan E, Singer Y. Adaptive subgradient methods for online learning and stochastic optimization. J Mach Learn Res. 2011;12(Jul): 2121–59.
23. Tieleman T, Hinton G. Lecture 6.5-rmsprop, coursera: Neural networks for machine learning. University of Toronto, Technical Report. 2012.
24. Srivastava N, Hinton G, Krizhevsky A, Sutskever I, Salakhutdinov R. Dropout: A simple way to prevent neural networks from overfitting. J Mach Learn Res. 2014;15(1):1929–58.
25. Ioffe S, Szegedy C. Batch normalization: Accelerating deep network training by reducing internal covariate shift. In: International Conference on Machine Learning; 2015. p. 448–56.
26. Stark C, Breitkreutz B-J, Reguly T, Boucher L, Breitkreutz A, Tyers M. Biogrid: a general repository for interaction datasets. Nucleic Acids Res. 2006;34(suppl_1):535–9.
27. Butte A-J, Kohane I-S. Mutual information relevance networks: functional genomic clustering using pairwise entropy measurements. In: Biocomputing 2000. World Scientific; 1999. p. 418–429.
28. Chua HN, Sung W-K, Wong L. Exploiting indirect neighbours and topological weight to predict protein function from protein–protein interactions. Bioinformatics. 2006;22(13):1623–30.
29. Abadi M, Agarwal A, Barham P, Brevdo E, Chen Z, Citro C, Corrado GS, Davis A, Dean J, Devin M, Ghemawat S, Goodfellow I, Harp A, Irving G, Isard M, Jia Y, Jozefowicz R, Kaiser L, Kudlur M, Levenberg J, Mané D., Monga R, Moore S, Murray D, Olah C, Schuster M, Shlens J, Steiner B, Sutskever I, Talwar K, Tucker P, Vanhoucke V, Vasudevan V, Viégas F., Vinyals O, Warden P, Wattenberg M, Wicke M, Yu Y, Zheng X. TensorFlow: Large-Scale Machine Learning on Heterogeneous Systems. Software available from tensorflow.org. 2015. https://www.tensorflow.org/ Accessed 21 Dec 2016.
30. Clevert D, Unterthiner T, Hochreiter S. Fast and accurate deep network learning by exponential linear units (elus). In: International Conference on Learning Representations; 2016.
31. Davis J, Goadrich M. The relationship between precision-recall and roc curves. In: Proceedings of the 23rd International Conference on Machine Learning. ACM; 2006. p. 233–240.
32. Maaten L. v. d., Hinton G. Visualizing data using t-sne. J Mach Learn Res. 2008;9(Nov):2579–605.
33. Mao F, Dam P, Chou J, Olman V, Xu Y. Door: a database for prokaryotic operons. Nucleic Acids Res. 2008;37(suppl_1):459–63.
34. Gustin MC, Albertyn J, Alexander M, Davenport K. Map kinase pathways in the yeastsaccharomyces cerevisiae. Microbiol Mol Biol Rev. 1998;62(4): 1264–300.
35. Miller JE, Zhang L, Jiang H, Li Y, Pugh BF, Reese JC. Genome-wide mapping of decay factor–mrna interactions in yeast identifies nutrient-responsive transcripts as targets of the deadenylase ccr4. G3: Genes, Genomes, Genet. 2018;8(1):315–30.
36. Liang J, Singh N, Carlson CR, Albuquerque CP, Corbett KD, Zhou H. Recruitment of a sumo isopeptidase to rdna stabilizes silencing complexes by opposing sumo targeted ubiquitin ligase activity. Genes Dev. 2017;31(8):802–15.
37. Babu M, Bundalovic-Torma C, Calmettes C, Phanse S, Zhang Q, Jiang Y, Minic Z, Kim S, Mehla J, Gagarinova A, et al. Global landscape of cell envelope protein complexes in escherichia coli. Nat Biotechnol. 2018;36(1):103.

Liu and Rajapakse *BMC Systems Biology* 2019, **13**(Suppl 2):37
https://doi.org/10.1186/s12918-019-0695-x

BMC Systems Biology

RESEARCH | Open Access

Fusing gene expressions and transitive protein-protein interactions for inference of gene regulatory networks

Wenting Liu[1,2] and Jagath C. Rajapakse[3*]

From The 17th Asia Pacific Bioinformatics Conference (APBC 2019)
Wuhan, China. 14–16 January 2019

Abstract

Background: Systematic fusion of multiple data sources for Gene Regulatory Networks (GRN) inference remains a key challenge in systems biology. We incorporate information from protein-protein interaction networks (PPIN) into the process of GRN inference from gene expression (GE) data. However, existing PPIN remain sparse and transitive protein interactions can help predict missing protein interactions. We therefore propose a systematic probabilistic framework on fusing GE data and transitive protein interaction data to coherently build GRN.

Results: We use a Gaussian Mixture Model (GMM) to soft-cluster GE data, allowing overlapping cluster memberships. Next, a heuristic method is proposed to extend sparse PPIN by incorporating transitive linkages. We then propose a novel way to score extended protein interactions by combining topological properties of PPIN and correlations of GE. Following this, GE data and extended PPIN are fused using a Gaussian Hidden Markov Model (GHMM) in order to identify gene regulatory pathways and refine interaction scores that are then used to constrain the GRN structure. We employ a Bayesian Gaussian Mixture (BGM) model to refine the GRN derived from GE data by using the structural priors derived from GHMM. Experiments on real yeast regulatory networks demonstrate both the feasibility of the extended PPIN in predicting transitive protein interactions and its effectiveness on improving the coverage and accuracy the proposed method of fusing PPIN and GE to build GRN.

Conclusion: The GE and PPIN fusion model outperforms both the state-of-the-art single data source models (CLR, GENIE3, TIGRESS) as well as existing fusion models under various constraints.

Keywords: Gene regulatory network (GRN), Gene expressions, Gaussian mixture model (GMM), Protein-protein interaction networks, Transitive protein-protein interactions

Background

Gene regulations describe the interactions among genes during cellular activity. Through regulation, genes orchestrate the level of synthesized mRNA and thereby control the expression of other genes and the rates at which proteins are produced, eventually deciding the state of the cell. Gene expression (GE) microarrays provide quantitative or semi-quantitative data on the cell state at a specific time and condition. By "reverse-engineering" GE data, regulatory interactions among genes can be identified and gene regulatory network can be mapped using computational methods [1, 2].

Vast majority of functional analysis approaches to modelling microarray GE data assume that genes with similar expression profiles have similar cellular functions [3–5]. A molecular pathway is a set of genes that activate together to achieve a specific task and thus share similar expression profiles. In this paper, we use a data-driven method - the model-based clustering - to model genes in distinct pathways. Specifically, we model each pathway

*Correspondence: asjagath@ntu.edu.sg
[3]School of Computer Engineering, Nanyang Technological University, Singapore, Singapore
Full list of author information is available at the end of the article

as a Gaussian model as it allows modelling correlations among gene expressions in a data-driven way. It is better suited for situations where the prior knowledge of the regulatory pathways is unknown. In addition, because genes naturally participate in more than one regulatory pathway, soft-clustering is used to allowed so that genes can have memberships in multiple pathways. Hence, we adopt the Gaussian mixture model (GMM) on GE data so that different regulatory pathways can be identified.

The rationale behind clustering is that co-expressed genes, i.e., genes in the same cluster are more likely to be functionally related and belong to the same cluster. However, regulatory processes of the genes in a cluster could not necessarily be direct as it could refer to an indirect regulation via proteins, metabolites, or ncRNAs. In cases where two interaction partners are transcription factors or where two proteins are in the same complex, the interactions are direct. In order to identify indirect regulations in GRN, evidences from multiple data sources should be used. For example, medical literature, protein-protein interaction (PPI) data, gene ontology, etc., have all been used to supplement wet lab data in the inference of GRN [6–9]. When more than one source are available, an essential step is to optimally combine evidences from multiple sources to derive a coherent GRN [10–12].

Since proteins are products of genes, protein interactions provide useful evidence for gene regulation. PPIN data have been fused with GE data for GRN inference in previous studies [13–17]. Most of these works considered only binary links of PPIN: if the link is consistent with the predicted edge from GE, the link is accepted as a true regulation. This approach throws away valuable information, so an accurate quantitative scoring scheme is needed to evaluate consistency between PPIN and gene regulation. On the other hand, existing PPINs are sparse and many real protein interactions are missing in current PPIN databases. Suggested by previous PPI prediction works [18, 19], there exist a large number of interactions between proteins in complexes, which have not yet been observed or recorded in current PPIN. We therefore propose a heuristic to quantitatively extend sparse PPIN by using transitive linkages. We then propose a novel way to score protein interactions by combining topological properties of extended PPIN and correlations of GE. Our experiments demonstrate that transitive protein interactions indeed play an important role in predicting protein interactions. We fuse the extended PPIN scores with GE data, using a Gaussian hidden Markov model (GHMM) to identify gene regulatory pathways, which are found to more consistent with PPIN than those produced by GMM. We further refine PPIN confidence scores by including gene interaction scores from GHMM, which makes the PPIN score more consistent with the existing GRN.

Since there exists no widely accepted model that universally fits GE data well [20–23], and different models capture different GE properties leading to different or complementary GRN structures [21], fusion of different models should lead to better GRN. The GHMM identifies regulatory pathways and obtain possible interacting genes by considering linear correlations between genes but misses conditional dependencies, i.e., non-linear relations, among genes in the same regulatory pathway. The Bayesian network (BN) model is good at capturing these conditional dependences but suffers from poor computational efficiency. We thus propose a systematic probabilistic framework that fuse these two models and derive coherent GRN closer to biological reality. Specifically, our framework takes a coarse-to-fine approach: GHMM generates regulatory pathways (i.e., a coarse GRN having high coverage) and obtains refined interaction scores, both of which are then used to constrain the GRN structure of a BN model (i.e., the Bayesian Gaussian mixture model). This generates GRN that are of good coverage and high precision. Furthermore, GRN structural constrains help greatly reduce the search space for BGM model, thereby reducing the overall computational complexity. Figure 1 shows the flow chart of our GRN inference process.

Methods

Soft-Clustering of GE by Gaussian mixture model

Gaussian mixture model

Since gene expression measurement can be viewed as an expression of every gene over all the possible pathways, we employ a Gaussian mixture model (GMM) [24] to describe gene expression (GE) data. For simplicity, we assume that the expression of genes in a pathway follows a Gaussian model, and gene expression data are generated by a finite mixture of underlying probability distribution, that is, by multivariate normal distributions. The key difference between our work and existing GMM hard clustering [13, 24] of GE data lies in our assumption that each gene can participate in multiple pathways. That is, we allow soft-clustering membership for genes so that they can participate in multiple pathways. When one gene participates multiple pathways, it can have soft-clustering membership values for different pathways, representing its contributions of gene expressions in the Gaussian model in different pathways.

Given a gene expression dataset, let $\mathcal{I} = \{i\}_{i=1}^{I}$ denote the set of genes, $\mathcal{L} = \{l\}_{l=1}^{L}$ be the set of pathways, and $\gamma_i \in \mathcal{L}$ denote the regulatory pathways in which gene i participates, i.e., the pathway assignment for gene i. The regulatory pathway assignments of gene set $\mathcal{I}$ is denoted by the set $\Gamma = \{\gamma_i\}_{i=1}^{I}$. The pathway assignment variable γ_i follows a multinomial distribution parameterized by vector $\Theta = \{\theta_l\}_{l=1}^{L}$, with assignment probabilities $p(\gamma_i = l) = \theta_l$, $\theta_l \in [0, 1]$ where $\sum_{l=1}^{L} \theta_l = 1$. We assume

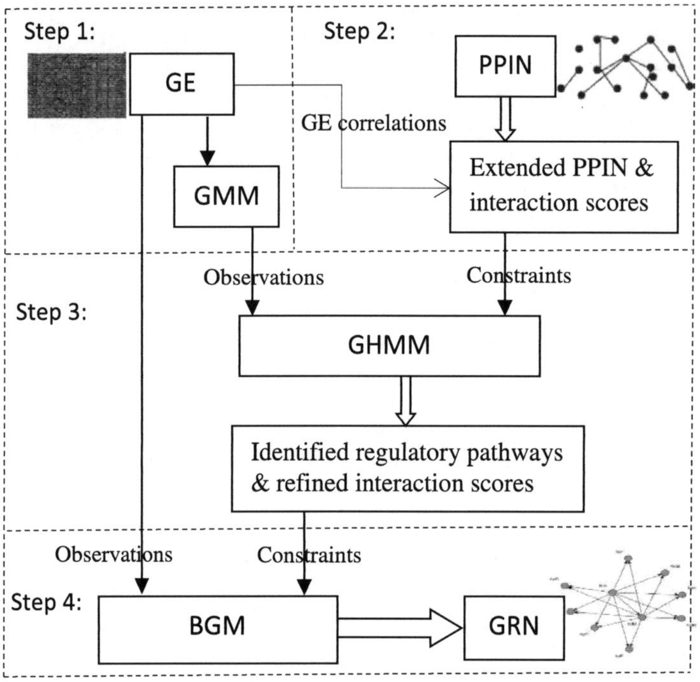

Fig. 1 Overall Model. Step 1: A Gaussian mixture model (GMM) is used to soft-cluster gene expression (GE) data. Step 2: A heuristic is proposed to quantitatively extend the sparse protein-protein interactions by using transitive linkages. A novel way is then proposed to score protein interactions by combining topological properties of extended protein-protein interaction network (PPIN) and GE correlations. Step 3: A Gaussian Hidden Markov Model (GHMM) is used to identify gene regulatory pathways and refine interaction scores, both of which are then used as structural priors to constrain the model of GRN. Step 4: Lastly, the GRN from GE is refined using a Bayesian Gaussian Mixture (BGM) model by including the structural priors derived from Step 3

that the set of weights $\Theta = \{\theta_l\}_{l=1}^{L}$ follow a Dirichlet distribution $p(\Theta) = Dir(\alpha_1, \ldots, \alpha_L)$ with Dirichlet weights having uniform priors.

Given gene expression data $\mathcal{X} = \{x_i\}_{i \in \mathcal{I}}$, each instance i has J continuous-valued attributes $x_i = \{x_{ij}\}_{j=1}^{J}$, where x_{ij} represents the gene expression value measured for gene i in experiment j out of a total of J experiments. We represent each pathway with a Gaussian model wherein the tuple representing the gene expression levels of each experiment (conditioned on a pathway) is a multivariate sample. The probability of observing a given tuple of gene expression levels conditioned on a pathway is $p(x_i | \gamma_i = l) \sim N(\mu_l, \Sigma_l)$, where μ_l is the mean vector and Σ_l is the $J \times J$ covariance matrix.

The pathway assignments are determined by $p(\Gamma | \mathcal{X}) \propto p(\mathcal{X}, \Gamma)$. Assuming independence between genes, the joint distribution over $\mathcal{X}$ and pathway assignments Γ is

$$p(\mathcal{X}, \Gamma) = \prod_{i=1}^{I} p(x_i, \gamma_i) = \prod_{i=1}^{I} \sum_{l=1}^{L} p(x_i, \gamma_i = l) \quad (1)$$

Learning parameters of the Gaussian mixture model

Expectation Maximization (EM) algorithm [25] is used to estimate the parameters of the joint likelihood in (1). The EM algorithm generates a sequence of parameter approximations that eventually maximizes the observed likelihood. The parameters are initialized to random values. The t-th iteration comprises the following two steps.

E-step: Compute the probability of gene $\{i : i = 1, \ldots, I\}$ conditioned on regulatory pathway $\{l : l = 1, \ldots, L\}$ by using current parameter estimates as:

$$p\left(x_i | \mu_l^t, \Sigma_l^t\right) \sim N\left(\mu_l^t, \Sigma_l^t\right);$$
$$\tau_{il}^t = p(x_i, \gamma_i = l) = \frac{\theta_l^t p\left(x_i | \mu_l^t, \Sigma_l^t\right)}{\sum_{l'=1}^{L} \theta_{l'}^t p\left(x_i | \mu_{l'}^t, \Sigma_{l'}^t\right)}. \quad (2)$$

This requires $I \times L$ computations.

M-step: Re-estimate the GMM parameters $\{\theta_l, \mu_l, \Sigma_l\}_{l=1}^{L}$ for all L pathways as:

$$\theta_l^{t+1} = \frac{1}{I} \sum_{i=1}^{I} \tau_{il}^t;$$
$$\mu_l^{t+1} = \frac{\sum_{i=1}^{I} \tau_{il}^t x_i}{\sum_{i=1}^{I} \tau_{il}^t}; \quad (3)$$
$$\Sigma_l^{t+1} = \frac{\sum_{i=1}^{I} \tau_{il}^t \left(x_i - \mu_l^{t+1}\right)\left(x_i - \mu_l^{t+1}\right)'}{\sum_{i=1}^{I} \tau_{il}^t}.$$

Once EM procedure converges, we accumulate the joint probability of each pair of genes (i, i') over all the pathways. This gives the pairwise interaction probability of two genes $G_{i,i'}$, which measures how likely both genes belong to the same pathway (assuming that genes are mutually independent):

$$G_{i,i'} = p(x_i, x_{i'}, y_i = y_{i'}) = \sum_{l=1}^{L} \tau_{il}\tau_{i'l}. \quad (4)$$

where $\tau_{il} = p(x_i, y_i = l)$ be the joint probability of gene i belonging to the l-th regulatory pathway. That is, from genes' pathway assignments matrix $\tau = \{\tau_{il}\}_{I\times L}$, we obtain the first version of GRN as a multiplication of τ and its transpose τ^T:

$$G = \tau \times \tau^T. \quad (5)$$

Note that $G = \{G_{i,i'}\}$ consists of the probabilities of gene pairs belonging to the same regulatory pathways. Each entry $G_{i,i'}$ denotes the interaction probability of gene i and i', estimated from GE data.

The number of mixture components L can be determined by the component-wise EM algorithm [25] automatically, instead of trying all possible $L \in [L_{min}, L_{max}]$ via the time-consuming EM algorithm. The idea is to make use of the non-increasing property of $p(\mathcal{X}, \Gamma|\mathcal{M}_L)$ with respect to L, implying that the minimum message length (MML) criterion $\text{MML}(\mathcal{X}, \Gamma, \mathcal{M}_L)$ also decreases with L. Starting from $L = L_{max}$, we run the EM algorithm and compute the MML. At the next iteration, L is decremented by eliminating the smallest or empty components (setting the smallest θ_l as 0). The E and M steps are repeated to compute an updated MML value at each iteration until $\text{MML}(\mathcal{X}, \Gamma, \mathcal{M}_L)$ converges. In this way, not all values of L need to be evaluated, saving unnecessary computations.

Structural inferences from PPIN

Implicit PPI derived from transitivity

Suggested by [18], there exist a large number of interactions between protein complex components that are not yet observed or recorded. Yamada et al. [19] showed that PPIN evolutionary properties, e.g., shortest path, clustering coefficient, give some clues for potential protein interactions. We therefore propose a heuristic to augment protein-interaction networks (PPIN) by assuming transitivity among known protein interactions. The rationale behind this heuristic is also the base for a lot of similarity-based approaches of predicting GRNs, e.g., the relevance networks based algorithms [26] assumed that "if A is similar to B, then A interacts with B". Hence, "if A is similar to B, and B is similar to C, then A is similar to C, implying that A interacts with C." As such, we infer implicit interactions based on the first-order transitivity assumptions of protein interactions, i.e., if A interacts with B and B interacts with C, we infer that A interacts with C. Suthram et al. [27] designed methods to assign confidence scores of predicted potential protein interactions from multiple data sources, i.e., by combining GE data, literature, and PPIN data. Inspired by their work, we assign confidence scores of implicit protein-protein interactions (PPI) based only on GE and PPIN data. We use a confidence scoring scheme similar to that of [27]. The novelty of our score lies in its combination of the shortest path score and the Markov clustering score (a graph property) instead of the clustering coefficient.

We extract the shortest path of length $d_{ii'}$ from protein/gene i to protein/gene i' in protein interaction network P whose gene set is $\mathcal{I}$ by using Dijkstra's shortest path algorithm [28]. Let $\mathcal{E}$ denote the set of shortest paths $d_{ii'} < \infty$. When the interaction dataset P is quite sparse, we choose a larger PPI dataset of the same species, whose gene set is a superset of $\mathcal{I}$. Use of a larger PPI dataset incorporates additional biological hints for the analysis. By choosing the shortest path as a potential PPI, we obtain a higher confidence in deriving inferred implicit interactions. We also avoid the need to consider many paths by introducing all the transitive PPIs. For example, we can avoid the hassle of considering protein interactions in a clique.

Scoring the shortest path

Implicit interactions are designed to have lower scores compared to explicit interactions; scores are assigned inversely proportional to the number of intermediaries, i.e., implicit interactions with a larger number of intermediaries have proportionately lower scores. The path length $d_{ii'}$ indicates the strength of the implicit association between gene i and i'. Let the confidence score $c_d \in [0, 1]$ be a non-increasing number series over path length d, which quantifies the implicit protein interaction between i and i'. Clearly, c_1 should be 1 for 1-hop connections. For $d \geq 2$, we set $c_d = \zeta^{d-1}$ where ζ is the probability of extending the path by one hop. We assume that each additional hop is independent of the previous hop.

Using the equivalence assumption, we deduce the closed form approximation for ζ as follows. Suppose there are paths of length $d = 2, \ldots, \infty$ between proteins i and i', then there should be a direct interaction between protein i and i', i.e., $\sum_{d=2}^{\infty} c_d = 1$. Thus, the solution to this infinite summation $\sum_{d=1}^{\infty} \zeta^d = \frac{\zeta}{1-\zeta} = 1$ is $\zeta = 0.5$. We therefore assign a confidence score of $c_d = 0.5^{d-1}$ to each shortest path of length d. This score decreases exponentially with the number of hops, e.g., $c_2 = 0.5, c_3 = 0.25, c_4 = 0.125$, etc.

For a gene pair (i, i') in $\mathcal{I}$ with a corresponding shortest transitive path length $d_{ii'}$, a confidence score of $c_{d_{ii'}}$ is assigned omitting $\sum_{pl>d_{ii'}}^{\infty} 0.5^{pl-1}$. In fact, this confidence score is a conservative estimate of the real confidence score because it assumes that there exists path lengths of length $d > pl$. In practice, there may exist only one path between two genes. The shortfall is at most $\sum_{pl=d>2}^{\infty} 0.5^{pl-1} < 0.5$. If there exist many paths of varying lengths (overwhelming evidence) between a genes pair, then the confidence score becomes an underestimate. On the other hand, if there exists only one path, which is also the shortest path, then the confidence score is an accurate depiction of the strength of regulation between two genes.

We use the method of [27] to estimate PPI scores by modeling protein interactions as a function of two random variables: (1) the implicit PPI confidence score based on the shortest path and (2) the Pearson correlation coefficient of expression measurements for the corresponding genes. Unlike the approach of [27], which learns the weight of each random variable from a training set of positive and negative examples, we simply use uniform weights. Our formulation involves only two variables: the implicit PPI confidence score and Pearson correlation coefficient derived from GE data. Let W with elements W_{ij} denote the graph with connection strengths computed from these two variables, then

$$W_{ij} = \rho_{ij} + c_{d_{ij}} = \rho_{ij} + 0.5^{d_{ij}-1}, \tag{6}$$

where ρ_{ij} is the correlation of gene i and j in the gene expression data, and d_{ij} is the shortest path length between gene/protein i and j in the PPIN.

Until now, W collected the evidence from GE correlation and transitive protein-protein interaction for predicting direct gene-gene interactions. In fact, there are other evidences for gene-gene interactions. Similar to [29] that used a random walk model to consider the incompleteness of current gene ontology (GO) or PPIN evidences, we propose a random walk model to allow collecting additional evidence in a random fashion for predicting gene-gene interactions.

Topological connectivity via random walk transitions

The random walk models including Markov clustering algorithm [30] and PageRank [31] have been successfully used to model the link structure of graphs. Likewise, we extract the topological structure of the extended PPIN by using a random walk model instead of modelling with a small-world clustering coefficient [27].

Given an undirected graph W, random walk transition matrix T is defined as

$$T = \begin{cases} c\frac{W_{ij}}{\sum_{j=1}^{I} W_{ij}} + \frac{1-c}{I} & \text{if } W_{ij} \neq 0, \\ \frac{1-c}{I} & \text{otherwise.} \end{cases} \tag{7}$$

where $c \in (0.5, 1)$ is the fusion parameter (a.k.a. the damping factor in PageRank, typically set to 0.85) that determines the probability of the next transition from one of the outgoing links versus the transition from going to any random link.

Enright et al. [30] has shown that transition matrix T converges quadratically to an equilibrium state representing the topological connectivity of the graph. The converged matrix, denoted by $\hat{T} = \lim_{k\to+\infty} T^k$, can be computed by Markov CLustering (MCL) algorithm [30]. In our context, the converged matrix of extended PPIN shows how likely protein pairs are related to one another.

Extended PPIN

Recall that the confidence score of how likely the predicted PPI occurs are inferred from the shortest path evidences from existing database, which can be treated as the confidence from biological knowledge. Here, the topological connectivities of the extended PPIN are inferred from the graph structure or the topological properties by random walk transitivity. We combine both of them to refine the confidence score of how likely all protein pairs will interact with each other in the PPIN as $C = W + W * T$.

Combining W from (6) and the transition matrix from (7), the final confidence PPI scores are arrived as follows.

$$C_{ij} = 0.5 \times W_{ij} + 0.5 \times \sum_{k=1}^{I} W_{ik}\hat{T}_{kj}, \tag{8}$$

where W_{ij} denotes the confidence score derived from the PPIN database and GE correlations and $\hat{T}_{kj}$ represents the converged random walk probability. Hence, $W \times \hat{T}$ yields the confidence on the link structures of the original network. The summation term actually compensates for the confidence score W_{ij} that could have been underestimated in the previous step using the shortest path.

To summarize, score C_{ij} starts from the values estimated from PPIN and GE correlations and then updated with converged values of a random walk model. Thus, the confidence score takes into account topological properties of GRN. The extended PPIN is derived from thresholding the C scores.

Gaussian hidden Markov model

When genes collaborate to achieve a specific task, the corresponding protein products generally interact [13]. The PPI serve as valuable hints to underlying regulations among genes in GRN pathways. To fuse PPI and GE data, we treat gene expressions as observations, pathways as hidden states, and protein interactions as transitions in a Gaussian Hidden Markov Model (GHMM) [24]. In the GHMM, gene expression is the observed variable, the pathway it belongs to is the hidden state, given a hidden state (pathway), the expression of genes in the pathway follows a Gaussian model. The are gene-gene interactions are treated as transitions between their corresponding hidden states (pathways): if they are in the same pathway, the gene-pathway assignments are reinforced; otherwise, they are penalized. In this way, the GHMM combines the extended PPIN in the gene-pathway assignments.

Hidden Markov random field (HMRF) models assume that the conditional distribution of a variable obeys Markov property, i.e., the probability of a variable only depends on the neighbouring variables (see [32] for a complete description of HMRF). In the present context, HMRF graph is represented by a set of nodes where node i represents observation x_i with hidden variable γ_i and the neighborhood graph is represented by constraints C_{ij}, each of which indicates the edge weight between node i and node j. Specifically, the extended PPIN with confidence score C is used as constraints (prior knowledge) to the GHMM, i.e., they are considered neighbourhood structural priors of the corresponding gene-gene associations. Figure 2 shows a sample GHMM model with PPI priors.

The prior probability of a particular cluster assignment Γ follows a Gibbs distribution [33] as $p(\Gamma|C) = \frac{1}{Z}\exp\left(\sum_{i=1}^{I}\sum_{j\neq i}^{I} -C_{ij}\delta(\gamma_j \neq \gamma_i)\right)$ where $\delta(\cdot)$ is the indicator function and $Z = \sum_{\Gamma} p(\Gamma|C)$ is the normalizing function, and C_{ij} is the PPIN confidence score between gene i and j. Exact inference of the posterior requires the complete evaluation of $p(\Gamma|\mathcal{X}, C) \propto p(\mathcal{X}, \Gamma)p(\Gamma|C)$, where $p(\mathcal{X}, \Gamma)$ follows the Gaussian mixture model defined in (1). The posterior [34] probabilities for the multivariate Gaussian case are approximated as

$$\tau_{il} = p(x_i, \gamma_i = l|C)$$
$$= \frac{\theta_l p(x_i|\mu_l, \Sigma_l)}{\sum_{l'}^{L} \theta_{l'} p(x_i|\mu_{l'}, \Sigma_{l'})} \exp\left(\sum_{j\neq i} -C_{ij}(1-\tau_{jl})\right), \quad (9)$$

where θ_l is the mixture weight and $p(x_i|\mu_l, \Sigma_l)$ is the probability density function of the component model.

Based on the above, we learn a GHMM by using component-wise EM algorithm of "Soft-Clustering of GE by Gaussian mixture model" section, from which we derive the GRN from the pathway assignments: $G = \tau \times \tau^T$. The gene interaction score is then refined as $R = 0.5 \times G + 0.5 \times C$ where C is the confidence score of PPIN in (8).

Bayesian Gaussian network model

The GHMM assumes that the genes in a pathway are mutually and conditionally independent and do not consider partial correlation of their mRNA expression levels. To model this inter-gene dependency in a pathway, we use a Bayesian network (BN), a directed acyclic graph with local conditional distributions. Here, nodes represent genes, proteins, and/or metabolites, while edges represent molecular interactions such as protein-DNA and protein-protein interactions, including indirect relationships like those from inferred PPI.

BN are prone to overfitting of noisy or sparse training data such as gene expression data. Overfitting could lead to a vastly incorrect graph structure of GRN. Furthermore, since protein levels are unobservable from microarray data, vast majority of BN models of GRN proposed so far only include mRNA levels of genes as nodes but do not include protein levels. Thus, one way to reduce overfitting the BN is to incorporate the prior knowledge extracted from PPI. We thus propose a novel scoring scheme to feed implicit protein interactions (derived from GHMM) as structural priors into BN, thereby enhancing the robustness of predicted GRN.

Another major limitation of BN is its exponentially growing solution space with respect to the size of the network and thus approximate solutions such as those using Monte Carlo Markov Chain (MCMC) or genetic

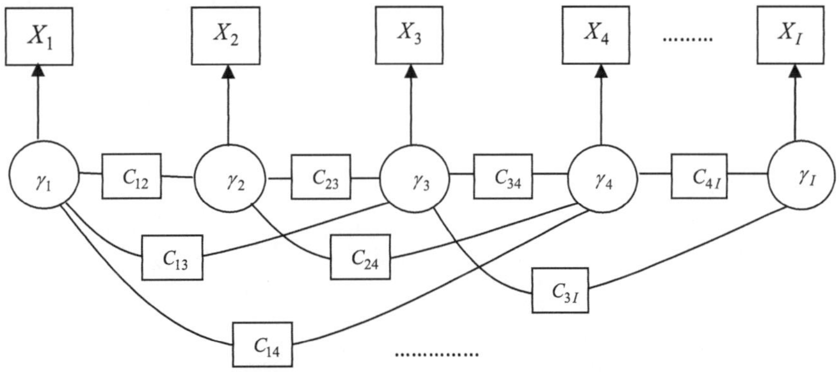

Fig. 2 A sample Gaussian Hidden Markov Model (GHMM) model. The gene expression observation of gene i is denoted by X_i, the circles denote hidden variables γ_i, and PPI confidence scores between genes i and j are denoted by C_{ij}

algorithms (GA) have been used. Such solutions are prone to errors and suboptimal. Since gene regulations exist among the genes in the same regulatory pathway, we constrain the parent set of each gene in the BN by first detecting the pathways, thereby greatly reducing the solution search space for subsequent computations. In a GRN, every gene appears in multiple pathways and therefore the relationship between genes in each pathway cannot be solved independently, i.e., by modelling each pathway as an independent BN. We thus use a Bayesian Gaussian Mixture (BGM) [35] that simply implements a BN with Gaussian Mixture observations. Here, observations for the Gaussian mixture reflect experimental observations, i.e., the experiments/observations are hard-clustered by using Gaussian Mixture Models. We thus use a MCMC inference method to learn the BGM model with structural priors derived from R scores in "Gaussian hidden Markov model" section, as shown in Fig. 1.

The BN is defined by a graph $\mathcal{G}$ with a family of conditional probability distributions and their parameters $\mathcal{Q} = \{q_i\}$, which together specify the joint distribution over the variables $p(\mathcal{X}|\mathcal{G}, \mathcal{Q})$. The joint distribution in a static BNs is factorized as $p(\mathcal{X}|\mathcal{G}, \mathcal{Q}) = \prod_{i=1}^{I} p(x_i|\pi_i, q_i)$ where each node x_i depends only on its parent nodes π_i. The parameter matrix $\mathcal{Q}$ is comprised of I vectors where each vector q_i specifies a local probability distribution.

If we assume a Gaussian Bayesian Network (GBN), the parameter vector $q_i = \{\mu_{\pi_i}, \sigma_{\pi_i}\}$ consists of the mean and standard deviation of the local probability distributions $p(x_i|\pi_i, q_i) \sim N(\mu_{\pi_i}, \sigma_{\pi_i})$. If we assume conditional independence among the J experiments, then $p(x_i|\pi_i, q_i) = \prod_{j=1}^{J} p(x_{ij}|\pi_i = x_{\pi_i,j}, q_i)$ where $p(x_{ij}|\pi_i = x_{\pi_i,j}, q_i) \sim N\left(\mu_{\pi_i,j}, \sigma^2_{\pi_i,j}\right)$. Assuming parameter independence, prior distribution $p(\mathcal{Q}|\mathcal{G})$ of the unknown parameters is expressed in terms of I local prior distributions: $p(\mathcal{Q}|\mathcal{G}) = \prod_{i=1}^{I} p(q_i|\mathcal{G}) = \prod_{i=1}^{I} p(q_i|\pi_i)$. The marginal likelihood $p(\mathcal{X}|\mathcal{G})$ is thus the integral over the parameter space:

$$\begin{aligned} p(\mathcal{X}|\mathcal{G}) &= \int p(\mathcal{X}, \mathcal{Q}|\mathcal{G}) d\mathcal{Q} \\ &= \int \left(\prod_{i=1}^{I} p(x_i|\pi_i, q_i) p(q_i|\pi_i) \right) d\mathcal{Q}, \end{aligned} \tag{10}$$

which can be rewritten as $p(\mathcal{X}|\mathcal{G}) = \prod_{i=1}^{I} \Psi(x_i, \pi_i)$, where $\Psi(x_i, \pi_i) = \int p(q_i|\pi_i) p(x_i|\pi_i, q_i) dq_i$.

GE data are generated under a variety of conditions that may include time-series experiments, i.e., each experiment may be a continuation of preceding experiments. Thus, it is infeasible to assume independence among individual experiments. Taking a step back, we assume the experiments are generated from Gaussian mixtures and adopt a Bayesian Gaussian mixture model to represent GE data.

Suppose there are K mixtures $\{D^{(k)}\}$ where each mixture $D^{(k)} = \{x_{\cdot,j}\}_{I*m}$ represents some m attributes of a gene set $\mathcal{I}$, and $\{\theta_k\}$ mixture weights of experiments in the BGM, then we express

$$\begin{aligned} p(\mathcal{X}|\mathcal{G}) &= \sum_{k=1}^{K} \theta_k p\left(D^{(k)}|\mathcal{G}\right) \\ &= \prod_{i=1}^{I} \sum_{k=1}^{K} \theta_k \Psi\left(D^{(k)}, x_i, \pi_i\right), \end{aligned} \tag{11}$$

where mixture weights $\{\theta_k\}$ are estimated with the EM algorithm, assuming independence among genes. Theoretical considerations and more details specifying the GBN parameters estimation can be found in [36].

Constraining the set of parent candidates

The GHMM derived in "Gaussian hidden Markov model" section can provide some structural constraints to BN learning. First, parents of genes should hail from the same regulatory pathway, i.e., if gene i belongs to some regulatory pathways, then other genes in these regulatory pathways could be its potential parents. In this manner, we generate a more accurate parent candidates set for each gene in addition to using the relationship defined in GRN. This is a nice middle ground between an exhaustive search through all the genes for potential parents and the limited set of GRN derived parents.

With the refined confidence score R assigned to every gene pair, the parent candidate set is limited to only top-k neighbours or top-N gene pairs as determined experimentally. One simple way to determine k and N is to use the average number extracted from a real GRN having a comparable size.

Incorporating neighborhood confidence scores

From [33], the prior probability of a particular structure $\mathcal{G}$ with constraints $R = \{R_{ij}\}$ follows a Gibbs distribution,

$$p(\mathcal{G}|R) = \frac{1}{Z} \exp\left(\sum_{i=1}^{I} \sum_{j \neq i}^{I} -R_{ij}\left(1 - \mathcal{G}_{ij}\right) \right) \tag{12}$$

where $Z = \sum_{\mathcal{G}} p(\mathcal{G}|R)$ is a normalizing function.

By using Markov networks framework to represent the correlations between neighbouring links, joint probability of data $\mathcal{X}$, given the BGM structure $\mathcal{G}$ and constraints R, can be written as $p(\mathcal{X}, \mathcal{G}|R) = p(\mathcal{X}|\mathcal{G}) p(\mathcal{G}|R)$. That is,

$$p(\mathcal{X}, \mathcal{G}|R) = \frac{1}{Z} \prod_{i=1}^{I} \left[\sum_{k=1}^{K} \theta_k \Psi\left(D^{(k)}, x_i, \pi_i\right) \right] \exp\left(\sum_{j \neq i}^{I} -R_{ji}(1 - \mathcal{G}_{ji}) \right) \tag{13}$$

Learning BGM with constraints via MCMC

Given GE data $\mathcal{X}$, fusing association scores R computed from GHMM and fixed parameters $\mathcal{Q}$, the structural posterior probability is written as

$$p(\mathcal{G}|\mathcal{X},\mathcal{Q},R) = \frac{p(\mathcal{X},\mathcal{G}|\mathcal{Q},R)}{p(\mathcal{X})} \propto p(\mathcal{X},\mathcal{G}|\mathcal{Q},R) \quad (14)$$

In the context of static BNs, different MCMC methods have been proposed for sampling Directed Acyclic Graphs (DAG) from the structural posterior distribution. We adopt the structural MCMC approach of [35] to sample our BGM structure $\mathcal{G}$ from posterior distributions $p(\mathcal{G}|\mathcal{X},\mathcal{Q},R)$. The idea is to give preference to the structures of higher posterior probability. The details of MCMC method to learn DAG from BGM are described in the supplementary materials. With model averaging [37], i.e., we run MCMC structural learning for a maximum number of iterations and each predicted edge in the GRN is assigned a confidence score that is measured by the number of occurrences of predicted edges among the generated graphs.

Experiments and results

Datasets

Our methods were tested on a dataset consists of 25 yeast genes, similar to those used by similar studies on fusion of GE and PPIN [6, 16]. The only difference is that we added five more genes (CDC28, CLB6, CLN3, FUS3, FKH2) that are highly connected to the 25 genes, based on current biological databases and literature. The 30 genes network involves in the cell-cycle regulation of yeast. A cell cycle is comprised of four phases: (i) Gap 1 (G1) phase - the checkpoint to ensure that the cell is ready for division, (ii) Synthesis (S) phase - involving DNA replication, (iii) Gap 2 (G2) phase - a checkpoint to ensure that the cell is ready to enter the next phase, and (iv) Mitotic (M) phase referring to cell division.

We collected data from several resources in order to construct a comprehensive target ground-truth network: (i) GeneNetWaver (http://wwwmgs.bionet.nsc.ru/mgs/gnw/genenet/) [38, 39] that includes 12873 transcriptional regulations among 4441 yeast genes where 62 transcriptional regulations are among our 30 cell cycle genes; (ii) 141 literature-reported regulatory relations among the 30 target genes manually collected from [40, 41], etc.; and (iii) Lee et al. [42] which proposed a integrated functional association score from mRNA expression data, PPIN, and literature mining edges of 5552 yeast genes. The linkage score showed good performance on independent benchmark datasets from KEGG (https://www.genome.jp/kegg/), STRING (https://string-db.org/), Gene Ontology (www.geneontology.org/), and experimentally-determined subcellular localization. We collected the "Int-Net" from [42] with a high likelihood of regulation score above 0.5, which included co-expression regulations from 717 experiments for yeast (divided into 27 experimental categories), protein-protein interaction experiments, and literature mining of edges. From this network, 166 linkages are among the 30 target genes. In total, the ground-truth network has 317 regulations among 30 cell cycle genes.

The GE data was obtained from [43] that contains 77 experiments collected over 8 yeast cell cycles by using four different synchronisation protocols. The PPI data for *Saccharomyces Cerevisiae* was downloaded from BioGRID (http://thebiogrid.org/), that contains 6263 proteins and 210,996 interactions.

For comparison, true positives (TP), false positives (FP), and false negatives (FN) of edges were computed by comparing the predicted pathways to the target ground truth network. Various performance metrics including Precision, Recall, and F1-score were evaluated. To compare with state-of-the-art methods of predicting GRN from GE data, such as CLR [26], GENIE3 [44], TIGRESS [45], we determined AUROC (The area under the receiver operating characteristic (ROC) curve) and AUPR (The area under the precision-recall (PR) curve) scores as defined in [44]. For these metrics, the best results that are significantly different (p-value < 0.05) from other methods are shown in bold. We implemented t-test to check whether the best result is significantly different from other methods; If the best result is not significantly different from the next best ones, we choose the best results as one group and the others as another group and then implemented the unpaired t-test for the two groups.

Feasibility of extending PPIN

In "Structural inferences from PPIN" section, we described a method to extend PPIN, using transitive protein interactions and assigned confidence scores C to the predicted protein interactions. We will demonstrate the feasibility of predicting protein interactions from this extended PPIN in this section.

To test how well extended PPIN (derived from C scores) recover an incomplete PPIN, we randomly selected 200 yeast genes and assign 3008 PPIs from BioGRID among them. We used 10-fold cross-validation on the PPINs of these 200 randomly selected genes, by randomly removing 10% edges in the target PPIN, deriving extended PPIN based on the remaining 90% edges, and then comparing the extended PPIN with the target PPIN on the missing 10% edges. We repeated 10-fold cross-validation experiments for 10 times to show the robustness of the method. The results demonstrated that the extended PPIN (when the cut-off threshold for C score is set as 0.5) effectively recover 346 out of the 383 removed edges on average. Further, our extended PPIN based only on 90% PPIN edges predicted the entire PPIN with an average F1-score of 43%, indicating the effectiveness of the method of extending PPIN (C score) and in its utility in predicting the missing PPIN.

As BioGrid may contain functional linkages predicted as protein-protein interactions, we also choose another PPIN dataset to show the reliability of the method of extending PPIN (C score) to predict real PPIs. The yeast PPIN data from http://interactome.dfci.harvard.edu/S_cerevisiae consists of the full set of physical interactions that occur in a physiologically relevant dynamic range between all its macromolecules, including protein-protein, DNA-protein, and RNA-protein interactions. By combing all the physical interactions, co-complex membership associations [46] and literature-curated interactions, [47] collected in "Yeast Interactome Datasets" from which we get 11995 interactions of 2234 yeast proteins where only 100 PPIs are found among the same randomly selected 200 yeast genes. The experiments of 10-fold cross-validation on these 100 PPIs showed that our extended PPIN (when the cut-off threshold for C score is set as 0.55) effectively recovers the removed edges on average. It also shows that the extended PPIN based only on 90% PPIN edges predict the entire PPIN with an average F1-score of 36.62% (average precision at 39.56% and average recall at 34.10%). This further validates the reliability of the method of extending PPIN (C score) to predict the real PPIs.

In the BGM with prior model, we added the prior information into the likelihood of gene expression data of the learned BN structure. If the GRN learned from GE data is consistent with prior information, the likelihood is reinforced; otherwise, the likelihood is penalized. Thus, the GRN from the fusion model achieves the maximum consistency between gene expression data and prior information. We choose the informative BioGRID PPINs as priors in our fusion model.

In order to show the effectiveness of predicting gene regulations from extended PPIN, we evaluated the extended PPIN alongside the original PPIN (denoted by "raw PPIN"). As mentioned in "Structural inferences from PPIN" section, we can extend PPIN either from a sub-network or from the global network where the global network is the PPIN for the whole genome of the same organism and the sub-network indicates the subnet PPIN defined by the genes in the target gene set. When the PPIN was extended from subnet information, we only consider transitive links among the target gene set. If the subnet PPIN is quite sparse, we choose the global network to make use of transitivity. However, extended PPIN from global information may introduce more noise especially when PPIN is spurious. We experimentally compared differences in the performance of extending PPIN from the sub-network (i.e., 30-gene PPIN for the benchmark) and the global network (i.e., the complete PPIN from 6263 yeast genes) on the 30-gene benchmark network. Table 1 shows the performance with three PPINs: raw PPIN, extended PPIN with subnet information, and extended PPIN with global information on predicting the GRN of the 30 benchmark genes. The cut-off thresholds to generate GRN from C score for the extended PPINs were set as 0.5 for a better trade-off between precision and recall as seen in Fig. 3.

Comparing two extended PPINs, we see that the extended PPIN with global information outperforms the prediction with only subnet information on recall, sacrificing some precision. This indicates that global information may introduce more incoherent genes to make prediction noisy. The observation that both local and global information achieve comparable performance shows that global information can complement subnet information from sparse PPIN data. Clearly, if not much protein interactions are lost in the target PPIN, extended PPIN with subnet information is a good choice to balance the precision and recall, as illustrated by the best F1 performance with extended PPIN from local information. We thus choose extended PPIN with subnet information for the 30-gene benchmark network in subsequent experiments.

Comparing extended PPIN with the raw PPIN, we see that the extended PPIN with subnet information yielded a notable 20% improvement on recall compared to raw PPIN at a hefty decrease of 12% in precision. This shows the feasibility of deriving regulations by using the transitivity present in the PPIN. It is reasonable that raw PPIN predicts GRN with the highest precision, indicating how much protein interactions in the current PPIN database are consistent with gene regulation data. The extended PPINs provide better recall/coverage on

Table 1 Performance of predicting GRN with different PPINs on 30 yeast genes ground-truth network

Method	*TP*	*FP*	*FN*	*Prec.(%)*	*Rec.(%)*	*F1(%)*	*AUROC*	*AUPR*
Raw PPIN	165	95	152	63.46	52.05	57.19	0.698	0.591
Extended PPIN $_{(subnet)}$	229	220	88	51.00	**72.24**	59.79	**0.713**	0.533
Extended PPIN $_{(global)}$	240	260	77	48.00	**75.71**	58.75	0.696	0.500

Best performance measures that are significantly different are shown in bold

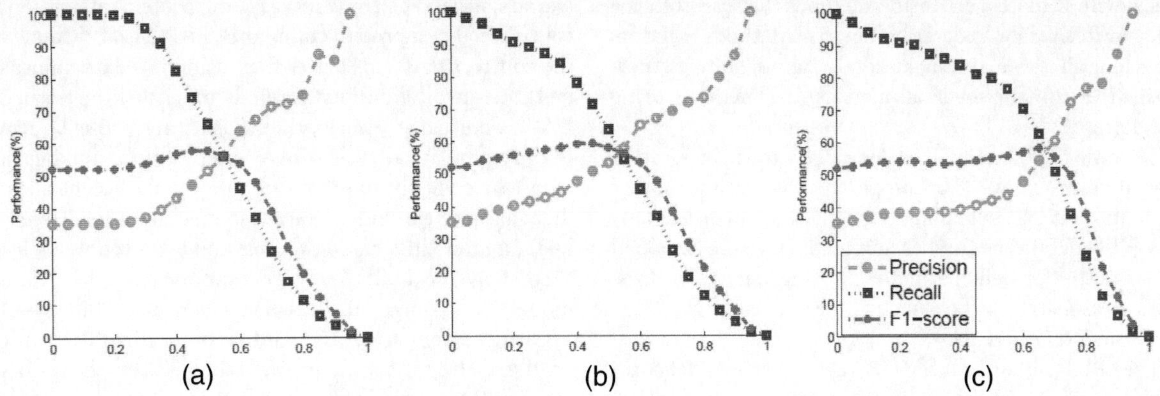

Fig. 3 Performance comparison of C and R scores from the subnet or the global information at various cut-off thresholds on 30 yeast genes ground-truth network. **a** C scores (global) **b** C scores (subnet) **c** R scores (subnet)

predicting GRN even with the subnet information, giving the best F1-score, indicating the incompleteness of the current PPIN database. The precision of extended PPIN can be improved by raising the C score threshold but the recall will then be decreased at the same time. Since our approach is a coarse-to-fine framework to predict GRN where the extended PPIN is used to limit candidate gene interaction set (coarse step), a higher coverage or recall is preferred. As such, we choose 0.5 as the cut-off threshold that gave us a reasonable recall. The high false positive rate from the extended PPIN can be reduced in the subsequent "refinement" step. As subsequent experiments demonstrate, our methods predict the final GRN with a high precision. Therefore, what follows next are demonstrated only with extended PPIN data.

Comparison of performances

We benchmarked the generated GRNs against the ground-truth networks. All the methods except the BGM generated networks with undirected edges. When comparing an undirected graph to a directed one, we considered an existence of a matched link regardless of its direction. For comparison, all the networks generated by the BGM were sampled and averaged for 10 models each of which was learned from 1000 iterations of MCMC structural moves. The performance of the BGMs were further improved by increasing the number of models by model averaging and the iterations of structural moves.

Since the generated graphs from C, R, and G assigned confidence scores for the edges, i.e., probabilities of edge existence, we chose a cut-off threshold and constructed the GRN, based on the scores: if the edge score is higher than this threshold, we confirm the existence of an edge; and otherwise, the edge was rejected. From experiments, as seen in Fig. 3, we found that the best cut-off thresholds (for C and R scores) that maximize the F1-score (which is also the best trade-off between precision and recall) of GRN tend to be clustered around 0.5. For GRN structure $\mathcal{G}$ from clustering or the BGM, scores came from the probability of genes belonging to the same pathway or the model averaging where both reflected how likely genes interact. We therefore intuitively fixed 0.5 as the threshold to determine interaction edges. Thus, for consistency, we set the cut-off thresholds for all our graphs to be 0.5. The confidence scores were normalized between 0.0 and 1.0 before averaging. The thresholds can also be experimentally fine-tuned to further improve the performances of our methods. The vast majority of existing works on prediction of GRN choose the cut-off of the confidence score as a trade-off between sensitivity and specificity of prediction [48].

Table 2 compares performances of building GRNs by various methods. The first four methods (CLR, GENIE3, TIGRESS, GMM) are recent state-of-the-art methods of predicting GRN from GE. We use MATLAB implementations of CLR [26], GENIE3 [44], and TIGRESS [45]. The other three methods (GHMM, R scores from GHMM,

Table 2 Performances of prediction of GRNs by various methods on 30 yeast genes ground-truth network

Method	*TP*	*FP*	*FN*	*Prec.(%)*	*Rec.(%)*	*F1(%)*	*AUROC*	*AUPR*
CLR	190	312	127	37.85	59.94	46.40	0.555	0.388
GENIE3	128	202	189	38.79	40.38	39.57	0.546	0.395
TIGRESS	140	207	177	40.35	44.16	42.17	0.546	0.392
GMM	172	266	145	39.27	54.26	45.56	0.583	0.412
GHMM	258	329	59	43.95	**81.39**	57.08	0.664	0.461
R scores (GHMM)	250	262	67	**48.83**	78.86	**60.31**	**0.705**	**0.501**
BGM (R scores)	202	237	115	46.01	63.72	53.44	0.627	0.446

Best performance measures that are significantly different are shown in bold

BGM with R scores) are fusion methods that use both the GE and PPIN data as described in the "Methods" section. As seen, all three fusion methods significantly outperformed (5–20% better in all metrics) the methods using GE data only.

To show the effect of extending PPIN by incorporating transitive edges, we also compared three fusion models with and without extending the PPIN (i.e., by using only raw PPIN data). The three fusion methods using extended PPIN outperformed the methods using only raw PPIN. This demonstrates the effectiveness of using extended PPIN in GRN inference.

The GRNs obtained using GHMM and refined R scores are undirected graphs, indicating gene interactions but not specifically the directional regulations. To figure out the directional dependencies among genes, a directional model such as BGM is needed. The BGM model based on R scores as constraints in fact gave the overall best performance and thus deserves further investigation.

BGM with different priors

Using the same set of parent candidates and the BGM model, we evaluated whether the BGM using refined confidence scores R can hold its own when benchmarked against other PPIN fusion methods. Specifically, we compared its performance with two well-known methods. Nariai et al. [14] tested if each added protein pair form a protein complex by considering the complex as a virtual node in the network. The method computes the principal component analysis (PCA) model of the protein pairs from original observations and checks if the protein complex contributes to a higher likelihood. If a higher likelihood is obtained, the complex (pair) is accepted and otherwise rejected. Imoto et al. [15] updated PPIN whenever the learned GRN produced different structure. Each inconsistent edge in the learned DAG is perturbed by either removing or reversing its direction. If a perturbation leads to a better likelihood, the perturbed PPIN is accepted otherwise it is rejected.

Both methods needed significantly more operations than the basic models and increased already substantial computational complexity. Further, they tend to converge to local optima given their greedy hill-climbing nature. Nariai's method can only increase the protein interactions contributed by a protein complex if the PCA projection of the complex performs better than using raw data. Imoto's method is overly-optimistic as it assumes that the learned DAG is correct. However, the MCMC method is known to only simulate a DAG; it may accept a large number of incorrect edges in one iteration and could accept all of them as updated PPIN in the subsequent iteration. Hence, updating the PPIN based on a learned DAG is unreliable.

To show the utility of each component in the fusion model, we compared the performance of the BGM by using priors from the GMM only, the PPIN only or the GHMM, as shown in Table 3. For the priors from PPIN, we also tested two existing fusion methods with the same priors. We described different components of our method as follows: "BGM (GMM)" denotes the BGM with parental constraints from GMM and predicts GRN only from GE data without PPIN data; "BGM (C scores)" denotes the BGM constrained with C scores from extended PPIN; "Nariai et al. (GHMM)" and "Imoto et al. (GHMM)" use parental constraints from the GHMM with extended PPIN to learn GRNs in BGM with corresponding fusion method; "BGM (R scores)" uses the extended PPIN in the GHMM fusion model and then R constraints for the BGM. For comparison, we also included a BGM without any constraints as the baseline.

The results in Table 3 show that the priors from GHMM significantly improve TPs (highest recall) with comparable F1 scores (higher coverage at the slight expense of precision) compared to the methods using priors only from the PPIN. This indicates that GE clustering help to predict more TPs but at a lower precision. It also demonstrates that the use of the GRN structure from the GMM and extended PPIN in the prediction of GRN. These results further validate the utility of each component in the fusion model. Results in Table 3 show that with same structural constraints, the fusion BGM (with R scores) beats both the BGM proposed by Nariai et al. and by Imoto et al. This means that the R scores are effective measure of the reliability of PPIN edges. Table 3 also shows the running-time of different algorithms implemented with MATLAB on a machine with an Intel Xeon E5-1620 3.6GHz CPU

Table 3 Performance of GRNs generated from the BGM with different priors on 30 yeast genes ground-truth network

Method	*TP*	*FP*	*FN*	*Prec.*(%)	*Rec.*(%)	*F1*(%)	*AUROC*	*AUPR*	*Time*(sec)
BGM	139	201	178	40.88	43.85	42.31	0.553	0.390	636
BGM (GMM)	185	293	132	38.70	58.36	46.54	0.555	0.383	606
BGM (C scores)	170	150	147	**53.13**	53.63	**53.38**	**0.653**	**0.490**	**518**
Nariai et al. (GHMM)	176	273	141	39.20	55.52	45.95	0.570	0.408	1174
Imoto et al. (GHMM)	184	293	133	38.57	58.04	46.35	0.565	0.401	1150
BGM (R scores)	202	237	115	46.01	**63.72**	**53.44**	0.627	0.446	608

Best performance measures that are significantly different are shown in bold

and 8GB RAM. As seen, the priors given by our confidence scores lead to computationally more efficient procedures.

Discussion & conclusion

We proposed an automated method to detect gene regulations by fusing GE data and PPI data. Gene expression data is first soft-clustered into regulatory pathways by a GMM where the number of regulatory pathways is automatically determined via a component-wise EM algorithm. Transitive protein interactions are derived using a novel confidence score and PPIN are refined and extended. The extended PPIN are then fused with the GMM derived from GE data, using a GHMM. This improves the biological relevance of the clustering results and also refines the confidence score of indirect gene regulations/protein interactions. Using refined PPI confidence scores, together with regulatory pathways obtained from the GHMM as structural constraints, a BGM model was used to capture direct gene regulations by using an effective and efficient MCMC procedure. Fusion of the BGM model with GHMM as structural priors generated more accurate GRN compared to those produced by the GHMM and the BGM models. We have experimentally shown that our procedures are more effective than two existing well-known methods for fusing GE and PPIN data.

Furthermore, the fusion framework reduces the overall time complexity of building GRN with Bayesian networks. The GHMM uses a component-wise EM algorithm to soft-cluster genes into pathways at a complexity $O\left(I^2 \times L^2\right)$ (where L denotes the number of components and I denotes the number of genes), indicating that the method scales well on large networks (large I). The BGM structural constraints, i.e., the parent of each node must itself be a node from the same regulatory pathway, help to significantly reduce the search space, i.e., from $O\left(I^I\right)$ to approximately $O\left((I/L)^I\right)$. In fact, for larger I, the reduction can even be more pronounced.

Our approach shows the importance of systematically fusing multiple sources of biological evidences in inferring useful and reliable GRN. In fact, our framework can be extended to fuse more than two data sources, such as gene ontologies, biological pathways (KEGG), transcription factors (known regulators), information mined from literature, and other multi-omics data [4, 5, 10–12]. Our methods may be easily extended by further considering dependencies among the experiments [35] and are worth investigating further.

Abbreviations

AUROC: The area under the receiver operating characteristic (ROC) curve; AUPR: The area under the precision-recall (PR) curve; BGM: Bayesian Gaussian mixture; EM: Expectation maximization; GHMM: Gaussian hidden Markov model; GRN: Gene regulatory networks; HMRF: Hidden Markov random field; MCL: Markov cLustering; MML: Minimum message length; MCMC: Monte Carlo Markov Chain; PPIN: Protein-protein interaction networks

Acknowledgements

Not applicable.

Funding

Publication of this article was sponsored by Tier-2 MOE2016-T2-1-029 grant by the Ministry of Education, Singapore.

Availability of data and materials

The MATLAB source code are available at https://gitlab.com/liuwt/fusionBGM.

About this supplement

This article has been published as part of *BMC Systems Biology Volume 13 Supplement 2, 2019: Selected articles from the 17th Asia Pacific Bioinformatics Conference (APBC 2019): systems biology*. The full contents of the supplement are available online at https://bmcsystbiol.biomedcentral.com/articles/supplements/volume-13-supplement-2.

Authors' contributions

WL developed and implemented the method. WL and JCR conceptualised the methods and wrote the manuscript. Both authors have read and approved the manuscript.

Ethics approval and consent to participate

Not applicable.

Consent for publication

Not applicable.

Competing interests

The authors declare that they have no competing interests.

Publisher's Note

Author details

[1]School of Public Health and Management, Hubei University of Medicine, Shiyan, Hubei, China. [2]Integrative Biology and Physiology, University of California, Los Angeles, Los Angeles, CA, USA. [3]School of Computer Engineering, Nanyang Technological University, Singapore, Singapore.

Published: 5 April 2019

References

1. Liu F, Zhang S-W, Guo W-F, Wei Z-G, Chen L. Inference of gene regulatory network based on local bayesian networks. PLoS Comput Biol. 2016;12(8):e1005024.
2. Liu Z-P. Reverse engineering of genome-wide gene regulatory networks from gene expression data. Curr genomics. 2015;16:3–22.
3. Langfelder P, Horvath S. Wgcna: an r package for weighted correlation network analysis. BMC Bioinformatics. 2008;9:559.
4. Chan TE, Stumpf MPH, Babtie AC. Gene regulatory network inference from single-cell data using multivariate information measures. Cell Syst. 2017;5:251–67.
5. Aibar S, González-Blas CB, Moerman T, Huynh-Thu VA, Imrichova H, Hulselmans G, Rambow F, Marine J-C, Geurts P, Aerts J, van den Oord J, Atak ZK, Wouters J, Aerts S. Scenic: Single-cell regulatory network inference and clustering. Nat Methods. 2017;14:1083–6.
6. Bernard A, Hartemink AJ. Informative structure priors: Joint learning of dynamic regulatory networks from multiple types of data. In: Pacific Symposium on Biocomputing. Hawaii: 2005. p. 459–470.
7. Wu M, Chan C. Learning transcriptional regulation on a genome scale: a theoretical analysis based on gene expression data. Brief Bioinform. 2011;13(2):150–61.
8. Wang L, Hou L, Qian M, Deng M. Integrating phosphorylation network with transcriptional network reveals novel functional relationships. PLoS ONE. 2012;7(3):33160.

9. Cheng C, Alexander R, Min R, et al. Understanding transcriptional regulation by integrative analysis of transcription factor binding data. Genome Res. 2012;22(9):1658–67.
10. Zarayeneh N, Ko E, Oh JH, Suh S, Liu C, Gao J, Kim D, Kang M. Integration of multi-omics data for integrative gene regulatory network inference. Int J Data Min Bioinform. 2017;185:223–39.
11. Banf M, Rhee SY. Enhancing gene regulatory network inference through data integration with markov random fields. Sci Rep. 2017;7:1–13.
12. Yugi K, Kubota H, Hatano A, Kuroda S. Trans-omics: How to reconstruct biochemical networks across multiple 'omic' layers. Trends Biotechnol. 2016;34:276–90.
13. Segal E, Wang H, Koller D. Discovering molecular pathways from protein interaction and gene expression data. In: International Conference on Intelligent Systems for Molecular Biology. Brisbane: 2003. p. 264–272.
14. Nariai N, Kim S, Imoto S. Using protein-protein interactions for refining gene networks estimated from microarray data by bayesian networks. In: Pacific Symposium on Biocomputing. Hawaii: 2004. p. 336–347.
15. Imoto S, Higuchi T, Goto T. Error tolerant model for incorporating biological knowledge with expression data in estimating gene networks. Stat Methodol. 2006;3(1):1–16.
16. Werhli AV, Husmeier D. Gene regulatory network reconstruction by bayesian integration of prior knowledge and/or different experimental conditions. J Bioinforma Comput Biol. 2008;6(3):543–72.
17. Li M, Wu X, Wang J, Pan Y, Towards the identification of protein complexes and functional modules by integrating ppi network and gene expression data. BMC Bioinformatics. 2012;13:109.
18. Chen Y, Xu D. Global protein function annotation through mining genome-scale data in yeast saccharomyces cerevisiae. Nucleic Acids Res. 2004;32(21):6414–24.
19. Yamada T, Bork P. Evolution of biomolecular networks — lessons from metabolic and protein interactions. Nat Rev Mol Cell Biol. 2009;10(11): 791–803.
20. Emmert-Streib F, Glazko GV, Altay G, Simoes RM. Statistical inference and reverse engineering of gene regulatory networks from observational expression data. Front Genet. 2012;3:8.
21. Bansal M, Belcastro V, Ambesi-Impiombato A. How to infer gene networks from expression profiles. Mol Syst Biol. 2007;3:78.
22. Kuffner R, Petri T, Tavakkolkhah P, Windhager L, Zimmer R. Inferring gene regulatory networks by anova. Bioinformatics. 2012;28(10):1376–82.
23. Zhang X, Zhao X-M, He K, Lu L, Cao Y, Liu J, Hao J-K, Liu Z-P, Chen L. Inferring gene regulatory networks from gene expression data by path consistency algorithm based on conditional mutual information. Bioinformatics. 2012;28(1):98–104.
24. Vignes M, Forbes F. Gene clustering via integrated markov models combining individual and pairwise features. IEEE Trans Comput Biol Bioinforma. 2009;6:260–70.
25. Figueiredo MAT, Jain AK. Unsupervised learning of finite mixture models. IEEE Trans Pattern Anal Mach Intell. 2002;24(3):381–96.
26. Faith JJ, Hayete B, Thaden JT, Mogno I, Wierzbowski J, Cottarel G, Kasif S, Collins JJ, Gardner TS. Large-scale mapping and validation of escherichia coli transcriptional regulation from a compendium of expression profiles. PLoS Biol. 2010;5(1):8.
27. Suthram S, Shlomi T, Ruppin E. A direct comparison of protein interaction confidence assignment schemes. BMC Bioinformatics. 2006;7:360.
28. Johnson DB. A note on dijkstra's shortest path algorithm. J Assoc Comput Mach (ACM). 1973;20(3):385–8.
29. Yang H, Nepusz T, Paccanaro A. Improving go semantic similarity measures by exploring the ontology beneath the terms and modelling uncertainty. Bioinformatics. 2012;28:1383–9.
30. Enright AJ, Dongen S, Ouzounis CA. An efficient algorithm for large-scale detection of protein families. Nucleic Acids Res. 2002;30(7):1575–84.
31. Page L, Brin S, Motwani R. The pagerank citation ranking: Bringing order to the web. Technical report, Stanford InfoLab. 1999.
32. Li SZ. Markov random field modeling in image analysis, 2nd edn. New York: Springer; 2001.
33. Chandgotia N. Generalisation of the Hammersley-Clifford theorem on bipartite graphs. Trans Amer Math Soc. 2017;369:7107–7137.
34. Lange T, Law MH, Jain AK, Buhmann JM. Learning with constrained and unlabelled data. In: IEEE Computer Society Conference on Computer Vision and Pattern Recognition. Beijing; 2005. p. 731–8.
35. Grzegorczyk M, Husmeier D, Edwards KD. Modeling non-stationary gene regulatory processes with a non-homogeneous bayesian network and the allocation sampler. Nucleic Acids Res. 2008;24(18):2071–8.
36. Grzegorczyk M. An introduction to gaussian bayesian networks. Methods Mol Biol. 2010;662:121–47.
37. Battle A, Jonikas M, Walter P. Automated identification of pathways from quantitative genetic interaction data. Mol Syst Biol. 2010;6:379.
38. Ananko EA. Genenet: a database on structure and functional organisation of gene networks. Nucleic Acids Res. 2002;30(1):398–401.
39. Schaffter T, Marbach D, Floreano D. Genenetweaver: In silico benchmark generation and performance profiling of network inference methods. Bioinformatics. 2011;27(16):2263–70.
40. Simon I, Barnett J, Hannett N, Harbison CT, Rinaldi NJ, Volkert TL, Wyrick JJ, Zeitlinger J, Gifford DK, Jaakkola TS, Young Ra. Serial regulation of transcriptional regulators in the yeast cell cycle. Cell. 2001;106(6):697–708.
41. Lee TI, Rinaldi NJ, Robert F, et al. Transcriptional regulatory networks in saccharomyces cerevisiae. Science. 2002;298(5594):799–804.
42. Lee I, Date SV, Adai AT. A probabilistic functional network of yeast genes. Science. 2004;306(5701):1555–8.
43. Spellman PT, Sherlock G, Zhang MQ. Comprehensive identification of cell cycle regulated genes of the yeast saccharomyces cerevisiae by microarray hybridization. Mol Biol Cell. 1998;9(12):3273–97.
44. Huynh-Thu VA, Irrthum A, Wehenkel L, Geurts P. Inferring regulatory networks from expression data using tree-based methods. PLoS ONE. 2010;5:9.
45. Haury A-C, Mordelet F, Vera-Licona P, Vert J-P. Tigress: Trustful inference of gene regulation using stability selection. BMC Syst Biol. 2012;6:145.
46. Collins SR, Kemmeren P, Zhao X-C, Greenblatt JF, Spencer F, Holstege FCP, Weissman JS, Krogan NJ. Toward a comprehensive atlas of the physical interactome of saccharomyces cerevisiae. Mol Cell Proteomics. 2007;6(3):439–50.
47. Reguly T, Breitkreutz A, Boucher L, Breitkreutz BJ, Hon GC, Myers CL, Parsons A, Friesen H, Oughtred R, Tong A, Stark C, Ho Y, Botstein D, Andrews B, Boone C, Troyanskya OG, Ideker T, Dolinski K, Batada NN, Tyers M. Comprehensive curation and analysis of global interaction networks in saccharomyces cerevisiae. J Biol. 2006;5(4):11.
48. Djebbari A, Quackenbush J. Seeded bayesian networks: Constructing genetic networks from microarray data. BMC Syst Biol. 2008;2:57.

Li et al. BMC Systems Biology 2019, 13(Suppl 2):26
https://doi.org/10.1186/s12918-019-0696-9

BMC Systems Biology

RESEARCH Open Access

FCMDAP: using miRNA family and cluster information to improve the prediction accuracy of disease related miRNAs

Xiaoying Li[1], Yaping Lin[1], Changlong Gu[1*] and Jialiang Yang[2]

From The 17th Asia Pacific Bioinformatics Conference (APBC 2019)
Wuhan, China. 14-16 January 2019

Abstract

Background: Biological experiments have confirmed the association between miRNAs and various diseases. However, such experiments are costly and time consuming. Computational methods help select potential disease-related miRNAs to improve the efficiency of biological experiments.

Methods: In this work, we develop a novel method using multiple types of data to calculate miRNA and disease similarity based on mutual information, and add miRNA family and cluster information to predict human disease-related miRNAs (FCMDAP). This method not only depends on known miRNA-diseases associations but also accurately measures miRNA and disease similarity and resolves the problem of overestimation. FCMDAP uses the k most similar neighbor recommendation algorithm to predict the association score between miRNA and disease. Information about miRNA cluster is also used to improve prediction accuracy.

Result: FCMDAP achieves an average AUC of 0.9165 based on leave-one-out cross validation. Results confirm the 100, 98 and 96% of the top 50 predicted miRNAs reported in case studies on colorectal, lung, and pancreatic neoplasms. FCMDAP also exhibits satisfactory performance in predicting diseases without any related miRNAs and miRNAs without any related diseases.

Conclusions: In this study, we present a computational method FCMDAP to improve the prediction accuracy of disease related miRNAs. FCMDAP could be an effective tool for further biological experiments.

Keywords: Disease-related miRNA, Leave-one-out cross validation, miRNA family information, miRNA cluster information, Nearest neighbor recommendation algorithm

Background

MicroRNAs (miRNAs) are small endogenous non-coding RNAs with length of about 22 nt and can regulate gene expression mainly through post-transcription [1]. The latest version of miRBase consists of 1881 human miRNAs, and most of them regulate more than 60% of human protein-coding genes. miRNAs regulate target genes through biological processes, such as cell growth, proliferation, differentiation and apoptosis. miRNAs play a critical role in the development of various diseases including cancers [2]. Takamizawa et al. [3] found that the expression level of let-7 decreases in lung neoplasms in vivo and in vitro, resulting in shortened post-operative survival of the patients. Moreover, let-7 is a potential therapeutic miRNA for prevention of tumorigenesis. Lung neoplasms are characterized by several key oncogene mutations, including p53, RAS, and MYC; some of which may be directly related to the decreased expression of let-7 and may be inhibited by introducing this miRNA [3]. miRNAs can be used as biomarkers to identify cancer tissure origin of unknown primary origin [4, 5]. Therefore, identification of disease-related miRNAs would benefit research on pathogenesis and diagnosis.

* Correspondence: guchanglong@hnu.edu.cn
[1]College of Computer Science and Electronic Engineering, Hunan University, Changsha 410082, Hunan, China
Full list of author information is available at the end of the article

Many disease-related miRNAs have been identified through biological experiments. Researchers have collected data from existing literature to build miRNA-related databases, such as miRBase [6], miRGen [7], miRTarBase [8], miRWalk [9], microRNA.org [10], miR-Cancer [11], HMDD [12], miR2Disease [13], dbDEMC [14], and PhenomiR [15]. These databases provide solid data foundation for study of miRNAs. However, methodologies for screening of miRNA-disease associations are costly and time consuming. In this regard, computational methods are used to predict miRNAs that are most likely associated with a disease and provide experimental targets for biological experiments to save cost and time.

Computational methods are classified into two main categories, namely, network-based methods and machine-learning-based methods [16]. Network-based methods predict unknown miRNA-disease associations by constructing different computational models using miRNAs and disease-related data resources to construct miRNA and disease similarity networks [17]; the obtained data are then combined with experimentally validated (or known) miRNA-disease networks. Jiang et al. [18] proposed a miRNA-prediction algorithm for the hypergeometric distribution scoring system, and the scores are ranked to select candidate disease- related miRNAs. Chen et al. [19] proposed WBSMDA method, which integrates the With-Score of miRNA and diseases similarity and the Between-Score of unknown miRNA-disease associations to predict potential miRNA-disease associations. However, the two methods make assumptions about probability distribution, and their prediction performances will be affected when the data resources are inconsistent with the assumptions. Xuan et al. [20] proposed HDMP method by considering weighted k most similar neighboring miRNAs and combining miRNA functional similarity to predict miRNAs associated with human diseases. RWRMDA [21] and MIDP [22] methods use random walk to calculate similarity of miRNAs and diseases. However, these methods cannot predict related miRNAs for diseases without any related miRNAs or new diseases (isolated diseases). Zou et al. [23] proposed KATZ to calculate the prediction score of different walking lengths between miRNAs and diseases through social network analysis. However, the performance of KATZ is poor because the known associations are sparse. KATZ also cannot predict related diseases for miRNAs without known related diseases or new miRNAs (isolated miRNAs). However, KATZ cannot be used to predict related miRNAs for isolated diseases.NCPMDA [24] develops network consistency projection to calculate potential miRNA–disease association score from miRNA and disease vector space projection scores. Li et al. [25] proposed a network similarity integration method (NSIM) for predicting potential miRNA-disease associations. However, NSIM are overly dependent on known miRNA-disease associations. HGIMDA [26] utilizes a heterogeneous graph iterative algorithm based on known miRNA–disease associations to predict miRNA–disease associations. However, HGIMDA is difficult to use in selecting parameters.

Machine learning-based methods aim to predict reliable miRNA-disease association by extracting effective features or solving specific optimization problems by using powerful machine-learning algorithms. Xu et al. [27] built a support vector machine (SVM) classifier by using four topological features based on the miRNA target-dysregulated network to predict potential miRNAs related to prostate cancer. The main disadvantage of Xu's method is the impossibility to obtain negative samples, thereby decreasing the prediction performance. Chen and Yan [28] proposed RLSMDA method that uses regularized least squares to predict miRNA-disease associations. This method is based on semi-supervised learning and avoid using negative samples but adjust parameters intricately. Li et al. [29] proposed MCMDA method using the matrix completion algorithm. Luo et al. [30] proposed CPTL method using the transduction learning collective prediction model to predict miRNA-disease associations. However, these methods cannot be applied to predict potential miRNAs for isolated diseases.

These above methods use only a single piece of information related to miRNAs or diseases, such as association of miRNAs and diseases verified by biological experiments, resulting in overestimation [31]. Therefore, researchers have investigated different types of miRNA- and disease-related a priori biological information to construct miRNA–disease associations through intermediaries. For example, Mørk et al. [32] developed a miRNA–protein–disease heterogeneity-related network, namely, miRPD, which uses protein-related associations as a bridge to link miRNAs and diseases. However, the prediction accuracy of miRPD is unsatisfactory because of its high false positive/negative rates. Xu et al. [33] used the network of interactions between miRNAs and target genes derived from matched miRNA and mRNA expression data and the network of interactions between specific miRNAs and diseases to sequence and identify miRNAs most likely associated with multiple diseases. Liu et al. [31] integrated miRNA-target gene and miRNA-lncRNA multiple data sources, established disease and miRNA similarity subnets, and predicted miRNA-disease associations in heterogeneous networks by using random walk with restart. Zeng et al. [34] used gene functional information, four main parameters of miRNAs and miRNA-disease associations to construct a bilayer networks. Then they used structural consistency as an indicator to estimate the link predictability of the bilayer networks, and used structural perturbation method (SPM) to predict potential miRNA-disease associations. SRMDAP [35] builds miRNA and disease similarity subnetworks by using the SimRank algorithm and density-based clustering recommender model based on known miRNA-mRNA interaction data, disease-gene data, and miRNA-disease association data. However, these methods lead to incomplete calculation of similarity and low prediction accuracy.

In our work, we propose a novel computational method, namely, FCMDAP, by using miRNA family and cluster information to improve the prediction accuracy of disease-related miRNAs. FCMDAP uses information entropy and mutual information (MI) to measure similarity between miRNAs based on miRNA–mRNA interaction and adds miRNA family information to reconstruct a miRNA similarity network. FCMDAP obtains functional similarity between diseases based on disease–gene interaction and semantic similarity between diseases based on disease directed acyclic graph (DAG). FCMDAP then integrates functional and semantic similarity to disease similarity. Based on the k-most similar neighboring recommendation algorithm, FCMDAP uses experimentally verified miRNA–disease association, miRNA similarity, and cluster information to predict potential miRNA–disease associations in miRNA space. FCMDAP also uses experimentally verified miRNA–disease association and disease similarity to predict potential miRNA–disease associations in disease space. The two predicted association scores are linearly integrated together. We implemented leave-one-out cross validation (LOOCV) and achieved AUC of 0.9165. Analysis of miRCancer, dbDEMC, or PhenomiR databases, confirmed the 50, 49, and 48 of top 50 predicted miRNAs in case studies of colorectal, lung, and pancreatic neoplasms, respectively. The average AUC values of FCMDAP to predict isolated diseases and miRNAs were 0.8417 and 0.8944, respectively. For isolated lung neoplasms, all of the top 50 predicted miRNAs were confirmed. For isolated hsa-mir-93, 9 of the top 10 diseases were confirmed. In conclusion, FCMDAP outperforms other methods.

Materials

Data

Data used in FCMDAP are obtained from five data sets:

(1) experimentally verified miRNA-disease related data from HMDD v2.0 database (http://www.cuilab.cn/hmdd, Jun-14-2014 Version) [12]. After filtering invalid data with disease name error or wrong miRNA name and removing redundant miRNA-disease associations, we obtained 5048 experimentally verified miRNA-disease associations including 475 miRNAs and 334 diseases as the benchmark dataset [see Additional file 1]. We use $M = \{m_1, m_2, \ \ , m_{nm}\}$ to represent the miRNA set and $D = \{d_1, d_2, \ \ , d_{nd}\}$ to represent the disease set, where nm is the number of miRNAs, and nd is the number of diseases. We also use the matrix AS to represent the known association of miRNAs and diseases. When miRNA i associates with disease j, $AS(i,j)$ is 1. Otherwise, $AS(i,j)$ is 0.

(2) experimentally verified miRNA-mRNA interactions from miRTarBase database (http://mirtarbase.mbc.nctu.edu.tw/, Release 6.0: Sept-15-2015) [36]. We use these data to measure functional similarity of miRNAs.

(3) experimentally verified disease-gene interaction from DisGeNET database (http://www.disgenet.org, Release 4.0: Oct-2016) [37]. We use these data to measure functional similarity of diseases.

(4) data on the relationship of various disease from the MeSH (http://www.nlm.nih.gov/, 2017 Version) descriptor of Category C, which are descripted as DAG. We use these data to measure semantic similarity of diseases.

(5) information of the family and cluster of human miRNAs from miRBase (http://www.mirbase.org, Release 21) [6]. We established the miRNA family information matrix FAM for the 475 miRNAs in the benchmark. $FAM(i,j) = 1$ if miRNA i and j are in the same family; otherwise, $FAM(i,j) = 0$. We also established the miRNA cluster information matrix CLU for 475 miRNAs. $CLU(i,j) = 1$ if the distance between miRNA i and j is less than 20 kb and we consider the two miRNAs in the same cluster; otherwise, $CLU(i,j) = 0$.

miRNA similarity network

Information entropy and mutual information (MI) are used to calculate similarity between miRNAs based on the set of mRNAs interacting with miRNAs.

In events set X, information entropy is a measure of the average information content that can be obtained if one of the events actually occurs [38]. This parameter can be defined as

$$H(X) = \sum_{x \in X} p(x) \ log \frac{1}{p(x)} = -\sum_{x \in X} p(x) \ log(p(x)) \quad (1)$$

where $p(x)$ is the probability of x.

For two discrete random variables X and Y, their MI can be described as

$$I(X;Y) = \sum_{x\in X}\sum_{y\in Y} p(x,y)\ log\frac{p(x,y)}{p(x)p(y)} \quad (2)$$

where $p(x)$ is the marginal probability distribution function of X, $p(y)$ is the marginal probability distribution function of Y, and $p(x, y)$ is the joint probability function of X and Y.

If the mRNAs set of miRNA A is $T_m^A = \{T_m^A(1), T_m^A(2), ..., T_m^A(ma)\}$, and the mRNAs set of miRNA B is $T_m^B = \{T_m^B(1), T_m^B(2), ..., T_m^B(mb)\}$ (where ma and mb are the target genes number of miRNA A and miRNA B, respectively), then information entropy of T_m^A can be calculated as

$$\begin{cases} H(T_m^A) = -\sum_{i=1}^{ma} p(T_m^A(i)) \log_2(p(T_m^A(i))) \\ p(T_m^A(i)) = n(T_m^A(i))/N \end{cases} \quad (3)$$

where N is the total number of the known miRNA–mRNA interactions in the dataset. $n(T_m^A(i))$ is the known number of interactions between the ith target gene in the target gene set of miRNA A and all miRNAs. $p(T_m^A(i))$ is the rate of the ith target gene in the target gene set of miRNA A with the known miRNA-mRNA interactions.

The similarity between miRNA A and miRNA B can use the normalized MI of T_m^A and T_m^B denoted as

$$SM(A,B) = \frac{2 * H(T_m^A \cap T_m^B)}{H(T_m^A) + H(T_m^B)} \quad (4)$$

where $H(T_m^A \cap T_m^B)$ is the information entropy of the intersection of T_m^A and T_m^B. When calculating the similarity of miRNA A and miRNA B, both of their information entropies and the common information entropies of their mRNAs are considered. Also, the frequency of occurrence of the target mRNAs are considered. It measures the similarity between miRNAs by MI according to the occurrence probability of target genes of miRNAs. The target gene with higher probability is more universal and carries less information, while the target gene with lower probability is more specific and carries more information. Obviously, the difference in target gene probability results in such a result. By comparing the similarity data, we find that the metric is determined by the above two factors, and the similarity between the two miRNAs can be appropriately measured.

Disease similarity network

In building disease similarity network, we first calculate the functional similarity of disease on the basis of disease-gene interaction dataset. We then calculate the semantic similarity of disease on the basis of disease DAG. Finally, we integrate both data into disease similarity to build a disease similarity network.

Disease functional similarity of known disease–gene interactions

If the interaction genes set of disease A is $T_d^A = \{T_d^A(1), T_d^A(2), ..., T_d^A(da)\}$, and $T_d^B = \{T_d^B(1), T_d^B(2), ..., T_d^B(db)\}$ is for disease B (where da and db are the target genes number of disease A and disease B, respectively), then the information entropy of T_d^A can be calculated as

$$\begin{cases} H(T_d^A) = -\sum_{i=1}^{da} p(T_d^A(i)) \log_2(p(T_d^A(i))) \\ p(T_d^A(i)) = n(T_d^A(i))/N \end{cases} \quad (5)$$

where N is the total number of known disease–gene interactions in the dataset, $n(T_d^A(i))$ is the known number of the interactions between the ith target gene in the target gene set of disease A and all diseases, and $p(T_d^A(i))$ is the rate of the ith target gene in the target gene set of disease A with known disease–gene interactions.

The functional similarity between disease A and disease B can use the normalized MI of T_d^A and T_d^B denoted as

$$SDF(A,B) = \frac{2 * H(T_d^A \cap T_d^B)}{H(T_d^A) + H(T_d^B)} \quad (6)$$

where $H(T_d^A)$ and $H(T_d^B)$ are the information entropies T_d^A and T_d^B of disease A and disease B, respectively. $H(T_d^A \cap T_d^B)$ is the information entropy of the intersection of T_d^A and T_d^B. When calculating the functional similarity of disease A and disease B, both the information entropy of the diseases and the common information entropy of their genes are considered.

Disease semantic similarity

Disease semantic similarity DD are built from disease DAG as reported in the literature [39].

$$DD(A,B) = \frac{\sum_{t\in T_A\cap T_B} D_A(t) + \sum_{t\in T_A\cap T_B} D_B(t)}{2 * \ min(DV(A), DV(B))} \quad (7)$$

where $DD(A, B)$ is the semantics similarity value between disease A and disease B in disease DAG. For the meaning of the symols, please refer to the literature [39].

Integrating disease similarity

We integrate disease functional similarity and semantic similarity to obtain disease similarity.

$$SD(A,B) = \gamma \cdot SDF(A,B) + (1-\gamma) \cdot DD(A,B) \quad (8)$$

where $\gamma \epsilon (0, 1)$ is the balance factor to tune the contribution level from disease function similarity and semantic similarity. The results are shown in Additional file 2.

miRNA similarity network reconstruction

miRNA family information is obtained from miRBase database. We establish the miRNA family information matrix *FAM* for 475 miRNAs in the benchmark dataset. $FAM(A, B) = 1$ if miRNA A and B are in the same family; otherwise, $FAM(A, B) = 0$. We recalculate the miRNA similarity by adding miRNA family information as follows

$$miRNAsim(A, B) = SM(A, B) * (1 + FAM(A, B)) \quad (9)$$

We then reconstruct the miRNA similarity network. The results are shown in Additional file 3.

FCMDAP prediction method

The flowchart of FCMDAP to predict disease-related miRNAs is shown in Fig. 1.

miRNA space score calculation

Calculating the recommendation score of neighboring miRNAs and disease

Wang et al. [39] proposed that miRNAs with the same similarity tend to be related to diseases with the same functions, and vice versa. In the miRNA space, the related score between miRNA and disease is associated with the correlation score of the neighbor nodes with the miRNA closest to the disease. Hence, if a similar neighbor of a miRNA is related to a disease, then the miRNA may be related to the disease. According to the collaborative recommendation algorithm, the association score of miRNA i and disease j is calculated based on the similarity scores of the top $k1$ nearest neighbor nodes of miRNA i and the association scores of these nodes and disease j. We normalize the association score of the top $k1$ most similar neighbor nodes of miRNA i and disease j by using the following:

$$s1(i, j) = \frac{\sum_{k=1}^{k1} SM1(i, k) \cdot AS(k, j)}{\sum_{k=1}^{k1} SM1(i, k)} \quad (10)$$

where $SM1$ is the row vector of each miRNA in the miRNA matrix *miRNAsim* and is sorted in descending order. Hence, miRNAs that are more similar will be ranked higher. $SM1(i, k)$ is a component of miRNA i and the kth closet similar neighbor nodes in the vector $SM1$. If miRNA k is related to disease j, then we calculate the sum of the related scores between miRNA i and miRNA k and divide the sum of the related scores of the top $k1$ similar neighbor nodes of miRNA i.

Calculating the prediction score in the same miRNA cluster

Baskerville S. and Bartel D.P. [40] found significant coexpression among the proximal pairs of miRNAs (< 50 kb). The closest miRNA cluster is usually expressed as a common regulatory unit of polycistronics, and intronic miRNAs are usually coexpressed with host genes, presenting

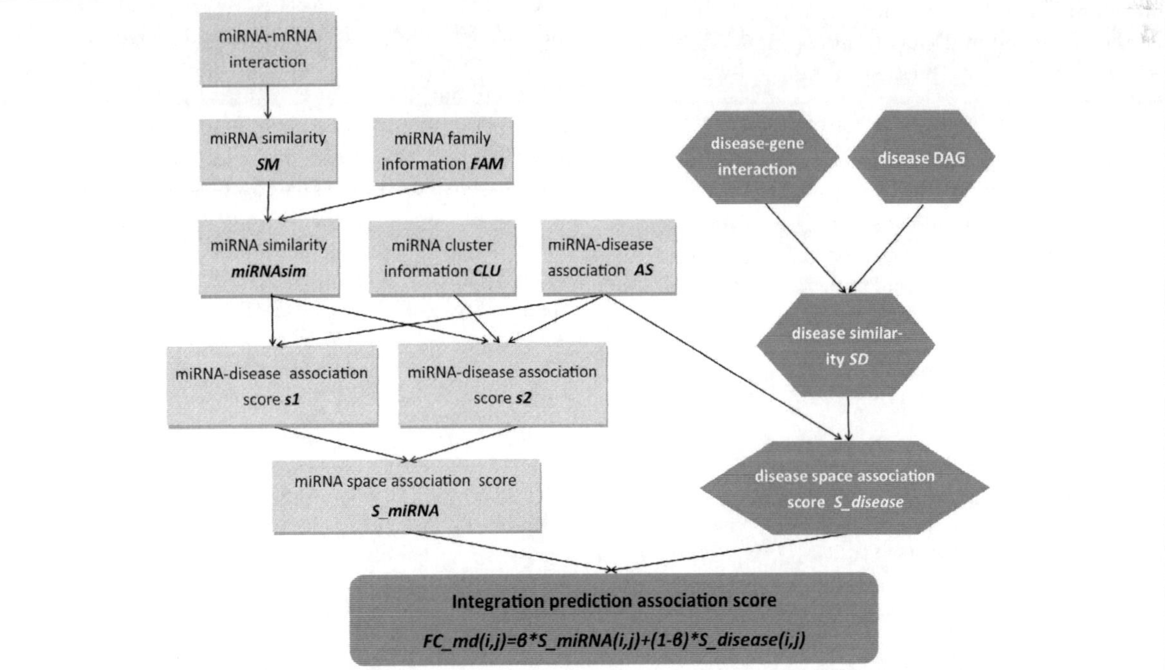

Fig. 1 The flowchart of FCMDAP

complex miRNA expression patterns. Lu et al. [41] performed statistical analysis and found that miRNAs in 46% of diseases have at least one neighboring member. For example, all of the 6 miRNAs (miR-17, miR-18a, miR-19a, miR-20a, miR-19b-1 and miR-92a-1) involved in hematopoietic malignancies are located in the miR-17 cluster. This result shows that neighboring miRNAs may be regulated by a common regulator under the same conditions and interactions, and their dysfunction may lead to the same disease. Wang et al. [39] confirmed that miRNAs are more likely to associate with the similar disease when clustered and located within 20 kb of genomic location. We downloaded the information of the location of human miRNAs in the genome from miRBase v.21, and clustered miRNAs are selected within a distance of 20 kb. A miRNA cluster matrix *CLU* is built for the 475 miRNAs in the benchmark dataset. Basing on the collaborative recommendation algorithm, we calculate the normalized related scores between miRNA i and disease j as

$$s2(i,j) = \frac{\sum_{k=1}^{n} SM2(i,k) \cdot AS(k,j)}{\sum_{k=1}^{n} SM2(i,k)} \tag{11}$$

where $SM2(i,k)$ is the similarity score of miRNA i and miRNA k in the same cluster, and n is the number of miRNAs in the same cluster as miRNA i. If miRNA k is related to disease j, then we add the similarity score $miRNAsim(i,k)$ of miRNA i and miRNA k and divide the sum of the similarity score of pairwise miRNAs in the same cluster as miRNA i. From the formula, we can find that the closer the miRNAs are in the same cluster with disease j, the closer the relation of miRNA i with disease j will be.

Integrating similarity score in miRNA space

In the miRNA space, the recommendation scores of miRNA–disease associations are calculated by integrating the score of top k similarity neighboring miRNAs of miRNA i and the recommendation score of miRNAs in the same cluster as miRNA i with disease j. The formula is as follows:

$$S\_miRNA(i,j) = \alpha * s1(i,j) + (1-\alpha) * s2(i,j) \tag{12}$$

where α is a tradeoff factor. Experiments show that FCMDAP gets the best performance when α is 0.5.

Calculating disease space score

In the disease space, we also use the k-nearest neighbor-based recommendation algorithm to calculate the predicted association score between disease and miRNA. If the k-nearest neighbor of a disease is related to a miRNA, then the disease is related to the miRNA.

According to the collaborative recommendation algorithm, for miRNA i with disease j, their recommendation score is calculated by the normalized similarity score between the $k2$-nearest neighbors of disease j and miRNA i. The formula is shown as follows

$$S\_disease(i,j) = \frac{\sum_{k=1}^{k2} AS(i,k) \cdot SD1(k,j)}{\sum_{k=1}^{k2} SD1(k,j)} \tag{13}$$

where $SD1$ is the column vector of all diseases in disease similarity matrix SD. These vectors are sorted in descending order, and the most similar disease is ranked as the highest. $SD1(k,j)$ represents the k-th component of the k-th nearest neighbor of disease j on the similarity column vector SD of disease j.

Calculating the final prediction score of disease-related miRNAs

The final prediction score of disease-related miRNAs of miRNA i with disease j is obtained by integrating the scores in miRNA space and disease space as follows

$$FC\_md(i,j) = \beta * S\_miRNA(i,j) + (1-\beta) * S\_disease(i,j) \tag{14}$$

where β is the factor used to balance the weight of two spaces. Experiments show that the optimal performance of FCMDAP can be obtained when the value of β is 0.8.

FCMDAP can predict isolated disease-related miRNAs and isolated miRNA-related diseases. Isolated disease-related miRNAs/miRNA-related diseases are miRNAs/diseases without any related diseases/miRNAs, such as newly discovered miRNAs/diseases. When we use FCMDAP to predict isolated disease-related miRNAs, all miRNAs related to disease j do not exist, leading to the prediction score $S\_miRNA(i,j)$ of 0. We calculate $S\_disease(i,j)$ from two parts, namely, similarity score between miRNA i and other diseases and similarity between diseases. Thus, FCMDAP can predict the association between isolated diseases and miRNAs. When we predict isolated miRNA-related disease, diseases related to miRNA i do not exist, leading $S\_disease(i,j)= 0$. We can calculate $S\_miRNA(i,j)$ from the relationship between other miRNA and disease j and the similarity between miRNAs to predict the association of miRNA i and disease j.

Results

Characteristics of the miRNA-disease association network

The benchmark data set include 5048 known miRNA–disease associations of 475 miRNAs and 334 diseases. The characteristics of these associations are shown in

Table 1. The average degree of diseases and miRNAs are 15.11 and 10.63, respectively.

Performance evaluation of FCMDAP

The LOOCV of known miRNA-disease associations is used to evaluate the performance of FCMDAP. For a given disease *d*, each known association of disease *d* is deleted in turn as a test sample, and the other known associations are used as training set. The remaining miRNAs without experimental evidence regarding their relation with disease *d* comprise the candidate miRNA set. The association prediction scores of these candidate miRNAs and diseases are calculated and ranked. If the rank exceeds a given threshold, then we consider FCMDAP to successfully predict the association of miRNA and disease. After changing the threshold, drawing the receiver operating characteristic (ROC) curve and calculating the area under the curve (AUC) value are conducted to evaluate prediction performance.

The ROC plots indicate the relationship between the true positive rate (TPR) and the false positive rate (FPR) at different thresholds. If TP, FP, TN, and FN represent true positive, false positive, true negative, and false negative, respectively, then TPR and FPR are calculated as

$$TPR = \frac{TP}{TP + FN} \tag{15}$$

and

$$FPR = \frac{FP}{TN + FP} \tag{16}$$

After one round of LOOCV, one association between miRNA and disease was excluded, and the prediction score was calculated by remaining associations. All these scores were sorted and a special ranking position was selected as threshold. TP and FP are the number of experimentally verified and unverified associations above the threshold, respectively. TN and FN are the number of unverified and verified associationas below the threshold, respectively.

Table 1 Global characteristic of the known miRNA-disease association network

Characteristic	Number
No. of miRNA-disease association	5048
No. of miRNAs	475
No. of diseases	334
Avg. degree of miRNAs	10.63
Avg. degree of diseases	15.11
Max degree of miRNAs	112
Min degree of miRNAs	1
Max degree of diseases	208
Min degree of diseases	1

We compared FCMDAP with SRMDAP, RLSMDA [28], KATZ [23], and Liu's method [31] in terms of prediction performance, AUC value, and ROC shapes on the benchmark data set. The values of the four parameters of FCMDAP are $\alpha = 0.5$, $\beta = 0.8$, $k1 = 50$, and $k2 = 30$. The optimal parameters of SRMDAP, RLSMDA, KATZ, and Liu's method are set as previously described. The comparison of the overall ROC curves and AUCs of all methods are shown in Fig. 2. The average AUC value of FCMDAP is 0.9165, which is 3.72, 5.81, 6.43, and 11.82% higher than those of SRMDAP, RLSMDA, KATZ and Liu's method, respecitively. When the FPR is lower than 0.2, the ROC of FCMDAP is more convex near the upper left corner, indicating that the prediction accuracy is higher. Therefore, FCMDAP shows higher prediction accuracy than the other methods.

To obtain reliable judgment, we tested 18 human diseases associated with at least 70 miRNAs. The results are shown in Table 2. Table 2 shows that FCMDAP obtained the highest AUC value of 0.8837 for pancreatic neoplasms and the lowest AUC value of 0.7572 for hepatocellular carcinoma. The average AUC value for the 18 diseases is 0.8195. The average AUC values for the 18 diseases obtained from SRMDAP, RLAMDA, KATA, and Liu's method are 0.8057, 0.6671, 0.6901, and 0.5178, respectively. The average AUC value obtained by FCMDAP is 1.38, 15.24, 12.94, and 30.17% higher than those of the four methods, respectively. Hence, FCMDAP exhibits better performance than SRMAPS, RLSMDA, KATA, and Liu's method.

Parameter effect

The five parameters in FCMDAP are α, β, γ, $k1$, and $k2$. We focus on miRNA space. In the miRNA space, α balances the tradeoff between the recommendation score from the neighboring miRNAs and the score from the miRNA cluster. β is the entire space balancing factor that sets different weights of recommendation scores from the miRNA and disease spaces. To obtain optimal parameters, we assign different values to α and β starting from 0.1 to calculate the recommendation scores of miRNA–disease association and evaluate the performance of FCMDAP by calculating AUC value. We repeat this work by increasing α and β in steps of 0.1 and calculating the AUC value until α and β are both 1. We obtain the best performance when $\alpha = 0.5$ and $\beta = 0.8$, and the AUC of FCMDAP is 0.9165. The results are shown in Fig. 3.

As shown in Fig. 3, the ordinate is the average AUC value, and the abscissa is the value at which β is magnified 10 times. Each curve in the figure represents the line connecting the points of the corresponding average

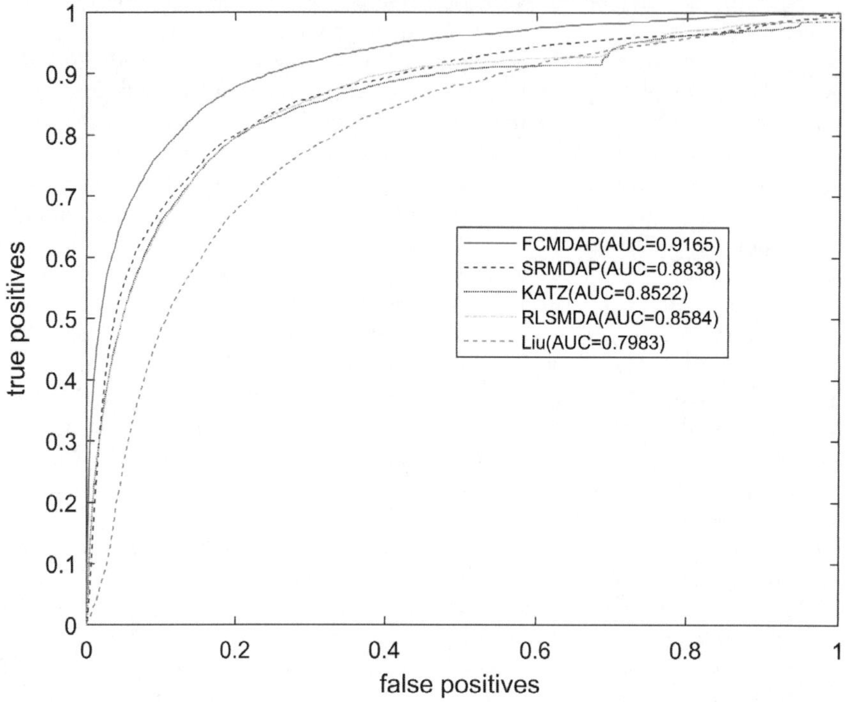

Fig. 2 The ROC curve and AUC value of FCMDAP and other compared methods

Table 2 AUC value of compared five methods for 18 diseases

Disease names	No. of related miRNAs	FCMDAP	SRMDAP	RLSMDA	KATZ	Liu's method
Carcinoma, Hepatocellular	208	0.7572	0.7639	0.6909	0.6881	0.4807
Breast Neoplasms	197	0.7733	0.7776	0.6814	0.6779	0.4147
Stomach Neoplasms	174	0.7658	0.7591	0.6635	0.6791	0.5498
Colorectal Neoplasms	143	0.7904	0.7929	0.6647	0.6895	0.4699
Melanoma	136	0.8300	0.7958	0.6584	0.6673	0.4804
Lung Neoplasms	128	0.8688	0.8874	0.7198	0.7675	0.5243
Heart Failure	120	0.7737	0.7538	0.6608	0.6622	0.5040
Prostatic Neoplasms	116	0.8185	0.8076	0.6704	0.7054	0.5440
Ovarian Neoplasms	112	0.8684	0.8732	0.7194	0.7705	0.5382
Carcinoma, Renal Cell	104	0.7878	0.7367	0.5815	0.6126	0.4932
Pancreatic Neoplasms	97	0.8837	0.8687	0.6829	0.7288	0.5355
Carcinoma, Non-Small-Cell Lung	94	0.8417	0.8322	0.6873	0.6981	0.5470
Glioblastoma	94	0.8383	0.7686	0.6421	0.6522	0.5644
Urinary Bladder Neoplasms	90	0.8214	0.7935	0.6231	0.6635	0.5475
Carcinoma, Squamous Cell	78	0.8640	0.8637	0.7179	0.7200	0.5398
Colonic Neoplasms	77	0.8278	0.8271	0.6582	0.6859	0.5490
Glioma	71	0.8679	0.8212	0.6727	0.7146	0.5591
Esophageal Neoplasms	70	0.7723	0.7789	0.6126	0.6383	0.4781
Average AUC value		0.8195	0.8057	0.6671	0.6901	0.5178

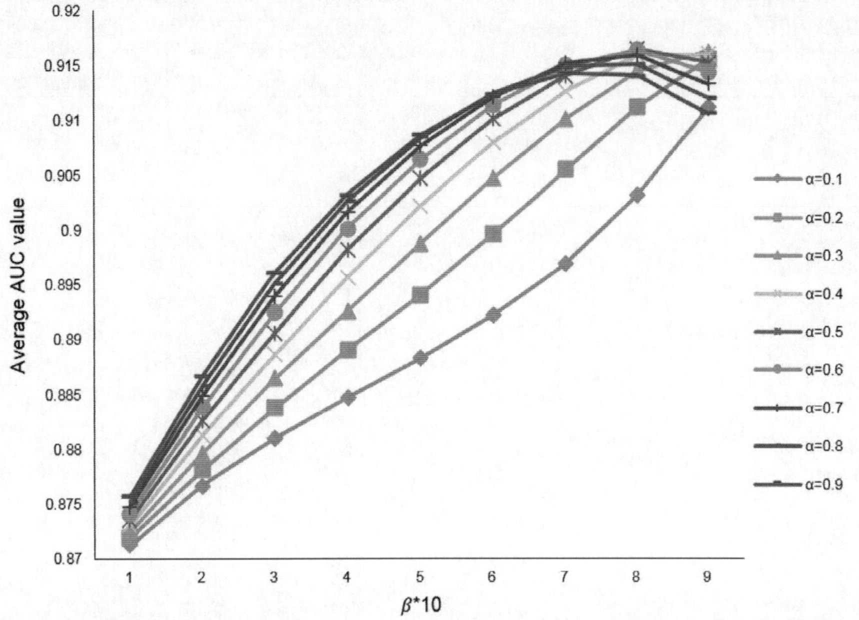

Fig. 3 Average AUCs value affected by α, β

AUC values when the same α value differs from the β value. The average AUC value varies from 0.8712 to 0.9165. When $\alpha = 0.1$, $\beta = 0.1$, the average AUC is the minimum value of 0.8712. When $\alpha = 0.5$, $\beta = 0.8$, the average AUC is the maximum value of 0.9165. The general trend is that the overall average AUC value increase with increasing α, β. γ denotes the balance factor in the disease similarity network based on disease functional similarity in disease–gene interactions and disease semantic similarity in disease DAG. $k1$ and $k2$ denotes the number of neighboring miRNAs and neighboring diseases in the recommendation algorithm, respectively. The values of γ, $k1$, and $k2$ are set as 0.5, 50, and 30, respectively, according to experience.

Case studies

Three important diseases (colorectal neoplasms, lung neoplasms, and pancreatic neoplasms) were selected to evaluate the performance of FCMDAP. The top 50 miRNA candidates of these three diseases were analyzed and verified using miRCancer (v. Oct. 2017), dbDEMC (v. 2.0), and PhenomiR (v. 2.0) databases and findings in the literature.

Colorectal neoplasms, the third most common cancer worldwide, severely affects the human health. In this regard, understanding colorectal-related miRNAs is important for diagnosis and prognosis of colorectal neoplasmsa. For example, patients with early colorectal neoplasms can be discriminated from healthy people by using serum miR-21, miR-29a, and miR-125b levels [42]. We used experimentally identified miRNA–disease associations as training samples to calculate the recommendation score of all candidate miRNAs through FCMDAP. We then ranked them in descending order and selected the top 50 miRNAs for verification. The top 50 candidate miRNAs and the corresponding evidence of their association with colorectal neoplasms are listed in Table 3. All the top 50 miRNAs were confirmed by analysis of miRCancer, dbDEMC, and PhenomiR databases.

Lung neoplasms is a malignant lung tumor caused by uncontrolled growth of lung tissue cells. Lung tumor cells can also rapidly spread from the lungs to other nearby tissues or other parts of the body. According to the World Health Organization's 2014 World Cancer Report [43], the number of patients with lung tumors worldwide reached 1.8 million in 2012. Lung neoplasms are the main cause of cancer-related death in men and women (other than breast neoplasms). In the United States, the 5-year survival rate for patients diagnosed with lung neoplasms is only 17.4%, which is lower than that in developing countries. Thus, effective methods for early diagnosis and treatment of lung neoplasms are important. Evidence indicates the important role of miRNAs in the pathogenesis, migration, and spread of lung neoplasms. For example, Takamizawa et al. [3] first found that the expression levels of let-7 are often reduced in lung neoplasms in vitro and in vivo in their study on 143 cases of lung neoplasms. The decrease in let-7 expression may affect the survival of patients that with lung neoplasms who were surgically treated. Johnson et al. [44] found that let-7 acts as a tumor

Table 3 The top 50 candidate miRNAs associated with colorectal neoplasms predicted by FCMDAP and the confirmation for their associations by miRCancer, PhenomiR or dbDEMC databases are listed here. All of them have been confirmed

Rank	miRNA	Evidence
1	hsa-mir-106b	miRCancer,dbDEMC, PhenomiR
2	hsa-mir-29b	miRCancer,dbDEMC, PhenomiR
3	hsa-mir-15a	miRCancer,dbDEMC, PhenomiR
4	hsa-mir-100	miRCancer,dbDEMC, PhenomiR
5	hsa-mir-192	miRCancer,dbDEMC, PhenomiR
6	hsa-mir-208b	dbDEMC
7	hsa-mir-24	miRCancer,dbDEMC, PhenomiR
8	hsa-let-7f	dbDEMC,PhenomiR
9	hsa-mir-101	miRCancer,dbDEMC, PhenomiR
10	hsa-let-7 g	dbDEMC,PhenomiR
11	hsa-mir-15b	miRCancer,dbDEMC, PhenomiR
12	hsa-mir-20b	dbDEMC
13	hsa-mir-193b	miRCancer,dbDEMC
14	hsa-mir-615	dbDEMC
15	hsa-mir-30c	dbDEMC,PhenomiR
16	hsa-mir-223	miRCancer,dbDEMC, PhenomiR
17	hsa-mir-130b	miRCancer,dbDEMC, PhenomiR
18	hsa-mir-296	miRCancer,dbDEMC, PhenomiR
19	hsa-mir-98	dbDEMC,PhenomiR
20	hsa-mir-125a	miRCancer,dbDEMC, PhenomiR
21	hsa-mir-29c	dbDEMC,PhenomiR
22	hsa-let-7d	dbDEMC,PhenomiR
23	hsa-mir-205	miRCancer,dbDEMC, PhenomiR
24	hsa-mir-23b	miRCancer,dbDEMC, PhenomiR
25	hsa-mir-10a	miRCancer,dbDEMC, PhenomiR
26	hsa-mir-128	miRCancer,dbDEMC, PhenomiR
27	hsa-mir-744	dbDEMC
28	hsa-mir-484	dbDEMC,PhenomiR
29	hsa-mir-32	miRCancer,dbDEMC, PhenomiR
30	hsa-mir-197	dbDEMC,PhenomiR
31	hsa-mir-151a	dbDEMC
32	hsa-mir-331	miRCancer,dbDEMC, PhenomiR
33	hsa-mir-138	miRCancer,dbDEMC, PhenomiR
34	hsa-mir-181d	dbDEMC
35	hsa-mir-449a	miRCancer,PhenomiR
36	hsa-mir-449c	dbDEMC
37	hsa-mir-326	miRCancer,dbDEMC, PhenomiR
38	hsa-mir-212	miRCancer,dbDEMC, PhenomiR
39	hsa-mir-196b	miRCancer,dbDEMC
40	hsa-mir-191	miRCancer,dbDEMC, PhenomiR
41	hsa-mir-30d	dbDEMC,PhenomiR
42	hsa-mir-214	miRCancer,dbDEMC, PhenomiR
43	hsa-mir-204	miRCancer,dbDEMC, PhenomiR
44	hsa-mir-99b	dbDEMC,PhenomiR
45	hsa-mir-449b	dbDEMC
46	hsa-mir-769	dbDEMC
47	hsa-mir-520 h	dbDEMC
48	hsa-mir-181c	dbDEMC,PhenomiR
49	hsa-mir-520 g	dbDEMC
50	hsa-mir-361	miRCancer,dbDEMC

Table 3 The top 50 candidate miRNAs associated with colorectal neoplasms predicted by FCMDAP and the confirmation for their associations by miRCancer, PhenomiR or dbDEMC databases are listed here. All of them have been confirmed *(Continued)*

suppressor in lung cells and negatively regulates the expression of the oncogene RAS. Hence, miRNAs can be used to develop drugs for treatment of lung tumors.

In our work, we used experimentally identified miRNA–disease associations as training samples to calculate recommendation scores of all candidate miRNAs based on FCMDAP. We then ranked them in descending order and selected the top 50 miRNAs for verification. The top 50 candidate miRNAs and the corresponding evidence of their association with colorectal neoplasms are listed in Table 4. Among these miRNAs, 48 miRNAs were confirmed in miRCancer, dbDEMC, and PhenomiR databases, and only two miRNAs (hsa-mir-520 g, hsa-mir-147a) were not confirmed. A recent study (PMID: 29033588) [45] showed that hsa-mir-147a is related to lung neoplasms. In this study, lncRNA HOXD-AS1 is specifically upregulated in non-small-cell lung cancer (NSCLC) tissues and promotes cancer cell growth by targeting miR-147a.

Pancreatic neoplasms are cellular masses caused by uncontrollable pancreatic cell proliferation. The most

Table 4 The top 50 candidate miRNAs associated with lung neoplasms predicted by FCMDAP and the confirmation for their associations by miRCancer, PhenomiR or dbDEMC databases are listed here. 49 of them have been confirmed

Rank	miRNA	Evidence
1	hsa-mir-429	dbDEMC,miRCancer
2	hsa-mir-141	dbDEMC,PhenomiR, miRCancer
3	hsa-mir-106b	dbDEMC,PhenomiR
4	hsa-mir-520 g	unconfirmed
5	hsa-mir-16	dbDEMC,PhenomiR, miRCancer
6	hsa-mir-215	dbDEMC,PhenomiR
7	hsa-mir-217	dbDEMC,PhenomiR, miRCancer
8	hsa-mir-376c	dbDEMC,PhenomiR
9	hsa-mir-181d	dbDEMC,PhenomiR
10	hsa-mir-20b	dbDEMC,PhenomiR
11	hsa-mir-15a	dbDEMC,PhenomiR, miRCancer
12	hsa-mir-195	dbDEMC,PhenomiR, miRCancer
13	hsa-mir-451a	dbDEMC
14	hsa-mir-99a	dbDEMC,PhenomiR
15	hsa-mir-193b	dbDEMC,PhenomiR
16	hsa-mir-130b	dbDEMC,PhenomiR, miRCancer
17	hsa-mir-194	dbDEMC,PhenomiR, miRCancer
18	hsa-mir-130a	dbDEMC,PhenomiR
19	hsa-mir-373	dbDEMC,PhenomiR
20	hsa-mir-15b	dbDEMC,PhenomiR, miRCancer
21	hsa-mir-10a	dbDEMC,PhenomiR
22	hsa-mir-378a	dbDEMC
23	hsa-mir-122	dbDEMC,PhenomiR
24	hsa-mir-449a	dbDEMC,PhenomiR, miRCancer
25	hsa-mir-148b	dbDEMC,PhenomiR, miRCancer
26	hsa-mir-449b	dbDEMC,PhenomiR, miRCancer
27	hsa-mir-204	dbDEMC,PhenomiR
28	hsa-mir-615	dbDEMC
29	hsa-mir-383	dbDEMC,PhenomiR, miRCancer
30	hsa-mir-340	dbDEMC,PhenomiR
31	hsa-mir-328	dbDEMC,PhenomiR
32	hsa-mir-151a	dbDEMC
33	hsa-mir-152	dbDEMC,PhenomiR
34	hsa-mir-153	dbDEMC,PhenomiR, miRCancer
35	hsa-mir-320a	dbDEMC,PhenomiR
36	hsa-mir-302d	dbDEMC,PhenomiR
37	hsa-mir-630	dbDEMC,miRCancer
38	hsa-mir-296	dbDEMC,PhenomiR
39	hsa-mir-139	dbDEMC,PhenomiR
40	hsa-mir-149	dbDEMC,PhenomiR
41	hsa-mir-423	dbDEMC,PhenomiR
42	hsa-mir-23b	dbDEMC,PhenomiR
43	hsa-mir-196b	dbDEMC,PhenomiR
44	hsa-mir-147a	PMID:29144017
45	hsa-mir-425	dbDEMC,PhenomiR
46	hsa-mir-99b	dbDEMC,PhenomiR, miRCancer
47	hsa-mir-324	dbDEMC,PhenomiR
48	hsa-mir-302c	dbDEMC,PhenomiR
49	hsa-mir-421	dbDEMC,PhenomiR
50	hsa-mir-484	dbDEMC,PhenomiR

Table 4 The top 50 candidate miRNAs associated with lung neoplasms predicted by FCMDAP and the confirmation for their associations by miRCancer, PhenomiR or dbDEMC databases are listed here. 49 of them have been confirmed *(Continued)*

common symptoms of pancreatic neoplasms include yellowing of the skin, abdominal or back pain, unexplained weight loss, and loss of appetite. Early pancreatic neoplasms are small and have no symptoms. Most pancreatic neoplasms are large when they are found and can metastasize to other parts of the body. According to reports, 411,600 people worldwide died of various pancreatic neoplasms in 2015. Pancreatic neoplasms most often occur in developed countries; that is, these malignancies rank as the fifth most common cancer in the UK and the fourth most common cancer in the United States [43, 46]. The prognosis of pancreatic neoplasms is very poor, with 25% survival rate for 1 year after diagnosis and 5% survival rate for 5 years. Thus, effective methods for early diagnosis, treatment, and prognosis of pancreatic neoplasms must be developed. At present, evidence supports the role of miRNA differential expression in the diagnosis, treatment, and prognosis of pancreatic neoplasms. For example, Sadakari et al. [47] found that the relative expression levels of miR-21 and miR-155 in tissues and pancreatic juice of patients with pancreatic ductal adenocarcinoma are significantly higher than those in patients with chronic pancreatitis; thus, miR-21 and miR-155 in pancreatic juice may be a potential biomarker for diagnosis of pancreatic ductal adenocarcinoma. Lodygin et al. [48] reported that the expression of miR-34a is silenced in several types of

Table 5 The top 50 candidate miRNAs associated with pancreatic neoplasms predicted by FCMDAP and the confirmation for their associations by miRCancer, PhenomiR or dbDEMC databases are listed here. 48 of them have been confirmed

Rank	miRNA	Evidence
1	hsa-mir-141	dbDEMC,PhenomiR, miRCancer
2	hsa-mir-29a	dbDEMC,PhenomiR, miRCancer
3	hsa-mir-181a	dbDEMC,PhenomiR, miRCancer
4	hsa-mir-29c	dbDEMC,PhenomiR, miRCancer
5	hsa-mir-19b	dbDEMC,PhenomiR
6	hsa-mir-93	dbDEMC,PhenomiR
7	hsa-mir-30a	dbDEMC,PhenomiR
8	hsa-mir-1	dbDEMC,PhenomiR
9	hsa-mir-98	dbDEMC,PhenomiR
10	hsa-mir-106b	dbDEMC,PhenomiR
11	hsa-mir-215	dbDEMC,PhenomiR, miRCancer
12	hsa-mir-520 g	dbDEMC
13	hsa-mir-7	dbDEMC,PhenomiR, miRCancer
14	hsa-mir-9	dbDEMC,PhenomiR
15	hsa-mir-195	dbDEMC,PhenomiR
16	hsa-mir-19a	dbDEMC,PhenomiR
17	hsa-mir-181d	dbDEMC,PhenomiR
18	hsa-mir-193b	dbDEMC
19	hsa-mir-125a	dbDEMC,PhenomiR
20	hsa-mir-135a	dbDEMC,PhenomiR
21	hsa-mir-205	PhenomiR,miRCancer
22	hsa-mir-26b	dbDEMC,PhenomiR
23	hsa-mir-138	dbDEMC,PhenomiR, miRCancer
24	hsa-mir-181c	dbDEMC,PhenomiR, miRCancer
25	hsa-mir-136	dbDEMC,PhenomiR
26	hsa-mir-133a	dbDEMC,PhenomiR, miRCancer
27	hsa-mir-320a	dbDEMC,PhenomiR, miRCancer
28	hsa-mir-20b	dbDEMC
29	hsa-mir-449a	dbDEMC
30	hsa-mir-615	dbDEMC,miRCancer
31	hsa-mir-140	dbDEMC,PhenomiR
32	hsa-mir-335	dbDEMC,PhenomiR, miRCancer
33	hsa-mir-378a	unconfirmed
34	hsa-mir-130b	dbDEMC,PhenomiR
35	hsa-mir-365a	unconfirmed
36	hsa-mir-423	dbDEMC,PhenomiR
37	hsa-mir-23b	dbDEMC,PhenomiR
38	hsa-mir-373	dbDEMC,PhenomiR, miRCancer
39	hsa-mir-149	dbDEMC,PhenomiR
40	hsa-mir-153	dbDEMC,PhenomiR
41	hsa-mir-30b	dbDEMC,PhenomiR
42	hsa-mir-27b	dbDEMC,PhenomiR
43	hsa-mir-22	dbDEMC,PhenomiR
44	hsa-mir-324	dbDEMC,PhenomiR
45	hsa-mir-185	dbDEMC,PhenomiR
46	hsa-mir-744	dbDEMC,miRCancer
47	hsa-mir-484	dbDEMC
48	hsa-mir-449b	dbDEMC
49	hsa-mir-328	dbDEMC,PhenomiR
50	hsa-mir-148b	dbDEMC,PhenomiR, miRCancer

Table 5 The top 50 candidate miRNAs associated with pancreatic neoplasms predicted by FCMDAP and the confirmation for their associations by miRCancer, PhenomiR or dbDEMC databases are listed here. 48 of them have been confirmed *(Continued)*

cancers, including pancreatic neoplasms, due to CpG methylation. By partially targeting CDK16, the re-expression of miR-34a in MiaPaC2 cell line with pancreatic neoplasms induces cellular senescence and cell cycle arrest. This observation indicates that miR-34a is a neoplasm suppressor gene, which is inactivated by CpG methylation and subsequent transcriptional silencing in various tumors, such as pancreatic neoplasms. Thus, miR-34a can be used as a therapeutic target for malignant neoplasms, such as pancreatic neoplasms.

In our work, we also calculated the recommendation score of all candidate miRNAs based on FCMDAP, ranked them in descending order, and selected the top 50 miRNAs for verification. The top 50 candidate miRNAs and the corresponding evidence of their associations with pancreatic neoplasms are listed in Table 5. Among the top 50 miRNAs, 48 miRNAs were confirmed in the miRCancer, dbDEMC, and PhenomiR databases, and only two miRNAs (miR-378a and miR-365a) were not confirmed.

Predicting isolated diseases and isolated miRNAs

FCMDAP can predict isolated disease-related miRNAs. In our work, we removed all experimentally verified disease-miRNA associations for a given disease and calculated the recommendation score by FCMDAP. We also ranked the miRNAs according to their recommendation scores. The average AUC of FCMDAP for predicting an isolated disease is 0.8417. For lung

Table 6 The top 50 miRNAs associated with isolated lung neoplasms predicted by FCMDAP and their evidence

Rank	miRNA	Evidence
1	hsa-mir-16	dbDEMC, PhenomiR, miRCancer
2	hsa-mir-15a	dbDEMC, PhenomiR, miRCancer
3	hsa-mir-195	dbDEMC, PhenomiR, miRCancer
4	hsa-mir-141	dbDEMC, PhenomiR, miRCancer
5	hsa-mir-106b	dbDEMC, PhenomiR
6	hsa-mir-429	dbDEMC, miRCancer
7	hsa-mir-296	dbDEMC, PhenomiR
8	hsa-mir-151a	dbDEMC
9	hsa-mir-122	dbDEMC, PhenomiR
10	hsa-mir-451a	dbDEMC
11	hsa-mir-130a	dbDEMC, PhenomiR
12	hsa-mir-378a	dbDEMC
13	hsa-mir-99a	dbDEMC, PhenomiR
14	hsa-mir-302b	dbDEMC, PhenomiR, miRCancer
15	hsa-mir-152	dbDEMC, PhenomiR
16	hsa-mir-193b	dbDEMC, PhenomiR
17	hsa-mir-708	dbDEMC
18	hsa-mir-625	dbDEMC
19	hsa-mir-204	dbDEMC, PhenomiR
20	hsa-mir-15b	dbDEMC, PhenomiR, miRCancer
21	hsa-mir-302c	dbDEMC, PhenomiR
22	hsa-mir-194	dbDEMC, PhenomiR, miRCancer
23	hsa-mir-320a	dbDEMC, PhenomiR
24	hsa-mir-449a	dbDEMC, PhenomiR, miRCancer
25	hsa-mir-149	dbDEMC, PhenomiR
26	hsa-mir-129	dbDEMC, PhenomiR, miRCancer
27	hsa-mir-20b	dbDEMC, PhenomiR
28	hsa-mir-139	dbDEMC, PhenomiR
29	hsa-mir-302a	dbDEMC, PhenomiR
30	hsa-mir-148b	dbDEMC, PhenomiR, miRCancer
31	hsa-mir-10a	dbDEMC, PhenomiR
32	hsa-mir-328	dbDEMC, PhenomiR
33	hsa-mir-215	dbDEMC, PhenomiR
34	hsa-mir-99b	dbDEMC, PhenomiR, miRCancer
35	hsa-mir-302d	dbDEMC, PhenomiR
36	hsa-mir-196b	dbDEMC, PhenomiR
37	hsa-mir-151b	dbDEMC
38	hsa-mir-373	dbDEMC, PhenomiR
39	hsa-mir-345	dbDEMC, PhenomiR
40	hsa-mir-449b	dbDEMC, PhenomiR, miRCancer
41	hsa-mir-452	dbDEMC, PhenomiR
42	hsa-mir-339	dbDEMC, PhenomiR
43	hsa-mir-367	dbDEMC, PhenomiR
44	hsa-mir-342	dbDEMC, PhenomiR
45	hsa-mir-130b	dbDEMC, PhenomiR, miRCancer
46	hsa-mir-211	dbDEMC, PhenomiR
47	hsa-mir-92b	dbDEMC, PhenomiR
48	hsa-mir-520c	dbDEMC
49	hsa-mir-520d	dbDEMC
50	hsa-mir-520a	dbDEMC

neoplasms, FCMDAP identifies the top 50 miRNAs related to lung neoplasms (Table 6). All of the top 50 miRNAs were confirmed by one or more databases (miRCancer, dbDEMC, or PhenomiR). Hence, FCMDAP exhibits satisfactory performance in predicting isolated diseases.

FCMDAP also shows satisfactory performance in predicting isolated miRNA-related diseases. In our work, we removed all disease association information for a given miRNA and calculated the recommendation score for all diseases for a given miRNA by using FCMDAP. We ranked these diseases and verified them in the databases. The average AUC of the FCMDAP to predict isolated miRNA is 0.8944. For hsa-mir-93, the top 10 related diseases predicted by FCMDAP are listed in Table 7. Among the 10 diseases, eight were confirmed to be related to hsa-mir-93 by dbDEMC or PhenomiR databases. Adrenocortical carcinoma, which ranked 8, was not confirmed by these two databases. Heart failure, which ranked 1, was confirmed to be related to hsa-mir-93 in the literature. Ke et al. [49] found that miR-93 is related to cardiomyocyte apoptosis, and miR-93 can prevent cardiomyocyte apoptosis induced by myocardial ischemia/reperfusion by inhibiting PI3K/AKT/PTEN signaling.

Discussion

In this work, we developed FCMDAP to predict human disease-related miRNAs. FCMDAP calculates the similarity between miRNAs by using mutual information based on the known miRNA-mRNA interaction information and adds the miRNA family information to

Table 7 The top 10 diseasesrelated with hsa-mir-93 predicted by FCMDAP and their evidences

Rank	miRNA	Evidence
1	Heart Failure	PMID:27119510
2	Colonic Neoplasms	PhenomiR, dbDEMC
3	Carcinoma, Squamous Cell	PhenomiR, dbDEMC
4	Leukemia, Lymphocytic, Chronic, B-Cell	PhenomiR
5	Mesothelioma	dbDEMC
6	Pancreatic Neoplasms	PhenomiR, dbDEMC
7	Hodgkin Disease	dbDEMC
8	Adrenocortical Carcinoma	unconfirmed
9	Glioblastoma	PhenomiR, dbDEMC
10	Leukemia, Myeloid, Acute	PhenomiR, dbDEMC

construct a miRNA space. FCMDAP integrates disease functional similarity based on the disease-gene interaction and disease semantic similarity based on the DAG from MeSH to construct a disease space. FCMDAP integrates the association scores between miRNA and disease from miRNA and disease spaces. The association scores between miRNA and disease are calculated based on the *k* most similar neighbor recommendation algorithm, and miRNA cluster information is added into miRNA space. Like NSIM and other method, FCMDAP also predict unknown associations by constructing miRNA network and disease network. However, in the process, the similarity calculation process of miRNA and disease are independent of each other. Multiple types of data including miRNA-mRNA interaction, miRNA family information, disease-gene interaction, DAG from MeSH to calculate miRNA similarity, and disease similarity are considered and the prediction does not only depend on the known miRNA–diseases associations, thereby improving the accuracy of similarity calculations. Using the *k* most similar neighbor recommendation algorithm and miRNA cluster information makes the prediction results more reasonable, and improves the predictive performance.

LOOCV and case research show that FCMDAP exhibits excellent performance in predicting miRNA-–disease associations. FCMDAP shows satisfactory performance in predicting diseases without any related miRNA information and miRNAs without any related disease information. The average AUC of FCMDAP for predicting isolated diseases and isolated miRNAs are 0.8417 and 0.8944, respectively. For isolated lung neoplasms, the prediction accuracy reached 100% in the top 50 predicted miRNAs. For the isolated hsa-mir-93, the prediction accuracy reached 90% in the top 10 diseases.

However, FCMDAP presents the following limitations. miRNA similarity can be further improved if other biomolecules that interact with miRNAs can be considered. As FCMDAP is developed on experimentally verified miRNA–disease associations, miRNA–disease associations can be experimentally verified, thereby improving the performance of FCMDAP.

Conclusion

In order to provide effective support for experimental research on miRNAs, we proposed a computational method FCMDAP to find potential disease-related miRNAs. FCMDAP exhibits excellent performance in predicting potential disease-related miRNAs. The FCMDAP could extend to study on other biomeolecular networks and help to decipher the study of complex human disease pathogenesis and diagnosis.

Additional files

Additional file 1: Known miRNA-disease associations. (XLSX 146 kb)
Additional file 2: Integrated disease similarity. (XLSX 1379 kb)
Additional file 3: Integrated miRNA similarity. (XLSX 1850 kb)

Abbreviations

AUC: Area under the curve; DAG: Disease directed acyclic; LOOCV: Leave-one-out cross validation; MI: Mutual information

Acknowledgements

Not applicable.

Funding

Publication costs were supported by National Natural Science Foundation of China (No. 61472127).

Availability of data and materials

All data generated or analysed during this study are included in this published article (and its supplementary information files).

About this supplement

This article has been published as part of BMC Systems Biology Volume 13 Supplement 2, 2019: Selected articles from the 17th Asia Pacific Bioinformatics Conference (APBC 2019): systems biology. The full contents of the supplement are available online at https://bmcsystbiol.biomedcentral.com/articles/supplements/volume-13-supplement-2.

Authors' contributions

CG and XL conceived of and designed the approach. XL carried out the experiments and wrote the manuscript. CG, YL and JY participated in revising the manuscript critically. All authors read and approved the final manuscript.

Ethics approval and consent to participate

Not applicable.

Consent for publication

Not applicable.

Competing interests

The authors declare that they have no competing interests.

Publisher's Note

Author details
[1]College of Computer Science and Electronic Engineering, Hunan University, Changsha 410082, Hunan, China. [2]Department of Genetics and Genomic Sciences, Icahn School of Medicine at Mount Sinai, New York, NY 10029, USA.

Published: 5 April 2019

References

1. Bartel DP. MicroRNAs. Genomics, biogenesis, mechanism, and function. Cell. 2004;116(2):281–97.
2. He L, Thomson JM, Hemann MT, Hernandomonge E, Mu D, Goodson S, Powers S, Cordoncardo C, Lowe SW, Hannon GJ. A microRNA polycistron as a potential human oncogene. Nature. 2005;435(7043):828–33.
3. Takamizawa J, Konishi H, Yanagisawa K, Tomida S, Osada H, Endoh H, Harano T, Yatabe Y, Nagino M, Nimura Y. Reduced expression of the let-7 microRNAs in human lung cancers in association with shortened postoperative survival. Cancer Res. 2004;64(11):3753–6.
4. Rosenfeld N, Aharonov R, Meiri E, Rosenwald S, Spector Y, Zepeniuk M, Benjamin H, Shabes N, Tabak S, Levy A, et al. MicroRNAs accurately identify cancer tissue origin. Nat Biotechnol. 2008;26:462.
5. Tang W, Wan S, Yang Z, Teschendorff AE, Zou Q. Tumor origin detection with tissue-specific miRNA and DNA methylation markers. Bioinformatics. 2018;34(3):398–406.
6. Kozomara A, Griffiths-Jones S. miRBase: annotating high confidence microRNAs using deep sequencing data. Nucleic Acids Res. 2013;42(D1): D68–73.
7. Megraw M, Sethupathy P, Corda B, Hatzigeorgiou AG. miRGen: a database for the study of animal microRNA genomic organization and function. Nucleic Acids Res. 2006;35(suppl_1):D149–55.
8. Hsu S-D, Lin F-M, Wu W-Y, Liang C, Huang W-C, Chan W-L, Tsai W-T, Chen G-Z, Lee C-J, Chiu C-M. miRTarBase: a database curates experimentally validated microRNA–target interactions. Nucleic Acids Res. 2010;39(suppl_1):D163–9.
9. Dweep H, Sticht C, Pandey P, Gretz N. miRWalk – database: prediction of possible miRNA binding sites by "walking" the genes of three genomes. J Biomed Inform. 2011;44(5):839–47.
10. Betel D, Wilson M, Gabow A, Marks DS, Sander C. The microRNA. Org resource: targets and expression. Nucleic Acids Res. 2008;36(suppl_1):D149–53.
11. Xie B, Ding Q, Han H, Wu D. miRCancer: a microRNA-cancer association database constructed by text mining on literature. Bioinformatics. 2013; 29(5):638–44.
12. Li Y, Qiu CX, Tu J, Geng B, Yang JC, Jiang TZ, Cui QH. HMDD v2.0: a database for experimentally supported human microRNA and disease associations. Nucleic Acids Res. 2014;42(D1):D1070–4.
13. Jiang Q, Wang Y, Hao Y, Juan L, Teng M, Zhang X, Li M, Wang G, Liu Y. miR2Disease: a manually curated database for microRNA deregulation in human disease. Nucleic Acids Res. 2009;37(Database):D98–104.
14. Yang Z, Wu LC, Wang AQ, Tang W, Zhao Y, Zhao HT, Teschendorff AE. dbDEMC 2.0: updated database of differentially expressed miRNAs in human cancers. Nucleic Acids Res. 2017;45(D1):D812–8.
15. Ruepp A, Kowarsch A, Schmidl D, Buggenthin F, Brauner B, Dunger I, Fobo G, Frishman G, Montrone C, Theis FJ. PhenomiR: a knowledgebase for microRNA expression in diseases and biological processes. Genome Biol. 2010;11(1):R6.
16. Zeng X, Zhang X, Zou Q. Integrative approaches for predicting microRNA function and prioritizing disease-related microRNA using biological interaction networks. Brief Bioinform. 2016;17(2):193–203.
17. Zou Q, Li J, Song L, Zeng X, Wang G. Similarity computation strategies in the microRNA-disease network: a survey. Brief Funct Genomics. 2015;15(1): 55–64.
18. Jiang QH, Hao YY, Wang GH, Juan LR, Zhang TJ, Teng MX, Liu YL, Wang YD. Prioritization of disease microRNAs through a human phenome-microRNAome network. BMC Syst Biol. 2010;4.
19. Chen X, Yan CC, Zhang X, You ZH, Deng LX, Liu Y, Zhang YD, Dai QH. WBSMDA: within and between score for MiRNA-disease association prediction. Sci Rep. 2016;6:21106.
20. Xuan P, Han K, Guo M, Guo Y, Li J, Ding J, Liu Y, Dai Q, Li J, Teng Z, et al. Prediction of microRNAs associated with human diseases based on weighted k most similar neighbors. PLoS One. 2013;8(8):e70204.
21. Chen X, Liu MX, Yan GY. RWRMDA: predicting novel human microRNA-disease associations. Mol BioSyst. 2012;8(10):2792–8.
22. Xuan P, Han K, Guo YH, Li J, Li X, Zhong YL, Zhang ZG, Ding J. Prediction of potential disease-associated microRNAs based on random walk. Bioinformatics. 2015;31(11):1805–15.
23. Zou Q, Li J, Hong Q, Lin Z, Wu Y, Shi H, Ju Y. Prediction of MicroRNA-disease associations based on social network analysis methods. Biomed Res Int. 2015;2015:810514.
24. Gu C, Liao B, Li X, Li K. Network consistency projection for human miRNA-disease associations inference. Sci Rep. 2016;6:36054.
25. Li XY, Lin YP, Gu CL. A network similarity integration method for predicting microRNA-disease associations. RSC Adv. 2017;7(51):32216–24.
26. Chen X, Yan CC, Zhang X, You ZH, Huang YA, Yan GY. HGIMDA: Heterogeneous graph inference for miRNA-disease association prediction. Oncotarget. 2016;7(40):65257.
27. Xu J, Li CX, Lv JY, Li YS, Xiao Y, Shao TT, Huo X, Li X, Zou Y, Han QL, et al. Prioritizing candidate disease miRNAs by topological features in the miRNA target-dysregulated network: case study of prostate cancer. Mol Cancer Ther. 2011;10(10):1857–66.
28. Chen X, Yan GY. Semi-supervised learning for potential human microRNA-disease associations inference. Sci Rep-Uk. 2014;4:5501.
29. Li JQ, Rong ZH, Chen X, Yan GY, You ZHMCMDA. Matrix completion for MiRNA-disease association prediction. Oncotarget. 2017;8(13):21187.
30. Luo J, Ding P, Liang C, Cao B, Chen X. Collective prediction of disease-associated miRNAs based on transduction learning. IEEE/ACM Trans Comput Biol Bioinform. 2017;14(6):1468–75.
31. Liu Y, Zeng X, He Z, Zou Q. Inferring microRNA-disease associations by random walk on a heterogeneous network with multiple data sources. IEEE/ACM Trans Comput Biol Bioinform. 2016.
32. Mork S, Pletscher-Frankild S, Caro AP, Gorodkin J, Jensen LJ. Protein-driven inference of miRNA-disease associations. Bioinformatics. 2014;30(3):392–7.
33. Xu C, Ping Y, Li X, Zhao H, Wang L, Fan H, Xiao Y, Li X. Prioritizing candidate disease miRNAs by integrating phenotype associations of multiple diseases with matched miRNA and mRNA expression profiles. Mol BioSyst. 2014; 10(11):2800–9.
34. Zeng X, Liu L, Lü L, Zou Q. Prediction of potential disease-associated microRNAs using structural perturbation method. Bioinformatics. 2018; 34(14):2425–32.
35. Li X, Lin Y, Gu C, Li Z. SRMDAP: SimRank and density-based clustering recommender model for miRNA-disease association prediction. Biomed Res Int. 2018;2018:11.
36. Chou CH, Chang NW, Shrestha S, Hsu SD, Lin YL, Lee WH, Yang CD, Hong HC, Wei TY, Tu SJ, et al. miRTarBase 2016: updates to the experimentally validated miRNA-target interactions database. Nucleic Acids Res. 2016; 44(D1):D239–47.
37. Pinero J, Queralt-Rosinach N, Bravo A, Deu-Pons J, Bauer-Mehren A, Baron M, Sanz F, Furlong LI. DisGeNET: a discovery platform for the dynamical exploration of human diseases and their genes. Database (Oxford). 2015: bav028.
38. Shannon CE. A mathematical theory of communication. ACM SIGMOBILE Mobile Computing and Communications Review. 2001;5(1):3–55.
39. Wang D, Wang JA, Lu M, Song F, Cui QH. Inferring the human microRNA functional similarity and functional network based on microRNA-associated diseases. Bioinformatics. 2010;26(13):1644–50.
40. Baskerville S, Bartel DP. Microarray profiling of microRNAs reveals frequent coexpression with neighboring miRNAs and host genes. RNA. 2005;11(3):241–7.
41. Lu M, Zhang Q, Deng M, Miao J, Guo Y, Gao W, Cui Q. An analysis of human microRNA and disease associations. PLoS One. 2008;3.
42. Yamada A, Horimatsu T, Okugawa Y, Nishida N, Honjo H, Ida H, Kou T, Kusaka T, Sasaki Y, Makato Y, et al. Serum miR-21, miR-29a and miR-125b are promising biomarkers for the early detection of colorectal neoplasia. Clin Cancer Res. 2015;21(18):4234–42.
43. McGuire S. World Cancer report 2014. Geneva, Switzerland: World Health Organization, International Agency for Research on Cancer, WHO press, 2015. Adv Nutr. 2016;7(2):418–9.
44. Johnson SM, Grosshans H, Shingara J, Byrom M, Jarvis R, Cheng A, Labourier E, Reinert KL, Brown D, Slack FJ. RAS is regulated by the let-7 microRNA family. Cell. 2005;120(5):635–47.
45. Wang Q, Jiang S, Song A, Hou S, Wu Q, Qi L, Gao X. HOXD-AS1 functions as an oncogenic ceRNA to promote NSCLC cell progression by sequestering miR-147a. OncoTargets Ther. 2017;10:4753–63.

46. Wang H, Naghavi M, Allen C, Barber RM, Bhutta ZA, Carter A, Casey DC, Charlson FJ, Chen AZ, Coates MM. Global, regional, and national life expectancy, all-cause mortality, and cause-specific mortality for 249 causes of death, 1980–2015: a systematic analysis for the global burden of disease study 2015. Lancet. 2016;388(10053):1459–544.
47. Sadakari Y, Ohtsuka T, Ohuchida K, Tsutsumi K, Takahata S, Nakamura M, Mizumoto K, Tanaka M. MicroRNA expression analyses in preoperative pancreatic juice samples of pancreatic ductal adenocarcinoma. JOP. 2010; 11(6):587–92.
48. Lodygin D, Tarasov V, Epanchintsev A, Berking C, Knyazeva T, Körner H, Knyazev P, Diebold J, Hermeking H. Inactivation of miR-34a by aberrant CpG methylation in multiple types of cancer. Cell Cycle. 2008;7(16):2591–600.
49. Ke Z-P, Xu P, Shi Y, Gao A-M. MicroRNA-93 inhibits ischemia-reperfusion induced cardiomyocyte apoptosis by targeting PTEN. Oncotarget. 2016; 7(20):28796.

Xue *et al. BMC Systems Biology* 2019, **13**(Suppl 2):34
https://doi.org/10.1186/s12918-019-0697-8

BMC Systems Biology

RESEARCH Open Access

Predicting disease-related phenotypes using an integrated phenotype similarity measurement based on HPO

Hansheng Xue[1,2], Jiajie Peng[1*] and Xuequn Shang[1*]

From The 17th Asia Pacific Bioinformatics Conference (APBC 2019)
Wuhan, China. 14–16 January 2019

Abstract

Background: Improving efficiency of disease diagnosis based on phenotype ontology is a critical yet challenging research area. Recently, Human Phenotype Ontology (HPO)-based semantic similarity has been affectively and widely used to identify causative genes and diseases. However, current phenotype similarity measurements just consider the annotations and hierarchy structure of HPO, neglecting the definition description of phenotype terms.

Results: In this paper, we propose a novel phenotype similarity measurement, termed as *DisPheno*, which adequately incorporates the definition of phenotype terms in addition to HPO structure and annotations to measure the similarity between phenotype terms. *DisPheno* also integrates phenotype term associations into phenotype-set similarity measurement using gene and disease annotations of phenotype terms.

Conclusions: Compared with five existing state-of-the-art methods, *DisPheno* shows great performance in HPO-based phenotype semantic similarity measurement and improves the efficiency of disease identification, especially on noisy patients dataset.

Keywords: Human phenotype ontology, Semantic similarity, Phenotype similarity

Background

With the high-speed development of next generation sequencing (NGS) techniques, large-scale biological and medical data is generated exponentially, which greatly contributes to Mendelian disease and cancer diagnosis [1–3]. However, it is still difficult to make accurate clinic diagnosis solely based on sequencing technologies, because of the complex and incomprehensible relationships between genetic variants and clinical phenotypes [4].

Some observable features of patients, such as behaviors and biomedical properties, are defined as patient phenotypes, which are usually determined by both genetic and environmental factors [5]. Currently, patient phenotypes are widely used to improve efficiency of disease diagnosis by analysing the complex relationships between clinic phenotypes and phenotypes of known diseases.

Human Phenotype Ontology (HPO) is a widely used ontology resource, which provides a standardized vocabulary of phenotypic abnormalities encountered in human disease [6]. HPO contains multiple types of information of phenotype, such as frequency modifier and definitions of phenotype terms. Besides, phenotype terms in HPO are organized as a directed acyclic graph (DAG) to describe the phenotypic characteristics and their relationships (An example is illustrated in Fig. 1). Based on HPO, researchers start to calculate phenotype similarity, which recently has been widely utilized to improve efficiency of disease diagnosis, and phenotype semantic similarity has become a rising research area [7, 8].

In phenotype semantic similarity area, previous researchers have proposed various HPO-based similarity

*Correspondence: jiajiepeng@nwpu.edu.cn; shang@nwpu.edu.cn
[1]School of Computer Science, Northwestern Polytechnical University, Xi'an, China
Full list of author information is available at the end of the article

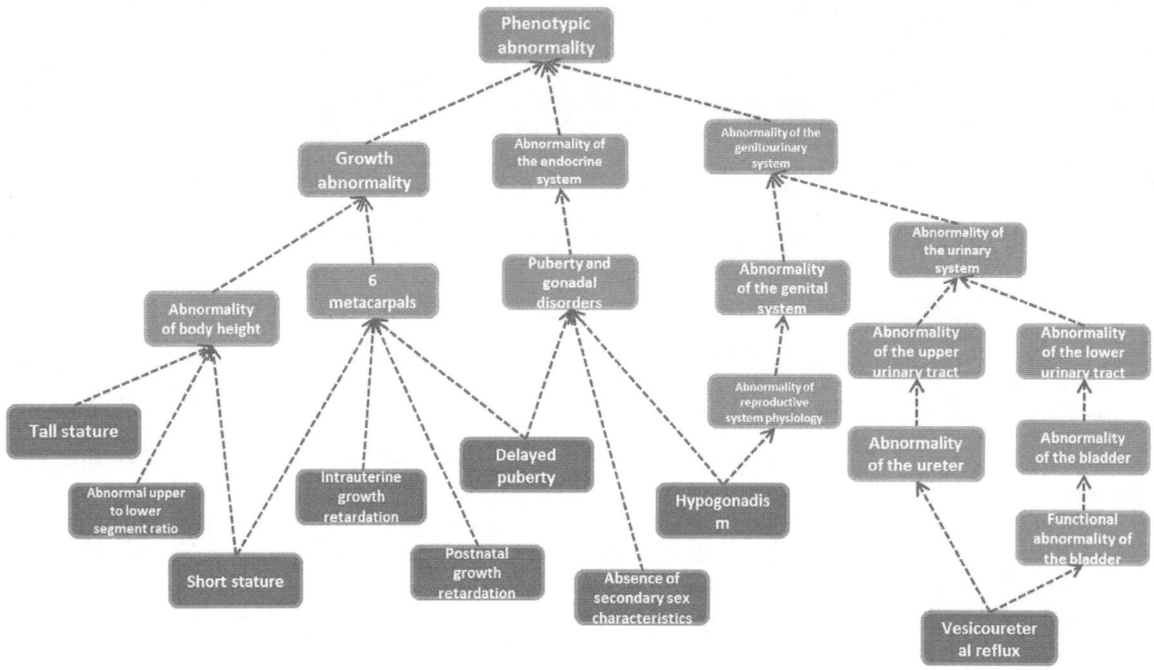

Fig. 1 An illustrative example of Human Phenotype Ontology (HPO). An example of Phenotypic abnormality (HP:0000118) forming a directed acyclic graph (DAG), in which nodes represent phenotype terms and edges represent "subclass of" relationships between phenotype terms

measurements. Most of existing semantic similarity measurements are based on Information Content (IC), such as Phenomizer [9], OWLSim [10] and PhenomeNet [11]. In detail, Phenomizer measures any two phenotype terms similarity based information content of phenotype ontology, which is similar as Masino et al. [12]. PhenomeNet and OWLSim extend simGIC [13] to calculate phenotype similarity of two phenotype sets. However, IC-based similarity measurements ignore the associated relationships of phenotype terms. Besides IC-based measurements, most existing measurements are similar to GO-based similarity measurements and neglect the unique topological structure of HPO [14–22]. And the main difference between HPO and GO is the biological knowledge representing by their structure. In the low-level of GO structure, sibling terms are often similar to each other. In contrast, sibling terms in the low-level of HPO structure are hard to prove that they have associations at the gene level or share any disease symptoms. For instance, phenotype terms "Split hand (HP:0001171)" and "Areflexia of upper limbs (HP:0012046)" are two leaf terms in HPO, but between them, there is no known gene-level associations nor shared disease symptoms [23].

Thus, it is essential to propose a novel and unique HPO-based semantic similarity measurement which designs for considering topological information of HPO. We designed a new path-constrained IC-based phenotype term semantic similarity measurement, termed as *PhenoSim*, which considers the unique DAG structure of HPO [23]. In addition, some practical online or offline tools have been developed for biological researchers, including HPOSim [24] and PhenoSimWeb [25]. HPOSim provides an offline R package, which implements seven common ontology-based similarity measurements, including Jiang [26], Lin [27], Wang [28] and Schlicker [29]. PhenoSimWeb is an easy-to-use online application which implements five phenotype measurements and provides an intuitive visualization interface.

Although above methods are widely used to calculate phenotype semantic similarity, none of them make the best of phenotype ontology information, such as definition description of phenotype term and phenotype annotation information. *PhenoSim* proposed a phenotype similarity measurement based on topological structure of HPO, but it neglects text description and association information of phenotype term. Current HPO-based methods adopt gene or disease annotations to represent information content of phenotype term. However, this method cannot describe phenotype term fully and accurately, since many annotations associated with a phenotype are still unknown [30–32]. Therefore, it is essential and necessary to explore a novel phenotype similarity measurement that make the best of phenotype ontology information, such as hierarchical structure, term annotation and text description of phenotype.

In this paper, we propose a novel phenotype similarity measurement, named *DisPheno*, which integrates hierarchy structure and phenotype term definition of HPO. Compared with existing methods, the main contributions of our work can be summarized as:

- To the best of our knowledge, *DisPheno* is the first HPO-phenotype similarity measurement integrating term annotation, hierarchical structure and text description.
- *DisPheno* applies Point-wise Mutual Information to calculate phenotype annotations and integrates into phenotype-set similarity measurement.
- The evaluation results show that *DisPheno* outperforms some state-of-the-art approaches.

Methods

In order to improve the performance of identifying disease-related phenotypes, we propose a novel phenotype similarity measurement, termed as *DisPheno*, which is a optimized method of a path-constrained information content-based similarity measurement. DisPheno mainly contains four steps. First, it annotates phenotype ontology information content using both genes and diseases. Second, it reconstructs topological structure of phenotype term using TF-IDF method [33]. Third, it computes semantic similarity between two phenotype term t_i and t_j considering information content(IC), distance between terms and DAG structure. Finally, it computes phenotype term associations using Point-wise Mutual Information (PMI) method [34] and calculates phenotype set similarity. The framework of *DisPheno* is shown in Fig. 2. and the detailed steps will be introduced as follows.

Step 1. Annotating phenotype ontology information content

Most of current phenotype similarity measurement are based on information content(IC), and the types of annotating phenotype term mainly contains gene annotation and disease annotation. Existing phenotype similarity measurement are annotated using gene or disease, and our method integrates these two types of annotations. In annotating part, we use a weighted coefficient w to adjust

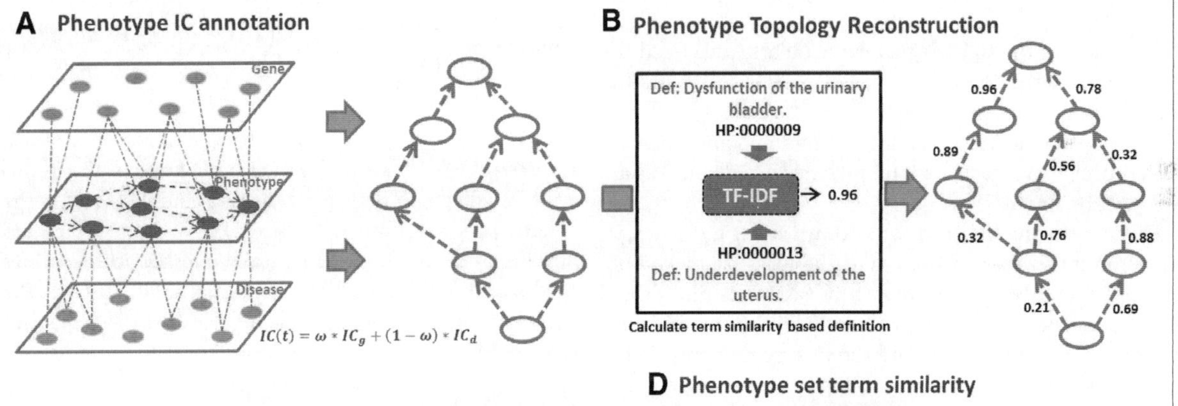

Fig. 2 The workflow of DisPheno. It mainly contains four parts: **a** Annotating phenotype ontology information content using both gene annotation and disease annotation; **b** Reconstructing topological structure of phenotype term by calculating phenotype term definition similarity using TF-IDF; **c** Measuring phenotype semantic similarity based on HPO by integrating term definition-similarity; **d** Calculating phenotype term association and set similarity by measuring phenotype term associations using Point-wise Mutual Information(PMI)

the ratio of two types of annotations. The IC of phenotype term t can be described as follows:

$$IC(t) = w * IC_{gene} + (1 - w) * IC_{disease}$$

$$IC_{gene}(t) = ln\left(\frac{G}{G_t}\right) \qquad IC_{disease}(t) = ln\left(\frac{D}{D_t}\right)$$

where $IC_{gene}(t)$ represents the information content of phenotype term t annotated by genes, G and G_t represent the size of genes annotated to the root and term t respectively ($IC_{disease}(t)$ is similar to gene annotation). Finally, we can comprehensively integrate the relationships between phenotypes and genes / diseases into the information content of phenotype terms.

Step 2. Reconstructing topological structure of phenotype term

Human Phenotype Ontology (HPO) provides a directed acyclic graph (DAG) to describe the phenotype term and associations. However, the edge of DAG has no weight and just indicate the hierarchical relationship. To further describe the relationship between phenotype terms, we try to turn original DAG into a weighted directed acyclic graph (WDAG). In our model, we calculate the cosine similarity between the definitions of phenotype terms using TF-IDF method and try to add weights for edges of original DAG.

To calculate the phenotype term similarity, we need to convert the term definition into vector by TF-IDF firstly. TF-IDF is short for term frequency-inverse document frequency, which is often used in data mining and information retrieval to measure the importance of a document in a collection or corpus [33].

Given a phenotype term definition $t = \{p_1, p_2, ..., p_n\}$, p_i represents a specify word, and the term frequency of p_i is $tf(p_i, t) = n_i/|t|$, where n_i represents the times that word p_i occurs in phenotype term definition t, and $|t|$ is the number of words in t. And the inverse document frequency of word p_i is $idf(p_i, T) = \log \frac{|T|}{|\{t \in T: p_i \in t\}|}$, where $|T|$ is the total number of phenotype term in the HPO corpus and $|\{t \in T : p_i \in t\}|$ is the number of phenotype term where the word p_i appears. Thus, the Term frequency-Inverse document frequency(TF-IDF) can be calculated as:

$$TF - IDF(p_i, t, T) = TF(p_i, t) * IDF(p_i, T)$$

After translating the phenotype term definitions into TF-IDF vectors by calculating the word TF-IDF scores, we can calculate the term similarity between pair-wise phenotype term using cosine similarity based on the TF-IDF vectors. Then, we can obtain a phenotype term similarity matrix $S \in R^{n*n}$, where n is the number of total phenotype terms. Finally, we add the phenotype term similarity into the DAG and we can reconstruct the unweighted directed acyclic graph into a weighted directed acyclic graph (WDAG). And the reconstructed WDAG will be used in the process of calculating phenotype term similarity.

Step 3. Measuring phenotype semantic similarity

Most phenotype similarity measurements are based on information content, they just consider the information content of most informative common ancestor or phenotype terms. They neglect the effects of hierarchy structure and text description of phenotype terms.

In our previous research, *PhenoSim* has proposed a path-constrained information content-based phenotype similarity measurement. The core idea is to consider the structural accessibility of phenotype terms. In detail, if there is a directed path between any two phenotype terms t_i and t_j in the hierarchy structure of HPO, we consider that these two terms are highly similar to each other and "reachable". Otherwise, these two phenotype terms are "unreachable" in the DAG structure of HPO.

Based on this measurement, we propose a novel method, termed as *DisPheno*, which considering the distance between term t_i and t_j and the pathway on the weighted directed acyclic graph. Thus, we define a novel phenotype-based similarity measurement as:

$$sim(t_i, t_j) = \begin{cases} WIC(t_{MICA}) * \left(1 - \frac{dist(t_i,t_j)}{mostDepth}\right) & reachable \\ 0 & otherwise \end{cases}$$

where $WIC(t_{MICA}) = \min(IC(t_i), IC(t_j)) * W(t_i, t_j)$, $(mostDepth - dist(t_i, t_j))/mostDepth$ implies the influences of distance between t_i and t_j, and $W(t_i, t_j)$ is the weight product from t_i to t_j among weighted directed acyclic graph of HPO. Specifically, *mostDepth* describes the longest path in the hierarchy structure of HPO, or the maximum number of edges that leaf node reaches the root node.

Step 4. Computing phenotype term association and set similarity

Before calculating the phenotype set similarity, we need to measure the association among all phenotype terms. Current phenotype set similarity measurements all adopt the average value of maximum phenotype term similarity between phenotype term and phenotype set as the phenotype set similarity. In our model, we introduce the phenotype association relationships and use Point-wise Mutual Information(PMI) to compute the phenotype term associations.

Assuming that if two term t_i and t_j belongs to same causative gene (or disease) in the gene-to-phenotype (or disease-to-phenotype) association file, we hold that term t_i and t_j are associated. The pair-wise association between phenotype terms can be calculated as:

$$PMI(t_i, t_j) = \log\left(\frac{p(t_i, t_j)}{p(t_i) * p(t_j)}\right)$$

where $p(t_i, t_j)$ is the probability that term t_i and t_j appear on the same gene or disease annotation set simultaneously, $p(t_i)$ and $p(t_j)$ are total probability of term t_i and t_j in the phenotype annotation set.

Given a patient and a candidate gene(or disease), the corresponding phenotype sets are T_p and T_c respectively. The phenotype set similarity between specific patient and candidate genes (or diseases) are the average value of pair-wise phenotype terms similarities between T_p and T_c:

$$Sim_{set}(T_p \rightarrow T_c) = \frac{1}{N_p} \sum_{t_i \in T_p} \max_{t_j \in T_c}(sim(t_i, t_j) * PMI(t_i, t_j))$$

$$Sim_{set}(T_c \rightarrow T_p) = \frac{1}{N_c} \sum_{t_j \in T_c} \max_{t_i \in T_p}(sim(t_j, t_i) * PMI(t_j, t_i))$$

where phenotype similarity $sim(t_i, t_j)$ is measured in previous step and N described the number of phenotype terms in set T. Due to the similarity score relies on the order of the phenotype-set and the above two equation are asymmetric, we use the following equation to eliminate the asymmetry affects. The symmetrical phenotype similarity measurement are described as:

$$Sim_{sym}(T_p, T_c) = \frac{1}{2}(Sim_{set}(T_p \rightarrow T_c) + Sim_{set}(T_c \rightarrow T_p))$$

Where Sim_{sym} is the average value of set similarities of two phenotype sets with different order. Phenotype term and set similarity measurement are the key of identifying true disease from candidate disease set. By modifying existing HPO-based similarity, we can further improve the efficiency of disease diagnosis.

Results

Data preparation

The experimental datasets were downloaded from Human Phenotype Ontology (HPO) official website (https://hpo.jax.org/), which contain 10,838 phenotype terms, 99,186 disease-to-phenotype annotations and 61,784 gene-to-phenotype annotations.

To evaluate the performance of our method, we used the patients that simulated in our previous work *PhenoSim*, which mainly contains "patients with known causative genes" and "patients with known diseases" two parts. Taking into account the clinical situation, we generated dataset with noise phenotype terms, named noisy dataset, and imprecision phenotype terms, named imprecision dataset. The optimal and noisy datasets used in this paper are same as our previous paper [35]. The details of simulating patients are described as follows.

Optimal dataset Each simulated patient was assigned one selected disease, and then we randomly added phenotype terms that selected disease associated with into this stimulated patient. In detail, if the randomly generated number was not greater than the known penetrance of the phenotype that disease associated with, this phenotype will be assigned to this simulated patient. The process was repeated for 100 times, then we obtained final optimal simulated patients.

Noisy dataset The noisy dataset is an extension of optimal, which considers the real clinic dataset. Before simulating noisy dataset, we firstly generated a noisy phenotype-set that much larger than the number of optimal phenotypes for every selected disease. The noise phenotype can be defined as the term which is not associated to this disease. After generating noisy phenotype-set, half number of noisy phenotype terms are selected and added into the phenotype set of simulated patients. Finally, we repeated this process for optimal patients and we generated the noisy simulated patients.

Noisy & Imprecision dataset Besides noisy phenotypes, clinical datasets usually contain imprecision phenotypes which attributes to the limitation of medical technology. The imprecision data is described as a kind of phenotype terms that one of their ancestors is associated with the disease d instead of the explicit phenotype term itself. In this noisy & imprecision dataset, we randomly selected half of the optimal terms and replaced them with one of their ancestors. Then we added noisy phenotype terms into the imprecision dataset, and the number of noise terms is half of the imprecision dataset. Finally, optimal, noisy and noisy & imprecision data all account for one-third of the whole dataset.

Performance evaluation on optimal dataset

We utilized the same evaluation criterion with PhenoSim to validate the prediction performance of *DisPheno* [12]. The main idea is to rank the candidate diseases of each simulated patient. We calculated the phenotype similarity value between the patient and each candidate diseases using *DisPheno*, then ranked all the candidate diseases in descending order by their similarity values. Higher the true disease's rank is, the better the algorithm's performance. Finally, we compared *DisPheno* with other five existing state-of-the-art measures on all the simulated datasets.

"Optimal patients with known causative gene" dataset contains 3300 simulated patients and each patient corresponds to one causative gene. We tested *DisPheno* and other five methods on this optimal dataset and compared the rank of true disease. Specifically, there is mapping relationship between causative genes and diseases. Because the HPO-based similarity measurements are usually used on disease diagnosis, we ranked the candidate diseases for each simulated patient instead of causative genes. In the cumulative rank distribution figure, we can find that *DisPheno* performed much better than the other methods (Fig. 3). 28.78% of true candidate diseases rank first using *DisPheno* which is the highest percentage among all methods. The percentage of rank among top-3 using *DisPheno* is 49.86%, while the ratio of other methods are 42.94% (PhenoSim), 35.69% (Masino), 28.55% (Lin), 30.69% (Jiang) and 28.42% (Schlicker) respectively. In addition, 60.43% candidate diseases rank among top-5 using *DisPheno* and it is 10.98% higher than *PhenoSim* (49.45%) which is the second best method.

"Optimal patients with known disease" dataset contains 24,000 simulated patients and each patient corresponds to one disease. We tested the performance of all six approaches on this optimal dataset (see Table 1). Although the percentage of top-1 using *DisPheno* (83.12%) is less than the ratio of Schlicker (96.36%), 99.10% of candidate diseases rank among top-3 which is the highest compared with other methods. Although top-1 percentage is not highest, *DisPheno* shows great performance on disease identification. In the clinical cancer diagnosis or disease prediction, it usually provides scientists with several top candidates instead of the single highest one.

In the optimal datasets, *DisPheno* performs better than other five methods. And it also shows great performance and latent capacity on predicting disease and disease diagnosis. Considering that clinical phenotype set often contains lots of noise data, we further validate the performance of *DisPheno* on the simulated patient with noisy phenotype terms.

Performance evaluation on noisy dataset

"Noisy patients with known causative gene" dataset contains noisy phenotypes which are not annotated phenotype terms of the causative gene. We applied *DisPheno* and other five measures on the noisy dataset. Our method performed the best in all the six measurements (Fig. 4). The ratio of true diseases rank among top-5 using *DisPheno* reaches the highest (57.38%), which is 11.20% higher than the second highest method PhenoSim (46.18%). The percentage of other methods perform on this dataset are 36.85% (Masino), 10.67% (Lin), 6.80% (Jiang) and 14.61% (Schlicker). *DisPheno* shows great performance on noisy patient with known causative gene, it indicates good application prospect on clinical diagnosis.

"Noisy patients with known disease" dataset contains noisy phenotypes which are not annotated phenotype terms of the disease. We applied *DisPheno* and other five approaches on the noisy dataset, and our method performed the best in all the six measurements (Fig. 5). On the noisy patients with known diseases, the performance of *DisPheno* is far superior than the other five

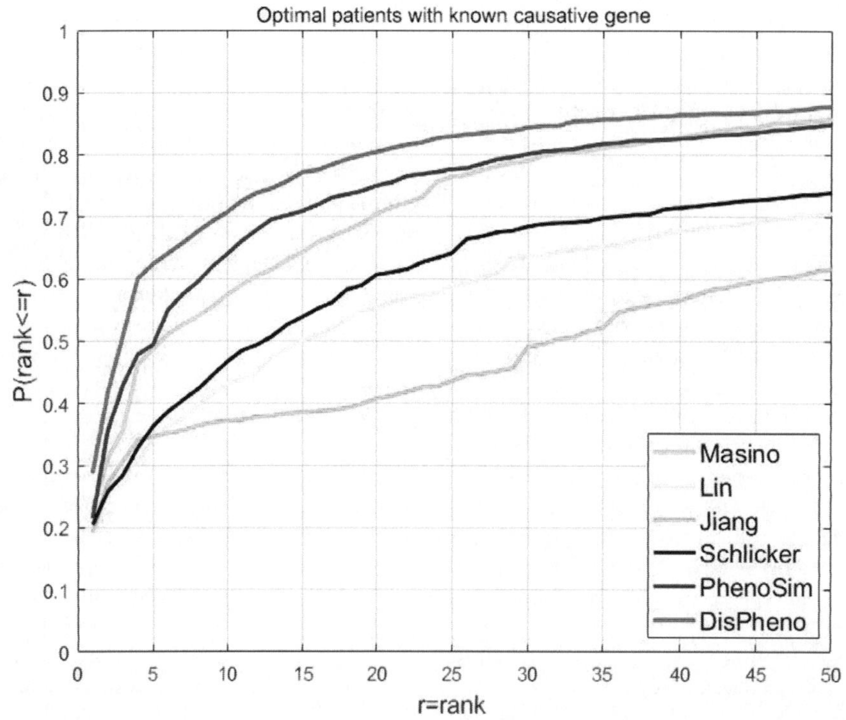

Fig. 3 Cumulative rank distribution of optimal patient dataset with the known causative gene. The x-axis is the rank threshold and the y-axis is the cumulative probability of true disease rank

Table 1 The percentage of cumulative rank distribution

Method	Top-1	Top-3	Top-5	Top-10
DisPheno	83.12%	99.10%	99.71%	99.87%
PhenoSim	79.50%	98.62%	99.45%	99.83%
Masino	82.48%	97.43%	98.63%	99.16%
Lin	95.68%	97.94%	98.63%	99.35%
Jiang	95.43%	98.17%	99.11%	99.69%
Schlicker	96.36%	98.31%	98.93%	99.53%

DisPheno was compared with other five methods on the optimal patient with the known disease

algorithms. 56.48% of candidate diseases rank the highest using *DisPheno*. Instead, the ratio of other five methods are 42.74% (PhenoSim), 20.04% (Masino), 0.5% (Lin), 0.32% (Jiang) and 1.82% (Schlicker). The second highest is PhenoSim, which is 13.74% less than *DisPheno*. The great gap shows the performance of our method in disease identification, especially on noisy simulated patient dataset.

Overall, *DisPheno* performs better than other five similarity measurements on the stimulated datset with noise phenotype terms, and it shows great robustness. It implies huge potential on clinical disease diagnosis.

Performance evaluation on noisy & imprecision dataset

Except noisy phenotype terms, clinical datasets often contains imprecision phenotypes. In this part, we performed *DisPheno* on the noisy and imprecision patient dataset with known disease to evaluate the performance respectively.

Compared with other five methods, *DisPheno* shows good and stable performance on simulated patients with noisy and imprecision phenotypes (see Table 2). The percentage of true disease rank among top-10 using *DisPheno* reaches 22.34%, which is much higher than others. It indicates that *DisPheno* would perform well on the clinical datasets and it shows great prospects on disease diagnosis.

Effects of parameters on DisPheno model

In this part, we test the various parameters on *DisPheno* model. In the first part of our model, we utilize both gene and disease annotations. We run *DisPheno* multiple times by varying the parameter w from 0.0 to 1.0 to test the performance of different weighted coefficients. Figure 6 shows that *DisPheno* achieves the best performance when the weighted coefficient is equal to 0.5 or 0.9.

Besides, we also run different parts of *DisPheno* to evaluate the contribution of different components in the model. Compared with previous algorithm *PhenoSim*, this novel model mainly adds four parts to improve the

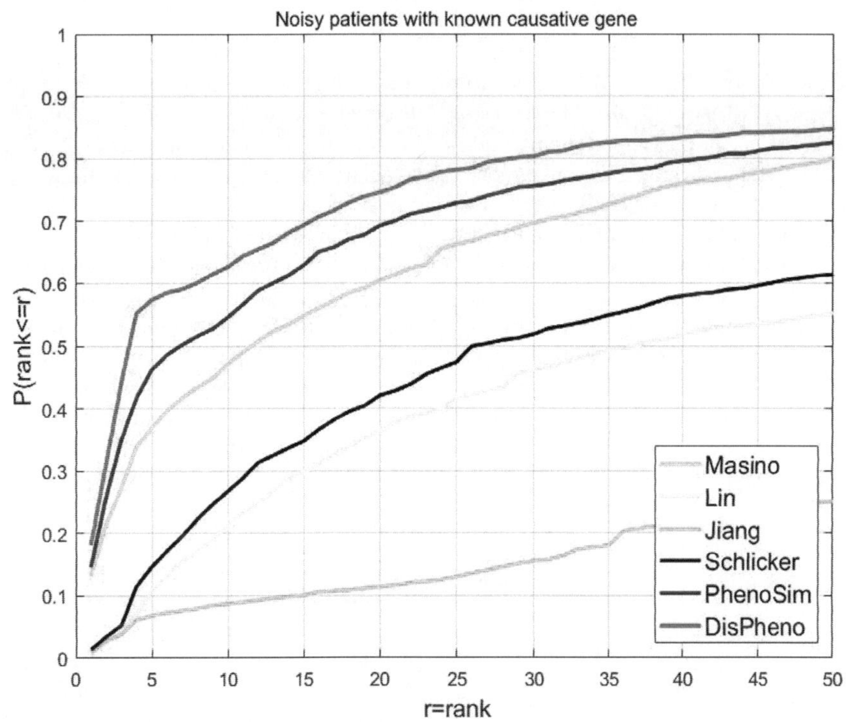

Fig. 4 Cumulative rank distribution of noisy patient dataset with the known causative gene. The x-axis is the rank threshold and the y-axis is the cumulative probability of true disease rank

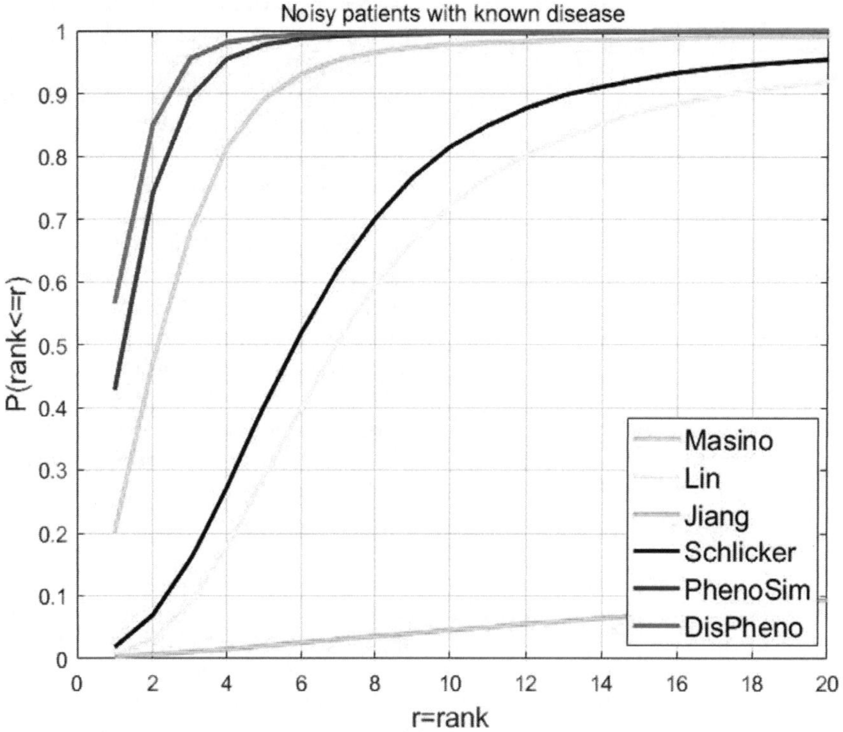

Fig. 5 Cumulative rank distribution of noisy patient dataset with the known disease. The x-axis is the rank threshold and the y-axis is the cumulative probability of true disease rank

performance of identifying true disease. First part is utilizing both gene and disease to annotate phenotype terms, named as *Anno*. Second part of our model mainly consider the effect of the distance between two phenotype terms, thus we add $(1 - dist(t_i, t_j)/mostDepth)$ in the process of calculating phenotype term similarity, named as *Depth*. Besides, we utilize TF-IDF and Cosine Similarity to measure the similarity between any two phenotype terms based on their definitions. We then add term definition similarities into phenotype topological structure, and convert original directed acyclic graph into a weighted directed acyclic graph. This part is named as *Weight*. In the part of calculating phenotype term similarity, we calculate PMI matrix to measure the association of phenotype terms. This step is named as *PMI*. We run our model with different single part to evaluate the performance of *DisPheno*. Figure 7 shows that each part of *DisPheno* contributes to improve the performance of identifying true disease from disease candidate sets. From this experimental results, we can find that the phenotype annotation method, distance between two phenotype, definition of phenotype term and association of phenotype sets are all critical to phenotype similarity measure and it could significantly improve the performance of disease diagnosis.

Table 2 The percentage of cumulative rank distribution

Method	Top-10	Top-20	Top-30	Top-40	Top-50
DisPheno	22.34%	36.00%	44.55%	52.55%	58.07%
PhenoSim	3.86%	11.76%	20.69%	29.03%	36.49%
Masino	7.12%	23.76%	38.21%	48.89%	56.57%
Lin	2.14%	8.16%	15.16%	21.77%	27.98%
Jiang	1.66%	2.57%	3.45%	4.32%	5.25%
Schlicker	1.89%	6.88%	13.67%	20.34%	26.78%

DisPheno was compared with other five methods on the noisy & imprecision patient with the known disease

Performance evaluation on gene and disease similarity

To further test the performance of *DisPheno*, we also apply our method on similarity measurement of gene and disease. Each gene or disease can be annotated by a set of phenotype terms. Therefore, gene or disease similarity measurement can be translated into a task of measuring phenotype set similarity. We run our method *DisPheno* on a gene set and a disease set. Both of the two sets contain 20 genes or diseases. We use venn diagram to show

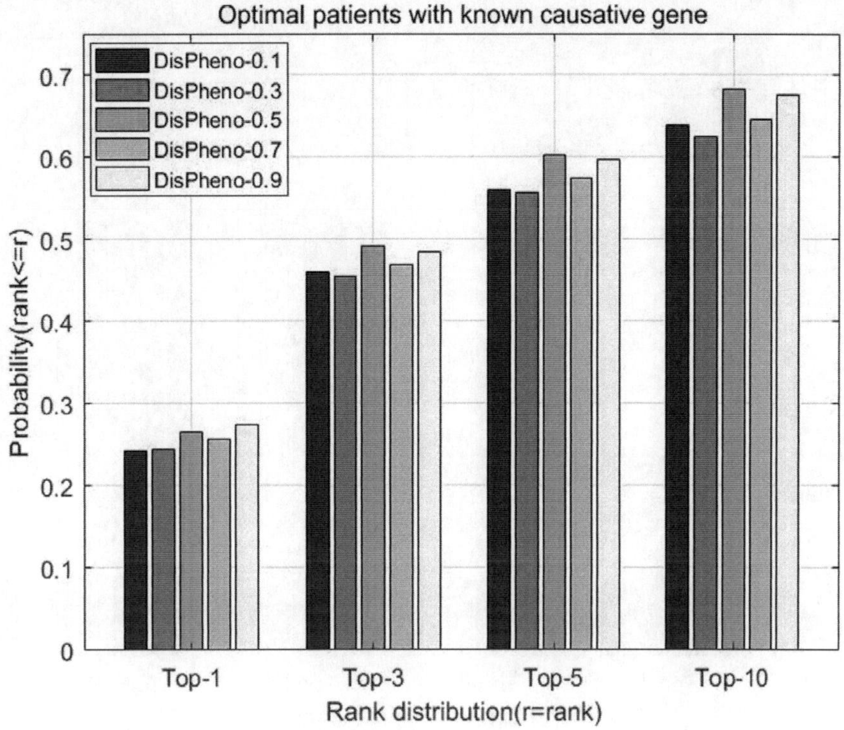

Fig. 6 The histogram of cumulative rank distribution with different weighted coefficient between gene and disease annotations. The x-axis is the four different top ranks and the y-axis is the cumulative probability of true disease rank. When the weighted coefficient is 0.5 or 0.9, DisPheno achieves better performance on the optimal patients with known causative gene dataset

the experimental results of five measurements (DisPheno, PhenoSim and other three methods randomly selected from Masino, Jiang, Lin and Schlicker). In detail, we firstly rank gene or disease pairwise similarities calculated by all five methods. Then, we calculate the intersection of top-20 gene pairs or disease pairs, and visualize the result by venn diagram.

The venn diagram (Fig. 8) shows that *DisPheno* is slightly better than other similarity measurements. We compare DisPheno and PhenoSim with other three methods which randomly selected from four phenotype similarity measurements. In the task of gene similarity calculation, the top-20 gene pairs of *DisPheno* are all part of others. In contract, *PhenoSim* contains 2 or 4 gene pairs which do not belong to any intersection. Similarity, *DisPheno* has fewer single disease-pairs than others in the task of measuring disease similarity.

Besides, we used the visualization tool of *PhenoSimWeb* to visualize the disease and gene set similarity [25]. PhenoSimWeb is an online application which can be used to calculated phenotype, gene and disease similarity. It also can predict disease and causative gene based on the input phenotype set. PhenoSimWeb contains other useful tools, such as text description translator and visualization interface. And the visualization interface of disease set similarity calculated by *DisPheno* is shown in Fig. 9. The main panel is the terms association network, where nodes represent disease terms and edges represent similarities between diseases. The upper left is the mini control panel, where you can adjust threshold and visual layout. The lower left part is the overall distribution of similarity scores. The upper right shows the neighborhood of selected disease term "OMIM:601894". This visualization webpage provides user a clear and convenient way to analysis the results of disease similarity.

PhenoSimWeb is an online phenotype similarity calculating and visualizing application, which currently contains five phenotype similarity measurements, including PhenoSim, Masino, Jiang, Lin and Schlicker. And in this paper, we propose a novel HPO-based phenotype similarity method. We will add our method *DisPheno* into the online tool *PhenoSimWeb* and enrich phenotype similarity measurement of this web application in the future.

Conclusions

The high-speed development of biological techniques such as next generation sequencing has greatly improved efficiency of cancer prediction and disease diagnosis.

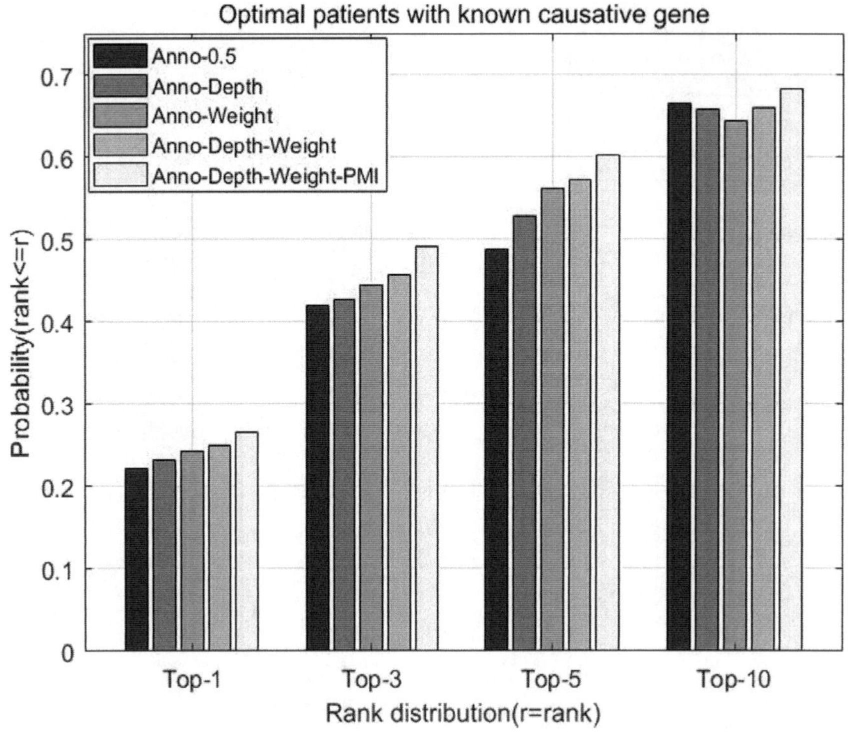

Fig. 7 The histogram of cumulative rank distribution with different parts of *DisPheno*. The x-axis is the four different top ranks and the y-axis is the cumulative probability of true disease rank. The navy blue bar is *PhenoSim* method with part *Anno* and the weighted coefficient is 0.5. The sky blue contains part *Depth* based on previous step. The green one contains part *Weight*. The orange one combines part *Depth* and *Weight*. The yellow bar is the method *DisPheno* which performs better than others

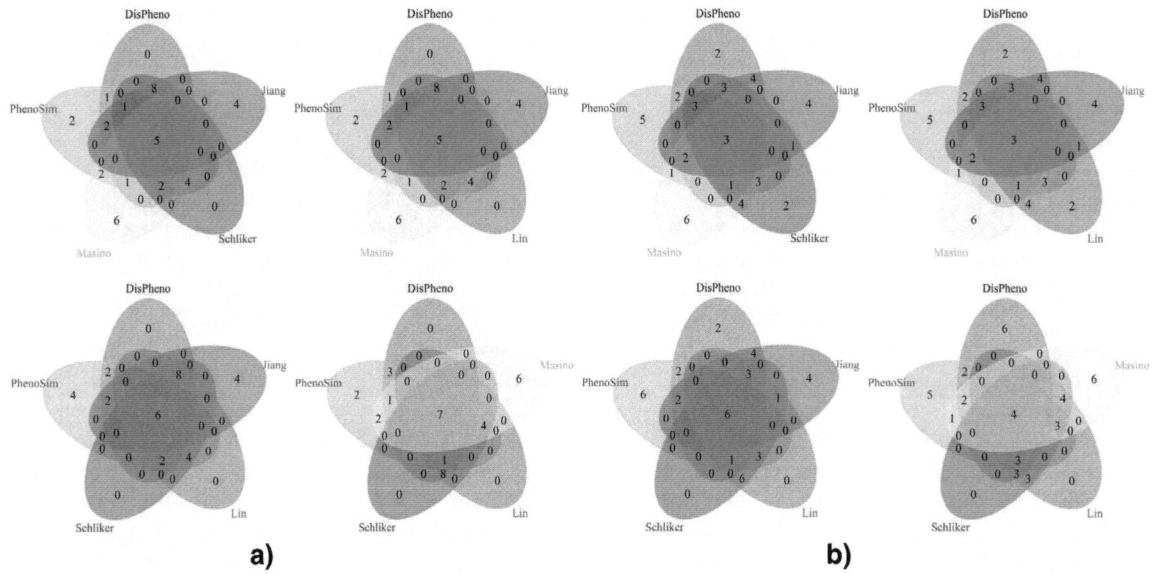

Fig. 8 The venn diagram of top-20 gene and disease pairwise similarity. The blue and green are *DisPheno* and *PhenoSim*. The purple, tomato, yellow and red are *Schlicker*, *Lin*, *Masino* and *Jiang* respectively. From the intersection of venn figure, *DisPheno* performs better than other methods on task of gene and disease similarity measurement. For instance, the upper-left venn diagram shows that there are 5 pairwise genes are included in all methods' results. All top-20 pairwise genes of DisPheno are contained by others. In contrast, there are 2 (PhenoSim), 4 (Jiang) and 6 (Masino) pairwise genes not belongs to any intersections. **a** Gene **b** Disease

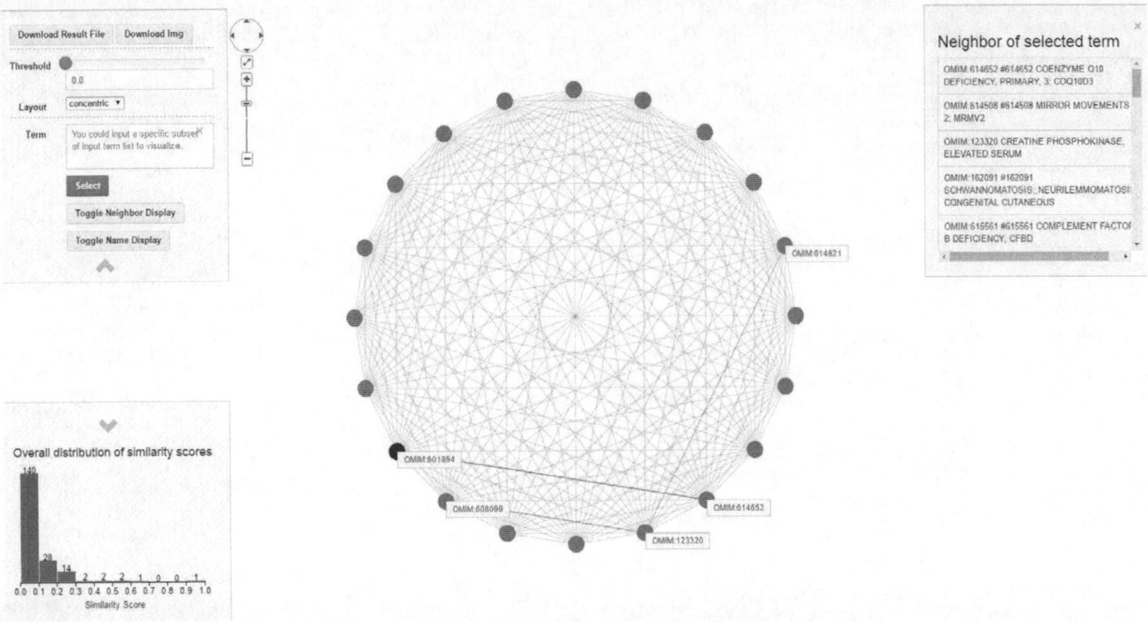

Fig. 9 Disease similarity using visualization interface of *PhenoSimWeb*. We ran *DisPheno* on a disease dataset and calculated the pair-wise similarity of these diseases. Then, we visualized the results using the visualization interface of PhenoSimWeb. The main panel is the diseases association network, where nodes represent disease terms and edges represent similarities between diseases. The upper left is the mini control panel, where you can adjust threshold and choose different visual layout. The lower left part is the overall distribution of similarity scores. The upper right shows the neighborhood of selected disease term "OMIM:601894"

However, intricate phenotype ontology and high genetic heterogeneity have stunted further improvement of disease identification. As an useful and powerful tool, HPO-based phenotype semantic similarity could fill this gap and accelerate the disease diagnosis effectively. In this paper, we proposed an unique and novel phenotype similarity measurement, called *DisPheno*, which integrates multiple types of information: hierarchical structure, phenotype term annotation and text description. Compared with existing five state-of-art methods on the optimal and noisy datasets, our method performs much better than the others. In summary, *DisPheno* accelerates the efficiency of disease identification significantly and it also shows greatly potentiality in practical clinical studies.

Abbreviations
DAG: Directed acyclic graph; HPO: Human phenotype ontology; IC: Information Content; PMI: Point-wise mutual information; TF-IDF: Term frequency-inverse document frequency

Acknowledgments
We thank all anonymous reviewers.

Funding
Publication of this article was sponsored by National Natural Science Foundation of China (No. 61702421, 61332014, 61772426), China Postdoctoral Science Foundation (No. 2017M610651), Fundamental Research Funds for the Central Universities (No. 3102018zy033), Top International University Visiting Program for Outstanding Young Scholars of Northwestern Polytechnical University.

Availability of data and materials
Not applicable.

About this supplement
This article has been published as part of *BMC Systems Biology Volume 13 Supplement 2, 2019: Selected articles from the 17th Asia Pacific Bioinformatics Conference (APBC 2019): systems biology*. The full contents of the supplement are available online at https://bmcsystbiol.biomedcentral.com/articles/supplements/volume-13-supplement-2.

Authors' contributions
JP and XS designed the algorithm framework; HX implemented the algorithm; JP and HX wrote this manuscript. All authors read and approved the final manuscript.

Ethics approval and consent to participate
Not applicable.

Consent for publication
Not applicable.

Competing interests
The authors declare that they have no competing interests.

Publisher's Note
Springer Nature remains neutral with regard to jurisdictional claims in published maps and institutional affiliations.

Published: 5 April 2019

References
1. De Ligt J, Willemsen MH, van Bon BW, Kleefstra T, Yntema HG, Kroes T, et al. Diagnostic exome sequencing in persons with severe intellectual disability. N Engl J Med. 2012;367(20):1921–9.

2. Yang Y, Muzny DM, Xia F, Niu Z, Person R, Ding Y, et al. Molecular findings among patients referred for clinical whole-exome sequencing. Jama. 2014;312(18):1870–9.
3. Study TDDD. Large-scale discovery of novel genetic causes of developmental disorders. Nature. 2015;519(7542):223–8.
4. Zemojtel T, Köhler S, Mackenroth L, Jäger M, Hecht J, Krawitz P, et al. Effective diagnosis of genetic disease by computational phenotype analysis of the disease-associated genome. Sci Transl Med. 2014;6(252): 252ra123.
5. Groza T, Köhler S, Moldenhauer D, Vasilevsky N, Baynam G, Zemojtel T, et al. The human phenotype ontology: semantic unification of common and rare disease. Am J Hum Genet. 2015;97(1):111–24.
6. Robinson PN, Köhler S, Bauer S, Seelow D, Horn D, Mundlos S. The Human Phenotype Ontology: a tool for annotating and analyzing human hereditary disease. Am J Hum Genet. 2008;83(5):610–5.
7. Peng J, Hui W, Shang X. Measuring phenotype-phenotype similarity through the interactome. BMC Bioinformatics. 2018;19(S5):114.
8. Peng J, Li Q, Shang X. Investigations on factors influencing HPO-based semantic similarity calculation. J Biomed Semant. 2017;8(1):34.
9. Köhler S, Schulz MH, Krawitz P, Bauer S, Dölken S, Ott CE, et al. Clinical diagnostics in human genetics with semantic similarity searches in ontologies. Am J Hum Genet. 2009;85(4):457–64.
10. Washington NL, Haendel MA, Mungall CJ, Ashburner M, Westerfield M, Lewis SE. Linking human diseases to animal models using ontology-based phenotype annotation. PLoS Biol. 2009;7(11):e1000247.
11. Hoehndorf R, Schofield PN, Gkoutos GV. PhenomeNET: a whole-phenome approach to disease gene discovery. Nucleic Acids Res. 2011;39(18):e119.
12. Masino AJ, Dechene ET, Dulik MC, Wilkens A, Spinner NB, Krantz ID, et al. Clinical phenotype-based gene prioritization: an initial study using semantic similarity and the human phenotype ontology. BMC Bioinformatics. 2014;15(1):1.
13. Pesquita C, Faria D, Bastos H, Ferreira AE, Falcao AO, Couto FM. Metrics for go based protein semantic similarity: a systematic evaluation. BMC Bioinformatics. 2008;9(5):4.
14. Peng J, Zhang X, Hui W, Lu J, Li Q, Liu S, et al. Improving the measurement of semantic similarity by combining gene ontology and co-functional network: a random walk based approach. BMC Syst Biol. 2018;12(1):18.
15. Peng J, Li H, Liu Y, Juan L, Jiang Q, Wang Y, et al. InteGO2: a web tool for measuring and visualizing gene semantic similarities using gene ontology. BMC Genomics. 2016;17(5):530.
16. Cheng L, Jiang Y, Wang Z, Shi H, Sun J, Yang H, et al. DisSim: an online system for exploring significant similar diseases and exhibiting potential therapeutic drugs. Sci Rep. 2016;6:30024.
17. Peng J, Uygun S, Kim T, Wang Y, Rhee SY, Chen J. Measuring semantic similarities by combining gene ontology annotations and gene co-function networks. BMC Bioinformatics. 2015;16(1):1.
18. Peng J, Wang H, Lu J, Hui W, Wang Y, Shang X. Identifying term relations cross different gene ontology categories. BMC Bioinformatics. 2017;18(16):573.
19. Teng Z, Guo M, Liu X, Dai Q, Wang C, Xuan P. Measuring gene functional similarity based on group-wise comparison of GO terms. Bioinformatics. 2013;29(11):1424–1432.
20. Caniza H, Romero AE, Heron S, Yang H, Devoto A, Frasca M, et al. GOssTo: a stand-alone application and a web tool for calculating semantic similarities on the Gene Ontology. Bioinformatics. 2014;30(15):2235–6.
21. Garla VN, Brandt C. Semantic similarity in the biomedical domain: an evaluation across knowledge sources. BMC Bioinformatics. 2012;13(1):261.
22. Batet M, Sánchez D, Valls A. An ontology-based measure to compute semantic similarity in biomedicine. J Biomed Inform. 2011;44(1):118–25.
23. Peng J, Xue H, Shao Y, Shang X, Wang Y, Chen J. A novel method to measure the semantic similarity of HPO terms. Int J Data Min Bioinform. 2017;17(2):173–88.
24. Deng Y, Gao L, Wang B, Guo X. HPOSim: an R package for phenotypic similarity measure and enrichment analysis based on the human phenotype ontology. PloS ONE. 2015;10(2):e0115692.
25. Peng J, Xue H, Hui W, Lu J, Chen B, Jiang Q, Shang X, Wang Y. An online tool for measuring and visualizing phenotype similarities using hpo. BMC Genomics. 2018;19(S6):571.
26. Jiang JJ, Conrath DW. Semantic similarity based on corpus statistics and lexical taxonomy. 1997709008. arXiv preprint cmp-lg/9.
27. Lin D. An information-theoretic definition of similarity. In: Proceedings of the Fifteenth International Conference on Machine Learning, ICML '98. San Francisco: Morgan Kaufmann Publishers Inc.; 1998. p. 296–304.
28. Wang JZ, Du Z, Payattakool R, Philip SY, Chen CF. A new method to measure the semantic similarity of GO terms. Bioinformatics. 2007;23(10): 1274–81.
29. Schlicker A, Domingues FS, Rahnenführer J, Lengauer T. A new measure for functional similarity of gene products based on Gene Ontology. BMC Bioinformatics. 2006;7(1):1.
30. Garcelon N, Neuraz A, Salomon R, Bahi-Buisson N, Amiel J, Picard C, et al. Next generation phenotyping using narrative reports in a rare disease clinical data warehouse. Orphanet J Rare Dis. 2018;13(1):85.
31. Malone BM, Perkins AD, Bridges SM. Integrating phenotype and gene expression data for predicting gene function. BMC Bioinformatics. 2009;10 Suppl 11(Suppl 11):S20.
32. Kumar AA, Van LL, Alaerts M, Ardeshirdavani A, Moreau Y, Laukens K, et al. pBRIT: Gene Prioritization by Correlating Functional and Phenotypic Annotations Through Integrative Data Fusion. Bioinformatics. 2018;34(13): 2254–2262.
33. Jing LP, Huang HK, Shi HB. Improved feature selection approach TFIDF in text mining. In: Proceedings. International Conference on Machine Learning and Cybernetics vol. 2. Beijing: IEEE; 2002. p. 944–6.
34. Church KW, Hanks P. Word association noms, Mutual Information, and lexicography. Comput Linguis. 1990;16(1):76–83.
35. Peng J, Xue H, Shao Y, Shang X, Wang Y, Chen J. Measuring phenotype semantic similarity using Human Phenotype Ontology. In: bioinformatics and biomedicine (BIBM), 2016 IEEE international conference on Shenzhen. IEEE; 2016. p. 763–6.

Liu *et al. BMC Systems Biology* 2019, **13**(Suppl 2):29
https://doi.org/10.1186/s12918-019-0698-7

BMC Systems Biology

RESEARCH Open Access

Anti-TNF-α treatment-related pathways and biomarkers revealed by transcriptome analysis in Chinese psoriasis patients

Lunfei Liu[1†], Wenting Liu[2,3†], Yuxin Zheng[4], Jisu Chen[1], Jiong Zhou[1], Huatuo Dai[1], Suiqing Cai[1], Jianjun Liu[3], Min Zheng[1] and Yunqing Ren[1*]

From The 17th Asia Pacific Bioinformatics Conference (APBC 2019)
Wuhan, China. 14-16 January 2019

Abstract

Background: Anti-tumor necrosis factor alpha (TNF-α) therapy has made a significant impact on treating psoriasis. Despite these agents being designed to block TNF-α activity, their mechanism of action in the remission of psoriasis is still not fully understood at the molecular level.

Results: To better understand the molecular mechanisms of Anti-TNF-α therapy, we analysed the global gene expression profile (using mRNA microarray) in peripheral blood mononuclear cells (PBMCs) that were collected from 6 psoriasis patients before and 12 weeks after the treatment of etanercept. First, we identified 176 differentially expressed genes (DEGs) before and after treatment by using paired t-test. Then, we constructed the gene co-expression modules by weighted correlation network analysis (WGCNA), and 22 co-expression modules were found to be significantly correlated with treatment response. Of these 176 DEGs, 79 DEGs (M_DEGs) were the members of these 22 co-expression modules. Of the 287 GO functional processes and pathways that were enriched for these 79 M_DEGs, we identified 30 pathways whose overall gene expression activities were significantly correlated with treatment response. Of the original 176 DEGs, 19 (GO_DEGs) were found to be the members of these 30 pathways, whose expression profiles showed clear discrimination before and after treatment. As expected, of the biological processes and functionalities implicated by these 30 treatment response-related pathways, the inflammation and immune response was the top pathway in response to etanercept treatment, and some known TNF-α related pathways, such as molting cycle process, hair cycle process, skin epidermis development, regulation of hair follicle development, were implicated. Furthermore, additional novel pathways were also suggested, such as heparan sulfate proteoglycan metabolic process, vascular endothelial growth factor production, whose transcriptional regulation may mediate the response to etanercept treatment.

Conclusion: Through global gene expression analysis in PBMC of psoriasis patient and subsequent co-expression module based pathway analyses, we have identified a group of functionally coherent and differentially expressed genes (DEGs) and related pathways, which has not only provided new biological insight about the molecular mechanism of anti-TNF-α treatment, but also identified several genes whose expression profiles can be used as potential biomarkers for anti-TNF-α treatment response in psoriasis.

Keywords: Psoriasis, Anti-tumor necrosis factor alpha (TNF-α) treatment, Etanercept, Peripheral blood mononuclear cells (PBMCs)

*Correspondence: yqren@zju.edu.cn
†Lunfei Liu and Wenting Liu contributed equally to this work
[1]Department of Dermatology, Second Affiliated Hospital, Zhejiang University School of Medicine, Hangzhou, China
Full list of author information is available at the end of the article

Background

Psoriasis is a common immune-mediated chronic inflammatory skin disease affecting up to 3% of the world's population [1]. It is characterized by the marked hyperproliferation of keratinocytes, infiltration of immune cells and vascular remodeling in the skin. Approximately a third of patients have a moderate-to-severe form of the disease, and have joint involvement, which has severely negative effect in the patient quality of life.

The immunopathogenesis of psoriasis is thought to be a complex interplay between T cells, dentric cells and keratinocytes, which generates the inflammatory process in the skin [2]. Biological agents (Etanercept, Infliximab and adalimumab) targeting TNF-α have showed high efficacy in psoriasis therapy, which consists with the pivotal role of TNF-α in the pathogenesis of psoriasis. As a pro-inflammatory cytokine, TNF-α possesses multiple activities in the development and maintenance of psoriasis, including recruiting T cells to the skin and increasing the proliferation of keratinocytes. It has been shown that TNF-α level significantly increased in all layers of the epidermis, dermal blood vessels in lesional skin, in the blood serum, and the synovial fluid of patients with psoriatic arthritis comparing with healthy controls [3] .

Inhibition of TNF-α is known to neutralizes the biological function of this cytokine by blocking its interaction with the cell surface TNF-a receptors, but the anti-inflammatory mechanisms of anti-TNF-α are still not fully understood. Recently, a few studies have been conducted to elucidate the mechanism of action of these anti-TNF agents in the resolution of psoriasis, focusing on different gene expression in psoriatic lesion and peripheral blood.

Skin samples from patients with anti-TNF-a treatment have been shown to downregulate Th17 [4, 5] and IL12/23 [6] -driven immune response in lesional psoriatic skin. Johnston et al. [7] further provided evidence that early responses of psoriasis to etanercept may be due to decrease tissue responsiveness to IL-17A as a result of suppressed IL-17RC expression in keratinocytes. In addition, the association with clinical remission induced by anti-TNF treatment was also described between inhibition of IL-20 subfamily [8] , CCR7/CCL1 [9] axis and p38 mitogen-activated protein kinase (MAPK) activity [10] in skin lesions. In another study, Bosè et al [11] reported a dual role of anti-TNF-α therapy by inhibiting the expression of Th17/Th1 cytokine and early inflammatory genes in the skin whilst enhancing TCR-mediated Th17/Th1 cell activation in peripheral blood. Chow et al. [12] also demonstrated the different effects of adalimumab on gene expression, finding that pathways modulating the differentiation and proliferation of epidermal keratinocytes were changed in the skin, whereas pathways involved in hematopoetic cell lineage and immune response were found to be differentially expressed in peripheral blood.

Overall, these findings demonstrated a domino effect model of anti-TNF-α action in improving psoriasis, which need to further clarify at a systemic level especially using a standardized protocol for method analysis. In the current study, we performed the transcriptome analysis of blood samples from psoriatic patients treated with etanercept to characterize gene pathways activated by treatment as well as treatment response-related expression biomarkers, aiming to obtain better mechanistic insight for the further improvement of psoriasis treatment.

Results

Sample description

Six adult patients with moderate to severe psoriasis were treated with etanercept 50 mg subcutaneously twice weekly for 12 weeks from a clinical trial. The patients were all responders, as assessed by the change in PASI75 at 12 wk. The time period of 12 wk was chosen according to the consensus statement on biological agents for the treatment of psoriasis. The demographics and clinical information of the patients were outlined in Table 1.

Differentially-expressed gene (DEG) analysis

34585 probes with detected signal p-value < 0.05, after background correction and quantile normalization, the probe-level expression is transformed into gene-level expression by using median expression of probes for the same gene symbol. After performing quality control (see materials and methods), 19777 genes were determined to be expressed and retained for differential expression analysis. A global expression analysis of all the 19777 genes was performed by using principal component analysis (PCA). No major batch effect was observed among the samples. One sample seemed to be more distant from others as "potential outliers", but was kept for further analyses (Fig. 1). An unsupervised cluster analysis was also performed by using all the expressed genes. The heatmap of the cluster analysis indicated that the two samples (0W and 12W) from each individual were mostly clustered together, but no separation was observed between the samples collected at the baseline and the ones collected 12 weeks after treatment (Fig. 1). Although not showing

Table 1 Characteristics of psoriasis patients included in this study

Patient ID	sex	Age	BMI	PASI	PASI12W	%PASI decrease
1	Male	64	21.6	21.6	1.2	94.4
2	Male	55	27.2	32.2	3.7	88.5
3	Male	23	20.1	24.3	2.8	88.5
4	Male	51	22	18.7	2.5	86.6
5	Male	61	27.3	43.6	6	86.2
6	Male	35	21.6	19.9	3	84.9

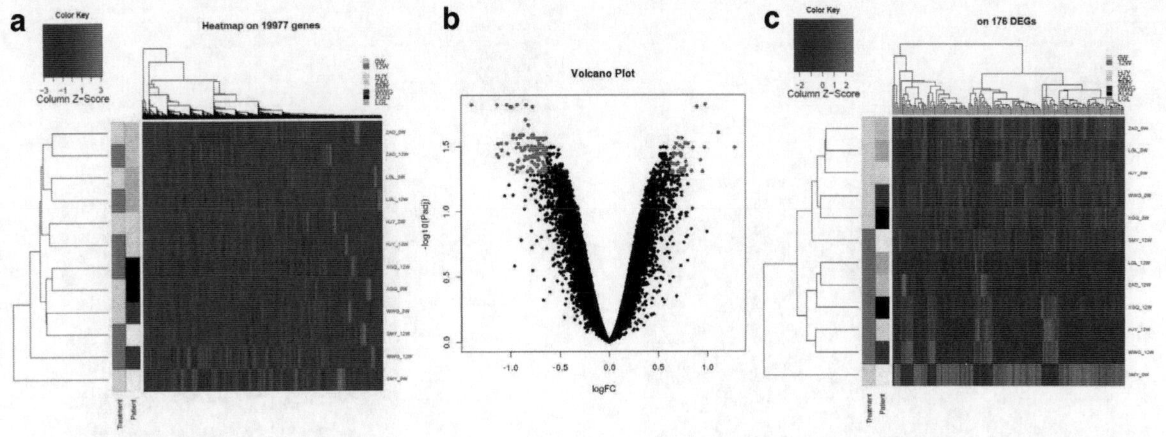

Fig. 1 **a** The heatmap on all expressed genes; **b** The volcano plot of moderated Paired-T test, the red dots are 176 differential expressed genes DEGs with at least 1.55-fold-change between two groups of samples, before and after treatment, and BH adjusted p-value<0.05; **c** the heatmap on 176 DEGs

evidence for strong batch effect, the global expression pattern mainly showed the individual difference instead of treatment effect (before and after).

We then performed differential expression analysis by carrying out pair-wise comparison analysis between the 0W and 12W samples of individual patients using paired t-test. We identified 176 differential expressed genes (DEGs) (Additional file 1) between baseline and 12-week after treatment with adjusted p-value<0.05 and 1.55-Fold Change, of which 45 were up-regulated and 131 were down-regulated after 12-week treatment (Fig. 1). An unsupervised cluster analysis was performed by using the expression profiles of these 176 DEGs. As expected, most of the 0W and 12W samples could be clustered into different groups and thus classified correctly between the baseline and after treatment. It shows that SMY_0W is out of the two groups, which is also suggested as "potential outliers" by the global expression analysis above. In the two groups of other samples, the same patient after treatment, "SMY_12W", is misclassified into the group before treatment. These misclassification results implicate that the individual gene signatures may be incomplete and inaccurate, therefore need to be further improved.

Gene co-expression module and module DEG analysis

To further identify the DEGs with coherent or related molecular functions, we investigated and grouped the genes/transcripts with correlated expressions into co-expression modules using the weighed gene co-expression network analysis (WGCNA). After identifying co-expression modules, we then evaluated the correlation of these co-expression modules with treatment response. To do so, the expression profiles of all the genes within each module were first 'summarized' through PCA analysis, and the first vector of PCA (capturing the largest portion of expression variation of the module) was then evaluated for its correlation with treatment. We found 13 modules that are significantly positive-correlated with treatment (correlation coefficient > 0.5, p-value < 0.05); and 9 modules that are significantly negative-correlated with treatment (correlation coefficient < -0.5, p-value < 0.05) (Additional file 2). These treatment correlated modules are likely enriched for the functionally related genes whose expression are correlated with treatment, and DEGs within these modules are more likely (than other DEGs) to be the ones that are involved in etanercept therapy response.

Among the 176 DEGs, only 79 were within these treatment correlated co-expression modules and referred to as "Module DEGs (M_DEGs)". As expected, the unsupervised clustering analysis using these 79 M_DEGs provided better performance than the analyses by using 176 DEGs, by perfectly separating all the 0W samples from all the 12W ones (Fig. 2).

GO enrichment analysis and GO_DEG

Last, we performed GO enrichment analysis to identify the functional processes and gene pathways that are enriched for the 79 M_DEGs. To further narrow down the pathways that play an important role in treatment response, we evaluated the correlation of the overall gene expression activity of each pathway with treatment response. About 30 pathways were found to be significantly correlated with treatment response (p-value < 0.01), and these pathways are likely to be the ones that are involved in treatment response (Table 2). Of these 30 pathways, 4 pathways were related to immune response and inflammation (positive regulation of CD4+ $\alpha\beta$ T cell

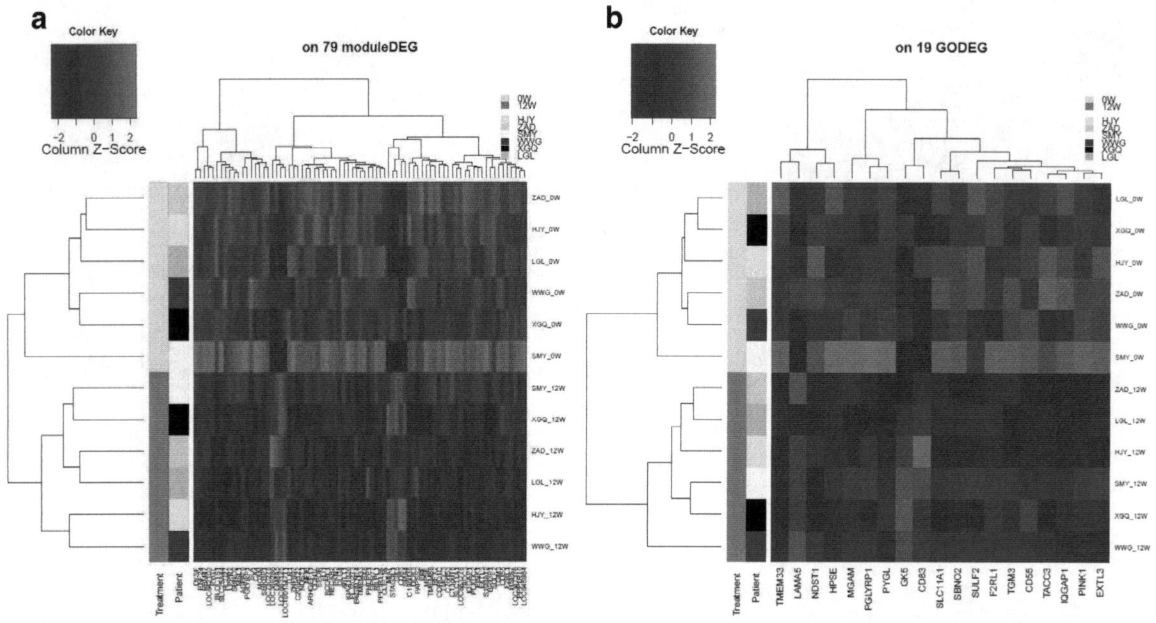

Fig. 2 a The heatmap on 79 module DEGs in the significantly treatment response correlated modules; **b** The heatmap on 19 final gene markers (GO_DEG)

activation, myeloid cell activation involved in immune response, regulation of CD4-positive, alpha-beta T cell activation, regulation of natural killer cell differentiation), 8 were involved in polysaccharide metabolic process, 5 were involved on the skin epidermal and hair follicle development, 4 pathways were enriched to positive regulation vascular endothelial growth factor (VEGF) production, positive regulation of protein kinase B signalling, glomerulus development and chondrocyte differentiation accordingly. Interestingly, only 19 M_DEGs were within these 30 pathways, and their expression profiles can achieve the same performance as the 79 M_DEGs in term of distinguishing the samples before the treatment (baseline 0W) from the ones after the treatment (12W) (Fig. 2) and can potentially be used as biomarkers for etanercept treatment.

Discussion

We performed gene expression analysis in the blood samples from psoriatic patients treated with etanercept. By comparing the samples collected at the baseline (before the treatment) and the ones collected 12 weeks after the treatment, we were able to identify 176 DEGs whose expression changes are correlated with treatment response. In addition, we integrated the results of deferential expression analysis with the co-expression analysis and further identified 79 DEGs (of the 176 DEGs) that shared coherent molecular functionalities (as suggested by their co-expression patterns). With coherent functionalities and significant correlation with treatment response, these 79 DEGs likely play an important role in etanercept treatment by mediating therapeutic response through gene expression alteration. As a result, the expression profile of these genes can clearly distinguish the blood samples collected before the treatment from the ones collected after treatment (12 weeks). Furthermore, by focusing on the GO pathways whose expression profile was correlated with treatment response, we could identify a small group of 19 DEGs whose expression profile can provide perfect separation between the blood samples collected before and after the treatment. These genes can be applied as expression biomarkers for markers of the anti-TNF-α treatment response in psoriatic patients.

In the current study, we pursued the integration of differential gene expression and co-expression analysis to identify the genes and pathways that mediate treatment response. The integration is implemented to reduce the false positives from single view and find the consistent evidence. As co-expression modules cluster the genes with similar expression patterns, these genes are more likely (than others) to share coherent functions. Using these co-expression modules as functional filters, we were able to prioritize the DEGs with coherent functions. As we demonstrated, 79 co-expression modules related DEGs performed a better job than the total 176 DEGs in term of separating the samples before and after the treatment,

Table 2 The top enriched GO pathways (pathway size <100; enrichment *p*-value < 0.1) which also significantly correlated with treatment response (|*correlation*| > 0.7; correlation *p*-value < 0.01)

GO Pathway	Treatment response correlation	GO-Enrichment *P*value	M_DEG
Heparan sulfate proteoglycan metabolic process	0.81	2.50E-06	EXTL3, HPSE, NDST1, SULF2
Cellular carbohydrate catabolic process	0.86	0.00135	GK5, MGAM, PYGL
Positive regulation of CD4-positive, alpha-beta T cell activation	0.90	0.00268	CD55, CD83
Hair follicle development	0.80	0.00273	HPSE, LAMA5, TGM3
Molting cycle process	0.80	0.00273	HPSE, LAMA5, TGM3
Hair cycle process	0.80	0.00273	HPSE, LAMA5, TGM3
Skin epidermis development	0.80	0.00283	HPSE, LAMA5, TGM3
Positive regulation vascular endothelial growth factor production	0.74	0.00317	HPSE, SULF2
Glucan catabolic process	0.88	0.00369	MGAM, PYGL
Cellular polysaccharide catabolic process	0.88	0.00397	MGAM, PYGL
Regulation of vascular endothelial growth factor production	0.79	0.00425	HPSE, SULF2
Polysaccharide catabolic process	0.88	0.00454	MGAM, PYGL
Vascular endothelial growth factor production	0.76	0.00485	HPSE, SULF2
Proteoglycan biosynthetic process	0.76	0.01521	EXTL3, NDST1
Myeloid cell activation involved in immune response	0.78	0.01734	F2RL1, SBNO2
Glycosaminoglycan catabolic process	0.76	0.01789	HPSE, PGLYRP1
Cellular glucan metabolic process	0.75	0.03104	MGAM, PYGL
Glucan metabolic process	0.75	0.03104	MGAM, PYGL
Glomerular epithelial cell development	0.88	0.03312	IQGAP1
Positive regulation of mitochondrial fission	0.79	0.03312	PINK1
Manganese ion transport	0.73	0.03961	SLC11A1
Regulation of hair follicle development	0.78	0.04284	HPSE
Cellular polysaccharide metabolic process	0.72	0.04295	MGAM, PYGL
Nucleus localization	0.76	0.07144	TACC3
Glycogen catabolic process	0.88	0.08388	PYGL
Peptide cross-linking	0.74	0.09616	TGM3
Cell proliferation in forebrain	0.85	0.09616	TACC3
Regulation of CD4-positive, alpha-beta T cell activation	-0.90	0.00485	CD55, CD83
Aminoglycan catabolic process	-0.75	0.02074	HPSE, PGLYRP1
Regulation of oxidative phosphorylation	-0.76	0.03312	PINK1
Nuclear pore organization	-0.80	0.03637	TMEM33
Regulation of natural killer cell differentiation	-0.71	0.03961	PGLYRP1
Pore complex assembly	-0.72	0.03961	TMEM33
Glycoprotein catabolic process	-0.77	0.04284	HPSE

suggesting that the functional coherence is a good constraint that can be used to narrow down genes whose expression alterations molecularly mediate the therapeutic response to anti-TNF-α treatment.

Using the 79 M_DEGs as seed, we were able to identify the pathways that are enriched for the genes and further identified the enriched pathways whose expression activities are correlated with anti-TNF-α treatment. Many of these pathways TNF-α related ones, such as skin epidermis development and molting cycle process and immune related processes, such as positive regulation of CD4-positive, alpha-beta T cell activation, myeloid cell activation involved in immune response and regulation of natural killer cell differentiation. These are well expected due to inhibition effect of the treatment on TNF-α activity.

In addition, our analyses also identified some novel pathways and/or functional processes, such as polysaccharide catabolic process and regulation VEGF production. Among the 6 pathways of polysaccharide catabolic process, the heparan sulfate proteoglycan (HSPG), belong to a heterogeneous family of the proteoglycans, are important components of the basement membrane (BM) of various tissues. Differential expression of HSPG has been shown in psoriatic lesions (comparing with normal skin) as well as chronically inflamed synovium from arthritis patients (comparing to normal individuals) [13, 14]. Several studies suggested their role in adhesion, cell-extracellular matrix interactions, migration, keratinocyte proliferation and differentiation, inflammation, and wound healing. There was also evidence that HSPG was involved in angiogenesis that might be related to the interaction of the HSPG with VEGF. Angiogenesis has a crucial role in the pathogenesis of psoriasis. As the most potent angiogenic factor, VEGF is directly implicated in influencing both skin lesions and systemic involvement in psoriasis and is a potential target for the treatment of psoriasis. It was significantly up-regulated in psoriatic skin lesions as compared with uninvolved skin in psoriasis patients and normal controls. Besides, Serum level of VEGF was significantly elevated in patients and correlated with clinical severity of psoriasis [15]. In addition, systemic anti-VEGF treatment strongly reduces skin inflammation in a mouse model of psoriasis [16] . Therefore, it is likely that a kind of antipsoriatic mechanism of TNF-α inhibitors is inhibition of angiogenesis by preventing the TNF-induced upregulation of VEGF(by reducing the synthesis of VEGF). Previous studies had also showed VEGF down-regulation in psoriatic skin lesions and in cutaneous mesenchymal stem cells from psoriasis patients responding to the anti-TNF therapy [9, 17]. In our study, we identified 3 pathways involved in VEGF production increase the expression activity after 12 weeks of anti-TNF-α treatment, which implied the dual role of anti-TNF-α therapy including inhibition of inflammation and angiogenesis. Although novel and molecularly interesting, these pathways/functional processes do need to be further investigated regarding to their involvement in mediating the treatment effect of etanercept on psoriasis.

Furthermore, we found 19 GO_DEGs within these 30 pathways. Interestingly, these 19 DEGs perform as well as 79 M_DEGs on discriminating the samples before and after the treatment. These genes can therefore be applied as expression biomarkers for the anti-TNF-α treatment response in psoriatic patients. Among them, CD83 and SLC11A1 are enriched in several pathways related to immune response and inflammation. CD83 is a marker for mature/activated dendritic cells, which are a major source of TNF in psoriasis lesions and provide signals for direct intralesional T cell activation. We found that CD83 were up-regulated after 12-week treatment, which is not with the finding from the previous study by Gottlieb et al. [18] where decreased CD83 mRNA expression was observed in psoriatic skin lesions after 12-week treatment of etanercept. Solute Carrier Family 11A Member 1 (SLC11A1), also known as natural resistance-associated macrophage protein 1(NRAMP1), has an immunoregulatory role in macrophage activation and the Th1/Th2 balance of the adaptive immune response to infection. Previous studies have showed that genetic variants in SLC11A1 gene are associated with the susceptibility to autoimmune diseases such as rheumatoid arthritis and multiple sclerosis, and infectious diseases including tuberculosis and leprosy [19]. So, it suggested these genes might be novel causal genes in related to immune response of etanercept treatment in psoriasis. We'll collect more clinical samples for further validation study on these novel findings.

Conclusions

In summary, through an integrated analysis of differential expression and co-expression, we have explored the application of functional coherence (co-expression in the current study) in transcriptome analysis and identified both known and novel anti-TNF-α treatment based novel function. Our result suggests that is a good constraint that can be used to narrow down genes whose expression alterations molecularly mediate the therapeutic response to anti-TNF-α treatment.

Methods

Sample description

Six adult patients with moderate to severe psoriasis who were going to be treated with etanercept 50 mg subcutaneously twice weekly were recruited for this study. Patients were excluded if they had already received any systemic or local medications 4 weeks before inclusion. Before and during treatment, patients were evaluated clinically with Psoriasis Area Severity Index (PASI) score. The time period of 12 wk was chosen according to the consensus statement on biological agents for the treatment of psoriasis. The patients were all responders, as assessed by the change in PASI75 at 12 wk. The demographics and clinical information of the patients were outlined in Table 1. All patients provided signed informed consent for participating in this study. The study was approved by the local institutional ethics review board, the 2nd affiliated hospital, Zhejiang University school of medicine.

RNA microarray data collection

The blood samples before etanercept treatment (0W) and 12 weeks (12W) after treatment of drug etanercept were collected from 6 psoriasis patients. Total RNA was extracted and purified from these 12 blood samples by using QIAGEN® QIAamp RNA Blood Mini Kit according to the manufacturer's instructions. After QC analysis, 500ng total RNA was amplified using Illumina® TotalPrep™ RNA Amplification Kit (Ambion). Amplified

RNA was hybridized overnight with HumanHT-12 v4 Expression BeadChips (Illumina) according to manufacturer's protocols. The slides were scanned by iSCAN. The Illumina GenomeStudio software generates the bead summary data with a single signal intensity value for each probe.

Background correction and normalization

The background correction was done in GenomeStudio by subtracting the mean signal of negative control probes [20] in a particular array from each bead summary value in that array. To avoid distortion of the results by noise due to background signal rather than specific probe-target hybridization, the specificity of individual probe signals was estimated using the detection *p*-value, which is the probability of seeing a certain signal level without probe-target hybridization. All the probes returning a detection *p*-value > 0.05 in both the control group and the case group were eliminated from further analysis. The quantile normalization was implemented by using "neqc" function in R package "limma" [21].

Gene-level differential expression test

As we focused on gene-level expression analysis, the expression level of a particular gene was calculated by using the median expression of all the probes within this gene. The low-expressed genes are filtered out by R package "genefilter" to make the remaining genes are those expressed at least 3 samples among the 12 samples.

Assuming the independence among genes, we used a linear model with sample-pair effects for each gene to fit the expression data by "lmFit" from "limma" [21] . After fitting the linear additive model, the limma moderated paired t-test (the moderated t statistic is modifying t-statistic by adding small constant to the pooled standard deviation to avoid divisions by an extremely small variance estimate, suitable for small sample size with small variance on some genes) of differential expression for each probe was implemented by "eBayes". The empirical Bayes analysis borrows the information across probes to improve power for small sample size by shrinking the probe-wise sample variances towards a common value [22] .

Treatment response correlated co-expression modules from WGCNA

Correlated expressions of different genes (co-expression) suggest shared molecular functions/gene pathways. To capture such global co-expression (functional coherent) modules from gene expression profiles, we performed weighted gene co-expression network analysis (WGCNA) by using the R package, "WGCNA" [23], which identify co-regulated gene sets based on correlation of gene pairs and assign sets of co-regulated genes to co-expression modules via hierarchical clustering. The minimum module size is set as 30.

We further attempted to further narrow down the co-expression modules whose expression activities are associated with the treatment response. First, we performed the PCA analysis using the expression data of all the genes within a particular co-expression module, and the overall expression activity of a module (eigengene expression) was estimated by using the value of the first principle component of PCA analysis. Then, we calculated the Pearson correlation of the eigengene expression of each module with treatment by coding the samples before treatment (0W) as "1" and the ones after treatment (12W) as "2" and estimated the significance of correlation (p vlaue) using student t-test. The significant treatment-correlated modules are the modules whose absolute correlations with traits are above 0.5 and corresponding *p*-value<0.05, that is, significantly correlated with the traits. Note that the positive correlation with "12W" means up-regulation of overall expression activity with treatment, and the negative correlation means down-regulation of overall expression activity with treatment.

Treatment-correlated top pathways that are enriched for module DEGs

The GO pathways enrichment was implement by an R package,"topGO" [24], where hypergeometric enrichment *p*-value is reported. We didn't report the FDR adjusted *p*-value here as it's too conservative to get very few GO pathways as significant [24]. To not omit the interesting GO pathways, we first used enrichment *p*-value to filter the GO pathways, and then used the pathway correlation with treatment traits to reduce the false positive discovery rate.

The specific GO pathways whose pathway size (the number of the gene members in the pathway) is less than 100, were enriched from the module refined DEGs (M_DEGs), and the significant ones with enrichment *p*-value less than 0.1 were selected. For these significant enriched pathways, their expression correlation with treatment traits were also investigated by the same procedures as what we did for the co-expression module in the above section. The significantly treatment traits correlated GO pathways (Pearson correlation above 0.7 or below −0.7; Pearson correlation *p*-value less than 0.01) were selected as treatment response correlated pathways, which were thus most likely the ones that mediate the anti-TNF-α treatment response through transcriptional alteration. Lastly, we identified the DEGs within these correlated GO pathways as GO_DEGs. The whole work flow of the data analyses procedures is shown in Fig. 3.

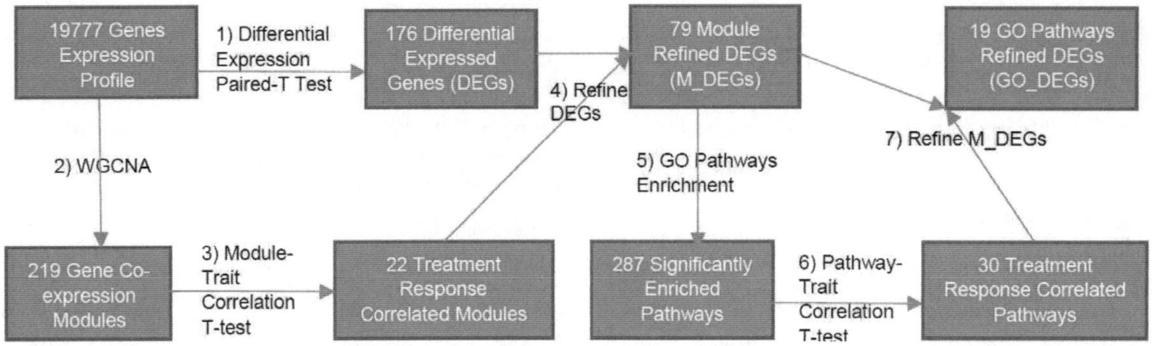

Fig. 3 The whole analysis procedures: 1) 176 DEGs are obtained by paired-t test of 19777 genes expression of patients before and after treatment; 2) 219 gene co-expression modules are obtained by Weighted Gene Co-expression Network Analysis (WGCNA) [23], 3) 22 gene co-expression modules are significant correlated with treatment response traits; 4) 79 DEGs are refined by the 22 treatment response correlated modules as M_DEGs; 5) 287 GO pathways are significantly enriched (pathway size < 100 and p-value < 0.1) from 79 M_DEGs; 6) 30 GO pathways gene expression activity are significantly correlated with treatment response traits ($|correlation| > 0.7$; correlation p-value < 0.01); 7) 19 M_DEGs are refined by the 30 treatment response correlated GO pathways as GO_DEGs

Additional files

Additional file 1: Additional Table 1. 176 differential expressed genes (DEGs) between baseline and 12-week after treatment with adjusted p-value<0.05 and 1.55-Fold Change by using paired t-test. (XLSX 20 kb)

Additional file 2: Additional Table 2. The details of co-expression modules generated from WGCNA, where 13 modules that are significantly positive-correlated with treatment (correlation coefficient > 0.5, p-value < 0.05); and 9 modules that are significantly negative-correlated with treatment (correlation coefficient < -0.5, p-value < 0.05). (XLSX 375 kb)

Abbreviations
DEG: Differentially-expressed gene; PCA: Principal components analysis; PBMC: Peripheral blood mononuclear cell; TNF: Tumor necrosis factor; WGCNA: Weighed gene co-expression network analysis

Acknowledgements
Not applicable.

Funding
Publication of this article was supported by Agency for Science, Technology and Research (A*STAR), Singapore; Biomedical Research Council (Strategic Positioning Fund SPF2014/001; and the National Natural Science Foundation of China (81573040).

Availability of data and materials
The data and materials related to this work are available upon request.

About this supplement
This article has been published as part of *BMC Systems Biology Volume 13 Supplement 2, 2019: Selected articles from the 17th Asia Pacific Bioinformatics Conference (APBC 2019): systems biology*. The full contents of the supplement are available online at https://bmcsystbiol.biomedcentral.com/articles/supplements/volume-13-supplement-2.

Authors' contributions
YR and MZ conceived and designed the experiments. LL, JC, and SC conducted sample recruitment. WL undertook data processing and statistical analysis. YZ, JZ and HD performed the experiments. YR, LL, WL and JL interpreted the results. WL, YR and JL wrote the paper. All authors read and approved the final manuscript.

Ethics approval and consent to participate
Not applicable.

Consent for publication
Not applicable.

Competing interests
The authors declare that they have no competing interests.

Publisher's Note
Springer Nature remains neutral with regard to jurisdictional claims in published maps and institutional affiliations.

Author details
[1]Department of Dermatology, Second Affiliated Hospital, Zhejiang University School of Medicine, Hangzhou, China. [2]School of Public Health and Management, Hubei University of Medicine, Shiyan, Hubei, China. [3]Human Genetics, Genome Institute of Singapore, Singapore, Singapore. [4]Second Clinical Medical College, Zhejiang Chinese Medical University, Hangzhou, China.

Published: 5 April 2019

References
1. Crow JM. Psoriasis uncovered. Nature. 2012;492:S50–1.
2. Kim J, Krueger JG. The immunopathogenesis of psoriasis. Dermatol Clin. 2015;33:13–23.
3. Schottelius AJ, Moldawer LL, Dinarello CA, Asadullah K, Sterry W, Edwards III CK. Biology of tumor necrosis factor-alpha- implications for psoriasis. Exp Dermatol. 2004;13:193–222.
4. Zaba LC, Suarez-Farinas M, Fuentes-Duculan J, Nograles KE, Guttman-Yassky E, Cardinale I, Lowes MA, Krueger JG. Effective treatment of psoriasis with etanercept is linked to suppression of IL-17 signaling, not immediate response TNF genes. J Allergy Clin Immunol. 2009;124:1022–30.
5. Balato A, Schiattarella M, Di Caprio R, Lembo S, Mattii M, Balato N, Ayala F. Effects of adalimumab therapy in adult subjects with moderate-to-severe psoriasis on Th17 pathway. J Eur Acad Dermatol Venereol. 2014;28:1016–24.
6. Brunner PM, Koszik F, Reininger B, Kalb ML, Bauer W, Stingl G. Infliximab induces downregulation of the IL-12/IL-23 axis in 6-sulfo-lacnac (slan)+ dendritic cells and macrophages. J Allergy Clin Immunol. 2013;132:1184–93.
7. Johnston A, Guzman AM, Swindell WR, Wang F, Kang S, Gudjonsson JE. Early tissue responses in psoriasis to the antitumour necrosis factor-alpha biologic etanercept suggest reduced interleukin-17 receptor expression and signalling. Br J Dermatol. 2014;171:97–107.
8. Wang F, Smith N, Maier L, Xia W, Hammerberg C, Chubb H, Chen C, Riblett M, Johnston A, Gudjonsson JE, et al. Etanercept suppresses

regenerative hyperplasia in psoriasis by acutely downregulating epidermal expression of interleukin (IL)-19, IL-20 and IL-24. Br J Dermatol. 2012;167:92–102.
9. Bose F, Petti L, Diani M, Moscheni C, Molteni S, Altomare A, Rossi RL, Talarico D, Fontana R, Russo V, et al. Inhibition of ccr7/ccl19 axis in lesional skin is a critical event for clinical remission induced by tnf blockade in patients with psoriasis. Am J Pathol. 2013;183:413–21.
10. Johansen C, Vinter H, Soegaard-Madsen L, Olsen LR, Steiniche T, Iversen L, Kragballe K. Preferential inhibition of the mRna expression of p38 mitogen-activated protein kinase regulated cytokines in psoriatic skin by anti-TNFα therapy. Br J Dermatol. 2010;163:1194–204.
11. Bose F, Raeli L, Garutti C, Frigerio E, Cozzi A, Crimi M, Caprioli F, Scavelli R, Altomare G, Geginat J, et al. Dual role of anti-TNF therapy: enhancement of TCR-mediated T cell activation in peripheral blood and inhibition of inflammation in target tissues. Clin Immunol. 2011;139: 164–76.
12. Chow M, Lai K, Ahn R, Gupta R, Arron S, Liao W. Effect of adalimumab on gene expression profiles of psoriatic skin and blood. J Drugs Dermatol. 2016;15:988–94.
13. Patterson AM, Cartwright A, David G, Fitzgerald O, Bresnihan B, Ashton BA, Middleton J. Differential expression of syndecans and glypicans in chronically inflamed synovium. Ann Rheum Dis. 2008;67:592–601.
14. Seyger MM, van den Born J, Schalkwijk J, van de Kerkhof PC, de Jong EM. Altered distribution of heparan sulfate proteoglycans in psoriasis. Acta Derm Venereol. 1997;77:105–9.
15. Bhushan M, McLaughlin B, Weiss JB, Griffiths CE. Levels of endothelial cell stimulating angiogenesis factor and vascular endothelial growth factor are elevated in psoriasis. Br J Dermatol. 1999;141:1054–60.
16. Schonthaler HB, Huggenberger R, Wculek SK, Detmar M, Wagner EF. Systemic anti-VEGF treatment strongly reduces skin inflammation in a mouse model of psoriasis. Proc Natl Acad Sci. 2009;106:21264–9.
17. Campanati A, Orciani M, Ganzetti G, Consales V, Di Primio R, Offidani A. The effect of etanercept on vascular endothelial growth factor production by cutaneous mesenchymal stem cells from patients with psoriasis. J Int Med Res. 2016;44:6–9.
18. Gottlieb AB, Chamian F, Masud S, Cardinale I, Abello MV, Lowes MA, Chen F, Magliocco M, Krueger JG. TNF inhibition rapidly down-regulates multiple proinflammatory pathways in psoriasis plaques. J Immunol. 2005;175:2721–9.
19. Awomoyi AA. The human solute carrier family 11 member 1 protein (SLC11A1): linking infections, autoimmunity and cancer? FEMS Immunol Med Microbiol. 2007;49:324–9.
20. Shi W, Oshlack A, Smyth GK. Optimizing the noise versus bias trade-off for illumina whole genome expression beadchips. Nucleic Acids Res. 2010;38:e204.
21. Ritchie ME, Phipson B, Wu D, Hu Y, Law CW, Shi W, Smyth GK. limma powers differential expression analyses for RNA-sequencing and microarray studies. Nucleic Acids Res. 2015;43:e47.
22. Smyth GK. Linear models and empirical bayes methods for assessing differential expression in microarray experiments. Stat Appl Genet Mol Biol. 2004;3:Article 3.
23. Langfelder P, Horvath S. WGCNA, an R package for weighted correlation network analysis. BMC Bioinformatics. 2008;9:559.
24. Alexa A, Rahnenfuhrer J. Gene set enrichment analysis with topGo. 2018. (www.bioconductor.org).

Sun et al. BMC Systems Biology 2019, 13(Suppl 2):28
https://doi.org/10.1186/s12918-019-0699-6

BMC Systems Biology

RESEARCH Open Access

A fast and efficient count-based matrix factorization method for detecting cell types from single-cell RNAseq data

Shiquan Sun[1,2,3,4], Yabo Chen[1], Yang Liu[1] and Xuequn Shang[1,2*]

From The 17th Asia Pacific Bioinformatics Conference (APBC 2019)
Wuhan, China. 14–16 January 2019

Abstract

Background: Single-cell RNA sequencing (scRNAseq) data always involves various unwanted variables, which would be able to mask the true signal to identify cell-types. More efficient way of dealing with this issue is to extract low dimension information from high dimensional gene expression data to represent cell-type structure. In the past two years, several powerful matrix factorization tools were developed for scRNAseq data, such as NMF, ZIFA, pCMF and ZINB-WaVE. But the existing approaches either are unable to directly model the raw count of scRNAseq data or are really time-consuming when handling a large number of cells (e.g. $n > 500$).

Results: In this paper, we developed a fast and efficient count-based matrix factorization method (single-cell negative binomial matrix factorization, scNBMF) based on the TensorFlow framework to infer the low dimensional structure of cell types. To make our method scalable, we conducted a series of experiments on three public scRNAseq data sets, brain, embryonic stem, and pancreatic islet. The experimental results show that scNBMF is more powerful to detect cell types and 10 - 100 folds faster than the scRNAseq bespoke tools.

Conclusions: In this paper, we proposed a fast and efficient count-based matrix factorization method, scNBMF, which is more powerful for detecting cell type purposes. A series of experiments were performed on three public scRNAseq data sets. The results show that scNBMF is a more powerful tool in large-scale scRNAseq data analysis. scNBMF was implemented in R and Python, and the source code are freely available at https://github.com/sqsun.

Keywords: Single-cell RNA sequencing, Matrix factorization, Read count, Deep learning

Background

Single-cell RNA-sequencing (scRNAseq) analysis plays an important role in investigating tumour evolution, and is more powerful to characterize the intra-tumor cellular heterogeneity [1, 2]. Compared with traditional RNA sequencing (i.e. bulk RNAseq) which measures the specific gene expression level within a cell population, scRNAseq quantifies the specific gene expression level within only an individual cell [3, 4]. scRNAseq is more likely to understand the detailed biological processes of cell developmental trajectories and cell-to-cell heterogeneity, providing us fresh insights into cell composition, dynamic cell states, and regulatory mechanisms [5–8].

However, there are still several big challenges we have to carefully deal with before analyzing scRNAseq data [9, 10]. The first challenge is that the scRNAseq data is easy to involve some unwanted variables [11, 12], e.g. batch effects, confounding factors, etc. Moreover, the scRNAseq data set has their own characterizes, such as gene expression matrix is extremely sparse because of the quite small number of mRNAs represented in each cell [13]; current sequencing technologies, e.g. CEL-Seq2 [14] and

*Correspondence: shang@nwpu.edu.cn
[1]School of Computer Science, Northwestern Polytechnical University, 710129 Xi'an, Shaanxi, People's Republic of China
[2]Key Laboratory of Big Data Storage and Management, Northwestern Polytechnical University, Ministry of Industry and Information Technology, 710129 Xi'an, Shaanxi, People's Republic of China
Full list of author information is available at the end of the article

Drop-seq [15], etc, do not have enough power to quantify the actual concentration of mRNAs (i.e. well-known "dropout events") [16]; the heavy amplifications may result into strong amplification bias [17]; cell cycle state, cell size or other unknown factors may contribute to cell-cell heterogeneity even within the same cell type [18].

The second important feature of the scRNAseq data set is of count nature [19]. In most RNA sequencing studies, the number of reads mapped to a given gene or isoform is often used as an intuitive estimate of its expression level. To account for the count nature of the RNA sequencing data, and the resulting mean-variance dependence, most statistical methods were developed using discrete distributions in differential expression analysis, i.e., PQLseq [20], edgeR/DESeq [21, 22], and MACAU [23]. Therefore, a nature choice of analyzing scRNAseq data is to develop count-based dimensionality reduction methods. Although several dimensionality reduction techniques have been already applied to scRNAseq data analysis, such as principal component analysis (PCA) [24]; independent components analysis (ICA) [25], and diffusion map [26]; partial least squares (PLS) [27, 28]; nonnegative matrix factorization (or factor analysis) [29, 30], gene expression levels are inherently quantified by counts, i.e., count nature of scRNAseq data [31, 32].

Therefore, developing the bespoke scRNAseq dimensionality reduction method has been triggered within the last two years. The first factor analysis method, ZIFA, is trying to model the drop-out events via the zero-inflated model, but the method does not take into account the count nature of the data [33]; pCMF is trying to build sparse Gamma-Poisson factor model within the Bayesian framework, but such method does not include the covariates [34]; ZINB-WaVE is trying to involve both gene-level and sample-level covariates via a hierarchical model, but the method is really time-consuming when sample size is large [35, 36].

Here, in this paper, we propose a fast and efficient count-based matrix factorization method that utilizes the negative binomial distribution to account for the over-dispersion problem of the count nature of scRNAseq data, single-cell Negative Binomial-based Matrix Factorization, scNBMF. The reason of choosing negative binomial model instead of zero-inflated negative binomial model is that not only the most scRNAseq data sets do not show much technical contribution to zero-inflation (Fig. 1a), but also can largely reduce the computation burden in estimating drop-out parameters for each gene. With the stochastic optimization method Adam [37] implemented within TensorFlow framework, scNBMF is roughly 10 – 100 times faster than the existing count-based matrix factorization methods, such as pCMF and ZINB-WaVE. To make the proposed method scalable, we apply scNBMF to analyze three publicly available scRNAseq datasets. The

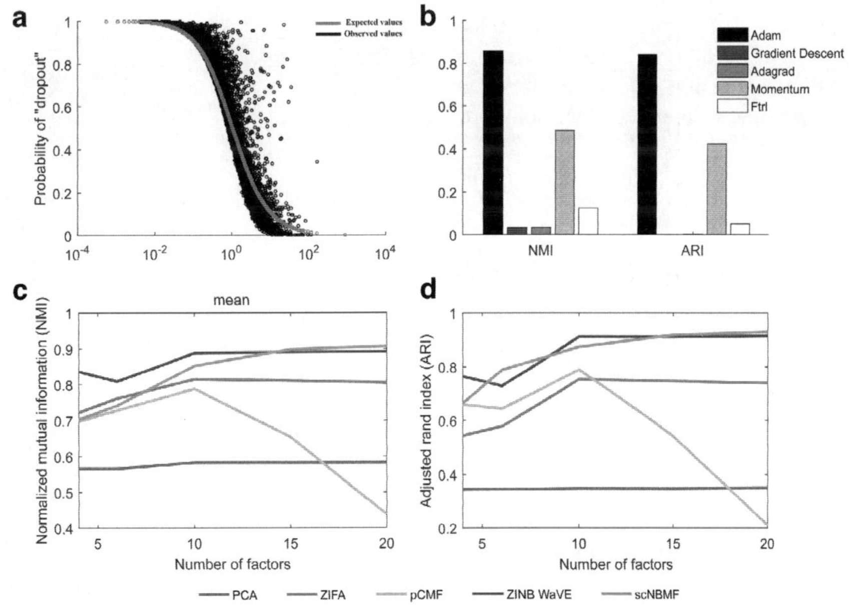

Fig. 1 A simple example to show the parameter effect or optimizer effect of NMI and ARI in scRNA-seq data on clustering. **a** This figure shows the relationship between mean gene expression levels and dropout rates. The black line indicates observed value, which is computed by the number of unexpressed cells divided by the number of cells; The red line represents expected value, which is calculated by negative binomial distribution with mean gene expression levels and dispersion parameter $\psi(\psi = mean(\psi_i))$ **b** This figure shows how optimizers affect the performance of different methods on NMI and ARI. **c-d** These two figure indicate how the number of factors affect the NMI and ARI, respectively

results demonstrate that scNBMF is more efficient and powerful than other matrix factorization methods.

Materials and methods

scNBMF: model and algorithm

scNBMF is to fit the logarithm likelihood function of negative binomial model-based matrix factorization. Given n cells and p genes, we denote Y as a gene expression matrix, and its element y_{ij} is the count of gene i and cell j. To account for the over-dispersion problem, we model the gene expression level y_{ij} as a random variable following the negative binomial distribution with parameters μ_{ij} and ϕ_i, i.e.,

$$y_{ij} \sim NB(\mu_{ij}, \phi_i)$$

where the rate parameter μ_{ij} denotes the mean expression level for gene i and cell j; the parameter ϕ_i represents variance of gene expression, typically means gene-specific over-dispersion; NB is the negative binomial distribution, i.e.

$$\Pr_{NB}(y_{ij}|\mu_{ij}, \phi_i) = \binom{y_{ij} + \phi_i - 1}{y_{ij}} \left(\frac{\mu_{ij}}{\mu_{ij} + \phi_i}\right)^{y_{ij}} \left(\frac{\phi_i}{\mu_{ij} + \phi_i}\right)^{\phi_i}.$$

For the rate parameter μ_{ij}, we consider the following regression model

$$log(\mu_{ij}) = log(N_j) + \sum\nolimits_{k=1}^{K} W_{ik}X_{kj}.$$

where N_j is the total read count for the individual cell j (a.k.a read depth or coverage); W_{ik} is the loadings while H_{kj} is the factors represents the coordinates of the cells, which can be used to identify cell type purpose; K is the pre-defined number of components; When all $\phi_i \to 0$, the negative binomial distribution will reduce to the standard Poisson distribution.

Therefore, the log-likelihood function for gene i and cell j is

$$\begin{aligned}\mathcal{L}_{NB}(\mu, \phi|Y) &= \sum_{i=1}^{p}\sum_{j=1}^{n} log\Pr_{NB}\left(y_{ij}|\mu_{ij}, \phi_i\right) \\ &= \sum_{i=1}^{p}\sum_{j=1}^{n} y_{ij}log(\mu_{ij}) + \phi_i log(\phi_i) \\ &\quad - (y_{ij} + \phi_i)log(\mu_{ij} + \phi_i) \\ &\quad + log\binom{y_{ij} + \phi_i - 1}{y_{ij}}.\end{aligned}$$

where μ denotes the mean gene expression matrix and its element $\mu_{ij} = e^{log(N_j)+\sum_{k=1}^{K} W_{ik}X_{kj}}$; ϕ is a p-vector, and its element ϕ_i represents the over-dispersion parameter for gene i.

To make our model more interpretation for the biological applications, we introduce a sparse penalty (LASSO) on loading matrix W since some genes are expressed while some are not in real-world biological processes. Therefore, the objective function of optimization problem becomes

$$\mathcal{L} = \mathcal{L}_{NB}(\mu, \phi|Y) + \lambda \sum_{i=1}^{p} \| W_i \|_1$$

where $\| \cdot \|_1$ is a l_1-norm (i.e. LASSO penalty); λ denotes the penalty parameter.

In the above model, we are interested in extracting the factor matrix H for detecting the cell type purposes. We first estimate the dispersion parameter ϕ_i) for each gene via edgeR [21] with default parameter settings, then fit the above model using Adam optimizer within TensorFlow. For deep learning model, we set the learning rate of the network as 0.001 and maximum iteration as 18000.

Compared methods and evaluations

To make scNBMF scalable, we compared seven existing methods, i.e. PCA, Nimfa, NMFEM, tSNE, ZIFA, pCMF, and ZINB-WaVE, in the experiments. Since PCA and ZIFA are only for normalized gene expression data, we normalized raw count data following previous recommendations [38]. Typically, we transformed the count data using base 2 and pseudo count 1.0, i.e., $log_2(Y + 1.0)$, into continuous data. The performance of each method was evaluated by the normalized mutual information (NMI), defined in [39]

$$NMI(L_e, L) = \frac{\sum_{k=1}^{K}\sum_{t=1}^{K_e} \frac{n_{kt}}{n} log\left(\frac{n_{kt}}{n}\right) - \sum_{k=1}^{K} \frac{n_k}{n} log\left(\frac{n_k}{n}\right) - \sum_{t=1}^{K_e} \frac{n_t}{n} log\left(\frac{n_t}{n}\right)}{\sqrt{\sum_{k=1}^{K} \frac{n_k}{n} log\left(\frac{n_k}{n}\right) * \sum_{t=1}^{K_e} \frac{n_t}{n} log\left(\frac{n_t}{n}\right)}}. \quad (1)$$

and the adjusted rand index (ARI), defined in [40]

$$ARI(L_e, L) = \frac{\sum_{kt}\binom{n_{kt}}{2} - \left(\sum_k \binom{n_k}{2} \sum_t \binom{n_t}{2}\right) / \binom{n}{2}}{\frac{1}{2}\left(\sum_k \binom{n_k}{2} + \sum_t \binom{n_t}{2}\right) - \left(\sum_k \binom{n_k}{2} \sum_t \binom{n_t}{2}\right) / \binom{n}{2}}.$$

where L_e and L are the predicted cluster labels and the true labels, respectively; K_e and K are the predicted cluster number and the true cluster number, respectively; n_k denotes the number of cells assigned to a specific cluster k $(k = 1, 2, \cdots, K)$; similarly n_t denotes the number of cells assigned to cluster $t(t = 1, 2, \cdots, K_e)$; n_{kt} represents the number of cells shared between cluster k and t; and n is the total number of cells.

Public scRNAseq data sets

Three publicly available scRNAseq data sets were collected from three studies:

- The first scRNAseq data set was collected from human brain [41]. There are 420 cells in eight cell types after excluded hybrid cells including, fetal quiescent cells (110 cells), fetal replicating cells (25 cells), astrocytes cells (62 cells), neuron cells (131 cells), endothelial (20 cells) and oligodendrocyte cells (38 cells) microglia cells(16 cells), and (OPCs, 16 cells), and remain 16,619 genes to test after filtering out the lowly expressed genes. The original data was downloaded from the data repository Gene Expression Omnibus (GEO; GSE67835);
- The second scRNAseq data set was collected from human pancreatic islet [42]. There are 60 cells in six cell types after excluding undefined cells including alpha cells (18 cells), delta cells (2 cells), pp cells (9 cells), duct cells (8 cells), beta cells (12 cells) and acinar cells (11 cells) ,and 116,414 genes to test after filtering out the lowly expressed genes. The original data was downloaded from the data repository Gene Expression Omnibus (GEO; GSE73727);
- The third scRNAseq data set was collected from the human embryonic stem [43]. There are 1018 cells which belong to seven known cell subpopulations that include neuronal progenitor cells (NPCs, 173 cells), definitive endoderm derivative cells (DEDs), endothelial cells (ECs, 105 cells), trophoblast-like cells (TBs, 69 cells), undifferentiated H1(212 cells) and H9(162 cells) ESCs, and fore-skin fibroblasts (HFFs, 159 cells), and contains 17,027 genes to test after filtering step. The original data was downloaded from the data repository Gene Expression Omnibus (GEO; GSE75748).

Results

Model selection

Our first set of experiments is to select the optimization method for the log-likelihood function of negative binomial matrix factorization model. Without loss of generality, we choose the human brain scRNAseq data set. Five optimization methods were compared to optimize the neural networks, i.e., Adam, gradient descent, Adagrad, Momentum and Ftrl. The results show that the Adam significantly outperforms other optimization methods regardless of what criteria we choose (Fig. 1b). Specifically, for NMI, Adam, gradient descent, Adagrad, Momentum, and Ftrl achieve 0.8579, 0.0341, 0.0348, 0.4859, and 0.1251, respectively. Therefore, in the following experiments, we will choose the Adam method to optimize the neural networks.

Our second set of experiments is to select the number of factors in the low dimensional structure of cell types. Without loss of generality, we still choose the human brain scRNAseq data set. We varied the number of factors ($k = 4$, 6, 10, 15, and 20). The results demonstrate that the number of factors does not impact PCA (Fig. 1c and d; bule line). The other four methods show an increasing pattern when the number of factors varied from 4 to 20 (Fig. 1c and d). Therefore, we choose the top 20 factors in the following experiments.

Public scRNAseq data sets

Our third set of experiments is to apply scNBMF to three scRNAseq real data sets, human brain, human pancreas islet, and human embryonic stem. The cell type information of the three data sets were reported by the original studies. For the comparison, we compared seven other methods, PCA, Nimfa, NMFEM, tSNE, ZIFA, pCMF and ZINB-WaVE. For the evaluation, we extracted the low dimensional structure with top 10 factors, and used k-means clustering method in an unsupervised manner, repeated 100 times to test how well each method can recover the cell type assignments on NMI and ARI in the studies.

The first biological data application is performed on the human brain scRNAseq data set. Figure 2 demonstrates the comparison results of tSNE with respect to seven compared clustering methods. scNBMF shows the clearly cell type patterns with the annotated cell type (Fig. 1h). Also, we carried out the same analysis using PCA (Fig. 2a), Nimfa (Fig. 2b), NMFEM (Fig. 2c), tSNE (Fig. 2d), ZIFA (Fig. 2e), pCMF (Fig. 2f), and ZINB-WaVE (Fig. 2g). For NMI and ARI, scNBMF outperforms the other methods. Specifically, for NMI criterion, PCA, Nimfa, NMFEM, tSNE, ZIFA, pCMF, ZINB-WaVE and scNBMF achieve, 0.582, 0.494, 0.456, 0.712, 0.797, 0.787, 0.892, and 0.901, respectively (Fig. 2i and Table 1); while for ARI criterion, PCA, Nimfa, NMFEM, tSNE, ZIFA, pCMF, ZINB-WaVE and scNBMF achieve, 0.339, 0.258, 0.264, 0.544, 0.721, 0.788, 0.916, and 0.933, respectively (Fig. 2i and Table 1).

The second biological data application is to investigate the character of human pancreas islet scRNAseq data set. This data set has a smaller number of cells - only 60 cells in six cell types. Since all methods do not have enough power to detect the cell type clustering patterns, we did not show the tSNE plots for this data set. For NMI and ARI, tSNE shows the highest performance, while scNBMF achieves the second best performance (Table 1). Specifically, tSNE achieves 0.973 and 0.652 on NMI and ARI, respectively; while scNBMF is 0.716 and 0.472 on NMI and ARI respectively.

The third biological data application is to investigate lineage-specific transcriptomic features at single-cell resolution. To elucidate the distinctions between different lineages, we performed eight matrix factorization methods, i.e., PCA (Fig. 3a), Nimfa (Fig. 3b), NMFEM (Fig. 3c), tSNE (Fig. 3d), ZIFA (Fig. 3e), pCMF (Fig. 3f), ZINB-WaVE (Fig. 3g), and scNBMF (Fig. 3h). scNBMF demonstrates more clearly their respective cell-type patterns

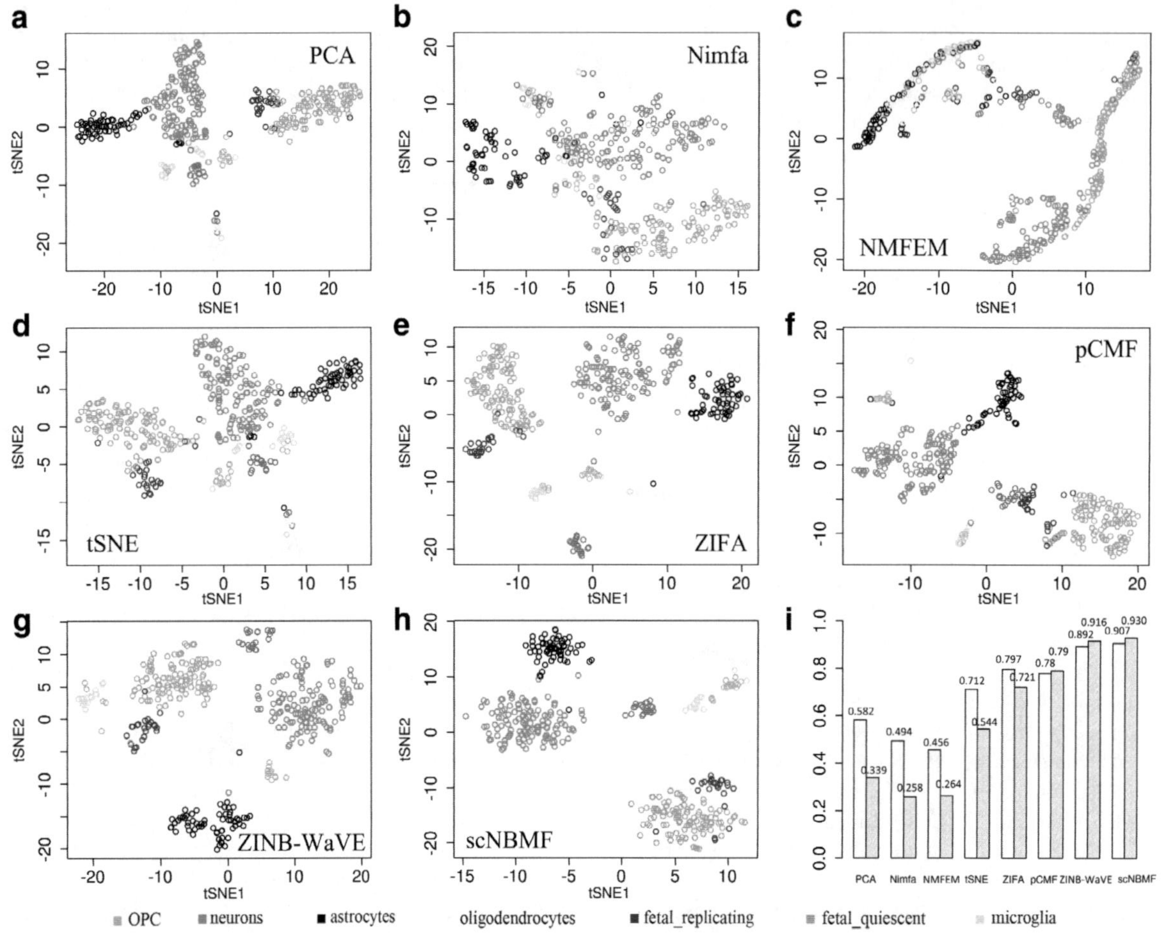

Fig. 2 Performance evaluation on human brain scRNA-seq data. In this data set there are 420 cells in eight different cell types after the exclusion of hybrid cells. Each kind of color represent a kind of cell type. **a-h** These eight figures display the clustering output of two dimension of tSNE using eight matrix factorization methods(PCA, Nimfa, NMFEM, tSNE, ZIFA, pCMF, ZINB-WaVE, and scNBMF). **f** This figure shows NMI and ARI values which are from eight compared methods

Table 1 Clustering comparison of the matrix factorization-based methods in terms of Normalized Mutual information (NMI) and Adjusted Random Index (ARI)

Method	Brain		Embryo		Pancreas	
	NMI	ARI	NMI	ARI	NMI	ARI
PCA	0.582	0.339	0.366	0.187	0.630	0.368
Nimfa	0.494	0.258	0.414	0.173	0.456	0.114
NMFEM	0.456	0.264	0.741	0.614	0.435	0.175
tSNE	0.712	0.544	0.658	0.538	**0.793**	**0.652**
ZIFA	0.797	0.721	0.888	0.748	0.641	0.429
pCMF	0.787	0.788	0.822	0.659	0.547	0.334
ZINB-WaVE	0.892	0.916	0.888	0.721	0.518	0.342
scNBMF	**0.901**	**0.933**	**0.908**	**0.763**	0.716	0.472

The number with bold indicates the best performance method and the number with grey represents the second best performance method

compared with other methods. The cell type H1 and H9 show the tight overlapping pattern to indicate the relative homogeneity of human ES cells, such results are also consistence with the previous results [43]. For NMI and ARI, scNBMF outperforms other methods (Fig. 3i and Table 1). Specifically, for NMI, PCA, Nimfa, NMFEM, tSNE, ZIFA, pCMF, ZINB-WaVE and scNBMF achieve, 0.366, 0.414, 0.741, 0.658, 0.888, 0.822, 0.888, and 0.908, respectively; For ARI, PCA, Nimfa, NMFEM, tSNE, ZIFA, pCMF, ZINB-WaVE and scNBMF achieve, 0.187, 0.173, 0.614, 0.538, 0.748, 0.659, 0.721, and 0.763, respectively.

Computation time

The last set of experiments is to compare the computation time of PCA, Nimfa, NMFEM, tSNE, ZIFA, pCMF, and ZINB-WaVE. Without loss of generality, we use human brain data set to show the computation time

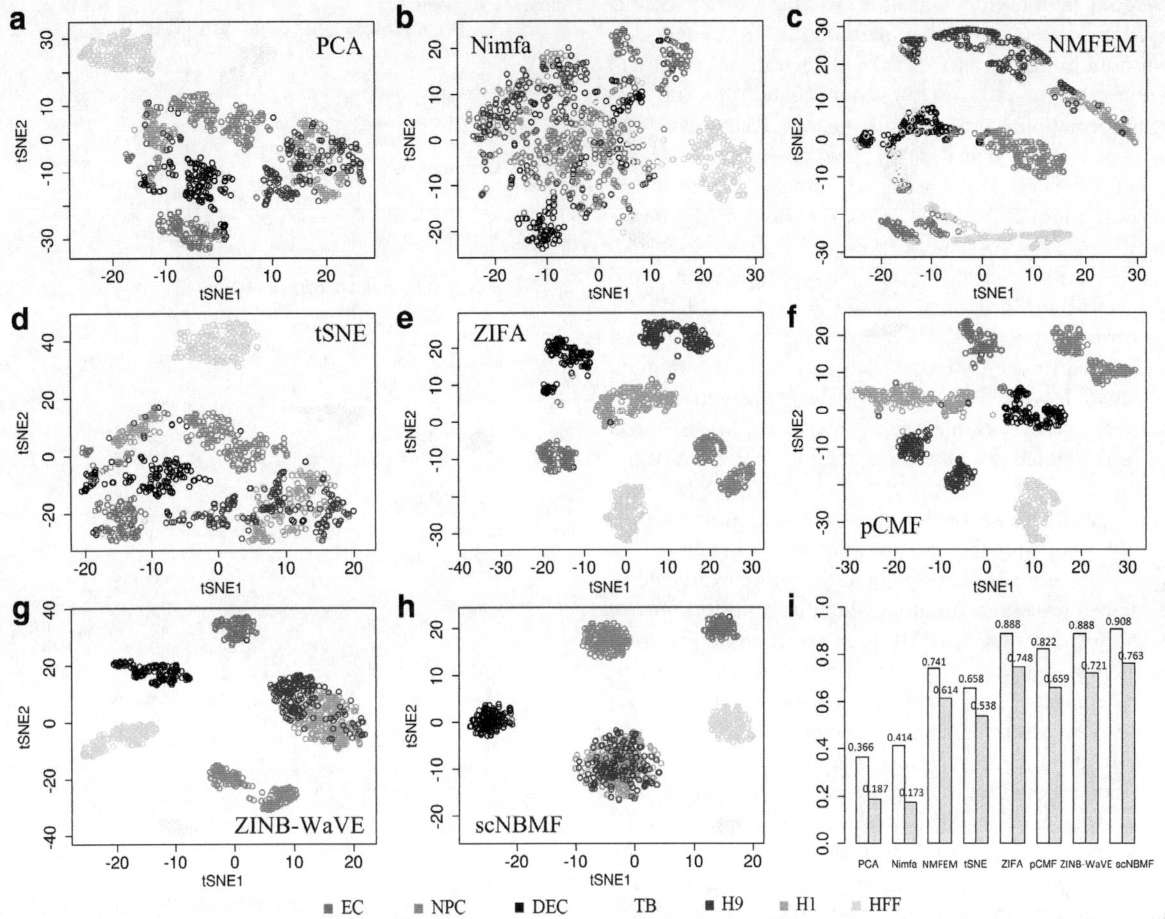

Fig. 3 Performance evaluation on human embryonic stem scRNA-seq data set, which contains 1018 cells in seven cell types. Different colors also represent different cell types. **a-h** These five figure display the clustering output of two dimension of tSNE using five matrix factorization methods(PCA, Nimfa, NMFEM, tSNE, ZIFA, pCMF, ZINB-WaVE, and scNBMF). **f** This figure shows NMI and ARI values which are from eight compared methods

of the compared methods (Table 2). Nimfa, NMFEM, ZIFA, pCMF, and ZINB-WaVE are the bespoke scRNAseq methods. Compared with the count-based methods, ZINB-WaVE and pCMF, scNBMF is roughly 100 folds faster than ZINB-WaVE, and 10 folds faster than pCMF. Even comparing the non-count based methods, ZIFA, Nimfa, and NMFEM, scNBMF is still the fastest method.

Conclusion

With rapid developing sequencing technology, a large amount of scRNAseq data sets is easily obtained via different sources. Therefore, computation time is one of these big issues for downstream analysis. On the other hand, scRNAseq data have their own characterizes, i.e., count nature, noisy, and sparsity, etc. These have been

Table 2 Computation times (second) of the matrix factorization-based methods on human brain scRNAseq data set, *k* represents the number of factors

Method	*k*=2	*k*=4	*k*=6	*k*=10	*k*=15	*k*=20
PCA	11.54	11.55	11.70	11.35	11.37	11.59
Nimfa	639.15	1990.66	2260.13	2490.05	2705.42	2924.87
NMFEM	1471.39	1628.2	1913.11	2248.18	2659.23	3027.5
tSNE	1.85	14.41	32.11	56.01	77.20	101.25
ZIFA	5331.25	5831.04	6347.08	6987.52	7338.26	7722.33
pCMF	12391.6	13517.12	14260.26	15111.55	15978.44	17158.42
ZINB-WaVE	71053.1	79402.17	90118.3	101072.9	115379.7	126575.2
scNBMF	456.12	478.90	541.31	717.88	1053.22	1563.75

triggered the development of a fast and efficient count-based matrix factorization method. In this paper, we proposed a count-based matrix factorization (scNBMF) method to model the raw count data, prevent losing information from normalizing raw count data. On three public biological scRNAseq data sets, scNBMF provides powerful performance compared with other seven methods in terms of NMI, ARI, and computation time.

Zero-inflated distribution is more appropriate method to account for dropouts, e.g. ZIFA and ZINB-WaVE. In current study, we did not consider the zero-inflated model because the tested data sets do not show too much dropouts. However, this is a necessary step in analyzing some scRNAseq data sets. Therefore, we will add the zero-inflated distribution in the future version of the scNBMF.

Biologically, if we incorporate all genes in scRNAseq data analysis, probably it would be able to involve some unwanted variables because not all genes are expressed in biological processes. An interesting direction to improve the performance of scNBMF is to select some informative genes first, this step can largely reduce unwanted variables, and exclude some redundancy genes [44, 45] in the downstream analysis. In addition, because gene expression levels are highly affected by other gene specific annotations, such as GC-content, gene length, and chromatin states [46]. If some interesting variables in the statistical model, such as "drop-out" parameter, can be inferred by annotation information, the method probably will significantly improve the power of detecting cell types from scRNAseq data.

Abbreviations

ARI: Adjusted rand index; DESeq: Differential expression; edgeR: Empirical analysis of digital gene expression data in R; ICA: Independent components analysis; MACAU: mixed model association for count data via data augmentation; NMI: Normalized mutual information; PCA: Principal component analysis; pCMF: Probabilistic count matrix factorization; PLS: Partial least squares; PQLseq: Penalized quasi-likelihood; scNBMF: Single-cell negative binomial matrix factorization; scRNAseq: Single-cell RNA sequencing; tSNE: t-distributed stochastic neighbor embedding; ZIFA: Zero-inflated factor analysis; ZINB-WaVE: Zero-inflated negative binomial-based wanted variation extraction;

Acknowledgements

No applicable.

Funding

Publication of this artical was sponsored by the Top International University Visiting Program for Outstanding Young scholars of Northwestern Polytechnical University; Fundamental Research Funds for the Central Universities (Grant: 3102017OQD098); Natural Science Foundation of China (NFSC; Grants: 61772426 and 61332014).

Availability of data and materials

scNBMF was implemented by R and Python, and the source code are freely available at https://github.com/sqsun. The three publicly scRNAseq datasets are available at https://www.ncbi.nlm.nih.gov/geo/query/acc.cgi?acc=GSE67835 https://www.ncbi.nlm.nih.gov/geo/query/acc.cgi?acc=GSE73727 https://www.ncbi.nlm.nih.gov/geo/query/acc.cgi?acc=GSE75748

About this supplement

This article has been published as part of *BMC Systems Biology Volume 13 Supplement 2, 2019: Selected articles from the 17th Asia Pacific Bioinformatics Conference (APBC 2019): systems biology*. The full contents of the supplement are available online at https://bmcsystbiol.biomedcentral.com/articles/supplements/volume-13-supplement-2.

Authors' contributions

SS, YC, YL and XS conceived and wrote the manuscript. SS and YC implemented the software and analyzed the data. All authors read and approved the final manuscript.

Ethics approval and consent to participate

No applicable.

Consent for publication

No applicable.

Competing interests

The authors declare that they have no competing interests.

Publisher's Note

Springer Nature remains neutral with regard to jurisdictional claims in published maps and institutional affiliations.

Author details

[1]School of Computer Science, Northwestern Polytechnical University, 710129 Xi'an, Shaanxi, People's Republic of China. [2]Key Laboratory of Big Data Storage and Management, Northwestern Polytechnical University, Ministry of Industry and Information Technology, 710129 Xi'an, Shaanxi, People's Republic of China. [3]Centre for Multidisciplinary Convergence Computing (CMCC), School of Computer Science, Northwestern Polytechnical University, 710129 Xi'an, Shaanxi, People's Republic of China. [4]Department of Biostatistics, University of Michigan, MI 48109 Ann Arbor, USA.

Published: 5 April 2019

References

1. Alexander J, et al. Utility of Single-Cell Genomics in Diagnostic Evaluation of Prostate Cancer. Cancer Res. 2018;78:348–58.
2. Love JC. Single-cell sequencing in cancer genomics. Cancer Res. 2015;75: IA14.
3. Conesa A, et al. A survey of best practices for RNA-seq data analysis. Genome Biol. 2016;17:13.
4. Vieth B, et al. powsimR: power analysis for bulk and single cell RNA-seq experiments. Bioinformatics. 2017;33:3486–8.
5. Buettner F, et al. Computational analysis of cell-to-cell heterogeneity in single-cell RNA-sequencing data reveals hidden subpopulations of cells. Nat Biotechnol. 2015;33:155–60.
6. Jiang L, et al. GiniClust: detecting rare cell types from single-cell gene expression data with Gini index. Genome Biol. 2016;17:144.
7. Kiselev VY, et al. SC3: consensus clustering of single-cell RNA-seq data. Nat Methods. 2017;14:483–6.
8. Lonnberg T, et al. Single-cell RNA-seq and computational analysis using temporal mixture modeling resolves T(H)1/T-FH fate bifurcation in malaria. Sci Immunol. 2017;2:eaal2192.
9. Wills QF, Mead AJ. Application of single-cell genomics in cancer: promise and challenges. Hum Mol Genet. 2015;24:R74–R84.
10. Yuan GC, et al. Challenges and emerging directions in single-cell analysis. Genome Biol. 2017;18:84.
11. Ding B, et al. Normalization and noise reduction for single cell RNA-seq experiments. Bioinformatics. 2015;31:2225–7.
12. Vallejos CA, et al. Normalizing single-cell RNA sequencing data: challenges and opportunities. Nat Methods. 2017;14:565–71.
13. Li WV, Li JYJ. An accurate and robust imputation method scImpute for single-cell RNA-seq data. Nat Commun. 2018;9:997.
14. Hashimshony T, et al. CEL-Seq2: sensitive highly-multiplexed single-cell RNA-Seq. Genome Biol. 2016;17:77.
15. Macosko EZ, et al. Highly Parallel Genome-wide Expression Profiling of Individual Cells Using Nanoliter Droplets. Cell. 2015;161:1202–14.

16. Ziegenhain C, et al. Comparative Analysis of Single-Cell RNA Sequencing Methods. Mol Cell. 2017;65:631–43.
17. Brennecke P, et al. Accounting for technical noise in single-cell RNA-seq experiments. Nat Methods. 2013;10:1093–5.
18. McDavid A, Finak G, Gottardo R. The contribution of cell cycle to heterogeneity in single-cell RNA-seq data. Nat Biotechnol. 2016;34:591–3.
19. Wu AR, Neff NF, Kalisky T, et al. Quantitative assessment of single-cell rna-sequencing methods. Nat Methods. 2014;11:41–6.
20. Sun S, Zhu J, Mozaffari S, Ober C, Chen M, Zhou X. Heritability estimation and differential analysis of count data with generalized linear mixed models in genomic sequencing studies. Bioinformatics. 2018. https://doi.org/10.1093/bioinformatics/bty644.
21. Robinson MD, McCarthy DJ, Smyth GK. edgeR: a Bioconductor package for differential expression analysis of digital gene expression data. Bioinformatics. 2010;26:139–40.
22. Anders S, Huber W. Differential expression analysis for sequence count data. Genome Biol. 2010;11:R106.
23. Sun S, Hood M, Scott L, Peng Q, Mukherjee S, Tung J, Zhou X. Differential expression analysis for RNAseq using Poisson mixed models. Nucleic Acids Res. 2017;e106:45.
24. Zurauskiene J, Yau C. pcaReduce: hierarchical clustering of single cell transcriptional profiles. BMC Bioinforma. 2016;17:140.
25. Trapnell C, et al. The dynamics and regulators of cell fate decisions are revealed by pseudotemporal ordering of single cells. Nat Biotechnol. 2014;32:381–U251.
26. Haghverdi L, Buettner F, Theis FJ. Diffusion maps for high-dimensional single-cell analysis of differentiation data. Bioinformatics. 2015;31: 2989–98.
27. Chen MJ, Zhou X. Controlling for Confounding Effects in Single Cell RNA Sequencing Studies Using both Control and Target Genes. Sci Rep. 2017;7:13587.
28. Sun SQ, Peng QK, Shakoor A. A Kernel-Based Multivariate Feature Selection Method for Microarray Data Classification. Plos ONE. 2014;9: e102541.
29. Shao CX, Hofer T. Robust classification of single-cell transcriptome data by nonnegative matrix factorization. Bioinformatics. 2017;33:235–42.
30. Zhu X, et al. Detecting heterogeneity in single-cell RNA-Seq data by non-negative matrix factorization. Peerj. 2017;e2888:5.
31. Miao Z, et al. DEsingle for detecting three types of differential expression in single-cell RNA-seq data. Bioinformatics. 2018;34:3223–4.
32. Streets AM, Huang YY. How deep is enough in single-cell RNA-seq. Nat Biotechnol. 2014;32:1005–6.
33. Pierson E, Yau C. ZIFA: Dimensionality reduction for zero-inflated single-cell gene expression analysis. Genome Biol. 2015;16:241.
34. Durif G, et al. Probabilistic Count Matrix Factorization for Single Cell Expression Data Analysis. BioRxiv; 2017.
35. Risso D, et al. A general and flexible method for signal extraction from single-cell RNA-seq data. Nat Commun. 2018;9:284.
36. Van den Berge K, et al. Observation weights unlock bulk RNA-seq tools for zero inflation and single-cell applications. Genome Biol. 2018;19:24.
37. Kingma DP, Ba J. Adam: A Method for Stochastic Optimization. the 3rd International Conference for Learning Representations. San Diego; 2015.
38. Lin PJ, Troup M, Ho JWK. CIDR: Ultrafast and accurate clustering through imputation for single-cell RNA-seq data. Genome Biol. 2017;18:59.
39. Ghosh J, Acharya A. Cluster ensembles. Adv Rev. 2011;4:305–15.
40. Hubert L, Arabie P. Comparing partitions. J Classif. 1985;2:193–218.
41. Darmanis S, et al. A survey of human brain transcriptome diversity at the single cell level. P Natl Acad Sci USA. 2015;112:7285–90.
42. Li J, et al. Single-cell transcriptomes reveal characteristic features of human pancreatic islet cell types. Embo Rep. 2016;17:178–87.
43. Chu LF, et al. Single-cell RNA-seq reveals novel regulators of human embryonic stem cell differentiation to definitive endoderm. Genome Biol. 2016;17:173.
44. Feng Z, Wang Y. Elf: extract landmark features by optimizing topology maintenance, redundancy, and specificity. IEEE ACM T Comput Bl. 2018;99:1.
45. Sun S, Peng Q, Zhang X. Global feature selection from microarray data using Lagrange multipliers. Knowl-Based Syst. 2016;110:267–74.
46. Sun S, Sun X, Zheng Y. Higher-order partial least squares for predicting gene expression levels from chromatin states. BMC Bioinforma. 2018;19: 113.

Tian et al. BMC Genomics 2019, 20(Suppl 2):192
https://doi.org/10.1186/s12864-019-5488-5

BMC Genomics

RESEARCH Open Access

MRCNN: a deep learning model for regression of genome-wide DNA methylation

Qi Tian[1], Jianxiao Zou[1], Jianxiong Tang[1], Yuan Fang[1], Zhongli Yu[1] and Shicai Fan[1,2*]

From The 17th Asia Pacific Bioinformatics Conference (APBC 2019)
Wuhan, China. 14-16 January 2019

Abstract

Background: Determination of genome-wide DNA methylation is significant for both basic research and drug development. As a key epigenetic modification, this biochemical process can modulate gene expression to influence the cell differentiation which can possibly lead to cancer. Due to the involuted biochemical mechanism of DNA methylation, obtaining a precise prediction is a considerably tough challenge. Existing approaches have yielded good predictions, but the methods either need to combine plenty of features and prerequisites or deal with only hypermethylation and hypomethylation.

Results: In this paper, we propose a deep learning method for prediction of the genome-wide DNA methylation, in which the Methylation Regression is implemented by Convolutional Neural Networks (MRCNN). Through minimizing the continuous loss function, experiments show that our model is convergent and more precise than the state-of-art method (DeepCpG) according to results of the evaluation. MRCNN also achieves the discovery of de novo motifs by analysis of features from the training process.

Conclusions: Genome-wide DNA methylation could be evaluated based on the corresponding local DNA sequences of target CpG loci. With the autonomous learning pattern of deep learning, MRCNN enables accurate predictions of genome-wide DNA methylation status without predefined features and discovers some de novo methylation-related motifs that match known motifs by extracting sequence patterns.

Keywords: Genome-wide DNA methylation, Convolutional neuro networks, Regression

Background

The process of DNA methylation is the selective addition of a methyl group to cytosine to form 5-cytosine under the action of DNA methyltransferase (Dnmt). DNA methylation primarily occurs symmetrically at the cytosine residues that are followed by guanine (CpG) on both DNA strands, and 70–80% of the CpG dinucleotides are methylated in the mammalian genomes [1]. The methylation status of cytosines in CpGs influences gene expression, chromatin structure and stability; and plays a vital role in the regulation of cellular processes including host defense against endogenous parasitic sequences, embryonic development, transcription, X-chromosome inactivation, and genomic imprinting, as well as possibly playing a role in learning and memory [2–5].

Determining the level of genome-wide methylation is the basis for further research. Recent technological advances have enabled DNA methylation assay and analysis at the molecular level [6–9], and high-throughput bisulfite sequencing is widely used to measure cytosine methylation at the single-base resolution in eukaryotes, including whole-genome bisulfite sequencing (WGBS) and Infinium 450 k/850 k. As the gold standard for genome-wide methylation determination, systems-level analysis of genomic methylation patterns associated with

* Correspondence: shicaifan@uestc.edu.cn
[1]School of Automation Engineering, University of Electronic Science and Technology of China, Chengdu, Sichuan, China
[2]Center for Informational Biology, University of Electronic Science and Technology of China, Chengdu, Sichuan, China

gene expression and chromatin structure can be achieved with WGBS [4, 5]. However, this method is not only expensive, but also constrained by bisulfite-converted genomes' lower sequence complexity and reduced GC content [3]. Apart from the above issues, the unstable environment and different platforms make the situation more formidable.

Therefore, computational prediction of CpG site-specific methylation levels is critical to enable genome-wide analysis [6], and forecasting through probabilistic models and machine learning methods has already received extensive attention [7]. As has been reported, gene methylation in normal tissues is mainly concentrated in the coding region lacking CpG; conversely, although the density of CpG islands in the promoter region is high, the gene remains unmethylated. Owing to this, some typical methods focus on the predicting methylation patterns of specific genomic regions, such as CGIs [10–16]. Other methods assume that the methylation status is encoded as a binary variable, which means that a CpG site is either methylated or unmethylated [14–19]. In addition, most of the methods need to combine a large amount of information, like knowledge of predefined features [6, 11, 13–16, 18]. Considering the number of methylation sites is large (usually tens of millions), the corresponding features for prediction are not easily accessible, which leads to large amount of manual annotation and preprocessing must be implemented before obtaining the final prediction.

Here, we report MRCNN, a computational method based on convolution neural networks for prediction of genome-wide DNA methylation states at CpG-site resolution [20, 21]. MRCNN leverages associations between DNA sequence patterns and methylation levels, using 2D-array-convolution to tackle the sequence patterns and characterize the target CpG site methylation. On the one hand, MRCNN does not need any knowledge of predefined features, because it's a deep learning method with end-to-end learning patterns. On the other hand, by using a continuous loss function to perform parameter calculations, a continuous value prediction of the methylation level can be achieved. We found that a series of convolution operations could extract DNA sequence patterns for our prediction and could yield substantially more accurate predictions of methylation from several different data sets. In addition, some de novo motifs are discovered from the filters of the convolution layer.

Methods

Data and encoding

We downloaded the whole genome bisulfite sequencing (WGBS) data (GEO, GSM432685) of H1 ESC from the GEO database for training and validation. The methylation level of each CpG locus is represented as a methylation ratio, varying from 0 to 1. The ratio is used as the network prediction target value, while the weights between the nodes in the network are optimized by minimizing the error between the predicted value and the target value. For independent testing, we chose genome-wide methylation data from multiple series of GEO databases, including the same series of H1 ESC (GEO, GSM432686) and different series of brain white matter, lung tissue, and colon tissue datasets (GEO, GSE52271). The DNA sequences selected were from the UCSC hg19 file, GRCh37 (Genome Reference Consortium Human Reference 37) with GenBank assembly accession number GCA_000001405.1.

In contrast to other traditional prediction tools with predefined features, our method exclusively takes the raw sequence as input. Given a DNA sequence, a fragment of 400 bps centered at the assayed methylation site was extracted. We choose the window size of 400 (without counting the target site and including each 200 bps DNA fragment upstream and downstream), with consideration for the potential workload of the calculation. Prior to conducting MRCNN training, these fragments needed to be encoded to convert the bases A, T, C, and G in the original sequence into matrices that could be input to the network. The strategy we select was one-hot encoding with the following rules: A = [0,0,0,1]; T = [1, 0, 0, 0]; C = [0, 1, 0, 0] and G = [0, 0, 1, 0]. After preprocessing, a matrix of 400*4 size could be generated for each target CpG site, in which every row represented a base (A, T, C, G) and the columns assembled the whole original fragment.

MRCNN

Deep learning is widely used in the field of image recognition due to its end-to-end mode, by which the convolutional neural network achieves good results with its specific partial connection. However, there is a lack of knowledge on how to construct a deep learning model that could be applied to the regression of methylation levels. As we know, a typical convolutional network is generally a convolution layer adjacent to a pooling layer, alternating in turn and finally output by a fully connected layer, such as VGG Net [22]. We were more concerned about solving the regression problem itself, and after tried many structures, we eventually found that, for the prediction of methylation sites, the required structure has its own unique characteristics. On the one hand, we must consider the complete coding information of single base. On the other hand, the method needs to implement efficient feature extraction to improve the prediction results. The final deep learning architecture of MRCNN is shown in Fig. 1.

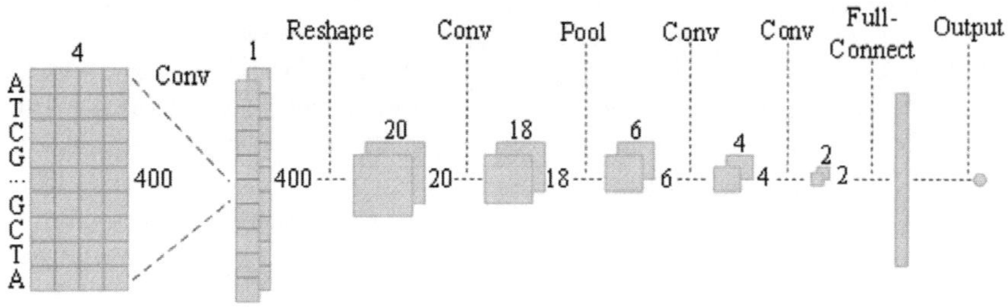

Fig. 1 The deep-learning architecture of MRCNN. The input layer is a matrix of one-hot coding for the DNA fragment centered at the methylation site, and the first convolution layer helps extract the information of each base. Then, it is reshaped as a 2D tensor for the following operations, and the convolution and pooling operations obtain higher-level sequence feature, while the next two convolution layers overcome the side effects of the saturated zone. Finally, the tensor is expanded by the full-connection layer, and the output node gives the prediction value

The first layer of the MRCNN is a single convolutional layer, which is mainly employed to extract single nitrogenous base information from the 400*4 input matrix. Because each base is a 1*4 independent code, the size of the convolution kernel can only be 1*4. This makes it possible to ensure that every base's information is entered into the network while the 16 feature maps are generated. In the design of the first layer, we choose not to adopt the pooling operation because the convolution of the first layer was essentially the synthesis of coding information, that is, ensuring each base's encoded information could be read completely by the network. For the input matrix $s_{n,\ x,\ y}$,

$$L_{n,1} = \sum_{x=1}^{400} \sum_{y=1}^{4} s_{n,x,y} w_{x,y}^{f,1} + b^{f,1}$$

Here, $w_{x,y}^{f,1}$ is the parameter or weight of the convolutional filter f for this layer, and $b^{f,\ 1}$ is the corresponding bias. Then, the output of the first layer $L_{n,\ 1}$ for each CpG site is a 400*1 tensor with 16 channels. To extract the information contained in the DNA sequence pattern, the output tensor is reshaped into a 20*20 tensor before being input into the next layer, which is advantageous for subsequent 2D-array-convolution and pooling operations. Here, each row of tensor $L_{n,\ 1}$ represents the synthesis information of every single base, then it is restructured following the original queue of bases while the shape is changed to 20*20.

The second and third layer are the traditional convolution and pooling layers. The size of the convolution kernel is 3*3, the pooling method is max pooling, and the step sizes are 1*1 and 3*3. Through this layer, higher-level sequence features can be extracted.

$$L_{n,2} = Relu\left(\sum_{x=1}^{20} \sum_{y=1}^{20} L_{n,1} w_{x,y}^{f,2} + b^{f,2}\right)$$

$$L_{n,3} = max_{3i \le x, 3i \le y}\left(L_{i,n,2}\right)$$

The Relu activation function sets negative values to zero, such that $L_{n,\ 2}$ corresponds to the evidence that the motif represented by $w_{x,y}^{f,2}$ occurs at the corresponding position. Nonoverlapping pooling is implemented to decrease the dimensions of the input tensor and, hence, the number of model parameters.

The next two layers are both single-convolution layers with the same size and step size as the second layer's convolution kernel. The convolution of the first layer and these two layers is linear convolution operation, with no pooling layer connection or activation function. The main purpose is to improve the effect of the convolution and nonlinear activation function, which results in part of the input falling into the saturated zone, with corresponding weights not being able to be updated. Finally, the tensor obtained by the last layer is expanded through the fully connected layer. A drop-out function is introduced for possible overfitting in training and then the methylation level could be obtained via the output layer. For the loss function in the training process, we chose the Mean Square Error (MSE) function for measurement, which is a classic solution to the problem of regression:

$$\mathrm{MSE}\left(Y, Y^0\right) = \frac{\sum_{i=1}^{n}\left(Y - Y^0\right)^2}{n}$$

where Y represents the predicted value of methylation and Y^0represents the true methylation level. Since the final predicted value is continuous, it may be more than 1 or less than 0, and we have incorporated this uniformly. For a prediction value greater than 1, the value is taken as 1, and a prediction value less than 0 is taken as 0.

Model construction and evaluation

For all training processes and evaluations, we used a holdout validation. First, for construction of the model, we selected nearly 10 million sites from WGBS for training. Since all chromosome numbers are disrupted, it is not necessary to consider the difference among different chromosomes, which is more conducive to the discovery of the genome-wide DNA methylation patterns.

Approximately 2 million CpG sites were randomly selected from the remaining sites as the validation set to help the network fine-tune the parameters. For testing the model, we randomly divided the sites in the test data set into a few copies to generate multiple independent test subsets. The division of the test set was based on two aspects, one being the original methylation level and the other being whether the region where the site is located belonged to the CpG islands. Details will be explained in the Results section. This also helps reduce the accidental errors in the model testing process, which is equivalent to a number of completely different test sets, as the training and test sites are completely different in origin. In general, we fitted the model on the training set, optimized the hyperparameters on the validation set, and performed the final model evaluation and comparison on the test sets.

To illustrate the model performance, we compared MRCNN with DeepCpG [7]. DeepCpG is the most state-of-art tool for genome-wide hypermethylation and hypomethylation prediction using deep learning. With a modular design, it uses a one-dimensional convolution DNA module and a bidirectional gated recurrent network of CpG module to achieve prediction. In addition, to compare the effect of network structural difference on the results, we also trained a simple CNN network as a baseline method. The specific structure of this network was an input layer, convolution layer 1, pooling layer 1, convolution layer 2, pooling layer 2, a fully connected layer, and an output layer. For simple CNN, we chose the same loss function and activation function to ensure univariate element during the experiments.

On the basis of the above, in order to analyze the sequence features extracted during the training of the model, we visualized the weight matrix of the convolutional filters by reverse decoding from weight assignment and corresponding raw tensor input. Specifically, the products of the first convolutional layer shared four types of weights, which corresponded to the original encoding of the four bases, so that the base sequence could be assigned according to the input, and then the weights of the different sequences could be reassigned according to the size of the filter weights. Motifs could be generated from MEME 5.0.1 by inputting the weighted sequences [23], and these de novo motifs were matched to annotated motifs given by Tomtom 5.0.1 [24]. Matches, where an FDR less than 0.05 was considered significant. All training and testing were implemented on our server with 128 G memory and 2 Nvidia 1080 graphics cards.

Evaluation metrics

We quantitatively evaluated the predictive performance from regression and classification. For regression, we chose the root mean square error (RMSE) and mean absolute error (MAE),

$$\mathrm{RMSE}(Y, Y^0) = \sqrt{\frac{\sum_{i=1}^{n}(Y-Y^0)^2}{n}}$$

$$\mathrm{MAE}(Y, Y^0) = \frac{1}{n}\sum_{i=1}^{n}|Y-Y^0|$$

where Y represents the predicted value of the methylation level and Y^0represents the true value.

For classification evaluation, we chose the sensitivity (SE), specificity (SP), classification accuracy (ACC) and area under the receiver operating characteristic curve (AUC). Here, TN, TP, FN and FP represented the number of true-negatives, true-positives, false-negatives and false-positives, respectively.

$$\mathrm{SE} = \frac{TP}{TP + FN} \quad \mathrm{SP} = \frac{TN}{TN + FP}$$

$$\mathrm{ACC} = \frac{TP + TN}{TP + FN + TN + FP}$$

Results

To evaluate the model prediction performance, we considered the two aspects, consisted of regression errors and binary classification performance. For regression errors, the model predictions of hypermethylation, hypomethylation and intermediate methylation status were compared to analyze the predictive properties of MRCNN for CpG methylation regression. These three states were grouped by different cutoff values of the methylation rate. Analysis of the classification performance was implemented by comparing the classification metrics of sites from the CpG islands and non-CpG islands among different models, which could be more comprehensive because of the difference in methylation patterns on distinct regions of the genome. Predictions results from other tissues were used to further analyze the robustness of MRCNN for more complicated methylation mechanisms. In addition, we also analyzed the filters from the model training process, and verified the validity of the sequence feature extraction, and obtained related de novo motifs.

Regression error

Here, to demonstrate the predictive ability for different methylation states, we distinguished successive methylation values in the raw data by different cutoff values. Most of the previous studies were focus on predictions of hypermethylation and hypomethylation, thus we also evaluated model performance based on predictions of the two states. However, in addition to this, in order to objectively evaluate the regression prediction, we added the evaluation for prediction of the intermediate methylation status. Specifically, if the original methylation label value was greater than 0.9, it was classified as "hyper", and if it was less than 0.1, it was classified as "hypo". The intermediate methylation status expressed as "mid" was defined by an original value greater than 0.4 but less than 0.6. Three different groups were formed and then regression results were evaluated by calculating the errors between the true and predicted values.

The different regression results of the three groups confirmed our previous expectation that MRCNN plays different roles in learning hypermethylation (hyper), hypomethylation (hypo) and intermediate methylation (mid) statuses. A comparison can be concluded from the boxplot in Fig. 2. For sites with significantly high methylation status, MRCNN was able to achieve smaller errors and obtain more satisfactory predictions compared with hypo and mid groups. On one hand, there were more sites with hypermethylation on genomes during training, on the other hand, potential more complex methylation mechanisms made prediction of hypo and mid methylation more difficult. In terms of the overall regression results, MRCNN achieved good results. First, maximum error for a single site prediction was approximately 0.5, and the prediction error distribution showed high accuracy of the predictions as most of the errors were concentrated around 0.1 for all test sites, see in Additional file 1. The RMSE and MAE of the three groups were calculated as follows: hyper: RMSE = 0.146806, MAE = 0.129885; hypo: RMSE = 0.23837, MAE = 0.207714; mid: RMSE = 0.281514, MAE = 0.268643. As seen from the RMSE and MAE values, the overall error was acceptable and would not produce a case in which a hyper site was predicted to be hypo, a hyper site was predicted to be mid, etc.

Classification performance

Considering that most previous studies on methylation were based on CpG islands [4], the evaluation of the classification performance was implemented for loci from CpG islands and non-CpG islands. Additionally, we compared MRCNN to DeepCpG for analysis of the classification ability for methylation under different deep-learning architectures and brought in the simple CNN model as the baseline method.

Since our label values and prediction results were continuous, we selected 0.5 as the cutoff value to divide the state of methylation into positive (> 0.5) and negative (≤0.5) samples. Via holdout validation ("Methods"), all methods were trained and tested on distinct methylation sites. In particular, these sites were previously grouped, with part of them from CpG islands and the rest from non-CpG islands. CpG islands are short CpG-rich regions of DNA which are often associated with the transcription start sites of genes. There are differences in methylation patterns between CpG islands and non-CpG islands, so we chose SE, SP, ACC and AUC to quantify the prediction performance of different models. The results of the classification comparison were shown in Fig. 3. The results showed that the overall prediction of MRCNN was better than that of DeepCpG, while the result of DeepCpG was better than that of the baseline model, CNN. It is worth mentioning that MRCNN achieved an accuracy of 93.2% and an AUC of 0.96 (t-test; P-value = 3.27×10^{-19}) on sites from CpG islands

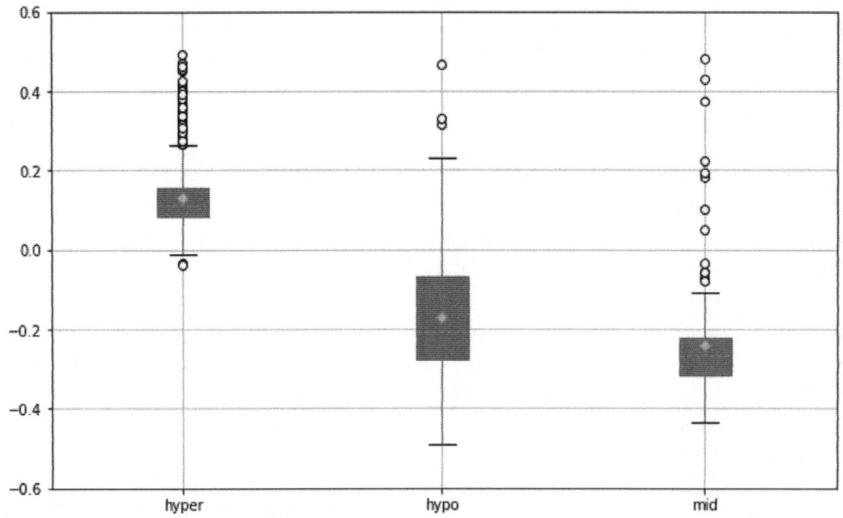

Fig. 2 MRCNN achieved regression of the whole genome methylation. The box diagrams depict the distribution of the prediction errors of the three groups of sites. The yellow diamonds represent the mean points and the green dotted lines represent the median lines. The points outside the upper and lower boundary lines are the outliers

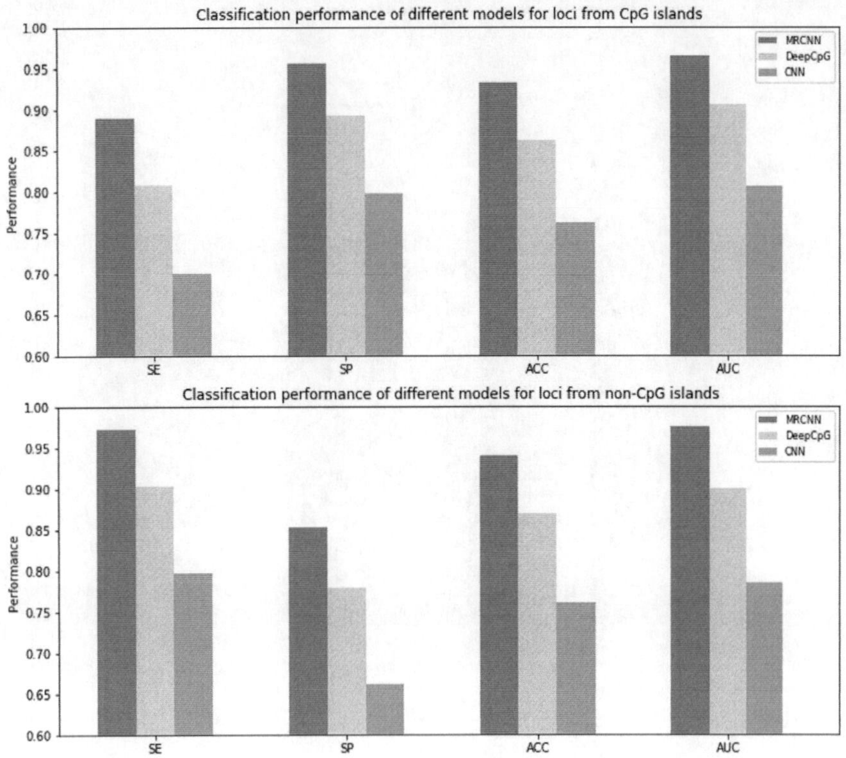

Fig. 3 MRCNN obtained better classification performance than DeepCpG and the baseline method, simple CNN. Different deep learning architectures lead to different effects in extracting features, which in turn affects the classification results for the test sets. The difference between the SE and SP between CpG islands and non-CpG islands reveals distinct methylation patterns in different regions of the genomes

and an accuracy of 93.8% and an AUC of 0.97 (t-test; P-value = 2.65×10^{-19}) on sites from non-CpG islands. To fully compare the classification performance of the three models, we also selected several sets of loci from the whole genome with different sizes for testing. The results were shown in Additional file 2.

We can see that even a general simple CNN model had a certain ability to describe the relationship between DNA sequences and CG sites after training and achieved an accuracy of more than 70% and an AUC of approximately 80%. However, there was still a gap compared to the well-designed MRCNN and DeepCpG. On one hand, we can see the powerful feature extraction capability of deep convolutional networks. On the other hand, we can conclude that a customized deep learning model for a specific problem is able to truly utilize its capability. In addition, we also find that in the prediction of sites from CpG islands, the SE is less than the SP, while this situation is exactly the opposite for sites from non-CpG islands. A significant reason for this is that CpG islands are enriched with sites of hypomethylation (more negative samples), while non-CpG islands are predominantly hypermethylated (more positive samples). This illustrates the effect of the different methylation patterns of CpG islands and non-CpG islands on feature extraction during model training.

We also considered the effect of different cell and tissue types on the prediction of MRCNN. Based on this, test was performed on several other tissue types of methylation data. Since the data for training the model come from the normal stem cells of human body, we compared the performance of predicting the methylation level of another three tissues. The test loci come from normal brain white matter, lung tissue, and colon tissue, which were randomly distributed on CpG islands and non-CpG islands for the consideration of genome-wide methylation prediction. The results of the classification performances were shown in Fig. 4. Precisely speaking, the prediction result from the H1 ESC was slightly better than the other three cell types, but the difference was very tiny, and the prediction of hypomethylation in lung tissue was better than that of H1 ESC (with higher SP). MRCNN got an AUC of 0.91 (t-test; *P*-value = 1.87 × 10–19) for brain white matter data, an AUC of 0.925 (t-test; P-value = 2.21 × 10–19) for normal lung tissue data and an AUC of 0.915 (t-test; P-value = 4.19 × 10–19) for normal colon tissue data.

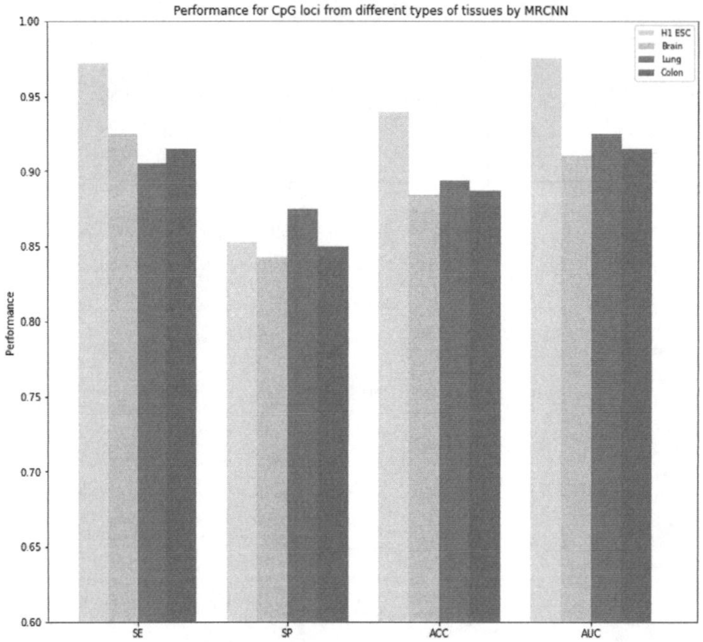

Fig. 4 MRCNN predicted methylation for different types of tissues. The H1 ESC was used as the control data, and the other three data were taken from the normal brain white matter, lung and colon tissue. Although MRCNN was trained on H1 ESC data, it still obtained high accuracy and performance when used to predict methylation levels of other types of tissues. The results showed that MRCNN had a certain robustness to more complicated methylation problems

Although MRCNN was trained based on human stem cells, we can see from the experimental results that the performance of MRCNN was still good on other tissue methylation data and further demonstrated the effectiveness of MRCNN as a universal predictive tool for genome-wide methylation. For more cautious consideration, we also evaluated the prediction of MRCNN in the cancerous phenotypes of the three tissues, and the results were shown in Additional file 3. Overall, MRCNN achieved satisfactory predictions for different types of cells and tissues, indicating that the model had considerable adaptability in face of more complex methylation mechanisms and confirmed the original intention of designing a universal genome-wide methylation prediction tool.

Feature analysis and motifs finding

To explore the extraction of DNA sequence pattern information during the training process, we also analyzed the feature maps from the network. In particular, we analyzed the learned filters of the first convolutional layer. First, we evaluated the ability of these filters to distinguish between hyper and hypo methylation states by visualizing the generated representations with t-SNE [25]. We compared the representation of the learned filters with the original input tensor representation and found that the learned filters were more able to distinguish the methylation level of the sites and explain the feature extraction by MRCNN. The t-SNE plot was shown in Fig. 5. The original feature could not distinguish the hyper and hypo methylation states quite well, while after the convolutional feature extraction, it could be roughly separated and would be sufficient to demonstrate the validity of the convolution operation. So, we can infer that the feature extraction was finished during the training and thus produced good prediction result.

These filters also recognize DNA sequence motifs similarly to conventional position weight matrices and can be visualized as sequence logos [7]. The discovered sequence motifs associated with DNA methylation are from the online motif-based sequence analysis tools MEME [23] (version 5.0.1). We submitted these de novo motifs into Tomtom [24] (version 5.0.1) to find similar known DNA motifs by searching public databases. This may contribute to our deeper knowledge of methylation and DNA sequences. Part of the motifs and their matches were shown in the Fig. 6. The top three motifs were from hypomethylation related sequences (with methylation rate < 0.1), the middle two motifs were from sequences with a methylation rate between 0.4 and 0.6, and the last ten motifs were from hypermethylation related sequences (with methylation rate > 0.9). It was interesting that, as intuitively seen from the logo, the hypermethylated corresponding motif tended to have

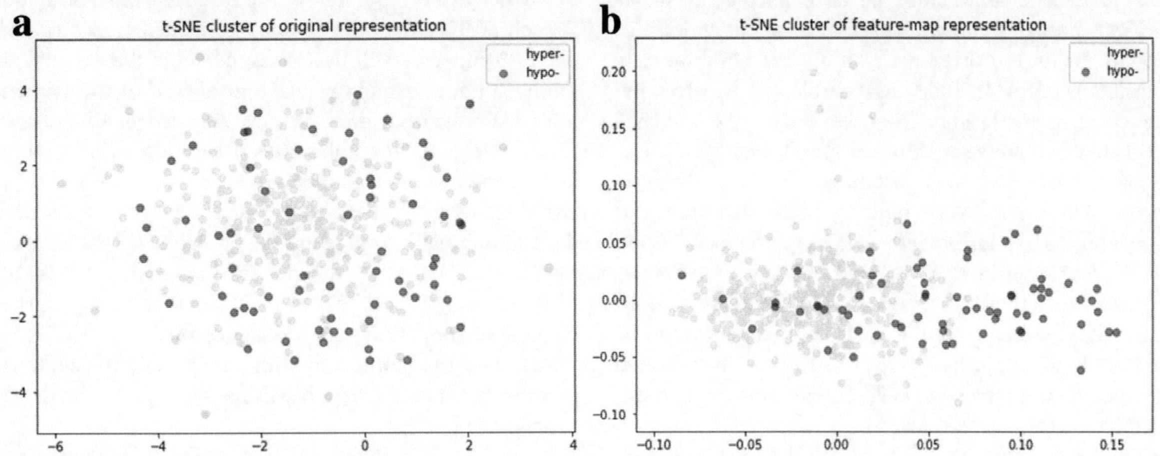

Fig. 5 Clustering results for hypermethylation and hypomethylation loci of the original features and the learned filters of the first convolutional layer. **a** t-SNE plot of the original input tensor representation. From the plot, we cannot apparently distinguish between hypermethylation and hypomethylation. **b** t-SNE plot of the learned feature map representation. Hypermethylation and hypomethylation are generally grouped

Motifs	de novo motifs logo	Match From Tomtom	Class & Family	*p*-value
hypo_1	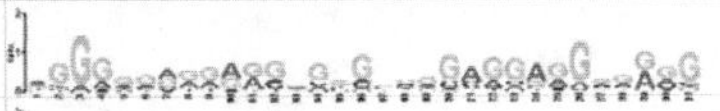	ZNF263	C2H2 zinc finger factors	2.98E-08
hypo_2		SP1	Three-zinc finger Krüppel-related factors	1.83E-05
hypo_3		POU3F3	Pou domain - N-terminal to homeobox domain, Homeobox domain	8.73E-05
mid_1		PLAG1	More than 3 adjacent zinc finger factors	2.07E-04
mid_2		NFIC::TLX1	SMAD/NF-1 DNA-binding domain factors::Homeo domain factors	3.09E-04
hyper_1		KLF4	Three-zinc finger Kruppel-related factors	3.45E-04
hyper_2		TEAD2	TEF-1-related factors	4.44E-04
hyper_3		PAX9	PAX9_DBD ('Paired box' domain, Homeodomain-like domain)	4.47E-04
hyper_4		PAX1	Paired domain only	7.01E-04
hyper_5		USF2	bHLH-ZIP factors	7.03E-04
hyper_6		VDR	Thyroid hormone receptor-related factors (NR1)	7.93E-04
hyper_7		HOXD8_DBD	Homeobox domain	9.93E-04
hyper_8		HOXA13	HOXA13_full (Hox protein A13 N terminal, Homeobox domain)	1.79E-03
hyper_9		ZBTB49	Zinc-finger of C2H2 type	2.85E-03
hyper_10		TEAD1	TEAD1_full (TEA/ATTS domain family)	4.96E-03

Fig. 6 Discovered sequence motifs associated with DNA methylation. The first column is the number of de novo motifs and the second column is the motif logo generated by MEME. The third column is the known motifs matched by Tomtom, and the next forth column is the corresponding class and family representing the biological factor species of the known motifs. The *p*-value is defined as the probability that a random motif of the same width as the target would have an optimal alignment with a match score as good as or better than the target's. Tomtom estimates the p-value using a null model consisting of sampling motif columns from all the columns in the set of target motifs

major bases of one certain type at a specific site, while there was no particularly obvious trend in the motif corresponding to the hypomethylation and intermediate methylation status. In addition, regardless of hypermethylation or hypomethylation sites, several of the matched known motifs were related to zinc finger factors, suggesting that it might play an important role in the methylation process. There have been reports in the literature that methylation is associated with zinc finger factors [26]. In this research, Carvin et al. pointed out that the selective targeting of methylation by zinc-finger proteins demonstrated that binding of distinct classes of factors could be monitored in living cells. Other matched motifs indicated more potential methylation background knowledge in different biological mechanisms.

Discussion

Through multilayer convolution learning, MRCNN achieves methylation prediction for CpG loci at single-base resolution, and thanks to the continuous MSE loss function, the method enables continuous value regression. Although the 2D-convolution approach can exploit more comprehensive DNA sequence features for methylation prediction, one potential problem is that when the sequence is fixed, the predicted results do not change. Thus, we not only tested the model on the same series of tissues, but also on different types of tissues. The results showed that the impact of this problem was tiny, and the prediction results did not produce obvious fluctuations. The most important point is that MRCNN achieves methylation prediction of CpG loci by local DNA sequences, not only does it overcome the cumbersome pre-processing, but the prediction results cover all ranges of methylation values. In addition, we also find the corresponding de novo motifs through the DNA sequence pattern extracted during the training process. The significance of MRCNN is more focused on the realization of a universal model for predicting the genome-wide methylation level of sites by the local DNA sequences. We can use deep learning to fit any problem we care about through large-scale data training, but more methods and attempts would be needed for problems with complex mechanisms and broad backgrounds, such as methylation. Here, MRCNN, as a general genome-wide methylation prediction tool, does not solve all methylation-related problems, it has its own scope of application, but builds a bridge between DNA sequences and methylation at CpG loci.

The application of deep learning methods in bioinformatics has become a hot phenomenon, especially in exploring the implicit relationships and nonlinear mapping of large-scale data. Nowadays, it is possible to analyze methylation at all levels with the massive data generated by high-throughput detection technology. Thus, the combination of the two aspects gives us an opportunity to further explore the methylation mechanism. Although we can analyze the phenomenon of methylation through deep learning, we still need to understand the essence of methylation, and only by making full use of the known information can we discover new knowledge. One of the future efforts is to combine deep learning with existing known methylated biological backgrounds to construct a comprehensive analytical model to achieve deeper understanding of this epigenetic phenomenon. In addition, the learned features during the training process of MRCNN may reflect the normal and abnormal patterns of methylation, which are worthy of further study. This research will be combined with cancer data for analysis in subsequent work, which will be a key point for us to expand the model.

Conclusion

In this paper, we propose a novel deep learning model based on convolutional neural networks for predicting DNA methylation at single-CpG-site precision using local DNA sequence. The specially designed network structure makes it a universal model for predicting genome-wide methylation of CpG loci. The extraction of DNA sequence features is achieved by multistep 2D-array-convolution, and the MSE loss function is minimized to achieve regression of the methylation values. Based on extensive training data, MRCNN achieves accurate methylation predictions and exhibits stability in prediction methylation of different types of tissues. We also further demonstrate the discovery of de novo motifs by analyzing the learned filters of the convolutional layer, and some of these motifs have been reported playing an important role in the regulation of methylation.

Additional file

Additional file 1: Additional figures. Distribution of the differences between the predicted value and the true value of all sites. (PDF 62 kb)

Additional file 2: Additional figures. Comparsion of the comprehensive classification performance metrics including ACC and AUC on the different size of test subsets. (PDF 64 kb)

Additional file 3: Additional figures. Comparison of the classification performances in three cancerous tissues. (PDF 56 kb)

Abbreviations

CNN: Convolutional Neural Network; GEO: gene expression omnibus; NCBI: National Center for Biotechnology Information; WGBS: whole genome bisulfite sequencing

Acknowledgements

Not applicable.

Funding

This project was supported by the National Natural Science Foundation of China under No. 61503061 and 61872063, the Fundamental Research Funds for the Central Universities under No. ZYGX2016J102. Publication of this article was sponsored by the National Natural Science Foundation of China (No. 61872063).

Availability of data and materials

The dataset is available at https://www.ncbi.nlm.nih.gov/geo/query/acc.cgi?acc=GSM432685
https://www.ncbi.nlm.nih.gov/geo/query/acc.cgi?acc=GSM432686
https://www.ncbi.nlm.nih.gov/geo/query/acc.cgi?acc=GSE52271
ftp://ftp.ncbi.nlm.nih.gov/genomes/archive/old_genbank/Eukaryotes/vertebrates_mammals/Homo_sapiens/GRCh37/Primary_Assembly/assembled_chromosomes/FASTA/
MRCNN is provided as an open-source tool available at https://github.com/TQBio/MRCNN.git

About this supplement

This article has been published as part of *BMC Genomics Volume 20 Supplement 2, 2019: Selected articles from the 17th Asia Pacific Bioinformatics Conference (APBC 2019): genomics*. The full contents of the supplement are available online at https://bmcgenomics.biomedcentral.com/articles/supplements/volume-20-supplement-2.

Authors' contributions

QT, JZ and SF designed the experiments carried out in the study, QT, JZ, JT, YF and ZY performed the analysis, edited the manuscript. JZ and SF led the research and reviewed the manuscript. All authors read and approved the manuscript.

Ethics approval and consent to participate

Not applicable.

Consent for publication

Not applicable.

Competing interests

The authors declare that they have no competing interests.

Publisher's Note

Published: 10 April 2019

References

1. Ma BS, Wilker EH, Willis-Owen SAG, Byun HM, Wong KCC, Motta V, Baccarelli AA, Schwartz J, Cookson WOCM, Khabbaz K, et al. Predicting DNA methylation level across human tissues. Nucleic Acids Res. 2014;42(6):3515–28.
2. Robertson KD. DNA methylation and human disease. Nat Rev Genet. 2005;6(8):597–610.
3. Suzuki MM, Bird A. DNA methylation landscapes: provocative insights from epigenomics. Nat Rev Genet. 2008;9(6):465–76.
4. Laird PW. Principles and challenges of genome-wide DNA methylation analysis. Nat Rev Genet. 2010;11(3):191–203.
5. Jones PA. Functions of DNA methylation: islands, start sites, gene bodies and beyond. Nat Rev Genet. 2012;13(7):484–92.
6. Zhang WW, Spector TD, Deloukas P, Bell JT, Engelhardt BE. Predicting genome-wide DNA methylation using methylation marks, genomic position, and DNA regulatory elements. Genome Biol. 2015;16:14.
7. Angermueller C, Lee HJ, Reik W, Stegle O. DeepCpG: accurate prediction of single-cell DNA methylation states using deep learning. Genome Biol. 2017;18:67.
8. Fan SC, Li CZ, Ai RZ, Wang MC, Firestein GS, Wang W. Computationally expanding infinium HumanMethylation450 BeadChip array data to reveal distinct DNA methylation patterns of rheumatoid arthritis. Bioinformatics. 2016;32(12):1773–8.
9. Fan SC, Huang K, Ai RZ, Wang MC, Wang W. Predicting CpG methylation levels by integrating Infinium HumanMethylation450 BeadChip array data. Genomics. 2016;107(4):132–7.
10. Cedar H, Bergman Y. Linking DNA methylation and histone modification: patterns and paradigms. Nat Rev Genet. 2009;10(5):295–304.
11. Irizarry RA, Ladd-Acosta C, Wen B, Wu ZJ, Montano C, Onyango P, Cui HM, Gabo K, Rongione M, Webster M, et al. The human colon cancer methylome shows similar hypo- and hypermethylation at conserved tissue-specific CpG island shores. Nat Genet. 2009;41(2):178–86.
12. Zheng H, Wu HW, Li JP, Jiang SW. CpGIMethPred: computational model for predicting methylation status of CpG islands in human genome. BMC Med Genet. 2013;6(Suppl 1):S13.
13. Previti C, Harari O, Zwir I, del Val C. Profile analysis and prediction of tissue-specific CpG island methylation classes. Bmc Bioinformatics. 2009;10:116.
14. Bock C, Paulsen M, Tierling S, Mikeska T, Lengauer T, Walter J. CpG island methylation in human lymphocytes is highly correlated with DNA sequence, repeats, and predicted DNA structure. PLoS Genet. 2006;2(3):243–52.
15. Fang F, Fan SC, Zhang XG, Zhang MQ. Predicting methylation status of CpG islands in the human brain. Bioinformatics. 2006;22(18):2204–9.
16. Fan SC, Zhang MQ, Zhang XG. Histone methylation marks play important roles in predicting the methylation status of CpG islands. Biochem Bioph Res Co. 2008;374(3):559–64.
17. Bhasin M, Zhang H, Reinherz EL, Reche PA. Prediction of methylated CpGs in DNA sequences using a support vector machine. FEBS Lett. 2005;579(20):4302–8.
18. Kim S, Li M, Paik H, Nephew K, Shi H, Kramer R, Xu D, Huang TH. Predicting DNA methylation susceptibility using CpG flanking sequences. Pacific Symposium on Biocomputing Pacific Symposium on Biocomputing. 2008:315–326.
19. Lu L, Lin K, Qian Z, Li H, Cai Y, Li YJJoBS, Engineering: Predicting DNA methylation status using word composition 2010, 3(07):672.
20. Krizhevsky A, Sutskever I, Hinton GE. ImageNet classification with deep convolutional neural networks. Commun ACM. 2017;60(6):84–90.
21. LeCun Y, Bengio Y, Hinton G. Deep learning. Nature. 2015;521(7553):436–44.
22. Simonyan K, AJapa Z. Very deep convolutional networks for large-scale image recognition; 2014.
23. Bailey TL, Elkan C. Fitting a mixture model by expectation maximization to discover motifs in bipolymers; 1994.
24. Gupta S, Stamatoyannopoulos JA, Bailey TL, Noble WS. Quantifying similarity between motifs. Genome Biol. 2007;8(2):R24.
25. van der Maaten L, Hinton G. Visualizing Data using t-SNE. J Mach Learn Res. 2008;9:2579–605.
26. Carvin CD, Parr RD, Kladde MP. Site-selective in vivo targeting of cytosine-5 DNA methylation by zinc-finger proteins. Nucleic Acids Res. 2003;31(22):6493–501.

Yin et al. BMC Genomics 2019, 20(Suppl 2):193
https://doi.org/10.1186/s12864-019-5489-4

BMC Genomics

RESEARCH Open Access

DeepHistone: a deep learning approach to predicting histone modifications

Qijin Yin, Mengmeng Wu, Qiao Liu, Hairong Lv* and Rui Jiang*

From The 17th Asia Pacific Bioinformatics Conference (APBC 2019)
Wuhan, China. 14-16 January 2019

Abstract

Motivation: Quantitative detection of histone modifications has emerged in the recent years as a major means for understanding such biological processes as chromosome packaging, transcriptional activation, and DNA damage. However, high-throughput experimental techniques such as ChIP-seq are usually expensive and time-consuming, prohibiting the establishment of a histone modification landscape for hundreds of cell types across dozens of histone markers. These disadvantages have been appealing for computational methods to complement experimental approaches towards large-scale analysis of histone modifications.

Results: We proposed a deep learning framework to integrate sequence information and chromatin accessibility data for the accurate prediction of modification sites specific to different histone markers. Our method, named DeepHistone, outperformed several baseline methods in a series of comprehensive validation experiments, not only within an epigenome but also across epigenomes. Besides, sequence signatures automatically extracted by our method was consistent with known transcription factor binding sites, thereby giving insights into regulatory signatures of histone modifications. As an application, our method was shown to be able to distinguish functional single nucleotide polymorphisms from their nearby genetic variants, thereby having the potential to be used for exploring functional implications of putative disease-associated genetic variants.

Conclusions: DeepHistone demonstrated the possibility of using a deep learning framework to integrate DNA sequence and experimental data for predicting epigenomic signals. With the state-of-the-art performance, DeepHistone was expected to shed light on a variety of epigenomic studies. DeepHistone is freely available in https://github.com/QijinYin/DeepHistone.

Keywords: Histone modification, Chromatin accessibility, Deep learning, Sequence analysis, Genetic variation

Background

Histone modifications, as covalent post-translational modifications (PTMs) to histone proteins, have been recognized as one of the major driving forces alters chromatin structures since the early 1960s [1]. Enabled by such innovative techniques as X-ray crystallography, it has been gradually clear that the modification of histone amino (N)-terminal tails would affect inter-nucleosomal interactions, alter the overall chromatin structure or recruit histone modifiers, and eventually impact gene expression [2]. It has also been known that histone modifications, including methylation, acetylation, phosphorylation, ubiquitylation and sumoylation, act in a variety of biological processes such as chromosome packaging [3, 4], transcriptional activation and inactivation [5–7], as well as DNA damage and repair [8]. Therefore, quantitative detection of histone modifications would provide useful information for not only a better understanding towards epigenetic regulation of cellular processes but also the development of drugs targeting on histone modifying enzymes [9].

Histone modifications are mainly profiled by such high-throughput experimental techniques as chromatin

* Correspondence: lvhairong@tsinghua.edu.cn; ruijiang@tsinghua.edu.cn
MOE Key Laboratory of Bioinformatics; Bioinformatics Division, Beijing National Laboratory for Information Science and Technology, Tsinghua University, Beijing 100084, China

immunoprecipitation followed by sequencing (ChIP-seq) [10]. For example, Barski et al. generated high-resolution maps for the genome-wide distribution of 20 histone lysine and arginine methylations and identified typical patterns of histone methylations exhibited at promoters, insulators, enhancers, and transcribed regions [11]. Whole-genome profiling of DNA regulatory elements, their relationship to target genes, their properties of histone modifications, and their features of chromatin accessibility, were conducted by the Encyclopaedia of DNA Elements (ENCODE) project [12]. Even larger scale global maps of regulatory elements in 111 reference human epigenomes, together with chromatin accessibility and gene expression information, were established by the Roadmap Epigenomics Consortium [13]. These abundant resources provided new insights into the function of histone modification and chromatin organization in genome, demonstrated the central role of epigenomic information for understanding gene regulation and cellular differentiation, and opened a door towards deciphering mechanisms of human disease.

Nevertheless, it is still too expensive and time-consuming to establish a landscape of histone modifications purely relying on biological experiments, due to the large number of cell types and known histone markers. It is, therefore, reasonable to take advantage of computational methods to predict histone modifications, complementing experimental approaches and facilitating the understanding of DNA signatures and modifications that contribute to gene expression. Towards this objective, Benveniste et al. designed a logistic regression model to predict histone modifications from transcription factor-binding profiles and recapitulated the importance of interactions between transcription factors and chromatin-modifying enzymes to gene expression [14]. Karlic et al. elucidated the correlation between histone modification levels and gene expression and designed a linear regression model to predict gene expression relying on a small number of histone modifications [15].

In the recent years, deep learning has been successfully incorporated into a variety of bioinformatics studies. For example, Alipanahi et al. proposed a convolutional neural network (CNN) named DeepBind to predict binding proteins and showed higher prediction power than traditional classifiers [16]. Zhou and Troyanskaya designed a model called DeepSEA to learn DNA regulatory signatures via a CNN from epigenomic data [17]. Quang and Xue combined a CNN and a bi-directional long short-term memory network to predict functions of DNA sequences and named their method DanQ [18]. Min et al. proposed a deep CNN model called DeepEnhancer to predict enhancers purely from DNA sequences [19]. Liu et al. designed a hybrid neural network to predict chromatin accessibility from sequence [20]. Min et al. further developed a representation learning formulation to embed *k*-mers into a low dimension space and then used the resulting vectors to predict chromatin accessibility via a deep neural network [21]. The success of these methods suggests that deep learning is a powerful technique in genomic studies. However, all these methods rely purely on DNA sequence information, which apparently lacks the power of making predictions in a cell line-specific manner, because DNA sequences are identical in different cell lines. To overcome this limitation, hybrid deep learning methods have been proposed and shown visible improvement in specific research by combining sequence information and biological experimental data. For instance, a recently proposed method named DeepTACT combined DNA sequences and chromatin accessibility to predict high-resolution chromatin contacts from promoter capture Hi-C data and achieved state-of-the-art performance [22].

Motivated by the above understanding, we purposed a deep learning approach named DeepHistone to predict histone modification by integrating DNA sequence information and chromatin accessibility data. The rationale for our method is to capture regulatory signatures from DNA sequences, while taking advantage of the compact relationship between histone modifications and chromatin accessibility to further improve the prediction performance. Through a serial of comprehensive validation experiments, we demonstrated that DeepHistone is superior to several baseline methods in the prediction of modification sites specific to different histone markers, not only within an epigenome but also across epigenomes. Besides, we illustrated that sequence signatures automatically extracted by our deep learning model was consistent with known transcription factor binding sites. As a potential application, we finally showed the possibility of our method in distinguishing functional single nucleotide polymorphisms (SNPs) from their nearby genetic variants.

Materials and methods

Data sources

We downloaded peak files of 7 histone modification markers for 21 human epigenomes from the Roadmap Epigenomics Project [13]. As shown in Table 1, the 7 markers, including H3K4me3, H3K4me1, H3K36me3, H3K27me3, H3K9me3, H3K27ac, and H3K9ac, are regarded as the most important markers that have been verified to be associated with such specific functional regions as enhancers and promoters in the genome [23]. The criterion for selecting an epigenome is that ChIP-seq assays should be performed for all the 7 markers for the tissue or cell line corresponding to the epigenome.

Given a marker, an epigenome, and the peak file of the corresponding ChIP-seq experiment, we used a window of 200 bp to scan the whole human genome (hg19) with step 200 bp and regarded a window that had at least 100 bp overlap with a peak as a histone modification site.

Table 1 The 15 epigenomes used in this research

Epigenome ID (EID)	Anatomy	Histone marker							
		H3K4me1	H3K4me3	H3K27me3	H3K36me3	H3K9me3	H3K9ac	H3K27ac	Total
E003	ESC	342,503	213,546	231,823	424,644	322,656	155,558	236,273	1,460,537
E005	ESC_DERIVED	433,887	166,189	107,589	456,459	287,790	71,753	258,471	1,349,880
E006	ESC_DERIVED	462,960	163,290	130,314	487,545	278,273	168,264	289,090	1,547,640
E007	ESC_DERIVED	263,177	159,076	69,494	192,670	149,941	294,433	118,545	957,424
E008	ESC	147,743	208,835	80,768	336,585	234,898	168,324	129,238	1,050,986
E017	LUNG	548,594	191,906	506,666	682,735	531,686	273,512	382,596	2,302,996
E114	LUNG	651,428	362,931	483,416	585,379	216,237	381,391	430,496	1,979,536
E117	CERVIX	593,890	285,181	406,167	347,919	95,082	260,910	347,656	1,533,734
E118	LIVER	541,823	271,863	205,199	340,891	246,440	220,457	278,768	1,451,201
E119	BREAST	563,758	192,740	260,776	242,863	376,845	232,169	366,810	1,566,171
E121	MUSCLE	591,321	244,747	708,930	483,319	156,381	356,341	486,568	2,110,799
E122	VASCULAR	582,593	211,344	404,019	317,600	252,718	260,503	360,372	1,617,712
E124	BLOOD	839,651	431,883	597,823	1,098,084	570,919	268,232	655,859	2,981,094
E125	BRAIN	627,409	258,248	321,184	400,306	178,567	359,857	424,867	1,629,166
E127	SKIN	708,918	289,273	466,659	513,074	186,065	302,827	416,314	1,900,958

7,626,807 modification sites specific to the 7 histone markers were identified from the 15 epigenomes

Applying this procedure to every marker and every epigenome and discarding epigenomes (total 6 epigenomes) that had only a small number of modification sites (< 50,000) for some histone markers, we identified a total of 7,626,807 sites in the human genome from 15 epigenomes, as detailed in Table 1.

For an epigenome, we further downloaded corresponding DNase-seq peak files from Roadmap. For a genomic position in a peak, we assigned the fold enrichment score of the peak, calculated by the standard pipeline of Roadmap [13], to the position, as its openness score to quantify the status of chromatin accessibility. For other genomic positions, we regarded their openness scores as zeros. By doing this for every epigenome, we obtained an openness score that was specific to the epigenome for every genomic position.

Design of DeepHistone

We designed a deep neural network model, named DeepHistone, to predict whether a DNA fragment in an epigenome is a site for the 7 histone markers. To achieve this aim, we first extended the input fragment upstream and downstream to obtain a region of 1000 bp centred at the fragment and then fed the resulting region to our model, which consists of three modules: a DNA module, a DNase module, and a Joint module, as illustrated in Fig. 1.

The DNA module, designed as a customized densely connected convolutional neural network [24], extracts sequence information for the input region. For this purpose, a one-hot encoding strategy is used to convert the sequence of the input region into a binary matrix. An initial convolutional layer is then adopted to scan the matrix for sequence patterns, i.e., motifs. The resulting patterns are further fed to two densely connected convolutional blocks connected in a tandem way by a convolutional layer and a pooling layer for extracting high-level features. These features, after passing through a convolutional layer and a pooling layer, are eventually fed to the joint module for the classification task. A densely connected convolutional block consists of three convolutional layers. Mediated by a batch normalization operation and a ReLU activation function, the first two layers connect to not only the subsequent layer but also all latter layers. The densely connected architecture is adopted here because recent advances in deep learning have shown that such an architecture can effectively overcome the vanishing gradient problem, strengthen feature propagation, utilize parameters more efficiently, and avoid the overfitting problem [24]. These shortcomings are common in a classical convolutional neural network, especially on tasks with small dataset. Detailed parameter settings of the DNA module are shown in Fig. 1.

The DNase module extracts chromatin accessibility information for an input region. This module has the identical architecture as the DNA module, except that an initial one-dimensional convolutional layer is used to deal with openness scores of positions in the region at the beginning.

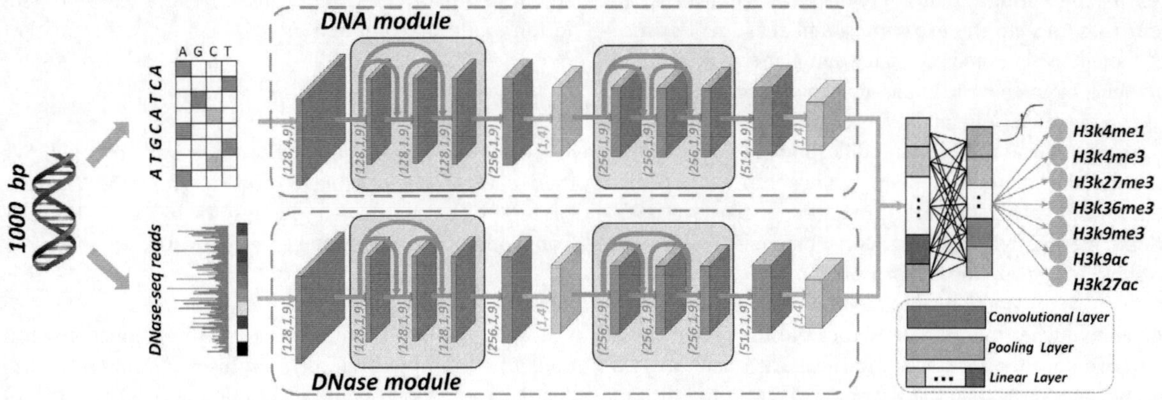

Fig. 1 Diagram of DeepHistone. The deep neural network model consists of three modules: a DNA module, a DNase module, and a joint module. The DNA module extracts sequence information via a densely connected convolutional neural network. The DNase module deals with chromatin accessibility information using the similar architecture. The joint module combines both sequence and chromatin accessibility features to distinguish histone modification sites of a marker from those of other markers. In figure, (128,4,9) indicates there are 128 kernels, each of width 4 and length 9 in a convolutional layer and (1,4) indicates kernel width and length are set to 1 and 4 respectively in a pooling layer. Others have similar meaning

The joint module integrates features extracted by the DNA and DNase modules to produce classification results. To achieve this objective, features extracted by these two modules are concatenated and fed to a feed-forward neural network, which uses 7 sigmoid functions to predict in parallel probabilities that a region is a site for the 7 histone modification markers. Note that multiple sigmoid functions instead of a softmax function are adopted because in reality the events that a site belongs to the markers are not mutually exclusive. In other words, a site can belong to multiple markers simultaneously.

We implemented DeepHistone in Python using Pytorch [25]. The high-performance NVIDIA GeForce GTX 1080Ti GPU was used to accelerate the computation. The cross entropy loss was used as the optimal function in model training, measuring the similarity between a true distribution p and the prediction probability q, as:

$$H(p,q) = \mathrm{E}_p\left[\frac{1}{\log q_x}\right]$$

Adam [26] was used to accelerate backpropagation with default parameters, except that the initial learning rate is set to 0.001. An early stopping strategy was used to reduce the training time.

Baseline methods

We compared the performance of DeepHistone with three baseline methods, including DeepSEA [17], DanQ [18], and gkm-SVM [27], with parameters proposed by the respective authors. Briefly, DeepSEA used three convolutional layers, a fully connected layer, and a sigmoid output layer to distinguish epigenomic sites. DanQ used a convolutional layer, a bi-directional long short-term memory layer, a fully connected layer, and a sigmoid output layer to classify DNA sequences. Gkm-SVM represented a DNA sequence as a gapped k-mer vector and then resorted to the widely used support vector machine (SVM) to do binary classification. We also proposed two variations of our model, named "DeepHistone (DNA-only)" and "DeepHistone (DNase-only)". The former discards the DNase module and predicts histone modification markers using only DNA sequence information, and the later discards the DNA module and makes predictions using only chromatin accessibility data.

Validation method and evaluation criteria

We adopted 5-fold cross-validation experiments to validate the performance of a method in predicting histone modification sites. Briefly, from ChIP-seq peak files regarding the 15 epigenomes and 7 histone markers, we identified a total of 7,626,807 modification sites. Given one of the 15 epigenomes, we partitioned all the known sites into five parts of nearly equal size. Then, in each fold of the validation, we used four parts to train a model and tested its performance on the remaining part. This procedure was repeated five times to guarantee that each site had been tested once and only once. Note that gkm-SVM is very time-consuming when compared with a deep learning method that can be accelerated by hardware (e.g., GPU). Consequently, we had to sample at random only a small number (50,000) of modification sites in the validation experiments for gkm-SVM, in order to complete the experiments in reasonable time.

Although our method can simultaneously predict whether a DNA fragment in an epigenome is a site for the 7 histone markers, a fragment has only two status for a certain marker, being a histone modification site or not. This understanding allows us to evaluate the performance of our method using the traditional formulation of binary classification. Specifically, given a histone marker, at a certain threshold of the prediction probability, we calculated the sensitivity as the fraction of its modification sites assigned a probability higher than the threshold, and the specificity as the fraction of sites not relevant to the marker and assigned a probability lower than the threshold. Varying the threshold value from 0 to 1, we were able to draw a receiver operating characteristic (ROC) curve. The area under this curve was then calculated as a criterion called auROC. Considering that the number of none-relevant modification sites for a marker is typically much larger than that of true sites, we further calculated the recall and precision at a threshold, drew a precision-recall curve by varying the threshold value, and obtained the area under this curve as another criterion called auPRC.

The rationale for our method and validation design is conceptually equivalent to using modification sites specific to a histone marker as positive set and those not relevant to the marker as negative set to train a binary classification model for the marker. However, our design has two advantages. First, instead of training 7 models for the 7 markers separately, our method can simultaneously train a model for all the 7 markers, thereby saving computational time. Second, the selection of the negative set in our design is much more stringent than such strategies as selecting DNA fragments at random from the whole genome, because modification sites for different markers may have some similar properties, e.g., GC contents, the distance to a gene, etc.

Motif visualization

To interpret how DeepHistone captures DNA sequence patterns, we proposed the following strategy to demonstrate the relationship between known DNA binding motif and sequence patterns extracted by the first convolutional layer of DNA module. Following the literature [18, 20], we first generated a position weighting matrix (PWM) for each kernel in first convolutional layer of the DNA module by scanning along all the input sequences to find activated regions and then averaging over all the activated regions. Formally, a region $\mathbf{x}_i$ in an input sequence $\mathbf{s}$ was regarded as activated, if

$$sum(\mathbf{w}^k \cdot \mathbf{x}_i) \geq \alpha \times EAV$$

where $\mathbf{w}^k$ is the weight matrix of the k-th kernel, $\alpha \in (0, 1)$ a control coefficient, and EAV the extreme activation value of $\mathbf{s}$ defined as

$$EAV = \max^{f()}(sum(\mathbf{w}^k \cdot \mathbf{x}_i^T)|\forall \boldsymbol{x}_i \in \boldsymbol{s})$$

We set the length of a kernel to 9 and α to 0.9. We then compared extracted PWMs to the JASPAR database [28] and illustrated the results by using the tool TomTom [29] with q-value threshold 0.05.

Analysis of functional implications of haQTLs

We applied DeepHistone to explore functional implications of single nucleoid polymorphisms (SNPs) related to histone acetylation quantitative trait loci (haQTLs) identified in a lymphoblastoid epigenome by the histone H3 acetylated on lysine 27 (H3K27ac) marker [30]. Given a SNP, we identified the 1000 bp DNA sequence centred at the SNP position and predicted two probabilities, p^{ref} and p^{alt}, that indicate the degree that the reference and alteration sequences being a histone modification site for the H3K27ac marker, respectively. Following the literature [20], the absolute value of the different between the two predictions was then defined as the functional implication score, $\Delta p = |p^{alt} - p^{ref}|$ for the SNP.

Results

DeepHistone accurately predicts histone modification sites

We first conducted 5-fold cross-validation experiments to assess the performance of our method (see Materials and methods). As shown in Table 2, for a histone marker, the auROC score averaging over the 15 epigenomes is close to 0.9, indicating the effectiveness of our method in predicting modification sites specific to a histone marker. From Fig. 2 (a), we observe that for a histone marker, the auROC score for an epigenome is typically above 0.87, though different epigenomes show fluctuations, also supporting this conclusion. Moreover, the effectiveness of our method is further supported by auPRC scores shown in Table 3 and Fig. 2 (b).

We then compared the performance of our method with that of the baseline approaches. Considering that our method uses both sequence and chromatin accessibility information, while the other approaches only rely on DNA sequence, we discarded the DNase module and implemented a variation of our method called DeepHistone (DNA-only). From Table 2, we observe that the mean auROC score over the 15 epigenomes for a histone marker yielded by this model, though in general has a slight drop when compared with the that generated by the original model, i.e., DeepHistone (Standard), is obviously significantly higher than all the three baseline methods (DeepSEA, DanQ, and gkm-SVM). For example, for H3K4me1, the mean auROC of the 15

Table 2 Performance of different methods in terms of auROC scores

Methods	H3K4me1	H3K4me3	H3K27me3	H3K36me3	H3K9me3	H3K9ac	H3K27ac
DeepHistone (Standard)	0.9065 (0.0292)	0.9459 (0.0404)	0.8896 (0.0292)	0.8890 (0.0174)	0.8969 (0.0340)	0.9039 (0.0331)	0.9137 (0.0242)
DeepHistone (DNA-only)	0.8685 (0.0550)	0.9381 (0.0595)	0.8766 (0.0635)	0.8834 (0.0447)	0.8744 (0.0535)	0.8900 (0.1013)	0.8834 (0.0605)
DeepHistone (DNase-only)	0.8335 (0.0304)	0.8627 (0.0257)	0.7132 (0.0319)	0.7080 (0.0245)	0.7639 (0.0354)	0.8353 (0.0596)	0.8690 (0.0361)
DeepSEA	0.7828 (0.0276)	0.8987 (0.0269)	0.7829 (0.0296)	0.8167 (0.0200)	0.8341 (0.0306)	0.8330 (0.0620)	0.7934 (0.0335)
DanQ	0.7649 (0.0258)	0.8934 (0.0367)	0.7722 (0.0266)	0.8138 (0.0191)	0.8272 (0.0336)	0.8238 (0.0353)	0.7726 (0.0241)
gkm-SVM	0.6361 (0.0399)	0.8092 (0.0592)	0.6434 (0.0415)	0.6666 (0.0335)	0.7278 (0.0421)	0.7167 (0.0454)	0.6445 (0.0367)

Numbers in a cell are the mean (left) and standard deviation (right) of auROC scores over the 15 epigenomes

epigenomes for DeepHistone (Standard) is 0.9065 ± 0.0290, while those for DeepHistone (DNA-only), DeepSEA, DanQ and gkm-SVM are 0.8685 ± 0.0550, 0.7828 ± 0.0280, 0.7649 ± 0.0260, 0.6361 ± 0.0400, respectively. This observation suggests that our method, even when using sequence information alone, is still superior over the three baseline methods in predicting modification sites specific to a histone marker.

From Fig. 2, we further confirmed this observation. Also taking H3K4me1 as an example, the median auROC of the 15 epigenomes for DeepHistone (Standard) is 0.9152 in the box plot, while those for DeepHistone (DNA-only), DeepSEA, DanQ and gkm-SVM are 0.8922, 0.8200, 0.8058, 0.6804, respectively. We then conducted a one-sided paired-sample binomial exact test to access whether the auROC scores of the 15 epigenomes yielded by a method for a histone marker is higher than those generated by another. Results show that DeepHistone (Standard) is superior to DeepHistone (DNA-only) with significant p-values for H3K4me1, H3K4me3, H3K9ac,

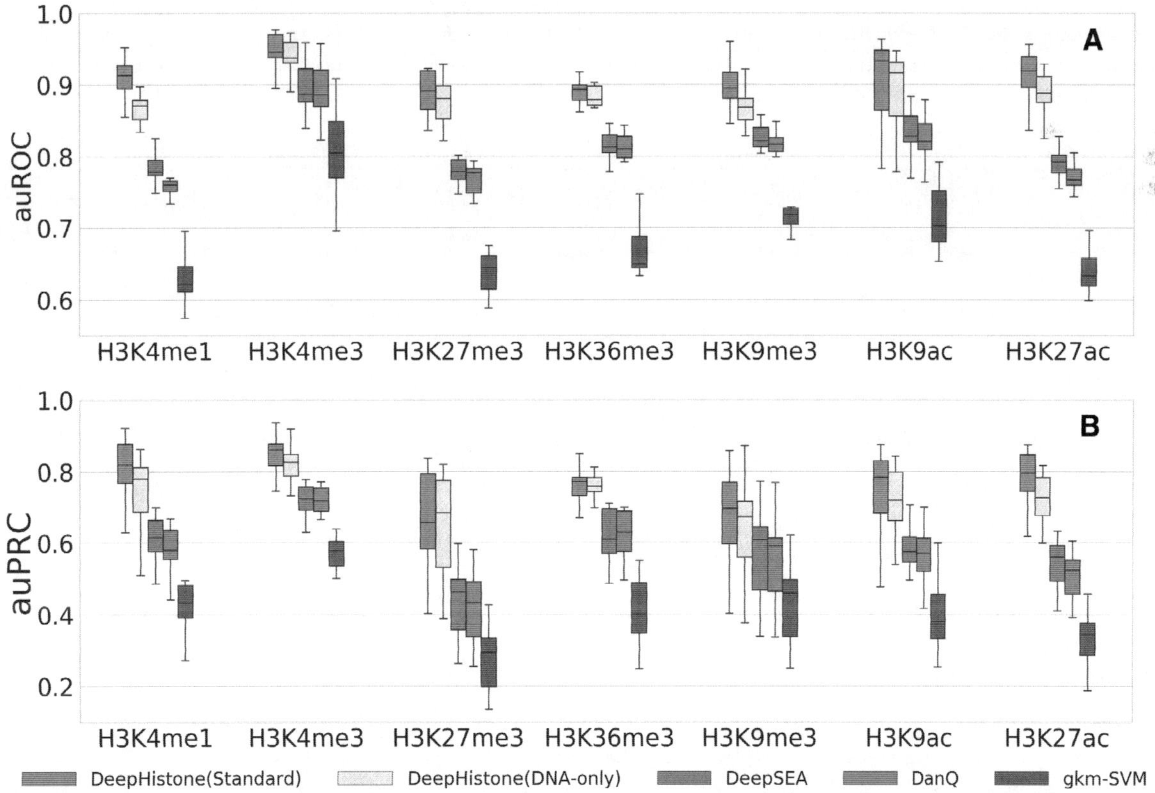

Fig. 2 Performance of different methods in terms of auROC (**a**) and auPRC (**b**). In 5-fold cross-validation experiments, DeepHistone (Standard) achieves higher performance than DeepHistone (DNA-only), which in turn outperforms the three baseline methods (DeepSEA, DanQ and gkm-SVM)

Table 3 Performance of different methods in terms of auPRC scores

Methods	H3K4me1	H3K4me3	H3K27me3	H3K36me3	H3K9me3	H3K9ac	H3K27ac
DeepHistone (Standard)	0.8116 (0.0751)	0.8432 (0.0801)	0.6655 (0.1032)	0.7551 (0.0688)	0.6779 (0.1329)	0.7271 (0.1044)	0.7714 (0.0822)
DeepHistone (DNA-only)	0.7429 (0.1268)	0.8208 (0.1348)	0.6543 (0.1382)	0.7493 (0.0754)	0.6427 (0.1424)	0.6888 (0.1835)	0.6990 (0.1535)
DeepHistone (DNase-only)	0.6664 (0.0985)	0.5962 (0.0549)	0.3219 (0.1459)	0.4043 (0.0776)	0.3489 (0.1401)	0.5701 (0.1465)	0.6673 (0.1179)
DeepSEA	0.6087 (0.0817)	0.7371 (0.0566)	0.4404 (0.1355)	0.6164 (0.0647)	0.5554 (0.1277)	0.5629 (0.1536)	0.5340 (0.1027)
DanQ	0.5805 (0.0783)	0.7303 (0.0737)	0.4233 (0.1046)	0.6178 (0.0756)	0.5420 (0.1379)	0.5502 (0.1034)	0.5015 (0.0859)
gkm-SVM	0.4237 (0.0722)	0.5858 (0.0981)	0.2774 (0.0945)	0.4069 (0.0898)	0.4221 (0.1208)	0.3889 (0.1187)	0.3340 (0.0700)

Numbers in a cell are the mean (left) and standard deviation (right) of auPRC scores over the 15 epigenomes

and H3K27ac (all p-values are equal to 3.052E-05) and marginal significant p-values for H3K9me3 (p-value = 3.693E-03) and H3K27me3 (p-value = 5.924E-02). For H3K36me3, these two methods show no apparent difference (p-value = 0.500). Furthermore, DeepHistone (DNA-only) is superior to all the three baseline methods for all the 7 histone markers (all p-values are equal to 3.052E-05). These results further support the conclusion that our method outperforms existing baseline approaches, even when using sequence information alone.

Contributions of the DNA and DNase modules

To evaluate contributions of the sequence information and chromatin accessibility data, we discarded the DNase and DNA module from our model, yielding two variations of our method, named DeepHistone (DNA-only) and DeepHistone (DNase-only), respectively. For each of the resulting model, we repeated the 5-fold cross-validation experiments and showed the results in Tables 2 and 3, and Fig. 3.

From Table 2, we first observe that both the sequence information and chromatin accessibility data have positive contributions to the final model, because the exclusion of either information leads to a drop in the performance. For example, for H3K4me1, the discard of the chromatin accessibility information results in an average auROC of 0.8685 over the 15 epigenomes, while the exclusion of the sequence information results in an average auROC of 0.8335 over the 15 epigenomes. Both

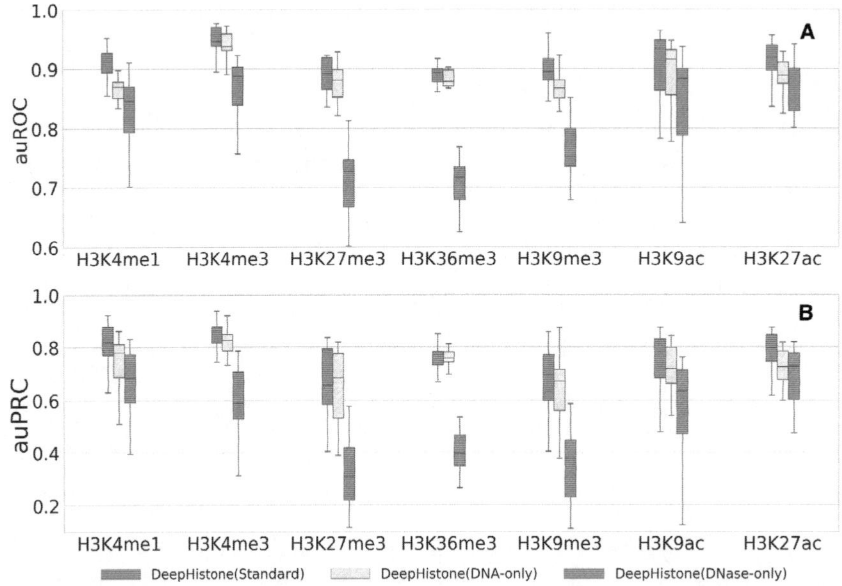

Fig. 3 Contribution of different channels of DeepHistone. In 5-fold cross-validation experiments, DeepHistone (Standard) achieves higher performance than DeepHistone (DNA-only) and DeepHistone (DNase-only) in terms of auROC (**a**) and auPRC (**b**) scores separately, indicating that both sequence and chromatin accessibility information have a positive contribution to our method

results are apparently lower than the full model, which yields an average auROC of 0.9065 over the 15 epigenomes.

We also notice that the sequence information contributes more to the final performance than the chromatin accessibility data, because the removal of the DNA module, i.e., DeepHistone (DNase-only), in general results in a larger drop in performance. To further confirmed this observation, we again conducted the aforementioned one-sided paired-sample binomial exact test to access whether the auROC scores of the 15 epigenomes yielded by DeepHistone (DNA-only) for a histone marker is higher than those generated by DeepHistone (DNase-only). Results show that the former is superior to the latter with significant *p*-values for H3K4me1, H3K4me3, H3K27me3, H3K36me3, H3K9me3 (all *p*-values are equal to 3.052E-05). For H3K9ac, the *p*-value is also significant as 4.883E-04. The only exception is H3K27ac, where the *p*-value (0.304) is not significant.

On one hand, it is not surprising that the sequence information contributes to the prediction of histone modification sites. Actually, this conclusion has been supported by abundant studies that demonstrate the effectiveness of such sequence patterns as transcription factor binding sites in the prediction of histone modification sites [14]. On the other hand, the effectiveness of chromatin accessibility information can also be explained by not only the relationship between histone methylation and DNA accessibility [31] but also the correlation between histone acetylation and chromatin status [32]. Moreover, we conjecture that chromatin accessibility information contributes less than sequence information might be due to the fact that DNase-seq data for all the 7 markers are identical for an epigenome, and signatures of chromatin accessibility may not be so strong as those of sequence.

DeepHistone predicts histone modification sites across epigenomes

Although the above cross-validation experiments demonstrated the success of our method in the prediction of modification sites specific to a histone marker, in reality it would be more meaningful to predict histone modification sites for an epigenome that has no biological experiment conducted. We therefore proposed the following collective scoring strategy to predict status (i.e., belonging to which histone markers) of modification sites for a novel epigenome.

Given a novel epigenome and a genomic region, we would like to predict whether this region was a modification site for a histone marker, with respect to the given epigenome. To achieve this objective, we resorted to a model trained on a known epigenome to predict a probability that indicated whether this region is a modification site for the same histone marker, and we averaged all such probabilities over all known epigenomes to obtain a final prediction probability. In this procedure, the input includes the DNA sequence of the region and the chromatin accessibility data specific to the novel epigenome.

We conducted a leave-one-out experiment to evaluate the performance of our method with this strategy. Specifically, in each validation run, we selected one of the 15 epigenomes and assumed that status of modification sites in this target epigenome is unknown. Then, we applied the collective scoring strategy to recover the status of these sites by making use of the remaining 14 epigenomes. Finally, we evaluated the performance of our method in terms of the auROC and auPRC scores by using the known status of the sites in the target epigenome as the gold standard. In implementation, we took advantage of the models trained in the aforementioned 5-fold cross-validation experiments and averaged the probabilities calculated by these 5 models to obtain the prediction probability with respect to a known epigenome.

We presented the results in Fig. 4, in which both auROC and auPRC scores were averaged over the 7 histone markers to make the presentation concise. From the figure, we can clearly see that the cross-epigenome prediction by DeepHistone is effective, in that for the 15 epigenomes, the auROCs are typically above 0.8, and the auPRCs are typically above 0.6. We also notice that the cross-epigenome prediction in general exhibits lower performance than self-prediction by DeepHistone (5-fold cross-validation). This is reasonable because an epigenome may have its specific sequence codes and chromatin accessibility patterns that might not be captured by the collective scoring strategy.

When compared with the three baseline methods, DeepHistone apparently achieves higher performance for all the 15 epigenomes. For example, the average auROCs for E003 (an embryonic stem cell line) are 0.8391, 0.7697, 0.7744, and 0.6711 for DeepHistone, DeepSEA, DanQ, and gkm-SVM, respectively. Actually, DeepHistone achieves higher auROC scores than all the three baseline methods for all the 15 epigenomes. As a result, a one-sided binomial exact test against the null hypothesis that the performance of DeepHistone across the 15 epigenomes is not different from a baseline method gives significant *p*-values for all the three methods (all *p*-values are equal to 3.052E-05). This conclusion is further supported when using auPRC as the evaluation criterion.

DeepHistone recovers TF binding motifs

To demonstrate sequence patterns automatically extracted by our method, we used the strategy described in Materials and methods to obtain sequence signatures (i.e.,

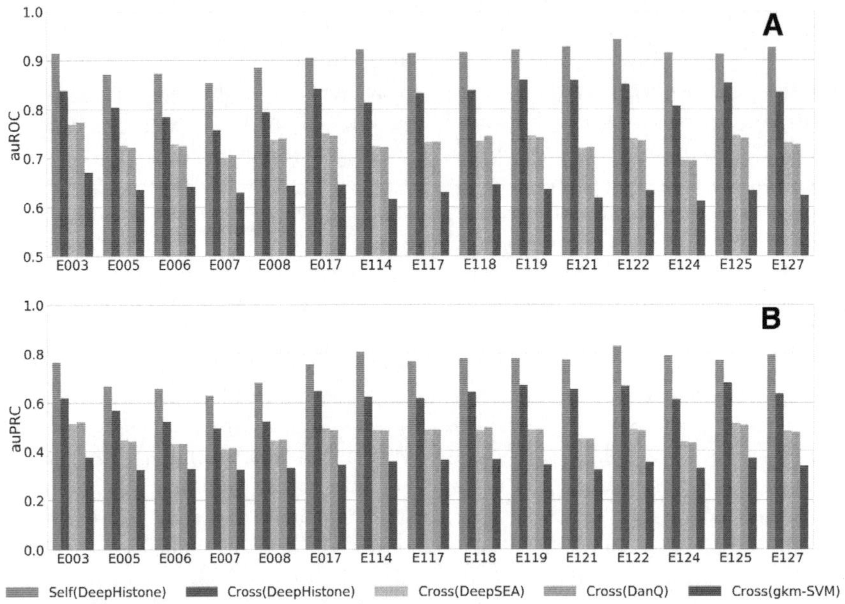

Fig. 4 Cross cell line prediction. Each bar indicates the mean of auROC (**a**) or auPRC (**b**) of 7 histone modification markers in a certain epigenome. "Self (DeepHistone)" indicates the performance of an epigenome predicting by a DeepHistone model trained in the same epigenome. "Cross (DeepHistone)" indicates the performance of cross epigenomes prediction using DeepHistone model. Others have similar meaning

PWMs) learned from the first convolutional layer of the DNA module with respect to an epigenome. We further identified putative sequence motifs by using the tool TomTom and match these PWMs to the JASPAR database. For each epigenome, we displayed the sequence logo of one of the matched motifs in Fig. 5.

In different carcinoma cell lines, DeepHistone recovered corresponding motifs to each cell line, which showed the sensitivity of DeepHistone. In the lung carcinoma cell line (E114), DeepHistone recovered E2F3, TFAP2C and GRHL2. It has been verified that the overexpression of E2F3 transcription factor promotes the development of lung cancer [33, 34]. TFAP2C has been previously shown to promote lung tumorigenesis and aggressiveness by upregulating of TGFBR1 [35, 36]. Different from E2F3 and TFAP2C, GRHL2 can suppress tumor metastasis by regulating of transcriptional activity of RhoG in lung cancer [37]. In HeLa-S3, the cervical carcinoma cell line (E117), PROX1 and NR2F6 were found by DeepHistone. The commitment of PROX1 positive cells is an early event in cervical neoplastic progression, and the expression of PROX1 is considered as evidence of an early lymphangiogenic switch [38]. The abnormal high expression of NR2F6 in early-stage cervical cancer predicts pelvic lymph node metastasis, tumor recurrence and poor prognosis and NR2F6 might be a potential therapeutic target of cervical cancer [39]. As for hepatocellular carcinoma cell line (E118), E2F8, GABBPA and SOX11 were recovered. It has been shown that E2F8 contributes to human hepatocellular carcinoma via regulating cell proliferation [40] and is considered as a potential therapeutic target of hepatocellular cancer [41]. GABBPA inhibits metastasis of hepatocellular carcinoma [42] and SOX11 is important in the regulation of hepatocellular carcinoma cell proliferation, migration and invasion [43]. Besides, DeepHistone recovers SREBF2, HOXA5 and ZNF24 in human umbilical vein endothelial primary (HUVEC) cell line (E122) and NKX6–1 in embryonic stem cell line (E008). Those recovered transcription factors are verified to play important roles in the corresponding cell line [44–48]. To sum up, DeepHistone has the ability to recover potential functional transcription factor corresponding to specific cell line.

DeepHistone explains functional implications of SNPs

Although genome-wide association studies (GWAS) have successfully identified thousands of single nucleotide polymorphisms (SNPs) associated with complex traits [49], most of these SNPs locate outside coding regions. The explanation of the functional implications of these SNPs has thus long been a critical task in genetic studies [50]. Recently, a new technique that combines a deep and long-read ChIP-seq assay on H3K27ac with a powerful statistical test has successfully enabled the identification of histone acetylation quantitative trait loci (haQTLs) related to a lymphoblastoid epigenome.

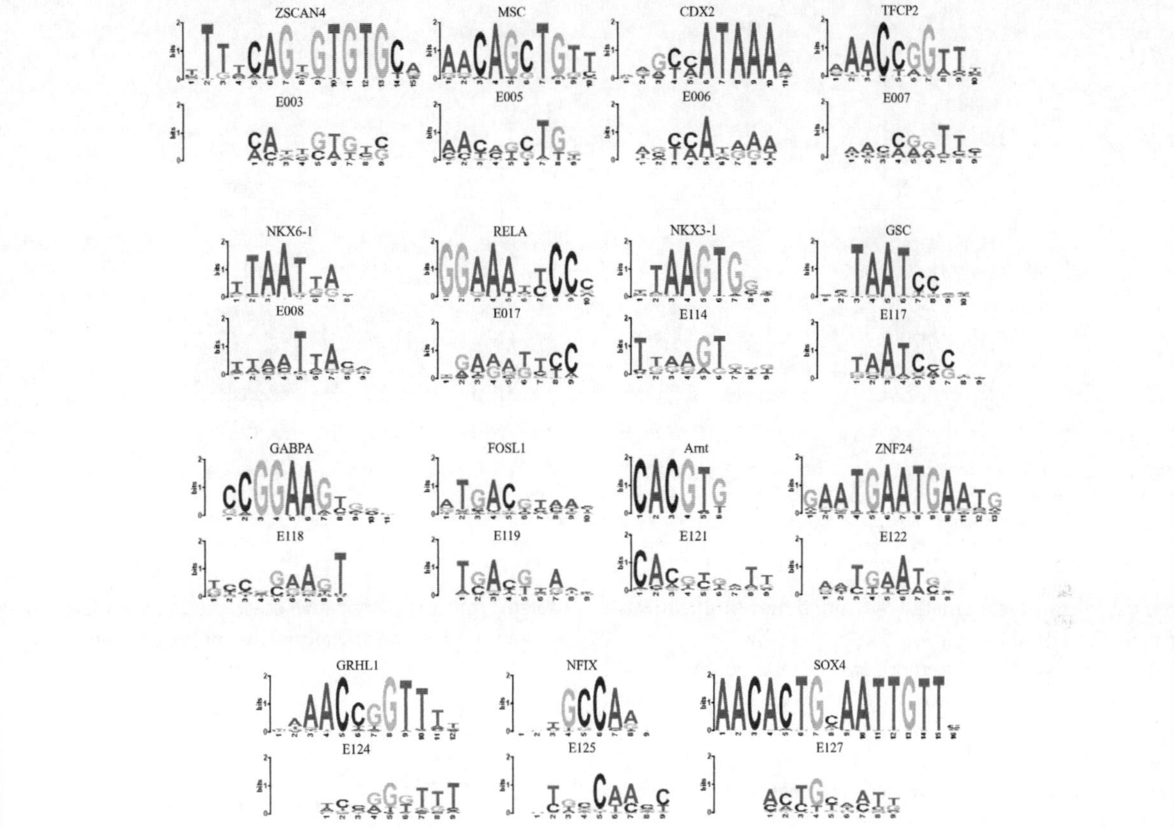

Fig. 5 Visualization of sequence patterns learned from DeepHistone. PWMs were learned from the first convolutional layer of the DNA module. In each subgraph, the upper logo is the motif from the JASPAR database, the bottom part is the PWM learned by DeepHistone

The identified SNPs exhibit highly predictive power in exploring mechanisms of autoimmune disease. We then applied DeepHistone to analyze these SNPs, demonstrating potential applications of our method.

From the literature [30], we identified a positive set that includes 7497 SNPs (haQTLs) specific to H3K27ac in the lymphoblastoid epigenome (E116) and appearing in the 1000 genomes project [51]. Meanwhile, we generated a negative control set that includes the same number of SNPs as the positive one by identifying for each haQTL a SNP that locates about 500 bp away, also from the 1000 genomes project. We then used the formulation detailed in Materials and methods to calculate functional implication scores for the identified SNPs and compared whether scores for positive SNPs are significantly different from those for negative ones. The results, as shown in Fig. 6, clearly show that the haQTLs tend to have higher functional implication scores than the control SNPs. A one-sided Wilcoxon rank sum test against the null hypothesis that the median score of these two sets of SNPs are identical yield a very significant *p*-value of 1.369E-140, strongly support the conclusion that haQTLs have higher functional implication scores. In other words, these SNPs are more likely to change the function of the lymphoblastoid epigenome, and thus are more likely to be responsible to a phenotype. We further generated other four control sets in which a SNP is required to be 1000, 1500, 2000 and 2500 bp away from a haQTL. The results, as shown in Fig. 6, give us a similar conclusion. All these results suggest that our method has the potential ability to discriminant SNPs responsible for a certain phenotype from their nearby genetic variants.

Conclusions and discussion

We have proposed a deep learning framework named DeepHistone to integrate DNA sequence information and chromatin accessibility data for predicting histone modification sites. Through comprehensive validation experiments regarding 7 histone markers and 15 epigenomes, we have shown that our approach is superior to several baseline methods in discriminating among modification sites specific to different histone markers, capable of making predictions across epigenomes, interpretable in extracted sequence features, and

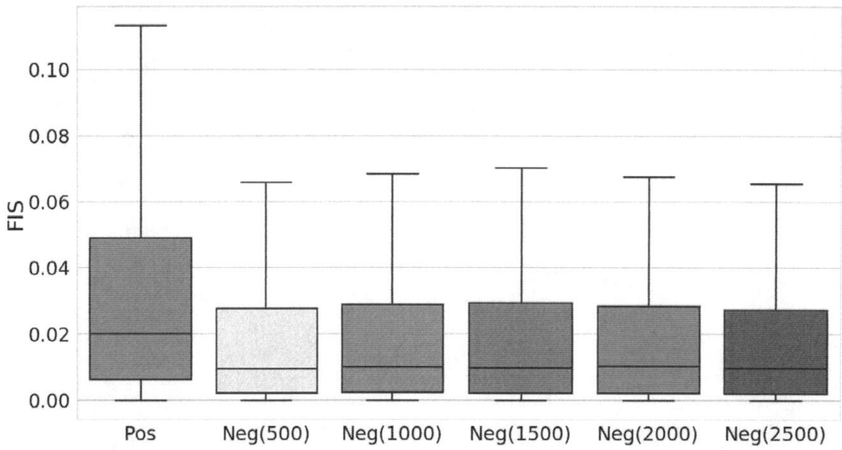

Fig. 6 Functional implication scores (FIS) of haQTLs and their nearby SNPs. "Pos" indicates haQTLs. "Neg(500)" indicates a negative dataset containing SNPs about 500 bp away from the haQTLs. Others have similar meaning. A significant difference of functional implication scores between positive and negative datasets suggested that DeepHistone can distinguish haQTLs against their nearby SNPs

applicable to the explanation of functional implications of genetic variants.

The success of our method can be attributed to the combination of the following facts. First, we have designed a novel deep neural network model with the incorporation of state-of-the-art techniques in the deep learning community. Particularly, the densely connected architecture effectively overcomes such problems as the vanishing gradient and overfitting, and greatly improves the prediction accuracy. Second, besides sequence information, we have also incorporated chromatin accessibility data into our model. These two types of information can then complement each other in our neural network model to capture subtle signals towards the accurate prediction of histone modification sites.

Certainly, our work can be further improved in several aspects. First, resorting to an embedding representation of DNA sequences instead of using the one-hot encoding may further improve the prediction accuracy [21]. Second, considering the sequential natural a DNA fragments, the incorporation of a recurrent neural network architecture, especially long short-term memory units, may further improve the performance of our method [18, 21]. Third, instead of scanning sequence motifs from the beginning using convolutional kernels, it is also possible to incorporate sequence patterns and design a hybrid network architecture [20]. Fourth, we used DNase-seq peaks from Roadmap to quantify chromatin accessibility. This treatment, though simple, may not be precise. Fifty, besides chromatin accessibility data, it is also worth to consider the integration of plenty of gene expression data. Finally, besides our current formulation of predicting for a certain epigenome putative modification sites specific to different histone markers, it will also be beneficial to formulate the problem from the perspective of predicting for a fixed histone marker putative sites for different epigenomes.

Abbreviations

auPRC: Area under the Precision-Recall curve; auROC: Area under the Receiver Operating Characteristic curve; CHIP-seq: Chromatin immunoprecipitation followed by sequencing; DNase-seq: DNase I hypersensitive sites sequencing; GWAS: Genome wide association study; haQTL: Histone acetylation quantitative trait loci; SNP: Single nucleoid polymorphism; TF: Transcription factor

Acknowledgements

We would like to thank Shaoming Song from Tsinghua University for helpful comments. We also appreciate detailed suggestions from anonymous reviewers who significantly help us improved the early version of this manuscript.

Funding

Publication charge for this work has been funded by the National Key Research and Development Program of China (No. 2018YFC0910404), the National Natural Science Foundation of China (Nos. 61873141, 61721003, 61573207, U1736210, 71871019 and 71471016, 31501081), and the Tsinghua-Fuzhou Institute for Data Technology. Rui Jiang is a RONG professor at the Institute for Data Science, Tsinghua University.

Availability of data and materials

Raw data are from the Roadmap Epigenomics Project, as well as the source code is freely available in https://github.com/QijinYin/DeepHistone.

About this supplement

This article has been published as part of *BMC Genomics Volume 20 Supplement 2, 2019: Selected articles from the 17th Asia Pacific Bioinformatics Conference (APBC 2019): genomics.* The full contents of the supplement are available online at https://bmcgenomics.biomedcentral.com/articles/supplements/volume-20-supplement-2.

Authors' contributions

HL and RJ provided guidance and planning for this project. QY and MW produced programs, analysed main results and wrote the manuscript. All authors read and approved the final manuscript.

Ethics approval and consent to participate

Not applicable.

Consent for publication
Not applicable.

Competing interests
The authors declare that they have no competing interests.

Publisher's Note

Springer Nature remains neutral with regard to jurisdictional claims in published maps and institutional affiliations.

Published: 10 April 2019

References

1. Allfrey VG, Faulkner R, Mirsky AE. Acetylation and methylation of histones and their possible role in the regulation of RNA synthesis. Proc Natl Acad Sci U S A. 1964;51(5):786–94.
2. Luger K, Mäder AW, Richmond RK, Sargent DF, Richmond TJ. Crystal structure of the nucleosome core particle at 2.8 Å resolution. Nature. 1997; 389(6648):251.
3. Peterson CL, Laniel MA. Histones and histone modifications. Curr Biol. 2004; 14(14):R546–51.
4. Bannister AJ, Kouzarides T. Regulation of chromatin by histone modifications. Cell Res. 2011;21(3):381–95.
5. Brehove M, Wang T, North J, Luo Y, Dreher SJ, Shimko JC, Ottesen JJ, Luger K, Poirier MG. Histone core phosphorylation regulates DNA accessibility. J Biol Chem. 2015;290(37):22612–21.
6. Cheung P, Allis CD, Sassone-Corsi P. Signaling to chromatin through histone modifications. Cell. 2000;103(2):263–71.
7. Binder H, Steiner L, Przybilla J, Rohlf T, Prohaska S, Galle J. Transcriptional regulation by histone modifications: towards a theory of chromatin re-organization during stem cell differentiation. Phys Biol. 2013;10(2):026006.
8. Kouzarides T. Chromatin modifications and their function. Cell. 2007;128(4): 693–705.
9. Kristeleit R, Stimson L, Workman P, Aherne W. Histone modification enzymes: novel targets for cancer drugs. Expert Opin Emerg Drugs. 2004; 9(1):135–54.
10. O'Geen H, Echipare L, Farnham PJ. Using ChIP-Seq technology to generate high-resolution profiles of histone modifications. Methods Mol Biol. 2011; 791:265–86.
11. Barski A, Cuddapah S, Cui K, Roh T-Y, Schones DE, Wang Z, Wei G, Chepelev I, Zhao K. High-resolution profiling of histone methylations in the human genome. Cell. 2007;129(4):823–37.
12. Consortium EP. The ENCODE (ENCyclopedia of DNA elements) project. Science. 2004;306(5696):636–40.
13. Roadmap Epigenomics C, Kundaje A, Meuleman W, Ernst J, Bilenky M, Yen A, Heravi-Moussavi A, Kheradpour P, Zhang Z, Wang J, et al. Integrative analysis of 111 reference human epigenomes. Nature. 2015;518(7539):317–30.
14. Benveniste D, Sonntag H-J, Sanguinetti G, Sproul D. Transcription factor binding predicts histone modifications in human cell lines. Proc Natl Acad Sci U S A. 2014;111(37):13367–72.
15. Karlic R, Chung HR, Lasserre J, Vlahovicek K, Vingron M. Histone modification levels are predictive for gene expression. Proc Natl Acad Sci U S A. 2010; 107(7):2926–31.
16. Alipanahi B, Delong A, Weirauch MT, Frey BJ. Predicting the sequence specificities of DNA- and RNA-binding proteins by deep learning. Nat Biotechnol. 2015;33(8):831–8.
17. Zhou J, Troyanskaya OG. Predicting effects of noncoding variants with deep learning-based sequence model. Nat Methods. 2015;12(10):931–4.
18. Quang D, Xie X. DanQ: a hybrid convolutional and recurrent deep neural network for quantifying the function of DNA sequences. Nucleic Acids Res. 2016;44(11):e107.
19. Min X, Zeng W, Chen S, Chen N, Chen T, Jiang R. Predicting enhancers with deep convolutional neural networks. BMC Bioinformatics. 2017;18(Suppl 13):478.
20. Liu Q, Xia F, Yin Q, Jiang R. Chromatin accessibility prediction via a hybrid deep convolutional neural network. Bioinformatics. 2018;34(5):732–8.
21. Min X, Zeng W, Chen N, Chen T, Jiang R. Chromatin accessibility prediction via convolutional long short-term memory networks with k-mer embedding. Bioinformatics. 2017;33(14):i92–i101.
22. Li W, Wong WH, Jiang R: DeepTACT: predicting high-resolution chromatin contacts via bootstrapping deep learning. bioRxiv 2018.
23. Brykczynska U, Hisano M, Erkek S, Ramos L, Oakeley EJ, Roloff TC, Beisel C, Schubeler D, Stadler MB, Peters AH. Repressive and active histone methylation mark distinct promoters in human and mouse spermatozoa. Nat Struct Mol Biol. 2010;17(6):679–87.
24. Huang G, Liu Z, Weinberger KQ: Densely connected convolutional networks. 2017 IEEE Conference on Computer Vision and Pattern Recognition (CVPR) 2017:2261–2269.
25. Paszke A, Gross S, Chintala S, Chanan G, Yang E, DeVito Z, Lin Z, Desmaison A, Antiga L, Lerer A. Automatic differentiation in pytorch; 2017.
26. Kingma DP, Ba J: Adam: A method for stochastic optimization. CoRR 2014, abs/1412.6980.
27. Lee D, Gorkin DU, Baker M, Strober BJ, Asoni AL, Mccallion AS, Beer MA. A method to predict the impact of regulatory variants from DNA sequence. Nat Genet. 2015;47(8):955–61.
28. Khan A, Fornes O, Stigliani A, Gheorghe M, Castro-Mondragon JA, van der Lee R, Bessy A, Cheneby J, Kulkarni SR, Tan G et al: JASPAR 2018: update of the open-access database of transcription factor binding profiles and its web framework. Nucleic Acids Res 2018, 46(D1):D260-D266.
29. Gupta S, Stamatoyannopoulos JA, Bailey TL, Noble WS. Quantifying similarity between motifs. Genome Biol. 2007;8(2):R24.
30. del Rosario RC, Poschmann J, Rouam SL, Png E, Khor CC, Hibberd ML, Prabhakar S. Sensitive detection of chromatin-altering polymorphisms reveals autoimmune disease mechanisms. Nat Methods. 2015;12(5):458–64.
31. Cui P, Li J, Sun B, Zhang M, Lian B, Li Y, Xie L. A quantitative analysis of the impact on chromatin accessibility by histone modifications and binding of transcription factors in DNase I hypersensitive sites. Biomed Res Int. 2013; 2013:914971.
32. Görisch SM, Wachsmuth M, Tóth KF, Lichter P, Rippe K. Histone acetylation increases chromatin accessibility. J Cell Sci. 2005;118(24):5825–34.
33. Cooper CS, Nicholson AG, Foster C, Dodson A, Edwards S, Fletcher A, Roe T, Clark J, Joshi A, Norman A. Nuclear overexpression of the E2F3 transcription factor in human lung cancer. Lung Cancer. 2006;54(2):155–62.
34. Trikha P, Sharma N, Pena C, Reyes A, Pécot T, Khurshid S, Rawahneh M, Moffitt J, Stephens JA, Fernandez SA, et al. E2f3 in tumor macrophages promotes lung metastasis. Oncogene. 2015;35:3636.
35. Kang J, Kim W, Lee S, Kwon D, Chun J, Son B, Kim E, Lee J, Youn H, Youn B. TFAP2C promotes lung tumorigenesis and aggressiveness through miR-183-and miR-33a-mediated cell cycle regulation. Oncogene. 2017;36(11):1585.
36. Kim W, Kim E, Lee S, Kim D, Chun J, Park KH, Youn H, Youn B. TFAP2C-mediated upregulation of TGFBR1 promotes lung tumorigenesis and epithelial–mesenchymal transition. Exp Mol Med. 2016;48(11):e273.
37. Pan X, Zhang R, Xie C, Gan M, Yao S, Yao Y, Jin J, Han T, Huang Y, Gong Y, et al. GRHL2 suppresses tumor metastasis via regulation of transcriptional activity of RhoG in non-small cell lung cancer. Am J Transl Res. 2017;9(9): 4217–26.
38. Cimpean AM, Mazuru V, Saptefrati L, Ceausu R, Raica M. Prox 1, VEGF-C and VEGFR3 expression during cervical neoplasia progression as evidence of an early lymphangiogenic switch. Histol Histopathol. 2012;27(10):1543.
39. Niu C, Sun X, Zhang W, Li H, Xu L, Li J, Xu B, Zhang Y. NR2F6 expression correlates with pelvic lymph node metastasis and poor prognosis in early-stage cervical Cancer. Int J Mol Sci. 2016;17(10):1694.
40. Deng Q, Wang Q, Zong WY, Zheng DL, Wen YX, Wang KS, Teng XM, Zhang X, Huang J, Han ZG. E2F8 contributes to human hepatocellular carcinoma via regulating cell proliferation. Cancer Res. 2010;70(2):782–91.
41. Lv Y, Xiao J, Liu J, Xing F. E2F8 is a potential therapeutic target for hepatocellular carcinoma. J Cancer. 2017;8(7):1205.
42. Zhang S, Zhang K, Ji P, Zheng X, Jin J, Feng M, Liu P. GABPA predicts prognosis and inhibits metastasis of hepatocellular carcinoma. BMC Cancer. 2017;17(1):380.
43. Wang Z, Li Z, Zhu J. Negative regulation of SOX11 in hepatocellular carcinoma. Int J Clin Exp Med. 2017;10(2):2809–17.
44. Xiao H, Lu M, Lin TY, Chen Z, Chen G, Wang WC, Marin T, Shentu TP, Wen L, Gongol B, et al. Sterol regulatory element binding protein 2 activation of NLRP3 inflammasome in endothelium mediates hemodynamic-induced atherosclerosis susceptibility. Circulation. 2013;128(6):632–42.
45. Zeng L, Liao H, Liu Y, Lee TS, Zhu M, Wang X, Stemerman MB, Zhu Y, Shyy JY. Sterol-responsive element-binding protein (SREBP) 2 down-regulates ATP-binding cassette transporter A1 in vascular endothelial cells: a novel role of SREBP in regulating cholesterol metabolism. J Biol Chem. 2004; 279(47):48801-7.

46. Fork C, Gu L, Hitzel J, Josipovic I, Hu J, SzeKa Wong M, Ponomareva Y, Albert M, Schmitz SU, Uchida S, et al. Epigenetic regulation of angiogenesis by JARID1B-induced repression of HOXA5. Arterioscler Thromb Vasc Biol. 2015;35(7):1645–52.
47. Jia D, Huang L, Bischoff J, Moses MA. The endogenous zinc finger transcription factor, ZNF24, modulates the angiogenic potential of human microvascular endothelial cells. FASEB J. 2015;29(4):1371–82.
48. Rezania A, Bruin JE, Xu J, Narayan K, Fox JK, O'Neil JJ, Kieffer TJ. Enrichment of human embryonic stem cell-derived NKX6.1-expressing pancreatic progenitor cells accelerates the maturation of insulin-secreting cells in vivo. Stem Cells. 2013;31(11):2432–42.
49. MacArthur J, Bowler E, Cerezo M, Gil L, Hall P, Hastings E, Junkins H, McMahon A, Milano A, Morales J. The new NHGRI-EBI catalog of published genome-wide association studies (GWAS catalog). Nucleic Acids Res. 2016; 45(D1):D896–901.
50. Wu M, Lin Z, Ma S, Chen T, Jiang R, Wong WH. Simultaneous inference of phenotype-associated genes and relevant tissues from GWAS data via Bayesian integration of multiple tissue-specific gene networks. J Mol Cell Biol. 2017;9(6):436–52.
51. Genomes Project C, Auton A, Brooks LD, Durbin RM, Garrison EP, Kang HM, Korbel JO, Marchini JL, McCarthy S, McVean GA, et al. A global reference for human genetic variation. Nature. 2015;526(7571):68–74.

Guzmán-Sáenz et al. BMC Genomics 2019, 20(Suppl 2):194
https://doi.org/10.1186/s12864-019-5490-y

BMC Genomics

RESEARCH Open Access

Signal enrichment with strain-level resolution in metagenomes using topological data analysis

Aldo Guzmán-Sáenz[1], Niina Haiminen[1], Saugata Basu[2] and Laxmi Parida[1*]

From The 17th Asia Pacific Bioinformatics Conference (APBC 2019)
Wuhan, China. 14-16 January 2019

Abstract

Background: A metagenome is a collection of genomes, usually in a micro-environment, and sequencing a metagenomic sample *en masse* is a powerful means for investigating the community of the constituent microorganisms. One of the challenges is in distinguishing between similar organisms due to rampant multiple possible assignments of sequencing reads, resulting in false positive identifications. We map the problem to a topological data analysis (TDA) framework that extracts information from the geometric structure of data. Here the structure is defined by multi-way relationships between the sequencing reads using a reference database.

Results: Based primarily on the patterns of co-mapping of the reads to multiple organisms in the reference database, we use two models: one a subcomplex of a Barycentric subdivision complex and the other a Čech complex. The Barycentric subcomplex allows a natural mapping of the reads along with their coverage of organisms while the Čech complex takes simply the number of reads into account to map the problem to homology computation. Using simulated genome mixtures we show not just enrichment of signal but also microbe identification with strain-level resolution.

Conclusions: In particular, in the most refractory of cases where alternative algorithms that exploit unique reads (i.e., mapped to unique organisms) fail, we show that the TDA approach continues to show consistent performance. The Čech model that uses less information is equally effective, suggesting that even partial information when augmented with the appropriate structure is quite powerful.

Keywords: Metagenomics, Topological data analysis, Multi-mapping reads, False positives

Background

A metagenome is a collection of genomes in a micro-environment, such as the gut of an animal, bottom of an ocean, or soil. This captures the influence of the immediate environment on the phenotype of an organism. For instance, one of the factors in the safety of our food supply chain is knowing the microbiome in the food [1]. The state of disease and health of a host has been shown to be related to the microbiomes in its gut [2]. The sturdiness or weakness of a plant is shown to be related to its soil microbiome [3]. It is turning out that these microorganisms are perhaps playing a much bigger role than earlier anticipated. The DNA technology to capture these organisms also has been disruptive in the area of microbiology, i.e., each organism does not need to be cultured individually before sequencing but the entire volume of samples can be put through the sequencing process en masse [4]. The obvious advantage is that the recalcitrant organisms that were resistant to being cultured no longer pose a problem, as long as careful sample processing is performed to avoid sequencing biases [5]. However, there are two major challenges with the sequencing approach.

*Correspondence: parida@us.ibm.com
[1]Computational Biology Center, IBM T. J. Watson Research Center, Yorktown Heights, NY, USA
Full list of author information is available at the end of the article

Firstly, the completeness and correctness of reference databases limits the power of detection. The database of reference sequences must systematically be updated when new genomes become available. Secondly, no matter how complete and accurate the databases are, there is the problem of correctly assigning sequencing reads when distinct organisms with very similar genomes are present. In this paper, we address the second challenge of accurately detecting the organisms present in a sample, given a database.

Characteristics of the short sequencing reads frequently results in them being mapped to multiple reference genomes in the database, even under very strict matching criteria. Thus when the database organisms are very similar to each other there is usually a substantial dearth of reads assigned to unique organisms. As a consequence, most solution pipelines yield mapping results that are riddled with false positives. In fact, in (microbial) simulation studies we find that often a large percentage of the predicted potential organisms, using standard pipelines from literature, are false positives. A recent benchmark study found the number of species reported for the same sample varying by orders of magnitude, depending on the classifier used [6]. Popular methods for metagenomic read classification include Kraken [7], CLARK [8], MetaPhlAn [9] and others.

Topological data analysis (TDA) is emerging as a promising approach for analyzing large genomic datasets for a variety of questions [10–12], with already demonstrated applications in evolutionary biology, cancer genomics, and analysis of complex diseases. TDA extracts information from the geometric structure of data; in our application the structure is defined by the relationships between sequencing reads and organisms in a reference database.

Based primarily on the patterns of co-mapping of reads to the organisms in a reference database, we use a subcomplex of a Barycentric subdivision complex to model the multi-way maps of reads, along with the extent of read coverage on the respective organisms. This subcomplex allows a natural mapping of our problem to homology computation and interpretation. We test this approach on the special scenario (dearth of uniquely mapped reads) and find that using an appropriate voting function on the typical bar diagrams of TDA, we can sort the organisms to separate true positives from false positives. Next, we test if a reduced information set, that is, only the number of reads without utilizing their lengths or coverage when combined with TDA is powerful enough to enrich for true positives. We observe success in this scenario as well, suggesting that the use of topology captures the non-obvious structure defined by the reads promiscuously mapping to multiple organisms. All the data used in the paper are simulated reads by necessity since it is the most reliable way of knowing the ground truth, to assess the results.

Methods

In this section we map the problem to topological data analysis, more specifically to bar diagrams arising from persistent homology.

Motivation and problem statement

Given a reference database and a collection of metagenomic reads from the sequencing process, we denote by Z the set of all possible organisms, resulting from a reads-to-organism-mapping (ROM) pipeline: $Z = \{X_1, X_2, ..., X_N\}$. We treat all the reads of the pipeline output equally, i.e., do not consider the extent of the match. The rationale is that these characteristics have already been used by the pipeline to produce the output. Hence we focus on the subsequent information gathered from the pipeline: the relationship between the reads an the organisms.

Consider the bipartite graph of reads and organisms, where an edge connects a read to the assigned organism, that captures the ROM relationships completely. But a much more natural fit is to a simplicial complex where a k-simplex, $k > 0$ (analogous to edge in a graph) is a read and the 0-simplex (analogous to a vertex in a graph) is an organism. Furthermore each of the the k-simplices are weighted, i.e., the weight is the number of reads that map to exactly these k organisms, which is not immediately apparent from the bipartite graph. The next question is whether the topological features of this naturally emerging simplicial complex can be exploited to solve the problem of false positives. In other words, is there some hidden characteristics of the ROM that separate the true from the false organisms.

In this paper, to make it more tractable, we simplify the problem as follows: Given the ROM pipeline output, obtain an order for the elements of Z as $X_{i_1}, X_{i_2}, ..., X_{i_N}$, such that the top of the list is enriched for true positives (TPs). In a practical scenario, a threshold can be used to snip away the bottom of the list to eliminate potential false positive candidates. Further, the retained true positives can be used to rescue and re-assign the reads that had been misassigned due to the false positives.

Model I: Barycentric subdivision

Each organism X_j has a collection of reads mapped to it. For each subset $Y = \{X_1, \ldots, X_k\} \in 2^{\mathcal{X}}$, we denote by S_Y the set of reads mapped to the organisms $X_1, X_2, ..., X_k$, and none other.

For instance, this implies that $S_{\{X_1,X_2\}} \cap S_{\{X_1,X_2,X_3\}} = \emptyset$. In general,

$$S_Y \cap S_{Y'} = \emptyset, \text{whenever} Y \neq Y'.$$

Definition 1 (area a_Y) *For $Y \in 2^{\mathcal{X}}$, the area a_Y is defined to be the sum total of all the lengths of the reads in S_Y.*

Definition 2 (depth $d_Y(X_j)$ for $Y \in 2^{\mathcal{X}}$ and $X_j \in Y$) *For $Y \in 2^{\mathcal{X}}$ and $X_j \in Y$ we denote the length of the genome of the organism X_j covered by the reads in S_Y by l_{Y,X_j}, and define*

$$d_Y(X_j) = \frac{1}{m^2}\left(\frac{a_Y}{l_{Y,X_j}}\right),$$

where $m = |Y|$.

This is motivated by the fact that if there is true overlap in the m organisms, then the depth d of coverage of each organism is magnified by some factor of m. Note that it is quite possible that for $Y = \{X_1, X_2, \ldots, X_m\}$, the values $d_Y(X_1), d_Y(X_2), d_Y(X_3), \ldots, d_Y(X_m)$ are all distinct.

Example 1 *Consider a set of 3 organisms $\mathcal{X} = \{A, B, C\}$. A graphical representation of the underlying mapped reads is shown in Fig. 1 and the values of d_Y for each set $Y \in 2^{\mathcal{X}}$ are given in Fig. 2 (top).*

The depth values were computed after simulating a small set of reads' placement on three reference genomes. For example, in this case no reads match both B and C (and not A), hence $d_{BC}(B) = d_{BC}(C) = 0$. However, some reads match all three organisms, hence $d_{ABC} > 0$. The values $d_A(A), d_B(B), d_C(C)$ indicate the depths of A, B, C based on their uniquely mapping reads.

Definition 3 ***(set Y^t for a $t \geq 0$)*** *For $Y \in 2^{\mathcal{X}}$ and $t \in \mathbb{R}, t \geq 0$, we denote*

$$Y^t = \{X_j \in Y \mid d_Y(X_j) \geq t\}.$$

Note that the sets $(X_i)_{1 \leq i \leq N}$ give a *cover* of the set of reads, while the sets $(S_Y)_{Y \in 2^{\mathcal{X}}}$ form a *partition* of the reads. This corresponds to the collection of organisms present at a time t, mapped to some stage of a filtration of a simplicial complex (see Definition 5).

Definition 4 *Given any finite set S, a simplicial complex with vertices in S, is a collection K of subsets of S, with the property that if $A \in K$, then for all $B \in 2^S$ with $B \subset A$, $B \in K$. A subcollection L of K is a subcomplex of K if L is a simplicial complex.*

In other words, K is closed under the operation of "taking subsets". This last property is crucial for the definition of the simplicial homology groups, $H_*(K)$, associated to a simplicial complex K.

Definition 5 *A finite filtered simplicial complex is a finite sequence $K_0 \subset K_1 \subset \cdots \subset K_N$ of simplicial complexes. In other words, K_i is a subcomplex of K_j for $i \leq j$. A filtration of a simplicial complex K is a filtered simplicial complex with $K_N = K$.*

This particular setup allows us to apply *persistent homology* to our problem.

To any finite simplicial complex K one can associate for each $j \geq 0$, a finite dimensional vector space $H_j(K, \mathbb{Q})$ (called the *j-th simplicial homology group of K*). Moreover, if we have a filtration of a simplicial complex K

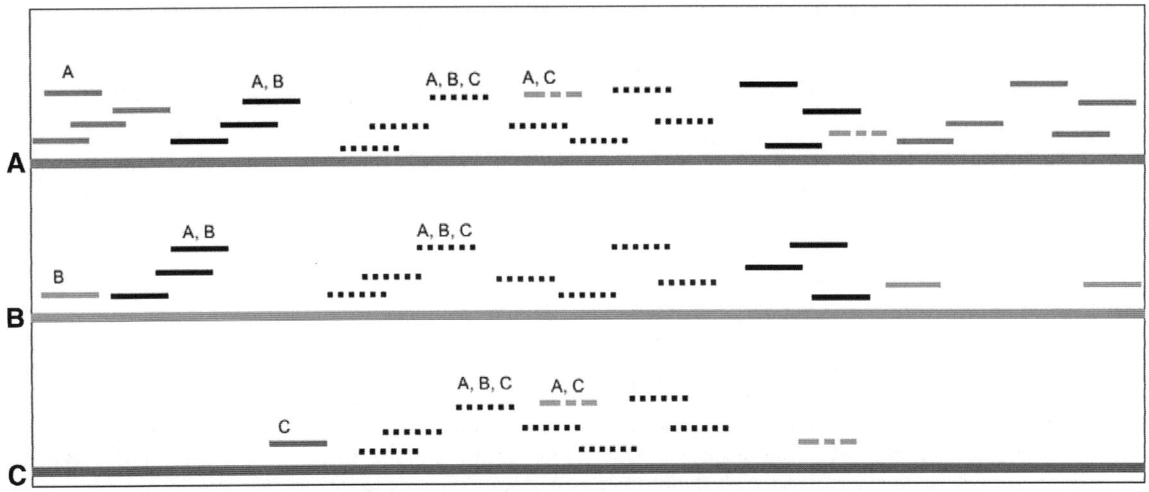

Fig. 1 A set of organisms {**a**, **b**, **c**} whose genomes are represented with a red, green, and blue line respectively. Reads mapped to the organisms are shown with shorter lines. The line colors and styles for the reads indicate unique and shared reads, according to the given labeling

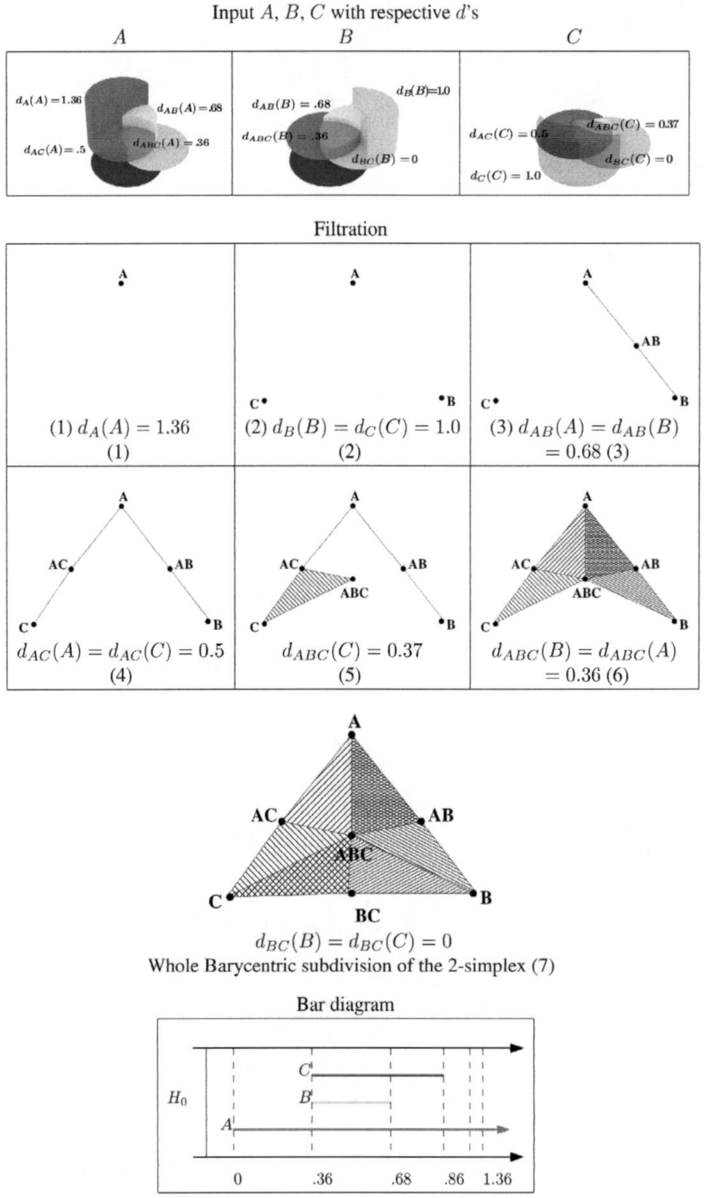

Fig. 2 Top: Values of depth *d* for the three organisms and read mapping shown in Fig. 1. Middle: Filtration of the Barycentric subdivision of the 2-simplex spanned by *A*, *B*, *C*. Bottom: The bar diagram in degree zero with organisms associated to bars

as above, then each inclusion $i_{s,t} : K_s \to K_t, s \leq t$ induces a *linear map*, $i_{s,t,*} : H_j(K_s, Q) \to H_j(K_t, Q)$. The images of these linear maps are usually called the *persistent homology groups* of the filtration, and their dimensions (i.e. the ranks of the maps $i_{s,t,*}$) determine the so called "bar diagram" associated to the filtration. Intuitively, the dimensions of the vector spaces $H_j(K, Q)$ (also called the *j*-th Betti number of the simplicial complex *K*) measure the number of independent *j*-dimensional cycles which are not boundaries of any $(j + 1)$-dimensional subcomplex of *K* (so called *j*-dimensional holes), and the *bar diagram* of the filtration is a record of the "times" of the births and deaths of these "homology classes", where we think of the sequence $(K_t)_{t \in [0,N]}$ as a complex growing with time *t*. Each bar in the bar diagram represents the interval in time in which a homology class *persisted.* We refer the reader to the book by Edelsbrunner and Harer [13] for further details about persistent homology and its use in Topological Data Analysis.

However, notice that for each $t \in \mathbb{R}, t \geq 0$, the set system $(Y^t)_{Y \in 2^{\mathcal{X}}}$ does not necessarily satisfy the condition of being closed under taking subsets – and hence, does not necessarily form a simplicial complex with $\mathcal{X}$ as the set of vertices.

Instead, we utilize the following construction (a simplicial subcomplex of a *barycentric subdivision*, detailed at the beginning of this section) that does produce a simplicial complex (in fact a filtration of complexes), and which is also naturally aligned with the various functions $d_Y(\cdot)$ defined earlier.

Intuitively, the simplices of the barycentric complex correspond to chains of subsets of $\mathcal{X}$. For example, if $X_1, X_2, X_3 \in \mathcal{X}$, then the sequence $\{X_1\} \subset \{X_1, X_2\} \subset \{X_1, X_2, X_3\}$ is a chain in the partially ordered set $2^{\mathcal{X}}$ (ordered by inclusion). Given $t \geq 0$, we include the chain $\{X_1\} \subset \{X_1, X_2\} \subset \{X_1, X_2, X_3\}$ in the barycentric complex iff $\{X_1\}^t \cap \{X_1, X_2\}^t \cap \{X_1, X_2, X_3\}^t \neq \emptyset$.

Barycentric subdivision and its subcomplex of interest

We now define more precisely the barycentric complex, and the subcomplex we will use.

Let $\Delta_{\mathcal{X}}$ denote the $(|\mathcal{X}| - 1)$-dimensional simplex with vertex set $\mathcal{X}$.

A sequence $\sigma = (Y_0, \ldots, Y_p)$, $Y_i \in 2^{\mathcal{X}}$ is called a *chain* of the poset $2^{\mathcal{X}}$ (ordered by inclusion) if $Y_0 \subsetneq Y_1 \subsetneq \cdots \subsetneq Y_p$. The first barycentric subdivision of $\Delta'_{\mathcal{X}}$ of $\Delta_{\mathcal{X}}$ (see Fig. 3) is then defined by

$$\Delta'_{\mathcal{X}} = \left\{ \mathbf{Y} = (Y_0, \ldots, Y_p) \mid \mathbf{Y} \text{is a chain in } 2^{\mathcal{X}} \right\}. \quad (1)$$

Now let $\mathcal{X}$ be a finite set, and let $\mathbf{d} = (d_Y : Y \to \mathbb{R})_{Y \in 2^{\mathcal{X}}}$ be a tuple of maps. For each $t \in \mathbb{R}$, we denote by $\Delta'_{\mathcal{X},\mathbf{d}}(t)$ the subcomplex of $\Delta'_{\mathcal{X}}$ defined as follows.

For $t \in \mathbb{R}$, and $\mathbf{Y} = (Y_0, \ldots, Y_p) \in \Delta'_{\mathcal{X}}$, let

$$\mathbf{Y}_\mathbf{d}(t) = \bigcap_{0 \leq i \leq p} Y_i^t = \bigcap_{0 \leq i \leq p} \{X \in Y_i \mid d_{Y_i}(X) \geq t\}, \quad (2)$$

and set

$$\Delta'_{\mathcal{X},\mathbf{d}}(t) = \left\{ \mathbf{Y} = (Y_0, \ldots, Y_p) \in \Delta'_{\mathcal{X}} \mid \mathbf{Y}_\mathbf{d}(t) \neq \emptyset \right\}. \quad (3)$$

We continue with Example 1 of three organisms A, B, C. In this case, the vertices of $\Delta'_{\mathcal{X}}$ are $\{A, B, C, AB, AC, BC, ABC\}$, where we write A for $\{A\}$, AB for $\{A, B\}$ and so on. The top part of Fig. 2 implies that we have only 7 different values for our functions d_Y. So we only need to specify the following collection of such $\Delta'_{\mathcal{X},\mathbf{d}}(t)$ (taking into account Remark 2 below):

$$\Delta'_{\mathcal{X},\mathbf{d}}(0) = \{[A]\}$$

$$\Delta'_{\mathcal{X},\mathbf{d}}(.36) = \Delta'_{\mathcal{X},\mathbf{d}}(0) \cup \{[B], [C]\}$$

$$\Delta'_{\mathcal{X},\mathbf{d}}(.68) = \Delta'_{\mathcal{X},\mathbf{d}}(.36) \cup \{[AB], [A, AB], [B, AB]\}$$

$$\Delta'_{\mathcal{X},\mathbf{d}}(.86) = \Delta'_{\mathcal{X},\mathbf{d}}(.68) \cup \{[AC], [A, AC], [C, AC]\}$$

$$\Delta'_{\mathcal{X},\mathbf{d}}(.99) = \Delta'_{\mathcal{X},\mathbf{d}}(.86) \cup \{[ABC], [C, ABC], [AC, ABC], [C, AC, ABC]\}$$

$$\Delta'_{\mathcal{X},\mathbf{d}}(1) = \Delta'_{\mathcal{X},\mathbf{d}}(.99) \cup \{[A, ABC], [B, ABC], [AB, ABC], [A, AB, ABC], [A, AC, ABC], [B, AB, ABC]\}$$

$$\Delta'_{\mathcal{X},\mathbf{d}}(1.36) = \Delta'_{\mathcal{X}}.$$

See Fig. 2 (middle) for a graphic representation of this filtered simplicial complex.

Fact 1 *$\Delta'_{\mathcal{X},\mathbf{d}}(t)$ is a simplicial complex with vertices in $2^{\mathcal{X}}$. Moreover, for any sequence $t_0 > t_1 > \ldots > t_N$,*

$$\Delta'_{\mathcal{X},\mathbf{d}}(t_0) \subset \Delta'_{\mathcal{X},\mathbf{d}}(t_1) \subset \cdots \subset \Delta'_{\mathcal{X},\mathbf{d}}(t_N).$$

Proof If $\mathbf{Y} = (Y_0, \ldots, Y_p) \in \Delta'_{\mathcal{X},\mathbf{d}}(t)$ is a chain, and $\mathbf{Z} = (Y_{i_0}, \ldots, Y_{i_q}), 0 \leq i_0 < \cdots < i_q \leq p$, is a subchain, then it is straightforward to check that

$$\mathbf{Y}_\mathbf{d}(t) \subset \mathbf{Z}_\mathbf{d}(t).$$

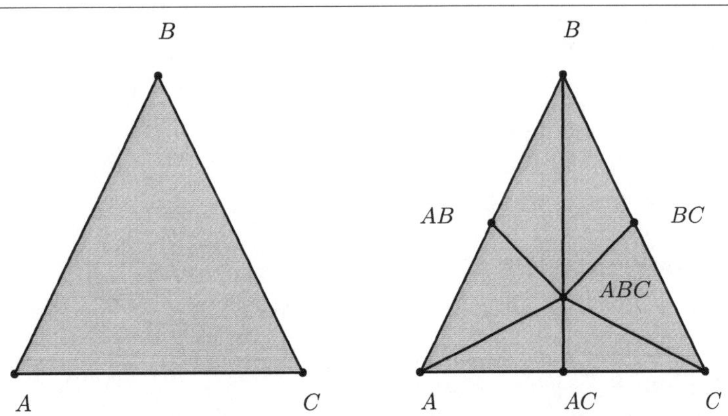

Fig. 3 Left: The 2-simplex $\Delta = 2^{\{A,B,C\}}$. Right: Its barycentric subdivision Δ'

Since $\mathbf{Y_d}(t)$ is not empty, $\mathbf{Z_d}(t)$ is not empty as well, so $\mathbf{Z} \in \Delta'_{\mathcal{X},\mathbf{d}}(t)$.

Now assume that $t_i > t_j$ for some $i, j \in \{0, \ldots, N\}$, and let $\mathbf{Y} \in \Delta'_{\mathcal{X},\mathbf{d}}(t_i)$. We have that if $\mathbf{Y_d}(t_i) \neq \emptyset$ then there exists some X such that for all Y_i in $\mathbf{Y}$ we have $d_{Y_i}(X) \geq t_i > t_j$, therefore $Y \in \Delta'_{\mathcal{X},\mathbf{d}}(t_j)$. □

Filtration

Armed with Fact 1 above, we have the following definition.

Definition 6 *Define the filtered simplicial complex*

$$\Delta'_{\mathcal{X},\mathbf{d}} = \bigcup_{t \in \mathbb{R}} \Delta'_{\mathcal{X},\mathbf{d}}(t). \tag{4}$$

This definition means that each of the simplicial complexes $\Delta'_{\mathcal{X},\mathbf{d}}(t)$ can be thought as a particular point in time of the filtration $\Delta'_{\mathcal{X},\mathbf{d}}$.

While the constructions so far deal with theoretical considerations, for actual computations we use the following remarks, which were in fact used to compute the filtration values for Example 1.

Remark 1 *Let* $\sigma = (Y_0, \ldots, Y_p) \in \Delta'_{\mathcal{X}}$, *and let*

$$\tau = \bigcap_{Y \in \sigma} Y.$$

Define t_0 *by*

$$t_0 = \max_{X \in \tau} \min_{Y \in \sigma} \{d_Y(X)\}. \tag{5}$$

Then t_0 *is the time of addition of* σ *to* $\Delta'_{\mathcal{X},\mathbf{d}}$.

Remark 2 *To make* $\Delta'_{\mathcal{X},\mathbf{d}}$ *covariant with respect to t, we use the change of variable* $t' = m - t$ *where m denotes the maximum value attained by the functions* d_Y.

From now on, unless stated otherwise, we will use the aforementioned change of variable.

Voting Scheme. Next we develop a voting scheme for all organisms that aims to solve the problem of ordering the organisms from true to false positives, using tools from algebraic topology. We refer the reader to Subsection 2.6 and Section 3 of [14] for the definition of persistent homology of a filtered simplicial complex, and a characterization in terms of $\mathcal{P}$-intervals respectively. In particular, using Corollary 3.1 of said article we have the following definition.

Definition 7 *Let K be a filtered simplicial complex and* $i \geq 0$ *an integer. We define the i-th degree bar diagram of K as the collection of* $\mathcal{P}$*-intervals associated to the i-th degree persistent homology of K.*

We now consider the persistent homology of the filtered complex $\Delta'_{\mathcal{X},\mathbf{d}}$. For $0 \leq i \leq |\mathcal{X}| - 1$ let $\mathcal{B}_i$ be a set of generators for the i-th degree persistent homology of $\Delta'_{\mathcal{X},\mathbf{d}}$, with fixed representing cycles, and $\mathcal{B} = \bigcup \mathcal{B}_i$. We can put these generators in a bijection with bars of the bar diagram of $\Delta'_{\mathcal{X},\mathbf{d}}$.

To help motivate the definition of the votings for each organism, we first list some of the components of the final formula. At each degree i, we assign to each organism X a collection of generators in $\mathcal{B}_i$ such that these generators witness X as a contributor to the features of our complex. For instance, for each X, consider all B such that X appears in its representing cycle (remember that such cycle is a linear combination of simplices, containing organisms as vertices). This gives rise to functions $F_i : \mathcal{X} \to 2^{\mathcal{B}_i}$.

For each generator $B \in \mathcal{B}$, let start(B) and end(B) denote its beginning and end as a bar. Consider now a strictly decreasing function $f : \mathbb{R}_{\geq 0} \to \mathbb{R}_{\geq 0}$ with $f(x) \to 0$ as $x \to \infty$; this function will modulate the contributions of each bar to the voting value of a given organism along the filtration value, while a decreasing function $g : \mathbb{R}_{\geq 0} \to \mathbb{R}_{\geq 0}$ will modulate the contributions along homology degree.

With all this in place, the vote $\nu(X)$ for organism X is computed as follows:

$$\nu(X) = \sum_i g(i) \left(\sum_{B \in F_i(X)} \big(f(\text{start}(B)) - f(\text{end}(B))\big) \right) \tag{6}$$

Examples of f, g include $f(t) = \frac{1}{t+1}$ and $g(i) = \frac{1}{i+1}$, which is what we used to obtain the results in this paper.

Continuing our example from Figs. 1and 2 (bottom) shows the resulting bar diagram with associated generators after application of the mentioned procedure on the depth values. In this case there is only non-trivial homology in 0-th degree. After applying the voting scheme described above on the bar diagram shown in Fig. 2, the votes are $\nu(A) = 1.00$, $\nu(B) = 0.14$, $\nu(C) = 0.20$.

Model II: Čech complex

Recall that $Z = \{X_1, X_2, ..., X_N\}$ is the set of all possible organisms, resulting from an ROM pipeline. Let $\left(Y = S_{X_1X_2..X_k}\right) \in 2^Z$ be the set of reads associated with organisms X_1, X_2, ..., and X_k. Note that for all $i, j, k, ...,$

$$|S_{X_i}| \geq |S_{X_iX_j}| \geq |S_{X_iX_jX_k}| \geq \ldots \tag{7}$$

The t (time) order filtration is $t_{\max}$ down to $t_{\min}$ in say steps of 1. A k-simplex on $X_1, X_2, ..., X_k, X_{k+1}$ at time t is introduced if $|S_{X_1X_2...X_kX_{k+1}}| \geq t$. Note that if the k-simplex on $X_1, X_2, ..., X_k, X_{k+1}$ belongs to the complex, then so does each $(k-1)$-simplex on $X_1, X_2, ..., X_k, X_{k+1}$ since, based on Eq. 7, for $1 \leq i \leq k$,

$$|S_{Y_1Y_2Y_3...Y_k}| \geq |S_{X_1X_2X_3...X_kX_{k+1}}|, \text{ where } Y_i \in \{X_1, X_2, X_3, ..., X_k, X_{k+1}\}.$$

Each node or organism X gets a true-positive score $v(X)$ as follows: In the bar diagram, let b be the h-simplex $= X_0X_1...X_h$ with bar length denoted as len(b). In this model, the vote $v(X)$ for organism X is computed as follows:

$$v(X) = \sum_h \left(\sum_{b \in H_h, X \in b} h \times \text{len}(b) \right) \tag{8}$$

A simple example with four organisms is shown in Fig. 4.

Results

We applied the model on simulated shotgun sequencing reads from a collection of 36 recently published *Salmonella* genomes [15], in an effort to study the applicability of the approach to strain-level detection. We simulated 150 bp paired end reads, 100,000 per each input genome, with dwgsim [16] (with parameters $y0, e0.005, E0.005, d500, s0, r0.001, R0.15, X0.3$) from the 36 genomes. The reads were mapped to a database consisting of the same set of genomes using bowtie2 [17] (very-sensitive-local mode, searching for up to 101 hits per read). Read simulation and mapping was performed with the Metagenomics Computation and Analytics Workbench (MCAW) [18].

The mapped reads were processed with custom scripts to prepare the Barycentric depth values and Čech set sizes for each set of organisms. Concordance of paired reads was checked (both reads of a pair mapping to the same organism), a random representative selected if the organism had several hits from the same read, and only the best

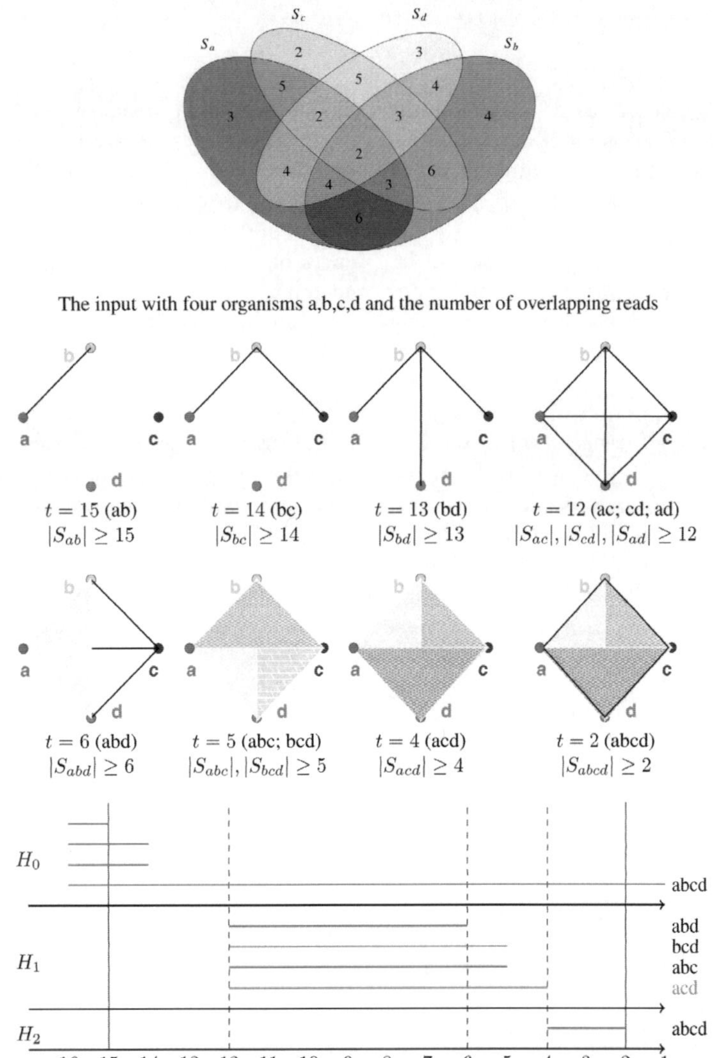

Fig. 4 Four organisms and the filtrations on the associated Čech complex with $t = 15$ down to $t = 2$

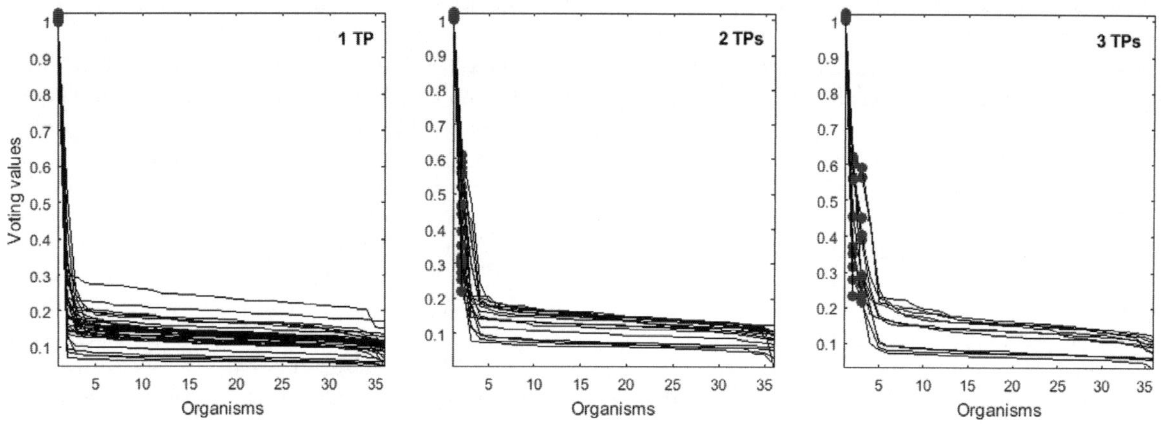

Fig. 5 Voting values (vertical axis) according to the Barycentric TDA method. The organisms (horizontal axis) are ordered according to their voting values from Eq. 6. Each line corresponds to a case where simulated reads from 1, 2, or 3 true positive (TP) input organisms were mixed together. The blue filled circles represent the voting values for the TPs

quality hits per read pair were used (based on sum of edit distances of reads in a pair). In this proof of concept we focused on the set of reads shared among 1 (unique reads), 2, or 3 organisms; the simulated data had 1–3 truly present organisms.

Indicative of the shared genome content between the closely related sequences, between only 2,087 to 92,176 of the 100,000 reads simulated per genome uniquely mapped to that genome after the process described above. Looking only at unique read counts per genome would certainly yield erroneous order of certain strains. For example, we observed a simulated mixture of 3 strains where two false positives (FPs) had more unique reads (2,596 and 2,427) than one of the true positives (TPs) (2,105). In this case, relying only on the number of unique reads would miss the truly present strain and falsely indicate the presence of missing ones. The observed unique reads for false positive organisms can arise from sequencing read errors, effects of the bioinformatics pipeline, and from subtle differences between the reference genomes and the observed genomes.

For analysis, the 36 *Salmonella* genomes were split into 18 non-overlapping sets of 2 strains each, and into 12 sets of 3 strains each. The TDA methodology, including the voting schemes described in Eqs. 6 and 8, was applied to the simulated reads from each set. The resulting voting values v for each organism were ordered from large (indicating truly present organisms) to small (indicating potential false positives).

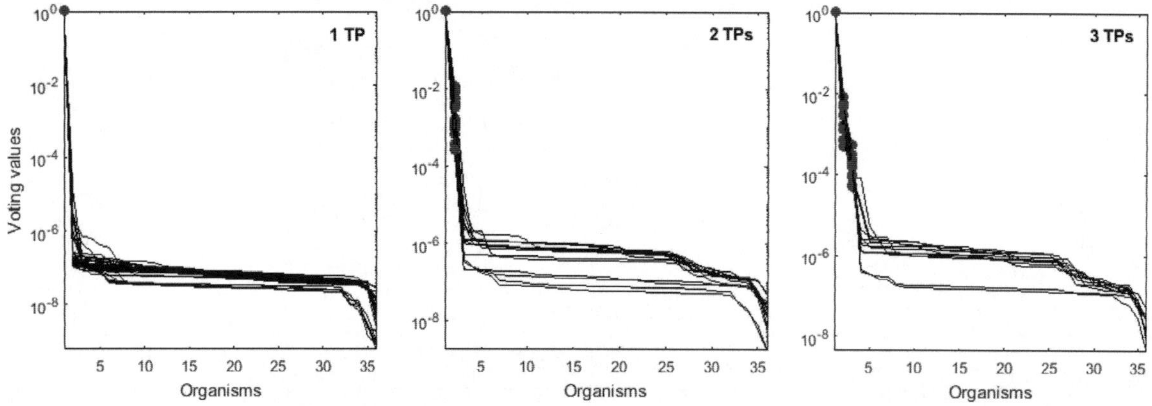

Fig. 6 Voting values (vertical axis) according to the Čech TDA method. The organisms (horizontal axis) are ordered according to their voting values from Eq. 8. Each line corresponds to a case where simulated reads from 1, 2, or 3 true positive (TP) input organisms were mixed together. The blue filled circles represent the voting values for the TPs

The voting values for simulated mixtures of 1, 2, or 3 *Salmonella* strains are visualized in Fig. 5 for the Barycentric approach and in Fig. 6 for the Čech approach. Ordering the organisms by voting values perfectly delineates the TPs from FPs with both methods. As a comparison, ordering the organisms by read coverage (computed by bedtools [19]) of the same reads as in the TDA input, the TPs could also be separated from FPs. However, as discussed earlier, using only the uniquely mapping reads would lead to false positives. We demonstrated that the TDA approach solves the problem of enriching the top of the list of voting values for truly present organisms.

Discussion and conclusions

In this proof of concept study we apply topological data analysis to the problem of separating signal from noise in the analysis of frequently multi-mapping metagenomic sequencing reads. Our approach is based on the construction of a particular subcomplex of a Barycentric subdivision complex, to rank-order the potential organisms and tease out the truly present ones.

The results from applying the approach on simulated genome mixtures show not just separation of signal from noise but also the potential for identifying microbes from metagenome samples, at strain level. We demonstrate the power of the TDA approach even in cases where alternative algorithms that exploit uniquely mapping reads fail. In fact, for the simple test cases the TPs bubble to the top from amongst a reference collection of highly related organisms, indicating promise of success for complicated real-life scenarios. The Čech model that uses less information is equally effective, suggesting that even partial information when augmented with the appropriate structure is quite powerful.

Additionally, the voting value curves show patterns of sharp decrease after the last true positive, suggesting automated calculation of cut-off thresholds. The same methodology could also be used to study higher taxonomic levels, e.g., separating true from false positive genera, families, etc. by modeling the read mappings across taxa.

Our next steps are to scale the implementation, and to apply it in the food safety as well as in the human health contexts. In both applications, a precise strain level assignment is of paramount importance.

Abbreviations

FP: False positive; ROM: Reads-to-organism mapping; TDA: Topological data analysis; TP: True positive

Acknowledgements

Not applicable.

Funding

Saugata Basu was partially supported by NSF Grant DMS-1620271. The publication charges were paid by the corresponding author's host institute.

Availability of data and materials

Not applicable.

About this supplement

This article has been published as part of *BMC Genomics Volume 20 Supplement 2, 2019: Selected articles from the 17th Asia Pacific Bioinformatics Conference (APBC 2019): genomics*. The full contents of the supplement are available online at https://bmcgenomics.biomedcentral.com/articles/supplements/volume-20-supplement-2.

Authors' contributions

LP conceived the model and designed the study. SB and AGS contributed towards the mathematical methodology. AGS designed and conducted the topological data analysis experiments. NH contributed to the methodology, and designed and conducted the metagenomic analysis. All authors contributed to writing the paper. All authors read and approved the final manuscript.

Ethics approval and consent to participate

Not applicable.

Consent for publication

Not applicable.

Competing interests

The authors declare that they have no competing interests.

Author details

[1]Computational Biology Center, IBM T. J. Watson Research Center, Yorktown Heights, NY, USA. [2]Department of Mathematics, Purdue University, West Lafayette, IN, USA.

Published: 10 April 2019

References

1. Kovac J, den Bakker H, Carroll LM, Wiedmann M. Precision food safety: A systems approach to food safety facilitated by genomics tools. Trends Anal Chem. 2017;96:52–61.
2. Hill-Burns EM, Debelius JW, Morton JT, et al. Parkinson's disease and Parkinson's disease medications have distinct signatures of the gut microbiome. Mov Disord. 2017;32:5.
3. Finkel OM, Castrillo G, Paredes SH, Gonzalez IS, Dangl JL. Understanding and exploiting plant beneficial microbes. Curr Opin Plant Biol. 2017;38: 155–63.
4. Forbes JD, Knox NC, Ronholm J, Pagotto F, Reimer A. Metagenomics: The Next Culture-Independent Game Changer. Front Microbiol. 2017;8:1069.
5. Costea PI, Zeller G, Sunagawa S, et al. Towards standards for human fecal sample processing in metagenomic studies. Nat Biotechnol. 2017;35:1069.
6. McIntyre ABR, Ounit R, Afshinnekoo E, et al. Comprehensive benchmarking and ensemble approaches for metagenomic classifiers. Genome Biol. 2017;18:182.
7. Wood DE, Salzberg SL. Kraken: ultrafast metagenomic sequence classification using exact alignments. Genome Biol. 2014;15:R46.
8. Ounit R, Wanamaker S, Close TJ, Lonardi S. CLARK: fast and accurate classification of metagenomic and genomic sequences using discriminative k-mers. BMC Genomics. 2015;16:236.
9. Segata N, Waldron L, Ballarini A, Narasimhan V, Jousson O, Huttenhower C. Metagenomic microbial community profiling using unique clade-specific marker genes. Nat Methods. 2012;9:8.
10. In: Przytyc ka T, editor. Research in Computational Molecular Biology. RECOMB 2015. Lecture Notes in Computer Science, vol 9029. Cham: Springer.
11. In: Natarajan R, Barua G, Patra MR, editors. Distributed Computing and Internet Technology. ICDCIT 2015. Lecture Notes in Computer Science, vol 8956. Cham: Springer.
12. Camara PG. Topological methods for genomics: Present and future directions. Curr Opin Syst Biol. 2017;1:95–101.

13. Edelsbrunner H, Harer JL. Computational Topology: An Introduction. Providence: American Mathematical Society; 2010.
14. Zomorodian A, Carlsson G. Computing Persistent Homology. Discrete Comput Geom. 2005;33:2.
15. Robertson J, Yoshida C, Gurnik S, Nash JHE. Completed Genome Sequences of Strains from 36 Serotypes of Salmonella. Genome Announc. 2018;6:3.
16. Homer N. DWGSIM: Whole Genome Simulator for Next-Generation Sequencing. 2010. https://github.com/nh13/DWGSIM. Accessed 23 Oct 2018.
17. Langmead B, Trapnell C, Pop M, Salzberg SL. Ultrafast and memory-efficient alignment of short DNA sequences to the human genome. Genome Biol. 2009;10:R25.
18. Edlund SB, Beck KL, Haiminen N, Parida LP, Storey DB, Weimer BC, Kaufman JH, Chambliss D. Design of the MCAW compute service for food safety bioinformatics. IBM J Res Dev. 2016;5(/6):60.
19. Quinlan AR, Hall IM. BEDTools: a flexible suite of utilities for comparing genomic features. Bioinformatics. 2010;26:6.

Biswas and Lai *BMC Genomics* 2019, **20**(Suppl 2):195
https://doi.org/10.1186/s12864-019-5491-x

BMC Genomics

RESEARCH

Open Access

A distance-type measure approach to the analysis of copy number variation in DNA sequencing data

Bipasa Biswas[1] and Yinglei Lai[2*]

From The 17th Asia Pacific Bioinformatics Conference (APBC 2019)
Wuhan, China. 14-16 January 2019

Abstract

Background: The next generation sequencing technology allows us to obtain a large amount of short DNA sequence (DNA-seq) reads at a genome-wide level. DNA-seq data have been increasingly collected during the recent years. Count-type data analysis is a widely used approach for DNA-seq data. However, the related data pre-processing is based on the moving window method, in which a window size need to be defined in order to obtain count-type data. Furthermore, useful information can be reduced after data pre-processing for count-type data.

Results: In this study, we propose to analyze DNA-seq data based on the related distance-type measure. Distances are measured in base pairs (bps) between two adjacent alignments of short reads mapped to a reference genome. Our experimental data based simulation study confirms the advantages of distance-type measure approach in both detection power and detection accuracy. Furthermore, we propose artificial censoring for the distance data so that distances larger than a given value are considered potential outliers. Our purpose is to simplify the pre-processing of DNA-seq data. Statistically, we consider a mixture of right censored geometric distributions to model the distance data. Additionally, to reduce the GC-content bias, we extend the mixture model to a mixture of generalized linear models (GLMs). The estimation of model can be achieved by the Newton-Raphson algorithm as well as the Expectation-Maximization (E-M) algorithm. We have conducted simulations to evaluate the performance of our approach. Based on the rank based inverse normal transformation of distance data, we can obtain the related *z*-values for a follow-up analysis. For an illustration, an application to the DNA-seq data from a pair of normal and tumor cell lines is presented with a change-point analysis of *z*-values to detect DNA copy number alterations.

Conclusion: Our distance-type measure approach is novel. It does not require either a fixed or a sliding window procedure for generating count-type data. Its advantages have been demonstrated by our simulation studies and its practical usefulness has been illustrated by an experimental data application.

Keywords: Genome-wide sequencing, DNA, Copy number variation, Distance-type measure, Geometric distribution, Mixture model

* Correspondence: ylai@gwu.edu
[2]Department of Statistics and Biostatistics Center, The George Washington University, Rome Hall, 7th Floor, 801, 22nd Street NW, Washington D.C 20052, USA
Full list of author information is available at the end of the article

Background

The next generation sequencing technology has advanced significantly during the recent decade. It allows us to obtain a large amount of short DNA sequence (DNA-seq) reads at a genome-wide level [1]. DNA-seq data have been increasingly collected during the recent years [2, 3]. Count-type data analysis is a widely used approach for DNA-seq data [4]. However, for data pre-processing, a moving window with the related window size need to be defined in order to obtain count-type data [5]. During the data pre-processing, given a window of genomic region, the observations within the region are summarized. However, useful information from the original data can be reduced after data pre-processing for count-type data.

DNA-seq has been widely used for the detection of copy number variation/alteration [1, 4]. Both copy number variation (CNV) analysis and copy number alteration (CNA) analysis are based on the assumption that the observed number of short reads of a local genomic region is proportional to the underlying copy number of DNA. CNV/CAN can also be discovered by other biomedical techniques such as fluorescent in situ hybridization, comparative genomic hybridization, array comparative genomic hybridization, and by virtual karyotyping with SNP arrays [2, 6, 7]. The next generation sequencing technology allows us to obtain millions of short DNA reads in a relatively short amount of time [8–10]. Digital karyotyping is a simple and powerful method [6]. Many statistical methods for CNV/CAN analysis are based on the count-type DNA-seq data [2, 3, 5, 11]. It is also necessary to consider GC content bias in the DNA-seq data [12]. The GC content can be calculated based on the Guanine (G) and Cytosine (C) bases in a reference genome. GC content bias refers to the dependence between the sequencing data and the related GC content.

Due to the limitations of current sequencing techniques, the sequencing data cannot be obtained for certain genomic regions. Then, the related count-type measure is simply zero. For a DNA sequence containing such a genomic region, a large gap without any short reads is observed. Then, the related distance-type measure can be very large (considered outliers statistically, see below for the definition of distance-type measure) and an artificial censoring can be considered.

In this study, we consider distance-type measure. Distances are measured in base pairs (bps) between two adjacent alignments of short reads mapped to a reference genome. Our distance-type measure approach is novel. It does not require either a fixed or a sliding window procedure for generating count-type data. Furthermore, we propose artificial censoring for the distance data so that distances larger than a given value are considered potential outliers. Our purpose is to simplify the pre-processing of DNA-seq data.

Statistically, we consider a mixture of right censored geometric distributions to model the distance data. Additionally, to reduce the GC-content bias, we extend the mixture model to a mixture of generalized linear models (GLMs). Our approach can be considered as an alternative to the method proposed by Shen and Zhang [11]. The estimation of model can be achieved by on the Newton-Raphson algorithm as well as the Expectation-Maximization (E-M) algorithm. Then, based on the rank based inverse normal transformation of distance data, we can obtain the related z-values for a follow-up analysis.

In the following, we first demonstrate how to conduct experimental data based simulation study to confirm the advantages of distance-type measure approach in both detection power and detection accuracy. Then, we introduce our statistical models and the related estimation procedure. The model performance can be demonstrated by simulation studies. The practical usefulness of our approach can be illustrated by an application to the DNA-seq data from a pair of normal and tumor cell lines [1] as well as a follow-up change-point analysis of z-values for the detection of DNA copy number alterations.

Methods

We consider a novel distance-type measure approach to the analysis of DNA-seq data. Distances are measured in base pairs (bps) between two adjacent alignments of short reads mapped to a reference genome. An illustration is shown in Fig. 1. A clear advantage is that the moving window method is no longer required for data pre-processing.

We first introduce how to compare the distance-type vs. count-type data so that more advantages of distance-type data can be confirmed. Then, we introduce a finite mixture [13, 14] of generalized linear models (GLMs) for modeling the distance between two adjacent reads with the adjustment to the GC-content. Artificial censoring of distance-type data can be considered so that potential outliers can be accommodated in the model. The basic distribution component for our method is the right censored geometric distribution. In the following, we briefly describe our method and the related estimation procedure. Their details are given in the Additional file 1.

Comparing distance-type vs. count-type data

The widely used coverage depth (or count-type) based approach can be briefly described as follows. For a genomic region, it is divided into non-overlapping smaller regions (windows or bins) and the number of sequencing reads is counted for each bin. For example, if a genomic region is divided into 100 bins, then 100 count-type data can be obtained from these bins. The original sequencing read data are then simplified to count-type data by

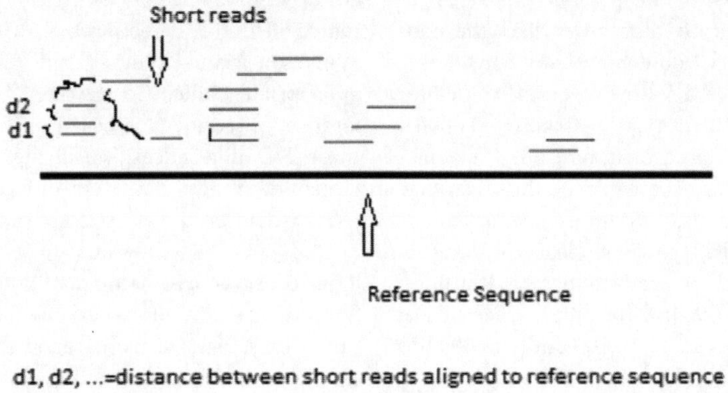

Fig. 1 Illustration of distance measure, which is measured in base pairs (bps) between the first aligned base pair of two adjacent alignments of short reads mapped to a reference sequence

this approach. (Our proposed distance based approach is to maintain the overall structure as well as the overall size of original sequencing read data.)

In this study, the DNA sequencing data were collected for the detection of copy number variations. A certain type of data (distance or count) is generated from the original sequencing read data (by the distance based approach or the coverage depth based approach, respectively). Between these two types of data, our choice is decided based on the statistical power for detecting changes. To demonstrate the advantage of our approach, we perform a simulation study to evaluate whether the distance based approach can be more statistically powerful than the coverage depth based approach. We consider a relatively simple scenario: if two genomic regions are known to be different in copy numbers, then which approach is more likely to detect the change? Furthermore, can accurate detection of change location be achieved when the distance-type data are used in the copy number variation/alteration analysis?

To address which type of data is more likely to detect the change (detection power), we consider the following simulation. Our simulation data were generated based on some real experimental sequencing data (instead of simulating data completely from some statistical distributions). We generated data based on the region with position 10-30Mbps from the normal sample (as presented in the section *Application to Sequencing Data*, but we used the original sequencing read data). In this region, we randomly selected two bins with length L for each bin. For each of the original sequencing reads from the second bin, we removed it randomly with 50% probability. (The purpose was to simulate a deletion of 50% copy number.) We kept all the original sequencing reads from the first bin. In this way, we simulated sequencing data for two genomic regions with different copy numbers. The distance-type data can be obtained as described at the beginning of Methods section. To obtain count-type data, we divided each region into bins with length W for each bin (and then counted the number of sequencing reads in each bin). To achieve an unbiased comparison between the distance-type data and the count-type data, we used the non-parametric Komogorov-Smirnov two-sample test to compare the empirical distributions of simulated data for two genomic regions. Notice that either the distance based approach or the coverage depth based approach is to generate a certain type of data (distance-type or count-type, respectively). It is necessary to compare which type of data can be more statistically powerful in the detection of changes. Our evaluation strategy does not depend on data type, as it does not require any parametric or statistical distribution assumptions.

We considered different values for L: 50,000, 200,000 and 1,000,000 bps for different lengths of copy number variations, as well as different values for W: 5000, 10,000 and 25,000 bps for different bin sizes. (We considered the range of bin size 5-25 kbps to avoid zero counts or very few count data.) For each setting, we repeated our simulation and analysis for 5000 times.

To address whether accurate detection of change location can be achieved (detection accuracy), we obtained more analysis results from the above simulation setting. In each simulation repetition, we considered the data for two genomic regions were connected. We evaluated how accurate the boundary (change) between two regions could be detected. As the data for two regions were ordered, the change location could be between any two consecutive data (for both distance-type and count-type). Therefore, we screened all the possible locations and considered the location with the lowest p-value as the identified change location. In this way, from each simulation repetition, we obtained the identified change location and its related p-value. Although the lowest p-value was considered, some cutoff values would still be usually considered in practice. As all the possible change locations were screened, a cutoff value lower than the conventional cutoff value 0.05 would be usually considered. Therefore, we considered the following cutoff values for the lowest p-values: 0.01, 0.001, 0.0001, 0.00001, 0.000001 and 0.0000001. Based on the results for detection power comparison (also see Table 1), the advantage of distance-type data was clear for short L and the advantage of count-type data became clear for long L. Therefore, we considered several moderate L values: 150,000, 300,000 and 450,000 bps for different lengths of copy number variations. 5000, 10,000 and 25,000 bps were still considered for W (bin sizes). For each setting, we still repeated our simulation and analysis for 5000 times.

For each simulation repetition, if the returned p-value was lower than the cutoff value, then we calculated the difference between the identified and true change location; otherwise, we considered the difference as no-detection (ND). Then, we would evaluate whether the overall difference based on the distance-type data was lower than the overall difference based on the count-type data. (Notice that a lower difference means a more accurate detection.) Statistically, the traditional Wilcoxon-Mann-Whitney rank sum test would be an appropriate choice (for testing "distance-type based difference" < "count-type based difference"). In the test, for these ND differences, we changed them to be a value larger than any observed differences as no-detection (ND) would be the worst scenario.

Table 1 Simulation based performance comparison between the distance-type data and the count-type data (for the detection power; W is the bin size)

p-value cutoff	Distance-type data	Count-type data		
		W = 25,000 bps	W = 10,000 bps	W = 5000 bps
Region length L = 50,000 bps				
0.05	28.9%	0	12.8%	16.1%
0.01	12.8%	0	0.1%	7.1%
0.001	3.4%	0	0	2.1%
0.0001	0.8%	0	0	0.3%
0.00001	0.1%	0	0	0
0.000001	0	0	0	0
Region length L = 200,000 bps				
0.05	73.8%	62.4%	68.2%	64.5%
0.01	59.4%	42.8%	50.3%	52.9%
0.001	40.5%	20.2%	30.5%	34.5%
0.0001	26.6%	0	14.3%	24.1%
0.00001	15.3%	0	8.0%	11.3%
0.000001	8.7%	0	1.6%	5.7%
Region length L = 1,000,000 bps				
0.05	97.6%	97.0%	96.8%	96.6%
0.01	94.7%	94.0%	94.1%	94.0%
0.001	89.7%	88.6%	90.1%	90.6%
0.0001	85.3%	83.0%	84.9%	85.9%
0.00001	80.7%	72.1%	80.2%	80.7%
0.000001	75.8%	62.8%	72.7%	75.7%

Right censored geometric distribution

Suppose Y is a random variable (e.g. distance-type measure) with a right censored geometric distribution with p as the probability of success. Notice that a geometric distribution can be considered as a model for the number of failures until first success. Let T be a given constant for the censoring value (e.g. the value for artificial right censoring). The indicator function $\delta = 1$ if $y \geq T$ and $\delta = 0$ if $y < T$. The probability distribution function of right censored geometric distribution is given by:

$$f(y) = \left(p(1-p)^y\right)^\delta \left((1-p)^{(T+1)}\right)^{(1-\delta)} \tag{1}$$

Mixture of Right Censored Geometric Distribution.

Based on the above right censored geometric distribution, a mixture of right censored geometric distribution can be described. Let $Y_1, \ldots, Y_n$ denote a random sample of size n. Each mixture component is a right censored geometric distribution given by Eq. (1) where p_j is the success probability for the j-th component. The probability distribution function of a mixture of g components is given by:

$$f(y, \pi_1, \pi_2, \ldots, \pi_g) = \sum\nolimits_{j=1}^{g} \pi_j f_j(y) \tag{2}$$

where

$$f_j(y) = \left(p_j\left(1-p_j\right)^y\right)^\delta \left(\left(1-p_j\right)^{(T+1)}\right)^{(1-\delta)}$$

$\pi_1, \pi_2, \ldots, \pi_g$ are the component proportions subject to the constraint $\sum_{j=1}^{g} \pi_j = 1$ and all $\pi_j \in [0, 1]$, all $p_j \in [0, 1]$. The indicator function $\delta = 1$ if $y \geq T$ and $\delta = 0$ if $y < T$. All the sample data are (artificially) censored at the same value T.

Mixture of right censured geometric distribution based GLMs

Based on the above mixture of right censored geometric distribution, a mixture of right censored geometric distribution based GLMs can be described. Eq. (2) can be extended to a finite mixture of g-component GLMs. Let $y_1, \ldots, y_n$ be n independent observations of the response variable. For each y_i, there is a covariate x_i. $\pi_1, \ldots, \pi_g$ are the component proportions as defined above. The mixture is given by:

$$f(y;x) = \sum_{j=1}^{g} \pi_j f_j(y;x) \quad (3)$$

where $f_j(y;x) = (p_j(x)(1-p_j(x))^y)^\delta((1-p_j(x))^{(T+1)})^{(1-\delta)}$. Notice that each p_j is a function of x. We first define $\mu_j = (1-p_j)/p_j$ and then we use a link function $\eta_j = \log(\mu_j) = \beta_{j0} + \beta_1 x$ for $j = 1, \ldots, g$. Here, β_1 is common for all different components since it represents the GC-content effect. Furthermore, the component proportions $\pi_1, \ldots, \pi_g$ do not depend on the covariate since they are considered as global parameters (not local). Therefore, the vector Ψ of parameters is given by $\Psi = (\pi_1, \ldots, \pi_{g-1}, \beta_{10}, \ldots, \beta_{g0}, \beta_1)$. The intercepts $(\beta_{10}, \ldots, \beta_{g0})$ are different for different distribution components.

Estimation procedure for mixture GLM based on geometric distribution

The log-likelihood based on Eq. (3) is given by
$l(\psi) = \log L(\psi) = \sum_{i=1}^{n} \log[\sum_{j=1}^{g} \pi_j f_j(y_i;x_i)]$.

The EM algorithm [15] can be used to obtain the maximum likelihood estimate of Ψ. For each y_i, missing component information can be considered. Accordingly, the vector $\{z_{ij}\}$ is introduced: $z_{ij} = 1$ if y_i belongs to the j-th component of the mixture model; $z_{ij} = 0$ otherwise ($j = 1, \ldots, g$; $i = 1, \ldots, n$). Then, the log-likelihood of "complete data" is given by:

$$l_C(\Psi) = \log L_C(\Psi) = \sum_{i=1}^{n}\sum_{j=1}^{g} z_{ij}\left\{ \log\pi_i + \log f_j(y_i;x_i)\right\}.$$

For the E-step, we calculate:
$\tau_{ij} = \mathbf{E}(z_{ij}|y) = \pi_j f_j(y_i;x_i)/[\sum_{h=1}^{g} \pi_h f_h(y_i;x_i).]$

For the M-step, we first calculate:
$\pi_j = \sum_{i=1}^{n} \tau_{ij}/n$.

The calculation for $(\beta_{10}, \ldots, \beta_{g0}, \beta_1)$ requires a numerical optimization procedure, which is used to solve the following equation system.

$$\sum_{i=1}^{n} \tau_{ij} \frac{\partial}{\partial \beta_{j0}} \log f_j(y_i;x_i) = 0 \text{ for } j = 1, \ldots, g$$

and

$$\sum_{i=1}^{n}\sum_{j=1}^{g} \tau_{ij} \frac{\partial}{\partial \beta_1} \log f_j(y_i;x_i) = 0$$

Based on the Newton-Ralphson method, we have the following iterative equation for $\boldsymbol{\beta} = (\beta_{10}, \ldots, \beta_{g0}, \beta_1)$:

$$\beta_{(r)} = \left[X^T W X\right]^{-1} X^T W z.$$

The details for this iterative equation as well as the EM algorithm are given in the Additional file 1.

Number of components

The number of components in the mixture model is determined by a likelihood ratio test (LRT) based approach. Consider two integers $g_0 < g_1$. We test the null hypothesis $H_0: g = g_0$ against $H_1: g = g_1$. We first use the observed data to fit a mixture model with g_0 components and a mixture model with g_1 components. A LRT score can be calculated for the observed data. Then, we can simulate data based on the fitted model with g_0 components (parametric bootstrap). For each set of simulated data, we can calculate the related bootstrapped LRT score. After B rounds of parametric bootstrap repetitions, the p-value of observed LRT score can be approximately calculated based on B bootstrapped LRT scores. (If p-value< 0.05, the null hypothesis can be rejected.)

Results

Distance-type data vs. count-type data

For the detection power based comparison, our evaluation results are summarized in Table 1. As different p-value cutoff values may be used in different scenarios of copy number variation analysis, we considered 0.05, 0.01, 0.001, 0.0001, 0.00001, and 0.000001. Given a cutoff value, a better approach should show a higher proportion (of p-values less than the cutoff value). When the regions are as long as 200 kbps, the advantage of distance based approach (distance-type data) can be clearly observed. It is more difficult to detect changes as the regions become shorter. However, the advantage of distance based approach (distance-type data) becomes even clearer when the regions are as short as 50 kbps. When the regions are as long as 1Mbps, two approaches perform overall similarly. From Table 1, it is also clear that the performance of coverage depth based approach (count-type data) depends on the choice of bin size. This has already been discussed in the literature. As our focus in this study is on the choice between distance-type data or count-type data, we prefer not to discuss the choice of bin size for count-type data.

For the detection accuracy based comparison, our evaluation results are summarized in Table 2. The proportion of the lowest p-value less than a given cutoff value is presented for each type of data. The proportion based on the distance-type data is always higher. Then, the one-sided p-value for each pair of comparison test is presented in Table 2. (Notice that, the relationship between one-sided and two-sided p-values for Wilcoxon-Mann-Whitney rank sum test is actually simple.) When L is relatively short as 150,000 bps, the detection accuracy based on the distance-type data is always better than the detection of accuracy based on the count-type data. When L is increased to 300,000 bps, the detection accuracy based on the distance-type data is

Table 2 Simulation based performance comparison between the distance-type data and the count-type data (for the detection accuracy evaluation; W is the bin size)

p-value cutoff	Distance-type based proportion	Count-type based proportion, and test p-value for distance-type data vs. count-type data		
		W = 25,000 bps	W = 10,000 bps	W = 5000 bps
Region length L = 150,000 bps				
0.01	68.6%	38.2%, $< 10^{-6}$	55.0%, 6×10^{-6}	57.0%, $< 10^{-6}$
0.001	42.6%	< 0.1%, $< 10^{-6}$	28.1%, $< 10^{-6}$	32.9%, $< 10^{-6}$
0.0001	24.7%	< 0.1%, $< 10^{-6}$	9.7%, $< 10^{-6}$	18.5%, $< 10^{-6}$
0.00001	13.0%	< 0.1%, $< 10^{-6}$	3.1%, $< 10^{-6}$	8.9%, $< 10^{-6}$
0.000001	6.3%	< 0.1%, $< 10^{-6}$	0.5%, $< 10^{-6}$	4.5%, 2×10^{-5}
0.0000001	3.0%	< 0.1%, $< 10^{-6}$	< 0.1%, $< 10^{-6}$	1.3%, $< 10^{-6}$
Region length L = 300,000 bps				
0.01	86.4%	73.8%, > 0.99	78.1%, 0.929	78.8%, 7×10^{-3}
0.001	70.3%	46.6%, $< 10^{-6}$	58.8%, $< 10^{-6}$	62.1%, 2×10^{-6}
0.0001	53.4%	26.9%, $< 10^{-6}$	41.4%, $< 10^{-6}$	47.5%, 3×10^{-5}
0.00001	39.9%	< 0.1%, $< 10^{-6}$	27.3%, $< 10^{-6}$	33.6%, $< 10^{-6}$
0.000001	29.1%	< 0.1%, $< 10^{-6}$	19.1%, $< 10^{-6}$	24.2%, $< 10^{-6}$
0.0000001	20.7%	< 0.1%, $< 10^{-6}$	10.1%, $< 10^{-6}$	17.6%, 3×10^{-4}
Region length L = 450,000 bps				
0.01	92.0%	85.4%, > 0.99	87.7%, > 0.99	87.6%, 0.214
0.001	81.0%	69.3%, > 0.99	75.2%, 0.493	76.2%, 9×10^{-3}
0.0001	69.1%	49.2%, $< 10^{-6}$	60.9%, 1×10^{-3}	64.3%, 3×10^{-3}
0.00001	57.5%	35.3%, $< 10^{-6}$	49.1%, 1×10^{-5}	52.6%, 6×10^{-4}
0.000001	48.0%	14.7%, $< 10^{-6}$	38.0%, $< 10^{-6}$	43.4%, 5×10^{-4}
0.0000001	38.9%	5.5%, $< 10^{-6}$	28.6%, $< 10^{-6}$	36.1%, 0.041

still clearly better when the cutoff value is 0.001 or lower. When L is relatively long as 450,000 bps, the detection accuracy based on the count-type data is better, but the detection accuracy based on the distance-type data is still better when the cutoff value is 0.0001 or lower. (Notice that, low cutoff values are necessarily considered in many practical situations.)

Estimation and testing performance for mixture of right censored geometric distributions

We performed a comprehensive simulation study to evaluate the estimation and also the testing performance of our proposed mixture of right censored geometric distribution. To understand estimation when artificial right censoring is applied we simulated a mixture of 3-component geometric with the parameters $\pi_1 = 0.3$, $\pi_2 = 0.5$, $\pi_3 = 0.2$ and $p_1 = 0.002$, $p_2 = 0.02$, $p_3 = 0.2$ (Fig. 2 and Additional file 2: Figures S1-S5). Another simulation was based on a mixture of 3-component geometric with the parameters $\pi_1 = 0.008$, $\pi_2 = 0.754$, $\pi_3 = 0.238$ and $p_1 = 0.999$, $p_2 = 0.011$, $p_3 = 0.0006$ (Fig. 3 and Additional file 2: Figure S6-S10). For estimation performance evaluation, we simulated 1000 times with 100, 250, 500, 1000 and 5000 samples each with no censoring, censoring at 250, 500 and 1000. We evaluated the bias, variance and root mean square error and used boxplots to summarize the findings (most figures provided as supplementary materials).

The standard deviation and the root mean square error (RMSE, Figs. 2 and 3) decreases with sample size and the censoring value. The RMSE combines both bias and variance. It shows a steady decline as the sample size increases. It also decreases with as the censoring value for observations increases. However for the second simulation, for $p_1 = 0.999$ we observed that the median consistently underestimated the true parameter value. Therefore, we need to be cautious when the true parameter value is close to 1. The inter-quartile range becomes narrower with increasing sample size and the artificial censoring value of the observations as well. Overall, both simulation studies support the use of right censored geometric distribution for DNA-seq data.

Furthermore, we have evaluated the effect of censoring on the performance of hypothesis testing. A data set can be simulated as above and we can test the hypothesis of 3 components vs. 4 based on 500 parametric bootstrap p-values. These steps can be repeated 500 times for each case (sample size and censoring value) to

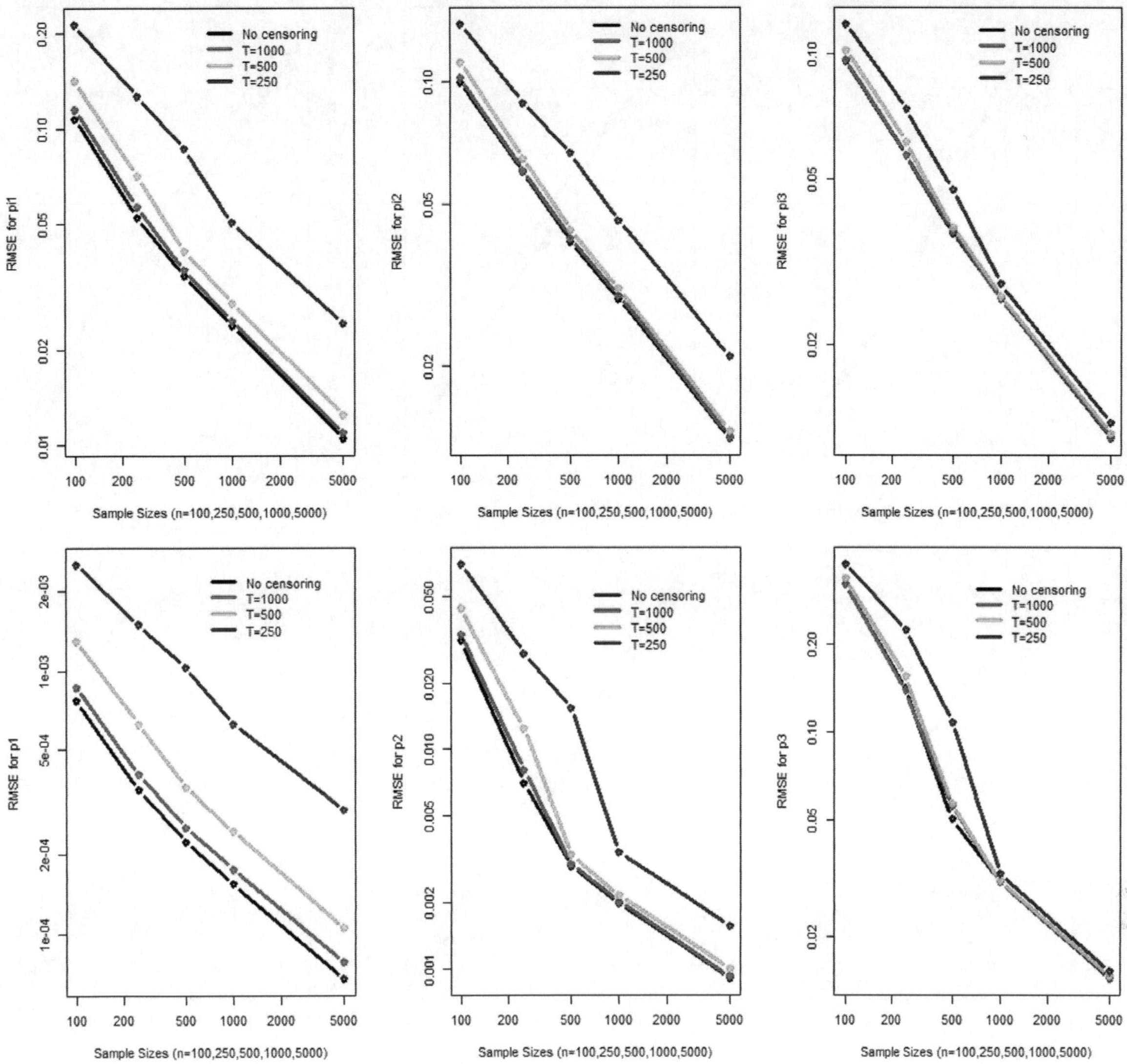

Fig. 2 RMSE of the parameter estimates ($\pi_1 = 0.3$, $\pi_2 = 0.5$, $\pi_3 = 0.2$ and $p_1 = 0.002$, $p_2 = 0.02$, $p_3 = 0.2$)

generate the empirical distribution of p-values. The empirical distribution plots for these p-values show uniform distribution like patterns (results not shown). Therefore, the hypothesis testing p-values can be appropriately calculated by the parametric bootstrap procedure.

Estimation and testing performance for mixture GLM where response variable has right censored geometric distributions

To understand estimation and testing performance when artificial right censoring is applied in the case of a mixture GLM, we simulated a mixture of 3-component GLM with the parameters $\pi_1 = 0.3$, $\pi_2 = 0.5$, $\pi_3 = 0.2$ and $\beta_{10} = 6.0$, $\beta_{20} = 3.6$, $\beta_{30} = 1, 1$, $\beta_1 = 2.0$. The choice of the component proportions are the same as above and the choice of β_{10}, β_{20}, β_{30} was based on setting $\beta_1 = 2.0$ and using $p_1 = 0.002$, $p_2 = 0.02$, $p_3 = 0.2$ as above (consistent with a previous simulation). For estimation performance evaluation, we simulated 1000 times with 10,000 samples each with no censoring (NC), censoring at 570, 937, and 1309 respectively (Additional file 2: Figure S11-S12). The censoring values were set such that approximately on average about 10, 5 and 2.5% of the data is censored (consistent with our previous simulations; also about 1–3% in our application). We generated the x-variable from the Uniform [0, 0.25] distribution. A second simulation was based on a mixture of 3-component geometric with the parameters $\pi_1 = 0.3$, $\pi_2 = 0.5$, $\pi_3 = 0.2$ and $\beta_{10} = 6.2$, $\beta_{20} = 3.9$, $\beta_{30} = 1.4$, $\beta_1 = 0.1$. The group proportions were set according to the same strategy as explained above (based on $\beta_1 = 0.1$). For

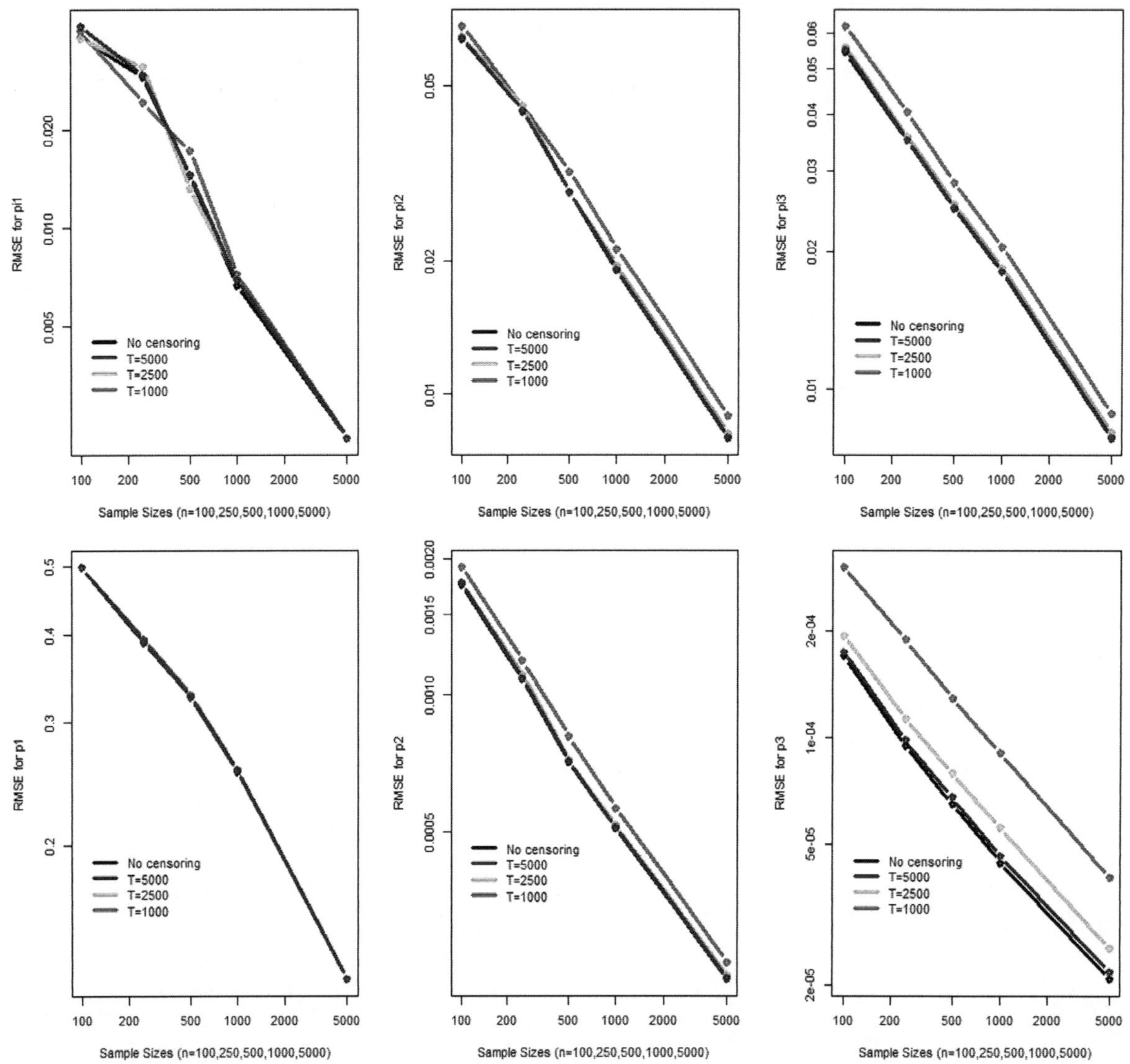

Fig. 3 RMSE of the parameter estimates ($\pi_1 = 0.008$, $\pi_2 = 0.754$, $\pi_3 = 0.238$ and $p_1 = 0.999$, $p_2 = 0.011$, $p_3 = 0.0006$)

estimation performance evaluation, we simulated 1000 times with 10,000 samples each with no censoring (NC), censoring at 1241, 895 and 549 (Additional file 2: Figure S13-S14). The censoring values were also set according to the same strategy as explained above. We evaluated the bias, variance and root mean square error (results not shown) and used boxplots to summarize the findings (most figures provided as supplementary materials).

The standard deviation and the root mean square error (RMSE) decreases with the increasing censoring value. The RMSE also decreases as the artificial censoring value for observations increases. The median gets closer to the true value with the increasing censoring value of the observations. The inter-quartile range gets narrower with increasing sample size and the censoring value of the observations as well. Overall, both simulation studies support the use of right censoring in mixture GLM.

Furthermore, we have evaluated the effect of censoring on the testing performance. A data set can be simulated as above and we can test the hypothesis of 3 components vs. 4 based on 100 parametric bootstrap p-values. These steps can be repeated 200 times for the first simulation scenario or 100 times for the second simulation scenario (also for each censoring value) to generate the empirical distribution of the *p*-values. (The number of parametric bootstraps was limited to 200 or 100 due to computational burden as this evaluation took much longer time in computing.) Additional file 2: Figure S15 (for the first simulation scenario) and Additional file 2: Figure S16 (for the second simulation scenario) show the empirical distribution plots for these p-values. Uniform distribution like patterns can be observed. Therefore, the hypothesis testing p-values can be appropriately calculated by the parametric bootstrap procedure.

Application to sequencing data

We considered an application to the DNA-seq data from Chiang et al. [1]. The data set was made publicly available and we selected a normal cell line and the related tumor cell line (HCC1954) for breast carcinoma. Furthermore, we focused on chromosome 9 for an illustration of our method. The value 5000 (bps) was used for the artificial right censoring (we observed at most 1–3% distance data greater than 5000 bps so most of the distance data were relatively short).

A 4-component mixture of right censored geometric distribution based generalized linear models was estimated for the normal sample and then for the tumor sample (Table 3). GC-content adjustment was considered. (We downloaded the 50Kbps-window based GC-content data from http://bioinfo-out.curie.fr/projects/freec .) Notice that, as the GC-content data were based on a reference genome, the model coefficient for GC-content was first estimated based on the normal sample data and then fixed in the model for tumor sample data (all the other model parameters were estimated separately for different sample data). This approach (pseudo-maximum likelihood estimation) was based on the work by Gong and Samaniego [16]. (About the number of mixture components, we could not perform the related hypothesis testing because the related computing could be not afforded. However, we conducted a preliminary data analysis based on the mixture of right censored geometric distributions without considering any covariates, in which we could performed hypothesis testing on the number of mixture components and four distinct components were confirmed. Therefore, we reasonably assumed 4-component for the mixture of GLMs.)

Table 3 Parameter estimates from mixture of GLM (censored at 5000 bps)

Parameter	Tumor	Normal
π_1	0.0065	0.0064
π_2	0.1802	0.0359
π_3	0.5414	0.5763
π_4	0.2719	0.3814
β_{10}	−1.106	−2.021
β_{20}	6.757	6.691
β_{30}	8.240	8.681
β_{40}	9.268	9.457
β_1	−4.428	−4.428

Inverse normal transformations (INTs) have been widely considered in practice. The purpose is to make the (transformed) data more similar to normal distributions. Both rank-based INTs and non-rank based INTs can be considered [17–26]. We considered the following INT. Based on our estimated model (for normal or tumor sample, see Table 3), the cumulative distribution function (CDF) for an observations y (not artificially right censored) was calculated by:

$$F(y;x) = \sum_{j=1}^{g} \pi_j F_j(y;x) \quad (4)$$

where F_j was the CDF based on f_j given in Eq. (3). Then, an inverse normal distribution function was applied. Through this INT, we obtain the distance-type measure related INT values.

The comparison between the distance data (log2 of distance+ 1) and the related INT values (inverse normal transformed values; not related to artificially right censored distance data) is shown in Fig. 4. Some uneven patterns can be visualized from the plots of distance data. These were not observed from the plots of z-values. Therefore, the GC-content adjustments have been well considered in our method.

Then, for the INT values related to these artificially right censored distance data, we simply assign a value 10 (larger than any INT values in Fig. 4). In Fig. 5, more details can be visualized for INT values. Notice that censored observations are highlighted by red vertical lines (there are 31 instances for the normal sample and 35 instances for the tumor sample). Some differences on the region around 200 K-38 M bps (chromosome 9) can be observed and we performed a change-point analysis for this region. (The normality was checked for INT values. We randomly selected regions with length 100 K bps for both the normal and tumor samples and generated the related quantile-quantile plots [Q-Q plots, Additional file 2: Figure S17-S18]. Then, we assumed that INT values were normally distributed, which was helpful to avoid the permutation based p-value calculation [27] in the next change-point analysis.)

For a change-point analysis of the above selected region (as an illustration), we considered a recursive combination approach [28]. For each round of recursion, at a given level α_C, any two adjacent blocks of INT values with two-sample test p-values larger than α_C were selected. Among them, the combination to generate the largest overall likelihood was applied. The recursion stopped when all the test p-values were less than α_C. Considering the large number of two-sample comparisons, we used 50 K bps bins to group INT values before the change-point analysis. We set $\alpha_C = 0.05/(m * (m - 1)/2)$, where m is the

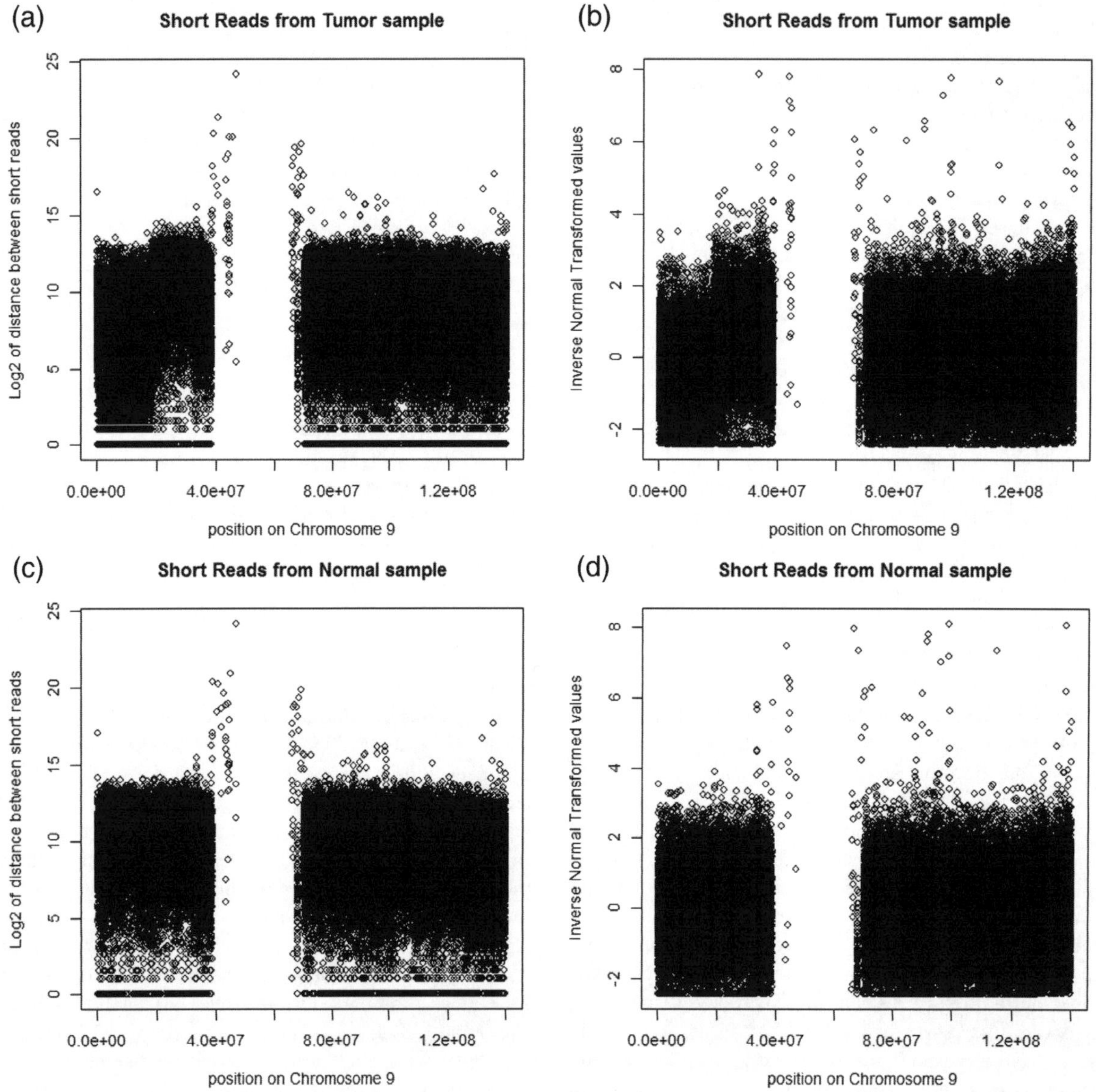

Fig. 4 Plots of $\log_2$ of the distance and INT of the cumulative distribution. (**a**) Plot of $\log_2$ of the distance between short reads for the tumor sample against the position on chromosome 9; (**b**) Plot of inverse normal transform of the cumulative distribution for the tumor sample versus the position on chromosome 9; (**c**) Plot of $\log_2$ of the distance between short reads for the normal sample against the position on chromosome 9; (**d**) Plot of inverse normal transform of the cumulative-distribution for the normal sample versus the position on chromosome 9

number of bins (767 for normal sample and 772 for tumor sample).

The change-point analysis results are shown in Fig. 6. For both normal and tumor samples, there is a change-point around 6-7 M bps. However, the magnitude of changes is small. This detection is unlikely to be biologically relevant. For the tumor sample, there is an additional change-point around 18-19 M bps, which implies a clear deletion (larger distance related to less DNA copy number). Cytobands have been widely considered in biomedical studies [29]. A search in UCSC data base (http://www.genome.ucsc.edu/cgi-bin/hgTracks) shows that the cytobands p13.2, p13.3, p21.1, p21.2, p21.3 and p22.1 are related to this detected region. To understand biological significance of this detected deletion, a literature search shows that this region has been investigated by many studies on the relationship between these cytobands and breast cancer [30, 31].

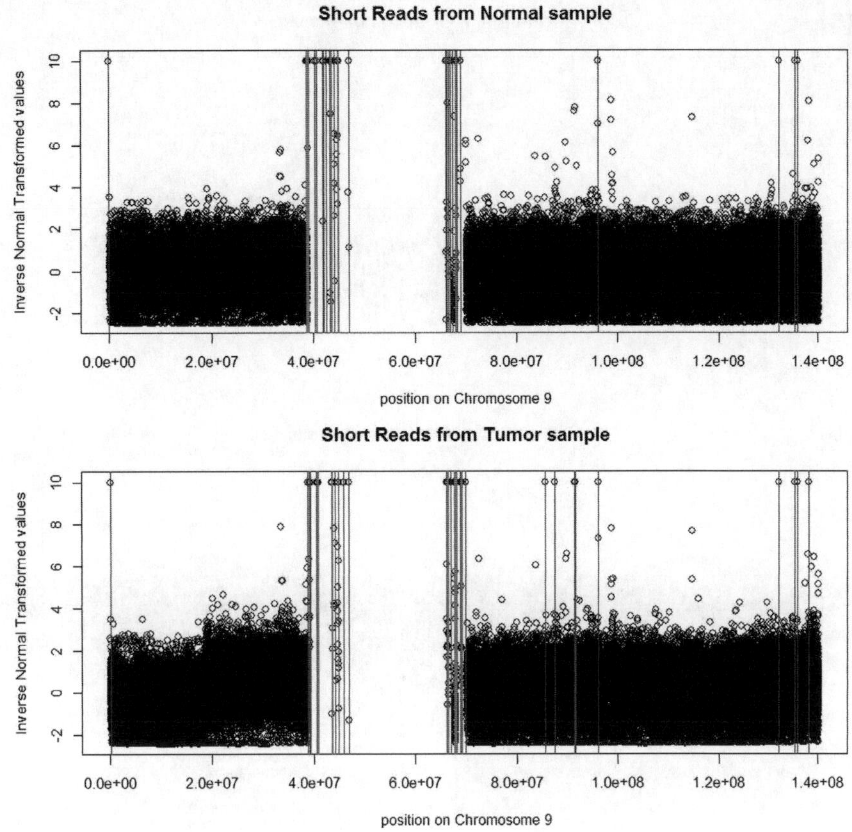

Fig. 5 Plots of INT of the cumulative distribution against the position on chromosome 9, represented as bps

Discussion

Count-type data analysis is a widely used approach for DNA-seq data. It requires either a fixed or a sliding window procedure for generating count-type data, and useful information can be reduced after this data pre-processing. In this study, we proposed to analyze DNA-seq data based on the related distance-type measure. Our experimental data based simulation study confirmed the advantages of distance-type measure approach in both detection power and detection accuracy. Furthermore, we proposed a mixture of generalized linear models for analyzing the recent DNA-seq data. Our model was based on the right censored geometric distribution. According to the mathematical details provided in the Additional file 1, our method is linear in term of the number of sequencing reads. Therefore, our method scales well. The validity and usefulness of this approach were demonstrated by the estimation and testing performances in our simulation study as well as an application to experimental data. The selection of the censoring value was based on that 1–3% of observed distance data were larger than 5000 bps. The optimization of the censoring value is an interesting research topic and we will investigate this in our future study. Our distance-based approach is novel with a clear advantage: it does not require either a fixed or a sliding window procedure for generating count-type data.

Similar as the exponential distribution (for continuous measurements), which is widely used to model waiting time processes, the geometric distribution (for discrete measurements) is also a memoryless probability distribution. The artificial right-censoring has been proposed to overcome the difficult in modeling extreme distance values. (For example, it is difficult to obtain sequencing reads near centromere regions.) Due to different biological properties in different genomic regions, which results in different distance distributions, different geometric components are necessary to model the data. As one geometric distribution does not well fit the distance data, a mixture of right-censored geometric distributions has been proposed. To understand the connection between a probability distribution component and its related underling biological properties, we need further statistical and biological investigations, which will be an interesting topic for our future research.

In our application, interesting change-points were observed from the tumor cell line data. The impact from

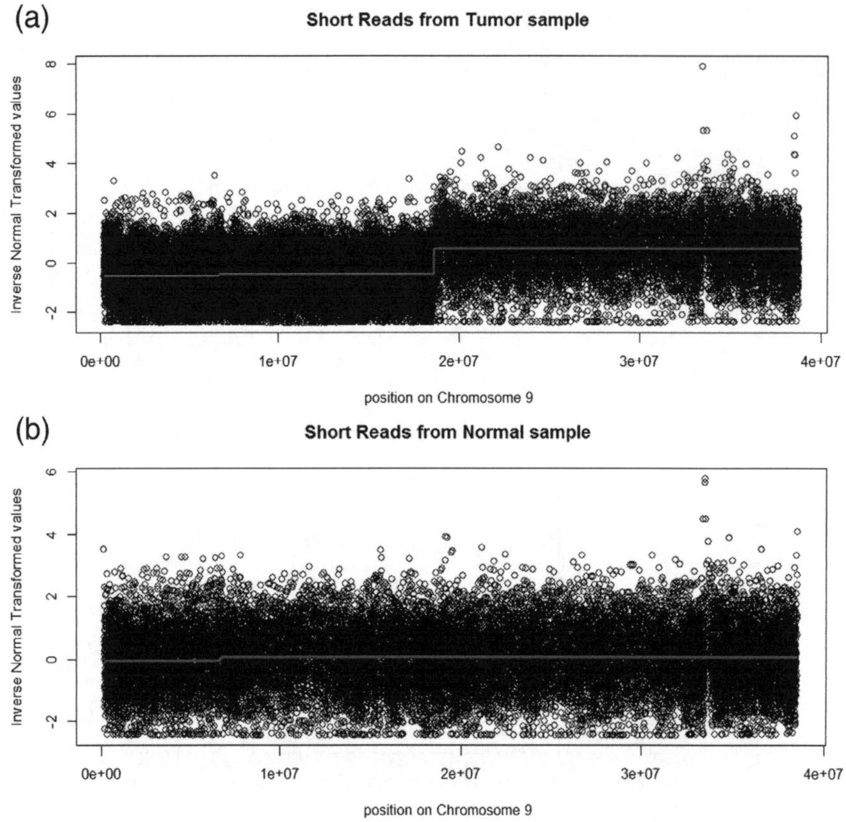

Fig. 6 INT of the cumulative distribution with the estimated mean represented by the red line. (**a**) Plot for the tumor sample where the red line is the estimated mean from the recursive combination algorithm; (**b**) plot for the normal sample where the red line is the estimated mean from the recursive combination algorithm

GC-content was considered in our analysis. In this study, we chose a set of relatively simple DNA-seq data for the purpose of illustration of our approach. Our approach can also be applied to paired-end DNA-seq data. In practice, either paired-end reads or single-end reads can be considered in sequencing experiments. If only the starting position of each paired-end read is considered, then the analysis scenario is equivalent to the analysis of single-end reads. However, this strategy may not well utilize the advantage of paired-end reads. (For example, longer regions can be covered by a paired-end read.) To consider the distance based approach and to utilize the advantage of paired-end reads, we will need to investigate which modifications or extensions are necessary to our method proposed in this study. This will be an interesting topic for our future research.

Conclusion

With the next generation sequencing technology, a large amount of short DNA sequence reads can be obtained at a genome-wide level. DNA-seq data have been increasingly used in different areas of biomedical studies. Although the count-type based approaches have been widely used for DNA-seq data analysis, a moving window based data pre-processing (or similar) is required with a pre-specified window size. Useful information can be reduced after the data pre-processing. In this study, we have developed a distance-type measure based method. Distances are measured in base pairs (bps) between two adjacent alignments of short reads mapped to a reference genome. This is a novel approach. Its advantages have been demonstrated by our simulation studies and its practical usefulness has been illustrated by an experimental data application.

Additional files

Additional file 1: Mathematical details. (PDF 271 kb)

Additional file 2: Supplemental figures. (PDF 1697 kb)

Abbreviations

CNA: copy number alteration; CNV: copy number variation; DNA-seq: short DNA sequence; E-M algorithm: Expectation-Maximization algorithm; GLM: generalized linear model; INT: Inverse normal transformation;

LRT: likelihood ratio test; Q-Q plot: quantile-quantile plot; RMSE: root mean square error

Acknowledgements
Not applicable.

Funding
This work was partially supported by the NIH grant GM-092963 (Y.Lai). The publication costs were funded by the Department of Statistics at The George Washington University. The U. S. Food and Drug Administration (FDA) and the sponsors (NIH and The George Washington University) had no role in the study design, data collection and analysis, decision to publish, or preparation of the manuscript.

Availability of data and materials
Not applicable.

About this supplement
This article has been published as part of BMC Genomics Volume 20 Supplement 2, 2019: Selected articles from the 17th Asia Pacific Bioinformatics Conference (APBC 2019): genomics. The full contents of the supplement are available online at https://bmcgenomics.biomedcentral.com/articles/supplements/volume-20-supplement-2.

Authors' contributions
Bipasa Biswas and Yinglei Lai conceived of the study, developed the methods, performed the statistical analysis, and drafted the manuscript. Both authors have read an approved the final manuscript.

Ethics approval and consent to participate
Not applicable.

Consent for publication
The authors agree the consent for publication.

Competing interests
The authors declare that they have no competing interests.

Publisher's Note

Author details
[1]Diagnostics Devices Branch 1, FDA/CDRH/OSB-DBS, White Oak Bldg #66, Room 2222, 10903 New Hampshire Avenue, Silver Spring, MD 20993, USA. [2]Department of Statistics and Biostatistics Center, The George Washington University, Rome Hall, 7th Floor, 801, 22nd Street NW, Washington D.C 20052, USA.

Published: 10 April 2019

References

1. Chiang DY, et al. High-resolution mapping of copy-number alterations with massively parallel sequencing. Nat Methods. 2009;6:99–103.
2. Miller CA, et al. ReadDepth: a parallel R package for detecting copy number alterations from short sequencing reads. PLoS One. 2011;6(1):e16327.
3. Xie R, et al. Detecting structural variations in the human genome using next-generation sequencing. Brief Funct Genomics. 2011;9:405–15.
4. Kim TM, et al. rSW-seq: algorithm for detection of copy number alterations in deep sequencing data. BMC Bioinformatics. 2010;11:432.
5. Krishnan NM, et al. COPS: a sensitive and accurate tool for detecting somatic copy number alterations using short-read sequence data from paired samples. PLoS One. 2012;7(10):e47812.
6. Korbel JO, et al. Paired-end mapping reveals extensive structural variation in the human genome. Science. 2007;318(5849):420–6.
7. Sudmant PH, et al. Diversity of human copy number variation and multicopy genes. Science. 2010;330(6004):641–6.
8. Wang TL, et al. Digital karyotyping. Proc Natl Acad Sci U S A. 2002;99:16156–61.
9. Leary RJ, et al. Digital karyotyping. Nat protocols. 2007;2:1973–86.
10. Morozova O, et al. Applications of next-generation sequencing technologies in functional genomics. Genomics. 2008;95:255–64.
11. Shen JJ, Zhang N. Change-point model on nonhomogeneous Poisson process with application in copy number profiling by next-generation DNA sequencing. Ann Appl Stat. 2012;6:476–96.
12. Benjamini Y, Speed TP. Summarizing and correcting the GC content bias in high-throughput sequencing. Nucleic Acids Res. 2012;40(10):e72.
13. Aitkin M, Rubin DB. Estimation and hypothesis testing in finite mixture models. Royal Statistical Society. 1985;47(1):67–75.
14. McLachlan G, Peel D. Finite Mixture Models. New York: Wiley; 2002.
15. Dempster AP, Laird NM, Rubin DB. Maximum likelihood from incomplete data via the EM algorithm. J R Stat Soc Ser B. 1977;39(1):1–38.
16. Gong G, Samaniego FJ. Pseudo maximum likelihood estimation: theory and applications. Ann Stat. 1981;9(4):861–9.
17. Ashton GC, Borecki IB. Further evidence for a gene influencing spatial ability. Behav Genet. 1987;17(3):243–56. https://doi.org/10.1007/BF01065504.
18. Silverman EK, et al. Biochemical intermediates in alpha 1-antitrypsin deficiency: residual family resemblance for total alpha 1-antitrypsin, oxidized alpha 1-antitrypsin, and immunoglobulin E after adjustment for the effect of the pi locus. Genet Epidemiol. 1990;7(2):137–49. https://doi.org/10.1002/gepi.1370070204.
19. Tzou GG, et al. Classification of beef calves as protein-deficient or thermally stressed by discriminant analysis of blood constituents. J Anim Sci. 1991;69:864–73.
20. Martin LJ, Crawford MH. Genetic and environmental components of thyroxine variation in Mennonites from Kansas and Nebraska. Hum Biol. 1998;70(4):745–60.
21. Basrak B, et al. Copulas in QTL mapping. Behav Genet. 2004;34(2):161–71. https://doi.org/10.1023/B:BEGE.0000013730.63991.ba.
22. Przybyla-Zawislak BD, et al. Identification of rat hippocampal mRNAs altered by the mitochondrial toxicant, 3-NPA. Ann N Y Acad Sci. 2005;1053:162–73. https://doi.org/10.1196/annals.1344.014.
23. Li M, et al. Quantitative trait linkage analysis using Gaussian copulas. Genetics. 2006;173(4):2317–27. https://doi.org/10.1534/genetics.105.054650.
24. Hicks BM, et al. Genes mediate the association between P3 amplitude and externalizing disorders. Psychophysiology. 2007;44(1):98–105. https://doi.org/10.1111/j.1469-8986.2006.00471.x.
25. Kraja AT, et al. Rheumatoid arthritis, item response theory, Blom transformation, and mixed models. BMC Proc. 2007;1(Suppl. 1):S116.
26. Knoll J, Ejeta G. Marker-assisted selection for early-season cold tolerance in sorghum: QTL validation across populations and environments. Theor Appl Genet. 2008;116(4):541–53. https://doi.org/10.1007/s00122-007-0689-8.
27. Lai Y. Change-point analysis of paired allele-specific copy number variation data. J Comput Biol. 2012;19(6):679–93.
28. Lai Y. On the adaptive partition approach to the detection of multiple change-points. PLoS One. 2011;6(5):e19754.
29. Furey TS, Haussler D. Integration of the cytogenetic map with the draft human genome sequence. Hum Mol Genet. 2003;12(9):1037–44.
30. Brenner AJ, Aldaz CM. Chromosome 9p allelic loss and p16/CDKN2 in breast cancer and evidence of p16 inactivation in immortal breast epithelial cells. Cancer Res. 1995;55(13):2892–5.
31. Nessling M, et al. Candidate genes in breast cancer revealed by microarray-based comparative genomic hybridization of archived tissue. Cancer Res. 2005;65(2):439–47.

Du *et al. BMC Genomics* 2019, **20**(Suppl 2):190
https://doi.org/10.1186/s12864-019-5475-x

BMC Genomics

RESEARCH

Open Access

Improving the sensitivity of long read overlap detection using grouped short *k*-mer matches

Nan Du[1], Jiao Chen[1] and Yanni Sun[2*]

From The 17th Asia Pacific Bioinformatics Conference (APBC 2019)
Wuhan, China. 14-16 January 2019

Abstract

Background: Single-molecule, real-time sequencing (SMRT) developed by Pacific BioSciences produces longer reads than second-generation sequencing technologies such as Illumina. The increased read length enables PacBio sequencing to close gaps in genome assembly, reveal structural variations, and characterize the intra-species variations. It also holds the promise to decipher the community structure in complex microbial communities because long reads help metagenomic assembly. One key step in genome assembly using long reads is to quickly identify reads forming overlaps. Because PacBio data has higher sequencing error rate and lower coverage than popular short read sequencing technologies (such as Illumina), efficient detection of true overlaps requires specially designed algorithms. In particular, there is still a need to improve the sensitivity of detecting small overlaps or overlaps with high error rates in both reads. Addressing this need will enable better assembly for metagenomic data produced by third-generation sequencing technologies.

Results: In this work, we designed and implemented an overlap detection program named GroupK, for third-generation sequencing reads based on grouped *k*-mer hits. While using *k*-mer hits for detecting reads' overlaps has been adopted by several existing programs, our method uses a group of short *k*-mer hits satisfying statistically derived distance constraints to increase the sensitivity of small overlap detection. Grouped *k*-mer hit was originally designed for homology search. We are the first to apply group hit for long read overlap detection. The experimental results of applying our pipeline to both simulated and real third-generation sequencing data showed that GroupK enables more sensitive overlap detection, especially for datasets of low sequencing coverage.

Conclusions: GroupK is best used for detecting small overlaps for third-generation sequencing data. It provides a useful supplementary tool to existing ones for more sensitive and accurate overlap detection. The source code is freely available at https://github.com/Strideradu/GroupK.

Keywords: Group hit criteria, Third-generation sequencing, Overlap detection, Metagenomics

*Correspondence: yannisun@cityu.edu.hk
[2]Electronic Engineering Department, City University of Hong Kong, Hong Kong SAR, China
Full list of author information is available at the end of the article

Background

The increased read length enables third-generation sequencing to close gaps in genome assembly [1, 2], reveal structural variations [3], and quantify gene isoforms with higher accuracy [4] in transcriptomic sequencing. In addition, using long reads holds promise in revealing the microbial community structure and deciphering intra-species variation for microbial communities [5, 6].

Genome assembly using third-generation sequencing data requires dedicated methods and tools. Existing genome assembly tools mainly utilize two types of graph models: overlap graph and de Bruijn graph. When the error rate is low, de Bruijn graph has the theoretical advantage that the graph size does not increase significantly with the sequencing coverage, which is usually high for Illumina datasets. For third-generation sequencing data, the high error rate and low coverage make the overlap graph a sensible choice for genome assembly [7]. A key step in constructing the overlap graph is to identify read pairs that share overlaps, which indicates that these reads are sequenced from the same loci in the underlying genome. Although there are a number of sequence alignment programs available for conducting overlap alignment [8, 9], a majority of them rely on dynamic programming and are too computationally expensive for high throughput sequencing data. Due to high error rates, existing short read overlap detection software using BWT (Burrows-Wheeler transform) or hash table [10, 11] cannot be directly applied to long reads.

Related work

Two strategies are currently being employed to detect overlaps for error-prone long reads. One strategy tries to correct sequencing errors in PacBio (Pacific Biosciences) and ONT (Oxford Nanopore) data before overlap detection. There exist a number of sequencing error correction tools [7, 12]. Some of them rely on hybrid sequencing, which requires preparation of at least two sequencing libraries and several types of sequencing runs and thus is not cost-effective for many applications. Others conduct error correction using long reads only. One representative method is described in Chin et al.'s hierarchical genome-assembly process HGAP [12], whose performance improves with higher read coverage. It is worth noting that for alignment-based error correction methods such as the one in HGAP, an important step is to identify reads that can be aligned quickly. Essentially, techniques used for overlap detection can be used for alignment detection as well.

The second category bypasses the difficulty of error correction and identifies overlaps using raw reads. Various approximate similarity search methods have been applied on PacBio and ONT data [13]. They generally follow seed-chain-align procedure [14]. Seed-based filtration step plays an essential role in controlling the trade-off between sensitivity and computational efficiency. Usually, these methods use short string matches as the filtration step. A short string or k-mer match requires exact matches of k consecutive characters between two sequences. Intuitively, overlapping reads tend to share more common k-mers than non-overlapping reads. Strategies that can quickly find the number of shared k-mers can thus be applied. In this section, we summarize the main strategies of several state-of-the-art overlap detection tools. We highlight the differences between our method and the existing ones in the following section.

MHAP [15], Minimap [14, 16], and DALIGNER [17] all use k-mer matches for identifying candidate overlapping pairs. Due to the high error rate, usually only short k-mers will be applied in order to achieve high sensitivity. However, identifying short k-mer matches between all pairs of reads is computationally expensive. Thus, the leading tools employed different data structures and algorithms for estimating k-mer-based similarity. MHAP converts long reads into sets of k-mers and sketches using minHash. Then the similarity between reads is estimated using the compact sketches. Minimap also uses a compact representation of the original reads by keeping minimizers rather than all possible k-mers of a read. Then collinear k-mers will be clustered and used for checking possible overlaps. DALIGNER directly sorts k-mers based on their positions and then utilizes merge sort to identify the number of shared k-mers. As the sorting is cache efficient, DALIGNER is practically very efficient.

BLASR [18] was initially designed for mapping PacBio reads to a reference genome. It is also widely used as an overlap detection tool for PacBio data. BLASR uses BWT and FM index to identify short k-mer matches and then clusters k-mer matches within a given distance range. The clustered k-mers are ranked based on a function of the k-mer frequency. Only highly ranked clusters will be kept for downstream analysis.

Being different from the above tools, GraphMap [19] uses spaced seeds that allow matches of non-consecutive characters. Spaced seeds were initially used in homology search for improving the trade-off between sensitivity and filtration efficiency [20–22]. In particular, spaced seeds containing the pattern "11*" have high sensitivity in capturing homologous protein-coding genes because of the codon structure. However, designing optimal spaced seeds (i.e., deciding the positions of the wildcard characters) is NP-hard [23, 24]. GraphMap empirically chooses two spaced seeds. Ideally, different sets of seeds may be designed for input data of different error profiles.

Overview of our work

The high error rates and also the different error profiles of PacBio and ONT data motivate us to use a more flexible seeding strategy called group hit criteria [25], which define a group of possibly overlapping *k*-mers satisfying statistically derived distance constrains. For brevity, we will call the *k*-mers set satisfying the group hit criteria as a "group seed". A group seed was initially proposed and used for homology search. Given the error profiles, such as the estimated indels and mismatch probabilities, thresholds for grouping short *k*-mers can be computed using the waiting time distribution and the one-dimensional random walk [25]. A group seed can effectively handle all types of errors and is ideal to detect small overlaps. With group seeds, we can achieve high sensitivity using short *k*-mers (e.g., 9-mer) while still maintaining a desirable specificity.

In this work, we employ group seeds for detecting overlapping long reads for genome assembly. Our implementation, named GroupK, provides a complementary tool to existing methods for detecting small overlaps or overlaps compounded by high error rates of both reads. This ability enables our tool a sensible choice for genome assembly in metagenomic data sequenced by third-generation sequencing platforms. As these community samples usually contain microorganisms with heterogeneous coverage, being able to identify small overlaps

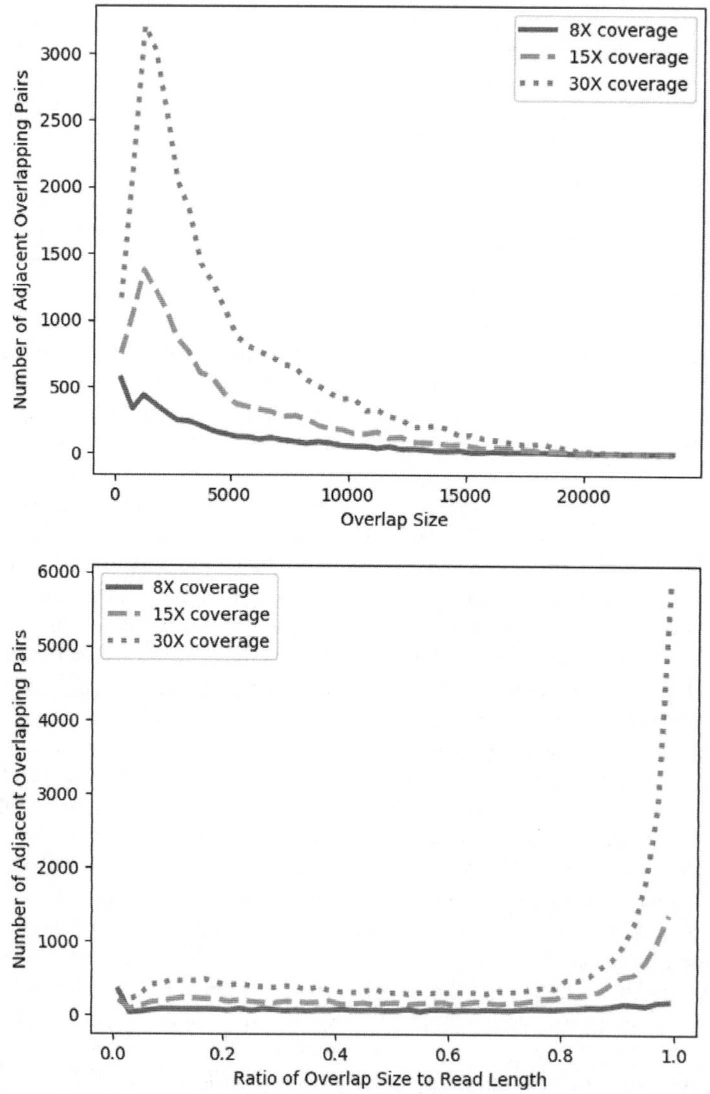

Fig. 1 Histograms of irreducible overlap sizes (top) and the ratio of overlap size to the read length (bottom) when comparing adjacent overlapping reads on simulated PacBio *E. coli* datasets given different coverages. The bin width for the overlap size is 500. For the 30X dataset, the average read length is 8366 and the number of reads is 16644. For the 15X dataset, the average read length is 8253 and the number of reads is 8436. For the 8X dataset, the average read length is 8414 and the number of reads is 4413

will be very important for reconstructing genomes of rare species.

Methods

GroupK is designed for improving the sensitivity of detecting small overlaps or overlaps with low identity. Currently, third-generation sequencing data still has high error rates. The overlapping regions formed by two error-prone long reads can have lower sequence identity than mapping a long read against a reference genome. Figure 1 presents the histogram of the overlap size and the corresponding ratio of overlap size to the read length between two adjacent reads. The reads are simulated using PBSIM [26] from *E. coli* with three different coverages. As we know the position of each simulated read in the genome, the overlap size can be easily decided. Note that a read can form overlaps with multiple reads sequenced from the same region. However, the two figures are generated using overlaps between two adjacent reads, which define an "irreducible" edge [27] in an overlap graph. Thus, these overlaps can decide the continuity of the final genome assembly. The figures show that there are still substantial regions with small overlaps. For example, there are 45.46%, 34.76%, and 31.19% of the overlaps shorter than the 50% of the read length for data with coverage of 8X, 15X, and 30X, respectively. It will be ideal to detect relatively small overlaps to fully take advantage of the long reads for generating more complete assemblies.

Pipeline

Identifying small overlaps is computationally difficult. Thus, we use a carefully designed hierarchical filtration strategy to distinguish true overlapping reads from non-overlapping ones. The pipeline of GroupK consists of three key steps: filtration, group seed matching, and chaining (Fig. 2). Filtration is used to reduce the search space by quickly identifying read pairs sharing a minimum number of k-mers. High insertion/deletion error rates tend to produce short k-mer matches on different diagonals. Thus, we adopt group seed matching to identify a group of short k-mer matches in close proximity. There are two types of distance constraints. 1) The distance (number of nucleotides) between the k-mers on x-axis and y-axis must be smaller than a given threshold; 2) the diagonal distance, which is the difference of the diagonals of two k-mer matches, must be within a given range. Chaining is used to estimate the final overlap region. Figure 3 shows that applying group seed can remove a large number of random k-mer hits while keeping the k-mer matches within the overlapping region. When $k = 15$, there are only two hits, and it is difficult to determine whether there is an overlap. When $k = 9$, there is clearly a chain formed by hits in the overlapping region. However, there are also a large number of random hits. With group seed matching criteria, most of the random 9-mer hits are filtered out. So the downstream analysis becomes more straightforward.

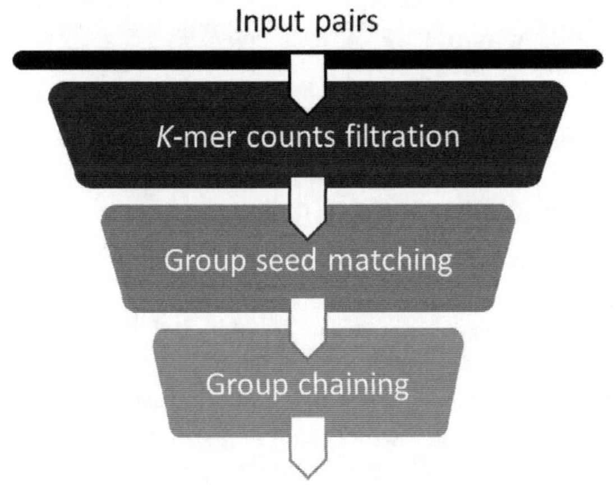

Fig. 2 The pipeline of GroupK

Estimate the expected number of *k*-mers for the filtration stage

In this section, we analyze the expected number of random k-mer hits between two reads and also k-mer hits in overlaps. The analysis will be used for determining the k-mer size and also other parameters for the filtration stage. Given two reads (S_1 and S_2) with length L and error rate ε, we want to determine how many k-mer hits we expect to find between S_1 and S_2.

We first consider the case that S_1 and S_2 are not related (no overlap). By assuming that the bases in S_1 and S_2 are randomly distributed, the expected number of random k-mer matches $E[X_r]$ is roughly:

$$E[X_r] = \left(\frac{1}{|\Sigma|}\right)^k \cdot L^2 \tag{1}$$

Note that this equation is different from the expected number of shared k-mers in MHAP [15] because we distinguish k-mer hits based on their locations rather than the k-mers themselves. Also, we assume that overlapping k-mers are independent.

In the second case of S_1 and S_2 forming an overlap, we estimate the expected number of k-mer matches in the overlap by first computing P_o, which is the probability of observing a k-mer match within the overlap. In the case of no sequencing error (i.e. $\varepsilon = 0$), the probability of observing a k-mer match at an aligned position in an overlap is simply 1.0. But in the practical case of $\varepsilon > 0$, we need to consider two scenarios in order to determine the probability of observing two identical characters at an aligned position in the overlap: 1) the characters from S_1 and S_2 are correct; 2) the two characters from S_1 and S_2 are errors

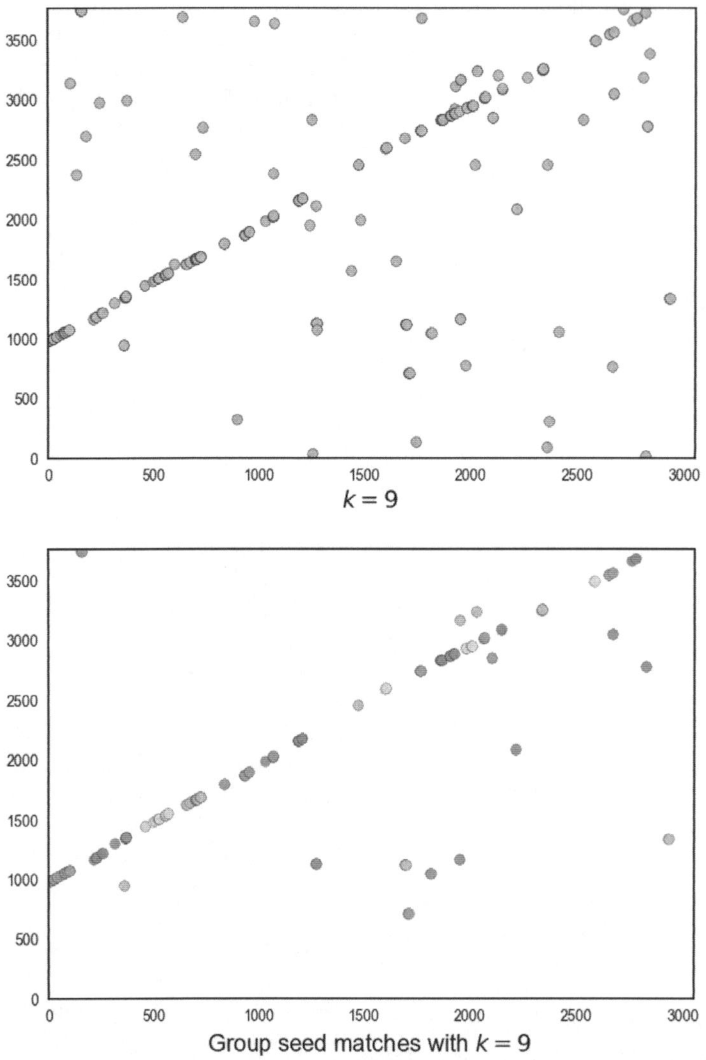

Fig. 3 The dot plot of k-mer hits and group seeds matches for one overlapping pair from the *E. coli* PacBio dataset used in Results. Each dot is a k-mer hit. The x-axis and y-axis show the locations of the hits on the reads. The overlap region is roughly from 0 to 2800 on the x-axis, and from 1000 to 3800 on the y-axis. Top: all 9-mer hits. Bottom: 9-mer hits that passed the group hit criteria. A group seed is represented by closely located dots of the same color

and are randomly substituted by the same character. The two cases are visualized in Fig. 4. So the probability is given by [15]:

$$P_o = \left[(1-\varepsilon)^2 + \varepsilon^2 \frac{1}{|\Sigma|-1}\right]^k \tag{2}$$

Considering both random k-mer matches and k-mer matches in an overlap, the expected number of shared k-mers between two overlapping reads is estimated by:

$$E[X_o] = P_o \cdot M + \left(\frac{1}{|\Sigma|}\right)^k \cdot L^2 \tag{3}$$

M is the size of the overlap. Note that the above equation slightly over-counts the number of k-mer hits in an overlap because the random k-mer hits inside the overlap may be counted twice with probability $P_o \cdot \left(\frac{1}{|\Sigma|}\right)^k$. (Also, we assume that the probabilities of substitution and insertion/deletion are on the same order and thus do not distinguish them in the above equation.)

As we are mainly interested in finding small overlaps or overlaps with low sequence identity, we plot the expected number of k-mer hits with the overlap size being 1/4 of the read size in Fig. 5. In order to plot the figure, we compute $E[X_o]$ and $E[X_r]$ using the read lengths from a 15X *E. coli* PacBio dataset (the data of our second experiment

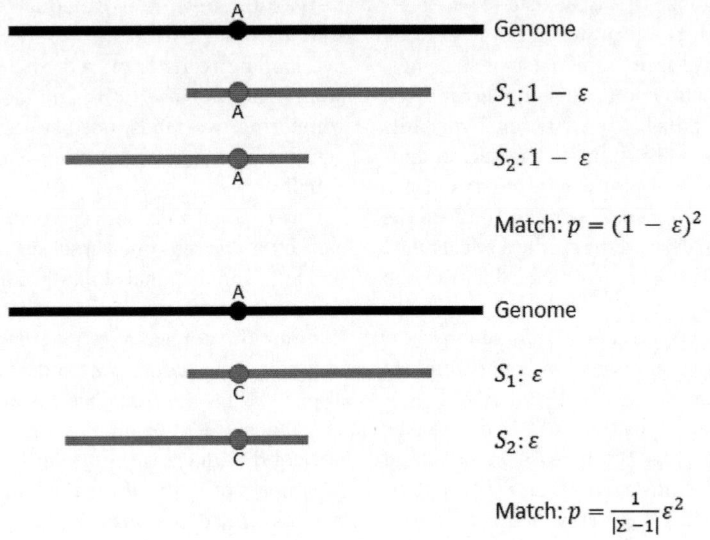

Fig. 4 Two cases contributed to the identical characters at an aligned position in the overlapping region of S_1 and S_2. S_1 and S_2 are two reads sequenced from the same region of the underlying genome and form an overlap. Top: both bases on S_1 and S_2 are correct, forming a match. Bottom: both bases on S_1 and S_2 are sequencing errors, and substituted by the same character

in Results). We only consider reads of length above 2,000. These figures allow us to choose the appropriate threshold for k-mer-counting based filtration. For example, Fig. 5 shows the expected number of 15-mers between reads of different error rates. $E[X_r]$ started as 4 when $\varepsilon = 0.15$. And, for larger ε, $E[X_r]$ is even smaller. Thus, our default filtration threshold is two 15-mers in order to ensure high filtration sensitivity. The implementation details of the k-mer counting stage can be found towards the end of the "Methods" section.

Group hit criteria

Sequencing errors tend to produce short k-mer matches. In addition, the insertion/deletion errors lead to k-mer

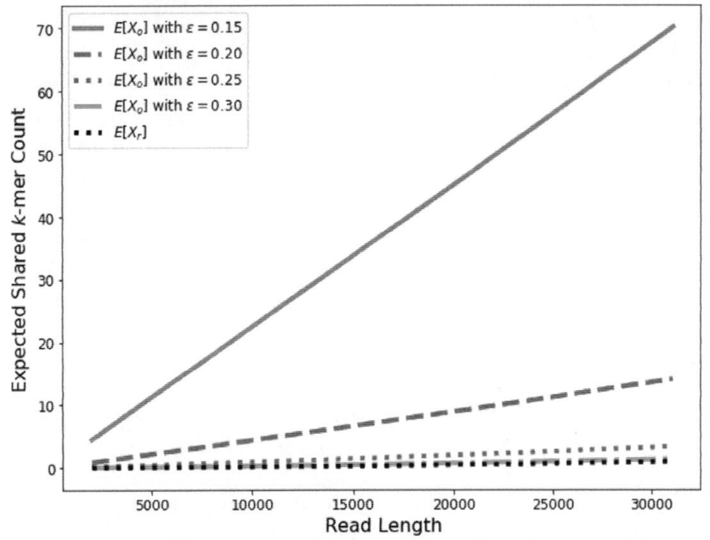

Fig. 5 The change of $E[X_o]$ and $E[X_r]$ (y-axis) with the increase of the read length (x-axis), which is obtained from a real PacBio dataset. The overlap size is set as the 1/4 of the read length as we focus on identifying the hard case of small overlaps. $k = 15$

matches on different diagonals. Thus, instead of using relatively long k-mers (such as 15 or 16-mers) as existing tools do, we use a group of short k-mers (such as 9-mer) to accommodate the high insertion or deletion error rates. A group seed is a set of possibly overlapping k-mer hits with statistically calculated constraints. The region containing group seeds is more likely to be inside an overlap than a single k-mer hit. Reference [25] first introduced the group hit criteria and also derived the method to calculate the criteria statistically. We apply their method for overlap detection.

Assume that we have two reads S_1 and S_2 of length m and n, respectively. The numbers of k-mers at different positions in S_1 and S_2 are $m - k + 1$ and $n - k + 1$, respectively. A k-mer hit at position (i,j) is defined by $S_1[i \ldots i+k-1] = S_2[j \ldots j+k-1]$, where $i \leq m-k+1$ and $j \leq n-k+1$. For two k-mer hits at (i_1, j_1) and (i_2, j_2), their inter-seed distance $D((i_1,j_1),(i_2,j_2))$ is the maximum of $|i_2 - i_1|$ and $|j_2 - j_1|$. The k-mer diagonal of a k-mer hit at (i,j), $d(i,j)$, is defined as $j - i$.

With these notations, the goal is to solve the following inequalities given confidence level $1 - \alpha$ defined by significance level α:

$$D((i_1,j_1),(i_2,j_2)) \leq \rho \tag{4}$$

$$|d(i_1,j_1) - d(i_2,j_2)| \leq \delta \tag{5}$$

ρ and δ are integers we need to define the group hit criteria. For example, when $\alpha = 0.05$, our goal is to derive ρ and δ so that with 95% chance the inter-seed distance and the diagonal shift between two k-mers in overlapping reads are at most ρ and δ, respectively.

Constraint on k-mer distance We followed the model described in the article [25]. Runs of head in n-independent Bernoulli trials are used to model k-mer matches, with the probability p for a match and $(1-p)$ for a mismatch. In this model, a k-mer match can be treated as k consecutive match runs with probability of p^k. From the waiting time distribution [28, 29], the probabilities of the inter-seed distance x of two k-mers in an overlap are:

$$\mathcal{P}[D_k = x] = \begin{cases} 0 & \text{for } 0 \leq x < k \\ p^k & \text{for } x = k \\ (1-p)p^k\left(1 - \sum_{i=0}^{x-k-1} \mathcal{P}[D_k = i]\right) & \text{for } x > k \end{cases} \tag{6}$$

With the confidence level $1 - \alpha$, ρ can be solved using following equation:

$$\mathcal{P}[D_k \leq \rho] = 1 - \alpha \tag{7}$$

In the actual implementation we use $\alpha = 0.05$. Thus, there is 95% chance that the inter-seed distance of the two seeds in an overlap is less than the ρ calculated using Eq. 7.

Constraint on k-mer diagonal distance The diagonal shift between two k-mers, $|d(i_1,j_1) - d(i_2,j_2)|$, in two overlapping reads is caused by insertions and deletions. Note that the insertions and deletions are defined by comparing two reads, not between a read to a reference genome. So the insertion rate and deletion rate are treated equally.

The diagonal shift between two k-mer hits can be modeled by a discrete one-dimensional random walk model [25, 30]. The diagonal shift starts from 0. Let the steps of the random walk be l. Assume that the insertion and deletion rate is q across the whole read. Thus, the probability of a diagonal change is q, and the probability of staying in place is $1 - 2q$. Also, we assume that in l steps, there are n_i inserted nucleotides (increase shift), n_d deleted nucleotides (decrease shift), and n_m matched nucleotides (no impact on shift). If the final diagonal shift is i, we have the following equations:

$$\begin{cases} n_i + n_d + n_m = l \\ n_i - n_d = i \end{cases} \tag{8}$$

Reference [25] calculated the probability of obtaining a diagonal shift i after l steps in the random walk. According to Eq. (8), we have $n_i = i + n_d$ and $n_m = l - (i + 2n_d)$. For a specific n_d, the probability of a random walk producing diagonal shift i can be calculated as the number of the possible paths $\binom{l}{i+2n_d} \cdot \binom{i+2n_d}{i+n_d}$, times the probability product of all insertions, deletions, and no shift change at each step $q^{n_d} q^{n_d+i}(1-2q)^{l-(i+2n_d)}$. To calculate the probability of generating a diagonal shift i given l, $P[i,l]$, we need to consider all possible values of n_d, which is from 0 (no deletion) to $(l-i)/2$ (no match). So we have:

$$\mathcal{P}[i,l] = \sum_{n_d=0}^{(l-i)/2} \binom{l}{i+2n_d} \cdot \binom{i+2n_d}{i+n_d} \cdot q^{n_d} q^{n_d+i} (1-2q)^{l-(i+2n_d)} \tag{9}$$

To calculate δ, we sum up the probabilities $\mathcal{P}[i,l]$ for $i = 0, \pm 1, \pm 2, \ldots, \pm l$ until we reach the level $1-\alpha$. We refer the reader to the original article [25] for a more detailed discussion of the model and a practical implementation using generating functions.

In our experiment, we set the sequencing accuracy $p = 0.85$ and the indel rate $q = 0.06$, as PacBio reads tend to have higher indel rates than substitution error rates. Users can adjust these parameters based on their data properties. The significance level α is 0.05 and k is 9. Using Eq. 6, we can obtain $\rho = 54$. Using Eq. 9 and ρ as an estimation of l, we can obtain $\delta = 5$ given $\rho = 54$. Thus, when $k = 9$, seeds with inter-seed distance $D((i_1,j_1),(i_2,j_2)) \leq 54$ and diagonal shift $|d(i_1,j_1) - d(i_2,j_2)| \leq 5$ are clustered in the same group. In addition, if k-mers (i_1,j_1) and (i_2,j_2) are in the same group and k-mers (i_2,j_2) and (i_3,j_3) are in the same group, we will cluster (i_1,j_1) and (i_3,j_3) as well.

Group chaining

With group hit criteria, GroupK can find short similar regions. To identify the overlapping region, we aim to find a chain of group seed matches that maximizes the number of matched bases. We used the modified sparse dynamic programming for chaining [31, 32].

After generating a chain of group seed matches, we need to determine whether this chaining defines an overlap. We develop two criteria for this purpose. First, we calculate the expected number of matched bases from the group hit criteria, assuming that the chain covers the possible overlapping region with length L_O (L_O can be estimated by the extension of both ends of the optimal chain). We used the following equation to calculate the expected number of the matched bases n_e:

$$n_e = \frac{1}{c} \cdot \frac{L_O}{\rho} \cdot k \quad (10)$$

Where c is a coefficient to control the criteria, k is the size of k-mer, ρ is the group hit criteria for the inter-k-mer distance. We only report the chaining result if the number of matched bases $n \geq n_e$. Second, we require that both reads have similar sizes inside the overlapping region.

In our experiment, we found that sometimes using the optimal chain generated from sparse dynamic programming may overestimate the overlap region, as shown in Fig. 6. This overestimation can jeopardize the sensitivity of detecting small overlaps. We fix this problem by only keeping the collinear group seeds, which are used to estimate the overlap size.

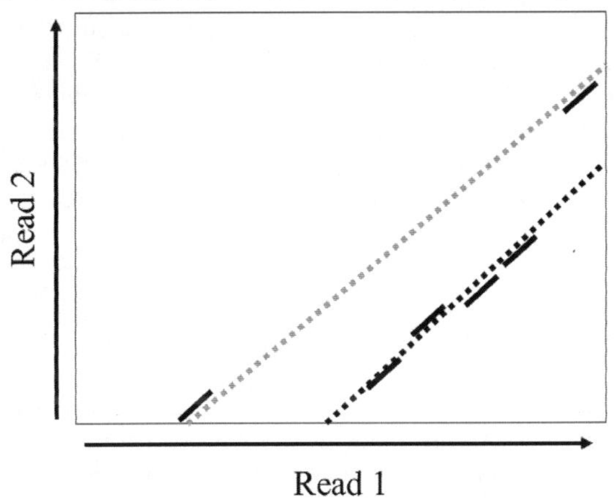

Fig. 6 An example of the overlap size estimation. The suffix of read 1 and the prefix of read 2 form an overlap. Each short solid line represents a group seed match in the optimal chain. The black dashed line indicates the true overlap alignment region between the two reads. The gray dashed line, which is formed by the two ending group seeds in the optimal chain, can overestimate the overlap size

Implementation details of the major components

Filtration by k-mer counts In the first step of our pipeline, we use k-mer counting-based filtration to remove large numbers of read pairs that are not likely sequenced from the same loci on the underlying genome. We implemented k-mer counting using a generalized suffix array and the derived longest common prefix (LCP) array. The generalized suffix array SA is created from the concatenated reads (delimited by special characters such as $) using a linear algorithm [33]. Then, we create the LCP using both the suffix array SA and the reversed suffix array SA' [34, 35]. Let the sequence of concatenated reads be T. Following the definition of the reversed suffix array, for a suffix starting at position $SA[i]$, we have $SA'[SA[i]] = i$. For each position i in the LCP, LCP[i] contains the size of the longest common prefix between $SA[i]$ and $SA[i-1]$. The key observation [33] for efficient computation of LCP[i] is: for a position j in T, if $LCP[SA'[j-1]]$ is L, $LCP[SA'[j]] \geq L-1$. The whole LCP array construction takes linear time to the size of T [33].

In order to count the shared k-mers between reads and also report read pairs passing the k-mer counting threshold, we use both LCP and an auxiliary data structure recording the read IDs (denoted as array *readID*). For a suffix starting at position $SA[i]$, its read ID is at $readID[i]$. The pseudocode of finding the number of shared k-mers can be found in Algorithm 1. In practice, we also count k-mers between a read and the other read's reverse complement.

Group seed match and chaining Any pair of reads that pass the above filtration stage will be used as input for finding group seed matches. All other pairs will be discarded. Currently, we are using the codes of YASS [36, 37] for finding the group seed matches. The program uses hashing table to find short exact matches and creates the groups of matches on the fly. It is our future work to re-implement group seed matching using more efficient indexing-based methods. Implementation of chaining algorithm has been modified from the program of global chaining algorithm in SeqAn library [38].

Time complexity analysis If we have N reads with average read length L, our text size is NL. Therefore, we have NL elements in the suffix array and the corresponding LCP array. For each suffix in the suffix array, suppose on average, it can form LCPs with m other suffixes with size above k, which is the size of k-mer used in the k-mer-count filtration steps. So the time complexity of finding all the shared k-mers for all possible reading pairs is $\mathcal{O}(mNL)$. If k is large enough (e.g., $k = 15$ and 11 in our experiment), we have $m \ll NL$, so the time complexity will be dominated by NL.

Algorithm 1 Counting k-mers between reads

Input: LCP array $LCP[1..n]$, read ID array $readID[0..n]$, k-mer size k, k-mer counts threshold τ

Output: Read pairs whose shared k-mer $counts \geq \tau$

```
Initialize a map counts[key][value] for recording k-mer counts
for i = 1 to n do
    readi ← readID[i]
    j ← i + 1
    L ← LCP[j]
    while L ≥ k and j ≤ n do
        readj ← readID[j]
        if readi < readj then
            key ← readi : readj
        end if
        if readj < readi then
            key ← readj : readi
        end if
        if counts[key] does not exist then
            counts[key] ← 1
        else
            counts[key]++
        end if
        j++
        L ← min (L, LCP[j])    ▷ min(L, LCP[j]) returns the minimum of L and LCP[j]
    end while
end for
for all key in counts do
    if counts[key] ≥ τ then
        output key    ▷ key contains read pair IDs passing the k-mer count threshold
    end if
end for
```

For N' read pairs that pass the filtration stage, let the average number of k-mer hits for each pair be q. Sorting the hits will need $\mathcal{O}(q \log q)$ and iterating through all hits to find groups is linear to q. For all the read pairs, the time complexity is $\mathcal{O}(N' q \log q)$.

Assuming finally we have r group seeds, the chaining procedure has complexity in $\mathcal{O}(r \log r)$ for each read pairs. For all the read pairs, the time complexity is $\mathcal{O}(N' r \log r)$. It is practically very fast because the number of group seed matches is very small compared to the original seed hits (indicated by Fig. 3).

Results

We focus on evaluating the sensitivity and precision of overlap detection. We applied GroupK to three PacBio datasets and one ONT dataset: a simulated PacBio RSII *E. coli* sequencing dataset, a real PacBio RSII *E. coli* sequencing dataset, a PacBio RSII human foot metagenomic sequencing dataset, and an ONT *E. coli* (SQK-MAP-006) dataset. For simulated *E. coli* dataset, we have the true sampling position for each read as our ground truth. For the real *E. coli* dataset and human foot dataset, we determine the ground truth via BLASR's [18] alignments against the reference genome.

We benchmarked GroupK's performance with Minimap [16], Minimap2 [14], DALIGNER [17], MHAP [15], and GraphMap [19]. Those tools and methods are representative overlap detection tools for long erroneous reads from PacBio or ONT [13]. All the detailed parameters can be found at the website listed in "Availability of data and materials". Our main metrics include: (1) sensitivity, which measures the ratio of the true overlaps identified by each program to the whole set of overlapping pairs; (2) precision, which quantifies the ratio of true overlap detected by each program to the total reported overlapping pairs; and (3) F1 score, which is the harmonic mean of sensitivity and precision. A reported overlapping pair is regarded as correct if it is also present in our ground truth. The detailed overlap region and overlap length were not considered in the current evaluation because these read pairs can go through a more accurate alignment program for generating the final overlap alignment. As we discussed in Background, our goal is to identify overlapping reads without using error correction. All tested tools will thus be applied to their raw data set.

Simulated *E. coli* dataset

We first evaluated the performance of our method on a simulated *E. coli* dataset. The dataset was generated using PBSIM [26] with *E. coli* K-12 MG1655 as the reference genome [39]. The length distribution and the quality profile were derived from real PacBio P6-C4 *E. coli* dataset [40]. The simulated dataset has 5620 reads, with average length of 8,344.78 bps and 14.5% average error rate (8.6% insertions, 4.4% deletions, and 1.4% substitutions).

From the report by PBSIM, we can obtain the exact locations where the simulated reads are sampled in the genome. This information provides us with the ground truth for reads' overlaps so that we can calculate the sensitivity and precision.

Following the pipeline we discussed in Methods, we first used k-mer-counting as the filtration stage. According to Eq. 1, Eq. 3, and Fig. 5, we discarded all read pairs with less than two 15-mer matches. The sensitivity of the filtration is 0.979 and only about 6% of read pairs are kept for downstream analysis.

We evaluated the performance of our tool by adjusting the group seed match criteria coefficient c, which is introduced in Methods. With the increase of c, sensitivity will become higher, and the precision will become lower. As shown in Fig. 7, GroupK can achieve 5 to 6% improvement

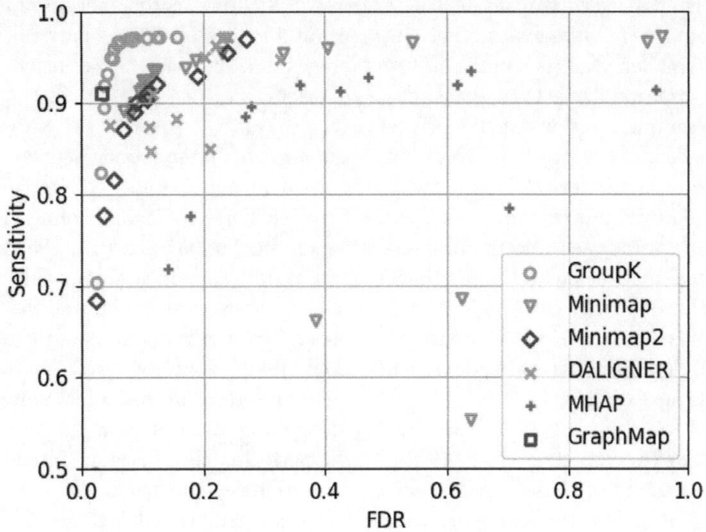

Fig. 7 The ROC-like plot using GroupK, Minimap, Minimap2, DALIGNER, MHAP, and GraphMap on the simulated PacBio *E. coli* dataset. The x-axis represents the false discovery rate (FDR = 1 − precision). Y-axis is the sensitivity (0.5 to 1)

on the sensitivity with similar precision to other overlap detection tools.

Running time and memory usage

We evaluated the running time and the peak memory usage of the tested tools in this experiment. We run all overlap detection tools with a single core of 2.4Ghz 14-core Intel Xeon E5-2680v4 CPU and 32 GB memory requested from the High-Performance Computing Center at Michigan State University. The performance is measured with the best F1 score. The results are reported in Table 1. For the memory usage, Minimap2 is the most efficient one but all others are comparable. GroupK is slower than other tools, partially because we use small *k*-mers. We found that the bottleneck of our program is the group matching stage, which accounts for about 1200 of 1871 s. By implementing a more efficient indexing-based method, we expect to reduce the running time of this stage. For example, we can speed up *k*-mer counting by adopting the method used in KMC [41].

Real PacBio *E. coli* dataset

After using the simulated dataset to evaluate our method's performance, we applied GroupK to a real PacBio RS II (P6-C4) *E. coli* dataset [40]. The coverage of the whole dataset is 150X. To test the performance of low coverage data, we sampled a 15X coverage dataset based on the read length distribution of the whole dataset. The dataset has 14,262 reads, with the average length of 4,882.09 bps and average error rate of 14.14% (error rate is estimated using quality score).

We applied BLASR to map the reads to the reference genome to estimate the ground truth (BLASR was run with parameters: minReadLength, 2000; maxScore, 1000; maxLCPLength, 16; minMatch, 12; m 4 and nCandidates/bestn set to 10× sequencing coverage). The mapping result from BLASR may contain short alignments due to the repeat in the genomes. So when we determine the ground truth, we only consider the alignments that cover at least 80% of the read. By removing short noisy alignments, we can make sure the BLASR alignments are close to the underlying ground truth.

We used the same filtration setup adopted in the simulated *E. coli* experiment. Among all overlap detection tools we tested, GroupK still achieved the highest sensitivity with comparable precision (Table 2). With slightly higher

Table 1 Computational performance on the simulated *E. coli* dataset

	GroupK	Minimap	Minimap2	DALIGNER	GraphMap	MHAP
Time (seconds)	1871	30	16	39	171	858
Memory (GB)	1.994	1.754	1.097	2.288	1.873	2.562

The computational performance of overlap detection using GroupK, Minimap, Minimap2, DALIGNER, MHAP, and GraphMap on the simulated *E. coli* dataset

Table 2 Overlap detection on the real *E. coli* dataset

	GroupK	Minimap	Minimap2	DALIGNER	GraphMap	MHAP
Best F1 score	**0.9311**	0.9037	0.8426	0.8340	0.7758	0.7023
Sensitivity	**0.9330**	0.8939	0.8778	0.9066	0.6741	0.7258
Precision	**0.9292**	**0.9138**	0.8101	0.7722	**0.9137**	0.6802

The performance of overlap detection using GroupK, Minimap, Minimap2, DALIGNER, MHAP, and GraphMap on the real PacBio RS II (P6-C4) *E. coli* dataset. Here we only report the experiment results with the highest F1 score for each tool precision, our sensitivity is 4% better than the next best tool, Minimap. Compared to the previous experiment, the difference in sensitivity is smaller. One reason lies in the construction of the ground truth dataset. In the simulated dataset, we used the sample positions of all reads to determine whether two reads form an overlap. Thus, that dataset can include reads with small overlaps or reads with higher error rates. In this dataset, our method discarded BLASR alignments with high error rates and the remaining alignments have higher similarities with the reference genome and thus produce fewer "hard cases". As Minimap is the second best tool for this dataset, we further analyzed the performance of GroupK and Minimap on read pairs of different overlap size in the next section.

Performance with different overlap size We divide all overlapping pairs into bins of width 500 based on the overlap size. For example, the first bin has the read pairs with overlap sizes from 0 to 499, and the second bin has read pairs with overlap sizes from 500 to 999, and so on. For each bin, we compared the sensitivity of GroupK and Minimap using parameters yielding the similar precision in Fig. 8. For Minimap, we showed the result with the highest F1 score. For GroupK, we selected a parameter so that it achieves similar precision to Minimap (F1 score: 0.9241, sensitivity: 0.9344, precision: 0.9140). According to Fig. 8, GroupK has much better sensitivity when the overlap size is less than 2000. As we showed in Fig. 1, there are a significant number of overlaps with overlap size smaller than 2000 even for 30X coverage. Being able to identify small overlaps allows us to generate more complete assemblies using long reads. This is particularly useful for low coverage data, such as what we usually have in metagenomic datasets.

Human foot metagenomic dataset

One of the major utilities of our tool is to identify overlaps between reads in metagenomic data that are sequenced using the PacBio platform. For complicated microbial communities, metagenomic data containing only short reads poses serious computational challenges for assembly and the downstream composition/functional analysis. Long reads hold the promise to produce more complete and accurate microbial genome assemblies for the metagenomic dataset. In this experiment, we evaluated the performance of overlap detection for a mock metagenomic dataset constructed from a real human foot dataset [5]. A particular challenge for this experiment is the low coverage of the component species in the metagenomic dataset, which could be caused by sequencing throughput and complexity of the sample.

The human foot sample was sequenced by linear PacBio RSII TdT (terminal deoxynucleotidyl transferase). Sequences that can be mapped to the human genome were removed as host-derived DNA. According to the Supplementary Materials of [5], there are about 1,000 bacteria and viruses in this metagenomic dataset. However, we cannot evaluate the performance of overlap detection on all the reads from the 1000 microbes because the coverages of many species are too low to yield meaningful

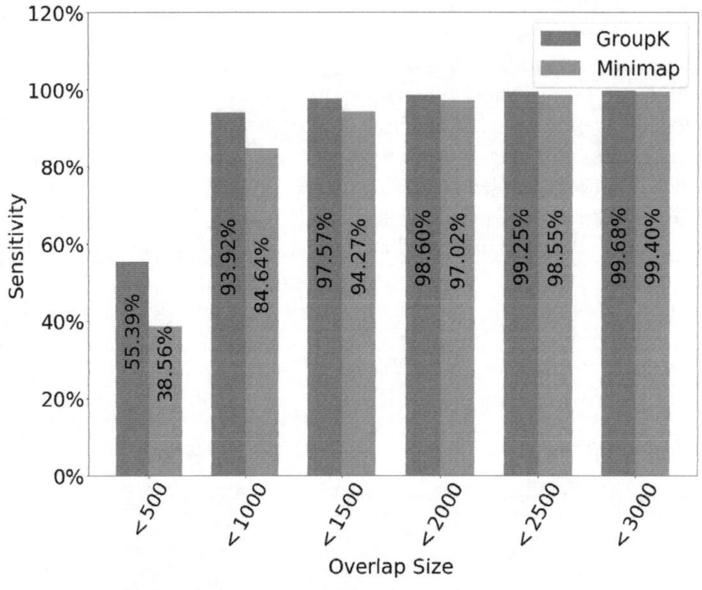

Fig. 8 Sensitivity of GroupK and Minimap for detecting overlaps of different size on PacBio *E. coli* dataset. The x-axis represents the overlap size. Y-axis is the corresponding sensitivity of the bin. The X-axis bin width is 500 and the figure only included the first 6 bins (i.e. up to overlap size 3000) as their sensitivity becomes more similar with the increase of the overlap size

overlaps. In order to construct the ground truth, we need to align the reads against species with known reference genomes and reasonable coverage. Thus, we only choose reads satisfying the following criteria: 1) the reads are sequenced from a species with known reference genome; and 2) the coverage of the species cannot be too small (e.g., >3X coverage). Based on these criteria, we keep the reads sequenced from three bacteria: *Corynebacterium aurimucosum* (6.3X Coverage), *Corynebacterium tuberculostearicum* (8.5X Coverage), and *Staphylococcus hominis* (3.2X Coverage). The reads are recruited via BLASR. The alignment positions are used to determine which reads form an overlap. Note that *Corynebacterium aurimucosum* and *Corynebacterium tuberculostearicum* belong to the same genus and may contribute to the false positive overlap detection due to their shared regions. The average length of the reads is 1696.25 bps, which is much shorter than the reads in the previous experiments.

As this dataset contains much shorter reads, the expected number of k-mer hits will change. Intuitively we need to use shorter k-mers to ensure high filtration sensitivity. Using the read length distribution, we calculated $E[X_r]$ (Eq. 1) and $E[X_o]$ (Eq. 3) and determined the k-mer counting-based filtration criteria. In this experiment, we only kept read pairs that share at least three 11-mers.

For this mock metagenomic dataset, GroupK yielded significantly better performance than other tools on metrics including F1 score, sensitivity, and precision (Table 3). Compared to other tools, GroupK can produce much higher sensitivity without sacrificing precision, leading to the higher F1 score. Besides evaluating the performance of various tools on all the reads from the three species, we also reported the performance of different overlap detection tools on each single bacteria dataset without mixing with other species (Table 3). In these tools, GraphMap has high specificity for all three with sacrifice of sensitivity. However, GroupK still achieves the best performance overall. The comparisons suggest that our method has great potential to detect overlaps for data with very low coverage (around 5X). This will enable better assembly for PacBio sequenced metagenomic data, which will become more available with the advances of long read sequencing technologies.

Table 3 Overlap detection on the human metagenomic dataset

	GroupK	Minimap	Minimap2	DALIGNER	GraphMap	MHAP
Total:						
Best F1 score	**0.9163**	0.8306	0.8776	0.6911	0.7188	0.7812
Sensitivity	**0.8954**	0.7802	0.8352	0.6012	0.5721	0.6803
Precision	0.9381	0.8880	0.9245	0.8027	**0.9666**	0.9174
C. aurimucosum:						
Best F1 score	**0.9512**	0.8072	0.9117	0.7432	0.7397	0.8545
Sensitivity	**0.9228**	0.6858	0.8467	0.6266	0.5892	0.8045
Precision	**0.9814**	**0.9806**	**0.9874**	0.9131	**0.9937**	0.9111
C. tuberculostearicum:						
Best F1 score	**0.9454**	0.8688	0.9050	0.8315	0.7276	0.7961
Sensitivity	**0.9105**	0.7958	0.8346	0.7150	0.5727	0.7355
Precision	**0.9830**	0.9567	**0.9884**	**0.9934**	**0.9977**	0.8675
S. hominis:						
Best F1 score	**0.9163**	0.7651	0.8024	0.8754	0.6343	0.6861
Sensitivity	**0.8733**	0.6887	0.6813	**0.7967**	0.4658	0.6348
Precision	0.9245	0.8606	0.9759	0.9713	**0.9938**	0.7464

The performance of overlap detection using GroupK, Minimap, Minimap2, DALIGNER, MHAP, and GraphMap on the mock metagenomic dataset. Here we only report the experimental results with the highest F1 score for each tool

Real ONT *E. coli* dataset

We also tested our method on one ONT dataset. We used downsampled 15X coverage 2D reads from the SQK-MAP-006 dataset as 2D reads provide higher quality than 1D reads. We followed the same pipelines we used for the real *E. coli* PacBio dataset. As 2D ONT reads have similar error rates to the PacBio reads, we expect that our tool can still achieve reasonable performance given the same setup for the PacBio dataset. Therefore, we used the same parameters as the ones we used for the PacBio *E. coli* dataset.

Among these tools, GraphMap was designed for ONT data, Minimap2 provides a specific setup for finding the overlap on ONT dataset. All other tools are not specifically designed for ONT datasets. GroupK achieves the best F1 score compared to other tools' default setup while keeping the highest sensitivity (Table 4). This result suggests that our strategy is robust with different types of long reads.

Discussion

Seeding is a key step for overlap detection because of the high error rate of long reads. Successful seeding strategies should balance the sensitivity and the specificity to

Table 4 Overlap detection on the ONT *E. coli* dataset

	GroupK	Minimap	Minimap2	DALIGNER	GraphMap	MHAP
F1 score	**0.9383**	**0.9310**	0.9090	0.8272	**0.9280**	0.8122
Sensitivity	**0.9597**	**0.9546**	0.9362	0.8730	0.8991	0.8871
Precision	0.9178	**0.9085**	0.8833	0.7860	**0.9589**	0.7490

The performance of overlap detection using GroupK, Minimap, Minimap2, DALIGNER, MHAP, and GraphMap on the real ONT SQK-MAP-006 *E. coli* dataset. Minimap2 uses the ava-ont setup, which is optimized for ONT data achieve the optimal performance. Popular seeding methods include maximal exact matches, spaced seeds, and gapped spaced seeds. However, to successfully find a hit between two reads, these methods still need either to find relatively long continuous exact matches (large *k*-mer) or to find inexact matches following certain error patterns (spaced seed). Compared to these methods, group seed matching is more flexible as it requires multiple short exact matches without specifying the error patterns. This flexibility leads to high sensitivity, and meanwhile the specificity is still guaranteed with the group seed match criteria.

Currently the group seed matching step based on hash table is the bottleneck of our overlap detection pipeline. A new method that can improve the running time efficiency of this step is needed to make the algorithm achieve the same speed as other faster overlap detection tools.

Conclusions

In this work, we developed an overlap detection tool for third-generation sequencing data. By adopting the group hit criteria to cluster a group of short *k*-mer hits that satisfy statistically derived distance constrains, our method can improve the sensitivity of overlap detection without sacrificing precision. Our experimental results have shown that for datasets with low sequencing coverage, our program can detect significantly more overlapping pairs while keeping high precision. One utility of our approach is to detect small overlaps between long reads of rare species in a microbial community.

Abbreviations

BWT: Burrows–wheeler transform; LCP: Longest common prefix; ONT: Oxford Nanopore; PacBio: Pacific biosciences; SMRT: Single molecule real time sequencing; TdT: Terminal deoxynucleotidyl transferase

Acknowledgements

Not applicable

Funding

Publication of this article was sponsored by NSF Grant IOS-1740874 and City University of Hong Kong.

Availability of data and materials

The source code of GroupK is available at https://github.com/Strideradu/GroupK.
The experiment details are at https://github.com/Strideradu/GroupK/tree/master/experiment.

About this supplement

This article has been published as part of *BMC Genomics Volume 20 Supplement 2, 2019: Selected articles from the 17th Asia Pacific Bioinformatics Conference (APBC 2019): genomics*. The full contents of the supplement are available online at https://bmcgenomics.biomedcentral.com/articles/supplements/volume-20-supplement-2.

Authors' contributions

YS proposed the original idea. YS and ND both designed the algorithm and the experiments. ND implemented the main algorithms, conducted the experiments, and analyzed the data. JC implemented the LCP array algorithm. All Authors wrote, read and approved the manuscript.

Ethics approval and consent to participate

Not applicable.

Consent for publication

Not applicable.

Competing interests

The authors declare that they have no competing interests.

Publisher's Note

Author details

[1]Department of Computer Science and Engineering, Michigan State University, 48824 East Lansing, MI, USA. [2]Electronic Engineering Department, City University of Hong Kong, Hong Kong SAR, China.

Published: 10 April 2019

References

1. Koren S, Harhay GP, Smith TP, Bono JL, Harhay DM, Mcvey SD, Radune D, Bergman NH, Phillippy AM. Reducing assembly complexity of microbial genomes with single-molecule sequencing. Genome Biol. 2013;14(9):101.
2. Conlan S, Thomas PJ, Deming C, Park M, Lau AF, Dekker JP, Snitkin ES, Clark TA, Luong K, Song Y, et al. Single-molecule sequencing to track plasmid diversity of hospital-associated carbapenemase-producing enterobacteriaceae. Sci Transl Med. 2014;6(254):254–126254126.
3. Chaisson MJ, Huddleston J, Dennis MY, Sudmant PH, Malig M, Hormozdiari F, Antonacci F, Surti U, Sandstrom R, Boitano M, et al. Resolving the complexity of the human genome using single-molecule sequencing. Nature. 2015;517(7536):608–11.
4. Tilgner H, Grubert F, Sharon D, Snyder MP. Defining a personal, allele-specific, and single-molecule long-read transcriptome. Proc Natl Acad Sci. 2014;111(27):9869–74.
5. Tsai Y-C, Conlan S, Deming C, Segre JA, Kong HH, Korlach J, Oh J, Program NCS, et al. Resolving the complexity of human skin metagenomes using single-molecule sequencing. MBio. 2016;7(1): 01948–15.
6. Giallonardo FD, Töpfer A, Rey M, Prabhakaran S, Duport Y, Leemann C, Schmutz S, Campbell NK, Joos B, Lecca MR, et al. Full-length haplotype reconstruction to infer the structure of heterogeneous virus populations. Nucleic Acids Res. 2014;42(14):115.
7. Koren S, Schatz MC, Walenz BP, Martin J, Howard JT, Ganapathy G, Wang Z, Rasko DA, McCombie WR, Jarvis ED, et al. Hybrid error correction and de novo assembly of single-molecule sequencing reads. Nat Biotechnol. 2012;30(7):693–700.
8. Altschul SF, Gish W, Miller W, Myers EW, Lipman DJ. Basic local alignment search tool. J Mol Biol. 1990;215(3):403–10.
9. Schwartz S, Kent WJ, Smit A, Zhang Z, Baertsch R, Hardison RC, Haussler D, Miller W. Human–mouse alignments with BLASTZ. Genome Res. 2003;13(1):103–7.
10. Simpson JT, Durbin R. Efficient construction of an assembly string graph using the FM-index. Bioinformatics. 2010;26(12):367–73.
11. Gonnella G, Kurtz S. Readjoiner: a fast and memory efficient string graph-based sequence assembler. BMC Bioinformatics. 2012;13(1):82.
12. Chin C-S, Alexander DH, Marks P, Klammer AA, Drake J, Heiner C, Clum A, Copeland A, Huddleston J, Eichler EE, et al. Nonhybrid, finished microbial genome assemblies from long-read SMRT sequencing data. Nat Methods. 2013;10(6):563–9.
13. Chu J, Mohamadi H, Warren RL, Yang C, Birol I. Innovations and challenges in detecting long read overlaps: an evaluation of the state-of-the-art. Bioinformatics. 2016;33(8):1261–70.
14. Li H. Minimap2: pairwise alignment for nucleotide sequences. Bioinformatics. 2018;34(18):3094–100.
15. Berlin K, Koren S, Chin C-S, Drake JP, Landolin JM, Phillippy AM. Assembling large genomes with single-molecule sequencing and locality-sensitive hashing. Nat Biotechnol. 2015;33(6):623–30.
16. Li H. Minimap and miniasm: fast mapping and de novo assembly for noisy long sequences. Bioinformatics. 2016;32(14):2103–10.

17. Myers G. Efficient local alignment discovery amongst noisy long reads. In: International Workshop on Algorithms in Bioinformatics. Berlin: Springer; 2014. p. 52–67.
18. Chaisson MJ, Tesler G. Mapping single molecule sequencing reads using basic local alignment with successive refinement (BLASR): application and theory. BMC Bioinformatics. 2012;13(1):238.
19. Sović I, Šikić M, Wilm A, Fenlon SN, Chen S, Nagarajan N. Fast and sensitive mapping of nanopore sequencing reads with GraphMap. Nat Commun. 2016;7:11307.
20. Buhler J, Keich U, Sun Y. Designing seeds for similarity search in genomic DNA. J Comput Syst Sci. 2005;70(3):342–63.
21. Sun Y, Buhler J. Designing multiple simultaneous seeds for DNA similarity search. J Comput Biol. 2005;12(6):847–61.
22. Ma B, Tromp J, Li M. PatternHunter: faster and more sensitive homology search. Bioinformatics. 2002;18(3):440–5.
23. Ma B, Li M. On the complexity of the spaced seeds. J Comput Syst Sci. 2007;73(7):1024–34.
24. Nicolas F, Rivals E. Hardness of optimal spaced seed design. J Comput Syst Sci. 2008;74(5):831–49.
25. Noé L, Kucherov G. Improved hit criteria for DNA local alignment. BMC Bioinformatics. 2004;5(1):149.
26. Ono Y, Asai K, Hamada Mq. PBSIM: PacBio reads simulator—toward accurate genome assembly. Bioinformatics. 2012;29(1):119–21.
27. Myers EW. The fragment assembly string graph. Bioinformatics. 2005;21(suppl_2):79–85.
28. Aki S, Kuboki H, Hirano K. On discrete distributions of orderk. Ann Inst Stat Math. 1984;36(1):431–40.
29. Benson G. Tandem repeats finder: a program to analyze DNA sequences. Nucleic Acids Res. 1999;27(2):573.
30. Feller W. An Introduction to Probability: Theory and Its Applications vol. 1. Hoboken, New Jersey, United States: Wiley; 2008.
31. Joseph D, Meidanis J, Tiwari P. Determining DNA sequence similarity using maximum independent set algorithms for interval graphs. In: Scandinavian Workshop on Algorithm Theory. Berlin: Springer; 1992. p. 326–37.
32. Gusfield D. Algorithms on Strings, Trees and Sequences: Computer Science and Computational Biology. Cambridge, England: Cambridge University Press; 1997.
33. Rajasekaran S, Nicolae M. An elegant algorithm for the construction of suffix arrays. J Discret Algoritm. 2014;27:21–8.
34. Kärkkäinen J, Sanders P. Simple linear work suffix array construction. In: International Colloquium on Automata, Languages, and Programming. Berlin: Springer; 2003. p. 943–55.
35. Kasai T, Lee G, Arimura H, Arikawa S, Park K. Linear-time longest-common-prefix computation in suffix arrays and its applications. In: Annual Symposium on Combinatorial Pattern Matching. Berlin: Springer; 2001. p. 181–192.
36. Noé L, Kucherov G. YASS: Similarity search in DNA sequences. Research Report. 2003;20. RR-485 2, INRIA.
37. Noé L, Kucherov G. YASS: enhancing the sensitivity of DNA similarity search. Nucleic Acids Res. 2005;33(suppl_2):540–3.
38. Döring A, Weese D, Rausch T, Reinert K. SeqAn an efficient, generic c++ library for sequence analysis. BMC Bioinformatics. 2008;9(1):11.
39. Hayashi K, Morooka N, Yamamoto Y, Fujita K, Isono K, Choi S, Ohtsubo E, Baba T, Wanner BL, Mori H, et al. Highly accurate genome sequences of Escherichia coli K-12 strains MG1655 and W3110. Mol Syst Biol. 2006;2(1)2006.0007.
40. PacificBiosciences. E. coli Bacterial Assembly Primary Analysis (Instrument Output) Data. 2016. https://github.com/PacificBiosciences/DevNet/wiki/E.-coli-Bacterial-Assembly. Accessed 13 May 2016.
41. Kokot M, Długosz M, Deorowicz S. KMC 3: counting and manipulating k-mer statistics. Bioinformatics. 2017;33(17):2759–61.

Abe et al. BMC Genomics 2019, 20(Suppl 2):191
https://doi.org/10.1186/s12864-019-5476-9

BMC Genomics

RESEARCH Open Access

ENIGMA: an enterotype-like unigram mixture model for microbial association analysis

Ko Abe[1], Masaaki Hirayama[2], Kinji Ohno[3] and Teppei Shimamura[4*]

From The 17th Asia Pacific Bioinformatics Conference (APBC 2019)
Wuhan, China. 14-16 January 2019

Abstract

Background: One of the major challenges in microbial studies is detecting associations between microbial communities and a specific disease. A specialized feature of microbiome count data is that intestinal bacterial communities form clusters called as "enterotype", which are characterized by differences in specific bacterial taxa, making it difficult to analyze these data under health and disease conditions. Traditional probabilistic modeling cannot distinguish between the bacterial differences derived from enterotype and those related to a specific disease.

Results: We propose a new probabilistic model, named as ENIGMA (Enterotype-like uNIGram mixture model for Microbial Association analysis), which can be used to address these problems. ENIGMA enabled simultaneous estimation of enterotype-like clusters characterized by the abundances of signature bacterial genera and the parameters of environmental effects associated with the disease.

Conclusion: In the simulation study, we evaluated the accuracy of parameter estimation. Furthermore, by analyzing the real-world data, we detected the bacteria related to Parkinson's disease. ENIGMA is implemented in R and is available from GitHub (https://github.com/abikoushi/enigma).

Keywords: Enterotype, Topic model, Unigram mixture, Bayesian inference, Metagenomics

Background

More than 100 trillion microbes live on and within human beings and form of complex microbial communities (microbiota). Most microbes cannot be cultured in laboratories, making it difficult to understand how individual microorganisms mediate vital microbiome-host interactions under health and disease conditions. However, recent important advances in high-throughput sequencing technology have enabled observation of the composition of these intestinal microbes. For each sample drawn from an ecosystem, the number of occurrences of each operational taxonomic units (OTUs) is measured and the resulting OTU abundance can be summarized at any level of the bacterial phylogeny. Discovering recurrent microbial compositional patterns that are related to a specific disease is a significant challenge, as individuals with the same disease typically harbor different microbial community structures.

Recent large-scale sequencing surveys of the human intestinal microbiome, such as the US NIH Human Microbiome Project (HMP) and the European Metagenomics of the Human Intestinal Tract project (MetaHIT), have revealed considerable variations in microbiota composition among individuals [1, 2]. Particularly, community clusters characterized by differences in the abundance of signature taxa, referred to as enterotypes, were first reported in humans [3]. Later, other studies identified enterotype-like clusters that may reflect features of the host-microbial physiology and homeostasis in different species [4, 5] or at different human body sites [6–9]. This

*Correspondence: shimamura@med.nagoya-u.ac.jp
[4]Division of Systems Biology, Nagoya University Graduate School of Medicine, 65 Tsurumai-Cho, Showa-Ku, 466-8550 Nagoya, Japan
Full list of author information is available at the end of the article

microbial stratification has motivated the development of methods for examining unknown clusters of microbial communities.

Probabilistic modeling of microbial metagenomics data often provides a powerful framework for characterizing the microbial community structures [10–12]. For example, Knights et al. [10] applied a Dirichlet prior to a single-level hierarchy and proposed a Bayesian approach for estimating the proportion of microbial communities. Holmes et al. [11] extended the Dirichlet prior to Dirichlet multinomial mixtures to facilitate clustering of microbiome samples. Shafiei et al. [12] proposed a hierarchical model for Bayesian inference of microbial communities (BioMiCo) to identify clusters of OTUs related to environmental factors of interest.

However, such models are not suitable for discovering enterotype-like clusters of microbial communities and associations between microbes and a specific disease for the following two reasons. First, the frameworks of Knights et al. [10] and Holmes et al. [11] do not explicitly address the association between the microbial compositional patterns and environmental depend on the interest. Second, the framework of Shafiei et al. [12] models the structure of each sample using a hierarchical mixture of multinomial distributions that are depends on the factors of interest. Individual host properties such as body mass index, age, or gender cannot explain the observed enterotypes [3]. Thus, such enterotype-like clusters that describes interindividual variability among humans do not always to directly affect host probabilities such as diseases ranging from localized gastroenterologic disorders to neurologic, respiratory, metabolic hepatic, and cardiovascular illnesses.

Here, we introduce a novel probabilistic model of a microbial community structures, named as ENIGMA (Enterotype-like uNIGram mixture model for Microbial Association analysis), to address these problems. ENIGMA includes the following contributions:

1. ENIGMA uses OTU abundances as input and models each sample by the underlying unigram mixture whose parameters are represented by unknown group effects and known effects of interest. The group effects are represented by baseline parameters that change with a latent group of microbial communities. One of the most important features of our model is that the group effects are independent of the effects of interest. This enables the separation of interindividual variability and fixed effects of the host properties related to disease risk.
2. ENIGMA is regarded as Bayesian learning for detecting associations between a community structure and factors of interest. Our model can be used to simultaneously learn how enterotype-like clusters of OTUs contribute to the microbial structure and how microbial compositional patterns may be related to known features of the sample.
3. We provide an efficient learning procedure for ENIGMA by using a Laplace approximation to integrate latent variables and estimate the evidence of the complete model and credible intervals of the parameters. The software package that implements ENIGMA in the R environment is available from https://github.com/abikoushi/enigma.

We describe our proposed framework and algorithm in the "Methods" section. We evaluate the performance of ENIGMA using simulated data in terms of its accuracy to estimate parameters and identify clusters in the "Simulation study" section. We apply ENIGMA to clinical metagenomics data and demonstrate how ENIGMA simultaneously identifies enterotype-like clusters and gut microbiota related to Parkinson's disease (PD) in the "Results on real data" section.

Methods

The key idea of ENIGMA is to adjust for the effects of the enterotype and evaluate the increases and decreases of bacterial abundance associated with environmental factors. Figure 1 shows a conceptual view of ENIGMA. The formal definition of the model is described in the following Mode section. Here we introduce several notations.

Suppose that we observe microbiome count data of K taxa for N samples with M individual host properties, (y_{nk}, x_{nm}) $(n = 1, \ldots, n; k = 1, \ldots, K; m = 1, \ldots, M)$ where $y_{nk} \in \mathbb{N}$ represents the abundance of the k-th taxa in the n-th sample and x_{nm} represents a binary variable such that $x_{nm} = 1$ if the n-th sample has the m-th host property and is otherwise $x_{nm} = 0$. Here the word taxa can represent any level of the bacterial phylogeny, e.g., species, genes, family, order, etc.

Model

Figure 2 shows a plate diagram of the proposed model for metagenome sequencing, where $\boldsymbol{y}_n$ is the read count vector of the n-th sample, $\boldsymbol{x}_n$ is the vector of the host properties of the n-th sample and $z_n \in \{1, \ldots, L\}$ is a latent class of the n-th sample. Our model is a simple extension of the unigram mixture model. We assume that each sample is generated from a multinomial distribution with the parameter vector $\boldsymbol{p}_n = (p_{n1}, \ldots, p_{nK})^\top$. The elements of $\boldsymbol{p}_n$ and p_{nk} $(k = 1, \ldots, K)$ are probabilities of the occurrence of the K taxa for the n-th sample. We also assume that p_{nk} can be influenced

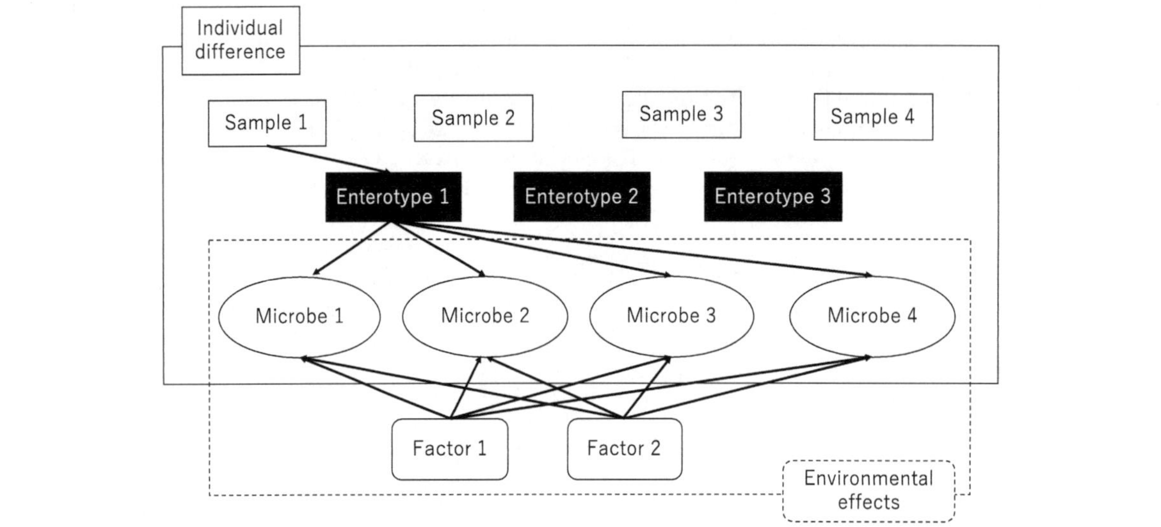

Fig. 1 A conceptual view of ENIGMA

independently by the environmental factor on the taxa that is common to all latent classes and the interindividual factor on the latent enterotype-like classes. More specifically, the generative process of ENIGMA is defined as follows:

$$\begin{aligned} \boldsymbol{y}_n | z_n, x_n, \boldsymbol{\beta} &\sim \text{Multinomial}\left(\boldsymbol{p}_n\right) \\ \boldsymbol{p}_n &= \text{softmax}\left(\boldsymbol{\gamma}_{z_n} + \boldsymbol{x}_n \boldsymbol{B}\right) \\ z_n | \boldsymbol{\pi} &\sim \text{Categorical}(\boldsymbol{\pi}) \\ \boldsymbol{\pi} | \boldsymbol{\alpha} &\sim \text{Dirichlet}(\boldsymbol{\alpha}) \\ \boldsymbol{\beta}_m &\sim \text{Normal}_K\left(O_K, \sigma^2 I_K\right) \\ \boldsymbol{\gamma}_l &\sim \text{Normal}_K\left(O_K, \tau^2 I_K\right) \end{aligned} \tag{1}$$

where $\boldsymbol{\gamma}_l$ is baseline parameter (K-dimensional vector) that changes with the latent class, $M \times K$ matrix $\boldsymbol{B} = (\beta_{mk})$ is effect of a environmental factor common to all enterotype-like clusters, $\boldsymbol{\beta}_m$ is a m-th row-vector of $\boldsymbol{B}$, $\boldsymbol{\pi} = (\pi_1, \ldots, \pi_L)$ is a mixing ratio of components, O_K is a K-dimensional zero matrix and I_K is K-dimensional identity matrix. Here, the softmax function is defined by $\text{softmax}(\boldsymbol{x}) = \frac{\exp(\boldsymbol{x})}{\sum_{k=1}^{K} \exp(x_k)}$ for a vector $\boldsymbol{x} = (x_1, \ldots, x_K)^\top$ using an element-wise exponential function and the probability function of categorical distribution is parameterized as $\Pr(z = l|\boldsymbol{\pi}) = \pi_l$, $l \in \{1, \ldots, L\}$. In a Bayesian approach, the prior distributions for $\boldsymbol{\pi}$, $\boldsymbol{\beta}$, and $\boldsymbol{\gamma}_l$ must be defiend. We set a prior based on the Dirichlet distribution for $\boldsymbol{\pi}$, and flat priors to the hyperparameters σ and τ for

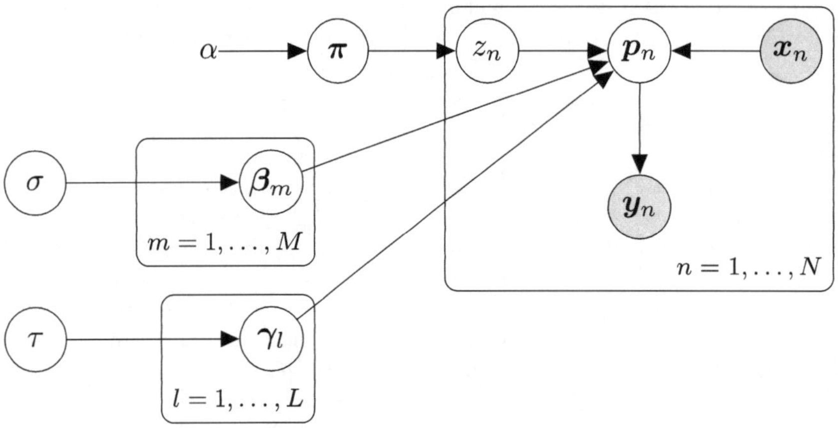

Fig. 2 Plate diagram of the model for ENIGMA $\boldsymbol{y}_n$ is affected from environmental factors $\boldsymbol{x}_n$ and latent variables $\boldsymbol{z}_n$

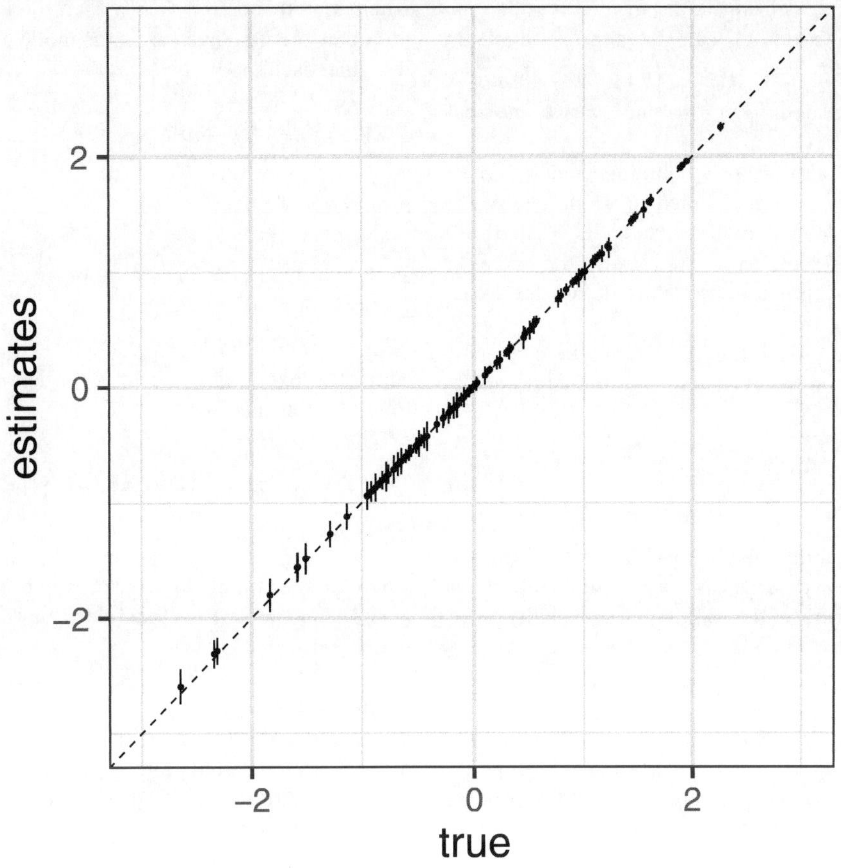

Fig. 3 Simulation result of $\boldsymbol{B}$ The comparison true $\boldsymbol{B}$ and the mean of $\hat{\boldsymbol{B}}$. The error bars indicates SE

$\boldsymbol{\beta}$ and $\boldsymbol{\gamma}$, respectively. For the convenience of later section, let $\boldsymbol{p}'_l = \mathrm{softmax}(\boldsymbol{\gamma}_l)$ be the probabilities of the occurrence of bacteria in the latent classes l.

Parameter estimation

Let us denote observed matrix by $\boldsymbol{Y} = (y_{nk})$, $\boldsymbol{X} = (x_{nm})$, the unknown parameters by $\boldsymbol{\theta} = (\boldsymbol{\alpha}, \boldsymbol{B}, \boldsymbol{\gamma}_1, \ldots, \boldsymbol{\gamma}_L, \sigma, \tau)$, and their prior by $\phi(\boldsymbol{\theta})$. The posterior distribution is represented as follows:

$$p(\boldsymbol{\theta}, \boldsymbol{z}|\boldsymbol{Y}) \propto \prod_{n=1}^{N} p(\boldsymbol{y}_n|z_n, \boldsymbol{x}_n, \boldsymbol{\theta}) p(z_n|\boldsymbol{\theta}) \phi(\boldsymbol{\theta}) \tag{2}$$

First, latent variable z_n must be marginalized. The likelihood is described by

$$\prod_{n=1}^{N} p(\boldsymbol{y}_n|\boldsymbol{x}_n, \boldsymbol{\theta}) = \prod_{n=1}^{N} \sum_{l=1}^{L} \pi_l p(\boldsymbol{y}_n|z_n = l, \boldsymbol{x}_n, \boldsymbol{\theta}). \tag{3}$$

The posterior distribution is proportional to the product of the likelihood and prior density:

$$p(\boldsymbol{\theta}|\boldsymbol{Y}) \propto \exp\left\{\sum_{n=1}^{N} \log p\left(\boldsymbol{y}_n|\boldsymbol{x}_n, \boldsymbol{\theta}\right) + \log \phi(\boldsymbol{\theta})\right\}$$

Let $\hat{\boldsymbol{\theta}}$ be the MAP estimator of $\boldsymbol{\theta}$, found by maximizing $\log p(\boldsymbol{\theta}, \boldsymbol{Y}, \boldsymbol{X})$.

We use a Laplace approximation [13] for parameter estimation. A Taylor expansion around $\hat{\boldsymbol{\theta}}$ gives

$$\log p(\boldsymbol{\theta}|\boldsymbol{Y}, \boldsymbol{X}) \approx \log p(\hat{\boldsymbol{\theta}}|\boldsymbol{Y}, \boldsymbol{X}) + \frac{1}{2}(\boldsymbol{\theta} - \hat{\boldsymbol{\theta}})^{\top} H(\hat{\boldsymbol{\theta}})(\boldsymbol{\theta} - \hat{\boldsymbol{\theta}}) \tag{4}$$

where $H(\hat{\boldsymbol{\theta}})$ is the Hessian of $\log p(\boldsymbol{\theta}|\boldsymbol{Y}, \boldsymbol{X})$ evaluated at $\hat{\boldsymbol{\theta}}$. Eq. 4 gives

$$p(\boldsymbol{\theta}|\boldsymbol{Y}, \boldsymbol{X}) \approx \frac{1}{C} \exp\left\{\frac{1}{2}(\boldsymbol{\theta} - \hat{\boldsymbol{\theta}})^{\top} H\left(\hat{\boldsymbol{\theta}}\right)\left(\boldsymbol{\theta} - \hat{\boldsymbol{\theta}}\right)\right\}$$

where C is a normalizing constant. This relationship shows that $p(\theta|\boldsymbol{Y}, \boldsymbol{X})$ can be approximated by the normal distribution $N\left(\hat{\boldsymbol{\theta}}, H^{-1}\left(\hat{\boldsymbol{\theta}}\right)\right)$. Credible intervals can be calculated from this multivariate normal distribution.

We used the stochastic programming language Stan (http://mc-stan.org/) for its implementation. The MAP estimators were obtained by the L-BFGS method. Credible intervals were computed from the using a Stan function to compute the Hessian at the MAP estimates.

After fitting the model, the enterotype-like cluster of each sample must be classified. The conditional probability of $z_n = l$ is

$$\Pr(z_n = l) = \frac{\pi_l p(y_n|\boldsymbol{\gamma}_l, \boldsymbol{\beta}, \boldsymbol{x}_n)}{\sum_{l=1}^{L} \pi_l p(y_n|\boldsymbol{\gamma}_l, \boldsymbol{\beta}, \boldsymbol{x}_n)}. \quad (5)$$

This is the probability that the n-th sample belongs to cluster l. Next, the n-th sample is then classified into the l-th cluster that maximizes the conditional probability given by Eq. 5.

Model Selection

We also examined whether or not the whole set rather than individual bacteria is related to the environmental factors of interest. We compared between the two models when $\boldsymbol{B} \neq \boldsymbol{0}$ and $\boldsymbol{B} = \boldsymbol{0}$. We used the log marginal likelihood as the goodness of fit for model comparison. The marginal likelihood is given by

$$P(\boldsymbol{Y}|\boldsymbol{X}) = \int p(\boldsymbol{Y}, \boldsymbol{\theta}|\boldsymbol{X})\, d\boldsymbol{\theta}. \quad (6)$$

From Eq. 4, we have

$$\int p(\boldsymbol{\theta}, Y|\boldsymbol{X})\, d\boldsymbol{\theta} \approx p\left(\hat{\boldsymbol{\theta}}|Y, X\right) \int \exp\left(\frac{1}{2}\left(\boldsymbol{\theta} - \hat{\boldsymbol{\theta}}\right)^{\top} H\left(\hat{\boldsymbol{\theta}}\right)\left(\boldsymbol{\theta} - \hat{\boldsymbol{\theta}}\right)\right) d\boldsymbol{\theta}. \quad (7)$$

Thus, the log marginal likelihood is approximated by the following formula:

$$\log P(Y|X) \approx \log p\left(Y|\hat{\boldsymbol{\theta}}, X\right) + \phi\left(\hat{\boldsymbol{\theta}}\right) + \frac{D}{2}\log 2\pi - \frac{1}{2}\log\left|H\left(\hat{\boldsymbol{\theta}}\right)\right| \quad (8)$$

where D is the number of free parameters. In model comparison, we choose the model showing larger log marginal likelihood.

Simulation study

To demonstrate the performance of ENIGMA, we conducted several simulation experiments. The synthetic data were naturally produced via our generative process

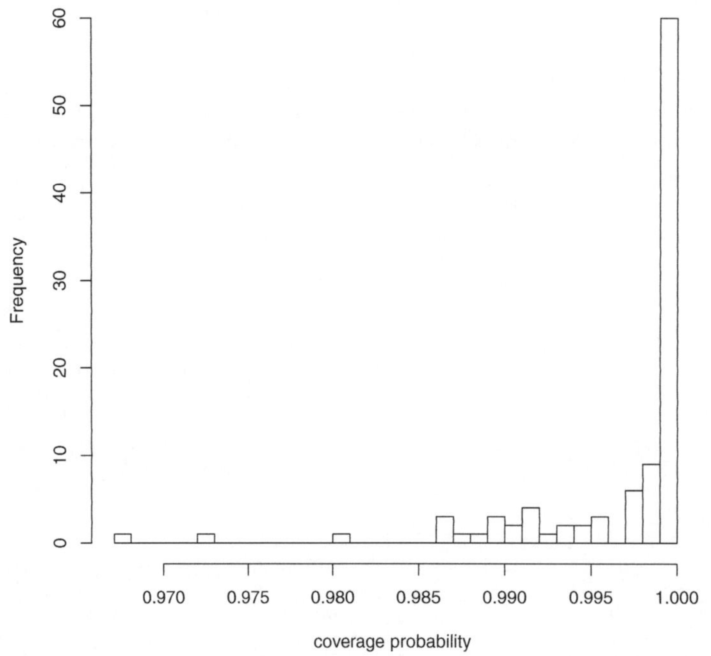

Fig. 4 Coverage probability of ***B***. The histogram of coverage probability of ***B***

given by Eq. 1. We set $M = 2000$, $L = 3$, $\pi_l = 1/3$, and $\boldsymbol{\alpha} = (1, 1, 1)^T$. We first generated $\boldsymbol{B}$ and $\boldsymbol{\gamma}_l$ from the standard normal distribution. The variables $\boldsymbol{x}_n$, z_n, and $\boldsymbol{y}_n$ are then sampled from the Bernoulli distribution with probability of 0.5, the categorical distribution, and the multinomial distribution, respectively. For the above parameter settings, we randomly generated a count dataset of 100 taxa for 100 samples for evaluation.

- **Coverage probability (CP)**: The coverage probability is the proportion of the time over which the interval contains the true value. A discrepancy between the coverage probability and the nominal coverage probability frequently occurs. When the actual coverage is greater than the nominal coverage, the interval is referred to as conservative. If the interval is conservative, there is no inconsistency in interpretation.
- **Bias**: The bias of $\boldsymbol{B}$ is defined by the difference between true value and estimated value $E[\hat{\boldsymbol{B}}] - \boldsymbol{B}$.
- **Standard error (SE)**: The standard error is the standard deviation from the estimate. A smaller standard error indicates the higher accuracy of estimation.
- **Root mean squared error (RMSE)**: The RMSE is defined by $\sqrt{E[\left(\hat{\boldsymbol{B}} - \boldsymbol{B}\right)^2]}$. A smaller RMSE indicates the higher accuracy of the estimation.
- **Accuracy**: The accuracy is the percentage of samples correctly classified into original group.

To calcurate these metrics, we detrmined that we calculated the sample means and standard deviations of $\hat{\boldsymbol{B}}$ and $\left(\hat{\boldsymbol{B}} - \boldsymbol{B}\right)^2$ from the 10,000 synthetic datasets.

Figure 3 shows a comparison of the true $\boldsymbol{B}$ and the mean and standard deviation of estimates $\hat{\boldsymbol{B}}$ obtained from the 10,000 simulations. We observed that the points were arranged diagonally, indicating that the estimator of ENIGMA was unbiased. We also calculated the proportion of the time for which the 95% credible interval contains the true value of $\boldsymbol{B}$. We found that this proportion was greater than nominal value of 0.95 for all $\boldsymbol{B}$ in Fig. 4. Table 1 shows the coverage probability (CP), bias, standard error (SE), and RMSE of $\hat{\boldsymbol{B}}$, respectively. We observed that the bias and standard error decreased when β_{mk} was large (i.e. the corresponding abundance was large). We also found that the accuracy of classification given by Eq. 5 was exactly 100%. Thus, these results indicate that ENIGMA can produce reasonable estimates.

Table 1 Coverage probability (CP), bias, standard error (SE), and RMSE of $\hat{\boldsymbol{B}}$

$\boldsymbol{\beta}$	CP	Bias	SE	RMSE	$\boldsymbol{\beta}$	CP	Bias	SE	RMSE
-3.40	0.97	0.08	0.15	0.17	-0.04	1.00	0.01	0.05	0.05
-2.65	0.97	0.06	0.15	0.16	-0.04	1.00	0.01	0.05	0.05
-2.34	0.99	0.04	0.12	0.13	-0.01	1.00	0.01	0.05	0.05
-2.32	0.99	0.03	0.12	0.12	0.01	1.00	0.01	0.04	0.04
-1.83	0.98	0.03	0.14	0.15	0.02	1.00	0.01	0.06	0.06
-1.59	0.99	0.02	0.13	0.13	0.02	1.00	0.01	0.04	0.05
-1.58	0.99	0.03	0.13	0.13	0.03	1.00	0.01	0.04	0.04
-1.51	0.99	0.02	0.14	0.14	0.10	1.00	-0.00	0.08	0.08
-1.51	0.99	0.02	0.13	0.13	0.13	1.00	0.01	0.03	0.03
-1.29	0.99	0.02	0.11	0.11	0.14	1.00	0.01	0.03	0.03
-1.14	0.99	0.01	0.11	0.11	0.21	1.00	0.01	0.06	0.06
-0.95	1.00	0.01	0.09	0.09	0.23	1.00	0.00	0.08	0.08
-0.95	0.99	0.01	0.12	0.12	0.29	1.00	0.01	0.04	0.04
-0.92	1.00	0.01	0.09	0.09	0.31	1.00	0.01	0.05	0.05
-0.88	0.99	0.01	0.12	0.12	0.32	1.00	0.00	0.08	0.08
-0.84	1.00	0.01	0.05	0.05	0.33	1.00	0.01	0.04	0.04
-0.82	1.00	0.01	0.08	0.08	0.44	0.99	-0.02	0.10	0.10
-0.78	0.99	0.01	0.13	0.13	0.46	1.00	0.01	0.05	0.05
-0.78	1.00	0.01	0.07	0.07	0.50	1.00	-0.01	0.08	0.08
-0.76	1.00	0.01	0.08	0.08	0.53	1.00	0.00	0.06	0.06
-0.72	0.99	0.00	0.12	0.12	0.54	1.00	-0.00	0.08	0.08
-0.68	1.00	0.01	0.10	0.10	0.55	1.00	0.01	0.04	0.04
-0.65	0.99	0.01	0.11	0.11	0.55	1.00	0.01	0.03	0.03
-0.65	0.99	0.01	0.11	0.11	0.56	1.00	0.01	0.05	0.05
-0.65	1.00	0.01	0.06	0.06	0.76	1.00	-0.00	0.07	0.07
-0.61	1.00	0.01	0.06	0.06	0.79	1.00	0.00	0.06	0.06
-0.58	1.00	0.01	0.06	0.06	0.84	1.00	0.00	0.05	0.05
-0.58	1.00	0.01	0.07	0.07	0.90	1.00	0.01	0.04	0.04
-0.56	1.00	0.01	0.05	0.05	0.93	1.00	0.00	0.05	0.05
-0.52	1.00	0.01	0.06	0.06	0.96	1.00	-0.01	0.08	0.08
-0.52	1.00	0.01	0.07	0.07	0.98	1.00	0.01	0.04	0.04
-0.51	1.00	0.01	0.04	0.05	1.01	1.00	-0.01	0.08	0.08
-0.50	1.00	0.01	0.05	0.05	1.08	1.00	0.00	0.05	0.06
-0.50	1.00	0.01	0.04	0.04	1.10	1.00	0.00	0.05	0.05
-0.49	0.99	0.00	0.11	0.11	1.13	1.00	0.01	0.04	0.04
-0.47	1.00	0.01	0.05	0.05	1.14	1.00	0.01	0.04	0.04
-0.45	1.00	0.01	0.09	0.09	1.16	1.00	-0.01	0.07	0.07
-0.42	0.99	-0.01	0.13	0.13	1.22	1.00	0.01	0.04	0.04
-0.33	1.00	0.01	0.07	0.07	1.23	1.00	-0.02	0.09	0.09
-0.28	1.00	0.00	0.09	0.09	1.43	1.00	0.00	0.04	0.04
-0.27	1.00	0.01	0.07	0.07	1.45	1.00	0.01	0.04	0.04
-0.23	1.00	0.00	0.09	0.09	1.47	1.00	0.00	0.04	0.04
-0.21	1.00	0.01	0.07	0.07	1.55	1.00	-0.01	0.07	0.08
-0.18	1.00	0.00	0.10	0.10	1.60	1.00	0.01	0.03	0.03
-0.15	0.99	-0.01	0.11	0.11	1.61	1.00	0.00	0.05	0.05
-0.11	1.00	0.01	0.06	0.06	1.89	1.00	0.01	0.03	0.03
-0.09	1.00	0.00	0.09	0.09	1.91	1.00	0.01	0.03	0.03
-0.05	1.00	0.01	0.04	0.04	1.95	1.00	0.01	0.02	0.02
-0.05	1.00	0.01	0.04	0.04	2.25	1.00	0.00	0.04	0.04

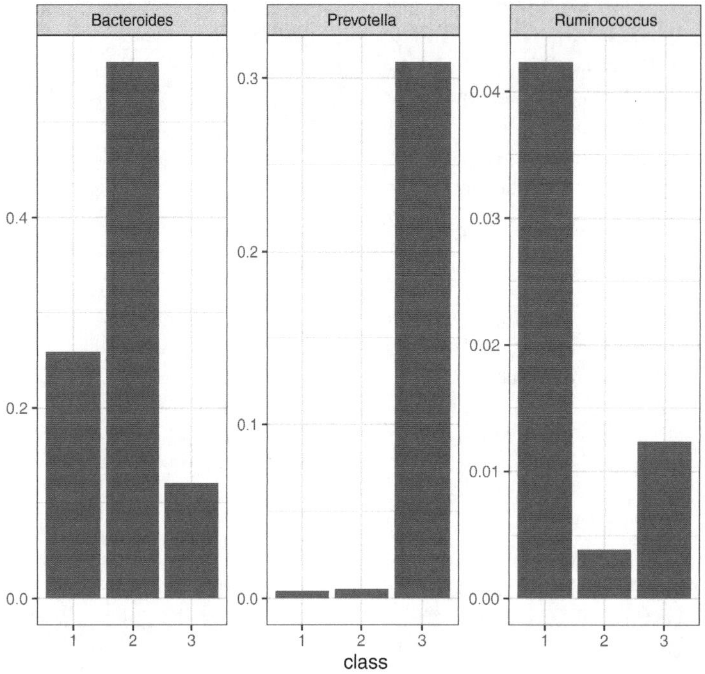

Fig. 5 Probability of occurrence in three bacteria

Results on real data

Arumugam et al. (2011)'s data

We demonstrated that the enterotype-like cluster can be estimated using the data of Arumugam et al. [3]. This data is $N = 33$, $K = 55$. The data of Arumugam et al. [3] does not disclose the total read count. Thus, We used the relative abundance multiplied by 10,000 as y_{nk}. Based on the result of Arumugam et al. [3], the number of latent classes in ENIGMA was chosen to be $L = 3$. We estimated the parameters using the ENIGMA and setting all $\beta_{mk} = 0$ in Eq. 1. We set the hyperparameters of Dirichlet prior $\boldsymbol{\alpha} = (1, \ldots, 1)^\top$, which is equivalent to a noninformative prior.

Arumugam et al. [3] showed that the enterotype is characterized by the differences in the abundance of *Bacteroides*, *Prevotella*, and *Ruminococcus*. Estimates of the probability of occurrence of those bacteria in three clusters are shown in the Fig. 5. Class 1 contains high-level *Ruminococcus*, class 2 contains high-level *Bacteroides*, and class 3 contains high-level *Prevotella*. This result is consistent with that of Arumugam et al. (2011) [3].

Parkinson's disease data

To validate the performance of ENIGMA in discovering clusters of microbial communities and associations between microbes and a specific disease, we applied ENIGMA to the real metagenomic sequencing data from Scheperjans et al. [14], Hill-Burns et al. [15], Heintz-Buschart et al. [16] and Hopfner et al. [17]. The data was analyzed by sequencing the bacterial 16S ribosomal RNA genes sampled from patients with Parkinson's disease (PD) and controls in Finland, USA, and Germany. Table 2 shows the summary statistics of the data. The OTUs were mapped to the SILVA taxonomic reference, version 132 (https://www.arb-silva.de/) and the abundances of family-level taxa were calculated.

To assess the optimal number of clusters, we used the Calinski-Harabasz (CH) Index. It is defined as:

$$\mathrm{CH}_l = \frac{\mathrm{BC}_l/(l-1)}{\mathrm{WC}_l/(n-l)} \tag{9}$$

Table 2 Data summary

	PD	CO
Finland	74	74
German	55	64
USA	207	139

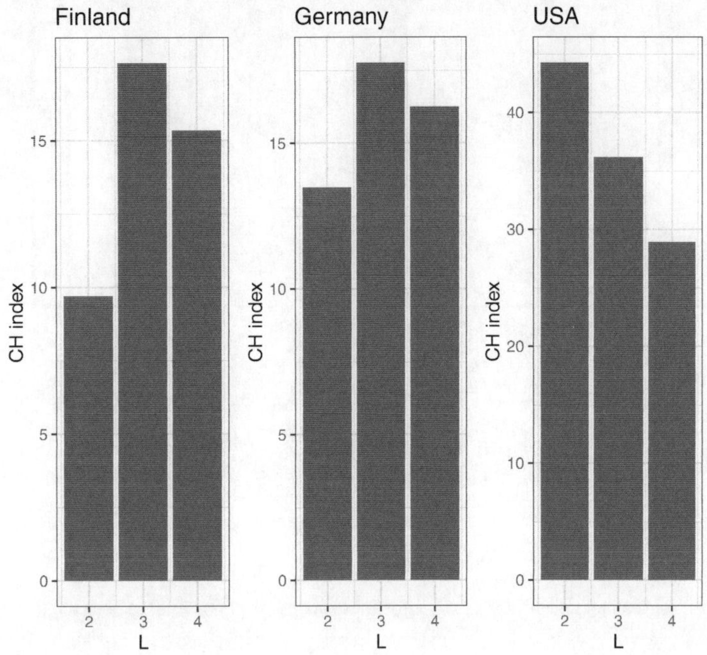

Fig. 6 The CH indexes in each country

where BC_l is the between-cluster sum of squares (i.e. the squared distances between all points i and j, for which i and j are not in the same cluster) and WC_l is the within-clusters sum of squares (i.e. the squared distances between all points i and j, for which i and j are in the same cluster). Here, we used Jensen-Shannon divergence (JSD) as the distance. The JSD between samples $\boldsymbol{a} = (a_1, \ldots, a_K)$ and $\boldsymbol{b} = (b_1, \ldots, b_K)$ is defined as follows:

$$\mathrm{JSD}(\boldsymbol{a}, \boldsymbol{b}) = \frac{1}{2}\left(\sum_{k=1}^{K} a_k \log(a_k/b_k)) + \sum_{k=1}^{K} b_k \log(b_k/a_k))\right). \quad (10)$$

When calculating the JSD, we used the normalized abundance obtained by dividing y_{nk} by the total read count, and 0 was replaced with pseudo count 10^{-6}. We chose the number of clusters L such that CH_l was maximal. To evaluate the CH Index, we use the function `index.G1()` from the R library clusterSim. The number of latent classes in ENIGMA was chosen to be $L = 3$ in Finland and Germany and $L = 2$ in USA by the CH indexes (Fig. 6).

First, we evaluated whether the model assumption was satisfied when using this data. According to Arumugam et al. (2011) [3], the gender of the host is not related with the enterotype. The genders of the subjects were published in the Finland study. We examined the relationship between gender and enterotype-like cluster using the data from Finland. Table 3 shows there was no correlation between them. We conducted a Chi-squared test for independence as shown in Table 3 and the p-value was 0.66.

We evaluated whether the model showing that bacteria were associated with PD is better than the model without the associations in terms of marginal likelihood. Marginal

Table 3 Cross-tabulation of gender and cluster

Class	1	2	3
Female	22	31	21
Male	21	27	26

Table 4 Comparison marginal likelihood

	Finland	Germany	USA
$\mathcal{M}_0$	-442734.62	-5913441.14	-3010279.35
$\mathcal{M}_1$	-355079.50	-3807297.76	-2063932.02

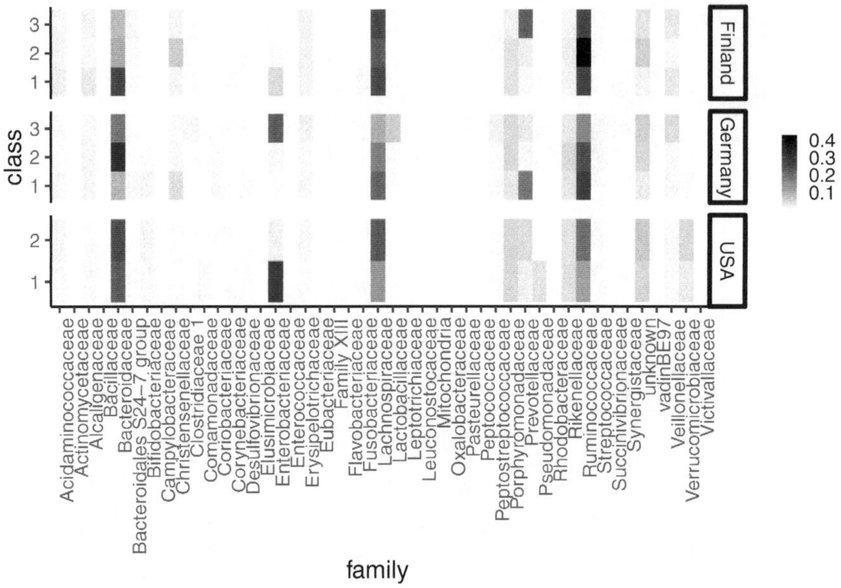

Fig. 7 Heatmap showing $(\hat{\boldsymbol{p}}'_l)$. These quantities correspond to the probabilities of the occurrences of bacteria for the three latent classes

likelihood represents the model evidence expressing the preference of the data for different models. Let $\mathcal{M}_1$ be the model which is described by Eq. 1 and $\mathcal{M}_0$ be the model setting all $\beta_{mk} = 0$ in Eq. 1. Table 4 shows that the marginal likelihood of $\mathcal{M}_1$ was greater than $\mathcal{M}_0$. It is preferrd to explain the data by considering the association between the microbiota and PD.

Figure 7 shows the estimated probabilities of the occurrences of bacteria for the three latent classes, $\boldsymbol{p}'_l$, $(l = 1, 2, 3)$. Bacteria detected in fewer three countries were removed. Arumugam et al. [3] showed that enterotype is characterized by the differences in the abundance of *Bacteroides, Prevotella,* and *Ruminococcus.* The results of ENIGMA showed the same tendency as the previous survey. Figure 8 shows the $(\hat{\boldsymbol{\gamma}}_l)'$ values and their credible intervals. The top three microbes in each enterotype-like cluster are shown in excerpts for this plot. According to the results of ENIGMA, the abundance of *Enterobacteriaceae* and *Lachnospiraceae* also differed greatly among clusters. Bacterial abundance differed between countries. In the USA, there was a high abundance of *Verrucomicrobiaceae,* while in Finland, few of these bacteria were detected. In contrast, Finland showed more *Prevotellaceae,* with fewer in the in USA it is less.

Table 5 shows the coefficients whose 95% credible intervals did not contain zero in more than two countries. The microbes with these coefficients indicates that the corresponding microbial composition patterns were significantly related to PD. We found that at the family levels, *Clostridiaceae, Comamonadaceae, Prevotellaceae, Actinomycetaceae, Bifidobacteriaceae, Enterococcaceae, Synergistaceae, Verrucomicrobiaceae* and *Victivallaceae,* the signs of the coefficients matched in all countries. These results are consistent with those of previous studies. Hill-Burns et al. [15] reported that patients with PD contained high levels of *Bifidobacteriaceae* and *Verrucomicrobiaceae.* Scheperjans et al. [14] reported PD patients contained high levels of *Verrucomicrobiaceae* and low levels of *Prevotellaceae.* Hopfner et al. reported that patients with PD have high levels of *Enterococcaceae.*

We compared ENIGMA to the Wilcoxon rank sum test, a classical methods for identifying bacteria related with an environmental factor of interest [16]. Table 6 shows the bacteria significantly related to PD with p-value < 0.05 in more than two countries. We observed that the bacteria detected by the Wilcoxon test were mostly included in those of ENIGMA (Table 5). Notably, all of the corrected p-values in Table 6 are larger than 0.05. This result shows that ENIGMA was superior to the Wilcoxon rank sum test in terms of identifying a larger number of associations between microbiota and PD.

Finally, we combined the results of ENIGMA to those of PICRUSt (version 1.1.3) [18] in order to

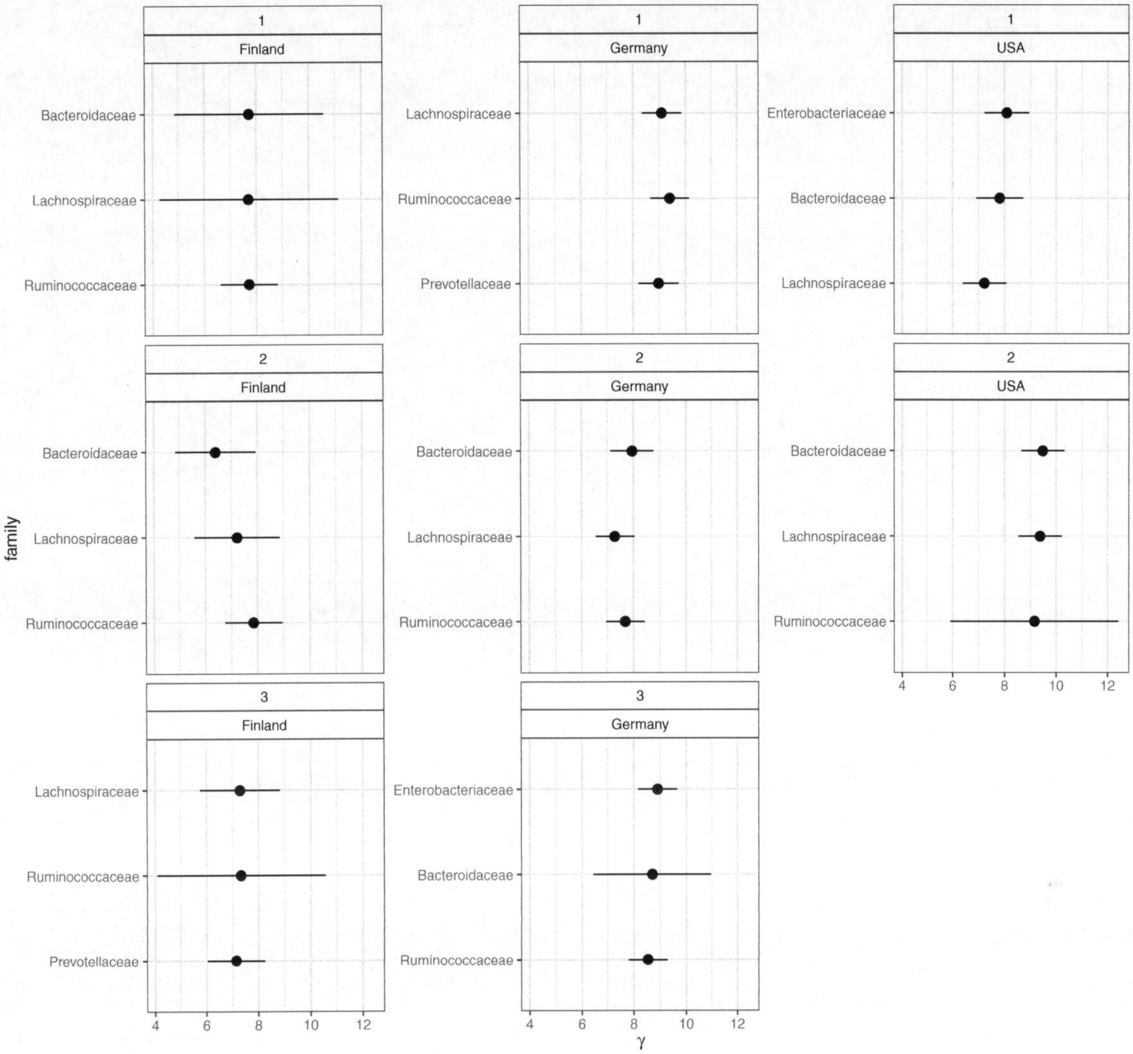

Fig. 8 Top three $(\hat{\gamma}_l)'$ in each country

evaluate which functions are related to PD. In the present study, PICRUSt was performed using the default settings. The Fig. 9 shows the functions exhibiting an increase and decrease from the median, which matched in all countries and clusters with respect to PD and control (CO). This result indicates that ENIGMA is a valuable tool for discovering new disease-related functions.

The analyses using real-world data thus show that ENIGMA can identify enterotype-like clusters and the associations between the gut microbiota and PD. Some of the results were strongly supported by those of previous studies.

Conclusion

We proposed a novel hierarchical Bayesian model, ENIGMA, for discovering the underlying microbial community structures and associations between microbiota and their environmental factors from microbial metagenome data. ENIGMA is based on a probabilistic model of a microbial community structures and supplied with labels for one or more environmental factors of interest for each sample. The structures of each sample are modeled by a multinomial distribution whose parameters are represented independently by group and environmental effects of each sample, which prevents mixing of individual differences and the effects of interest. This

Table 5 Bacteria significantly associated with PD in more than two countries

Family	Finland			Germany			USA		
	$\hat{\beta}$	Lower bound	Upper bound	$\hat{\beta}$	Lower bound	Upper bound	$\hat{\beta}$	Lower bound	Upper bound
Anaeroplasmataceae	-0.87	-1.28	-0.45	-1.69	-2.03	-1.35	-	-	-
Bacteroidales S24-7 group	-0.52	-0.93	-0.11	0.22	-0.12	0.56	-0.80	-1.16	-0.44
Bradyrhizobiaceae	-	-	-	-0.82	-1.17	-0.47	-1.44	-2.21	-0.66
Brevibacteriaceae	-	-	-	-1.02	-1.38	-0.66	-0.65	-1.05	-0.25
Brucellaceae	-	-	-	-1.69	-2.50	-0.87	-1.34	-1.75	-0.92
Clostridiaceae 1	-0.54	-0.96	-0.13	-0.08	-0.42	0.26	-0.52	-0.88	-0.16
Comamonadaceae	-0.85	-1.35	-0.35	-1.27	-1.61	-0.93	-0.21	-0.57	0.15
Elusimicrobiaceae	-4.17	-5.60	-2.74	-2.11	-2.54	-1.68	2.52	1.03	4.01
Intrasporangiaceae	-	-	-	-3.47	-4.86	-2.07	-3.00	-4.72	-1.28
Leuconostocaceae	-2.66	-4.30	-1.02	0.50	0.13	0.86	-1.74	-2.22	-1.25
Moraxellaceae	-	-	-	-1.58	-1.92	-1.24	-0.92	-1.28	-0.56
Pasteurellaceae	-1.62	-2.07	-1.17	0.30	-0.04	0.64	-1.88	-2.25	-1.51
Prevotellaceae	-2.46	-2.87	-2.05	-0.03	-0.37	0.30	-0.53	-0.89	-0.17
Rhodocyclaceae	-	-	-	-3.53	-4.93	-2.13	-0.75	-1.18	-0.32
Actinomycetaceae	0.11	-0.78	1.01	0.42	0.07	0.78	0.91	0.54	1.28
Bacillaceae	1.72	0.34	3.11	-2.35	-2.72	-1.99	0.80	0.43	1.17
Bdellovibrionaceae	-	-	-	1.43	0.40	2.46	3.07	1.78	4.36
Bifidobacteriaceae	1.34	0.82	1.86	0.54	0.20	0.88	0.01	-0.35	0.37
Campylobacteraceae	0.36	-0.31	1.03	4.90	4.48	5.33	0.83	0.46	1.21
Cytophagaceae	-	-	-	2.45	1.56	3.34	1.70	0.27	3.13
Enterococcaceae	3.87	2.70	5.05	0.74	0.40	1.08	0.09	-0.28	0.45
Lactobacillaceae	3.00	2.56	3.43	-0.51	-0.85	-0.18	1.73	1.36	2.09
Leptotrichiaceae	-0.90	-1.89	0.09	2.57	1.88	3.26	0.82	0.36	1.27
Methanobacteriaceae	-	-	-	0.93	0.59	1.27	0.67	0.30	1.04
Mitochondria	0.60	-1.27	2.46	0.73	0.11	1.36	1.57	0.95	2.20
Paenibacillaceae	-	-	-	2.19	1.28	3.10	1.71	1.30	2.12
Planococcaceae	-	-	-	1.06	0.72	1.41	3.26	2.67	3.85
Rhizobiaceae	-	-	-	0.64	0.24	1.03	1.52	1.08	1.95
Streptococcaceae	0.44	0.03	0.86	0.84	0.50	1.17	0.26	-0.10	0.62
Succinivibrionaceae	-0.32	-0.76	0.11	0.74	0.40	1.08	4.31	3.76	4.86
Synergistaceae	1.26	0.80	1.71	0.25	-0.10	0.61	1.44	1.06	1.82
Verrucomicrobiaceae	1.71	1.23	2.19	1.62	1.29	1.96	-0.06	-0.42	0.30
Victivallaceae	0.42	-0.00	0.85	0.68	0.34	1.02	0.93	0.54	1.32

The "-" notation indicates the bacteria undetected in that country

framework enables the model to simultaneously learn (*i*) how microbes contribute to an underlying community structures (cluster) and (*ii*) how microbial compositional patterns are explained by environmental factors of interest. The effectiveness of ENIGMA was evaluated through experiments involving both synthetic and read-world datasets. These newly discovered clusters and associations estimated using ENIGMA can provide insight into the the mechanisms of a microbial communities.

Table 6 *p*-Value of Wilcoxon test

	Finland	Germany	USA
Lachnospiraceae	0.009371	0.719014	0.002839
Lactobacillaceae	0.030404	0.077771	0.000002
Pasteurellaceae	0.006493	0.495315	0.004232
Prevotellaceae	0.001303	0.030892	0.194592

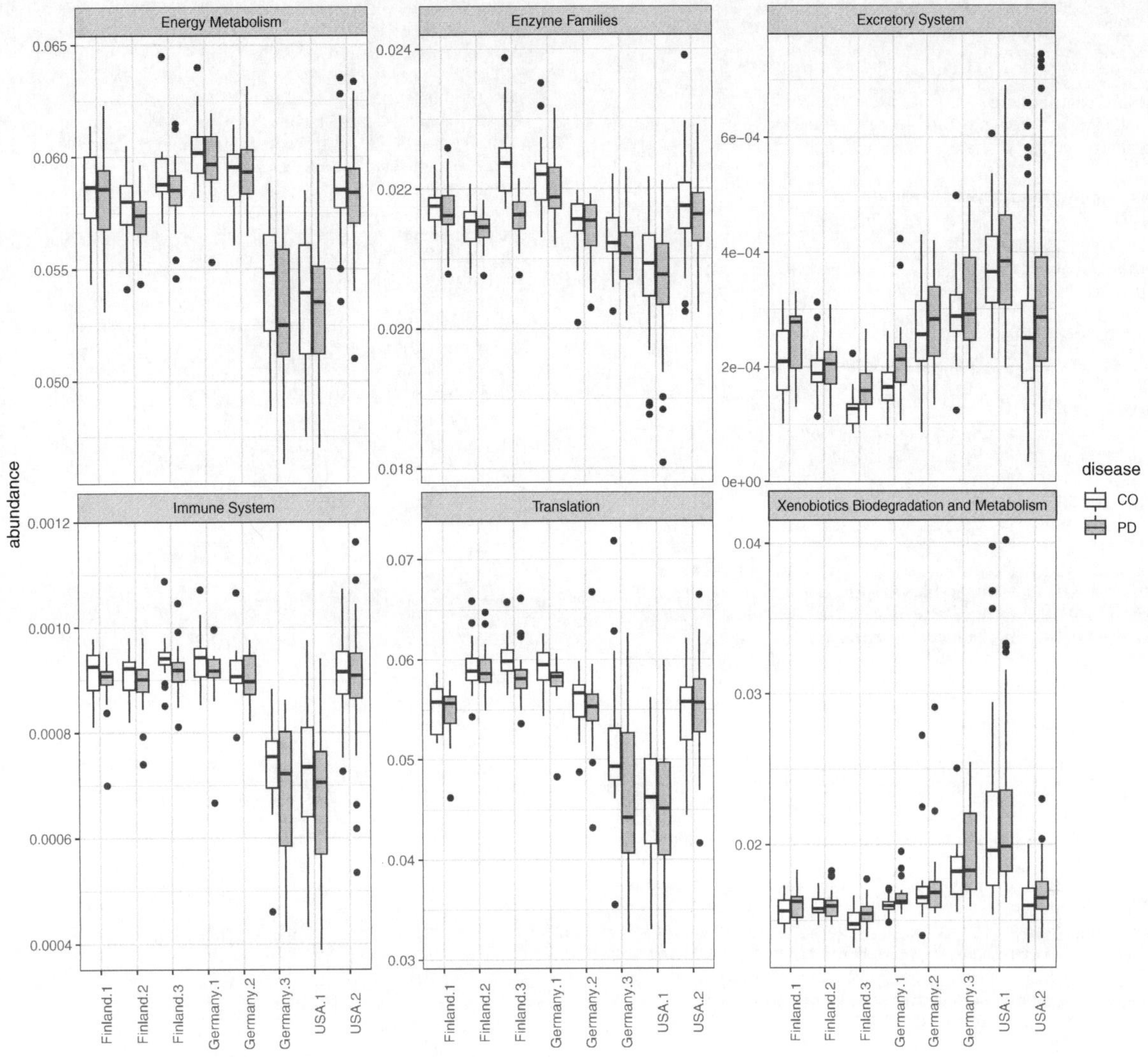

Fig. 9 Predicted functional trait abundance in each country and cluster

The major limitation of ENIGMA is its scalability and efficiency, as the number of the parameters in the model increase proportionally with the number of taxa when the number of environmental factors of interest is large. Further studies should focus on developing a scalable probabilistic model of microbial compositions to analyze underlying microbial structures with a large number of these effects by using sparse parameter estimation [19]. We are also interested in developing a dynamic probabilistic model similar to that reported by Blei and Lafferty [20] for analyzing time-varying bacteria compositions during disease progression.

Abbreviations
CH: Calinski-Harabasz; CO: Control; CP: Coverage probability; JSD: Jensen-Shannon divergence; MAP: Maximum a posteriori; OTU: Operational taxonomic units; PD: Parkinson's disease; RMSE: Root mean squared error; SE: Standard error

Acknowledgements
Not applicable.

Funding
This work was supported by Grants-in-Aid from the Ministry of Education, Culture, Sports, Science and Technology of Japan (MEXT); Ministry of Health, Labour and Welfare of Japan (MHLW); Japan Agency for Medical Research and Development (AMED), and the Hori Sciences and Arts Foundation. Publication of this article was sponsored by AMED CREST JP18gm1010002.

Availability of materials
ENIGMA is implemented with R and is available from GitHub (https://github.com/abikoushi/enigma).

About this supplement
This article has been published as part of *BMC Genomics Volume 20 Supplement 2, 2019: Selected articles from the 17th Asia Pacific Bioinformatics Conference*

(APBC 2019): genomics. The full contents of the supplement are available online at https://bmcgenomics.biomedcentral.com/articles/supplements/volume-20-supplement-2.

Authors' contributions
KA and TS designed the proposed algorithm. KO and MH designed the experiments. All authors have read and approved the final manuscript.

Ethics approval and consent to participate
Not applicable.

Consent for publication
Not applicable.

Competing interests
The authors declare that they have no competing interests.

Publisher's Note
Springer Nature remains neutral with regard to jurisdictional claims in published maps and institutional affiliations.

Author details
[1]Division of Systems Biology, Nagoya University Graduate School of Medicine, 65 Tsurumai-Cho, Showa-Ku, 466-8550 Nagoya, Japan. [2]School of Health Sciences, Nagoya University Graduate School of Medicine, 1-1-20 Daiko-Minami, Higashi-Ku, 461-8873 Nagoya, Japan. [3]Division of Neurogenetics, Center for Neurological Diseases and Cancer, Nagoya University Graduate School of Medicine, 65 Tsurumai-Cho, Showa-Ku, 466-8550 Nagoya, Japan. [4]Division of Systems Biology, Nagoya University Graduate School of Medicine, 65 Tsurumai-Cho, Showa-Ku, 466-8550 Nagoya, Japan.

Published: 10 April 2019

References
1. Huttenhower C, Gevers D, Knight R, Abubucker S, Badger JH, Chinwalla AT, et al. Human Mircrobiome Project Consortium. Structure, function and diversity of the healthy human microbiome. Nature. 2012;486:207–14.
2. Le Chatelier E, Nielsen T, Qin J, Prifti E, Hildebrand F, Falony G, et al. Richness of human gut microbiome correlates with metabolic markers. Nature. 2013;500:541-6.
3. Arumugam M, Raes J, Pelletier E, Le Paslier D, Yamada T, Mende DR, et al. Enterotypes of the human gut microbiome. Nature. 2011;473(7346):174.
4. Moeller AH, Degnan PH, Pusey AE, Wilson ML, Hahn BH, Ochman H. Chimpanzees and humans harbour compositionally similar gut enterotypes. Nat Commun. 3:1179.
5. Hildebrand F, Nguyen TL, Brinkman B, Yunta RG, Cauwe B, Vandenabeele P, et al. Inflammation-associated enterotypes, host genotype, cage and inter-individual effects drive gut microbiota variation in common laboratory mice. Genome Biol. 2013;14(1):R4.
6. Ravel J. Vaginal microbiome of reproductive-age women. Proc Natl Acad Sci USA. 2011;108.Supplement 1:4680-7.
7. Koren O, Knights D, Gonzalez A, Waldron L, Segata N, Huttenhower C, et al. A guide to enterotypes across the human body: meta-analysis of microbial community structures in human microbiome datasets. PLoS Comput Biol. 2013;9(1):e1002863.
8. Ding T, Schloss PD. Dynamics and associations of microbial community types across the human body. Nature. 2014;509:357–60.
9. Zhou Y, Mihindukulasuriya KA, Gao H, La Rose PS, Wylie KM, Martin JC, et al. Exploration of bacterial community classes in major human habitats. Genome Biol. 2014;15:R66.
10. Knights D1, Costello EK, Knight R. Supervised classification of human microbiota. FEMS Microbiol Rev. 2011;35(2):343–59.
11. Holmws I, Harris K, Quince C. Dirichlet multinomial mixtures: generative models for microbial metagenomics. PloS ONE. 2012;7(2):e30126.
12. Shafiei M, Dunn KA, Boon E, MacDonald SM, Walsh DA, Gu H, et al. BioMiCo: a supervised Bayesian model for inference of microbial community structure. Microbiome. 2015;3(1):8.
13. Bishop C. Pattern recognition and machine learning. New York: Springer-Verlag; 2006.
14. Scheperjans F, Aho V, Pereira PA, Koskinen K, Paulin L, Pekkonen E, et al. Gut microbiota are related to Parkinson's disease and clinical phenotype. Mov Disord. 2015;30(3):350–8.
15. Hill-Burns EM, Debelius JW, Morton JT, Wissemann WT, Lewis MR, Wallen ZD, et al. Parkinson's disease and Parkinson's disease medications have distinct signatures of the gut microbiome. Mov Disord. 2017;32(5):739–49.
16. Heintz-Buschart A, Pandey U, Wicke T, Sixel-Döring F, Janzen A, Sittig-Wiegand E, et al. The nasal and gut microbiome in Parkinson's disease and idiopathic rapid eye movement sleep behavior disorder. Mov Disord. 2018;33(1):88–98.
17. Hopfner F, Künstner A, Müller SH, Künzel S, Zeuner KE, Margraf NG, et al. Gut microbiota in Parkinson disease in a northern German cohort. Brain Res. 1667:41–5.
18. Langille MG, Zaneveld J, Caporaso JG, McDonald D, Knights D, Reyes JA, et al. Predictive functional profiling of microbial communities using 16S rRNA marker gene sequences. Nat Biotechnol. 2013;31(9):814–21.
19. Yang Y, Chen N, Chen T. mLDM: a new hierarchical Bayesian statistical model for sparse microbioal association discovery. bioRxiv. 2016;042630. https://doi.org/10.1101/042630.
20. Blei DM, Lafferty JD. Dynamic topic models. In: Proceedings of the 23rd international conference on Machine learning. New York: ACM; 2006. p. 113–20.

An *et al. BMC Genomics* 2019, **20**(Suppl 2):224
https://doi.org/10.1186/s12864-019-5477-8

BMC Genomics

RESEARCH

Open Access

TSEE: an elastic embedding method to visualize the dynamic gene expression patterns of time series single-cell RNA sequencing data

Shaokun An[1,3], Liang Ma[2*] and Lin Wan[1,3*]

From The 17th Asia Pacific Bioinformatics Conference (APBC 2019)
Wuhan, China. 14–16 January 2019

Abstract

Background: Time series single-cell RNA sequencing (scRNA-seq) data are emerging. However, the analysis of time series scRNA-seq data could be compromised by 1) distortion created by assorted sources of data collection and generation across time samples and 2) inheritance of cell-to-cell variations by stochastic dynamic patterns of gene expression. This calls for the development of an algorithm able to visualize time series scRNA-seq data in order to reveal latent structures and uncover dynamic transition processes.

Results: In this study, we propose an algorithm, termed time series elastic embedding (TSEE), by incorporating experimental temporal information into the elastic embedding (EE) method, in order to visualize time series scRNA-seq data. TSEE extends the EE algorithm by penalizing the proximal placement of latent points that correspond to data points otherwise separated by experimental time intervals. TSEE is herein used to visualize time series scRNA-seq datasets of embryonic developmental processed in human and zebrafish. We demonstrate that TSEE outperforms existing methods (e.g. PCA, tSNE and EE) in preserving local and global structures as well as enhancing the temporal resolution of samples. Meanwhile, TSEE reveals the dynamic oscillation patterns of gene expression waves during zebrafish embryogenesis.

Conclusions: TSEE can efficiently visualize time series scRNA-seq data by diluting the distortions of assorted sources of data variation across time stages and achieve the temporal resolution enhancement by preserving temporal order and structure. TSEE uncovers the subtle dynamic structures of gene expression patterns, facilitating further downstream dynamic modeling and analysis of gene expression processes. The computational framework of TSEE is generalizable by allowing the incorporation of other sources of information.

Keywords: Nonlinear dimensionality reduction, Elastic embedding, Visualization, Single-cell RNA sequencing, Time series, Cell fate decisions, Gene expression pattern, Oscillation, In-group proportion

*Correspondence: mal@big.ac.cn; lwan@amss.ac.cn
[1]Academy of Mathematics and Systems Science, Chinese Academy of Sciences, 100190 Beijing, China
[2]Beijing Institute of Genomics, Chinese Academy of Sciences, 100101 Beijing, China
Full list of author information is available at the end of the article

Background

Single-cell RNA sequencing (scRNA-seq) technology provides snapshots of transcriptomes at single-cell resolution, offering a comprehensive approach to study complex biological processes, such as cell fate decisions [1–3]. Given the challenges raised by high-dimensional gene expression profiles of large-scale heterogeneous cell populations at diverse stages during cell state-transitions, many computational methods have been proposed for the visualization, clustering, and reconstruction of scRNA-seq data (see [4], a recent review).

Cells used for a single-cell experiment are snapshot of heterogeneous dynamic populations. However, the dynamic range is usually limited, and only specific cell fate decisions can be analyzed when the samples are collected at a single stage or condition. Therefore, to extend the dynamic range and account for whole processes of cell development, time series scRNA-seq data are emerging. These time series data are generated by sampling single cells collected and sequenced at multiple time stages along the time course of cell developmental processes [5, 6]. Yet, by integrating time series analysis into scRNA-seq, new questions and challenges arise. Importantly, we know that scRNA-seq data across time stages are contaminated by assorted sources of variations during data collection and generation [7]. Moreover , the cell-to-cell gene expression is highly variable across the stochastic dynamics of gene transcription [8]. To address these issues, a handful computational time series scRNA-seq methods have been developed to (1) determine temporal trajectories [9], (2) reconstruct cell development landscapes based on the sophisticated mathematical tool of optimal transport [10], and (3) identify gene-gene interactions, as well as gene networks [8, 11].

However, for such methods to perform model reduction, computation and validation, they must incorporate dimensionality reduction and/or visualization of single-cell high-dimensional gene expression profiles. Accordingly, dimensionality reduction would ideally reveal the intrinsic structure of the data by representing both global structure by preserving the topology and geometry and local structure by preserving the neighborhood relationship. Among the numerous dimensionality reduction and visualization methods, principal component analysis (PCA) and t-distributed stochastic neighbor embedding (tSNE) are most widely used in the single-cell community to visualize data structures [4, 12]. However, PCA is designed to preserve linear structure based on eigendecomposition of a matrix into canonical form, and as such, it is incapable of handling nonlinear structures. While tSNE emphasizes neighborhood information to reveal the local cluster structures of the data, it tends to shatter trajectories and fails to preserve the global structures [12].

To address the limits of dimensionality reduction methods, like PCA and tSNE, we recently adopted the nonlinear dimensionality reduction algorithm, Elastic Embedding (EE) [13], to accurately visualize and reconstruct the embedded intrinsic latent space of cell development trajectory[14]. Both tSNE and EE embed high-dimensional data points into low-dimensional latent space by modeling the data points interactively with two terms. One term is an attractive force that attracts pairs of points towards each other, while the other term activates a repulsive force that simultaneously separates all pairs of points. A similar embedding idea was adopted in force-directed layout embedding [15], which has also been used in the visualization of scRNA-seq [16]. As an extension of tSNE, the EE algorithm penalizes placing latent points far from similar data points, as well as penalizes placing latent points from dissimilar data points close together [13], thereby preserving both local and global intrinsic data structures [17].

However, to the best of our knowledge, no dimensionality reduction and visualization methods can now incorporate temporal information of single-cell experiments into times series scRNA-seq data, i.e., time stages when samples are collected. Instead, scRNA-seq data from different time stages are combined as the input for methods like PCA and tSNE to obtain 2-dimensional visualization of the data structure [5, 6, 10]. TSEE can carefully incorporate experimental temporal information, resulting in significant improvement of temporal resolution of the cells on the 2-dimensional plane, as well as uncovering the subtle structures of dynamic gene expression patterns.

In this study, we propose a time series elastic embedding (TSEE) algorithm for dimensionality reduction and visualization of time series scRNA-seq data by incorporating the temporal information of the experiments. TSEE is an extension of EE by introducing an additional repulsive force term when pairs of data points are collected at distinct time stages (see Eq. 1). TSEE penalizes placing latent points in close proximity to data points otherwise separated by experimental time interval. For developmental processes, the rationale that underlies TSEE holds that data points sampled at the same time stage should be more similar than those sampled from adjacent time stages, while those data points sampled from adjacent time stages should be more similar than data points sampled from the nonadjacent time stages. In this way, TSEE preserves the temporal order and structure of time series data through the input of experimental temporal information. A similar motivation was successfully applied in the time-dependent community detection of time-varying networks [18].

In this paper, we first introduce the TSEE algorithm, and then provide an efficient numerical implementation of TSEE based on the partial-Hessian method developed by EE [19]. Next, we demonstrate the power of TSEE in the visualization of two datasets of time series scRNA-seq: human preimplantation embryo (hereinafter denoted as HPE) dataset [5] and early zebrafish embryonic development (hereinafter denoted as Zebrafish) dataset [6]. By establishing a new constraint term of temporal repulsive force, TSEE dilutes the distortions of the assorted sources of data variations across time stages and achieves temporal resolution enhancement. Compared to existing methods such as PCA, tSNE and EE, our TSEE shows superior ability to gain time resolution on 2-dimensional space by preserving local, global and temporal structures of time series scRNA-seq data. Furthermore, the visualization represented by TSEE uncovers the subtle patterns of dynamic gene expression by showing continuous patterns with regularity at the interface between samples from adjacent time stages. For example, TSEE reveals the oscillating waves of gene expression for *HER1*, *HER7*, *SOX2* and etc. along the time course, providing a solid foundation for downstream mathematical modeling and analysis. We also demonstrate robustness in the choice of TSEE.

Methods

Time series elastic embedding (TSEE)

Given the time series scRNA-seq dataset, the single-cell samples are collected at n time stages $\{t_1, t_2, \ldots, t_n\}$, and for each time stage $t_i (1 \le i \le n)$, a number of m_i single cells are sequenced with the corresponding gene expression vectors $y_1^{(t_i)}, \ldots, y_{m_i}^{(t_i)} \in \mathbb{R}^D$, where D is the number of genes for single cells. Thus, the single cell gene expression profile of a total number of $N = \sum_{i=1}^{n} m_i$ cells is contained in the data matrix

$$Y_{N\times D} = \left(y_1^{(t_1)}, \ldots, y_{m_1}^{(t_1)}, \ldots, \ldots, y_1^{(t_n)} \cdots, y_{m_n}^{(t_n)}\right)^T.$$

TSEE is an extension of EE, a nonlinear dimensionality reduction method [13] by incorporating the information of t_i, embedding the N data points in D-dimensional space into the latent d-dimensional ($d \ll D$) coordinates of

$$X_{N\times d} = \left(x_1^{(t_1)}, \ldots, x_{m_1}^{(t_1)}, \ldots, \ldots, x_1^{(t_n)} \cdots, x_{m_n}^{(t_n)}\right)^T.$$

In a manner that minimizes the pseudo potential energy function as

$$E[X;\lambda,\beta] = \sum_{n,m=1}^{N} w_{nm}^{P} \|x_n - x_m\|^2 + \lambda \sum_{n,m=1}^{N} \left(w_{nm}^{N} + \beta t_{nm}\right) \exp\left(-\|x_n - x_m\|^2\right), \quad (1)$$

where the weights $w_{nm}^P, w_{nm}^N, t_{nm}$ are defined as

$$\begin{aligned} w_{nm}^P &= N^+ w_{nm}^+ = N^+ \exp\left(\frac{-\|y_n - y_m\|^2}{2\sigma^2}\right), \\ w_{nm}^N &= N^- w_{nm}^- = N^- \|y_n - y_m\|, \\ t_{nm} &= N^- |t(n) - t(m)|. \end{aligned}$$

Among these, N^+ and N^- are normalization factors of weights for the two summation terms which, respectively, are $N^+ = \left(\sum_{n,m=1}^{N} w_{nm}^+\right)^{-1}$ and $N^- = \left(\sum_{n,m=1}^{N} w_{nm}^- + \beta |t(n) - t(m)|\right)^{-1}$; $|t(n) - t(m)|$ is the time interval between sample y_n and y_m. TSEE is different from EE by incorporating additional temporal information of t_{nm} in the model.

The pseudo potential energy function of TSEE has two terms on the right hand side of Eq. 1. The first term (*attractive* term) attracts pairs of data points towards each other, while the second term (*repulsive* term) separates all pairs of points. The weights $W^P = \left(w_{nm}^P\right)$ of the *attractive* term are defined based on the similarity between data points in high-dimensional space (original space) [20], enabling TSEE to place neighboring points in high-dimensional space still close to each other in low-dimensional embedding space. The weights of the *repulsive* term of TESS are composed of two parts. The weights $W^N = \left(w_{nm}^N\right)$ are the disparities between data points in high-dimensional space (original space), and $T^- = (t_{nm})$ are the time intervals between samples. Thus, in the constructed latent space, TSEE not only preserves the local and global structures of data hidden in original high-dimensional space, as EE does [14, 17], but also further preserves temporal structure of data by penalizing placing close together latent points that correspond to data points separated by experimental time intervals. Intuitively, the experimental temporal information of the experimental time of samples reflects, to some extent, cell development stages. In other words, the later a cell is collected in time, the more likely it will be in later developmental stages, and a longer experimental time interval between two cells typically indicates a longer distance between the cells, along the developmental process.

TSEE has two regularization parameters in Eq. 1. Parameter λ trades off the attractive and repulsive terms, while parameter β trades off the composite weights of data disparities in high-dimensional space and time intervals on the repulsive term. By fixing λ, a smaller β will have smaller effects by the temporal order/structure of samples, leading to less consideration of temporal information. If $\beta = 0$, TSEE degenerates into EE, and no temporal information will be incorporated. If fixing β, a smaller λ makes TSEE focus on local structures, determining $X_{N\times d}$ based on the attractive term, and when $\lambda = 0$, TSEE degenerates to Laplacian Eigenmap methods [13]. Choices and robustness analysis of both λ and β on embeddings in latent space are discussed in "Results" section. If either λ or β is too large, the result will either be distortion of local structures or loss of global structures in latent space.

The pseudocode of TSEE is in Algorithm 1.

Algorithm 1: TSEE algorithm

Input : single cell expression $Y_{N\times D} = (Y_1, \ldots, Y_N)^T$, time of single cells $T = (t_1, \ldots, t_N)$, λ and β, the default of which are both set to be 10

Output: low-dimensional embedding $X_{N\times d} = (X_1, \ldots, X_N)^T$

Step1: Normalize data

$$Y \leftarrow \frac{Y - \min_{i,j} y_{ij}}{\max_{i,j} y_{ij} - \min_{i,j} y_{ij}},$$

$$T \leftarrow \frac{T - \min_i t_i}{\max_i t_i - \min_i t_i}$$

Step2: Calculate two weight graphs W^P, W^N

$W^P \leftarrow$entropic affinities of $Y_1, \ldots, Y_N$

$W^- \leftarrow \{\|Y_i - Y_j\|\}_{N\times N}$

$T^- \leftarrow \{|t_i - t_j|\}_{N\times N}$

$W^N \leftarrow W^- + \beta T^-$

Normalize graphs:

$W^P \leftarrow \frac{1}{2}\left(W^P + \left(W^P\right)^T\right)$, *diagonal*$(W_P) \leftarrow 0$, $W^P \leftarrow \frac{W^P}{\sum_{i,j} w_{ij}^P}$

$W^N \leftarrow \frac{1}{2}\left(W^N + \left(W^N\right)^T\right)$, *diagonal*$(W_N) \leftarrow 0$, $W^N \leftarrow \frac{W^N}{\sum_{i,j} w_{ij}^N}$

Step3: Calculate low-dimensional embedding X

$$X \leftarrow \text{TSEE}\left(W^P, W^N, \lambda\right)$$

The numerical implementation of TSEE

In the numerical implementation of TSEE, we adopt the efficient computational method of partial-Hessian proposed in [19] to find $X_{N\times d}$ that minimizes the function E of TSEE (Eq. 1).

For a given symmetric weighted graph (matrix) $W = (w_{nm})$, the Laplacian matrix graph is defined as $L = D - W$, where $D = \text{diagonal}\left(\sum_{m=1}^{N} w_{nm}\right)$ is defined as degree matrix. L is positive semi-definite if W is non-negative. The gradient of objective function E of TSEE can be represented in terms of Laplacian as

$$\nabla E = 4LX,$$

where L is the Laplacian matrix of $W = (w_{nm})$ with

$$w_{nm} = w_{nm}^P - \lambda\left(w_{nm}^N + \beta t_{nm}\right)\exp\left(-||x_n - x_m||^2\right).$$

This optimization problem can be solved by an iterative approach in the form of

$$x_{k+1} = x_k + \alpha_k p_k,$$

where $\alpha_k > 0$ is the step size in each iteration that satisfies the linear search principles (e.g., Armijo rule), and $p_k \in \mathbb{R}^{ND}$ is the search direction. In each iteration, p_k is determined by

$$B_k p_k = -g_k,$$

where g_k is the gradient at k-th iteration, and B_k is a positive-definite matrix to achieve a descent direction satisfying $p_k^T g_k < 0$. The procedures iterate until certain conditions are satisfied, for example, the iteration terminates when the number of iterations achieves the given constraint, or the distance between x_k and x_{k+1} obtained by two adjacent iterations is less than a given threshold.

The partial-Hessian method relies on the *spectral direction* based on partial Hessian, the attractive Hessian $L^P \otimes I_d$, to strike the best compromise between deep descent and efficient computation, where L^P is the Laplacian matrix of W^P. In order to prevent the problem of singularity, a small $\mu_k I$ is added to the partial Hessian, ensuring positive definiteness of B_k, that is

$$B_k = L^P \otimes I_d + \mu_k I$$

and μ_k is set as $10^{-10}\min_{i,j}\left(L^P\right)_{i,j}$ in this study. Algorithm 2 shows the detailed steps of Partial-Hessian strategies. The authors of EE [19] demonstrated that the spectral direction obtained, as described above, can be rapidly computed, leading to global and fast convergence. Compared with EE, TSEE adds an additional repulsive factor in the objective function, which results in a simple modification of W and L with no effect on computational performance, making the complexity of TSEE comparable to that of EE. In terms of the large-scale Zebrafish dataset with sample size $\sim 40k$, the computational time of EE and TSEE is 38 min and 45 min, respectively.

Datasets

Two publicly available datasets of time series scRNA-seq are used in this study. The Zebrafish dataset [6] was obtained across 12 closely spaced stages of early zebrafish development spanning from high blastula stage (3.3 hours-postfertilization (hpf), just after transcription from the zygotic genome begins) to six-somite stage (12 hpf, shortly after the completion of gastrulation). The number of cells at each time stage ranges from 200 to 7162, comprising a total of 39,505 cells. The UMI count data can be accessed at NCBI GEO (accession no.GSE106587). The human preimplantation embryo (HPE) dataset [5] consists of individually isolated embryonic cells during the human preimplantation process, starting from the 8-cell stage at embryonic day 3 (E3) up to

Algorithm 2: TSEE: Partial-Hessian Strategy

Input : two weighted graphs W^P, W_N, λ
Output: low-dimensional embedding X
Step1: Calculate positive definite matrix based on Partial Hessian
$D^P \leftarrow diagonal\left(\sum_{j=1} w_{ij}^P\right), L^P \leftarrow D^P - W^P$
$B \leftarrow 4 \times \left(L^P + 10^{10} \min\left\{l_{ij}^P : l_{ij}^P > 0\right\} I_{N\times N}\right)$
$R \leftarrow$ upper triangular cholesky factorization of B
Step2: Search spectral directions
initialize:
$opts.maxit \leftarrow 100$, $count \leftarrow 1$
$Xold \leftarrow$random initial coordinates
$Eold \leftarrow$value of objective function at $Xold$
while $count \leq opts.maxit$ **do**
    $W \leftarrow \left\{w_{ij}^P - \lambda w_{ij}^N \exp\left(-\|Xold_i - Xold_j\|^2\right)\right\}_{N\times N}$
    $D \leftarrow diagonal\left(\sum_j w_{ij}\right)$
    $L \leftarrow D - W$
    $G \leftarrow 4LX$ # G is the gradient of objective function
    $P \leftarrow -\left(R^T\right)^{-1} R^{-1} G$ # P is the *spectral direction*
    line search on step size:
    $\alpha \leftarrow 1,\ \rho \leftarrow 0.8,\ c \leftarrow 0.1$
    $E \leftarrow$value of objective function at $Xold + \alpha P$
    $tmp \leftarrow \sum_{i,j} p_{ij} g_{ij}$
    while $E > Eold + \alpha \times c \times tmp$ **do**
        $\alpha \leftarrow \alpha \times \rho$
        $E \leftarrow$ value of objective function at $Xold + \alpha P$
    end
    $X \leftarrow Xold + \alpha P$
    $Xold \leftarrow X$
    $count \leftarrow count + 1$
end

the time point just prior to implantation at E7, consisting of a total of 1529 single cells. The data can be downloaded from https://www.ebi.ac.uk/arrayexpress/experiments/E-MTAB-3929/.

In addition, the scRNA-seq dataset of *Drosophila* embryos [21] is utilized here as a demonstration of the generalizability of TSEE for incorporating spatial information. The high-quality data of a total 1297 cells, as defined in [21], were selected. The data are used to reconstruct the spatial positions of cells in the precisely staged embryos. The 84 genes used as predictors of spatial positions of cells in the precisely staged embryos as in [21] are utilized as our spatial genes. The data can be obtained at https://shiny.mdc-berlin.de/DVEX/.

Preprocessing of time series scRNA-seq data

Prior to the implementation of TSEE, the time series scRNA-seq data are preprocessed as follows. First, we select out the most variable genes according to the Z-scores of their variations across all samples. Second, the selected gene expression profile is normalized with all values, subtracting the minimum in the matrix, followed by dividing all values by their maximum of the processed matrix, such that the values of the elements of the normalized profile matrix range from 0 to 1. The time points of samples are also normalized to range from 0 to 1 with the initial time stage set as 0 and the final time stage set as 1. Third, the PCA algorithm is utilized to select the top principal components which preserve most variations, and the dimension of components is determined following the procedure in [14]. Finally, the data of top PCA components are further normalized, as described in the second step.

Performance comparisons

We compare TSEE with PCA, tSNE and EE to evaluate their individual performance on visualizing data, preserving local and global structures, and revealing gene expression patterns.

The scatting plots of cells on the 2-dimensional embedded space are first displayed to compare their visualization results. A good visualization result should reveal the local and global structures, as well as uncover the latent specific gene expression patterns.

To measure the preservation of structures quantitatively, we adopt 1) the in-group proportion (*IGP*) [22] and 2) *IGP2*, a modification of *IGP*, to evaluate local structure preservation. In addition, we adopt the Pearson correlation coefficient (PCC) between experimental time and pseudotime obtained by DensityPath [14], to evaluate global structure preservation.

The *IGP*, defined as the proportion of samples in a group, the nearest neighbors of which are also in the same group [22], can be used to assess the accuracy of distinction of cells on 2-dimensional space. In this study, we use temporal information to define groups. For a given dataset $X = \{x_1, x_2, \ldots, x_N\}$, consisting of N samples which belong to n groups $U = \{1, 2, \ldots, u, \ldots, n\}$, we define $j^N = \arg\min_{k\neq j} \|x_j - x_k\|$ as the index of x_j's nearest sample, and we regard $Class_x(j)$ as the label for sample x_j. Based on the definition above, *IGP* of group u based on dataset X is defined as

$$IGP(u, X) = \frac{\#\left\{j|Class_x(j) = Class_x\left(j^N\right) = u\right\}}{\#\{j|Class_x(j) = u\}}.$$

The value of *IGP* ranges from 0 to 1, and a higher value indicates a better cell distinction between groups.

Because of the heterogeneity of cells, the cells at distinct time stages may still belong to the same cell type, which

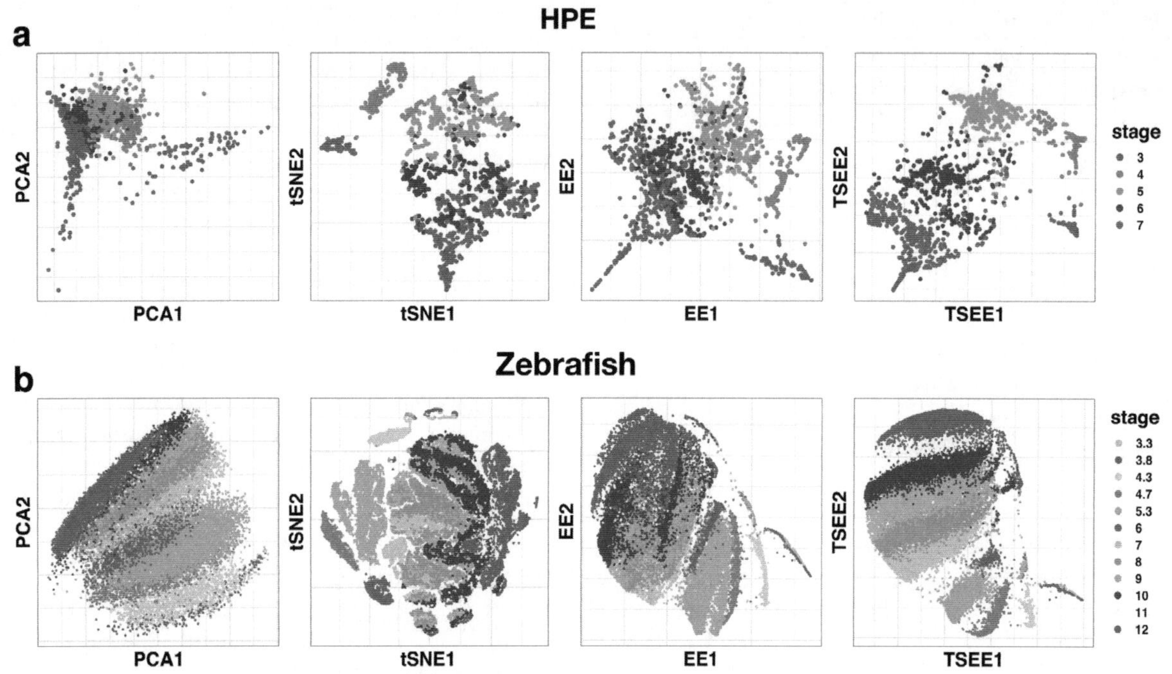

Fig. 1 Visualization results by PCA, tSNE, EE and TSEE. The four dimensionality reduction methods are applied to two datasets of time series scRNA-seq: **a** HPE and **b** Zebrafish. Each cell is colored according to the corresponding experimental time stage

means that *IGP* tends to underestimate the performance of cell-specific distinction. Consequently, we weaken the requirements of *IGP* and propose a new IGP score by defining the proportion of samples in a group, the nearest neighbors of which are also in the same, or adjacent group (adjacent time stages here). The so-called *IGP2* of the adapted index is formulated as

$$IGP2(u,X) = \frac{\#\{j||Class_x(j) - Class_x(j^N)| \leq 1, Class_x(j) = u\}}{\#\{j|Class_x(j) = u\}}. \quad (2)$$

In this study, the tSNE algorithm is implemented by the *Rtsne* function in the R package **Rtsne** with default parameters, and the EE algorithm is performed according to the code downloaded from http://faculty.ucmerced.edu/mcarreira-perpinan/software.html.

Results

We analyze two time series datasets: Zebrafish [6] and HPE [5] (see "Methods" section for details). To validate the performance of TSEE, we compare it with three other dimensionality reduction algorithms, PCA, tSNE and EE.

TSEE outperforms other dimensionality reduction methods in visualization and structure preservation

We apply the 4 dimensionality reduction algorithms noted above to HPE and Zebrafish datasets and demonstrate the embedding results on the 2-dimensional embedding space to visualize cell development structures. TSEE not only preserves the local and global data structures, but also has higher temporal resolution of cells with different time stages (Fig. 1).

More specifically, in the HPE dataset, PCA combines cells from E4 to E7 tightly, grouping cells into two groups visually, one mainly containing cells at E3 and the other

Table 1 Evaluation of local structure preservation based on averaged values of *IGP*

	PCA	tSNE	EE	TSEE
HPE	0.7747	0.8900	0.8466	0.9407
Zebrafish	0.4757	0.8799	0.7127	0.8943

Table 2 Evaluation of local structure preservation based on averaged values of *IGP2*

	PCA	tSNE	EE	TSEE
HPE	0.9661	0.9961	0.9936	0.9965
Zebrafish	0.8525	0.9930	0.9196	0.9954

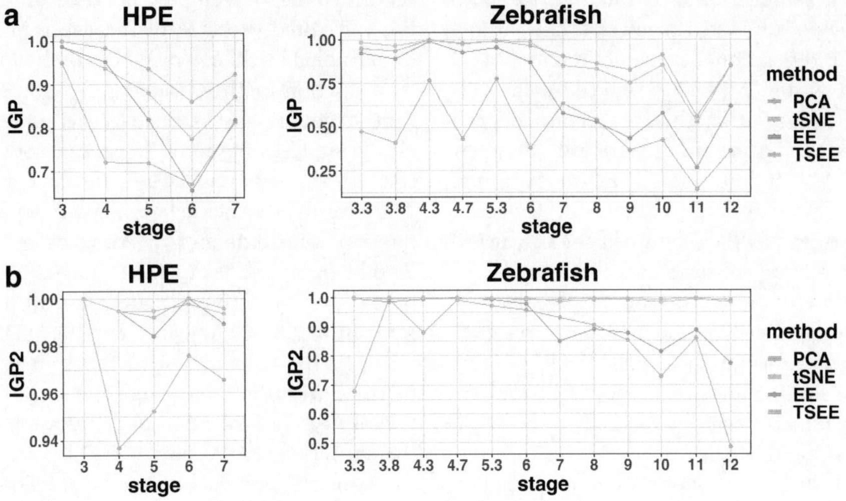

Fig. 2 Comparison of local structure preservation performance by *IGP* and *IGP2*. Performances in local structure preservation of PCA, tSNE, EE and TSEE is individually evaluated based on two metrics, **a** *IGP* and **b** *IGP2*, on the HPE and Zebrafish datasets

filled with cells from E4 to E7, which would affect downstream analysis, such as clustering. The embedded points with various stages obtained by tSNE and EE are both evenly distributed as a whole, but cells from E5 to E7 heavily overlap, with some E7 cells even falling into the area of E5 cells (Fig. 1a). For TSEE, cells at different time stages are well separated so that only cells at adjacent time stages are mixed in agreement with the heterogeneity of cells (Fig. 1a).

For the Zebrafish dataset, all four methods arrange cells along time with relatively large gaps arising in early developmental stages (Fig. 1b). However, PCA and EE results show that cells from 8 to 12 hpf are highly mixed, almost stacked together, even when two or more time intervals exist among them. Although clearly separating points with different time stages, tSNE shows distorted 2-dimensional temporal structures in that cells from 3.3 hpf to 4.3 hpf are distributed along time course, while cells at 4.7 hpf are separated with the cells before 4.7 hpf by cells at 5.3 hpf (Fig. 1b). In contrast, for TSEE, the cells from 3.3 to 6 hpf are obviously separated. Even though cells from 7 hpf to 12 hpf are not separated as distinctly as the cells before 6 hpf, significant gaps still exist between cells at adjacent time stages, thus showing the best performance on preserving structures along time (Fig. 1b).

To quantitatively evaluate the performance of TSEE, we apply *IGP* and *IGP2* scores to measure the distinction of cells, regarding time stages as standard of cell groups. TSEE always has highest averaged values of *IGP* and *IGP2* (see Tables 1 and 2), and its performance is similar to that of tSNE based on the two scores across all time stages in HPE and Zebrafish (Fig. 2), indicating that TSEE preserves the local structures and differentiates cells at various time stages better than the three other methods.

To evaluate the preservation of global and temporal structures by the four methods, we measure performance quantitatively by applying DensityPath [14], which has high accuracy in the reconstruction of cell development

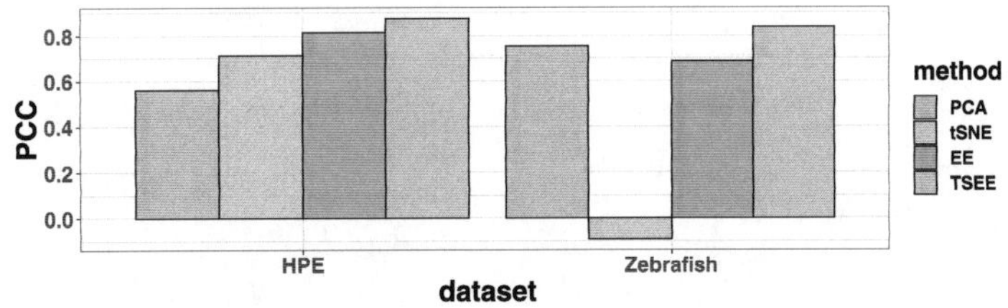

Fig. 3 Quantitative comparison of global structure preservation. Performance in global structure preservation of PCA, tSNE, EE and TSEE is evaluated based on PCC metric of pseudotime calculated by DensityPath and experimental time for the HPE and Zebrafish datasets

trajectory, as well as pseudotime calculation, for both datasets. Since our purpose is to investigate the performance of the four dimensionality reductions, we only apply DensityPath to dimensionality reduced data of 2-dimensional space by the four methods to reconstruct cell state-transition path and calculate pseudotime. After setting the root as 1472-th and 1-st cell in the HPE and Zebrafish datasets, respectively, the accuracy of calculated pseudotime is measured by PCC between the calculated pseudotime and experimental time of the cells. A larger value of PCC indicates more consistency in global structure preservation along time. We find that TSEE preserves temporal structure information best, while the PCC value of tSNE on the Zebrafish dataset is negative, indicating that the embedding of cells by tSNE distorts whole structures of data along time (Fig. 3).

The hierarchical clustering algorithm is also applied to measure the preservation of distance of time stage. The complete linkage method is adopted here based on centroids of samples at each time stages, where the calculated distance of the centroids is based on 2-dimensional Euclidean distances of the four dimensionality reduction methods separately. Although the adjacent time stages are always close to each other in hierarchical clustering trees based on the results of all four methods (Fig. 4), the clustering tree constructed on the basis of TSEE results shows the best consistency between the distance of centroids and the time order. Specifically, the lengths of leaf nodes (3.3, 3.8, 4.3, 4.7, 5.3, and 6) to the root of the clustering tree decrease with increasing time stage from 3.3 to 6 hpf, and then the lengths of leaf nodes (7, 8, 9, 10, 11, and 12) to the root of clustering tree increase with the increasing time stage from 7 to 12 hpf (Fig. 4). As an indication of cell development processes, the clustering tree result by TSEE indicates that the centroids of cell populations propagate along one direction from 3.3 to 6 hpf and then propagate along another direction from 7 to 12 hpf, which can be further supported by

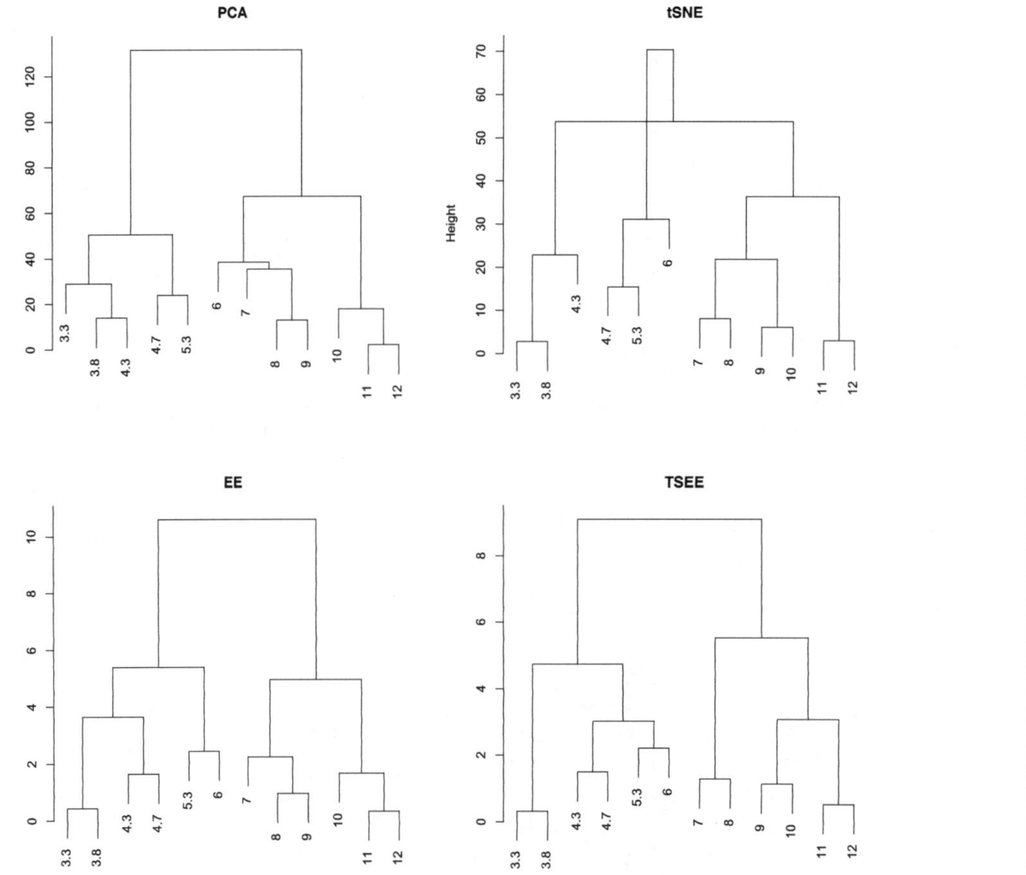

Fig. 4 Hierarchical cluster tree of centroids at various time stage, as obtained by PCA, tSNE, EE and TSEE. The centroids of samples with different time stages are submitted to hierarchical clustering to validate the preservation of time order for the Zebrafish dataset

the large gap between 6 and 7 hpf on the 2-dimensional space by tSNE, EE, and TSEE (Fig. 1b).

TSEE reveals subtle dynamic patterns of gene expression during zebrafish embryogenesis on 2-dimensional space

The large-scale Zebrafish dataset with closely spaced stages allows us to further explore dynamic patterns of gene expression during zebrafish embryogenesis.

The power of TSEE to allow visualization of structures that show dynamic patterns of gene expression can first be illustrated through the expression of marker genes *HER1* and *HER7* on the 2-dimensional space by PCA, tSNE, EE and TSEE in Zebrafish data (Fig. 5). *HER1* and *HER7*, for which the existence of oscillation expression patterns by negative feedback has been reported in presomitic mesoderm of zebrafish [23], clearly show oscillatory expression patterns along time stages on the 2-dimensional space of TSEE, indicating that TSEE successfully reveals the underlying dynamic patterns of gene expression. On the other hand, PCA only shows two wide strips of highly expressed genes on the whole plot, and tSNE gathers cells with highly expressed genes together without any oscillation pattern whatsoever. Meanwhile EE just shows a narrow region with oscillation expression pattern in early developmental stages, and the cells with high-level expression are stacked in the latter developmental stages, resulting in low temporal resolution. For TSEE, however, points on the 2-dimensional plane clearly demonstrate the oscillation patterns of gene expression with high resolution, validating the effectiveness of TSEE in preserving the intrinsic structures of data with high resolution, while, at the same time, allowing visualization of the underlying dynamic structures of genes.

We also explore the expression patterns of genes related to embryogenesis from the list provided by [6] on the 2-dimensional spaces by the four methods to examine the developmental patterns. Here, visualization by TSEE can also show clearer oscillation waves of gene expressions along time compared to the other methods (see Fig. 6 for four examples). PCA tends to distribute the cells with high-level gene expression uniformly, or accumulate them in the region of late developmental stages. The tSNE methods tends to gather together the cells with high gene expression, while small gaps occur among them, but this only shows meaningful clusters of cells without any evidence of patterns of gene expression. EE displays the products resulting from mixtures of cells at various time stages as the region of high-level gene expression, where genes are continuously expressed. The three methods place together the regions where the specific genes are highly expressed, failing to uncover the dynamic structures even if some special gene expression patterns exist among all cells. In TSEE, different patterns of gene expression can be revealed. As examples, *WNT8A*, an essential transcript for zebrafish axis development [24], displays an oscillatory gene expression pattern throughout

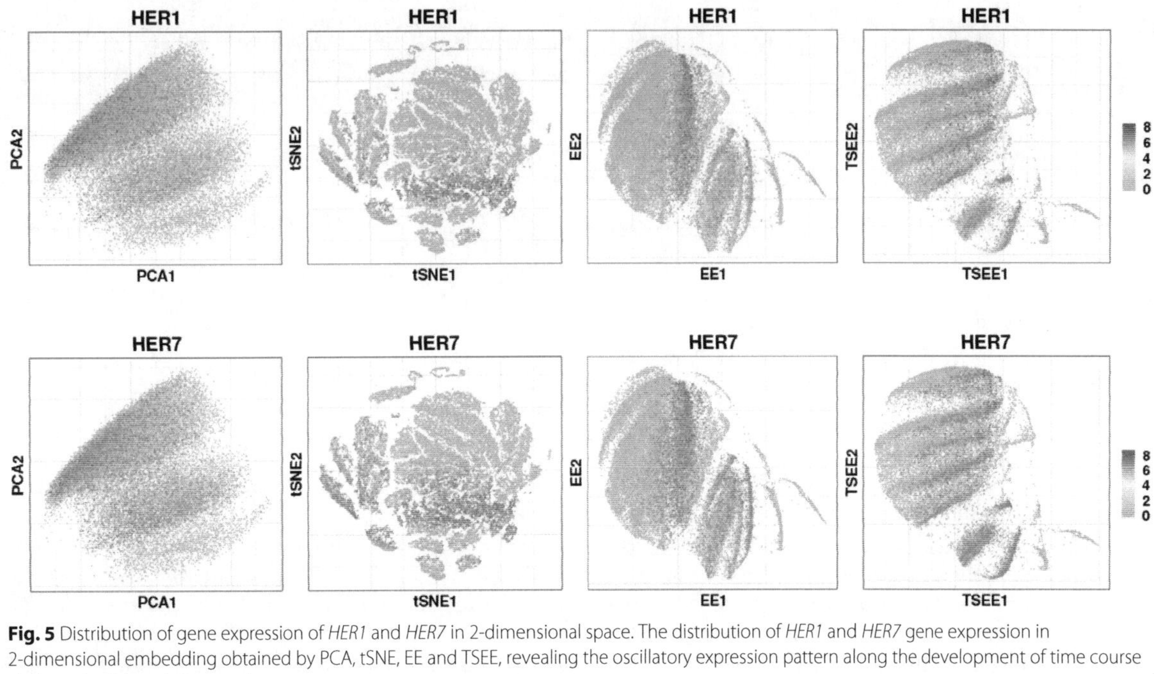

Fig. 5 Distribution of gene expression of *HER1* and *HER7* in 2-dimensional space. The distribution of *HER1* and *HER7* gene expression in 2-dimensional embedding obtained by PCA, tSNE, EE and TSEE, revealing the oscillatory expression pattern along the development of time course during zebrafish embryogenesis

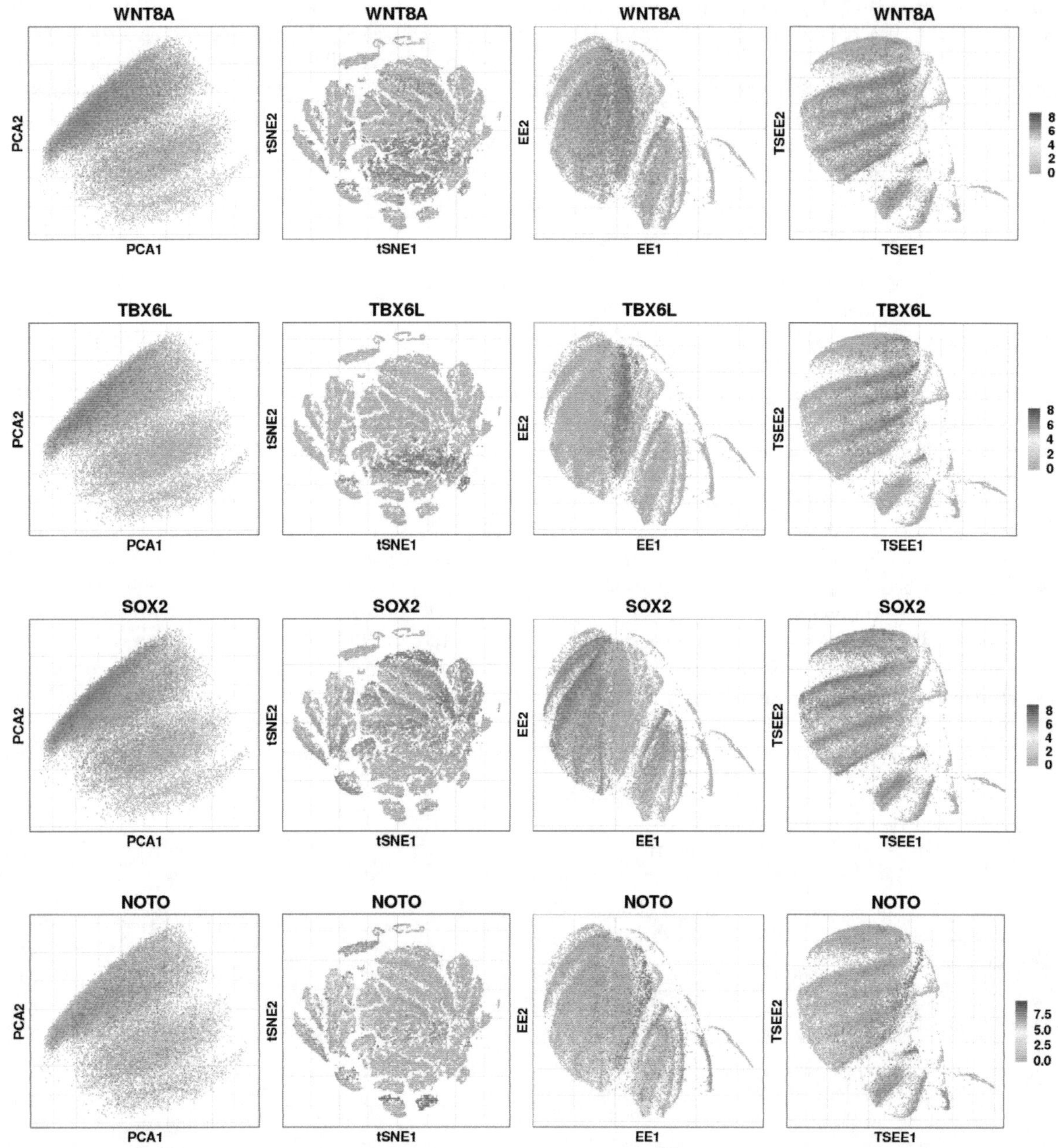

Fig. 6 Distribution of expression of four genes in 2-dimensional space. Distribution of expression of four genes in 2-dimensional embedding obtained by PCA, tSNE, EE and TSEE. Dynamic gene expression patterns are revealed by TSEE

all samples. *TBX6L*, which contributes to posterior paraxial mesoderm formation during zebrafish embryogenesis [25], reflects a periodically expressed pattern in the right region. *SOX2*, identified as an essential transcription factor to maintain self-renewal or pluripotency [26], reflects oscillation, as well, in the left region in the whole stages. Finally, *NOTO* displays a periodic gene expression pattern entirely, but fades away gradually along time. More genes with an oscillating pattern if expression can be found at https://github.com/ShaokunAn/TSEE.

Oscillating patterns of gene expression revealed by TSEE may be linked to the genome-scale oscillations in DNA methylation during exit from pluripotency [27]. Therefore, our TSEE results provide new insights and perspectives for subsequent analysis of dynamic transitions and regulation mechanisms of key genes.

Parameter choices and robustness analysis of TSEE

Two tuning parameters, λ and β, are found in the pseudo potential energy function E of TSEE (Eq. 1). The parameter choices of λ and β are critical to the embedding results by TSEE. In the implementation of TSEE, we set the default values of both λ and β equal to 10.

To test the robustness of the parameter choices, we first choose the values of each parameter separately in a wide range from 1 to 1000, while keeping the other parameter fixed at default, and calculate weighted mean of *IGP* scores for both HPE and Zebrafish datasets (Fig. 7). The weighted mean of *IGP* is the sum of *IGPs* at each time stage, weighting by the proportion of samples (cells) at each time stage.

When fixing $\beta = 10$, TSEE achieves the highest weighted mean of *IGP* at $\lambda = 100$ in both datasets (Fig. 7a,b). However, the large $\lambda(\geq 100)$ tends to separate samples into clusters on the 2-dimensional space of TSEE, breaking the structure into discontinuities (Fig. 8a,c), which is inappropriate for continuous embryonic development. Therefore, $\lambda = 10$ best balances the preservation of local and global structures.

When fixing $\lambda = 10$, the weighted mean of *IGP* increases sharply when increasing β from 1 to 10 and tends to be saturated after $\beta \geq 100$ (Fig. 7c,d). Although the global structures of cells on the 2-dimensional space of TSEE are stable by varying β from 1 to 1000, TSEE may sacrifice some local structures with large $\beta(\geq 100)$ since the repulsive terms are determined by the difference between time stages, while the dissimilarities based on gene expression are seldom considered. For example, the two corner sections at stage E7 and the one at stage E4 become less distinguishable as β grows in HPE data. Therefore, $\beta = 10$ is sufficient for the incorporation of temporal information data and trades off the weights between gene expression and time stage.

To determine the parameters more precisely, as well as analyze the robustness of TSEE, we further study the performance of TSEE when the parameters are tested in detail in the region from 1 to 100. The three metrics *IGP*, *IGP2* and PCC are employed here. Based on *IGP* and *IGP2*, TSEE has similar performance for various β and λ from 1 to 50 in the HPE and Zebrafish datasets, indicating that TSEE is quite robust to the change of parameters on preserving local structures. We calculate PCC for HPE and Zebrafish data to analyze the robustness to parameter choice, as well. Figure 9 demonstrates that TSEE is quite robust for the choices of β, as well as choices λ, when varying them separately in the range from 1 to 50.

Based on all the results above, we set the optimal default values of both β and λ to be equal to 10.

Discussion

In summary, we propose a novel visualization method, TSEE, for time series scRNA-seq data. To the best of our knowledge, in the single-cell community, TSEE is the first visualization method that considers additional temporal information.

TSEE enhances resolution of the map by correcting for unknown noise variation. The repulsive force based on time intervals of samples enables TSEE to align cells along time, preserving temporal structures, while the intrinsic structures remain well preserved owing to the incorporation of the attractive term and the repulsive term based on their distance in gene expression space. Figure 10 displays the correction of cells by TSEE. The samples colored in red are those for which the nearest neighboring points are at least two time intervals apart, indicating with high probability that they are misplaced in the 2-dimensional plane. The misplaced points by EE are distributed uniformly (Fig. 10a), but when displayed on the 2-dimensional space obtained by TSEE (Fig. 10b), they are aligned into their respective time stages, indicating that TSEE corrects the misplacement of samples by EE. Moreover, the misplaced cells by TSEE are generally located in the interface of two time stages (Fig. 10c), which makes sense since cells are typically heterogeneous, especially at late developmental stages. Because of the

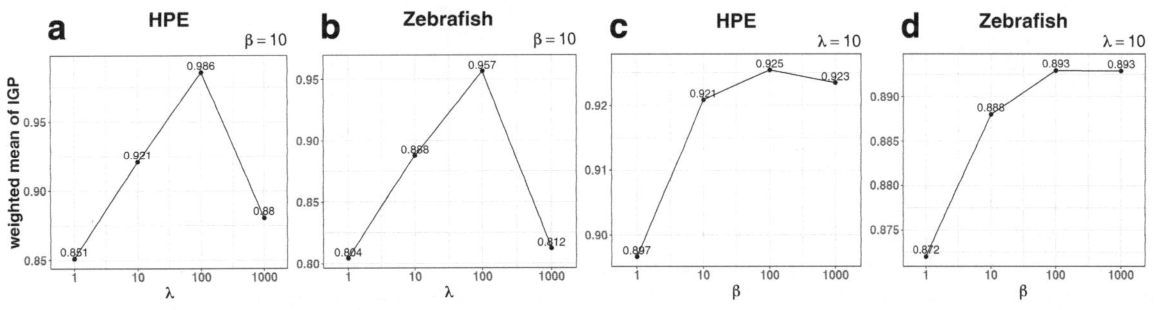

Fig. 7 Weighted mean of *IGP* under different parameter values. The y-axis is the weighted mean of *IGP* where the weights are the proportions of the number of cells at each time stage. The x-axis represents parameter choices (in logarithmic scale). (**a**) results of different λ for HPE dataset, (**b**) results of different λ for Zebrafish dataset, (**c**) results of different β for HPE dataset, (**d**) results of different β for Zebrafish dataset

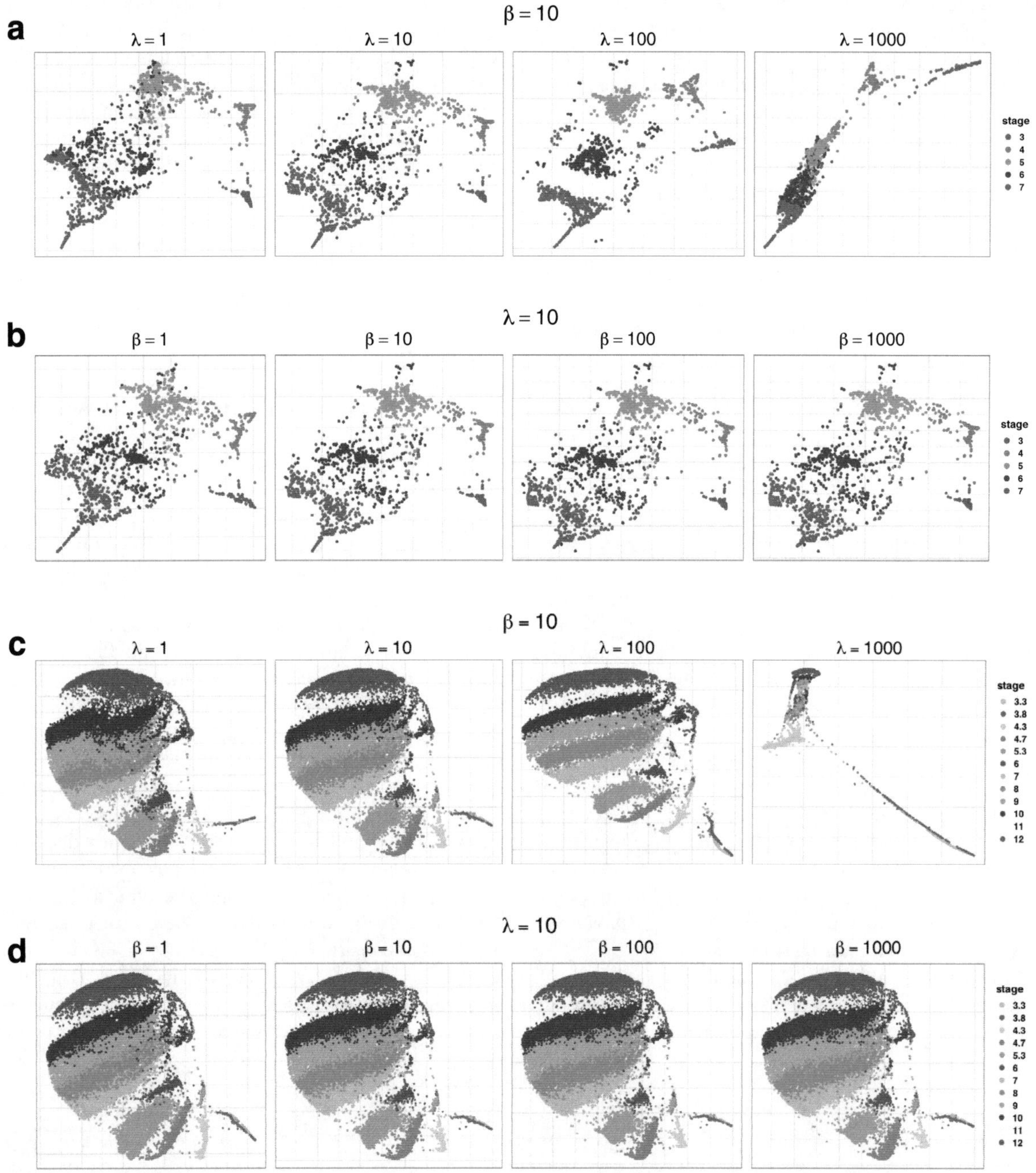

Fig. 8 The 2-dimensional embeddings of TSEE under different parameter settings. The visualization results of TSEE by varying one parameter, while the other is fixed at the default value of 10, are displayed on **a**, **b** for the HPE dataset and on **c**, **d** for the Zebrafish dataset

correction of cells along time, gene expression patterns can be revealed, such as the oscillation process of *HER1* and *HER7* in Zebrafish data, as demonstrated in the "Results" section. The increasing number of genes with oscillatory expression pattern discovered is further supported by the genome-scale oscillations in DNA methylation[27], and the uncovered subtle dynamic structures of time series scRNA-seq data, as, for example, through the use of TSEE, can be utilized for further analysis of gene expression.

The visualization tool of the forced-directed layout [15] utilizes the concept of attractive force and repulsive force, as well, to visualize network structures in low-dimensional space. The data tend to be collapsed along

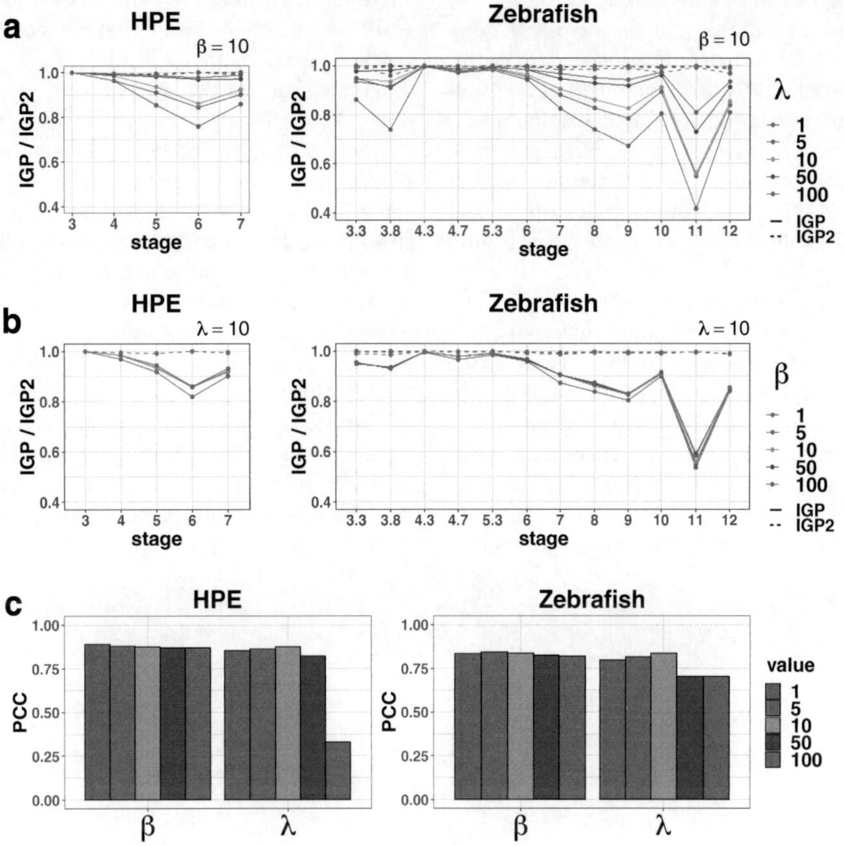

Fig. 9 Robustness analysis of parameters on two datasets. Robustness analysis is performed for λ and β on the two datasets. To evaluate preservation of local structure under each set of parameters, *IGP* and *IGP2* are used. **a** When β is set as default 10, *IGP* and *IGP2* display the effect by the change of λ. **b** When λ is set as default 10, *IGP* and *IGP2* display the effect by the change of β. **c** PCC is used to assess the preservation of global and temporal structures

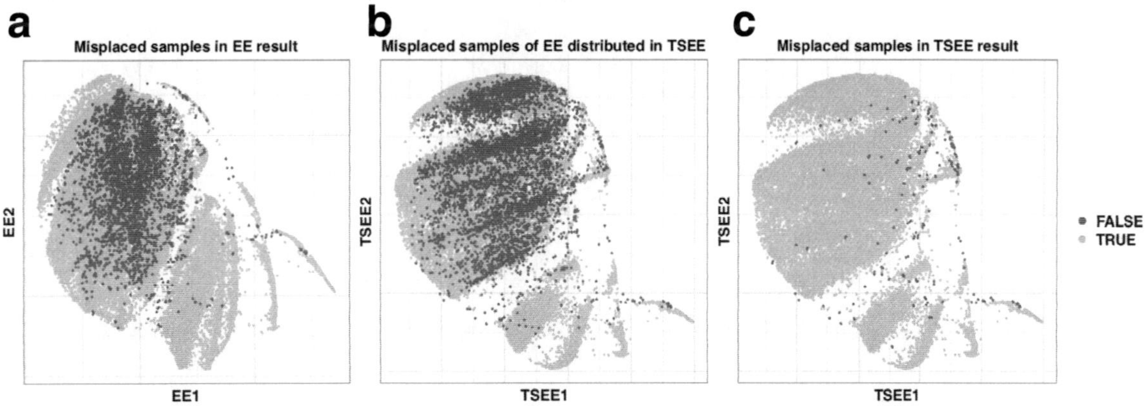

Fig. 10 Distribution of misplaced samples based on 2-dimensional embeddings. When one cell's nearest neighboring point is at least two time intervals away from it, the cell is considered to be misplaced with high probability and is colored in red in above figures. **a** shows the distribution of misplaced cells by EE on latent space of EE, **b** displays the distribution of misplaced cells by EE on latent space of TSEE, and **c** displays the distribution of misplaced cells by TSEE on latent space of TSEE

branches, covering cell-to-cell heterogeneity, as well as latent gene expression pattern, and the results in low-dimensional space generally need to be adjusted manually. For example, the marker *WNT8A* demonstrates an oscillating expression pattern in the TSEE visualization result, but the development tree result in the Supplementary File of [6], which was obtained by a hand-tuned force-direct layout merely shows a growing tendency along time, and the heterogeneity of cells is hidden because cells collapse along branches.

The computational framework of TSEE can also be potentially extended to incorporate other sources of information (e.g., spatial information) of scRNA-seq data. We utilize the scRNA-seq data of embryo cells from [21] to demonstrate the extended application based on spatial information of cells. A set of 84 marker genes are considered to uniquely classify almost every position within the embryo [21]. Therefore, we employ the information from the 84 marker genes to quantify the spatial distance of single cells, defining as $1 - \text{PCC}\left(x_i, x_j\right)$ between samples i and j, where x_i and x_j are the expression vectors of the 84 marker genes. Figure 11 displays the expression of three markers of dorsal ectoderm in 2-dimensional space, as obtained by tSNE, EE, and TSEE, respectively. Compared to the results of EE and tSNE, the highly expressed genes tend to locate along the boundary of the 2-dimensional space by the modified spatial TSEE, showing better consistency with the spatial patterns of the genes in the embryo. The accurate reconstruction of spatial location of single cells will be challenging, and this will be an interesting topic for our future study.

Conclusions

In this study, we propose an efficient algorithm, TSEE, for the visualization of time series scRNA-seq data. TSEE is an extension of EE by introducing an additional repulsive force term when pairs of data points are collected at distinct time stages, thereby balancing the effects from

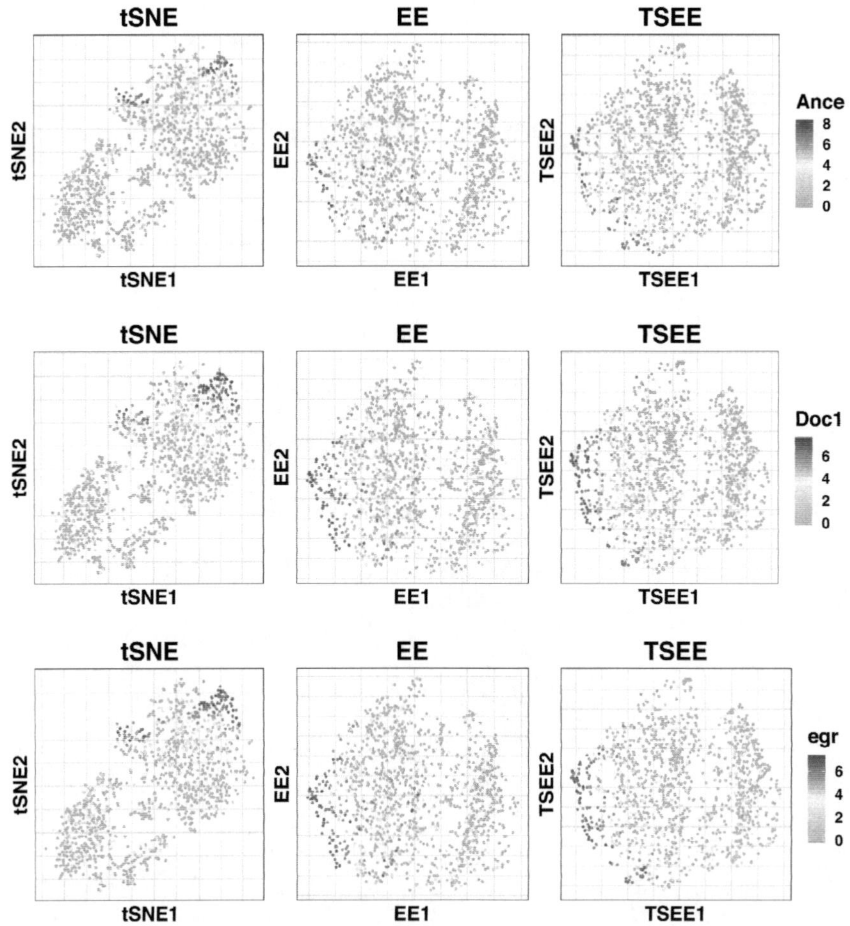

Fig. 11 Distribution of genes expressed in *Drosophila* embryo cells. The expression of *Ance*, *Doc1* and *egr* are displayed in 2-dimensional space, as obtained by tSNE, EE and modified spatial TSEE

disparities between samples in high-dimensional gene expression space.

To incorporate the temporal information of data, TSEE adds additional terms of the time intervals between samples into the repulsive terms to balance the effects from the disparities between samples in high-dimensional gene expression space. In this way, TSEE dilutes the distortions of the assorted sources of variations of the data across time stages and achieves temporal resolution enhancement by preserving temporal order and structure. In addition, TSEE uncovers the subtle dynamic structures of gene expression patterns, as exemplified by oscillating waves in our results, facilitating further downstream dynamic modeling and analysis of gene expression processes. The computational framework of TSEE is generalizable for the incorporation of other sources of information.

Abbreviations
EE: Elastic embedding; HPE: Human preimplantation embryo; IGP: In-group proportion; IGP2: A modification of in-group proportion; PCA: Principal component analysis; PCC: Pearson correlation coefficient; scRNA-seq: Single-cell RNA sequencing; TSEE: Time series elastic embedding; tSNE: t-distributed stochastic neighbor embedding

Acknowledgements
We are grateful to Xiangqi Bai and Ziwei Chen for their helpful discussions.

Funding
Publication of this article was sponsored by NSFC grants (Nos.11571349, 91630314, 81673833), the Strategic Priority Research Program of Chinese Academy of Sciences (CAS) (XDB13050000), the NCMIS of CAS, the LSC of CAS, and the Youth Innovation Promotion Association of CAS. LW would like to thank the Mathematical Biosciences Institute (MBI) at Ohio State University for partially supporting this research. MBI receives its funding through NSF grant DMS 1440386.

Availability of data and materials
The data used in the current study are all publicly available. HPE data can be obtained at https://www.ebi.ac.uk/arrayexpress/experiments/E-MTAB-3929/, while Zebrafish data in UMI count format can be assessed at NCBI GEO with accession no.GSE106587, and the scRNA-seq dataset of *Drosophila* embryos is available at https://shiny.mdc-berlin.de/DVEX/.
The TSEE code is available at https://github.com/ShaokunAn/TSEE, and the oscillation genes discovered by TSEE among the genes related to embryogenesis can be found at https://github.com/ShaokunAn/TSEE/tree/Additional-files.

About this supplement
This article has been published as part of BMC Genomics Volume 20 Supplement 2, 2019: Selected articles from the 17th Asia Pacific Bioinformatics Conference (APBC 2019): genomics. The full contents of the supplement are available online at https://bmcgenomics.biomedcentral.com/articles/supplements/volume-20-supplement-2.

Author's contributions
SA participated in method design and data analysis, wrote the program, and drafted the manuscript. LM conceived the project and participated in method design and writing. LW conceived and designed the study and participated in method design and writing. All authors read and approved the manuscript.

Ethics approval and consent to participate
Not applicable.

Consent for publication
Not applicable.

Competing interests
The authors declare that they have no competing interests.

Author details
[1]Academy of Mathematics and Systems Science, Chinese Academy of Sciences, 100190 Beijing, China. [2]Beijing Institute of Genomics, Chinese Academy of Sciences, 100101 Beijing, China. [3]University of Chinese Academy of Sciences, 100049 Beijing, China.

Published: 10 April 2019

References

1. Trapnell C, Cacchiarelli D, Grimsby J, Pokharel P, Li S, Morse M, Lennon NJ, Livak KJ, Mikkelsen TS, Rinn JL. The dynamics and regulators of cell fate decisions are revealed by pseudotemporal ordering of single cells. Nat Biotechnol. 2014;32(4):381–6. https://doi.org/10.1038/nbt.285.
2. Qiu X, Mao Q, Tang Y, Wang L, Chawla R, Pliner HA, Trapnell C. Reversed graph embedding resolves complex single-cell trajectories. Nat Methods. 2017;14(10):979–82. https://doi.org/10.1038/NMETH.440.
3. Setty M, Tadmor MD, Reich-Zeliger S, Ange O, Salame TM, Kathail P, Choi K, Bendall S, Friedman N, Pe'er D. Wishbone identifies bifurcating developmental trajectories from single-cell data. Nat Biotechnol. 2016;34(6):637–45. https://doi.org/10.1038/nbt.356.
4. Chen J, Renia L, Ginhoux F. Constructing cell lineages from single-cell transcriptomes. Mol Asp Med. 2018;59:95–113. https://doi.org/10.1016/j.mam.2017.10.00.
5. Petropoulos S, Edsgard D, Reinius B, Deng Q, Panula SP, Codeluppi S, Reyes AP, Linnarsson S, Sandberg R, Lanner F. Single-Cell RNA-Seq Reveals Lineage and X Chromosome Dynamics in Human Preimplantation Embryos. Cell. 2016;165(4):1012–26.
6. Farrell JA, Wang Y, Riesenfeld SJ, Shekhar K, Regev A, Schier AF. Single-cell reconstruction of developmental trajectories during zebrafish embryogenesis. Science. 2018;360(6392):979. https://doi.org/10.1126/science.aar313.
7. Butler A, Hoffman P, Smibert P, Papalexi E, Satija R. Integrating single-cell transcriptomic data across different conditions, technologies, and species. Nat Biotechnol. 2018;36(5):411. https://doi.org/10.1038/nbt.409.
8. Gao NP, Ud-Dean SMM, Gandrillon O, Gunawan R. SINCERITIES: inferring gene regulatory networks from time-stamped single cell transcriptional expression profiles. Bioinformatics. 2018;34(2):258–66. https://doi.org/10.1093/bioinformatics/btx57.
9. Rashid S, Kotton DN, Bar-Joseph Z. TASIC: determining branching models from time series single cell data. Bioinformatics. 2017;33(16):2504–12. https://doi.org/10.1093/bioinformatics/btx17.
10. Schiebinger G, Shu J, Tabaka M, Cleary B, Subramanian V, Solomon A, Liu S, Lin S, Berube P, Lee L, Chen J, Brumbaugh J, Rigollet P, Hochedlinger K, Jaenisch R, Regev A, Lander E. Reconstruction of developmental landscapes by optimal-transport analysis of single-cell gene expression sheds light on cellular reprogramming. bioRxiv. 2017. https://doi.org/10.1101/191056. https://www.biorxiv.org/content/early/2017/09/27/191056.full.pdf.
11. Ding J, Aronow BJ, Kaminski N, Kitzmiller J, Whitsett JA, Bar-Joseph Z. Reconstructing differentiation networks and their regulation from time series single-cell expression data. Genome Res. 2018;28(3):383–95. https://doi.org/10.1101/gr.225979.11.
12. Moon KR, van Dijk D, Wang Z, Burkhardt D, Chen W, van den Elzen A, Hirn MJ, Coifman RR, Ivanova NB, Wolf G, Krishnaswamy S. Visualizing transitions and structure for high dimensional data exploration. bioRxiv. 2017. https://doi.org/10.1101/120378.
13. Carreira-Perpiñán MÁ. The elastic embedding algorithm for dimensionality reduction. In: Proceedings of the 27th International Conference on Machine Learning (ICML-10), June 21-24, 2010, Haifa, Israel; 2010. p. 167–74. http://www.icml2010.org/papers/123.pdf.
14. Chen Z, An S, Bai X, Gong F, Ma L, Wan L. Densitypath: a level-set algorithm to visualize and reconstruct cell developmental trajectories for large-scale single-cell rnaseq data. bioRxiv. 2018. https://doi.org/10.1101/276311. https://www.biorxiv.org/content/early/2018/03/05/276311.full.pdf.
15. Jacomy M, Venturini T, Heymann S, Bastian M. ForceAtlas2, a Continuous Graph Layout Algorithm for Handy Network Visualization Designed for the Gephi Software. PLoS ONE. 2014;9(6):. https://doi.org/10.1371/journal.pone.009867.

16. Weinreb C, Wolock S, Klein AM. SPRING: a kinetic interface for visualizing high dimensional single-cell expression data. Bioinformatics. 2018;34(7): 1246–8. https://doi.org/10.1093/bioinformatics/btx79.
17. Wasserman L. Topological data analysis. Annu Rev Stat Appl. 2018;5(1): 501–32. https://doi.org/10.1146/annurev-statistics-031017-100045.
18. Mucha PJ, Richardson T, Macon K, Porter MA, Onnela J-P. Community Structure in Time-Dependent, Multiscale, and Multiplex Networks. Science. 2010;328(5980):876–8. https://doi.org/10.1126/science.118481.
19. Vladymyrov M, Carreira-Perpiñán MÁ. Partial-hessian strategies for fast learning of nonlinear embeddings. In: Proceedings of the 29th International Conference on Machine Learning (ICML-12). 2012. p. 345–352.
20. Vladymyrov M, Carreira-Perpinan M. Entropic affinities: Properties and efficient numerical computation. In: Dasgupta S, McAllester D, editors. Proceedings of the 30th International Conference on Machine Learning. Proceedings of Machine Learning Research, vol. 28. Atlanta, Georgia, USA: PMLR; 2013. p. 477–85. http://proceedings.mlr.press/v28/vladymyrov13.html.
21. Karaiskos N, Wahle P, Alles J, Boltengagen A, Ayoub S, Kipar C, Kocks C, Rajewsky N, Zinzen RP. The Drosophila embryo at single-cell transcriptome resolution. Science. 2017;358(6360):194–9. https://doi.org/10.1126/science.aan323.
22. Kapp AV, Tibshirani R. Are clusters found in one dataset present in another dataset?. Biostatistics. 2007;8(1):9–31. https://doi.org/10.1093/biostatistics/kxj02.
23. Kageyama R, Niwa Y, Isomura A, Gonzalez A, Harima Y. Oscillatory gene expression and somitogenesis. Wiley Interdiscip Rev Dev Biol. 2012;1(5): 629–41. https://doi.org/10.1002/wdev.4.
24. Wylie AD, Fleming J-AGW, Whitener AE, Lekven AC. Post-transcriptional regulation of wnt8a is essential to zebrafish axis development. Dev Biol. 2014;386(1):53–63. https://doi.org/10.1016/j.ydbio.2013.12.00.
25. Morrow ZT, Maxwell AM, Hoshijima K, Talbot JC, Grunwald DJ, Amacher SL. tbx6l and tbx16 are Redundantly Required for Posterior Paraxial Mesoderm Formation During Zebrafish Embryogenesis. Dev Dyn. 2017;246(10):759–69. https://doi.org/10.1002/DVDY.2454.
26. Hagey DW, Muhr J. Sox2 Acts in a Dose-Dependent Fashion to Regulate Proliferation of Cortical Progenitors. Cell Rep. 2014;9(5):1908–20. https://doi.org/10.1016/j.celrep.2014.11.01.
27. Rulands S, Lee HJ, Clark SJ, Angermueller C, Smallwood SA, Krueger F, Mohammed H, Dean W, Nichols J, Rugg-Gunn P, Kelsey G, Stegle O, Simons BD, Reik W. Genome-Scale Oscillations in DNA Methylation during Exit from Pluripotency. Cell Syst. 2018;7(1):63. https://doi.org/10.1016/j.cels.2018.06.01.

Hua and Zhang *BMC Genomics* 2019, **20**(Suppl 2):183
https://doi.org/10.1186/s12864-019-5467-x

BMC Genomics

RESEARCH Open Access

Estimating the total genome length of a metagenomic sample using k-mers

Kui Hua[1,2] and Xuegong Zhang[1,2,3*]

From The 17th Asia Pacific Bioinformatics Conference (APBC 2019)
Wuhan, China. 14-16 January 2019

Abstract

Background: Metagenomic sequencing is a powerful technology for studying the mixture of microbes or the microbiomes on human and in the environment. One basic task of analyzing metagenomic data is to identify the component genomes in the community. This task is challenging due to the complexity of microbiome composition, limited availability of known reference genomes, and usually insufficient sequencing coverage.

Results: As an initial step toward understanding the complete composition of a metagenomic sample, we studied the problem of estimating the total length of all distinct component genomes in a metagenomic sample. We showed that this problem can be solved by estimating the total number of distinct k-mers in all the metagenomic sequencing data. We proposed a method for this estimation based on the sequencing coverage distribution of observed k-mers, and introduced a k-mer redundancy index (*KRI*) to fill in the gap between the count of distinct k-mers and the total genome length. We showed the effectiveness of the proposed method on a set of carefully designed simulation data corresponding to multiple situations of true metagenomic data. Results on real data indicate that the uncaptured genomic information can vary dramatically across metagenomic samples, with the potential to mislead downstream analyses.

Conclusions: We proposed the question of how long the total genome length of all different species in a microbial community is and introduced a method to answer it.

Keywords: Metagenomics, Sequencing coverage, Distinct k-mers, Genome length

Background

It is now widely known that microbiomes or the ecological community of microbes living at a certain site of the human host such as the gut can play important roles in human health [1–5]. Metagenomic sequencing is a powerful technology for studying the microbiome by sequencing DNAs from all the genomes of its component microbes [5]. Since it is impossible to capture the full components of a microbiome, a 'metagenomic sample' is actually a subset of the target metagenome captured with the sequencing process, as a sample from a population in statistics [6]. The basic task of a metagenomic study is to read out the underlying information about the microbiome from the metagenomic sample.

For any genomic sequencing study, a fundamental property we need to consider is the sequencing coverage, which is the fraction of genomic materials that has been captured and sequenced. This, however, has been largely ignored in metagenomic studies [6]. The level of coverage of a metagenomic sample is of key importance for recovering the information about the microbiome. Variations caused by coverage differences between metagenomic samples can be wrongly attributed to biological reasons, resulting in misleading conclusions [6].

The question of estimating the coverage of a sequencing sample has been attracting researchers' attention since the beginning of human genome project. In 1988, Eric S. Lander and Michael S. Waterman introduced the famous Lander-Waterman theory to show how well a genome can

*Correspondence: zhangxg@tsinghua.edu.cn
[1] MOE Key Laboratory of Bioinformatics Division and Center for Synthetic & System Biology, BNRIST, 100084 Beijing, China
[2] Department of Automation, Tsinghua University, 100084 Beijing, China
Full list of author information is available at the end of the article

be recovered for a certain sequencing strategy [7]. It had played a key role in guiding the design and completion of the human genome project. Lander-Waterman theory was specially designed for single genomic sequencing projects. It is no longer suitable for most metagenomic data since the relative abundances of component genomes in a microbiome are very uneven and therefore the sequencing procedure violates the uniform distribution assumption [8]. This is also true for other types of sequencing projects like RNA-sequencing or ChIP-seq where distributions of components to be sequenced are uneven. Methods were therefore introduced to estimate the coverage or solve similar problems in such situations [8–12]. For example, Hooper et al. proposed a method to estimate the total number of genomic bins in a metagenome by assuming certain abundance distribution of the microbial composition [8]. Rodriguez et al. assessed the abundance-weighted coverage of a metagenomic sample by examining the redundancy among individual reads [10]. Daley and Smith introduced an empirical Bayesian method to predict the number of previously un-sequenced molecules that would be observed if additional reads were provided [9]. This method has been demonstrated powerful in different kinds of sequencing data such as ChIP-seq data and RNA-seq data, but its effectiveness on metagenomic data has not been studied.

For the genomic sequences that have been captured in a metagenomic sample, the basic information we want to get is what types of microbes are there at what abundances. This is referred to as taxonomy profiling. A straightforward way of taxonomy profiling is to map sequencing reads to reference genomes in known databases. Known microbial genomes only represent a small proportion of existing microbes. Even for the type of well-studied communities like human gut, it's typical that around 30%–60% of sequencing reads in a metagenomic sample could not be mapped to any known microbial genomes [13]. Furthermore, it has been observed that the fraction of unmapped reads can vary dramatically across different samples in the same study, say, ranging surprising from 2 to 96% [14]. This type of between-samples variation is lost when relative abundances are calculated based on mapped reads. Ignoring such loss of information can be misleading in downstream analyses [5].

Mainly because of the incomplete coverage and the existence of unmapped reads, the genomes that can be profiled from a metagenomic sample are only a part of all genomes that exist in the microbiome. It is therefore desirable to make estimations on the genomes that have been missed. Even if it is not possible to make accurate estimations on the number of missed genomes and their relative abundances, any educated guess about any properties of missed genomes can provide useful information for the comparison of samples based on known genomes. In this paper, we study the problem of estimating the total length of all distinct genomes in a metagenomic sample. If we can estimate this with reasonable accuracy, we will know a lot about the missed genomes by subtracting those known and mapped genomes from the total. This is the same question as estimating the actual coverage of the unknown targeting whole microbiome by the observed sequencing reads in the metagenomic sample. In preparation of this manuscript, a similar question has been studied in [15], but the method requires both long reads and short reads. For most cases where only short reads are available, we found that this question can be solved by solving the related question of estimating the number of distinct k-mers in the metagenome if we have infinite sequencing depth. A statistical model is introduced to predict the number of distinct k-mers in a metagenome that have not been included in the observed data. And we define a k-mer redundancy index (*KRI*) that helps to estimate the total genome length from total distinct k-mer count. Since the underlying truth is unknown in any real metagenomic data, we simulated a set of synthetic metagenomic datasets for different situations of microbial composition. Experiments on these data showed that the proposed method works well.

Methods

Problem statements

The problem we study is to estimate the total length of distinct genomes in a microbiome based on the metagenomic sequencing data. A more accurate statement of this problem in practice depends on the criteria for two genomes to be identified as distinct from each other. This is a complicated taxonomic question considering the wide existence of strains and substrains within each microbial species. To focus on the key mathematic problem behind the question, we simply assume that genomes from the same species are same while genomes from different species are distinct. We will give further discussion about this later in the "Estimating *KRI* of the distinct genome set" section.

Understanding DNA sequence as a collection of k-mers

A DNA sequence can be viewed as a collection of k-mers by breaking the sequence into nucleotide substrings of length k, as illustrated in Fig. 1a. From the k-mer perspective, we define total k-mer count (*TKC*), distinct k-mer count (*DKC*) and k-mer redundancy index (*KRI*) as three properties of a sequence. *TKC* is the number of all k-mers obtained when breaking a sequence into k-mers. *DKC* is the amount of distinct k-mers, i.e., the amount of remaining k-mers after removing all replicates of k-mers. *KRI* is defined as the ratio of *TKC* and *DKC*, which reflects the degree of repetition of k-mers in the sequence. The values of these three properties depend on the target sequence

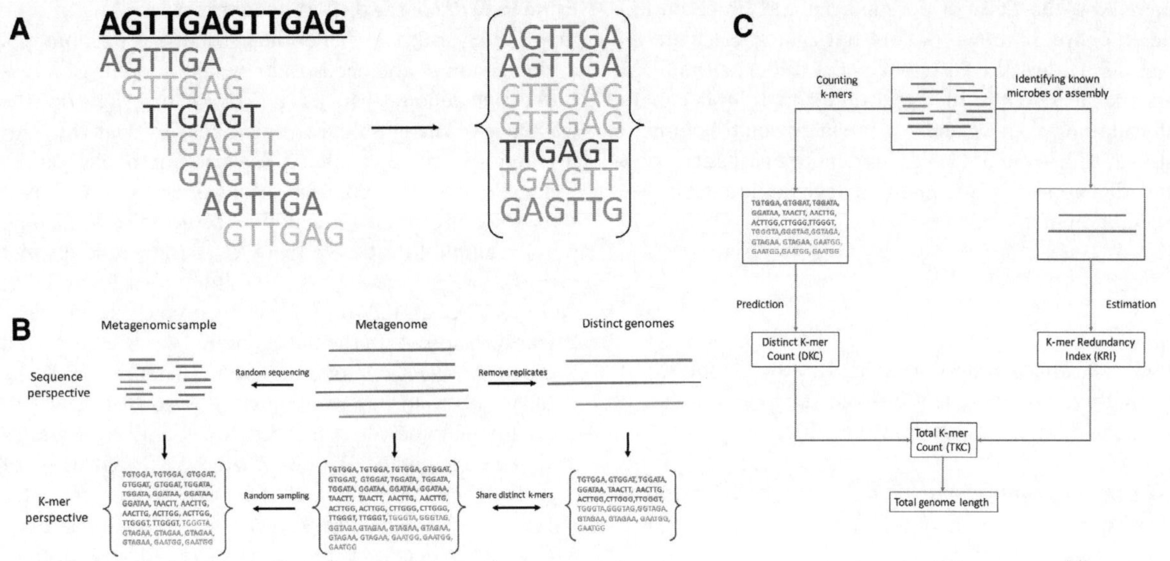

Fig. 1 Overview of the proposed method. **a** An illustration of understanding DNA sequence as a collection of k-mers. In this simple case, sequence length $L = 12$, $k = 6$ for the k-mer counting, *TKC*=$L - k + 1 = 7$, *DKC*=5, *KRI*=*TKC*/*DKC*=1.2. **b** Relationships between metagenome, metagenomic sample and the set of distinct genomes in the metagenome. **c** Workflow of the proposed method

and the selection of k-mer size (k). For a given k, any of the three properties can be obtained if the other two are provided. For example, $TKC = DKC * KRI$, which means *TKC* is achievable if we know *DKC* and *KRI* of a k-mer collection. Obviously, for a sequence of length L, $TKC = L - k + 1$, indicating that *TKC* can be roughly taken as the sequence length if $L \gg k$, which is satisfied when studying genomes using small k-mers. These simple mathematical relations form the basic idea of our work.

Similarly, a set of sequences can also be treated as a collection of k-mers by breaking every single sequence into k-mers. Therefore, a metagenomic sample, the metagenome and the set of distinct genomes in a metagenome can all be viewed as a collection of k-mers, respectively, as illustrated in Fig. 1b.

Overview of our solution

From the k-mer perspective, our aim of estimating total genome length of all distinct genomes in a metagenome is equivalent to estimating *TKC* of the set of distinct genomes (Fig. 1b). Since it is impossible to count *TKC* of the true metagenome from the metagenomic sample due to finite sequencing coverage and unknown genome composition, we predict *TKC* of the distinct genome set by estimating its *DKC* and *KRI* separately (Fig. 1c). A metagenome and the corresponding set of distinct genomes of all its components differ only in genome abundances, they share the same distinct k-mers and have equal *DKC*s. We estimate *DKC* of the metagenome from the observed metagenomic data by modeling the sequencing event as a Poisson sampling procedure. *KRI* of the distinct genome set can be estimated based on known genomes detected in the metagenomic sample. Finally, the total genome length, which is roughly equal to *TKC*, can be achieved simply by taking the product of *KRI* and *DKC*.

Predicting *DKC* of the metagenome

A metagenomic sample can be viewed as a subset of the metagenome obtained by random sampling, as illustrated in Fig. 1b. *DKC* of a metagenomic sample can be readily obtained by counting k-mers in the sequences, either from the original sequencing reads or from the assembled scaffolds. We need to estimate the number of k-mers in the metagenome that have not been covered in the metagenomic sample. The frequency that a given k-mer i is sequenced, denoted as x_i, can be modeled as a Poisson distribution with an unknown parameter λ_i. The probability that k-mer i will not been sequenced is $e^{-\lambda_i}$. We call these k-mers as uncaptured k-mers. Although the frequencies of k-mers overlapping with each other are dependent, such limited dependence can be well-approximated by assuming independence [16, 17]. Therefore, we further assume that λ_i independently and identically follow some unknown distribution $\mu(\lambda)$, the number of uncaptured k-mers is

$$N \int_0^\infty e^{-\lambda} d\mu(\lambda) \tag{1}$$

where N is the *DKC* of the metagenome. Since both N and $\mu(\lambda)$ are unknown, we are not able to calculate the value of (1) directly. Fortunately, the frequencies of captured k-mers in the metagenomic sample also contain information about N and $\mu(\lambda)$, which would help us to estimate the value of (1). Let n_j denote the number of k-mers that appear j times in the metagenomic sample. The expectation of n_j can be written as

$$E(n_j) = N \int_0^\infty e^{-\lambda}\lambda^j/j!\,\mathrm{d}\mu(\lambda) \tag{2}$$

If we take the observation n_j as its expectation $E(n_j)$, the mathematical problem of estimating the number of uncaptured k-mers can be formulated as:

Given observations $n_1, n_2, n_3, \ldots, n_M$, which follow the formula

$$n_j = N \int_0^\infty e^{-\lambda}\lambda^j/j!\,\mathrm{d}\mu(\lambda)$$

where N and $\mu(\lambda)$ are unknown. Find the value of

$$N \int_0^\infty e^{-\lambda}\mathrm{d}\mu(\lambda)$$

To solve this mathematical problem, let $\omega(\lambda) = N\lambda e^{-\lambda}$, $m_i = (i+1)!\,n_{i+1}$, the problem can be re-written as

Given observations $m_0, m_1, m_2, \ldots, m_{M-1}$, which follow the formula

$$m_j = \int_0^\infty \lambda^j \omega(\lambda)\mathrm{d}\mu(\lambda)$$

where $\omega(\lambda)$ and $\mu(\lambda)$ are unknown. Find the value of

$$\int_0^\infty \frac{1}{\lambda}\omega(\lambda)\mathrm{d}\mu(\lambda)$$

This is a special type of Gaussian quadrature problem that can be solved using the Golub-Welsch algorithm [9, 18]. The final estimation of (1) can be written as

$$N \int_0^\infty e^{-\lambda}\mathrm{d}\mu(\lambda) \approx \sum_{\mathrm{i}=1}^{\mathrm{M}} \frac{\alpha_\mathrm{i}}{\lambda_\mathrm{i}} \tag{3}$$

where α_i and λ_i are decided by the Golub-Welsch algorithm taking $m_0, m_1, m_2, \ldots, m_{M-1}$ as the input. *DKC* of the metagenome is finally achieved by adding this estimated uncaptured number of k-mers to *DKC* of the metagenomic sample. The variability and reliability of the estimation can be reflected by the confidence interval achieved by the bootstrap method.

Estimating *KRI* of the distinct genome set

To precisely estimate *KRI* of the set of distinct genomes of a metagenome, one needs to know all different genomes in the metagenome, which is usually unachievable due the existence of many unknown microbes. To deal with this problem, we reasoned that *KRI* of a genome set can be well estimated use only part of the genomes in it. Therefore, we can use known genomes detected in a metagenomic sample to estimate the *KRI* of the whole distinct genome set. In practice, we first apply MetaPhlan2 [19] and GOTTCHA [20] on the metagenomic data to identify known species in the metagenome. For each detected species, we select one of its reference genomes from the database [9] to form a genome set. An alternative way to form the genome set is to take the assembled scaffolds as detected genomes. We estimated the *KRI* of this set of detected genomes as the *KRI* of the distinct genome set.

The way of selecting known genomes to estimate *KRI* actually decides the criteria of identifying distinct genomes in our work. Since we select only one genome for each detected species to estimate the *KRI* of the set of distinct genomes, the estimation is restricted to species level, even if two strains of the same species were detected in the metagenomic sample. If we include genomes for all detected strains in the *KRI* estimation, the estimation will be at strain level.

Implementation of the method

We first adopt Pollux [21] to correct the sequencing error in the metagenomic samples. Counting all k-mers in a metagenomic sample can be computationally heavy. We employ jellyfish2 [22], one of the fastest k-mer counting approaches, for the k-mer counting step. We use the Golub-Welsch algorithm implemented in preseq [9, 17] to estimate the distinct k-mer count. MetaPlan2 [19] and GOTTCHA [20] are used to identify the known species from the metagenomic sample. Genomes for those known species are selected from existing database [23] to estimate the *KRI* for the whole community.

Simulated metagenomic datasets

Due to the complexity of real-world microbiome compositions, it is hard, if possible, to find real metagenomic data that have complete true answer of all components. To test the performance of our method, we simulated several microbial communities of different situations and generate synthetic metagenomic samples. We simulated communities with 10 species and 50 species as representatives of a simple case and a more complicated case. We used three types of composition abundance distributions to form microbial communities of low, medium and high complexities (LC, MC and HC) following the way of a previous simulation study [24]. LC, MC and HC are defined based on the number of dominant microbe who has a high relative abundance. LC has only one dominant microbe. MC has two or more dominant species. HC has no dominant species. The fraction of information captured by the metagenomic data is of key importance for estimating the total genome length. To reflect this property of a metagenomic sample, we define initial coverage as the fraction of distinct k-mers in the set of distinct genomes of the target community included in the sequencing data. For each community, metagenomic samples of different reads numbers were generated to simulate the situation of different sequencing depths and the initial coverages of the community. To check how robust the method is to random effect, we use three random seeds to generate samples for the same parameters. In total, 225 metagenomic samples with 10 species and 243 samples with 50 species were generated with an in-house simulation tool [25]. Beside the error-free samples, we also generated a set of metagenomic samples with sequencing errors for each community.

We did some simple simulations to show that *KRI* of a genome set can be estimated using part of all genomes. We simulated four metagenomes with 10, 50, 100 and 200 species, respectively. For each metagenome, we randomly select 60% of its component genomes as known ones to estimate the *KRI* of the whole metagenome. Although in real world, the known microbes are not randomly selected from the nature, the order in which they were known has nothing to do with their sequence contents. Therefore, we believe such random selection is reasonable.

Real metagenomic datasets

We select two datasets to conduct our method on. One dataset contains 65 oral metagenomic samples from Human Microbiome Project (HMP) [26] and the other consists of 145 human gut metagnomic samples, including 71 from normal people and 74 from type 2 diabetes patients [27].

Results

Results on simulated metagenomic datasets

We tested our method on all synthetic metagenomic samples. Fig. 2 shows how well the number of distinct k-mers (*DKC*) in a community can be estimated from

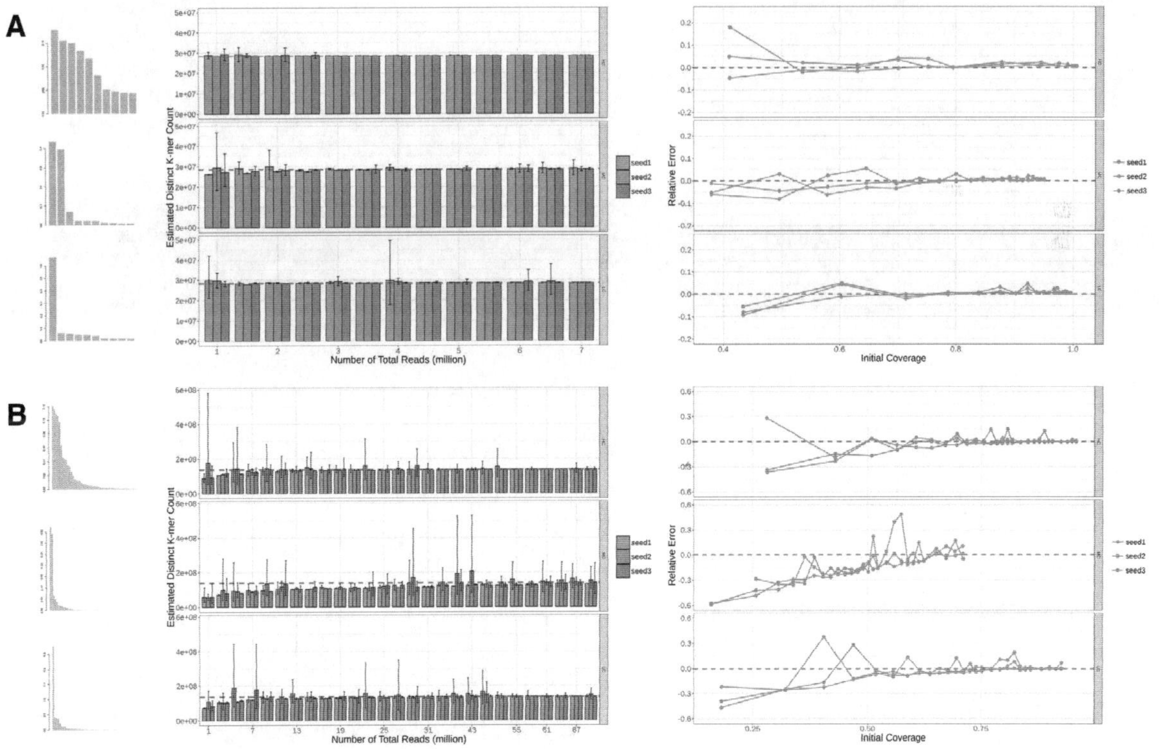

Fig. 2 Different microbial communities are simulated to test the performance of the proposed method. (**a**) Results for microbial communities with 10 species. The three histograms on the left show the abundance distributions of different simulated communities. The middle panel shows the estimation results of distinct k-mer count. Each bar represents an estimation result based on a synthetic metagenomic sample and the error bar shows the 95% bootstrap confidence interval of the estimation. The black dash line is the true distinct k-mer count. The right panel shows how the relative error goes as the initial coverage increases ($k = 20$). (**b**) The same as (**a**) except that the species number is 50. (Note that some of the samples with 10 species are not shown in the barplot, see Additional file 1: Figure S1 for all samples with 10 species)

a metagenomic sample. The whole figure contains two parts, showing results for communities with 10 species and 50 species, respectively. Each part consists of three panels, displayed from left to right. Further explanations about each panel are given in the figure caption. As expected, the overall prediction in samples with 10 species is better than in samples with 50 species. Communities with high complexity achieve best prediction accuracy among those three kinds of abundance distributions. This agrees with the intuition that the more even the abundance distribution is, the better the prediction will be. The performances on communities with medium complexity are the worst. This is because the two dominant species make up more than 70% of the community, which means that most of the reads are sequenced from them. Since less than 30% of the reads come from the rest of all species, only a small part of information about their genomes is reflected in the sequencing data, leading to the bad performance, especially when sequencing depth is low. We also show how the performance goes when the initial coverage increases. The performance is measured by relative error, defined as the difference between estimated value and the true value divided by the true value. In general, the performance gets better as the initial coverage increases. Another interesting observation is that, for most cases, Golub-Welsch algorithm gives a good estimation which trends to be no larger than the ground truth, and the corresponding bootstrap confidence interval is usually small. For the exaggerated estimations, Golub-Welsch algorithm is more likely to give a large bootstrap confidence interval. Therefore, Golub-Welsch algorithm provides a reliable estimation of the lower bound of DKC, as suggested in preseq [9].

Effects of K and sequencing errors

To see how the parameter k affects the results, We chose different k to do the estimation for a simulated metagenomic sample (50 species, high complexity, 25 million reads). Results show that the estimation is robust to the selection of k (Fig. 3c).

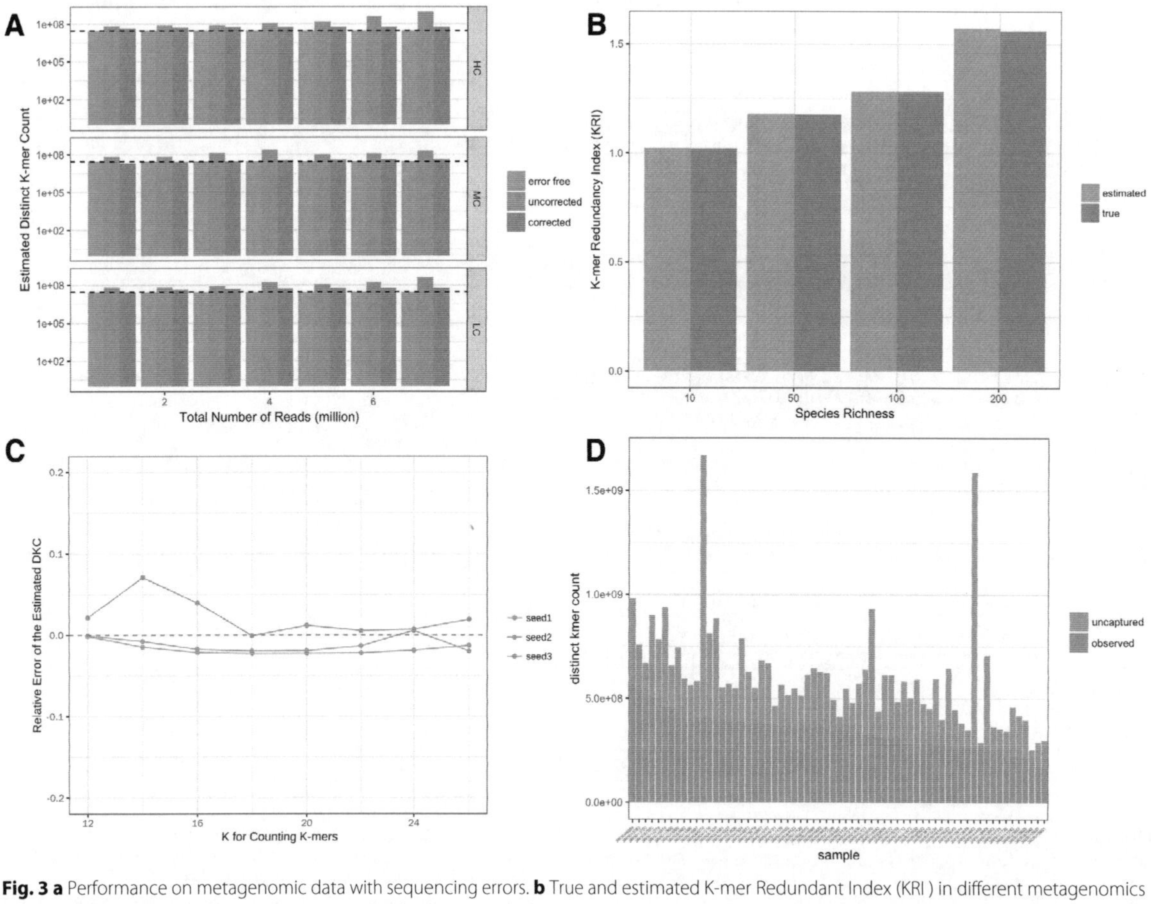

Fig. 3 a Performance on metagenomic data with sequencing errors. **b** True and estimated K-mer Redundant Index (KRI) in different metagenomics communities. About 60% of the species are randomly chosen as the known species to estimate the KRI of all species. **c** Results of different selections of K. Simulated metagenomic sample with 50 speices and high complexity of the abundance distribution was used. **d** Results on HMP Tongue Dorsum datasets

Despite the good performance on error-free sequencing data, the Golub-Welsch algorithm can given bad prediction when the sequencing data contains errors (Fig. 3a). Sequencing errors introduce novel k-mers that should not exist in the data. A higher fraction of low-count k-mers will be considered by the algorithm as the implication of more low-abundant microbes. Therefore, sequencing errors lead to exaggerated estimation of total distinct k-mers and this exaggeration grows as the sequencing depths increases (Fig. 3a, green bars). To solve this problem, we use Pollux [21] to correct the sequencing error before counting k-mers. Results on simulation data show that the performance can be under control after correcting the sequencing errors (Fig. 3a, blue bars).

Comparison between different methods

Besides Golub-Welsch algorithm, we also applied the major algorithm rational function approximation (RFA) in preseq on the simulated metagenomic samples with 50 species (Additional file 1: Figure S2) and compared its performance with Golub-Welsch algorithm. Both methods achieve a good performance and each present their own strength (Additional file 1: Figure S3). RFA outperforms Golub-Welsch algorithm in the median complexity communities (two species with a total relative abundance higher than 70%), indicating a stronger ability of extrapolation. For communities with high complexity or low complexity, Golub-Welsch algorithm makes stable and accurate results with only few exceptions. RFA also gives a good result, but with a slight trend to exaggerate the estimation.

Estimating KRI using known species

There's a gap between distinct k-mer count (*DKC*) and total genome length or *TKC*. We use *KRI* to bridge this gap as introduced above. For simulated metagenomic samples, GOTTCHA succesfully identified most species therefore led to a perfect estimation of KRI. We did some simple simulations to show that *KRI* of a genome set can be estimated using part of all genomes. In general, *KRI* of

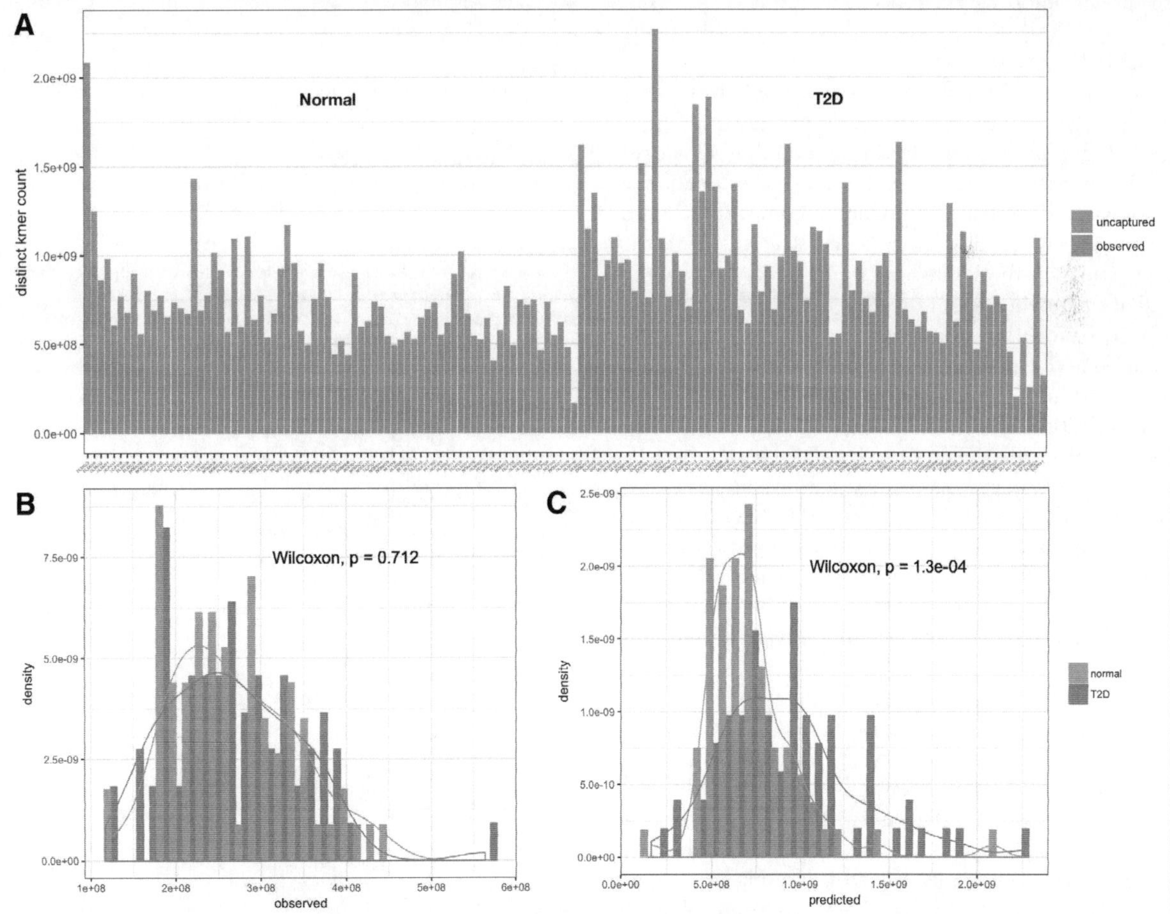

Fig. 4 Results on T2D metagenomic datasets. **a** Observed and estimated k-mer count. **b** Histogram and density of the observed distinct k-mer count. **c** Histogram and density of the predicted distinct k-mer count

the community increases as there are more species in the community, as shown in Fig. 3b. The result shows that *KRI* of a community can be well estimated use only part of the species, which demonstrates the feasibility of estimating *KRI* of a community based only on known species.

Results on real metagenomic datasets

We applied our method on the two selected datasets (Figs. 3d and 4). One general observation in the results is that, the number of uncaptured k-mers can differ a lot between samples, even when the observed k-mer counts are similar (Figs. 3d and 4a). Further comparison between normal samples and T2D samples shows that the predicted distinct k-mer counts present significant difference while observed k-mer counts do not (Fig. 4c and d). In the original study, it was reported that the difference of within-sample diversity (entropy of gene abundance) between normal group and T2D group is not significant [27]. Since the gene abundances were calculated based only on extracted sequence data, chances are that the significance had been masked by ignoring the difference in the 'unseen' information.

Conclusion and discussion

In this paper, we proposed the question of 'how long the total genome length of all different species in a microbial community is' and introduced a method to answer it. This is an important step toward the estimation of unknown and unseen component genomes in a microbiome. We invented a k-mer-based strategy to liberate the reliance on the limited microbial reference genomes so that unknown species can be included in the estimation. To explore the information that has not been directly captured in the metagenomic sample, we developed a statistical method to estimate the number of uncaptured k-mers. Distinct k-mer count was multiplied by the k-mer redundancy index (*KRI*), an index defined to reflect the repetition of k-mers and estimated from known species, to get the total genome length. Performance on the simulation data shows that the proposed method works well, and the precision of the estimation is mainly affected by factors including the sequencing error, the initial coverage of the community and the complexity of the microbial diversity.

Extracting information from the metagenomic data is the foundation of downstream analysis. The complex nature of microbial community and inadequate microbial diversity represented in existing databases make it challenging to extract the full information. A metagenomic sample can capture only part of the information about the microbial community due to its complexity, among which only part can be extracted due to the limited known references. Ignoring these 'uncaptured' and 'unknown' information can mislead downstream analyses. In the work of estimating total genome length, we adopted the reference-free strategy to include the 'unknown' information and a statistical model was employed to estimate the 'uncaptured' part so that the completeness of the extracted information can be pursued to the maximum. The experiments on simulated data showed the feasibility of the proposed method and results on real datasets revealed that downstream analyses may be biased if 'unseen' information is ignored. Further studies are needed in the future to explore ways by which the estimated total metagenome length can help to better extracting information about unknown or uncaptured species from the metagenomic data and comparing metagenome samples.

Additional file

Additional file 1: This file contains **Figure S1** – **Figure S3**. (PDFk 6194 kb)

Abbreviations

DKC: Distinct k-mer count; KRI: K-mer redundancy index; TKC: Total k-mer count

Acknowledgements

Not applicable.

Funding

The publication of this work was sponsored by the National Natural Science Foundation of China (61673231 and 61721003).

Availability of data and materials

K-mer count tables for all simulated datasets and real datasets can be found at https://github.com/stevenhuakui/Total-genome-length-data.

About this supplement

This article has been published as part of *BMC Genomics Volume 20 Supplement 2, 2019: Selected articles from the 17th Asia Pacific Bioinformatics Conference (APBC 2019): genomics*. The full contents of the supplement are available online at https://bmcgenomics.biomedcentral.com/articles/supplements/volume-20-supplement-2.

Authors' contributions

KH conceived the study, developed methodology, performed data analysis and wrote the manuscript. XZ conceived the study and wrote the manuscript. Both authors have read and approved the final manuscript.

Ethics approval and consent to participate

Not applicable.

Consent for publication

Not applicable.

Competing interests

The authors declare that they have no competing interests.

Author details

[1]MOE Key Laboratory of Bioinformatics Division and Center for Synthetic & System Biology, BNRIST, 100084 Beijing, China. [2]Department of Automation, Tsinghua University, 100084 Beijing, China. [3]School of Life Sciences, Tsinghua University, 100084 Beijing, China.

Published: 10 April 2019

References

1. Gordon JI. Honor thy gut symbionts redux. Science. 2012;336(6086): 1251–3.
2. Falony G, Wijmenga C, Raes J, et al. Population-level analysis of gut microbiome variation. Science. 2016;352(6285):560–4.
3. Zhernakova A, Wijmenga C, Fu J, et al. Population-based metagenomics analysis reveals markers for gut microbiome composition and diversity. Science. 2016;352(6285):565–9.
4. Cui H, Li Y, Zhang X. An overview of major metagenomic studies on human microbiomes in health and disease. Quant Biol. 2016;4(3):192–206.
5. Zhang X, Liu S, Cui H, Chen T. Reading the underlying information from massive metagenomic sequencing data. Proc IEEE. 2017;105(3):459–73.
6. Rodriguez RL, Konstantinidis KT. Estimating coverage in metagenomic data sets and why it matters. ISME J. 2014;8(11):2349–51.
7. Lander ES, Waterman MS. Genomic mapping by fingerprinting random clones: a mathematical analysis. Genomics. 1988;2(3):231–9.
8. Hooper SD, Dalevi D, Pati A, Mavromatis K, Ivanova NN, Kyrpides NC. Estimating dna coverage and abundance in metagenomes using a gamma approximation. Bioinformatics. 2010;26(3):295–301.
9. Daley T, Smith AD. Predicting the molecular complexity of sequencing libraries. Nat Methods. 2013;10(4):325–7.
10. Rodriguez RL, Konstantinidis KT. Nonpareil: a redundancy-based approach to assess the level of coverage in metagenomic datasets. Bioinformatics. 2014;30(5):629–35.
11. Tamames J, de la Pena S, de Lorenzo V. Cover: a priori estimation of coverage for metagenomic sequencing. Environ Microbiol Rep. 2012;4(3): 335–41.
12. Wendl MC, Kota K, Weinstock GM, Mitreva M. Coverage theories for metagenomic dna sequencing based on a generalization of stevens' theorem. J Math Biol. 2013;67(5):1141–61.
13. Segata N, Waldron L, Ballarini A, Narasimhan V, Jousson O, Huttenhower C. Metagenomic microbial community profiling using unique clade-specific marker genes. Nat Methods. 2012;9(8):811–4.
14. Oh J, Byrd AL, Deming C, Conlan S, Program NCS, Kong HH, Segre JA. Biogeography and individuality shape function in the human skin metagenome. Nature. 2014;514(7520):59–64.
15. Bankevich A, Pevzner PA. Joint analysis of long and short reads enables accurate estimates of microbiome complexity. Cell Syst. 2018;7(2):192–200.
16. Barbour AD, Chen LHY, Loh WL. Compound poisson approximation for nonnegative random-variables via stein method. Ann Probab. 1992;20(4): 1843–66.
17. Daley T, Smith AD. Modeling genome coverage in single-cell sequencing. Bioinformatics. 2014;30(22):3159–65.
18. Golub GH, Welsch JH. Calculation of gauss quadrature rules. Math Comput. 1969;23(106):221–30.
19. Truong DT, Franzosa EA, Tickle TL, Scholz M, Weingart G, Pasolli E, Tett A, Huttenhower C, Segata N. Metaphlan2 for enhanced metagenomic taxonomic profiling. Nat Methods. 2015;12(10):902–3.
20. Freitas TAK, Li P-E, Scholz MB, Chain PS. Accurate read-based metagenome characterization using a hierarchical suite of unique signatures. Nucleic Acids Res. 2015;43(10):69.
21. Marinier E, Brown DG, McConkey BJ. Pollux: platform independent error correction of single and mixed genomes. BMC Bioinformatics. 2015;16:10.
22. Marcais G, Kingsford C. A fast, lock-free approach for efficient parallel counting of occurrences of k-mers. Bioinformatics. 2011;27(6):764–70.
23. Pruitt KD, Tatusova T, Maglott DR. Ncbi reference sequences (refseq): a curated non-redundant sequence database of genomes, transcripts and proteins. Nucleic Acids Res. 2007;35(Database issue):61–5.
24. Mavromatis K, Hugenholtz P, Kyrpides NC, et al. Use of simulated data sets to evaluate the fidelity of metagenomic processing methods. Nat Methods. 2007;4(6):495–500.
25. Liu S, Hua K, Chen S, Zhang X. Comprehensive simulation of metagenomic sequencing data with non-uniform sampling distribution. Quant Biol. 2018;6(2):175–85.
26. Turnbaugh PJ, Ley RE, Hamady M, Fraser-Liggett CM, Knight R, Gordon JI. The human microbiome project. Nature. 2007;449(7164):804–10.
27. Qin J, Kristiansen K, Wang J, et al. A metagenome-wide association study of gut microbiota in type 2 diabetes. Nature. 2012;490(7418):55–60.

Ren *et al. BMC Genomics* 2019, **20**(Suppl 2):184
https://doi.org/10.1186/s12864-019-5468-9

BMC Genomics

RESEARCH **Open Access**

GPU accelerated sequence alignment with traceback for GATK HaplotypeCaller

Shanshan Ren, Nauman Ahmed, Koen Bertels and Zaid Al-Ars*

From The 17th Asia Pacific Bioinformatics Conference (APBC 2019)
Wuhan, China. 14–16 January 2019

Abstract

Background: Pairwise sequence alignment is widely used in many biological tools and applications. Existing GPU accelerated implementations mainly focus on calculating optimal alignment score and omit identifying the optimal alignment itself. In GATK HaplotypeCaller (HC), the semi-global pairwise sequence alignment with traceback has so far been difficult to accelerate effectively on GPUs.

Results: We first analyze the characteristics of the semi-global alignment with traceback in GATK HC and then propose a new algorithm that allows for retrieving the optimal alignment efficiently on GPUs. For the first stage, we choose intra-task parallelization model to calculate the position of the optimal alignment score and the backtracking matrix. Moreover, in the first stage, our GPU implementation also records the length of consecutive matches/mismatches in addition to lengths of consecutive insertions and deletions as in the CPU-based implementation. This helps efficiently retrieve the backtracking matrix to obtain the optimal alignment in the second stage.

Conclusions: Experimental results show that our alignment kernel with traceback is up to 80x and 14.14x faster than its CPU counterpart with synthetic datasets and real datasets, respectively. When integrated into GATK HC (alongside a GPU accelerated pair-HMMs forward kernel), the overall acceleration is 2.3x faster than the baseline GATK HC implementation, and 1.34x faster than the GATK HC implementation with the integrated GPU-based pair-HMMs forward algorithm. Although the methods proposed in this paper is to improve the performance of GATK HC, they can also be used in other pairwise alignments and applications.

Keywords: Semi-global alignment with traceback, Optimal alignment, GATK HaplotypeCaller (HC), GPUs

Background

NGS (Next Generation Sequencing) platforms offer the capacity to generate large amounts of DNA sequencing data in a short time and at a low cost. However, the analysis of the dramatic amounts of DNA sequencing data is still a computational challenge. Researchers have proposed many methods to improve the performance of the DNA sequencing data analysis tools and applications. One method is to execute these tools and applications on high performance computing architectures, such as supercomputers, clusters and even cloud environments. Another method is to use accelerators, such as GPUs and FPGAs, to accelerate the time-consuming kernels of these tools and applications to improve their performance.

GATK HaplotypeCaller (HC) is a popular variant caller, which is used to find the differences (or variants) between the sample DNA sequence compared with the reference sequence. Although GATK HC has higher accuracy of identifying variants compared with many other variant callers, its feasibility is limited by the long execution time needed for the analysis, which has proven to be difficult to optimize. This has driven researchers to improve its performance. Intel and IBM researchers both employ vector instructions to optimize the pair-HMMs forward algorithm [1, 2], which is the most time-consuming part of GATK HC, to reduce the total execution time. Ren et al. [3, 4] uses GPUs to accelerate the pair-HMMs forward algorithm in GATK HC, which achieved 1.71x speedup

*Correspondence: z.al-ars@tudelft.nl
Delft University of Technology, Mekelweg 4, 2628 CD Delft, The Netherlands

in single thread mode. After accelerating the pair-HMMs forward algorithm on GPUs, profiling of GATK HC shows that the semi-global pairwise sequence alignment accounts for around 34.5% of the overall execution time, making it the most time-consuming kernel in the application. This provides an opportunity to further improve the performance of GATK HC using GPU acceleration.

Pairwise sequence alignment, which includes global alignment, semi-global alignment and local alignment, is one of the commonly used techniques in many biological tools and applications. The global alignment and the semi-global alignment are calculated by the Needleman-Wunsch algorithm and the modified Needleman-Wunsch algorithm, respectively, while the local alignment is calculated by the Smith-Waterman algorithm. Although there are some differences existing in the three algorithms, the main framework of these algorithms is similar, which includes two stages: (1) a dynamic programming kernel to calculate the score matrices and find the optimal alignment score; (2) a traceback (or backtracking) kernel to find the optimal alignment.

Since three kinds of pairwise sequence alignment (global, semi-global and local) have the same framework and differ only in details, techniques of speeding up one can be applied to the other two with tiny modifications. Different kinds of high-performance platforms, especially accelerators, such as FPGAs [5, 6] and GPUs [7–16], are used to reduce their execution time.

There has been much research done to reduce the execution time of the three kinds of pairwise alignment on GPUs. There are two approaches to implement the first stage of the pairwise sequence alignment on GPUs (which is to calculate the optimal alignment score): inter-task parallelization model and intra-task parallelization model. The former is that each thread performs one alignment independently, such as [7] and [8]. The latter is that threads in a block cooperate to perform an alignment, such as [9]. If the pairwise sequence alignment is applied for sequence database scanning, aligning a query sequence with all database sequences for sequence similarity, a query profile and related data storage and access techniques are employed to reduce memory accesses on GPUs, such as [10] and [11]. In [11], alignments are performed in interleaved mode in order to amortize the cost of initiating each execution pass.

However, very few researchers implement the second stage on GPUs. The existing implementations can be classified into two groups. The implementations of the first group are based on backtracking matrices. Liu et al. [11] proposed to store the score matrices and backtrack the score matrices to obtain the optimal alignment. However, the method is not described clearly. gpu-pairAlign [12] proposed to store the alignment moves in four Boolean backtracking matrices during the first stage and retrieve the four Boolean backtracking matrices instead of the score matrices. This group of implementations obtain the optimal alignment in linear time, but the disadvantage is that their space complexity is quadratic. The implementations of the second group are based on the Myers-Miller algorithm. MSA-CUDA [13] developed a stack-based iterative implementation of the Myers-Miller algorithm [17] to retrieve the optimal alignment in linear space. SW# [14] proposed a modified Myers-Miller algorithm. CUDAlign 2.0 [15] combined the Myers-Miller and Smith-Waterman algorithm. Moreover, with several versions of incremental optimizations, CUDAlign 4.0 [16] is able to achieve the optimal alignment of chromosome-wide sequences using multiple GPUs. However, their approaches have quadratic time complexity, making them only suitable for the pairwise alignment of very long DNA and protein sequences.

In this paper, we provide an accelerated solution tailored to GATK HC which implements the semi-global pairwise sequence alignment with traceback on GPUs to further improve the performance. The contributions of this paper are as follows:

- We first analyze the characteristics of the semi-global alignment in GATK HC and then propose a GPU-based implementation of the semi-global alignment with traceback based on the analysis.
- During the first stage, we propose to record the length of consecutive match(es)/mismatch(es) and store the alignment moves in a special backtracking matrix.
- We also propose a new algorithm that allows for retrieving the optimal alignment efficiently on GPUs.
- We benchmark the results and show an overall speedup of GATK HC of about 2.3x over the non-accelerated version.

Although this paper proposes to improve the performance of GATK HC, the GPU-based implementation of the semi-global alignment with traceback can be used in other applications and tools. Moreover, since there are only small differences among the global alignment, semi-global alignment and local alignment, the methods proposed in this paper can also be applied to the global alignment and local alignment.

Methods

A brief overview of semi-global alignment

Semi-global alignment finds the overlap between two sequences. Insertion and deletions introduce gaps in the alignment. Gaps at the start or end of the sequences may be neglected. Hence, different types of semi-global alignments are possible between two sequences. Figure 1 shows an example of the type of the semi-global alignment performed in GATK HC.

```
R1:          AGTCTAG--CG-
R2:          -GAC--GAACGT
-------------------------
Alignment:   MMMDDMIIMM
```

Fig. 1 An example of an semi-global alignment of two sequences in GATK HC. R1 and R2 represent two sequences. Gaps ('-') at the start and end of two sequences are neglected. In the alignment, there are three kinds of operations indicating how R2 aligns with R1. 'M' indicates that a base in R2 aligns with a base in R1 (matches or mismatches); 'I' indicates that a base in R2 is not in R1; 'D' indicates that a base in R1 is not in R2

The pairwise sequence alignment is to find the optimal alignment between two sequences, which has the optimal alignment score. The modified Needleman-Wunsch algorithm with affine gap penalties to calculate the optimal alignment score of the semi-global alignment in GATK HC is defined as

Initialization:

$$M_{i,0} = 0 \quad (0 \le i \le m) \qquad M_{0,j} = 0 \quad (0 \le j \le n) \tag{1}$$

Recurrence:

$$M_{i,j} = max \begin{cases} M_{i-1,j-1} + sbt\left(R1[\,i\,], R2\left[j\right]\right) \\ D_{i,j} \\ I_{i,j} \end{cases}$$

$$D_{i,j} = max \begin{cases} D_{i-1,j} - \beta \\ M_{i-1,j} - \alpha \end{cases} \tag{2}$$

$$I_{i,j} = max \begin{cases} I_{i,j-1} - \beta \\ M_{i,j-1} - \alpha \end{cases}$$

Termination:

$$Result = max \begin{cases} \max_{\{1 \le i \le m\}} M_{i,n} \\ \max_{\{1 \le j \le n\}} M_{m,j} \end{cases} \tag{3}$$

where m and n are the length of R1 and R2, respectively. In these equations, $M_{i,j}$ represents the optimal alignment score of two subsequences $R1[\,1\,]...R1[\,i\,]$ and $R2[\,1\,]...R2\left[j\right]$, while $I_{i,j}$ and $D_{i,j}$ represent the optimal alignment score of two subsequences $R1[\,1\,]...R1[\,i\,]$ and $R2[\,1\,]...R2\left[j\right]$ with $R2\left[j\right]$ aligned to a gap and $R1[\,i\,]$ aligned to a gap, respectively. Here, the semi-global alignment uses an affine gap penalty model to calculate gap penalties, in which α and β are the gap open penalty and the gap extension penalty, respectively. *sbt* is the score of a match or mismatch. As shown by Eq. 1, the penalties of gaps at the start and end of two sequences are neglected. As shown by Eq. 3, the optimal alignment score of the semi-global alignment in GATK HC is the greatest value of the elements in the last row and the last column of the matrix M.

These equations indicate that the computation complexity of the modified Needleman-Wunsch algorithm is $O(mn)$, which makes the execution time increase quadratically with the sequence length. Usually, the algorithm is implemented using dynamic programming which solves three two dimensional matrices. According to the equations, $M_{i,j}$, $I_{i,j}$ and $D_{i,j}$ only depend on the up-left, up and left neighbor elements, which implies that the elements on the same anti-diagonal can be computed in parallel. Thus, a method employed by many researchers to reduce the execution time is to exploit this inherent parallelism in the algorithm.

If the alignment only needs to find the optimal alignment score of two sequences, the dynamic programming kernel can be calculated in linear space. Otherwise, the alignment with affine gap penalties generally uses three backtracking matrices to store the scores or alignment moves calculated by the dynamic programming kernel. The optimal alignment traceback starts from the position of the element with the optimal alignment score until reaching any element in the first row or the first column of the backtracking matrices, which is calculated in linear time. Figure 2 presents an example of backtracking an optimal alignment based on the score matrices.

Cigar format

In GATK HC, the goal is to get the optimal semi-global alignments, which are represented in the CIGAR format [18], and POS. CIGAR is a string including one or more

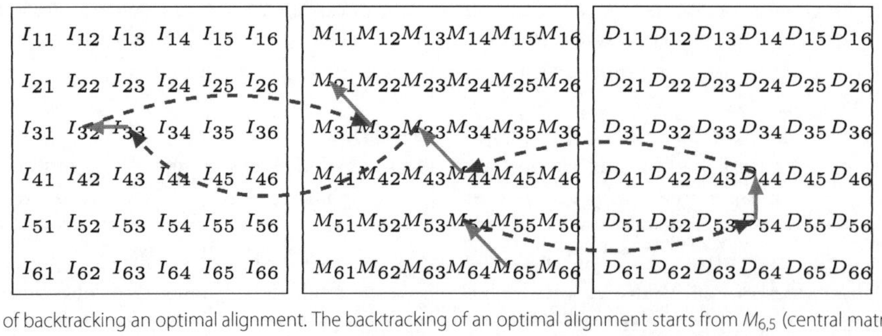

Fig. 2 An example of backtracking an optimal alignment. The backtracking of an optimal alignment starts from $M_{6,5}$ (central matrix in the figure), passes through $M_{5,4}$, jumps to $D_{5,4}$ (right matrix in the figure), passes through $D_{4,4}$, jumps to $M_{4,4}$, passes through $M_{3,3}$, jumps to $I_{3,3}$ (left matrix in the figure), passes through $I_{3,2}$, jumps to $M_{3,2}$ and ends at $M_{2,1}$. The optimal alignment retrieved is "MDMIM"

number-character pair(s). The character, including 'M', 'I', 'D', 'N', 'S', 'H', 'P', '=' and 'X', defines an operation explaining how the base in R2 aligns to the base in R1. Table 1 shows the CIGAR operations used in GATK HC. The number defines the length of the consecutive operations. POS is 0-based left most position of the first matching base of R1, which indicates the position of R1 where the alignment starts.

Take the alignment in Fig. 1 for example. The CIGAR representation of the alignment is "3M2D1M2I2M1S" and POS of the alignment is 1.

GPU architecture

Modern GPUs are widely used to accelerate computationally intensive algorithms. A GPU consists of thousands of small cores capable of executing one thread at a time. On NVIDIA GPUs, threads are grouped into *blocks* and these blocks are grouped into *grids*. Furthermore, consecutive threads in the same block are grouped into *warps*. The size of a warp is usually 32. The memory hierarchy includes registers, shared memory, global memory, cache and so on. Each thread is assigned a set of registers. The shared memory is accessed by all threads in a block. Using the shared memory, the threads in a block can exchange data at a very fast rate. The global memory is accessed by all the threads on the GPU. The latency of the global memory access is high since it resides on the device DRAM. If the data accessed by each thread in the same warp are stored at consecutive addresses, the global memory accesses of these threads can be coalesced. Usually, the width of one global memory access is 128 bytes. If the global memory accesses of threads in a warp are coalesced, there will be only one global memory access when the data accessed by each thread is not more than 4 bytes. Otherwise, there would be 32 sequential global memory accesses in the worst-case situation.

Semi-global alignment in GATK HC

Implementation of alignment in GATK HC

In GATK HC, the semi-global pairwise alignment is performed in two stages.

The implementation of the first stage is realized with a two-layer loop, which results in the $O(mn)$ computational complexity. The results of the first stage are two matrices: the score matrix *sw*, which stores matrix M, and the backtracking matrix *btrack*. In *btrack*, the value of each element can be classified into three kinds, which is defined as follows:

- > 0 - indicates a deletion and the length of the consecutive deletion(s) is the value of the element
- $= 0$ - indicates a match or mismatch and the length of the consecutive match(es)/mismatch(es) is increased by 1
- < 0 - indicates an insertion and the length of the consecutive insertion(s) is the absolute value of the element

The absolute values of the elements in the backtracking matrix are calculated by recording the length of the consecutive deletion(s) and consecutive insertion(s) when calculating the score matrix.

The implementation of the second stage is to calculate the optimal alignment in CIGAR format and POS. The score matrix sw is first used to find the optimal alignment score and the backtracking matrix *btrack* is then used to obtain the optimal alignment and POS. The optimal alignment is calculated in linear time. The backtracking matrix in GATK HC is helpful during backtracking. It is much easier to identify the next move compared with other methods since it does not need to jump among several backtracking matrices (shown in Fig. 1) or calculate the next move based on the current move [12]. Moreover, the lengths of the consecutive deletion(s) and consecutive insertion(s) are given by the element of the *btrack* matrix. However, the length of the consecutive match(es)/mismatch(es) is not given, which is increased by one instead.

Data analysis

In GATK HC, the semi-global alignment is performed in three situations:

1. Align the reference path with the dangling path to recover dangling branches for the local assembly.
2. Align the read with the assembled haplotype.
3. Align the assembled haplotype with the reference to decide whether the assembled haplotype satisfied the defined requirements.

We profiled GPU-based GATK HC [3] with a typical workload (Chromosome 10 of the whole human genome dataset G15512.HCC1954.1 [19]) to investigate which situation is most time-consuming. The profiling results in single-threaded mode are shown in Table 2, which specifies the relative execution time and the number of the semi-global alignments in each situation. As shown in the table, the execution time of all the semi-global alignment

Table 1 CIGAR operations used in GATK HC

Operation	Description
M	Match/mismatch
I	Insertion (gap in R1)
D	Deletion (gap in R2)
S	Soft clipping (base at the beginning or the end of R2 but not in R1)

Table 2 Execution time of the semi-global alignment in three situations of GATK HC

Situation	Number of alignments	Execution time
1	3529	0.03%
2	850376	14.58%
3	54802	19.89%
Total	908707	34.5%

accounts for 34.5% of the total execution time. Moreover, situation 2 and 3 consumes around 100% of the semi-global alignment execution time and the execution time in situation 1 is negligible. However, although the number of semi-global alignments in situation 2 is much larger than that in situation 3, the execution time in situation 2 is smaller than that in situation 3.

We then analyzed the lengths of the sequence pairs in situation 2 and 3. In situation 2, let R1 be the assembled haplotype and R2 be the read. In situation 3, let R1 be the assembled haplotype and R2 be the reference. Figure 3 shows a scatter plot of the lengths of the sequence pairs in these two situations. As shown in Fig. 3, the lengths of the sequence pairs in situation 2 (40∼350) are shorter than those in situation 3 (300∼520). Since the computation complexity is $O(mn)$, the execution time of each semi-global alignment in situation 3 is bigger than that in situation 2. This explains why the total execution time of situation 3 is bigger than that of situation 2, which is shown in Table 2. Moreover, in situation 2, the length of R2 (the read) is shorter than the length of R1 (the assembled haplotype).

In addition, we investigated the optimal alignments achieved in situation 2 and 3 and added up the number of M/I/D/S operations in each optimal alignment. Figure 4 shows that the number of M operations is the largest. Especially in situation 2, the number of M operations accounts for 99.65% of the total operations. Moreover, we found that most of M operations are consecutive in each optimal alignment. However, the length of the consecutive match(es)/mismatch(es) is increased by one during the optimal alignment retrieval.

Hence, although the computation complexity of the optimal alignment is linear, most of its execution time is used to calculate the length of the consecutive match(es)/mismatch(es).

We last studied the source code of GATK HC version 3.7 and found that the semi-global alignments in situation 2 and 3 can be grouped into many batches without big modifications of the source code. Each batch consists of many semi-global alignments of sequence pairs. The numbers of batches in situation 2 and 3 are 13,142 and 13,977, respectively. Figure 5 shows the number of sequence pairs of each batch in situation 2 and 3. The biggest number of sequence pairs in all the batches in situation 2 and 3 are 293 and 192, respectively. Furthermore, the majority of batches in situation 2 include 25 ∼125 of sequence pairs while the majority of batches in situation 3 include 1 ∼8 of sequence pairs.

Implementation on GPUs

The implementation of the semi-global pairwise alignment for GATK HC on GPUs is performed in two stages. In the first stage, it performs the modified Needleman-Wunsch algorithm in order to obtain the backtracking matrix and the position of the optimal alignment score. In the second stage, it retrieves the backtracking matrix in order to obtain the optimal alignment in CIGAR format and POS.

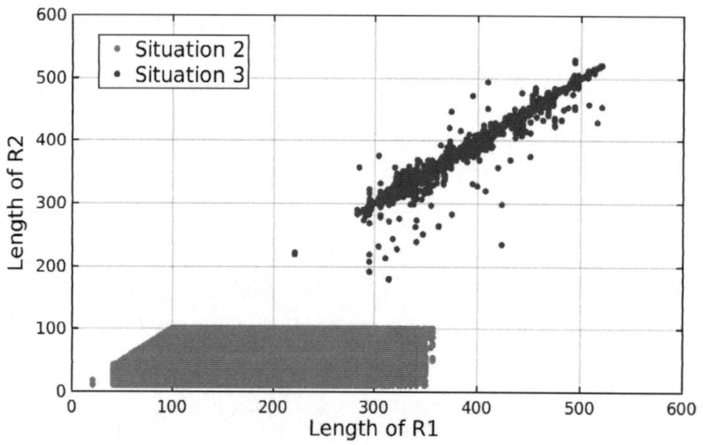

Fig. 3 Lengths of sequence pairs in situation 2 and 3. The lengths of the sequence pairs in situation 2 (40∼350) are shorter than those in situation 3 (300∼520)

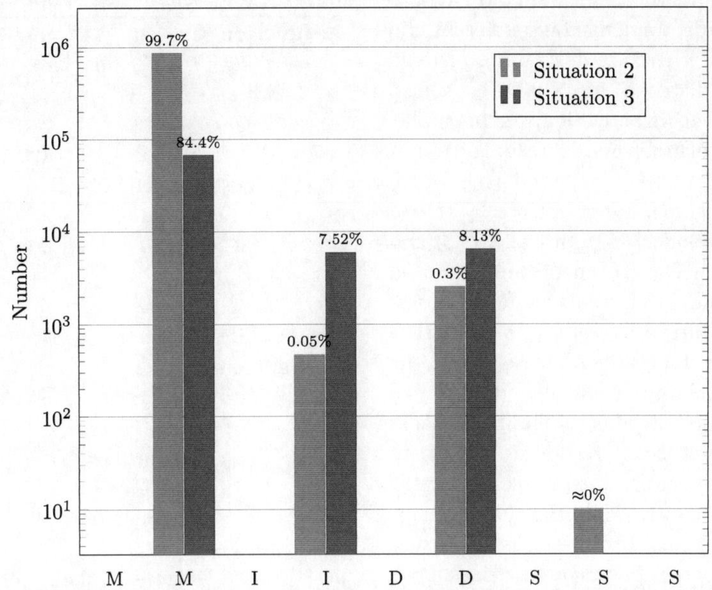

Fig. 4 Numbers of M/I/D/S in situation 2 and 3. The number of *M* operations is the largest. Especially in situation 2, the number of *M* operations accounts for 99.65% of the total operations

First stage implementation

Intra-task parallelization As mentioned in "Data analysis" subsection, the number of sequence pairs in each batch is less than 300. In order to effectively use the resources on GPUs, the intra-task parallelization model is employed to implement the modified Needleman-Wunsch algorithm on GPUs. For the implementation on GPUs, the elements on the same anti-diagonal of the score matrix M, I and D and backtracking matrix are calculated in parallel, reducing the computational complexity

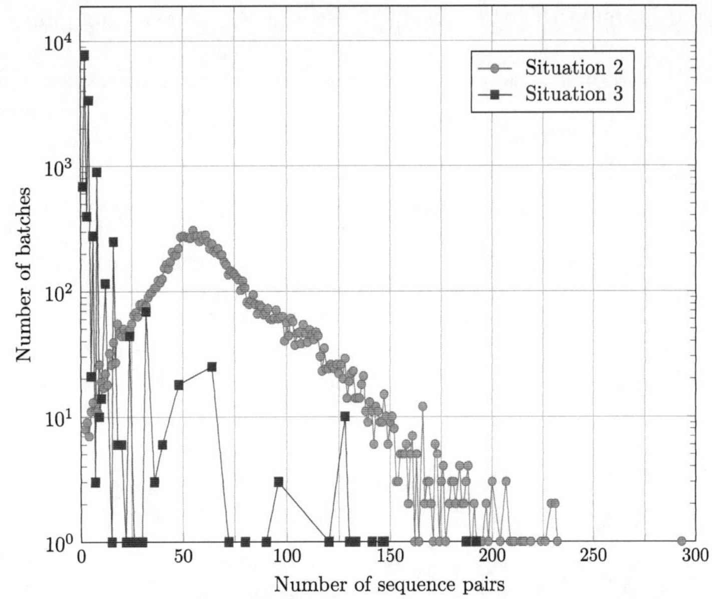

Fig. 5 Numbers of batches including different number of sequence pairs in situation 2 and 3. The biggest number of sequence pairs in all the batches in situation 2 and 3 are 293 and 192, respectively. Furthermore, the majority of batches in situation 2 include 25 ∼125 of sequence pairs while the majority of batches in situation 3 include 1 ∼8 of sequence pairs

to $O(m+n)$. Figure 6 shows the calculation of matrix M as an example to explain the implementation. Let R1 and R2 be the two sequences. There are in total 6 threads in the block and the size of matrix M is 6×6. At each step, the elements on an anti-diagonal are calculated in parallel and every element is calculated by one thread. For example, at step 5 (S5), $M_{5,1}$, $M_{4,2}$, $M_{3,3}$, $M_{2,4}$ and $M_{1,5}$, which are on the same anti-diagonal, are calculated by thread 0 (T0), thread 1 (T1), thread 2 (T2), thread 3 (T3) and thread 4 (T4), respectively. These elements are then used in the next step to calculate the elements on the next anti-diagonal. Moreover, each thread is responsible to calculate the elements in one column of matrix M. For example, the elements in the second column are calculated by thread 1 (T1). The goal of the first stage is to obtain the back-tracking matrix and the position of the optimal alignment score. Therefore, elements of the score matrix M, I and D are not stored. Instead, two vectors in the shared memory and three registers of each thread are used to store the intermediate results of the three score matrices. During the calculation of elements of the last column and the last row of matrix M, the optimal alignment score and its position are obtained. However, the drawback of the implementation is that some threads remain idle at the beginning or at the end of the calculation procedure, resulting in low thread utilization. If the length of R2 is smaller than the number of threads in a block, the execution is similar to Fig. 6 while some threads would remain idle during the whole calculation procedure. If the length of R2 is bigger than the number of threads in a block, there are two solutions to deal with it. One is to increase the size of a block until the number of threads in a block is equal to

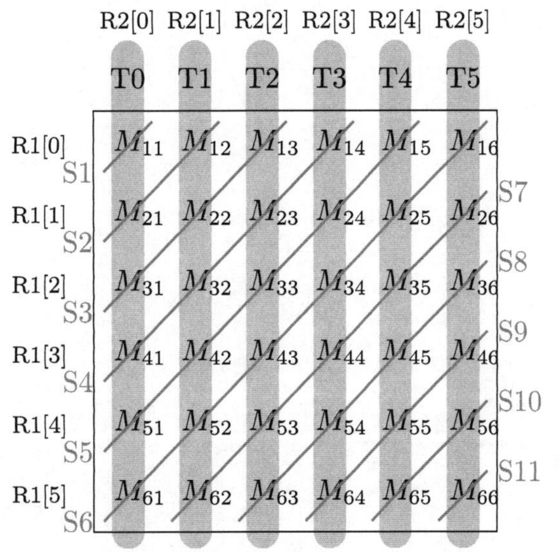

Fig. 6 Calculation of matrix M on GPUs

Algorithm 1 Pseudo code of optimal alignment retrieval

```
function CALCULATECIGAR(btrack[ ][ ], Cigar[ ], m, n, POS, P1, P2, max_col)
    state ←' N'                                  ▷ Initialization of state
    length ← 0
    if P2 > 0&&n − P2 > 0 then                   ▷ Check soft clippings at the end of R2
        Cigar[index].num = n − P2
        Cigar[index++].c =' S'
    end if
    i = P1, j = P2
    while i > 0&&j > 0 do                         ▷ Compute optimal alignment
        tp = B[(i − 1 + j − 1) * max_col + j − 1]
        if tp.x > 0 then
            new_state =' D'
            i = i − tp.x
            step_length = tp.x
        else
            if tp.x < 0 then
                new_state =' I'
                j = j − abs(tp.x)
                step_length = abs(tp.x)
            else
                new_state =' M'
                i = i − tp.y
                j = j − tp.y
                step_length = tp.y
            end if
        end if
        if state ==' N' then
            state = new_state
        end if
        if new_state == state then
            length+ = step_length
        else
            Cigar[index].num = length
            Cigar[index++].c = state
            length = step_length
            state = new_state
        end if
    end while
    Cigar[index].num = length
    Cigar[index++].c = state
    POS = i − 1
    if j > 0 then                                 ▷ Check soft clippings at the beginning of R2
        Cigar[index].num = j
        Cigar[index++].c =' S'
    end if
end function
```

or bigger than the length of R2. The other is to divide the calculation into several passes. In each pass, the execution is similar to Fig. 6. Three vectors in the global memory are used to store the intermediate results produced by the last thread of each pass, which would be used in the next pass. The advantage of the second solution is that it increases efficiency by reducing idle percentage of threads during the calculation procedure while its disadvantage is that it increases global memory accesses.

Recording the length of consecutive match(es)/mismatch(es) Besides recording the length of the consecutive deletion(es)/insertion(es), we also record the length of the consecutive match(es)/mismatch(es) in the first stage. The backtracking matrix on GPUs is stored in a *short*2 matrix. Each element of the matrix has two values, which are x and y. The value of x and y are defined as follows:

- $x > 0$ - indicates a deletion and the length of the consecutive deletion(s) is the value of the element
- $x = 0$ - indicates a match or mismatch and the length of the consecutive match(es)/mismatch(es) is y.
- $x < 0$ - indicates an insertion and the length of the consecutive insertion(s) is the absolute value of the element

The data type of x and y is *short*, of which the minimum value and maximum value are −32768 and 32767, respectively. The absolute values of the minimum value and maximum value are bigger than the theoretical maximum length of the consecutive operations, which is the length of R1 or R2. In order to calculate the backtracking matrix, a *short*2 vector in the shared memory and two registers of each thread are used.

Moreover, Since the backtracking matrix will be used in the next stage and the shared memory is not big enough to store it, the backtracking matrix is stored in the global memory. Similar to calculation of the matrix M shown in Fig. 6, elements of the backtracking matrix are calculated in anti-diagonal order. Thus, the backtracking matrices are stored in the diagonal-major data format (Fig. 7b), which is proposed in [20], instead of the row-major data format (Fig. 7a) to avoid non-coalesced global memory accesses of 32 threads in a warp and reduce global memory accesses.

Second stage implementation

In the second stage, we use the backtracking matrix *btrack* to obtain the optimal alignment and POS. Algorithm 1 presents the pseudo code of the optimal alignment retrieval on GPUs. $P1$ and $P2$ describe the position of the optimal alignment score. Algorithm 1 first checks whether there are soft clippings at the end of R2, and then computes the optimal alignment in a *while* loop. At the end, it checks whether there are soft clippings at the beginning of R2. The backtracking starts from $(P1, P2)$ and finishes when $i \leqslant 0$ or $j \leqslant 0$, which is calculated in linear time. POS is the value of $(i - 1)$ at the end of the *while* loop. In addition, the position of each element in the backtracking matrix is calculated by i, j and *max_col*, as shown in the 9th line in Algorithm 1. *max_col* is the column size of the maximum backtracking matrix of all sequence pairs.

The length of the deletion, insertion and match/-mismatch is given by the value of an element of the backtracking matrix, as shown in the 13th, 18th and 23rd line, respectively. This reduces the global memory accesses used to calculate the length of the operations.

Results

All the experiments are performed on IBM Power System S823L (82478-42L), which has 2 IBM Power8 processors (10 cores each) running at 3.6 GHz, 256 GB of DDR3 memory, and an NVIDIA Tesla K40 card. The NVIDIA Tesla K40 card has 2880 cores that run at up to 745 MHz and has a CUDA compute capability of 3.5.

We first compare the performance of the GPU-based semi-global alignment implementation with different techniques using the synthetic datasets. The synthetic datasets are created based on the output of Wgsim [21] with default parameters. We then compare the performance of GPU-based semi-global alignment implementation with gpu-pairAlign implementation using synthetic datasets. Next, we compare the performance of GPU-based semi-global alignment implementation with CPU-based implementation using synthetic and real datasets. We last integrate the GPU-based semi-global alignment implementation into GATK HC 3.7 and compare the overall performance.

Throughput is used as a performance metric of the first stage of the GPU-based implementation, which is measured by giga cell updates per second (GCUPS). Note that it is not fair to compare the throughput of the first stage of the semi-global alignment with traceback with that of the score-only alignments since the former needs to store backtracking matrices in the global memory.

Performance comparison of multi-pass

There are two solutions to implement the first stage of the semi-global alignment on GPUs if the length of R2 is bigger than the number of threads in a block. We realized these two solutions and used different synthetic datasets to compare the performance of the two solutions.

Figure 8 shows the performance of the two solutions with different synthetic datasets. There are 9 datasets each with a different length of R1/R2, namely: 64, 128, 192, 256, 320, 384, 448, 512 and 576. In each dataset, the lengths of

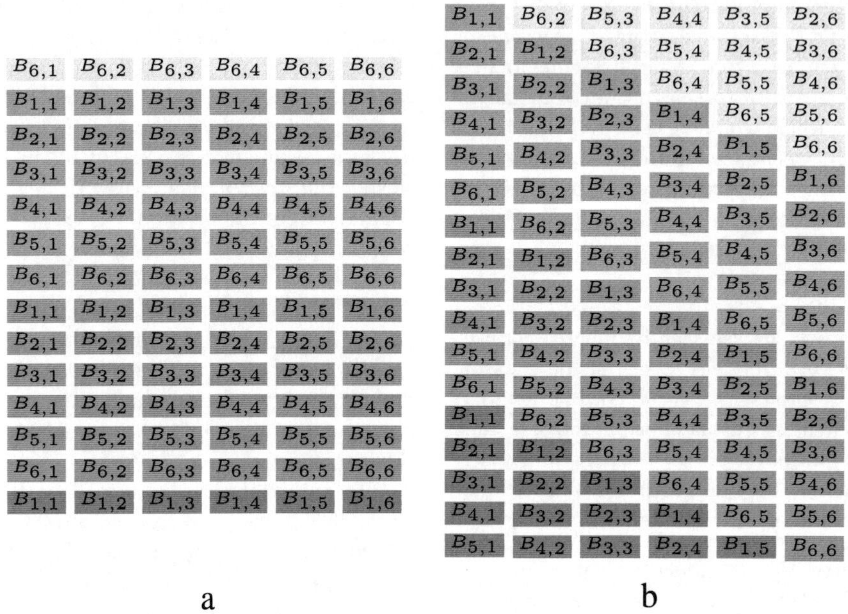

Fig. 7 Layout of the backtracking matrix in the global memory. Elements of different backtracking matrices are marked with different colors. **a** is the backtracking matrices stored in the row-major data format. **b** is the backtracking matrices stored in the diagonal-major data format

R1 and R2 are the same. The number of sequence pairs in the 9 datasets is 25, 100 and 1000.

For the first solution, which is to increase the block size, there are in total 9 implementations for 9 datasets. The differences of these implementations are the block size and the sizes of vectors in the shared memory which store the intermediate results. For the second solution, which employs multi-pass, there is 1 implementation with block size of 128.

As shown by Fig. 8, the throughput of the first solution is higher than that of the second solution when the number of sequence pairs of the datasets is 25 and 100.

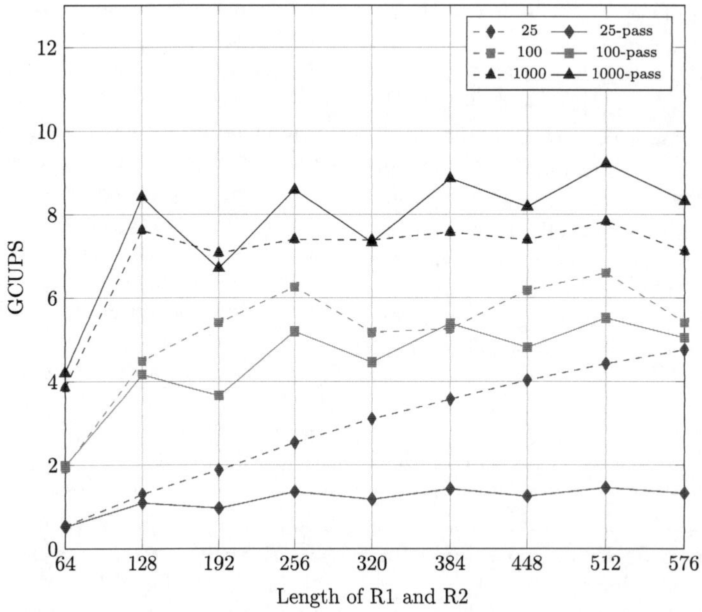

Fig. 8 Performance comparison of implementations for two solutions on synthetic datasets

However, when the number of sequence pairs of the datasets is 1000, the throughput of the second solution is higher in most cases. This is because the efficiency of the implementations for the first solution is smaller than that of the implementation for the second solution and the advantage of the second solution overweighs its disadvantage when the number of sequence pairs of the dataset is big. Thus, we can choose the implementation of these two solutions based on the number of sequence pairs of the dataset.

Performance comparison of recording match/mismatch lengths

In this section, we analyze the impact of recording the length of consecutive matches/mismatches on the performance of the second stage of the alignment on GPUs. We realized two implementations. The first implementation is our approach shown in Algorithm 1 in which the length of consecutive matches/mismatches is recorded in the backtrack matrix. The second implementation is similar to Algorithm 1 except that the length of M is increased by one and the coordinates (i, j) of M are decreased by 1. The backtracking matrices are produced by 9 implementations for the first solution using 9 synthetic datasets. Here, the synthetic datasets are not based on the output of Wgsim since we consider the best case, in which only many M operations exist in the optimal alignment. The lengths of R1/R2 in the 9 synthetic datasets are 64, 128, 192, 256, 320, 384, 448, 512 and 576. The number of sequence pairs in the 9 datasets is 100.

Figure 9 shows the execution time of the two implementations. The implementation which records match/mismatch lengths is faster. Moreover, its execution time remains nearly constant with increasing length of R1 and R2 as it only requires a single global memory access per R1 and R2 pair. The execution time of the implementation without recording match/mismatch lengths increases linearly with the length of R1 and R2. This is because the number of global memory accesses increases linearly with the number of M operations in the optimal alignment, which in turn increases linearly with the length of R1 and R2.

Performance comparison with gpu-pairAlign

As mentioned in "Background" section, there are two methods to implement the second stage on GPUs: the method based on the Myers-Miller algorithm and the method based on backtracking matrices. The method based on the Myers-Miller algorithm is only suitable for the pairwise alignment of very long DNA and protein sequences. Thus, we compared our implementation with gpu-pairAlign [12], which uses backtracking matrices to obtain the optimal alignments. gpu-pairAlign is designed to perform alignment of every given sequence pair on GPUs, especially for protein sequence pairs. It includes algorithms for global alignment, semi-global alignment and local alignment. We compare with its semi-global

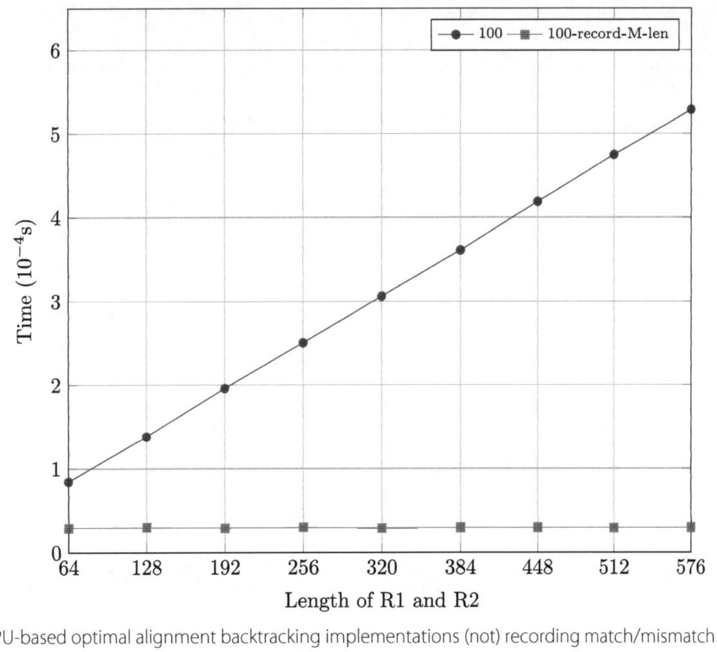

Fig. 9 Execution time of GPU-based optimal alignment backtracking implementations (not) recording match/mismatch lengths on synthetic datasets

alignment algorithm. The semi-global alignment algorithm of gpu-pairAlign is also performed in two stages: the optimal alignment score and the backtracking matrices are computed in the first stage; the backtracking is performed in the second stage.

There are two main differences between the gpu-pairAlign implementation and our implementation: (1) In the first stage, our implementation employs the intra-task parallelization model, while the gpu-pairAlign implementation employs the inter-task parallelization model; (2) The backtracking matrix of our implementation is a short2 matrix, elements of which are the length of consecutive deletion(es), insertion(es) and match(es)/mismatch(es), while the backtracking matrices of the gpu-pairAlign implementation are four Boolean matrices, elements of which indicate the proper direction of backtracking moves.

We modified the gpu-pairAlign implementation to make it to deal with the data produced by GATK HC: (1) Since the input data of our implementation is a set of sequence pairs instead of a set of sequences, the way in which the gpu-pairAlign implementation handles input data is modified; (2) Integer arrays are used to store the intermediate results instead of short arrays since the intermediate results are bigger than the maximum value of the short data type; (3) The alignments are modified to be represented using the CIGAR format and POS.

We first used the synthetic datasets described in "Performance comparison of multi-pass" subsection to compare the performance of the first stage of the two implementations, which is shown in Fig. 10. The performance of gpu-pairAlign implementation is much smaller than our implementation. The main reason is that when the size of the synthetic datasets is small, the resource on the GPU cannot be fully utilized for the inter-task parallelization model. The second reason is that the intermediate results are stored in integer arrays, which increases the size of shared memory of each block and the number of global memory accesses.

We then compared the performance of the second stage of the two implementations using the synthetic datasets described in "Performance comparison of recording match/mismatch lengths" subsection, which is shown in Fig. 11. The execution time of the second stage of our implementation remains nearly constant when the length of R1 and R2 increases, while the execution time of the second stage of the gpu-pairAlign implementation increases linearly with the length of R1 and R2. Although the gpu-pairAlign implementation reduces the global memory space by using four Boolean matrices, it still needs to calculate each move one by one, which is avoided in our implementation through storing the length of consecutive deletion(es), insertion(es) and match(es)/mismatch(es).

Performance comparison with CPU-based implementation

In this section, we compare the performance of our GPU-based semi-global alignment with traceback implementation with the CPU-based implementation using synthetic and real datasets. We used the first solution which increases the block size when the length of R2 is bigger than the block size and records the length

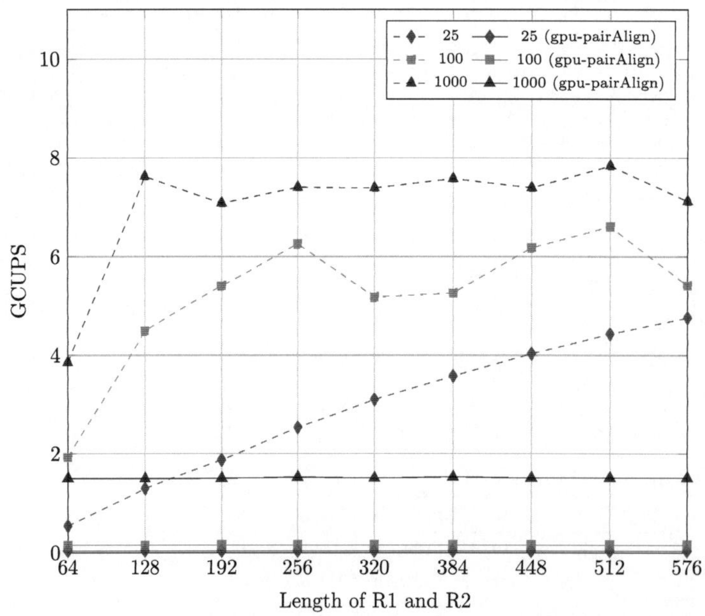

Fig. 10 Performance comparison of the first stage of two implementations on synthetic datasets

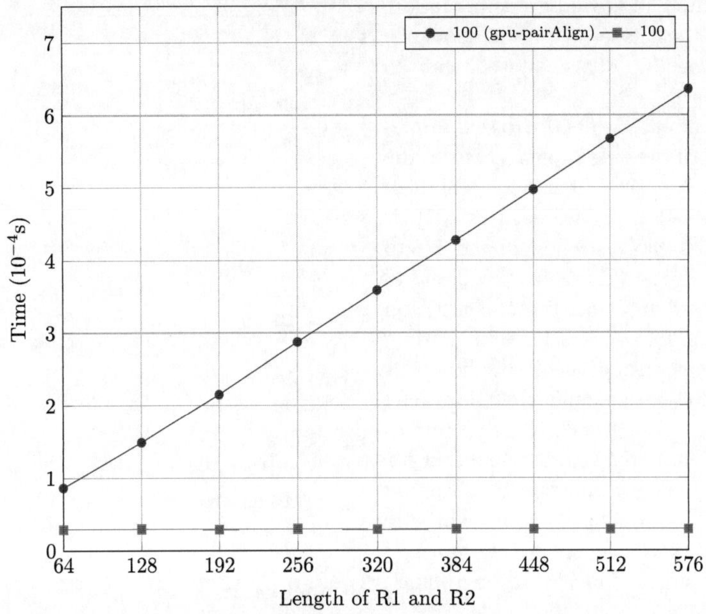

Fig. 11 Execution time of the second stage of two implementations on synthetic datasets

of consecutive matches/mismatches. The CPU-based implementation is written in the C++ programming language and compiled with gcc O3 optimization, running on one Power8 core. The real datasets are produced by using a typical workload (Chromosome 10 of the whole human genome dataset G15512.HCC1954.1).

Figure 12 shows the speedup of the GPU-based implementations compared with the CPU-based implementation using the synthetic datasets described in "Performance comparison of multi-pass" subsection. There are in total 9 GPU-based implementations for 9 datasets, block size of which are 64, 128, 192, 256, 320, 384, 448, 512 and 576. The GPU-based implementations

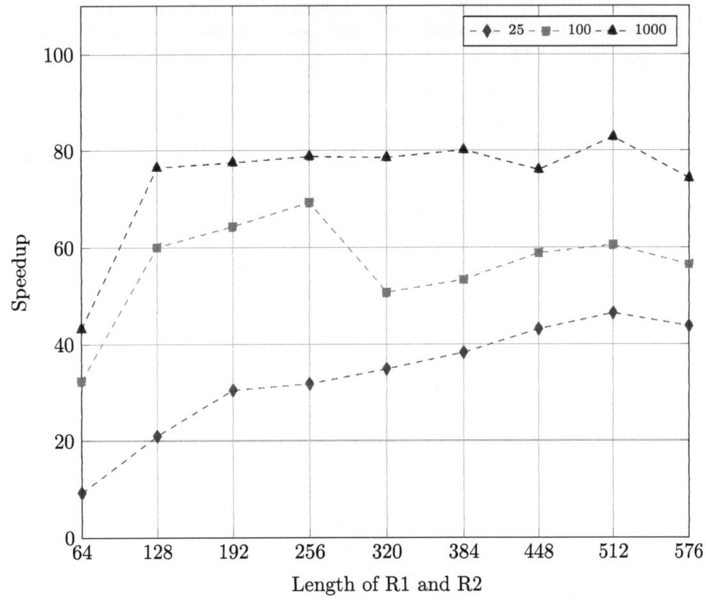

Fig. 12 Speedup of the GPU-based implementations compared with CPU-based implementation on synthetic datasets

is up to 80x faster than the CPU-based implementation. Moreover, the speedup of the datasets with 1000 sequence pairs is bigger than the speedup of the datasets with 25 and 100 sequence pairs.

Table 3 shows the execution time of GPU-based implementations with the real datasets. As shown by Fig. 3, the length of R2 in situation 2 is 40∼120 and the length of R2 in situation 3 is 300∼520. Thus, we used two GPU-based implementations with block size of 128 and 576 to execute the real datasets produced in situation 2 and 3, respectively. The GPU-based implementation of situation 2 is 14.14x faster than the CPU-based implementation, while the GPU-based implementation of situation 3 is 4.89x faster than the CPU-based implementation. The throughput of the first stage of the GPU-based implementation for situation 2 is 1.86 GCUPS, while that for situation 3 is 0.64 GCUPS. The throughput of situation 3 is much smaller than the throughput for the synthetic datasets with size 25. This is because the number of sequence pairs of batches in situation 3 is extremely small (1 ∼ 8 in most cases).

Integration into GATK HC

The two GPU-based implementations with block size of 128 and 576 are integrated into GATK 3.7 to accelerate the semi-global alignment with traceback of situation 2 and situation 3, respectively. The GATK HC implementation with both GPU-based pair-HMMs forward algorithm and GPU-based semi-global alignment with traceback is compared with other two GATK HC implementations: GATK HC (referred to as baseline), which is downloaded from the GATK website, and GATK HC with only GPU-based pair-HMMs forward algorithm. The dataset is Chromosome 10 of the whole human genome dataset (G15512.HCC1954.1). All the GATK HC implementations are performed in single thread mode.

Table 4 shows the overall execution time of these three implementations. The implementation with both GPU-based pair-HMMs forward algorithm and GPU-based semi-global alignment with traceback is 2.30x faster than the baseline implementation. Moreover, it is 1.34x faster than the implementation with only GPU-based pair-HMMs forward algorithm.

Note that the number of sequence pairs of each batch produced by GATK HC is small, leading to under utilization of the GPU resources. It is better to launch multiple GATK HC processes at the same time to fully utilize the GPU resources.

Table 3 Performance of GPU-based implementations on real datasets

	Throughput (GCUPS)	GPU (sec)	CPU (sec)	Speedup
Stage 1 of S2	1.86	2.32	43.93	18.94x
Stage 2 of S2	-	0.14	0.62	4.43x
Overall of S2	-	3.15	44.55	14.14x
Stage 1 of S3	0.64	10.20	53.29	5.22x
Stage 2 of S3	-	0.09	0.17	1.89x
Overall of S3	-	10.93	53.46	4.89x

S2 and S3 stand for situation 2 and 3, respectively. "Overall of S2" and "Overall of S3" represent the overall GPU execution time? CPU execution time and speedup of situation 2 and 3, respectively

Table 4 Execution time of GATK HC implementations

	Total time (s)	Speedup
Baseline	8034.05	-
GPU (only pair-HMMs)	4687.08	1.71x
GPU (pair-HMMs + semi-global alignment with traceback)	3490.70	2.30x

Conclusion

This paper presents an implementation of the semi-global alignment with traceback on GPUs to improve the performance of GATK HC. Semi-global alignment with traceback has two stages: in the first stage, a backtracking matrix is computed; in the second stage, the optimal alignment is calculated using the backtracking matrix. Based on the characteristics of the semi-global alignment with traceback in GATK HC, the intra-task parallelization model is chosen. The first stage of our GPU implementation is up to 18.94x faster than CPU. Moreover, our GPU implementation also records the length of consecutive matches/mismatches in addition to lengths of consecutive insertions and deletions as in the CPU implementation. This helps to reduce global memory accesses and provides a speedup of up to 4.43x in the second stage. Experimental results show that our alignment kernel with traceback is up to 80x and 14.14x faster than its CPU counterpart with synthetic datasets and real datasets, respectively. The GATK HC implementation with both GPU-based pair-HMMs forward algorithm and GPU-based semi-global alignment with traceback is 2.30x faster than the baseline GATK HC. It is 1.34x faster than the GATK HC implementation with only GPU-based pair-HMMs forward algorithm.

Abbreviations

HC: HaplotypeCaller; NGS: Next Generation Sequencing

Acknowledgements

The authors wish to thank the Texas Advanced Computing Center (TACC) at the University of Texas at Austin and IBM for the giving access to the IBM Power8 machines used in this paper.

Funding

This work was supported by CSC (Chinese Scholarship Council) grant and Delft University of Technology. Publication of this article was sponsored by Delft University of Technology.

Availability of data and materials
The algorithm generated in this manuscript as well as all input datasets are publicly available on a publicly available repository: https://github.com/ShanshanRen/semi-global-alignment-with-traceback.

About this supplement
This article has been published as part of BMC Genomics Volume 20 Supplement 2, 2019: Selected articles from the 17th Asia Pacific Bioinformatics Conference (APBC 2019): genomics. The full contents of the supplement are available online at https://bmcgenomics.biomedcentral.com/articles/supplements/volume-20-supplement-2.

Authors' contributions
SR designed and performed the experiments, analyzed the data, and wrote the manuscript. All the authors jointly developed the structure and arguments for the paper, made critical revisions and approved final version.

Ethics approval and consent to participate
Not applicable.

Consent for publication
Not Applicable.

Competing interests
The authors declare that they have no competing interests.

Published: 10 April 2019

References

1. Takeshi O, Yinhe C, Kathy T. Performance optimization of Broad Institute GATK Best Practices on IBM reference architecture for healthcare and life sciences. IBM Systems Technical White Paper. 2017. https://www.ibm.com/downloads/cas/LY1OY9XJ.
2. Proffitt A. Broad, Intel Announce Speed Improvements to GATK Powered by Intel Optimizations. Bio-IT World. 2014. http://www.bio-itworld.com/2014/3/20/broad-intel-announce-speed-improvements-gatk-powered-by-intel-optimizations.html.
3. Ren S, Bertels K, Al-Ars Z. GPU-Accelerated GATK HaplotypeCaller with Load-Balanced Multi-Process Optimization. In: IEEE International Conference on Bioinformatics and Bioengineering; 2017. p. 497–502.
4. Ren S, Bertels K, Al-Ars Z. Efficient Acceleration of the Pair-HMMs Forward Algorithm for GATK HaplotypeCaller on Graphics Processing Units. Evol Bioinform Online. 2018;14:1176934318760543.
5. Li IT, Shum W, Truong AK. 160-fold acceleration of the Smith-Waterman algorithm using a field programmable gate array (FPGA). BMC Bioinforma. 2007;8(1):1–7.
6. Benkrid K, Liu Y, Benkrid AS. A Highly Parameterized and Efficient FPGA-Based Skeleton for Pairwise Biological Sequence Alignment. IEEE Trans Very Large Scale Integr Syst. 2009;17(4):561–70.
7. Hasan L, Kentie M, Al-Ars Z. DOPA: GPU-based protein alignment using database and memory access optimizations. BMC Res Notes. 2011;4(1):261.
8. Ahmed N, Mushtaq H, Bertels K, Al-Ars Z. GPU accelerated API for alignment of genomics sequencing data. In: IEEE International Conference on Bioinformatics and Biomedicine; 2017. p. 510–5.
9. Maskell DL, Liu Y, Bertil S. CUDASW++: optimizing Smith-Waterman sequence database searches for CUDA-enabled graphics processing units. BMC Res Notes. 2009;2(1):73.
10. Liu Y, Schmidt B, Maskell DL. CUDASW++2.0: enhanced Smith-Waterman protein database search on CUDA-enabled GPUs based on SIMT and virtualized SIMD abstractions. BMC Res Notes. 2010;3(1):93.
11. Liu Y, Huang W, Johnson J, Vaidya S. GPU Accelerated Smith-Waterman. In: International Conference on Computational Science; 2006. p. 188–95.
12. Blazewicz J, Frohmberg W, Kierzynka M, Pesch E, Wojciechowski P. Protein alignment algorithms with an efficient backtracking routine on multiple GPUs. BMC Bioinforma. 2011;12(1):181.
13. Liu Y, Schmidt B, Maskell DL. MSA-CUDA: Multiple Sequence Alignment on Graphics Processing Units with CUDA. In: IEEE International Conference on Application-Specific Systems, Architectures and Processors; 2009. p. 121–8.
14. Korpar M, Sikic M. SW#-GPU-enabled exact alignments on genome scale. Bioinformatics. 2013;29(19):2494–5.
15. de O Sandes EF, de Melo ACMA. Smith-Waterman Alignment of Huge Sequences with GPU in Linear Space. In: 2011 IEEE International Parallel Distributed Processing Symposium; 2011. p. 1199–211.
16. Sandes EFO, Miranda G, Martorell X, Ayguade E, Teodoro G, Melo ACMA. CUDAlign 4.0: Incremental Speculative Traceback for Exact Chromosome-Wide Alignment in GPU Clusters. IEEE Trans Parallel Distrib Syst. 2016;27(10):2838–50.
17. Myers EW, Miller W. Optimal alignments in linear space. Comput Appl Biosci Cabios. 1988;4(1):11–7.
18. Li H, Handsaker B, Wysoker A, Fennell T, Ruan J, Homer N, et al. The Sequence Alignment/Map format and SAMtools. Bioinformatics. 2009;25(16):2078–9.
19. TCGA Mutation Calling Benchmark 4 Files. https://gdc.cancer.gov/resources-tcga-users/tcga-mutation-calling-benchmark-4-files. G15512.HCC1954.1.
20. Xiao S, Aji AM, Feng W. On the Robust Mapping of Dynamic Programming onto a Graphics Processing Unit. In: International Conference on Parallel and Distributed Systems; 2009. p. 26–33.
21. Wgsim. https://github.com/lh3/wgsim.

Ai et al. BMC Genomics 2019, 20(Suppl 2):185
https://doi.org/10.1186/s12864-019-5469-8

BMC Genomics

RESEARCH

Open Access

Explore mediated co-varying dynamics in microbial community using integrated local similarity and liquid association analysis

Dongmei Ai[1], Xiaoxin Li[1], Hongfei Pan[1], Jiamin Chen[2], Jacob A. Cram[3*] and Li C. Xia[2*]

From The 17th Asia Pacific Bioinformatics Conference (APBC 2019)
Wuhan, China. 14-16 January 2019

Abstract

Background: Discovering the key microbial species and environmental factors of microbial community and characterizing their relationships with other members are critical to ecosystem studies. The microbial co-occurrence patterns across a variety of environmental settings have been extensively characterized. However, previous studies were limited by their restriction toward pairwise relationships, while there was ample evidence of third-party mediated co-occurrence in microbial communities.

Methods: We implemented and applied the triplet-based liquid association analysis in combination with the local similarity analysis procedure to microbial ecology data. We developed an intuitive scheme to visualize those complex triplet associations along with pairwise correlations. Using a time series from the marine microbial ecosystem as example, we identified pairs of operational taxonomic units (OTUs) where the strength of their associations appeared to relate to the values of a third "mediator" variable. These "mediator" variables appear to modulate the associations between pairs of bacteria.

Results: Using this analysis, we were able to assess the OTUs' ability to regulate its functional partners in the community, typically not manifested in the pairwise correlation patterns. For example, we identified *Flavobacteria* as a multifaceted player in the marine microbial ecosystem, and its clades were involved in mediating other OTU pairs. By contrast, *SAR11* clades were not active mediators of the community, despite being abundant and highly correlated with other OTUs. Our results suggested that *Flavobacteria* are more likely to respond to situations where particles and unusual sources of dissolved organic material are prevalent, such as after a plankton bloom. On the other hand, *SAR11*s are oligotrophic chemoheterotrophs with inflexible metabolisms, and their relationships with other organisms may be less governed by environmental or biological factors.

Conclusions: By integrating liquid association with local similarity analysis to explore the mediated co-varying dynamics, we presented a novel perspective and a useful toolkit to analyze and interpret time series data from microbial community. Our augmented association network analysis is thus more representative of the true underlying dynamic structure of the microbial community. The analytic software in this study was implemented as new functionalities of the ELSA (Extended local similarity analysis) tool, which is available for free download (http://bitbucket.org/charade/elsa).

Keywords: Liquid association, Local similarity analysis, Microbial ecology, Time series data, Three-way association

* Correspondence: jcram@umces.edu; l.c.xia@stanford.edu
[3]Center for Environmental Science, University of Maryland, Cambridge, MA 21613, USA
[2]Department of Medicine, Stanford University School of Medicine, 269 Campus Dr., Stanford, CA 94305, USA
Full list of author information is available at the end of the article

Background

Marine ecosystem dynamics are likely controlled by environmental conditions and by the interactions between individual microbial taxa in the environment [1, 2]. Therefore, the study of the interplay among microbial species and the factors that control them are critical to understand and predict ecosystems' functions and their response to environmental changes. But how to apply theoretical approaches to predict species interactions in microbial communities from proximal data such as taxon abundances is still a key challenge in microbial ecology [3, 4]. Studies of microbial co-occurrence patterns [5–7] have led to novel insights into the ecology and habitats of different bacterial species. For instance, correlation network analyses have shown that related organisms, such as different ecotypes of *SAR11* respond to different environmental conditions and co-occur with different bacterial, archaeal and protistan species [7, 8]. Such analyses have also elucidated the differences in the interactions between grazers and viruses with bacteria [9, 10], and have suggested that changes in surface environments can have an effect on bacteria deep in the ocean [11, 12].

One limitation of previous correlation-based methods, such as *CoNet, MENAP, SparCC*, and *CCLasso* [13–16] is that they examine only pairwise correlations between organisms. It is however observed that some relationships between organisms are mediated by third variables [17–19]. For example, in marine ecosystems, mixotrophic eukaryotes feed on bacteria under some conditions (such as low sunlight or high prey density) but are photosynthetic under others. The trophic mode of these mixotrophs, is thus influenced by environmental conditions; in turn the way the mixotrophs interact with their environment is influenced by their trophic mode. Thus, a three variable mediation relationship could exist where sunlight levels (through trophic mode) affect the relationship between the mixotroph and its prey. Many bacteria are mixotrophic as well [20–22], and may similarly have interactions with their surroundings that are mediated by the variables that determine the trophic mode.

Other associations that likely include more than two variables would include symbioses that are favorable under only certain conditions. A well-known example is the symbioses between nitrogen fixing bacteria and their eukaryotic hosts [23–25], which are likely only prevalent in nitrogen-limited environments. Syntrophic relationships likewise would only exist under conditions favorable to that relationship, as would predation, competition and any other interaction that could occur between two organisms but might be mediated by third variables.

To examine how co-occurrence might be mediated by biological or environmental variables we employed liquid association (LA) analysis [26, 27] in succession to co-occurrence network analysis by the local similarity analysis (LSA) [6, 28, 29]. The rationale here is that LSA acts as a filtering mechanism to limit our three way analysis to only those variables that were at least sometimes associated with each other. Importantly, local associations, measured by ELSA [28–30], can include patterns that are present and robust over some parts of a time-series data set while weak or absent in other parts of that time-series. A natural question is whether the time-series are associated with a third variable. This question can then be measured by LA.

Our newly implemented Extended Liquid Association (ELA) analysis pipeline integrates both co-occurrence and mediation analysis to identify both pairwise and third party mediated associations. The ELA software is available as subroutines of the ELSA package (http://bitbucket.org/charade/elsa/). The tool will enable researchers to address the co-occurrence phenomenon with additional perspective of changing pairwise species-species or species-environment interaction with regard to a third species or environmental factor, or a varying ecosystem indicator. In this study, we identified multi-party interactions in inter-species and species-environment interaction using the *San Pedro Ocean Time-series (SPOT)* microbial community data by *Cram* et al. [11], which profiles the marine ecosystem near Los Angeles coast [7] using *Automated Ribosomal Intergenic Spacer Analysis (ARISA)* [31]. We also discussed the limitations of current analysis and possible future developments of ELA.

Methods

Analysis scheme

Our entire process, from data collection through visualization was outlined in Fig. 1 and is detailed in the following subsections. Briefly, we first collected the combined *SPOT* dataset (as detailed in ***Data Collection***) including both operational taxonomic unit (OTU) abundance levels and environmental factors co-measured. We then imputed the missing values and applied rank-based normalization to the dataset (as detailed in ***Data Normalization***). The normalized data was analyzed by the ELSA software to discover significant local pairwise associations (as detailed in ***Local Similarity Analysis***); these are associations that are present over part, though not necessarily all of the time series. Finally, we analyzed these significant local associations with the Extended Liquid Association routines within ELSA to reveal third party meditations of these pairwise interactions (as detailed in ***Liquid Association Analysis***). We finally described a novel visualization scheme to represent mediated association triplets in *Cytoscape* [32], and discussed their implications (as detailed in ***Visualization***).

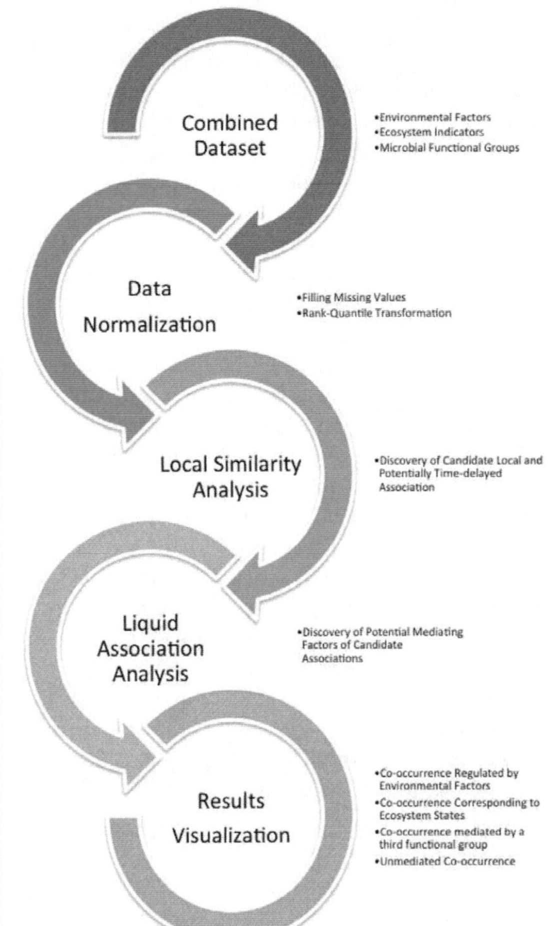

Fig. 1 The flowchart of integrated co-occurrence (Local Similarity) and mediation (Liquid Association) analysis pipeline

Our new bioinformatics tool - *Extended Liquid Association (ELA)* implements every step described in this analysis. This tool integrates two established and validated methods: Liquid Association analysis by *Li* et al [26] and the Local Similarity Analysis by *Ruan* et al. [6] and *Xia* et al. [28]. We chose to combine these tools because they behave synergistically with each other -- ELSA can uniquely identify local time-series associations and LA can elucidate those associations that are modulated by a third variable. Furthermore, the two tools, LSA and LA have been individually benchmarked extensively, comparing with other tools by independent studies [33, 34]. Indeed, in a comparison of many tools for correlation based network analysis, *Weiss* et al. [13] recently identified LSA as the best tool for correlation-based ecological time series data analysis.

Data collection

Our data were from the *San Pedro Ocean Time-series* project, which are publicly available from the *BCO-DMO (Biological & Chemical Oceanography Data Management Office)* websites: <http://www.bco-dmo.org/dataset/537137> (physical and chemical data) and <http://www.bco-dmo.org/dataset/535915> (biological data). The generation of these data has been described previously [35]. The samples were collected approximately monthly from 5-m depth from August 2000 through January 2011. Environmental parameters such as temperature and salinity measurements were measured in situ. Nitrate, nitrite, phosphate and silicate concentrations were measured by auto-analyzer. Bacterial heterotrophic productivity was measured through thymidine and leucine incorporation. Bacterial and viral concentrations were measured by SYBR (Synergy Brands, Inc) green epifluorescence microscopy. We also obtained satellite estimates chlorophyll-A concentration, and surface productivity. Other environmental variables included day length, virus to bacteria ratios, the excess phosphate concentration over Redfield ratios, and cell doubling time.

Bacterial community composition was determined by the *ARISA* fingerprinting method in conjunction with clone libraries. The *ARISA* data we used were a set of 423 of operational taxonomic units with putative taxonomic identities. The final data set is a 423 by 120 data matrix, where each row is an *ARISA* profile and each column is a time point (each entry value of an *ARISA* profile stands for the percentage of that operational taxon unit (OTU) in the community at the time. A 35 by 120 data matrix of supplementary biotic and abiotic environmental factors and ecosystem status indicators was combined with the *ARISA* data, in which, each row is one environmental factor and each column is the measure of that environmental factor at one time point, to obtain the final merged 458 by 120 raw data matrix.

Data normalization

We took the approach as in *Li* et al. [26] to normalize the relative abundance for *ARISA* datasets. In detail, to accommodate possible nonlinear associations and the variation of scales within the raw data, we apply the following approach to normalize the raw dataset before our analysis. We use X_i to denote the raw data of the *i*-th time spot of a variable. Assuming we have n samples in OTU profile, first, we take

$$R_k = \text{ rank of } X_k \text{ in } \{X_1, X_2, ..., X_n\} \tag{1}$$

Then, we take:

$$Z_k = \Phi^{-1}\left[\frac{R_k}{n+1}\right], \tag{2}$$

where Φ is the cumulative distribution function of the standard normal distribution. We will take $Z = Z_{[1:n]}$

obtained through the above procedure as the normalization of X.

A flowchart for incorporating Liquid Association (LA) with Local Similarity Analysis (LSA) was shown in Fig. 1. First, we used LSA to find candidate local and without time-delayed associations between factors X and Y. We filtered the results based on p-value and q-value. Then, given the significant LSA factors X and Y, we computed LA score to screen all environmental/OTU factors to discover potential mediating factor Z. Next, a permutation test for liquid association was performed and the results were filtered based on p-value and q-value to remove insignificant triplets. Finally, we used the software Cytoscape to visualize the resulted association network.

Local similarity analysis

In the next step, Local similarity analysis (LSA) was used to screen locally co-occurring pairs for later third-party mediation analysis. The LSA technique has been fully characterized in previous literature and has been successfully applied to the study of the co-occurrence networks within marine ecosystems [6, 7, 28, 29]. In this study, LSA was applied to find local associations with or without time delays. LSA was carried out with the ELSA software package [29] with parameters *'-d 0'* and *'-p theo'*. The *'-d 0'* flag in this case indicates to the algorithm that we only wanted to identify unlagged associations -- that is, associations of zero time delay. This parameter was chosen because we were only interested in synchronized co-occurrences and thus accepted only local similarity associations without delays. We also specified, with the *'-p theo'* flag that we wanted to estimate the p-value with the faster theoretical approximation [28] rather than the slower permutation based approach.

To correct for multiple testing, we use the false discovery rate (*FDR*) or q-value (Q). This q-value is defined as the fraction of false positives if a given association is declared as significant [36]. Resulting pairs with "$P \leq 0.001$", "$Q \leq 0.05$" and with an association segment spanning more than 50% of the total sampled time units were selected as correlated pairs. A cut-off for Local Similarity (LS) Score was determined at 0.28 by those p-value and q-value cut-offs, which could be interpreted as an analog of correlation coefficient of the same sample size.

Liquid association analysis

Next, we applied liquid association (LA) analysis to the each of the correlated pairs of variables to find their mediating factors. Liquid association analysis was originally developed by *K.C. Li et. al.* for characterizing the internal evolution of expression pattern of a pair of genes (X,Y) depending on a 'scouting' (mediator) gene Z [26]. In the microbial ecology setting, we analyze ecological functional factors instead of genes. Suppose X, Y, and Z are ecological functional factors and their measurements are standard normal distributed (after a normalization procedure) random variables with mean 0 and variance 1. The liquid association score (LA score) of X and Y with respect to$Z$, as denoted by LA (X; $Y| Z$), is defined as LA(X; $Y| Z$) = E(XYZ).

We estimate the LA score using the average product of three properly normalized factors:

$$\mathrm{LA}(X;Y|Z) = \frac{1}{m}\sum_{i=1}^{m} X_i Y_i Z_i \quad (3)$$

It can be seen that LA(X; $Y| Z$) = LA(Y; $X| Z$) = LA (Z; $Y| X$). Therefore, the LA score is a measurement of dependent association among the three functional factors; however, the score does not tell which two of the three are associated, which other is the mediating the relationship. In our analysis, the LSA approach supplements as the first step to identify the correlated pairs and then we used the identified pairs to scout for mediating factors.

To see whether the LA score computed is statistically significant, a permutation test was performed. To do this, after we computed the true LA score for X and Y with respect to Z, we randomly permuted the sequences of Z, and computed again a permuted LA score and compared it to the true LA score to see whether it was more extreme. The procedure was repeated many times ($N = 1000$) and the p-value (P) was calculated as the fraction that the permuted LA scores were higher than the true LA score. We used "$P \leq 0.001$" and "$Q \leq 0.05$" to control for statistical significance and multiple testing.

As shown in Fig. 2, There are four types of mediated correlation relationship, i.e., liquid association types, among factors X, Y and Z: (A) High Z level enhances the positive correlation between X and Y (Fig. 2a); (B) Low Z level enhances the negative correlation between X and Y (Fig. 2b); (C) Low Z level enhances the positive correlation between X and Y (Fig. 2c); (D) High Z level enhances the negative correlation between X and Y (Fig. 2d). In all cases, the correlation between X and Y is mediated by the level of Z. Those liquid association scenarios include cases when factors X and Y are correlated in one direction when Z is in one state, and X and Y cease to correlate or even to correlate reversely when the state of Z changes [26].

For example, the scenario a high Z level enhances the positive correlation between X and Y is shown in Fig. 2a. In the upper panel of Fig. 2a, factors X and Y were displayed in a scatter plot and the corresponding status of Z was both shape- and color-coded in the scattered points. A green circle stands for a low level of Z while an orange square stands for elevated level. The mediated

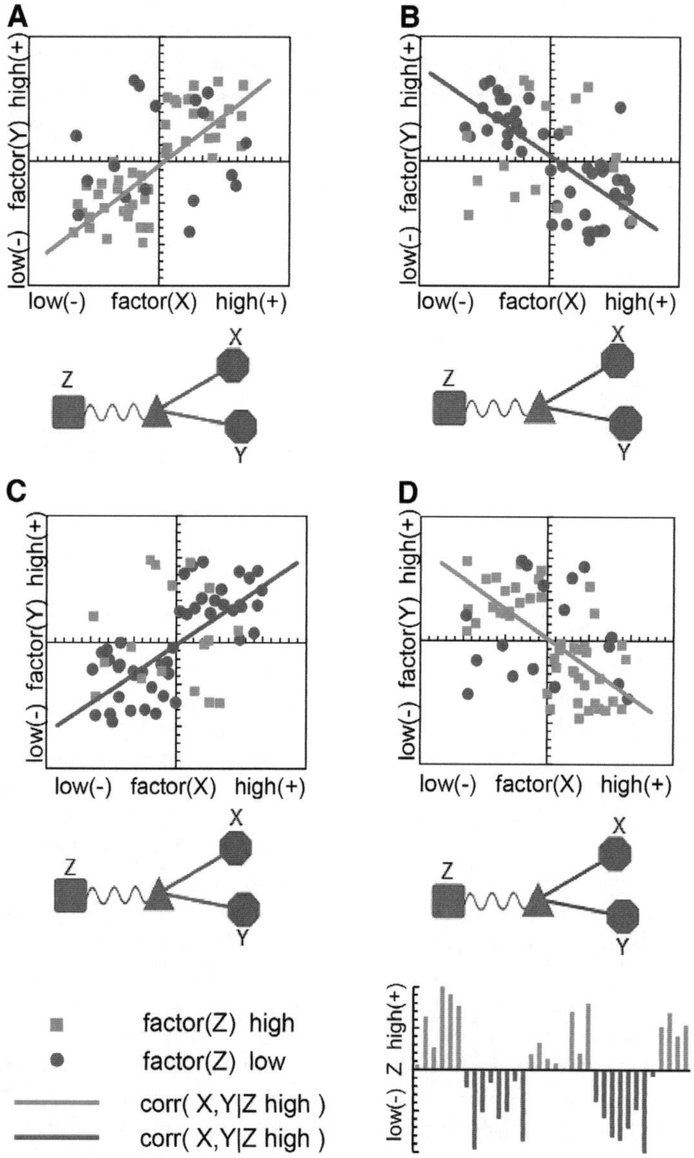

Fig. 2 Mediated correlation and example Cytoscape diagrams for all liquid association types of factors *X*, *Y* and *Z*. **a** High *Z* level enhances the positive correlation between *X* and *Y*; **b** Low *Z* level enhances the negative correlation between *X* and *Y*; **c** Low *Z* level enhances the positive correlation between *X* and *Y*; **d** High *Z* level enhances the negative correlation between *X* and *Y*

co-occurrence is represented by an orange-colored regression line applied to the scatter points of high level of *Z* only. We designed the graphic element as in the lower panel of Fig. 2a to illustrate this relationship for visualizing the co-occurrence networks using Cytoscape. Those three factors: *X*, *Y* and *Z*, were interconnected via a gray triangle to denote this three-way association. The wavy line connecting the triangular with the factor *Z* stands for mediation and the solid lines connecting the triangular with *X* and *Y* stands for the mediated correlation. We use red to indicate positive correlations and blue for negative correlations. Similarly, one can interpret the other types of liquid associations as depicted in the other panels of Fig. 2.

Visualization

To illustrate those complex triplet relationships in Cytoscape, we created a representation diagrams for each type: Green squares represent environmental factors or ecosystem status indicators. Violet octagons represent functional groups (OTUs). Gray triangles represent the three-way liquid association. Solid lines indicate local similarity associations - red solid lines indicate positive correlations while blue solid lines indicate the negative. Wavy lines indicate third-party mediation - red wavy lines indicate that the association is strong when the mediating variable is high, while the blue wavy lines indicate that the association is strong when the mediating variable is low (see Fig. 2).

We loaded all liquid association and local similarity connections into Cytoscape [32], along with metadata of bacterial and environmental factors. With Cytoscape we illustrated all significant three-way associations that involved environmental parameters. Those associations were all statistically significant and their pairwise *LS* scores were higher than 0.28. We went on to investigate associations that involved nodes of special interest. We first investigated pairwise interactions between operational taxonomic units that were regulated by environmental parameters. We also investigated associations that involved at least one OTU from the *SAR11* cluster, as well as associations that contained at least one OTU from the class *Flavobacteria*.

Results

Co-occurrence mediated by environmental factors

Bacterial abundance

Inspection of the liquid association interactions between pairs of OTUs that were mediated individual environmental parameters identified a variety of three-way interactions. Bacterial abundance (*Bact*) appeared to predict correlations of seven pairs of operational taxonomic units (Fig. 3). That is, seven pairs of positively and negatively correlated OTUs showed strongest correlations in their relative abundance in some cases when total bacterial abundance was high, and in other cases when bacterial abundance was low. For instance, the correlation between *AEGEAN_676.9* and *OTU522.8* is positive and was at its strongest when the total bacterial abundance was high. Conversely, as shown in the figure, the correlation between *AEGEAN_676.9* and *Prochl_HL (I)_828.8* was positive and at its strongest when total bacterial abundance was low.

In total, five of those seven pairs of bacteria (*SAR11_682.4, SAR11_703.7, AEGEAN_653.1, AEGEAN_679.4, AEGEAN_676.9*) contained at least one OTU from the *SAR11* cluster. Furthermore, six of the seven pairs were linked by a common OTU (*Bact*). That is, there were two OTUs (*OCS155_418.5* and *OM43_836.8*) that were correlated (positively or negatively) with another OTU when total bacterial abundance was high but correlated with a different OTU when the bacterial abundance was low. Overall it suggested that an alternating pattern exists for those common OTUs as they change interacting partners and types as the total bacteria abundance rise or drops, for example, when *Bact* is high *OM43* may compete with *Formos* but when *Bact* is low, *OM43* instead cooperates with *AEGEAN*.

Other environmental factors

There were many other environmental parameters that also appeared to be in liquid association with correlated OTU pairs, including silicate (12 connections), viral abundance (5 connections), salinity (6 connections), bacterial productivity as measured by thymidine incorporation (4 connections), the rate of change in day length (spring vs fall) (4 connections), phosphate concentration, and chlorophyll-A concentration (3 connections) (see Fig. 4).

For example, chlorophyll-A concentration (*Chol_A_Sat*) was a critical mediator of alphaproteobacteria and actinobacteria clades. When it was high, it enhanced the cooperation between *SAR11_703.7* and *AEGEAN169_653.1* clades - both were alphaproteobacteria, to promote carbon oxidization. This positive triplet association indicated likely oxidative bacteria bloom after a major increase of

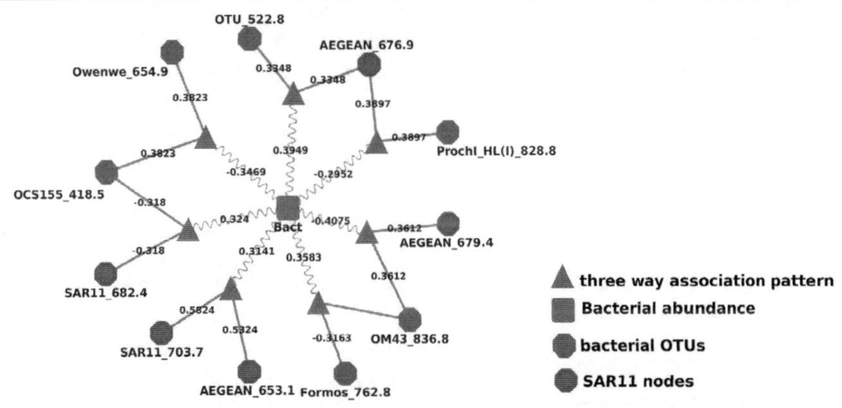

Fig. 3 A sub-network of local and liquid associations in which OTU correlations were mediated by the total bacterial abundance (*Bact*), an environmental factor

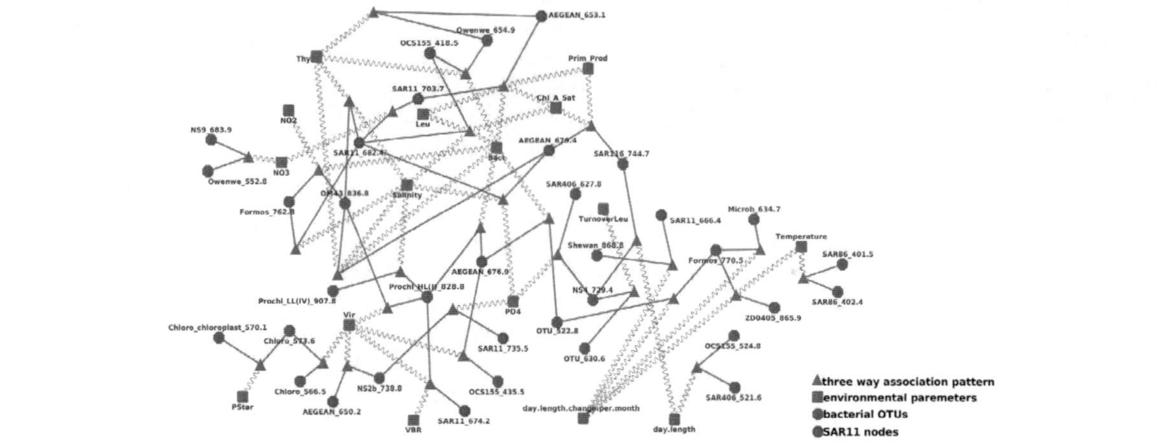

Fig. 4 A comprehensive network of local and liquid associations in which OTU correlations were mediated by environmental factors

marine productivity from photosynthetic plankton as indicated by high *Chol_A_Sat* level. Such alphaproteobacterial cooperativity however was inhibited by competition from OTU *OCS155_418.5*, which was a predominantly actinobacteria clade and was also involved in degrading organic matters, as we could observe an enhanced negative correlation between *SAR11_682.4* and *OCS155_418.5* when *Chol_A_Sat* concentration was high. A nature explanation for the observation was alphaproteobacteria and actinobacteria's competing biochemical nature to degrade organic matters. Interestingly, when *Chol_A_Sat* was low, it also enhanced the cooperation between two other alphaproteobacterial OTUs (*SAR216_744.7* and *AEGEAN_679.4*). This partner switch suggested subclones of alphaproteobacterium clades could rise or cease to dominate clade activity depending on changing environmental status.

Co-occurrence mediated by other operational taxonomic units

SAR11

Due to its unique metabolism and ecological role [37, 38], as well as its abundance and temporal variability in the surface of the san pedro channel [6] and globally [39], we focused on three-way associations in which any of the three nodes were from the *SAR11* clade (*Pelagibacterales*). It was evident that while many *SAR11*s were correlated with other operational taxonomic units, only OTU *SAR11_735.5* appeared to be involved in (several) three-way interactions with cutoffs of a liquid association effect larger than 0.8 (Fig. 5).

In the sub-network formed by the five correlating OTUs (Fig. 6), *SAR11_735.5* has a special status. First, *SAR11_735.5* and *OTU_545.8* (an OTU of unknown taxonomy) were positively correlated. They were connected by red solid lines and their correlations were mediated (gray triangle) by three OTUs (*OCS155_871.5*, *PAUG3G_600.6*, and *OTU_522.8*). When the abundance of those three mediating OTUs were high, the strength of the positive correlation between *SAR11_735.5* and *OTU_545.8* was stronger, and those positive three-way interactions were represented by red wavy lines. Further, the correlations of four pairs of *SAR11* OTUs (*OTU_545.8* and *OCS155_871.5*, *OTU_545.8* and *PAUC34_600.6*, *OCS155_871.5* and *OTU_522.8*, *PAUC34_600.6* and *OTU_522.8*) were mediated by *SAR11_735.5*. When the abundance of *SAR11_735.5* was high, the strength of these four correlations were stronger. Notably, all correlations between pairs of OTUs and three-way interactions in this network were all positive, implying these OTUs benefited from each other's boom and their relationships belonged to mutualism. This subgraph illustrated a complex and largely feed-forward trophic network that whose dynamics depended on the activity of *SAR11* clades.

Flavobacteria

The genus that appeared to be involved in the most three-way interactions was Flavobacteria with four nodes appearing in a tight cluster with other OTUs, most of which were liquid-associated (Fig. 7). These associations suggested that the bacteria in the cluster occasionally co-occurred together. At other times the Flavobacteria and their cooperatives were not co-occurring, but they were not necessarily all absent.

There were 31 liquid association patterns in the (Fig. 7). All the Flavobacteria OTUs, which were highlighted by bold outlines, were involved in triple associations. *OTU_522.8* was a special OTU (which was located at the left of Fig. 7) in this cluster. It was involved in more than 30 three-way association patterns. For instance, *OTU_522.8* was positively correlated with *OTU_545.8*,

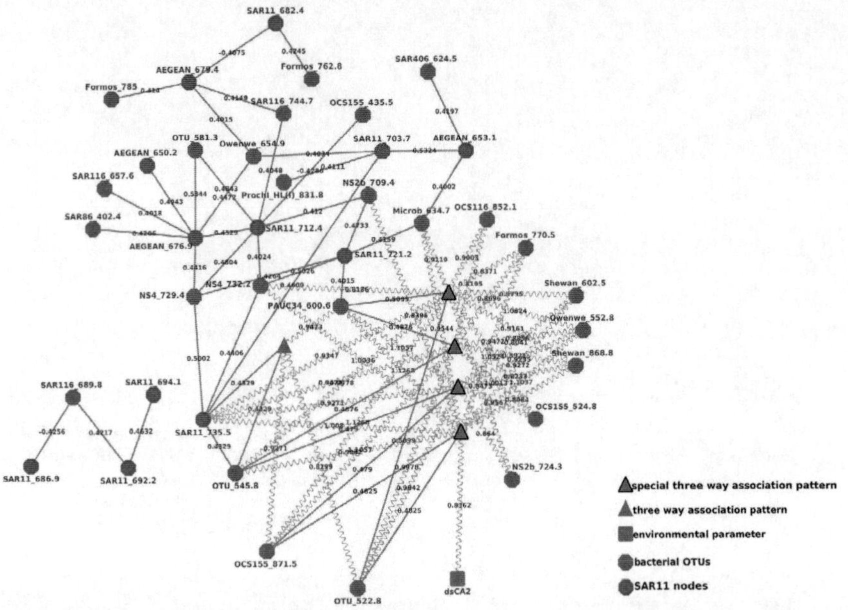

Fig. 5 A sub-network showing OTUs mediated and/or correlated to/by *SAR11* OTUs. SAR11 nodes were highlighted by bold outlines. All *SAR11* in the dataset and their significant liquid and local associations were shown

and their correlation was mediated by six OTUs, which included *Formos_785* and *NS9_683.9*. The red solid and wavy lines indicated that the relationship of those OTUs was mutualism. The correlation between *OTU_522.8* and *OCS155_871.5* was mediated by *OCS155_435.5*, the blue wavy line indicated that when the abundance of *OCS155_435.5* was high, the positive correlation between *OTU_522.8* and *OCS155_871.5* was weaker, which demonstrates a competitive relationship between *OCS155_435.5*, *OTU_522.8* and *OCS155_871.5*.

Discussion

The interactions between microbial species are complex and often non-linear. Some of those interactions are regulated by factors within the microbial community itself [40], such as the levels of key organic matter producers.

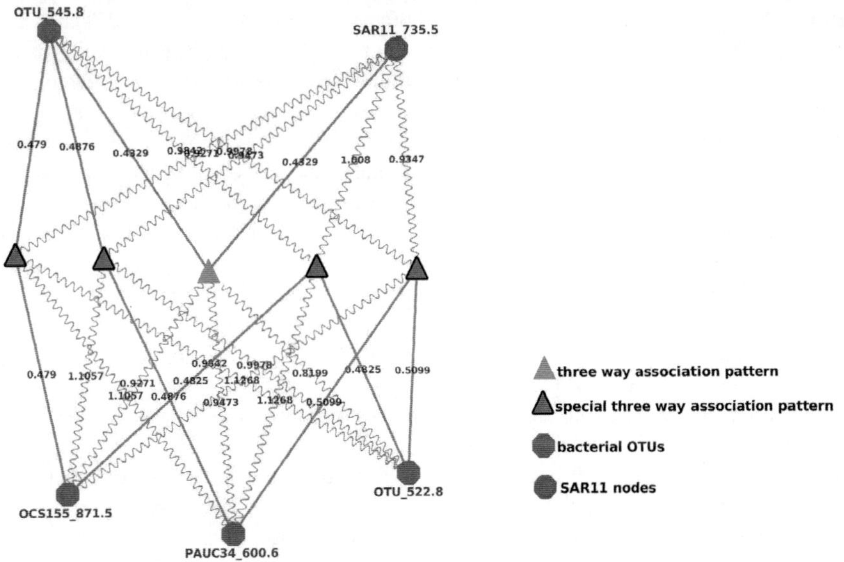

Fig. 6 A sub-network of four bacterial OTUs that were liquid-associated with *SAR11* and *SAR11_735.5*

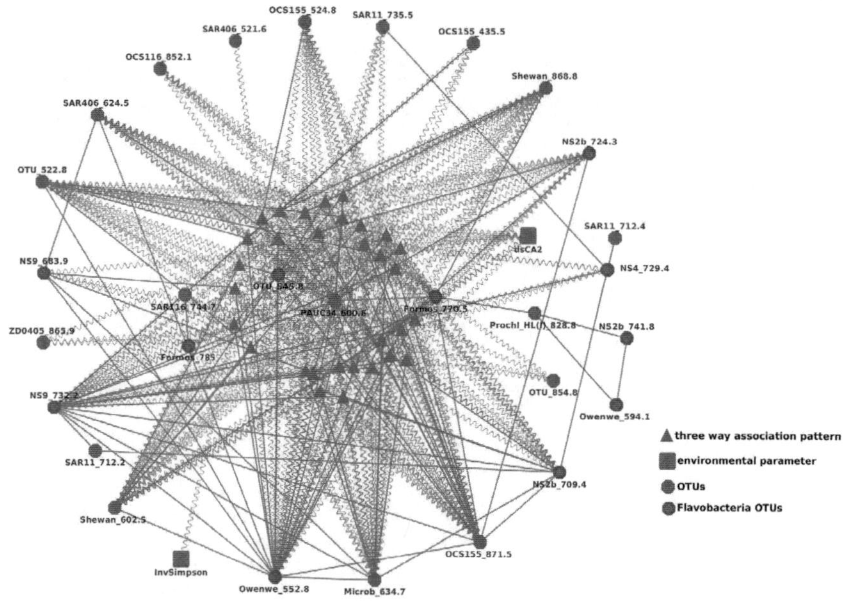

Fig. 7 A sub-network showing OTUs mediated and/or correlated to/by *Flavobacteria*. *Flavobacteria* nodes were highlighted by bold outlines

Some of them are however mediated by environmental factors, such as temperature, oxygen, among others. For example, water column mixing can disturb microbial communities by disrupting the physical–chemical gradients created by thermal stratification known to define niches for microorganisms [41–45]. Liquid association is thus promising in revealing and elucidating these complex non-linear co-mediated associations.

Environment

Analysis of the SPOT dataset suggested that there were a series of correlations that occurred most strongly only under certain conditions. We observed that several biotic and environmental parameters, especially bacterial and viral abundance, silicate concentrations and salinity appeared to predict a number of correlations between bacteria. Liquid associations involving these parameters suggest that not only do bacteria interact with each other and environmental parameters, but furthermore that the nature of the relationships between many bacteria are likely affected by their environment.

Liquid associations in which bacterial abundance mediated the relationships between pairs of OTUs could indicate symbioses or competitive interactions between species that depend upon either the overall density of the bacterial community. Denser microbial communities, for instance are likely to have higher encounter rates between bacteria pair of bacteria which could potentially increase opportunities for symbioses, antagonistic interactions, viral exchange and other interactions. Alternatively, some unmeasured parameter, such as predation [46]. For instance, the *AEGEAN_676.9* bacteria (which is a SAR11 relative [47], is positively associated with *OTU_522.8* under high bacterial abundance, and with *Prochlorococcus High Light Strain 828.8* under low bacterial abundance. SAR11 is known to require metabolic by products of other organisms, and one could expect that it gathers these metabolites from different organisms under different conditions.

Viruses are believed to have complex effects on bacterial communities, and this study suggests a number of interactions that may be mediated either by the viruses themselves or an unmeasured factor that relates to viral abundance [48, 49]. Silicate concentrations likely affect the protest community with diatoms utilizing silicate and dinoflagellates taking advantage of low silicate conditions; these eukaryotes likely interact with bacteria by producing chemicals, consuming resources, through predation and possibly through symbioses [50]. These different conditions might impact the kinds of relationships bacteria could have. For instance, the several bacteria that work together to break down a compound produced by one source might show up as a three-way positive interaction. Meanwhile two bacteria that compete to colonize a diatom symbiotic host or diatom carcass might show up as a three-way negative interaction with diatoms.

The parameter in our dataset that appeared to predict the associations between the most of OTUs was the number of days that had elapsed since the beginning of our data set. This suggests that a number of correlations that could be found in the first part of our data set were not seen in the later years of the data set and vice versa. This suggests a long-term temporal shift in the kinds of associations in the bacterial community.

OTUs

There were numerous cases of interactions between trios of bacteria and noticed that certain groups appeared to be associated with these three way interactions but not others. We observed notably different patterns between the liquid associations that contained at least one *SAR11* OTU and those liquid associations that contained at least one *Flavobacterial* OTU. *SAR11*s, with the exception of one OTU (*SAR11_735.5*), appear to not be found in three-way liquid associations in this dataset, meaning that while they may correlate with other bacteria, the magnitude of this correlation is rarely associated with the relative abundance of a third OTU. This lack of three way associations suggests that *SAR11* bacteria themselves do not influence associations between other kinds of bacteria. On the other hand, out of ten *Flavobacterial* OTUs that correlated with other bacteria, eight of these appeared to be part of liquid associations. The overall pattern for *Flavobacteria* is that they were part of a cluster of positive liquid and local associations. This suggests that there was a set of conditions in which *Flavobacteria* and related OTUs co-occurred positively. However, under other conditions, these OTUs were un-related to each other. Likely these *Flavobacteria* and their associates were all responding to some unmeasured environmental situation such as a bloom. That is, when the unmeasured condition was happening, one set of associations predominated, and when it was not happening another set predominated. This manifests as three way associations between OTUs. Another possibility is that communities may structure themselves in different ways, with different sorts of exchanges between organisms. Under one structure, we see one set of patterns and under another we see a different set of patterns. This structure could be driven by (unmeasured) environmental variability, could be a process shaped the interplay between organisms themselves. As no environmental parameters that we measure appeared to associate with this cohort, we can only speculate at this time what likely caused this association.

This difference in patterns between microorganisms suggests different trends in the ecological associations between organisms of different genetic makeup. *Flavobacteria* are known for being particle associated and able to break down complex molecules and might respond to situations in which particles and unusual sources of dissolved organic material are prevalent, such as after a plankton bloom [51]. *SAR11*s on the other hand are known as oligotrophic chemoheterotrophs with inflexible metabolisms. Accordingly, their relationships with other organisms may be less governed by environmental or biological factors.

Caveats

It should be noted that the collected data and obtained results are only associations (not causality), and need be further verified their validity by one by one OTU experimentation. Furthermore, the validity of ELA depends on the goodness of community fingerprint data, such as number of samples, accurate abundance of each species et.al. In the future, by applying ELA. In the future, by applying liquid association analysis to an integrated community big dataset, it is possible to find resilient associations are common to all communities as well as associations unique to certain community types. This can help us better understand the ecological dynamics with regard to community differences.

Conclusions

Multi-party correlations have been vastly revealed between biological entities ranging from ecosystems to genes [52]. Our approach of coupling local association with liquid association allows additional insight by first identifying statistical patterns, and then using liquid association to identify whether the strength of those patterns is modulated by additional variables. This approach examines many associations at once, and is thus inherently hypothesis generating. It allows for an initial exploration of patterns in a microbial data-set that go beyond pairwise correlations to three way correlations. In our analysis of the San Pedro Ocean time series, this approach has shown us for the first time that associations between organisms appear to change from the beginning to the end of the data set, that bacterial and viral total abundance appears to modulate the interactions between relative abundances of individual OTUs, and that OTU associations appear to be modulated both by environmental parameters and by other OTUs. These patterns provide a starting point for future observational approaches to explore whether these sorts of three-way patterns are robust across ecosystems, and for experimental and modeling approaches to explore the mechanistic underpinnings of these sorts of associations. We think this approach will be valuable in the analysis of any multivariate time-series data sets. We encourage groups to use the freely available software which has been added as an extension to the ELSA software package [28, 29] whose source code can be found at http://bitbucket.org/charade/elsa.

Abbreviations

ARISA: Automated ribosomal intergenic spacer analysis; ELSA: Extended local similarity analysis; FDR: False discovery rate; LA: Liquid association; LSA: Local similarity analysis; OTU: Operational taxonomic unit; SPOT: San Pedro Ocean Time Series

Acknowledgements
This study would not have been possible without the support of Drs. Fengzhu Sun and Jed Fuhrman, and many of their lab members at the University of Southern California. L.C.X. wants to thank Dr. Nancy Zhang at University of Pennsylvania and Dr. Hanlee Ji at Stanford University for their support and helpful discussions.

Funding
This work was supported by grants from the National Natural Science Foundation of China (61873027, 61370131). Publication charge of this article was sponsored by NSFC (61370131). L.C.X. was supported by the United States National Institute of Health (HG006137-07), the American Cancer Society (132922-PF-18-184-01-TBG), and the fund for Innovation in Cancer Informatics (ICI).

Availability of data and materials
All data generated or analyzed during this study are included in this published article.

About this supplement
This article has been published as part of *BMC Genomics Volume 20 Supplement 2, 2019: Selected articles from the 17th Asia Pacific Bioinformatics Conference (APBC 2019): genomics.* The full contents of the supplement are available online at https://bmcgenomics.biomedcentral.com/articles/supplements/volume-20-supplement-2.

Authors' contributions
D.A., J.A.C. and L.C.X. conceived and designed the experiments; D.A., X.X.L. and H. P. performed the experiments and analyzed the data; and D.A., J.C., J.A.C. and L.C.X. wrote the paper. All the authors have read and approved the final manuscript.

Ethics approval and consent to participate
Not applicable.

Consent for publication
Not applicable.

Competing interests
The authors declare that they have no competing interests.

Publisher's Note

Author details
[1]School of Mathematics and Physics, University of Science and Technology Beijing, Xueyuan Road, Haidian District, Beijing 100001, China. [2]Department of Medicine, Stanford University School of Medicine, 269 Campus Dr., Stanford, CA 94305, USA. [3]Center for Environmental Science, University of Maryland, Cambridge, MA 21613, USA.

Published: 10 April 2019

References

1. DeLong EF, Karl DM. Genomic perspectives in microbial oceanography. Nature. 2005;437:336.
2. Fuhrman JA, Cram JA, Needham DM. Marine microbial community dynamics and their ecological interpretation. Nat Rev Microbiol. 2015;13:133.
3. Chapin Iii FS, Zavaleta ES, Eviner VT, Naylor RL, Vitousek PM, Reynolds HL, Hooper DU, Lavorel S, Sala OE, Hobbie SE, et al. Consequences of changing biodiversity. Nature. 2000;405:234.
4. Widder S, Allen RJ, Pfeiffer T, Curtis TP, Wiuf C, Sloan WT, Cordero OX, Brown SP, Momeni B, Shou W, et al. Challenges in microbial ecology: building predictive understanding of community function and dynamics. ISME J. 2016;10:2557.
5. Bálint M, Bahram M, Eren AM, Faust K, Fuhrman JA, Lindahl B, O'Hara RB, M Ö, Sogin ML, Unterseher M. Millions of reads, thousands of taxa: microbial community structure and associations analyzed via marker genes. FEMS Microbiol Rev. 2016;40(5):686–700.
6. Ruan Q, Dutta D, Schwalbach MS, Steele JA, Fuhrman JA, Sun F. Local similarity analysis reveals unique associations among marine bacterioplankton species and environmental factors. Bioinformatics. 2006;22(20):2532–8.
7. Steele JA, Countway PD, Xia L, Vigil PD, Beman JM, Kim DY, Chow C-ET, Sachdeva R, Jones AC, Schwalbach MS, et al. Marine bacterial, archaeal and protistan association networks reveal ecological linkages. ISME J. 2011;5:1414.
8. Muyzer G. Marine microbial systems ecology: microbial networks in the sea. In: Stal LJ, Cretoiu MS, editors. The marine microbiome: an untapped source of biodiversity and biotechnological potential. Cham: Springer International Publishing; 2016. p. 335–44.
9. Chow C-ET, Kim DY, Sachdeva R, Caron DA, Fuhrman JA. Top-down controls on bacterial community structure: microbial network analysis of bacteria, T4-like viruses and protists. ISME J. 2013;8:816.
10. CJ A, PA E, FJ A. Dilution reveals how viral lysis and grazing shape microbial communities. Limnol Oceanogr. 2016;61(3):889–905.
11. Cram JA, Xia LC, Needham DM, Sachdeva R, Sun F, Fuhrman JA. Cross-depth analysis of marine bacterial networks suggests downward propagation of temporal changes. ISME J. 2015;9:2573.
12. Cram JA, Chow C-ET, Sachdeva R, Needham DM, Parada AE, Steele JA, Fuhrman JA. Seasonal and interannual variability of the marine bacterioplankton community throughout the water column over ten years. ISME J. 2014;9:563.
13. Weiss S, Van Treuren W, Lozupone C, Faust K, Friedman J, Deng Y, Xia LC, Xu ZZ, Ursell L, Alm EJ. Correlation detection strategies in microbial data sets vary widely in sensitivity and precision. ISME J. 2016;10(7):1669.
14. Deng Y, Jiang Y-H, Yang Y, He Z, Luo F, Zhou J. Molecular ecological network analyses. BMC Bioinformatics. 2012;13(1):113.
15. Friedman J, Alm EJ. Inferring correlation networks from genomic survey data. PLoS Comput Biol. 2012;8(9):e1002687.
16. Fang H, Huang C, Zhao H, Deng M. CCLasso: correlation inference for compositional data through lasso. Bioinformatics. 2015;31(19):3172–80.
17. Ho YY, Parmigiani G, Louis TA, Cope LM. Modeling liquid association. Biometrics. 2011;67(1):133–41.
18. Narisawa N, Haruta S, Arai H, Ishii M, Igarashi Y. Coexistence of antibiotic-producing and antibiotic-sensitive bacteria in biofilms is mediated by resistant bacteria. Appl Environ Microbiol. 2008;74(12):3887–94.
19. Schulz-Bohm K, Geisen S, Wubs EJ, Song C, de Boer W, Garbeva P. The prey's scent–volatile organic compound mediated interactions between soil bacteria and their protist predators. ISME J. 2017;11(3):817.
20. Eiler A. Evidence for the ubiquity of mixotrophic bacteria in the upper ocean: implications and consequences. Appl Environ Microbiol. 2006;72(12):7431–7.
21. Moore LR. More mixotrophy in the marine microbial mix. Proc Natl Acad Sci U S A. 2013;110(21):8323.
22. Stoecker DK, Hansen PJ, Caron DA, Mitra A. Mixotrophy in the marine plankton. Annu Rev Mar Sci. 2016;9(1):311.
23. Foster RA, Kuypers MMM, Vagner T, Paerl RW, Musat N, Zehr JP. Nitrogen fixation and transfer in open ocean diatom–cyanobacterial symbioses. ISME J. 2011;5(9):1484–93.
24. Zehr JP. Nitrogen fixation by marine cyanobacteria. Trends Microbiol. 2011;19(4):162–73.
25. Remigi P, Zhu J, Young JP, Masson-Boivin C. Symbiosis within Symbiosis: evolving nitrogen-fixing legume symbionts. Trends Microbiol. 2016;24(1):63–75.
26. Li K-C. Genome-wide coexpression dynamics: theory and application. Proc Natl Acad Sci. 2002;99(26):16875.
27. Wang L, Liu S, Ding Y, S-s Y, Ho Y-Y, Tseng GC. Meta-analytic framework for liquid association. Bioinformatics. 2017;33(14):2140–7.
28. Xia LC, Ai D, Cram J, Fuhrman JA, Sun F. Efficient statistical significance approximation for local similarity analysis of high-throughput time series data. Bioinformatics. 2013;29(2):230–7.
29. Xia LC, Steele JA, Cram JA, Cardon ZG, Simmons SL, Vallino JJ, Fuhrman JA, Sun F. Extended local similarity analysis (eLSA) of microbial community and other time series data with replicates. BMC Syst Biol. 2011;5(Suppl 2):S15.
30. Xia LC, Ai D, Cram JA, Liang X, Fuhrman JA, Sun F. Statistical significance approximation in local trend analysis of high-throughput time-series data using the theory of Markov chains. BMC Bioinformatics. 2015;16:301.
31. Brown MV, Schwalbach MS, Hewson I, Fuhrman JA. Coupling 16S-ITS rDNA clone libraries and automated ribosomal intergenic spacer analysis to show

marine microbial diversity: development and application to a time series. Environ Microbiol. 2005;7(9):1466–79.
32. Shannon P, Markiel A, Ozier O, Baliga NS, Wang JT, Ramage D, Amin N, Schwikowski B, Ideker T. Cytoscape: a software environment for integrated models of biomolecular interaction networks. Genome Res. 2003;13(11): 2498–504.
33. Duren Z, Wang Y. A systematic method to identify modulation of transcriptional regulation via chromatin activity reveals regulatory network during mESC differentiation. Scientific reports 6, 22656.
34. Gunderson T, Ho YY. An efficient algorithm to explore liquid association on a genome-wide scale. BMC Bioinformatics. 2014;15:371.
35. Chow C-ET, Sachdeva R, Cram JA, Steele JA, Needham DM, Patel A, Parada AE, Fuhrman JA. Temporal variability and coherence of euphotic zone bacterial communities over a decade in the Southern California bight. ISME J. 2013;7(12):2259–73.
36. Storey JD, Tibshirani R. Statistical significance for genomewide studies. Proc Natl Acad Sci U S A. 2003;100(16):9440–5.
37. Carini P, Campbell EO, Morré J, Sanudo-Wilhelmy SA, Thrash JC, Bennett SE, Temperton B, Begley T, Giovannoni SJ. Discovery of a SAR11 growth requirement for thiamin's pyrimidine precursor and its distribution in the Sargasso Sea. ISME J. 2014;8(8):1727.
38. Tripp HJ. The unique metabolism of SAR11 aquatic bacteria. J Microbiol. 2013;51(2):147–53.
39. James TM, Sundareshwar PV, Christopher TN, Bjorn K, Cahoon DR. Responses of coastal wetlands to rising sea level. Ecology. 2002;83(10):2869–77.
40. Faust K, Lahti L, Gonze D, de Vos WM, Raes J. Metagenomics meets time series analysis: unraveling microbial community dynamics. Curr Opin Microbiol. 2015;25:56–66.
41. Heaney SI, Talling JF. Dynamic aspects of dinoflagellate distribution patterns in a small productive lake. J Ecol. 1980;68(1):75–94.
42. Vincent WF, Neale PJ, Richerson PJ. Photoinhibition: algal responses to bright light during diel stratification and mixing in a tropical alpine lake. J Phycol. 1984;20(2):201–11.
43. Ovreås L, Forney L, Daae FL, Torsvik V. Distribution of bacterioplankton in meromictic Lake Saelenvannet, as determined by denaturing gradient gel electrophoresis of PCR-amplified gene fragments coding for 16S rRNA. Appl Environ Microbiol. 1997;63(9):3367–73.
44. Cytryn E, Minz D, Oremland RS, Cohen Y. Distribution and diversity of archaea corresponding to the limnological cycle of a hypersaline stratified lake (solar lake, Sinai, Egypt). Appl Environ Microbiol. 2000;66(8):3269–76.
45. Fenchel T, Finlay B. Oxygen and the spatial structure of microbial Communities. Biol Rev Camb Philos Soc. 2008;83:553–69.
46. Pernthaler J, Posch T, Simek K, Vrba J, Amann R, Psenner R. Contrasting bacterial strategies to coexist with a flagellate predator in an experimental microbial assemblage. Appl Environ Microbiol. 1997;63(2):596–601.
47. Alonso-Sáez L, Balagué V, Sà EL, Sánchez O, González JM, Pinhassi J, et al. Seasonality in bacterial diversity in north-west Mediterranean coastal waters: assessment through clone libraries, fingerprinting and FISH. FEMS Microbiol Ecol. 2007;60:98–112.
48. Fuhrman JA, Schwalbach M. Viral influence on aquatic bacterial communities. Biol Bull. 2003;204(2):192–5.
49. Michael SS, Ian H, Jed AF. Viral effects on bacterial community composition in marine plankton microcosms. Aquat Microb Ecol. 2004;34(2):117–27.
50. Zhang S, Liu H, Ke Y, Li B. Effect of the silica content of diatoms on protozoan grazing. Front Mar Sci. 2017;4:202.
51. Carlson CA, Giorgio PAD, Herndl GJ. Microbes and the dissipation of energy and respiration: from cells to ecosystems. Oceanography. 2007;20(2):89.
52. Coutinho FH, Gregoracci GB, Walter JM, Thompson CC, Thompson FL. Metagenomics sheds light on the ecology of marine microbes and their viruses. Trends Microbiol. 2018;26(11):955–65.

Zhang *et al. BMC Genomics* 2019, **20**(Suppl 2):186
https://doi.org/10.1186/s12864-019-5470-2

BMC Genomics

RESEARCH

Open Access

Large-scale 3D chromatin reconstruction from chromosomal contacts

Yanlin Zhang[1], Weiwei Liu[1], Yu Lin[2], Yen Kaow Ng[3] and Shuaicheng Li[1*]

From The 17th Asia Pacific Bioinformatics Conference (APBC 2019)
Wuhan, China. 14–16 January 2019

Abstract

Background: Recent advances in genome analysis have established that chromatin has preferred 3D conformations, which bring distant loci into contact. Identifying these contacts is important for us to understand possible interactions between these loci. This has motivated the creation of the Hi-C technology, which detects long-range chromosomal interactions. Distance geometry-based algorithms, such as ChromSDE and ShRec3D, have been able to utilize Hi-C data to infer 3D chromosomal structures. However, these algorithms, being matrix-based, are space- and time-consuming on very large datasets. A human genome of 100 kilobase resolution would involve ~30,000 loci, requiring gigabytes just in storing the matrices.

Results: We propose a succinct representation of the distance matrices which tremendously reduces the space requirement. We give a complete solution, called SuperRec, for the inference of chromosomal structures from Hi-C data, through iterative solving the large-scale weighted multidimensional scaling problem.

Conclusions: SuperRec runs faster than earlier systems without compromising on result accuracy. The SuperRec package can be obtained from http://www.cs.cityu.edu.hk/~shuaicli/SuperRec.

Keywords: Hi-C, 3D chromosome structure, Multidimensional scaling, Chromosome conformation capture, 3D genome

Backgound

Genome-wide sequencing studies, such as the Human Genome Project (HGP) [1, 2], have deciphered the genomic sequences of humans. We are now in a position to reconstruct the 3D structure of the genome, that is, the conformations of the chromosomes within the nucleus. This will further our understanding of chromosomal interactions.

Recent discoveries through imaging analysis revealed that, while chromosome conformations may vary from cell to cell, they are not random [3, 4]. Hotspots of interactions and transcriptions are unevenly distributed; transcriptionally inactive segments prefer locations such as on nuclear periphery, around nucleoli, or at nuclear substructures [5–16]. These observations all point to the fact that gene expressions are highly associated with the chromatin structure.

However, since imaging techniques are not yet able to achieve high enough resolutions for genome-wide studies, researchers have sought to reconstruct the chromatin structure from knowledge of the interactions between genomic loci [17–23]. Valuable insights on gene regulations, genome translocations, copy number variations, genome stability, *etc.*, have been derived, owing to the success of these methods [24–31]. Among the technologies for capturing interaction information, one called Hi-C has been used prominently. The method produces a matrix, called a *contact map*, which stores the normalized frequencies between all pairs of genome loci (also referred to in the literature as *bins*, *regions* or *windows*) at some resolution.

*Correspondence: shuaicli@cityu.edu.hk
[1] Department of Computer Science, City University of Hong Kong, 83 Tat Chee Avenue, Kowloon, Hong Kong SAR
Full list of author information is available at the end of the article

With these advances, it is reasonable to anticipate that the community will amass a very large collection of chromosome interaction data in the near future. These data are expected to be collected under many different conditions as well as resolutions, and for a variety of genomes, large and small. Their processing and analysis will present tremendous challenges to bioinformaticians [32].

A number of methods have been proposed for chromosome structure inferences from the contact maps. They can either infer a mean structure from the contact map [18, 19, 33–37] or solve multiple structures [28, 38–42]. Most of these methods operate on distance matrices, which consume very large amounts of memory for genomes with high resolution.

For our method, we assume the availability of high resolution data, which gives us more information at the expense of a larger problem size. We anticipate that data will be at the level of kilobase pairs (kbp). Existing strategies have difficulties with such data. In particular, distance matrix-based methods require very large amounts of memory to work; a resolution of one kbp for the human genome would require terabytes of memory. Processing time presents another issue.

The fastest method currently has a time complexity of $O\left(n^3\right)$ [36], rendering them inefficient for data of large sizes. In spite of the challenge, genome-scale 3D chromatin reconstruction has nonetheless been performed in at least two studies [43, 44]. In the study by Diament el al., a sparse contact matrix was generated with only a sampled portion of the Hi-C matrix, and the reconstruction was performed with a reduced set of constraints. This method becomes inefficient when the number of loci is increased [43]. In the study by Segal et al., [44] existing single chromosome structures were incorporated to form a whole genome structure. Since the method is dependent on existing tools such as ChromSDE for the inference of single chromosome structures, they are constrained by the efficiency of those tools.

In this work, we propose a *progressive* multidimensional scaling (MDS) approach for structure reconstruction from Hi-C data. We introduce a succinct representation of the distance matrix to reduce space consumption. The proposed approach progressively infer the coordinates to allow more flexible control of runtime. On the benchmark dataset which consists of simulated data of 100 to 30,000 loci, our approach (implemented as a program called SuperRec) performed 5 to 435 times faster compared to ShRec3D. In particular, it demonstrated a speedup of more than 400 times in reconstructing a structure of 30,000 loci, a length sufficient for us to analyze the longest human chromosome at a resolution of 10 kbp. When accessed with normalized root-mean-square deviation (RMSD) (Additional file 1: S1), Spearman's rank correlation coefficient (SRCC) and Pearson correlation coefficient (PCC), we found no loss of accuracy in the results obtained by SuperRec.

Methods

This section presents our method in detail. First, we model the sequence as a continuous linear polymer. We show how the contacts from Hi-C can be transformed and represented succinctly in the form of a distance matrix. The reconstruction then works by assigning coordinates to chromosome loci progressively. After all the coordinates are assigned, we refine and sharpen the coordinates iteratively through local search.

Structure modeling

In our algorithm, the chromosome is modelled as a continuous linear polymer, composed of many chromosomal loci. For example, human chromosome 1 can be modelled as ~25,000 bins at a resolution (number of base pairs per bin) of 10 kbp (kilo base pairs). We use each bin to represent a locus within the chromosome. During the structure reconstruction step, the determination of each locus's location is simplified to that of determining the 3D position of the locus's centroid. Although this model omits the local structures at each locus, it is the most widely accepted representation and is considered the most accurate model achievable by the resolution of current 3C-based techniques [35, 36, 41]. The 3D positions of all the loci are referred to as a *structure* or a *configuration*. In this work, we let n denote the number of loci.

Converting contact frequency to distance

To infer the 3D structures from Hi-C experimental data, the pairwise distances between loci are first computed. Most of the Hi-C protocols are cell population-based experiments; they provide the average contact frequencies across different cells. Every pair of loci i and j are associated with a number of m replicates by accumulating different structures, and the normalized contact frequency f_{ij} can be inferred from Hi-C dataset. Many structure inference methods assume a power-law relationship between contact frequency and 3D distance, allowing a contact $\left(f_{ij}\right)$ to be converted into a corresponding distance $\left(d_{ij}\right)$ through the equation $d_{ij} = 1/f_{ij}^{\alpha}$. The power-law coefficient α varies across datasets and needs to be estimated using other techniques. The special case where $\alpha = 1$ is called an inverse frequency (IF). Given a fixed α and the inferred structure X, the goodness of fit is calculated as $\sum_{f_{ij}>0}\left(f_{ij}^{'} - f_{ij}\right)^2$ where $f_{ij}^{'} = 1/d(X)_{ij}^{1/\alpha}$. Similar to ChromSDE, we assume that $0.1 \leqslant \alpha \leqslant 3$ (a range covering most α in previous studies) and use a golden section search to find the correct α through minimizing the goodness function. We face several challenges with this approach. First, Hi-C can only capture 2.5% of the contact loci, with large variations in the captured frequencies; these contacts are moreover only reliable for the physically close loci. The power-law relation $d_{ij} = 1/f_{ij}^{\alpha}$ also results in infinite distances for low frequency contact pairs. While this can be remedied by enforcing an upper bound on the distances, a criteria for deciding the upper bound would be difficult to derive. Furthermore, the distances converted with a power-law relationship are not metric, rendering many computations that work for metric relations unusable.

Recently, Lesne el al. solved these problems elegantly using the shortest-path method in graph theory [36]. They modeled the contact matrix as a connected graph, in which a vertex represents a locus and an edge is associated with a distance as the inversion contact frequency of the corresponding locus pair. The final distance between loci i and j is modified with the shortest-path distance within the graph. Not only is the shortest-path distance metric and represents a tighter estimation of locus distance, the approach also mitigates the problem due to low frequency contact pairs.

Computing shortest-path distance is, however, both time- and space-consuming since it requires the computation of all-pairs shortest-paths. Hence we propose a method to approximate this distance. We randomly choose ℓ ($\ell \ll n$) loci as *pivots*, and denote this set of loci as P. After that, we compute the single source shortest-path distances with each pivot locus as the source to all the n loci. Denote the shortest distance from pivot p to a vertex v as $d_p(v)$. With these shortest-path distances from the pivots we approximate the remaining shortest-path distances. Given a pair of loci i, j, if i or j is a pivot, we can obtain their shortest distance from the computed shortest-path distances. Otherwise, we use $\min_{p \in P} d_p(i) + d_p(j)$ to approximate the shortest distance between i and j. By increasing the number of pivots, the approximate shortest-path distance can be made arbitrarily close to the true shortest-path distance.

To reduce the space consumption, while computing the shortest-path distances, we adopt an adjacent list representation for the distance matrix derived from Hi-C dataset. Also, the approximated distances are not stored; we merely store the ℓ sets of shortest-path distances from the pivots — the approximated distances are computed on the fly. This data structure reduces the space complexity from $O(n^2)$ to O(ln+e) (e is the number of significant Hi-C contacts) to store a distance matrix with $n \times n$ dimensions. See Additional file 1: Figure S8 for a comparision of run-time memory usage.

Assigning coordinates progressively

The structure reconstruction problem is often formulated as: Given the pairwise distances (with errors) of all loci, to find a 3D configuration X for those loci which satisfy the distance constraints. Denote the distance between loci i and j inferred from contact information as $\hat{d}_{ij}$, and denote the Euclidean distance between loci i and j in a configuration X as $d_{ij}(X)$. The problem can be solved by minimizing the following objective function:

$$\min \sum \sum_{i \le j \le n} \left(d_{ij}(X) - \hat{d}_{ij} \right)^2 \tag{1}$$

which can be solved by multidimensional scaling (MDS) [45]. Such an approach has been utilized by several groups to reconstruct the chromatin structures [18, 24, 36, 42]. It performs well on Euclidean distance with small error. However, the distances inferred from contact information suffer from large errors, which are especially significant in the larger distances due to the underlying mechanism of Hi-C. This prompts us to adapt the formulation to a weighted one:

$$\min \sum \sum_{i \le j \le n} w_{ij} \left(d_{ij}(X) - \hat{d}_{ij} \right)^2 \tag{2}$$

where a weight w_{ij} can be assigned to each loci pair i and j according to their distance, d_{ij}, allowing us to give higher weights to closer pairs. Similar to the earlier problem, this problem has a solution through the use of WMDS (weighted multidimensional scaling) [46].

However, the use of MDS and WMDS remains time- and space-consuming, and will not scale on problems of larger sizes. To solve this, we propose a progressive solution for the MDS and WMDS problem, namely, iMDS (iterative MDS) and sMDS (scalable MDS). Both methods relay on conducting MDS on subsets of loci.

Our proposed approaches are based on the following insight: A distance matrix of size $n \times n$ contains $\binom{n}{2}$ variables. On the other hand, the inferred structures contain $3n - 6$ free variables. That is, the distance matrix contains information that may be considered redundant, which can be potentially discarded without affecting the quality of the inferred structure. Hence, our approach will reduce these redundant values, which we expect to lower the chances of errors. Besides this, our approach will also naturally allow the assignment of larger weights to distances that are more reliable. These will become apparent in the subsequent subsections.

Iterative MDS (iMDS) for structure polish

In the same way as performed in SC-MDS [47], we randomly split the set of loci into k overlapping subsets, $s_1, \ldots, s_i, \ldots, s_k$, with small intersections I_i between s_i and s_{i+1}. Then, classical multidimensional scaling [48] is performed on each subset to obtain the local coordinates of each locus. The subsets of loci are then combined by first selecting s_1 as the reference, and then combining each s_{i+1} to s_i iteratively until all the subsets are combined. Our combination method differs from SC-MDS in that we incorporate the reflection of objects, which we now describe.

Local structures recombination To combine s_{i+1} into s_i in iMDS, we use I_i as a set of anchors. Denote the local coordinates for I_i in s_i and s_{i+1} as P_I and P'_I, respectively. We want to superimpose P'_I onto P_I with a rigid transformation (a translation T and a rotation R) such as to minimize the RMSD. This problem is known to be solvable in linear time [49]. In order to solve the reflection problem, P'_I and its mirror were both superimposed to P_I. In the case that the RMSD between the mirror of P'_I and P_I is smaller than that between P'_I and P_I, the coordinates in s_{i+1} are replaced with those in the mirror of s_{i+1}.

$$RMSD = \min \sum_{i=1}^{N} \sqrt{\left\| p_i' - (Rp_i + T) \right\|^2} \quad (3)$$

After obtaining T and R, we apply them to the coordinates of s_{i+1} such that s_{i+1} and s_i would have the same frame of reference. This integrates s_{i+1} into s_i. In addition, we average P_I and the transformed coordinates of P'_I by R and T to update the coordinates of I_i.

Successive subsetting and combination As discussed in [47], grouping only neighboring loci is not as beneficial as grouping both close and distant loci. Hence, we randomly split all loci into k overlapping subsets. In order to combine two 3D sets successfully, we need to select at least 4 points not lie on the same plane (points on the same plane cannot identity a structure in 3D space), additionaly, a small number of loci may lead to a poor estimation of the rotation and translation matrix. Hence, we need to choose enough loci in I_i. In practice, we set the number of loci in I_i as 50, which is large enough for successive combinations in a three dimensional space. The performance of random subsetting and intersection is described in Additional file 1: S2.

Due to Hi-C's mechanism, the distances $\hat{d}$ measured have different degrees of reliability, which may aversely affect MDS and iMDS. This situation worsens when noise is elevated. To address this, we propose a scalable MDS to polish the structure obtained by iMDS.

Incorporating different distances

The structure from iMDS is further improved to better agree with the data from Hi-C experiments. In Eq. 2, a weight w_{ij} was introduced for each distance $\hat{d}_{ij}$ from Hi-C. Since shorter distances are more reliable than the longer ones, we set $w_{ij} = \hat{d}_{ij}^{-2}$ [42] to decrease the influence from the longer distances. We found this weighting scheme to be more robust than other schemes such as $w_{ij} = 1$ and $w_{ij} = 1/\hat{d}_{ij}$ [35, 36] (Additional file 1: S3).

However, computing WMDS for our framework remains infeasible for large data sets due to time and space complexities. We propose a scalable MDS approach here to address this issue. As far as we know, our approach is among the very few that are currently available for large-scale WMDS.

Scalable MDS

Our proposed scalable MDS is an iterative procedure which employs WMDS as a subroutine. At each iteration, we permute the loci randomly, and partition them by the permuted order into sets of size k each. Then, we apply WMDS to each resultant set, S say. Denote the loci coordinates in S before executing WMDS and after as P_S and P'_S respectively. We apply RMSD to discover the optimal superposition that maps each locus in P'_S to itself in P_S. Then, we update the coordinates for the locus to its average values from P_S and P'_S after the superposition. After we iterate through every set S, we restart the process for another round. This is repeated until the coordinates of the loci converge. The performance of random subsetting and iteration is described in Additional file 1: S2 and the sensitivity analysis of parameter settings is described in Additional file 1: S4.

WMDS is an iterative algorithm, the time complexity in each iteration is $O(n^3)$. In comparsion, the time complexity of sMDS for updating all loci once is $O(nk^2)$. Though sMDS need more iterations than WMDS (10 times more in practice), sMDS is faster.

Simulated contact maps

Binary contact map

We used known 3D structures as the basis for our simulation. For a given 3D structure, we constructed a binary matrix to store the pairwise contact information, with the top k nearest pairs of loci set to 1 and others set to 0.

Poisson distribution model

For a given 3D structure with pairwise distance d_{ij}, We generated the $M(i,j)$ entry of a simulated contact map M as a Poisson-distributed random number n_{ij}. The parameter λ_{ij} is defined as $\lambda_{ij} = \beta / d_{ij}^{\alpha}$ based on the power-law conversion of distance and contact frequency. β is a tunable parameter to control the signal coverage of the contact map.

Data preparation

In silico genome structure

The in silico nucleus with 1 and 16 chromosome(s) (Additional file 1: Table S1) used to test our software were generated using a polymer model. The coordinates were taken from the Langevin dynamics simulation after it has reached thermal equilibrium. After that, we constructed Poisson distribution-based contact maps and binary contact maps for the in silico structures. In addition to the contact matrices corresponding to in silico chromosomes, we also included contact maps generated from Poisson distribution model at different signal coverage levels of the regular helix structure used as benchmark dataset by Zou et al. [37] as well as the structure of chromosome 2 reconstructed from a real Hi-C dataset (mESC).

Hi-C experimental dataset

In this study, we used Hi-C data from two cell lines mESC and GM12878. These raw sequence data are transformed into contact map with Juicer [50]. The matrix balancing method described in [51] was used to normalize the contact matrix in order to remove biases in Hi-C dataset.

Results and discussion

We implemented our proposed approach in a package named *SuperRec*. We compared SuperRec with public publicly available softwares on a ubuntu 16.04 server equipped with two Intel(R) Xeon(R) E5-2620 CPUs, and 256 GB memory. All softwares were executed with suggested configurations, default setting were used when there is no recommended configuration.

The approximated shortest-path distances are reliable

We first assessed the quality of our approximation of the shortest-path distance. Figure 1 visualizes the Hi-C dataset SRX764938 (GM12878) [52] of human. Initially, there are many distances of small values derived from power-law conversion that are indistinguishable from each other (Fig. 1a). After using the shortest-path distances to refine the power-law converted distances, the local distances along the genome became significantly closer than the longer-range ones as expected (Fig. 1b).

An effective solution here is to use the shortest-path distance refinement approach to infer the spatial distances. However, the approach is time- and space-consuming, requiring time cubic to the number of loci; the very large number of loci to compute for renders it infeasible.

To overcome this we approximated the shortest-path distances through the use of pivots as described earlier. To assess the effects of the accuracy loss, we attempted to reconstruct the 3D configuration from the approximated distances of an in silico dataset with 10,000 loci. We repeated this for a range of different number of pivots from 1 to 10,000 to examine how having more pivots would affect the result; the case of using 10,000 pivots is the same as when the actual shortest-path distances are used.

We first assessed the accuracy loss due to approximation through two parameters: *Approximation Ratio (AR)* and *Exact Matching Rate (EMR)*. The approximation ratio is defined to be the ratio between the actual all-pair shortest-path length and the approximated all-pair shortest-path length. We obtained favorable AR values of 0.93 for 100 pivots, and 0.98 for 500 pivots. With 1000 pivots, the approximation ratio became more than 0.99 (Fig. 2a). The exact matching rate is defined to be the rate of the approximated shortest paths lengths that are equal to the corresponding actual shortest-path length. We obtained EMR of 0.85, 0.93 and 0.98 with 500, 1000, and 20,00 pivots, respectively (Fig. 2b). Additionaly, both EMR and AR show very small variances in our repeated analysis (Additional file 1: Figure S9).

We next examined if the quality of our reconstruction would be affected when the approximated shortest-path distances are used instead of the actual shortest-path distances. We conducted experiments on the same data with 10,000 loci, and with varying number of pivots from 1 to 10,000. The PCC between pairwise distance calculated from reconstructed and true structures was found to be 0.89 when 50 pivots are used. These values converge to 0.95 when more than 100 pivots are used. Similarly, the

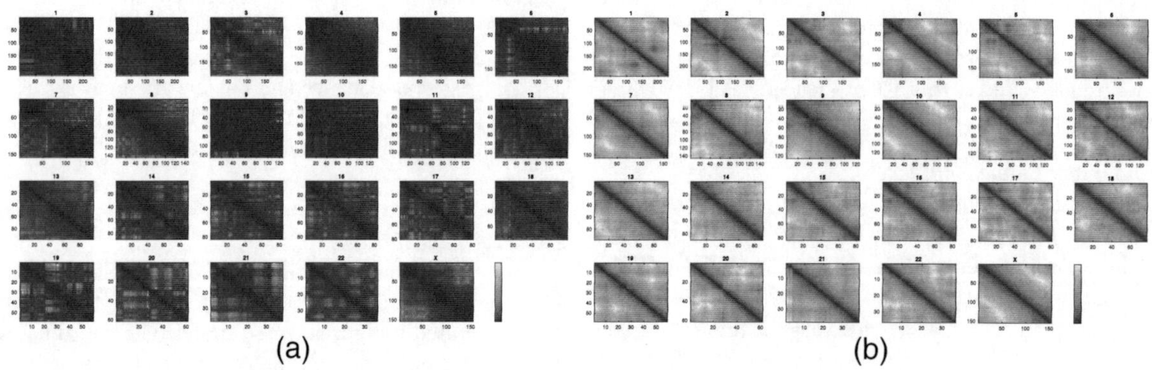

Fig. 1 Pairwise distances within each chromosome: **a** distances converted with power-law conversion, **b** power-law converted distances with the refinement of all-pair shortest-path distance

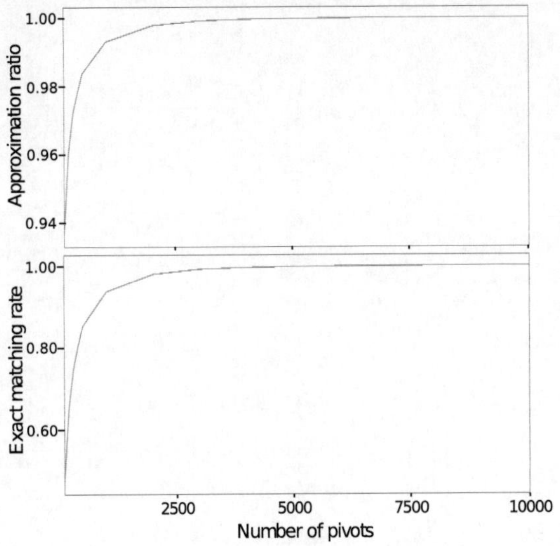

Fig. 2 Accuracy of approximate shortest-path distance: Approimation ratio calculated as the ratio of sum of all-pair shortest-path distance; Number of matched pairs of distance between approximate shortest-path distance and real shortest-path distance

SRCC between pairwise distances calculated from reconstructed and real structures converge to 0.965 when more than 100 pivots are used. The normalized RMSD between reconstructed and real structure varies from 0.17 to 0.27 throughout the experiment (Fig. 3); these values are considered small since they are the aggregate of 10,000 loci.

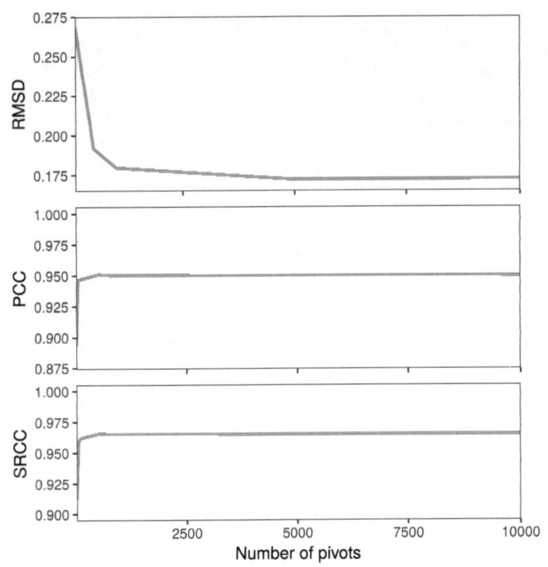

Fig. 3 Accuracy analysis of a reconstructed structure using various numbers of pivots to compute the shortest-path distance. For cases with more than 500 pivots, the difference between structures from accurate shortest-path distances and those from approximated shortest-path distances is negligible. Normalized RMSD, PCC as well as SRCC measurements are consistent

These show that there is no significant degradation of result from the use of the approximated shortest-path distances. In our experiment, SuperRec performs well when using 10% or more loci as pivots (Additional file 1: Figures S6, S7), and we suggest using at least 10% loci as pivots when using SuperRec.

SuperRec is fast

To compare the speed of SuperRec with existing methods, we simulated chromosomal structures with different numbers of loci up to 30,000 (Additional file 1: Table S1). The corresponding binary contact information were inferred with in silico structures as described in Method. These binary contact data were further analyzed with SuperRec using 10% loci as pivots as well as ChromSDE, HSA [37], and ShRec3D. Figure 4 shows the computation time plotted against the number of loci. SuperRec achieves significant improvements when handling thousands of loci. For large dataset with 2000 loci and more, SuperRec performed between 5 to 435 times faster than its alternatives. In one instance with 30,000 loci, SuperRec took only 43 min, whereas ShRec3D required more than 13 days. ChromSDE and HSA would require a few hours to complete on cases with more than 1000 loci Hence, we stopped the comparison with ChromSDE and HSA on the larger cases.

SuperRec is accurate

Whereas we have used only synthetic data in our speed benchmark test, the quality assessment of our algorithm was performed using both synthetic data and actual Hi-C experimental data. We first performed the reconstruction from the synthetic contact matrix used (Additional file 1: Table S1) using HSA, ChromSDE, ShRec3D, and SuperRec. To achieve a fair and comprehensive comparison, three different measurements were calculated: (1) normalized RMSD of each pair of reconstructed structure and original structure, (2) PCC and (3) SRCC between the original and the reconstructed distances (Fig. 5). On Poisson distribution-based contact map datasets, all algorithms achieved comparable and accurate results under all three kinds of measurement, consistently reporting correlation coefficients greater than 0.90. The corresponding normalized RMSDs reported are small compared to the number of loci, except for HSA with 400 and 500 loci. On binary contact maps datasets, SuperRec and ShRec3D achieved comparable and accurate results under all three kinds of measurement, consistently reporting correlation coefficients greater than 0.90. The corresponding normalized RMSDs reported are also small compared to the number of loci. In contrast, the performance of HSA and ChromSDE is dissatisfactory due to their inability to handle binary contact map. This shows that SuperRec is as accurate as these state-of-the-art algorithms.

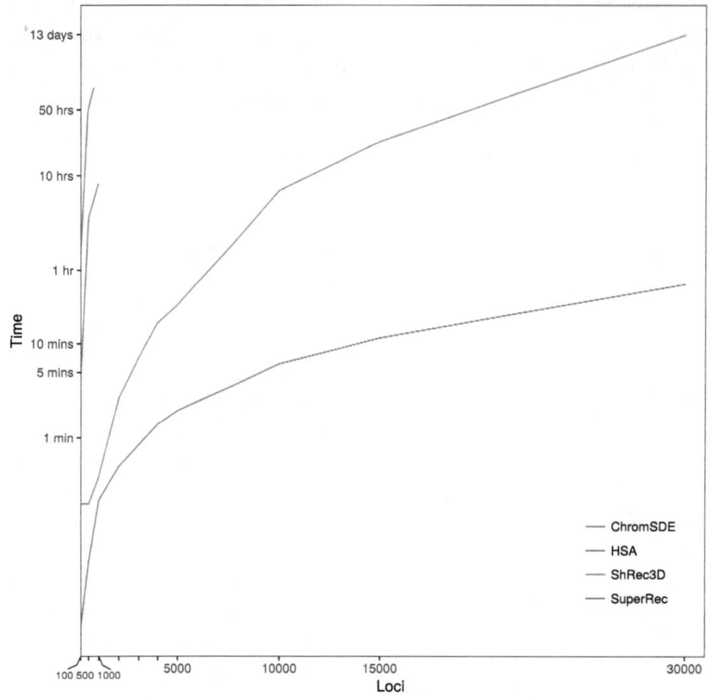

Fig. 4 Time usage for reconstructing chromosomal structures

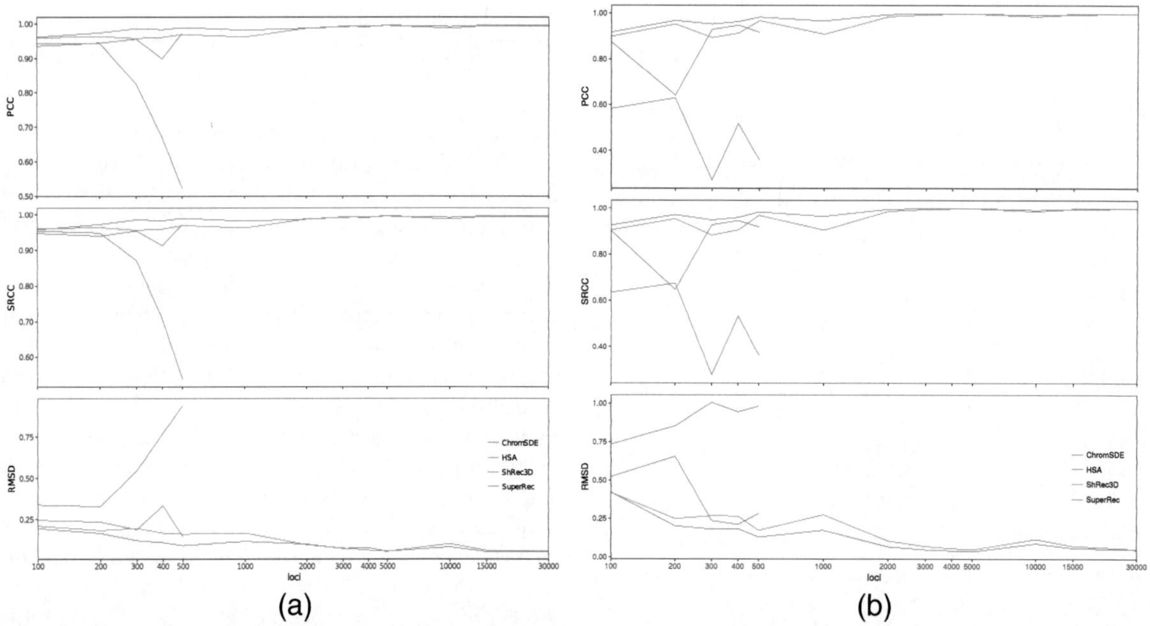

Fig. 5 Accuracy measurement by normalized RMSD, PCC and SRCC vs different numbers of loci. For structures with large number of loci, the SRCC and PCC is almost one, which indicated the reconstructed structures are close to the original structures. We stopped ChromSDE, and HSA computation beyond 500 loci due to the high runtime: **a** Poisson distribution-based contact maps; **b** Binary contact maps

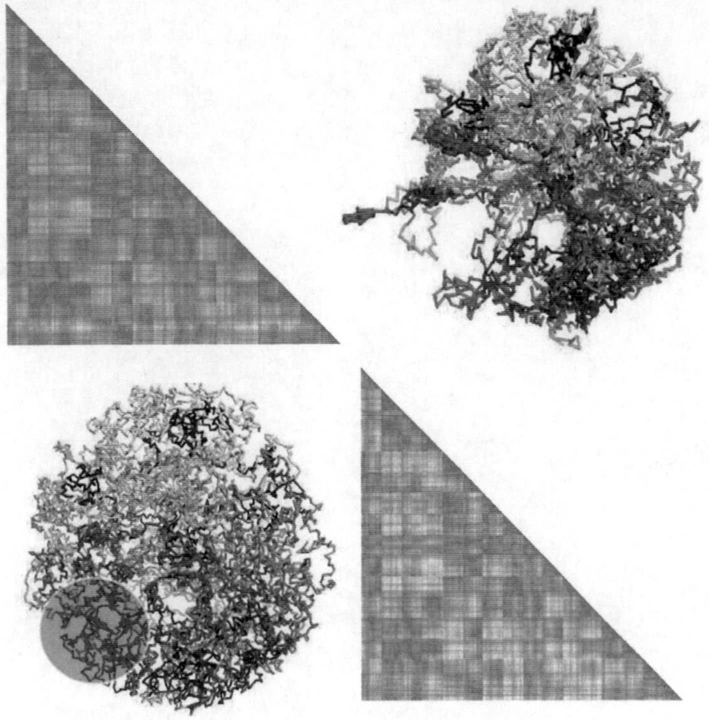

Fig. 6 The upper right and bottom left are two structures with 16 chromosomes and 10,000 loci: the upper right corresponding to a reconstructed structure by SuperRec with 1000 pivots, while the bottom left is the corresponding original structure. The upper left and bottom right heatmaps are plotted using pairwise distances inferred from reconstructed and real structures respectively

Figure 6 shows two structure: an in silico structure with 10,000 loci, and the structure reconstructed by SuperRec. The two structures are highly similar except for the purple highlighted region. The discrepancy is likely due to the relative sparsity of loci in the highlighted region, which resulted in the loci of that region to have fewer contacts with others, thus complicating the inference. On the other hand, we note that the polymer connectivity is preserved in our reconstructed structure (Fig. 7).

In addition to the comparison based on synthetic chromatin structures, we also reconstructed the structure of the regular helix from contact matrices at different signal coverage levels with HSA, ShRec3D, ChromSDE as well as SuperRec. Evaluating by the normalized RMSD between reconstructed structure and real structure, we find our method to be effective and robust (Additional file 1: Table S2).

We also performed the analysis with an *in situ* Hi-C dataset (GM12878) [52] of human at 1MB resolution. The chromosome (Fig. 8) reconstructions were performed with the usage of intra-contact matrices. Since the underlying structure of the Hi-C dataset is unknown, to evaluate the accuracy of the reconstructed structures we compared the pairwise distances from the reconstructions with the all pair shortest-path distances calculated from

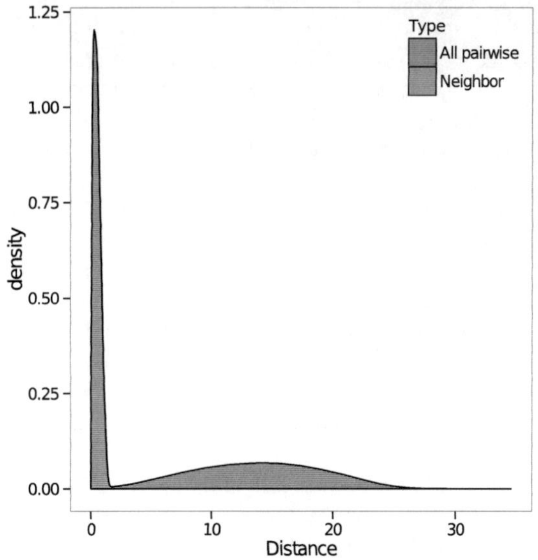

Fig. 7 Polymer connectivity: Histogram of distances between neighboring loci along chromosomal sequence (blue) and all pairwise distances (red) computed from a reconstructed configuration of 10,000 loci

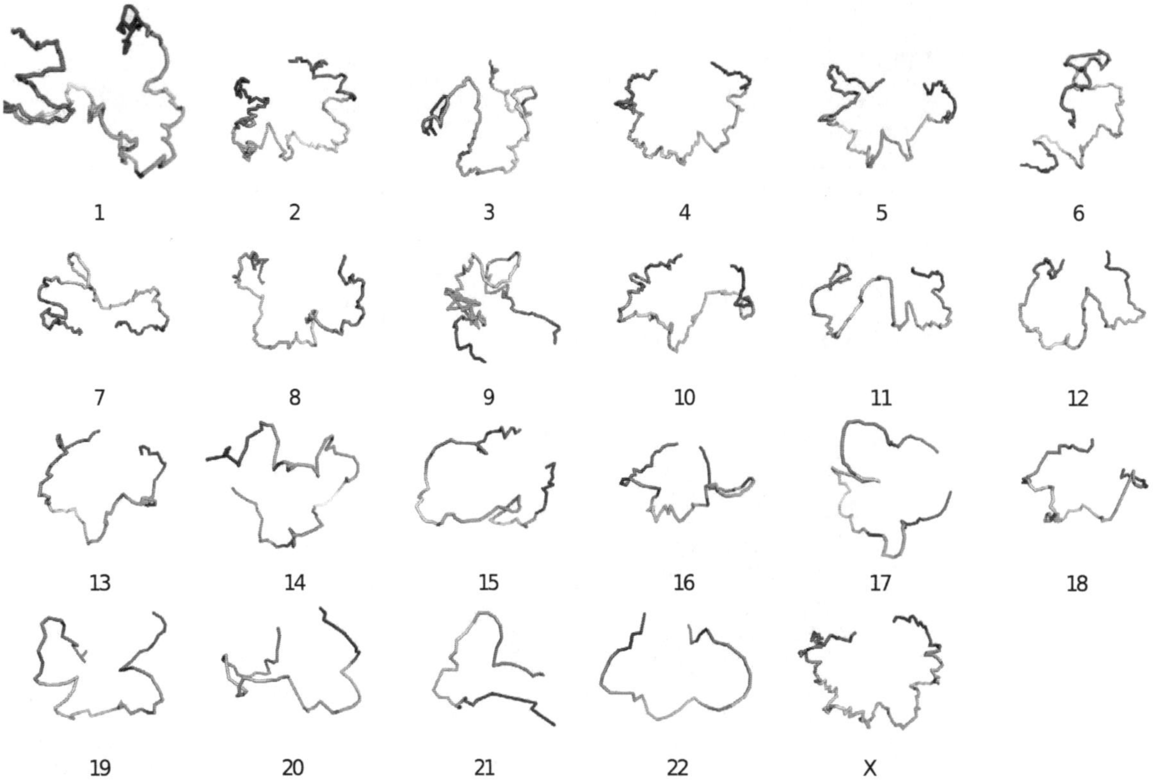

Fig. 8 3D visualization of human autosomes and X chromosome based on Hi-C data (SRX764938) [52] at 1MB resolution, only intra-contact frequency account for those reconstructions

contact frequency. The similarity is then expressed using correlation coefficients (Table 1). ShRec3D and SuperRec achieved similar performances when handling chromosome level reconstruction. At genome-wide level of number of loci, SuperRec slightly outperformed ShRec3D.

Validations and comparisons using FISH data

We also compared the methods' performance in inferring 3D chromatin structures using known distances derived from public 3D-FISH data for the cell lines mESC [53]. We selected six pairs of genomic loci from chromosome 2 or chromosome 11 for validation, with distances derived from FISH probes at 40-kb resolution. We inferred structures of chromosome 2 and chromosome 11 with HSA, BACH, ShRec3D and SuperRec with Hi-C contact data at 40-kb resolution. Single-track HSA, multi-track HSA and BACH were executed with the raw contact maps of NcoI and HindIII. ShRec3D and SuperRec were executed with the normalized contact maps of NcoI and HindIII. In addition, a modifed version of ShRec3D with a fixed α for distance conversion was included in our analysis. PCCs between the predicted distances from reconstructed structures and distance from 3D-FISH were calculated. Since each FISH locus spans two neighboring loci in the structures derived from Hi-C dataset, different combinations of neighbor loci at the two ends of FISH probed pair were used to compute the distances from 3D structures, and a range of PCCs for each FISH data set were obtained. The PCCs of SuperRec and multi-track HSA are most robust and significantly higher than those of other approaches (Fig. 9).

Application to sparse contact map

We carried out reconstructions using SuperRec on simulated contact maps of chromosome 2 (mouse) at different signal coverages (from 10 to 90%), and 10 simulated contact maps were generated at each signal coverage. SuperRec works well for both sparse and dense contact map when accessed with PCC and SRCC between distances from reconstructed structure and true structure. PCC ranges between 0.60 and 0.75, SRCC ranges between 0.65 and 0.80 at 10% signal coverage. The correlations increased with increasingly high signal coverage, with both PCC and SRCC approaching 0.9 when signal coverage reaches above 50 (Fig. 10). This demonstrates SuperRec's ability in handling Hi-C contact maps from low to high coverage.

Table 1 Accuracy measurement of SuperRec, ShRec3D and HSA based on real Hi-C data using Spearman's rank/Pearson's coefficient between distances calculated from reconstructions and all-pair shortest-path distances computed from contact information

Correlation	Method	Chromosome																							
		1	2	3	4	5	6	7	8	9	10	11	12	13	14	15	16	17	18	19	20	21	22	X	All
Spearman	SuperRec	0.981	0.986	0.973	0.942	0.971	0.964	0.989	0.993	0.997	0.981	0.987	0.979	0.991	0.992	0.976	0.976	0.973	0.980	0.979	0.974	0.974	0.970	0.982	0.880
	ShRec3D	0.981	0.983	0.984	0.988	0.985	0.989	0.990	0.987	0.998	0.991	0.977	0.981	0.985	0.991	0.986	0.986	0.974	0.983	0.986	0.982	0.992	0.979	0.990	0.851
	HSA	0.799	0.916	0.916	0.740	0.857	0.884	0.866	0.952	0.904	0.904	0.892	0.801	0.733	0.891	0.880	0.873	0.759	0.808	0.747	0.841	0.934	0.828	0.892	-
Pearson	SuperRec	0.982	0.985	0.974	0.947	0.974	0.960	0.987	0.993	0.998	0.982	0.989	0.983	0.989	0.992	0.980	0.980	0.978	0.985	0.982	0.972	0.970	0.978	0.981	0.896
	ShRec3D	0.981	0.978	0.983	0.976	0.977	0.985	0.986	0.988	0.998	0.983	0.983	0.980	0.984	0.990	0.984	0.984	0.974	0.984	0.985	0.980	0.991	0.977	0.979	0.857
	HSA	0.820	0.915	0.904	0.719	0.812	0.831	0.835	0.935	0.909	0.897	0.879	0.813	0.757	0.893	0.851	0.890	0.790	0.813	0.752	0.786	0.914	0.763	0.836	-

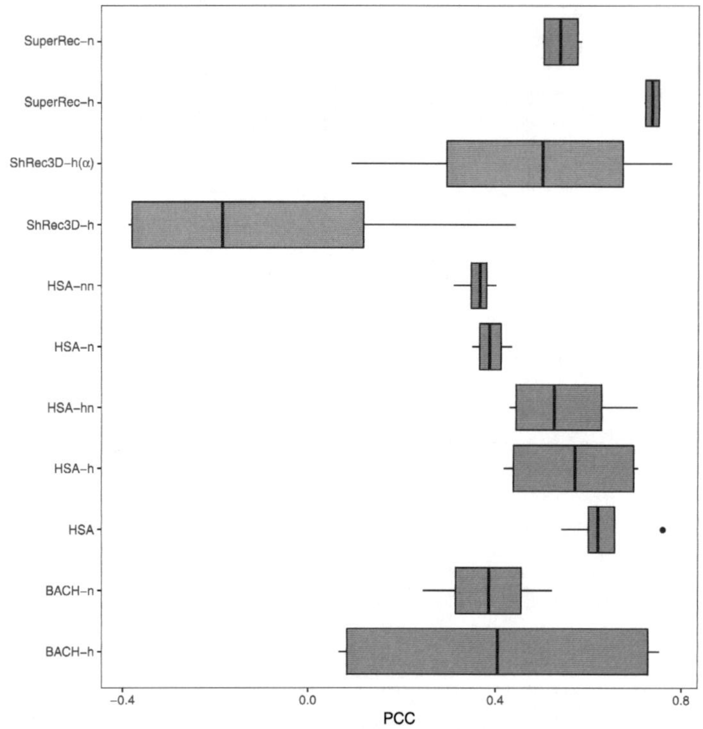

Fig. 9 Box plots of PCCs between FISH measured distance and predicted distances by different Hi-C dataset of mESC at 40-kb resolution. *BACH-h* BACH with contact map of HindIII, *BACH-n* BACH with contact map of NcoI, *HSA* multi-track HSA, *HSA-h* HSA with contact map of HindIII, *HSA-nn* HSA with normalized contact map of NcoI, *HSA-hn* HSA with normalized contact map of HindIII, *HSA-n* HSA with contact map of NcoI, *ShRec3D-h* ShRec3D with normalized contact map of HindIII, *ShRec3D-h(α)* modified ShRec3D with given α on normalized contact map of HindIII, *SuperRec-h* SuperRec with normalized contact map of HindIII, and *SuperRec-n* SuperRec with normalized contact map of NcoI

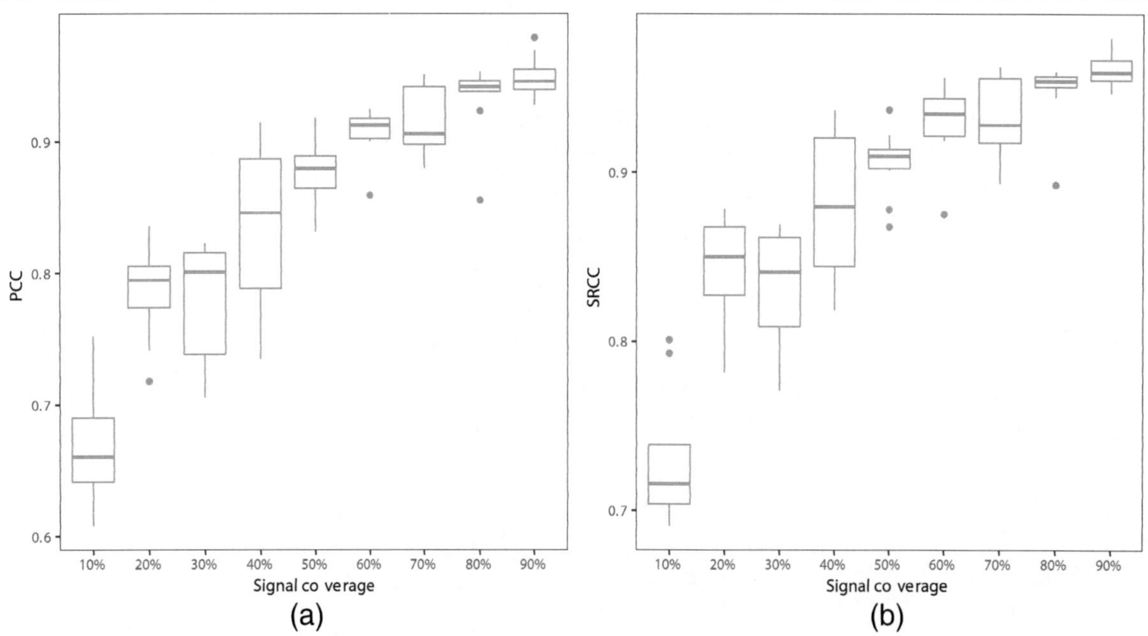

Fig. 10 Box plots of PCC (**a**) and SRCC (**b**) between true distances and predicted distances by simulated contact maps at different signal coverage

Conclusion

In this study, we devised a novel method for 3D chromatin reconstruction from chromosomal contacts, and implemented it into a complete software solution called *SuperRec*. We tested SuperRec on both synthetic and real Hi-C datasets. SuperRec achieved significant improvements in the analysis of longest human chromosome, completing the reconstruction at a resolution of 10 kbp within hours without loss of accuracy in the results.

Additional file

Additional file 1: Large Scale 3D Chromatin Reconstruction From Chromosomal Contacts - Supplementary materials. This file contains the supplementary text and figures mentioned in the text. (PDF 6926 kb)

Abbreviations

3C: Chromosome conformation capture; AR: Approximation ratio; EMR: Exact matching rate; FISH: Fluorescence in situ hybridization; IF: Inverse frequency; iMDS: Iterative multi-dimensional scaling; kbp: Kilobase pairs; MDS: Multi-dimensional scaling; PCC: Pearson correlation coefficient; RMSD: Root mean square deviation; sMDS: Scalable multi-dimensional scaling; SRCC: Spearman's rank correlation coefficient; WMDS: Weighted multi-dimensional scaling

Acknowledgements

Not applicable.

Funding

The work described in this manuscript and the publication costs for this manuscript were supported by a grant of GRF Project from RGC General Research Fund [9042181 (CityU 11203115)].

Availability of data and materials

The GM12878 dataset analysed during the current study are available at GEO repository, https://www.ncbi.nlm.nih.gov/geo/query/acc.cgi?acc=GSE63525. The in silico and mESC datasets used in this study can be downloaded from here: http://www.cs.cityu.edu.hk/~shuaicli/SuperRec/.

About this supplement

This article has been published as part of BMC Genomics Volume 20 Supplement 2, 2019: Selected articles from the 17th Asia Pacific Bioinformatics Conference (APBC 2019): genomics. The full contents of the supplement are available online at https://bmcgenomics.biomedcentral.com/articles/supplements/volume-20-supplement-2.

Authors' contributions

SL conceived the project. SL and YZ designed the algorithm. YZ implemented the algorithm. YZ and WL performed the analyses. YZ, WL and SL evaluated the results, and wrote the manuscript. All authors reviewed the manuscript. All authors read and approved the final manuscript.

Ethics approval and consent to participate

Not applicable.

Consent for publication

Not applicable.

Competing interests

The authors declare that they have no competing interests.

Publisher's Note

Springer Nature remains neutral with regard to jurisdictional claims in published maps and institutional affiliations.

Author details

[1]Department of Computer Science, City University of Hong Kong, 83 Tat Chee Avenue, Kowloon, Hong Kong SAR. [2]Research School of Computer Science, the Australian National University, Canberra, Australia. [3]Department of Computer Science, Faculty of Information and Communication Technology, Universiti Tunku Abdul Rahman, Kampar, Malaysia.

Published: 10 April 2019

References

1. Lander ES, Linton LM, Birren B, Nusbaum C, Zody MC, Baldwin J, Devon K, Dewar K, Doyle M, FitzHugh W, et al. Initial sequencing and analysis of the human genome. Nature. 2001;409(6822):860–921.
2. Venter JC, Adams MD, Myers EW, Li PW, Mural RJ, Sutton GG, Smith HO, Yandell M, Evans CA, Holt RA, et al. The sequence of the human genome. Science. 2001;291(5507):1304–51.
3. Boyle S, Rodesch MJ, Halvensleben HA, Jeddeloh JA, Bickmore WA. Fluorescence in situ hybridization with high-complexity repeat-free oligonucleotide probes generated by massively parallel synthesis. Chromosom Res. 2011;19(7):901–9.
4. Muller I, Boyle S, Singer RH, Bickmore WA, Chubb JR. Stable morphology, but dynamic internal reorganisation, of interphase human chromosomes in living cells. PLoS ONE. 2010;5(7):11560.
5. Bantignies F, Roure V, Comet I, Leblanc B, Schuettengruber B, Bonnet J, Tixier V, Mas A, Cavalli G. Polycomb-dependent regulatory contacts between distant hox loci in drosophila. Cell. 2011;144(2):214–26.
6. Branco MR, Pombo A. Intermingling of chromosome territories in interphase suggests role in translocations and transcription-dependent associations. PLoS Biol. 2006;4(5):138.
7. Brown JM, Green J, das Neves RP, Wallace HA, Smith AJ, Hughes J, Gray N, Taylor S, Wood WG, Higgs DR, et al. Association between active genes occurs at nuclear speckles and is modulated by chromatin environment. J Cell Biol. 2008;182(6):1083–97.
8. Cremer T, Cremer C. Chromosome territories, nuclear architecture and gene regulation in mammalian cells. Nat Rev Genet. 2001;2(4):292–301.
9. Fraser P, Bickmore W. Nuclear organization of the genome and the potential for gene regulation. Nature. 2007;447(7143):413–7.
10. Guelen L, Pagie L, Brasset E, Meuleman W, Faza MB, Talhout W, Eussen BH, de Klein A, Wessels L, de Laat W, et al. Domain organization of human chromosomes revealed by mapping of nuclear lamina interactions. Nature. 2008;453(7197):948–51.
11. Iborra FJ, Pombo A, Jackson DA, Cook PR. Active rna polymerases are localized within discrete transcription "factories' in human nuclei. J Cell Sci. 1996;109(6):1427–36.
12. Németh A, Conesa A, Santoyo-Lopez J, Medina I, Montaner D, Péterfia B, Solovei I, Cremer T, Dopazo J, Langst G. Initial genomics of the human nucleolus. PLoS Genet. 2010;6(3):1000889.
13. Pirrotta V, Li H-B. A view of nuclear polycomb bodies. Curr Opin Genet Dev. 2012;22(2):101–9.
14. Schoenfelder S, Sexton T, Chakalova L, Cope NF, Horton A, Andrews S, Kurukuti S, Mitchell JA, Umlauf D, Dimitrova DS, et al. Preferential associations between co-regulated genes reveal a transcriptional interactome in erythroid cells. Nat Genet. 2010;42(1):53–61.
15. Tolhuis B, Blom M, Kerkhoven RM, Pagie L, Teunissen H, Nieuwland M, Simonis M, de Laat W, van Lohuizen M, van Steensel B. Interactions among polycomb domains are guided by chromosome architecture. PLoS Genet. 2011;7(3):1001343.
16. van Koningsbruggen S, Gierliński M, Schofield P, Martin D, Barton GJ, Ariyurek Y, den Dunnen JT, Lamond AI. High-resolution whole-genome sequencing reveals that specific chromatin domains from most human chromosomes associate with nucleoli. Mol Biol Cell. 2010;21(21):3735–48.
17. Van Berkum NL, Lieberman-Aiden E, Williams L, Imakaev M, Gnirke A, Mirny LA, Dekker J, Lander ES. Hi-c: a method to study the three-dimensional architecture of genomes. J Vis Exp. 2010;39:e1869. https://doi.org/10.3791/1869.
18. Duan Z, Andronescu M, Schutz K, McIlwain S, Kim YJ, Lee C, Shendure J, Fields S, Blau CA, Noble WS. A three-dimensional model of the yeast genome. Nature. 2010;465(7296):363–7.
19. Lieberman-Aiden E, van Berkum NL, Williams L, Imakaev M, Ragoczy T, Telling A, Amit I, Lajoie BR, Sabo PJ, Dorschner MO, et al. Comprehensive mapping of long-range interactions reveals folding principles of the human genome. Science. 2009;326(5950):289–93.

20. Rodley C, Bertels F, Jones B, O'sullivan J. Global identification of yeast chromosome interactions using genome conformation capture. Fungal Genet Biol. 2009;46(11):879–86.
21. Sexton T, Yaffe E, Kenigsberg E, Bantignies F, Leblanc B, Hoichman M, Parrinello H, Tanay A, Cavalli G. Three-dimensional folding and functional organization principles of the drosophila genome. Cell. 2012;148(3): 458–72.
22. Jin F, Li Y, Dixon JR, Selvaraj S, Ye Z, Lee AY, Yen C-A, Schmitt AD, Espinoza CA, Ren B. A high-resolution map of the three-dimensional chromatin interactome in human cells. Nature. 2013;503(7475):290–4.
23. Nagano T, Lubling Y, Stevens TJ, Schoenfelder S, Yaffe E, Dean W, Laue ED, Tanay A, Fraser P. Single-cell hi-c reveals cell-to-cell variability in chromosome structure. Nature. 2013;502(7469):59–64.
24. Ay F, Bunnik EM, Varoquaux N, Bol SM, Prudhomme J, Vert J-P, Noble WS, Le Roch KG. Three-dimensional modeling of the p. falciparum genome during the erythrocytic cycle reveals a strong connection between genome architecture and gene expression. Genome Res. 2014;24(6): 974–88.
25. De S, Michor F. Dna replication timing and long-range dna interactions predict mutational landscapes of cancer genomes. Nat Biotechnol. 2011;29(12):1103–8.
26. Dixon JR, Selvaraj S, Yue F, Kim A, Li Y, Shen Y, Hu M, Liu JS, Ren B. Topological domains in mammalian genomes identified by analysis of chromatin interactions. Nature. 2012;485(7398):376–80.
27. Homouz D, Kudlicki AS. The 3d organization of the yeast genome correlates with co-expression and reflects functional relations between genes. PloS ONE. 2013;8(1):54699.
28. Kalhor R, Tjong H, Jayathilaka N, Alber F, Chen L. Genome architectures revealed by tethered chromosome conformation capture and population-based modeling. Nat Biotechnol. 2012;30(1):90–8.
29. Lemieux JE, Kyes SA, Otto TD, Feller AI, Eastman RT, Pinches RA, Berriman M, Su X-z, Newbold CI. Genome-wide profiling of chromosome interactions in plasmodium falciparum characterizes nuclear architecture and reconfigurations associated with antigenic variation. Mol Microbiol. 2013;90(3):519–37.
30. Ryba T, Hiratani I, Lu J, Itoh M, Kulik M, Zhang J, Schulz TC, Robins AJ, Dalton S, Gilbert DM. Evolutionarily conserved replication timing profiles predict long-range chromatin interactions and distinguish closely related cell types. Genome Res. 2010;20(6):761–70.
31. Shen Y, Yue F, McCleary DF, Ye Z, Edsall L, Kuan S, Wagner U, Dixon J, Lee L, Lobanenkov VV, et al. A map of the cis-regulatory sequences in the mouse genome. Nature. 2012;488(7409):116–20.
32. Gibcus JH, Dekker J. The hierarchy of the 3d genome. Mol Cell. 2013;49(5):773–82.
33. Baù D, Sanyal A, Lajoie BR, Capriotti E, Byron M, Lawrence JB, Dekker J, Marti-Renom MA. The three-dimensional folding of the α-globin gene domain reveals formation of chromatin globules. Nat Struct Mol Biol. 2011;18(1):107–14.
34. Ben-Elazar S, Yakhini Z, Yanai I. Spatial localization of co-regulated genes exceeds genomic gene clustering in the saccharomyces cerevisiae genome. Nucleic Acids Res. 2013;41(4):2191–201.
35. Zhang Z, Li G, Toh K-C, Sung W-K. 3d chromosome modeling with semi-definite programming and hi-c data. J Comput Biol. 2013;20(11): 831–46.
36. Lesne A, Riposo J, Roger P, Cournac A, Mozziconacci J. 3d genome reconstruction from chromosomal contacts. Nat Methods. 2014;11(11): 1141–3.
37. Zou C, Zhang Y, Ouyang Z. Hsa: integrating multi-track hi-c data for genome-scale reconstruction of 3d chromatin structure. Genome Biol. 2016;17(1):40.
38. Hu M, Deng K, Qin Z, Dixon J, Selvaraj S, Fang J, Ren B, Liu JS. Bayesian inference of spatial organizations of chromosomes. PLoS Comput Biol. 2013;9(1):1002893.
39. Rousseau M, Fraser J, Ferraiuolo MA, Dostie J, Blanchette M. Three-dimensional modeling of chromatin structure from interaction frequency data using markov chain monte carlo sampling. BMC Bioinformatics. 2011;12(1):414.
40. Tjong H, Gong K, Chen L, Alber F. Physical tethering and volume exclusion determine higher-order genome organization in budding yeast. Genome Res. 2012;22(7):1295–305.
41. Trieu T, Cheng J. Large-scale reconstruction of 3d structures of human chromosomes from chromosomal contact data. Nucleic Acids Res. 2014;42(7):52.
42. Varoquaux N, Ay F, Noble WS, Vert J-P. A statistical approach for inferring the 3d structure of the genome. Bioinformatics. 2014;30(12):26–33.
43. Diament A, Tuller T. Improving 3d genome reconstructions using orthologous and functional constraints. PLoS Comput Biol. 2015;11(5): 1004298.
44. Segal MR, Bengtsson HL. Reconstruction of 3d genome architecture via a two-stage algorithm. BMC Bioinformatics. 2015;16(1):373.
45. Kruskal JB. Multidimensional scaling by optimizing goodness of fit to a nonmetric hypothesis. Psychometrika. 1964;29(1):1–27.
46. Borg I, Groenen P. Modern multidimensional scaling: theory and applications. J Educ Meas. 2003;40(3):277–80.
47. Tzeng J, Lu HH, Li W-H. Multidimensional scaling for large genomic data sets. BMC Bioinformatics. 2008;9(1):179.
48. Torgerson WS. Multidimensional scaling: I. theory and method. Psychometrika. 1952;17(4):401–19.
49. Arun KS, Huang TS, Blostein SD. Least-squares fitting of two 3-d point sets. Pattern Anal Mach Intell, IEEE Trans. 1987;PAMI-9(5):698–700.
50. Durand NC, Shamim MS, Machol I, Rao SS, Huntley MH, Lander ES, Aiden EL. Juicer provides a one-click system for analyzing loop-resolution hi-c experiments. Cell Syst. 2016;3(1):95–8.
51. Knight PA, Ruiz D. A fast algorithm for matrix balancing. IMA J Numer Anal. 2013;33(3):1029–47.
52. Rao SS, Huntley MH, Durand NC, Stamenova EK, Bochkov ID, Robinson JT, Sanborn AL, Machol I, Omer AD, Lander ES, et al. A 3d map of the human genome at kilobase resolution reveals principles of chromatin looping. Cell. 2014;159(7):1665–80.
53. Eskeland R, Leeb M, Grimes GR, Kress C, Boyle S, Sproul D, Gilbert N, Fan Y, Skoultchi AI, Wutz A, et al. Ring1b compacts chromatin structure and represses gene expression independent of histone ubiquitination. Mol Cell. 2010;38(3):452–64.

Yang et al. BMC Genomics 2019, 20(Suppl 2):187
https://doi.org/10.1186/s12864-019-5471-1

BMC Genomics

RESEARCH Open Access

Meta-network: optimized species-species network analysis for microbial communities

Pengshuo Yang[1†], Shaojun Yu[1†], Lin Cheng[2] and Kang Ning[1*]

From The 17th Asia Pacific Bioinformatics Conference (APBC 2019)
Wuhan, China. 14-16 January 2019

Abstract

Background: The explosive growth of microbiome data provides ample opportunities to gain a better understanding of the microbes and their interactions in microbial communities. Given these massive data, optimized data mining methods become important and necessary to perform deep and comprehensive analysis. Among the various priorities for microbiome data mining, the examination of species-species co-occurrence patterns becomes one of the key themes in urgent need.

Results: Hence, in this work, we propose the Meta-Network framework to lucubrate the microbial communities. Rooted in loose definitions of network (two species co-exist in a certain samples rather than all samples) as well as association rule mining (mining more complex forms of correlations like indirect correlation and mutual information), this framework outperforms other methods in restoring the microbial communities, based on two cohorts of microbial communities: (a) the loose definition strategy is capable to generate more reasonable relationships among species in the species-species co-occurrence network; (b) important species-species co-occurrence patterns could not be identified by other existing approaches, but could successfully generated by association rule mining.

Conclusions: Results have shown that the species-species co-occurrence network we generated are much more informative than those based on traditional methods. Meta-Network has consistently constructed more meaningful networks with biologically important clusters, hubs, and provides a general approach towards deciphering the species-species co-occurrence networks.

Keywords: Microbial network, Data-mining, Network analysis, Associate-rule mining

Background

Network-based approaches are gaining momentum as one of the most helpful tools for the analysis of microbial community structure. They offer new methodological and biological insights to investigate species interactions. Many microorganisms co-exist by interacting with each other and effectively exert various functions [1]. In addition, due to currently insufficient understanding of the community structure, the mounting volume of metagenomics data limits the traditional network analysis to recover the real relationships in bacterial community [2]. Hence finding out intricate yet important associations (for example, to explore the cyclical process of a substance or element in a bacterial community [3]) becomes increasingly challenging for traditional methods.

To address these limitations, we propose Meta-Network to establish the species-species co-occurrence network. The loose definition method is introduced first to recover more correlations before correlation calculation. Then we utilize the FS-Weight and PCA-PMI (Part Mutual Information adjusted by Path Consistency Algorithm) methods to explore the indirect and non-linear correlations, respectively. To investigate the optimized species-species co-occurrence network, systematic evaluation is investigated to discover

* Correspondence: ningkang@hust.edu.cn
†Pengshuo Yang and Shaojun Yu contributed equally to this work.
[1]Key Laboratory of Molecular Biophysics of the Ministry of Education, College of Life Science and Technology, Huazhong University of Science and Technology, Wuhan 430074, Hubei, China
Full list of author information is available at the end of the article

meaningful biological implications in the species-species co-occurrence network.

Method

Construction workflow of species-species co-occurrence network

While direct and linear correlations were calculated by Pearson and Spearman [4], limitations of these algorithms are developed in quantifying the relationship among species pairs: First, the correlations are calculated when pairwise species exist in all samples, which fail to recover the microbial communities owing to the sparse distribution of species. Second, Pearson and Spearman algorithms only capture direct and linear correlations for pairwise species, but cannot detect complex ecology correlations patterns [5]. In response to these observations, we applied several association rule mining algorithms to uncover complex correlations among species (for example, indirect correlation and non-linear correlations) (Fig. 1, a). First, we propose to recover correlations filtered by traditional methods, termed as the loose definition method (Fig. 1, a.2). Second, While direct relationships have been reported in many existing works in the context of intricate bacterial community, more complex correlations can be detected using association rule mining [6]. Hence we develop two methods to detect complex forms of correlations like indirect correlations (FS-Weight) and non-linear correlations (PCA-PMI) (Fig. 1, a.3, and a.4).

Loose definition of species-species relationship (Fig. 1, a.1)

To measure the number of co-exist samples for pairwise species, we introduce the co-occurrence probability (Fig. 1, a.1). To be precise, we first convert the original abundance matrix into a presence-absence matrix, then we calculate the ratio of co-exist samples to all the samples as co-occurrence probability for each pairwise species (Fig. 2, a and b). When co-occurrence probability reaches above a user-defined threshold, the

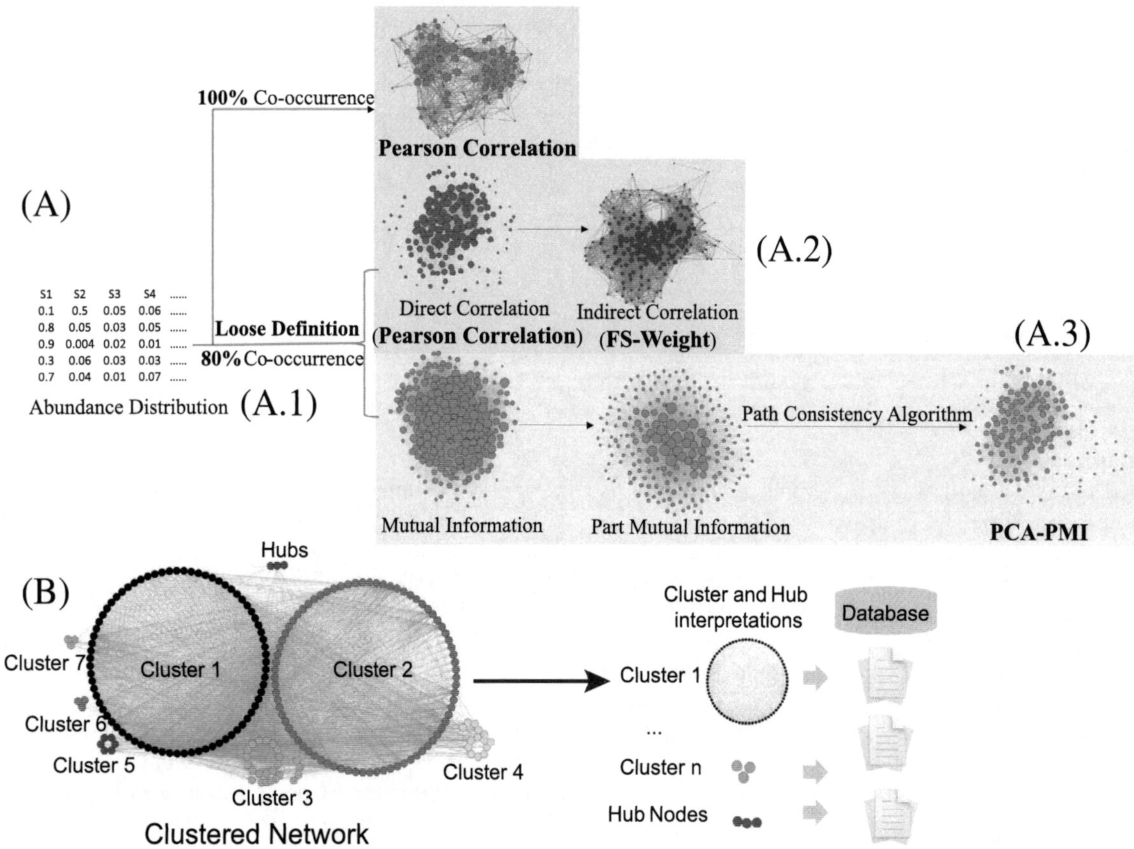

Fig. 1 Network establishment and cluster analysis for species-species co-occurrence network. **a** Network construction. Based on the loose definition method (A.1), networks were constructed by two approaches. First, the direct correlations were detected and a graph-based algorithm (FS-Weight) was applied to detect indirect correlations (A.2). Second, the part mutual information was calculate, followed by a few rounds of adjustment based on Path Consistency Algorithm (PCA) until the network has no correlations added or removed (A.3). **b** Clustering analysis. Density MCODE clustering algorithm is applied to categorize the network into functionally or taxonomically enriched units. Hub nodes represent species connected to many other species and usually play key roles in the community. Databases curation and literature mining are used to annotate the clusters and hub nodes

correlation for this pairwise species will be calculated (80% is used here as default, Fig. 2, c). It is worth mentioning that some genera pairs exist in a few number of samples but yield abnormally high correlation from Pearson and Spearman algorithms. However, no known literature and functionality can support this kind of correlation. Noise like this was filtered out based on the sample quantity set of loose definition.

Indirect relationships in the network (Fig. 1, a.2)

While direct relationships have been reported in many existing works in the context of intricate bacterial community, it is possible to detect more complex correlations using association rule mining [7]. In bacterial community, bacteria which do not interact but share interaction partners may play roles in the same functional pathway [8]. Hence, in our context, the FS-Weight method is applied to detect the indirect correlations (Fig. 1, a.2). FS-Weight measures the overlap between the pairwise species, and is originally designed to estimate association between direct and indirect correlations based on network structure [9].

For pairwise species, FS-Weight value between them was calculated in two steps. First, the direct correlations were calculated using a user-defined threshold for edge value. In this work, Pearson coefficient correlation was selected and the thresholds was set as 0.5 on genus level and 0.7 on OTU level based on previous research [10]. Second, the network constructed by FS-Weight was applied to filter out less reliable correlations and to add meaningful indirect relationships. FS-Weight threshold was set to 0.7 on OTU level and 0.5 on genus level based on the comparison between different thresholds. The benchmark result for FS-Weight on genus level was shown at Additional file 1: Figure S1, A and B.

Nonlinear associations (Fig. 1, a.3)

In microbial communities, nonlinear relationships play important roles [11]. Thus, to explore potentially nonlinear

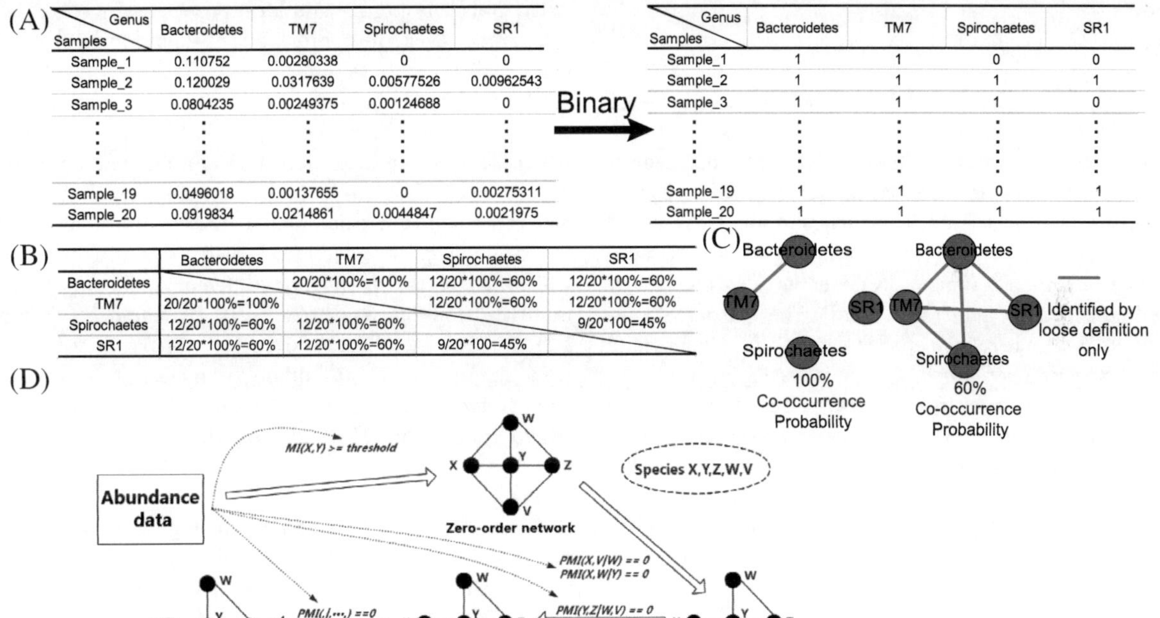

Fig. 2 New methods for detecting potential correlations. **a** The presence–absence matrix is the binary form of original abundance matrix. In this matrix, columns present samples, rows present species, and entries are the presence (1) or absence (0) of a species. Samples are organized as a presence–absence matrix in which i = 1 to R rows and j = 1 to C columns. The entry a_{ij} in the matrix represents the presence (1) or absence (0) of species j in sample i. **b** Co-occurrence probability matrix for the sample data. In loose definition method, co-occurrence means two species in the same sample have both non-zero abundance. The co-occurrence probability of two species is defined as co-occurrence times / total sample count. **c** Network establishment based on different co-occurrence probability thresholds. Traditional network construction workflow only calculates correlations between species between 100% co-occurrence probability (left network), which is too strict for sparse distribution. More correlations are detected between co-occurrence probability set to 60% (right network, correlations detected by loose definition only were colored in red). **d** An overview of the PCA-PMI process. Species abundance distribution was calculated by the mutual information method and then the zero-order network, which contains all the mutual information, was constructed. PCA-PMI process was applied to adjust the zero-order network until no correlations changed

correlations, we adopt PCA-PMI method to explore non--linear correlations in microbial [12] (Fig. 2, d). PCA-PMI method calculate the partial information to measure the linear and non-linear relationship for each pairwise correlations in microbial community. PMI is defined as follows: assuming X and Y represent two one-dimensional variables representing abundance distribution in all the samples for species A and B respectively, and Z means an n-2 dimensional vector (n-2 > 0) representing other species abundance distribution in all the samples, the PMI between species x and y given indirect neighbor z is defined as below:

$$\mathrm{PMI}(x,\ y|z) = \sum\nolimits_{x,y,z} p(x, y, z) \log \frac{p(x, y|z)}{p^*(x|z)p^*(y|z)} \quad (1)$$

Where the Part independence of species x and y given indirect neighbor z is defined as:

$$p^*(y|z) = \sum\nolimits_x p(y|z, x)p(x) \quad (2)$$

$$p^*(x|z) = \sum\nolimits_y p(x|z, y)p(y) \quad (3)$$

In order-one PCA-PMI, only one species was considered as indirect neighbor between two species, and the Path Consistency Algorithm (PCA) [13] was applied to adjust the correlation distribution using a user-defined correlation threshold (0.02 as default). This threshold was set based on the comparison between different thresholds. The benchmark result for FS-Weight on genus level was shown at Additional file 1: Figure S1, C and D. The sparse matrix is defined as a matrix owing a large number of nodes and sparse species distribution, thus it is feasible to use PCA method to construct the species-species co-occurrence network [14]. After all linear and non-linear correlations calculated, the network was updated by detecting higher orders of PMI (more species were set as indirect neighbor for two species). Higher orders of PMI are calculated to check the correlation reliability iteratively until no more edges are changed.

Identification of clusters and hubs in the network

Detecting co-occurrence patterns and their biological significance (Fig. 1, b) supply an important material to understand the microbial network. First, we perform cluster calculation and hub nodes detection based on density clustering. We apply the MCODE cluster algorithm to detect the potential clusters [15]. Aligning cluster members against taxonomical or functional annotation databases can predict potential cluster functions and taxonomical compositions.

To identify hub nodes, we calculate the most connected nodes as candidates for hub nodes. These nodes are selected to perform Kruskal-Wallis test (test differences in network distribution before and after deleting hub node) to decide whether they are hub nodes. Finally, we interpret the clusters and hub nodes based on known taxonomical and functional databases and literatures.

Systematic evaluation methods

In order to compare Pearson, loose definition, FS-Weight, PCA-PMI algorithm, the correlation coefficient calculated by loose definition which is implemented by the loose definition plus Pearson method.

First, we calculate the global network alignment by MAGNA++ [16] based on genetic algorithm. The alignment quality is measured by edge correctness (EC), induced conserved structure (ICS) and symmetric substructure score (S3) for each pair of networks. Then, after all nodes compared based on Jaccard index, we construct a similarity tree measured by the global similarity between any two networks been compared.

Secondly, these four algorithms were compared based on their topological structures including network similarity, global properties and local properties by compNet software [17]. Furthermore, we analyze the local properties for all taxon based on the R package igraph (http://igraph.org/r/), including coreness, degree, eigenvector centrality, eccentricity, betweeness and degree centrality.

Thirdly, we examined the correlation distribution using the same set of species as target. Based on top 100 abundant genera in gut microbiome datasets, four networks are constructed to investigate their edge distributions, which reflect the correlation difference. Then network motif distribution was investigated by mfinder (version 1.2) [18]. The motif size is set to 4, the query random network size set as 100, and other parameters are set as default. Network motif distribution was calculated in all sub networks and ranked by their frequency.

Results and discussions

Network construction based on human gut datasets

We applied the Meta-Network on a dataset of health young Chinese as our representation of human gut (MGP15838 in MG-RAST database) [19]. It consists of 314 healthy young adults, covering 20 rural and urban cohorts from 7 ethnic groups and 9 provinces throughout China. 5,102,015 high-quality sequences were generated and QIIME (version 1.91) [20] was applied to process this dataset. Finally, we obtained in total 24, 125 microbial OTUs. On genus level, we identified in total 2124 genera in which 102 genera process the relative abundance above 0.1%.

We compared networks constructed by all four algorithms (Pearson, Loose Definition, FS-Weight and PCA-PMI) on genus level. First, we calculated the global properties and illustrated in Fig. 3, a. Based on Pearson algorithm, 31 genera and 172 correlations were detected. Based on loose definition method, 29 genera and 287 correlations were detected. Network constructed by FS-Weight method detected 38 nodes and 252 edges. In network constructed by PCA-PMI method, 44 nodes and 235 edges were detected. Based on loose definition algorithm, the network removed 2 genera and 30 correlations, in which 81.16% correlations are among genera counted for average relative abundance less than 0.1%. On the other hand, 145 correlations are added among the genera which dominate in the microbial community (average abundance over 0.1%). This result indicates the importance roles of high-abundance species in microbial community, which is consistence with previous researches [21]. Furthermore, comparison between network constructed by loose definition method and FS-Weight and PCA-PMI methods shows that more complex forms of correlations have been detected by Meta-Network analysis. In our analysis workflow, both FS-Weight and PCA-PMI are calculated only based on Pearson Correlation coefficient. It would be interesting to see how it performs for network construction based on other pair-wise correlation analysis method under the Meta-Network workflow. Hence we applied the same workflow to construct the networks based on CCLasso [5], and the analysis results were provided in Additional file 1.

Network comparison based on human gut datasets

We performed a systematic evaluation to compare the network constructed by Pearson, loose definition, FS-Weight and PCA-PMI algorithms. Global network alignment was carried to compare the network constructed by these four algorithms on OTU level (Fig. 3, a). The alignment between networks constructed by Pearson correlation (100% co-occurrence probability) and loose definition method (80% co-occurrence probability) measured the co-occurrence probability optimization. Owing to a low EC and S3 score (0.63, 0.579, respectively), the two networks had a low match. Network constructed by

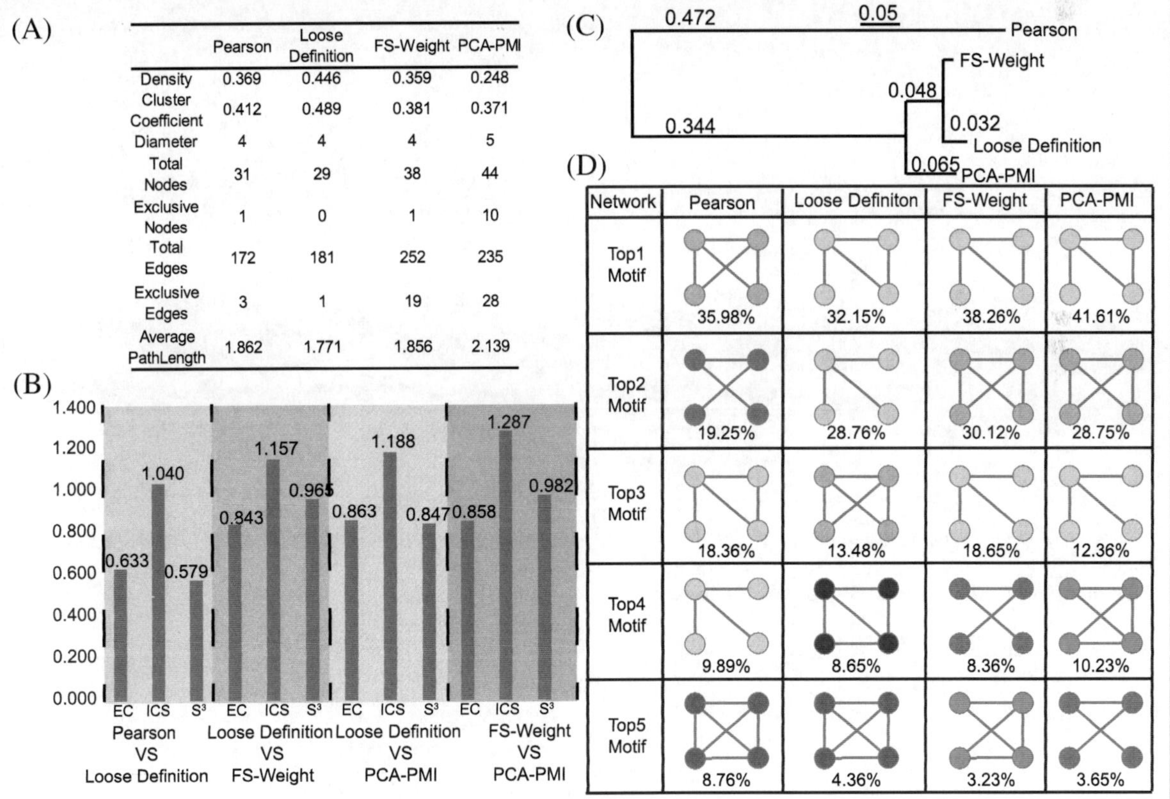

	Pearson	Loose Definition	FS-Weight	PCA-PMI
Density	0.369	0.446	0.359	0.248
Cluster Coefficient	0.412	0.489	0.381	0.371
Diameter	4	4	4	5
Total Nodes	31	29	38	44
Exclusive Nodes	1	0	1	10
Total Edges	172	181	252	235
Exclusive Edges	3	1	19	28
Average PathLength	1.862	1.771	1.856	2.139

Fig. 3 Network constructed and comparison among networks constructed by Pearson, Loose definition, FS-Weight and PCA-PMI methods on genus level. **a** Global properties for the four methods. The properties including network density, cluster coefficient, total nodes and edges, specific nodes and edges for each network and average path length were calculated. **b** Global network alignment result. Vertical axis represents the different alignment quality score. Horizontal axis represents the different pairwise alignments results. The quality score is ordered by edge correctness (EC), induced conserved structure (ICS) and symmetric substructure score (S^3). **c** Network similarity tree constructed by Jaccard Index. FS-Weight algorithm was chosen as root. Based on different construction process, network constructed by FS-Weight and PCA-PMI has similar structures. **d** Motif distributions of four network construction methods. The four-node motif ranked with top 5 frequencies were illustrated. The percentage represented the proportion of motif in the entire 4-node-network motif, and each color indicated one specific motif in different network construction method

FS-Weight and PCA-PMI methods had the highest quality score (S3 score: 0.982), indicating that the network constructed by FS-Weight and PCA-PMI methods had the highest similarity. Network alignment results indicate that: First, both two association rule mining methods show a low match to network constructed by Pearson method. Second, two different methods show a high similarity.

Based on the similarity tree based on Jaccard index, network constructed by Pearson method had a low similarity with other three networks (Fig. 3, b). Network constructed by loose-definition and FS-Weight had the shortest distance because FS-Weight employed loose-definition abundance as input to find indirect correlations and PCA-PMI based network had a similar structural composition compared to network constructed by FS-Weight method.

Based on their motif distributions, we compared these four networks to reflect homology relationship (Fig. 3, c). Network constructed by loose definition and FS-Weight method present similar network motif distributions, and this result suggested a high homology [22]. Motifs detected in Pearson method had a different distribution with the others (Fig. 3, d).

Correlation distribution and subnetwork comparison based on human gut datasets

To investigate the edge distribution, four algorithms were compared based on the top 100 genera (Fig. 4, a). In network constructed by Pearson and Spearman algorithms, 52 and 63 correlations were identified among low abundance genera (illustrated as small node size). However, in network constructed by FS-Weight and PCA-PMI methods, most correlations were identified among the nodes counted as high abundance (over 0.1%). Furthermore, in view of the importance of abundance and function in microbial community, we can further speculate that these genera play important role in the gut microbiome [23].

In the network constructed by Pearson and Spearman algorithm (Fig. 4, b), Genera *Lactococcus* and *Granulicatella* processed low abundance (average abundance 1.25e-5,

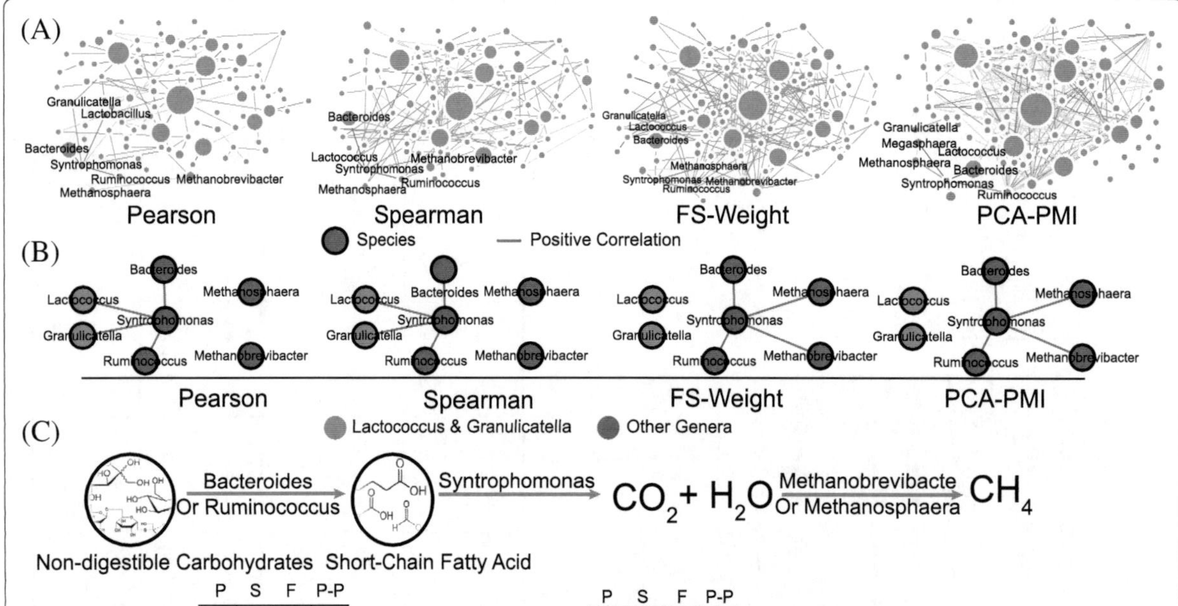

Fig. 4 Subnetwork comparison for the network constructed by Pearson, Spearman, FS-Weight and PCA-PMI methods for human gut microbiome. **a** Network generated based on four methods based on top 100 gut genera. Genera whose average abundance ranked top 100 are selected, based on which the four methods were compared. **b** Correlation of genus in the same subnetwork. Pearson and Spearman method detect non-related correlations which could be filtered out by loose definition method. Based on indirect neighbor (FS-Weight) and non-linear correlations (PCA-PMI) calculations, previously hidden correlations (between *Syntrophomonas* and *Methanosphaera*, for example) could be discovered. In (**a**) and (**b**), each node represents a genus and the node size represented the average abundance in all samples. Edges indicates the correlations and red edges represent the positive correlations and green edges represent the negative correlations. **c** The possible interaction mechanism in the subnetwork for (**b**). Non-digestible carbohydrates could be degraded by *Bacteroides* and *Ruminococcus* to produce the short-chain fatty acid. Further reactions are generated between *Syntrophomonas* and methanogens to transfer the short-chain fatty acid into the methane. **d** Correlation distribution for four methods in non-digestible carbohydrate pathway. We calculated the correlations between *Syntrophomonas* and other genera (*Bacteroides, Ruminococcus, Methanosphaera, Methanobrevibacter*) detected by different methods (P: Pearson, S: Spearman, F: FS-Weight, P-P: PCA-PMI), and a tick represented the correlation was detected, and no detection was indicated by a cross

2.58e-4, respectively) and existed in 15 and 14 samples, respectively. In addition, a strong and significant positive correlation were detected between them (correlation coefficient 0.915, 0.823, respectively, both *p*-value < 0.01). No convincing literature and reported function (*Lactocuccus*: commonly identified as produce lactic acid, *Granulicatella*: commonly identified as potential pathogenic bacteria) could prove their relationships and both genera [19, 20]. These two correlations might be noise correlations and do not take part in this non-digestible carbohydrates degradation pathway.

More importantly, FS-Weight and PCA-PMI methods are capable to discover correlations undetected by the Pearson and Spearman algorithms (Fig. 4, d). For example, genus *Syntrophomonas* is reported as a fatty-acidoxidizing genus which cooperates with methanogens such as genus *Methanosphaera* and *Methanobrevibacter* [24]. However, Pearson method failed to detect these correlations, probably due to the large heterogeneity (different kinds of methanogens in gut samples) in human gut dataset. On the contrary, the correlations between *Syntrophomonas* and methanogens could be clearly identified by FS-Weight and PCA-PMI methods (FS-Weight correlation: 0.873, 0.926; PCA-PMI correlation: 0.912, 0.915). The interaction mechanism was illustrated in Fig. 4, c. First, genera such as *Bacteroides* and *Ruminococcus* degrade the non-digestible carbohydrates into the short-chain fatty acid [25]. Second, the *Syntrophomonas* and methanogens cooperate to transfer the short-chain fatty acid into methane and energy. Combining the network analysis results with literature review, we speculate that *Bacteroides* and *Ruminococcus* play the role, as indirect neighbor, between the *Syntrophomonas* and methanogens.

In network constructed by CCLasso, the indirect correlation pattern could also be found, which could again prove the indirect pattern and advantage of FS-Weight and PCA-PMI. The result is provided in Additional file 1.

Network construction and analysis based on the Tara oceans datasets

We applied the same workflow to *Tara* Oceans Project (PRJEB1787, also known as Project ERP001736 on EBI Metagenomics Portal). The processed nucleotide sequences from 245 experimental runs were publicly available, of which the volume exceeded 1.3 TB. Parallel-Meta (version 3.0) [26] was adopted to process this dataset to calculate the taxonomical abundance distribution on genus and OTU levels. Based on the Pearson algorithm, 154 nodes and 4289 edges were detected, and 6 clusters were identified by the MCODE cluster algorithm (Fig. 5, a). The largest cluster was mainly composed with the members of

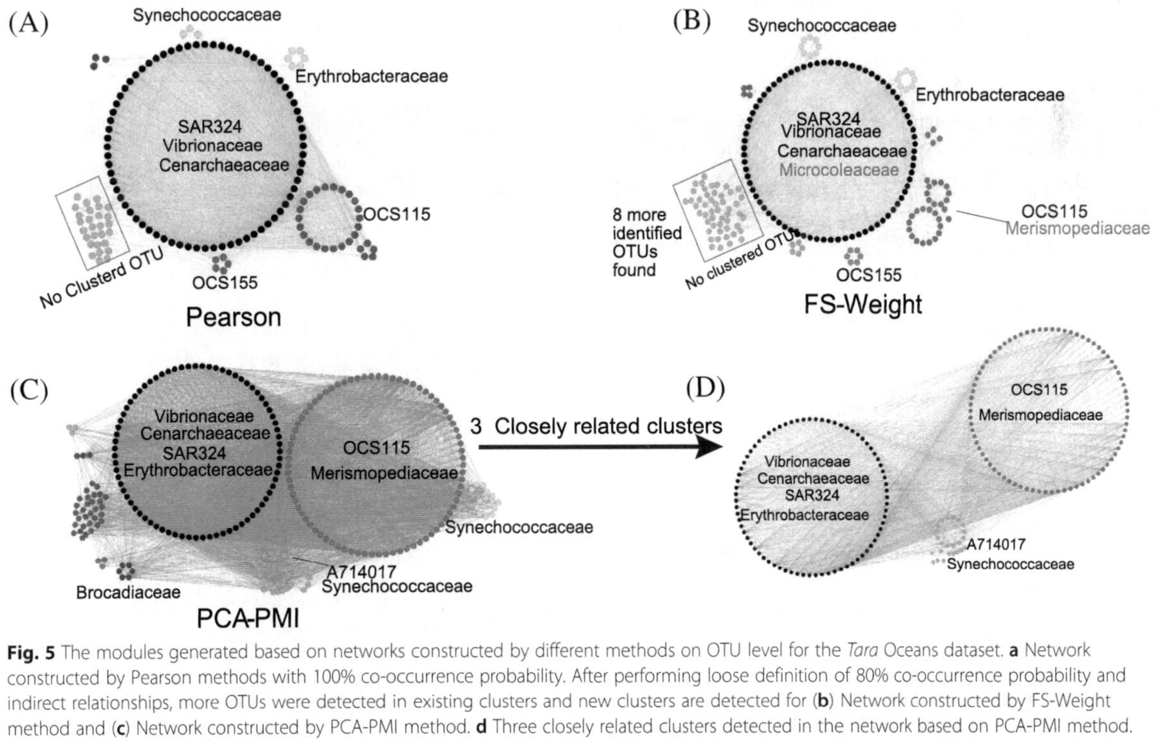

Fig. 5 The modules generated based on networks constructed by different methods on OTU level for the *Tara* Oceans dataset. **a** Network constructed by Pearson methods with 100% co-occurrence probability. After performing loose definition of 80% co-occurrence probability and indirect relationships, more OTUs were detected in existing clusters and new clusters are detected for (**b**) Network constructed by FS-Weight method and (**c**) Network constructed by PCA-PMI method. **d** Three closely related clusters detected in the network based on PCA-PMI method. In all four panels, each cluster was composed by *phylogenetic*ally close species are labeled in the same color. Based on the calculation of identified OTUs on the family level, the most abundant members of every clusters are annotated

phylum *Proteobacteria*, which was confirmed as the largest phylum in the ocean bacterial community [27].

Figure 5, b illustrated the network constructed by FS-Weight algorithm, in which 190 nodes and 8135 edges were detected, including 3910 indirect correlations. Ten clusters were calculated and 6 of them were composed with similar taxonomical composition compared to the network constructed by Pearson method. New cluster members were identified as common members in the ocean bacterial community [28]. The network constructed by PCA-PMI method was composed with 237 nodes and 5861 edges, with a clustering coefficient of 0.666 (Fig. 5, c). *Nitrosopumilus* and *Candidatus_Scalindua* were detected only in network constructed by PCA-PMI method and were identified as common members in the ocean environment [29].

These results have again, proved that the Meta-Network workflow has the ability to discover more complex formats of correlations. Moreover, Networks constructed by two different methods (FS-Weight and PCA-PMI) shared many identical network members and clusters, indicating that both methods portrayed the complex formats of correlations in microbial communities.

Conclusion

Species-species co-occurrence network is becoming one of the emerging fronts for microbiome research, largely due to the important ecological patterns it could reveal. However, previous methods are limited by only detecting the direct correlations among pairwise species. This work presents the Meta-Network method that optimized the construction, analysis and interpretation of the species-species co-occurrence network, focusing on three critical processes for species-species correlation analysis: we first employ loose definition to recover correlations missed by strict co-occurrence probability before calculating correlations. Based on these candidate correlations, we also develop two association rule mining methods for the recurrence of the real bacterial community network: a graph-based method FS-Weight to detect indirect correlations, PCA-PMI method to detect indirect correlations. Therefore, we believe that Meta-Network provides a general approach towards deciphering the species-species co-occurrence networks.

This method could be improved in several ways: in line with current gene expression network inferences, we can also improve its high-level associations (i.e., associations between two clusters of species). A complete species-species network, which provides the full picture of the regulation profile in the community, should also include viruses. All of these will be considered in our future work.

Additional file

Additional file 1: Figure S1. Threshold selection for FS-Weight and PCA-PMI method. (A) Global network properties for networks constructed by FS-Weight method. (DOCX 354 kb)

Abbreviations

OTU: Operational Taxonomic Unit; PCA: Path Consistency Algorithm; PMI: Part Mutual Information

Acknowledgements

We thank Torsten Juelich for linguistic assistance during the preparation of this manuscript.

Funding

This work is partially supported by National Science Foundation of China grant 31871334 and 31671374, Ministry of Science and Technology's high-tech (863) grant 2018YFC0910502, and Sino-German Research Center grant GZ878. Publication of this article was sponsored by Ministry of Science and Technology's high-tech (863) grant 2018YFC0910502.

Availability of data and materials

The source code and documentation are accessible at http://www.microbioinformatics.org/software/Meta-Network.htm.

About this supplement

This article has been published as part of *BMC Genomics Volume 20 Supplement 2, 2019: Selected articles from the 17th Asia Pacific Bioinformatics Conference (APBC 2019): genomics*. The full contents of the supplement are available online at https://bmcgenomics.biomedcentral.com/articles/supplements/volume-20-supplement-2.

Authors' contributions

The whole study was designed by KN. PSY and SJY collected samples and analyzed the data. PSY, CL, KN explained the results, PSY and KN wrote the initial draft of the manuscript, and all revised the manuscript. All of the authors have read and approved the final manuscript.

Ethics approval and consent to participate

Not applicable.

Consent for publication

Not applicable.

Competing interests

The authors declare no competing financial interests.

Publisher's Note

Author details

[1]Key Laboratory of Molecular Biophysics of the Ministry of Education, College of Life Science and Technology, Huazhong University of Science and Technology, Wuhan 430074, Hubei, China. [2]Department of Engineering, Trinity College, 300 Summit Street, Hartford, CT 06106, USA.

Published: 10 April 2019

References

1. Hurwitz BL, Westveld AH, Brum JR, Sullivan MB. Modeling ecological drivers in marine viral communities using comparative metagenomics and network analyses. Proc Natl Acad Sci U S A. 2014;111:10714–9.
2. Weiss S, Van Treuren W, Lozupone C, Faust K, Friedman J, Deng Y, Xia LC, Xu ZZ, Ursell L, Alm EJ, et al. Correlation detection strategies in microbial data sets vary widely in sensitivity and precision. ISME J. 2016;10:1669–81.
3. Vasquez-Cardenas D, van de Vossenberg J, Polerecky L, Malkin SY, Schauer R, Hidalgo-Martinez S, Confurius V, Middelburg JJ, Meysman FJ,

Boschker HT. Microbial carbon metabolism associated with electrogenic Sulphur oxidation in coastal sediments. ISME J. 2015;9:1966–78.
4. de Winter JC, Gosling SD, Potter J. Comparing the Pearson and Spearman correlation coefficients across distributions and sample sizes: a tutorial using simulations and empirical data. Psychol Methods. 2016; 21:273–90.
5. Fang H, Huang C, Zhao H, Deng M. CCLasso: correlation inference for compositional data through lasso. Bioinformatics. 2015;31:3172–80.
6. Zhao J, Zhou Y, Zhang X, Chen L. Part mutual information for quantifying direct associations in networks. Proc Natl Acad Sci U S A. 2016;113:5130–5.
7. Price MN, Deutschbauer AM, Skerker JM, Wetmore KM, Ruths T, Mar JS, Kuehl JV, Shao W, Arkin AP. Indirect and suboptimal control of gene expression is widespread in bacteria. Mol Syst Biol. 2013;9:660.
8. Lammel DR, Barth G, Ovaskainen O, Cruz LM, Zanatta JA, Ryo M, de Souza EM, Pedrosa FO. Direct and indirect effects of a pH gradient bring insights into the mechanisms driving prokaryotic community structures. Microbiome. 2018;6:106.
9. Chua HN, Sung WK, Wong L. Exploiting indirect neighbours and topological weight to predict protein function from protein-protein interactions. Bioinformatics. 2006;22:1623–30.
10. Barberan A, Bates ST, Casamayor EO, Fierer N. Using network analysis to explore co-occurrence patterns in soil microbial communities. ISME J. 2012; 6:343–51.
11. Chaffron S, Rehrauer H, Pernthaler J, von Mering C. A global network of coexisting microbes from environmental and whole-genome sequence data. Genome Res. 2010;20:947–59.
12. Marino S, Baxter NT, Huffnagle GB, Petrosino JF, Schloss PD. Mathematical modeling of primary succession of murine intestinal microbiota. Proc Natl Acad Sci U S A. 2014;111:439–44.
13. Zhou C, Zhang SW, Liu F. An ensemble method for reconstructing gene regulatory network with jackknife resampling and arithmetic mean fusion. Int J Data Min Bioinform. 2015;12:328–42.
14. Zhang X, Zhao XM, He K, Lu L, Cao Y, Liu J, Hao JK, Liu ZP, Chen L. Inferring gene regulatory networks from gene expression data by path consistency algorithm based on conditional mutual information. Bioinformatics. 2012;28: 98–104.
15. Wang J, Zhong J, Chen G, Li M, Wu FX, Pan Y. ClusterViz: a Cytoscape APP for cluster analysis of biological network. IEEE/ACM Trans Comput Biol Bioinform. 2015;12:815–22.
16. Vijayan V, Saraph V, Milenkovic T. MAGNA++: maximizing accuracy in global network alignment via both node and edge conservation. Bioinformatics. 2015;31:2409–11.
17. Kuntal BK, Dutta A, Mande SS. CompNet: a GUI based tool for comparison of multiple biological interaction networks. BMC Bioinformatics. 2016;17:185.
18. Milo R, Shen-Orr S, Itzkovitz S, Kashtan N, Chklovskii D, Alon U. Network motifs: simple building blocks of complex networks. Science. 2002;298:824–7.
19. Zhang J, Guo Z, Xue Z, Sun Z, Zhang M, Wang L, Wang G, Wang F, Xu J, Cao H, et al. A phylo-functional core of gut microbiota in healthy young Chinese cohorts across lifestyles, geography and ethnicities. ISME J. 2015;9: 1979–90.
20. Kuczynski J, Stombaugh J, Walters WA, Gonzalez A, Caporaso JG, Knight R. Using QIIME to analyze 16S rRNA gene sequences from microbial communities. Curr Protoc Microbiol. 2012, Chapter 1:Unit;27:1E–5.
21. Rivett DW, Bell T. Abundance determines the functional role of bacterial phylotypes in complex communities. Nat Microbiol. 2018;3:767–72.
22. Alon U. Network motifs: theory and experimental approaches. Nat Rev Genet. 2007;8:450–61.
23. Si J, You HJ, Yu J, Sung J, Ko G. Prevotella as a hub for vaginal microbiota under the influence of host genetics and their association with obesity. Cell Host Microbe. 2017;21:97–105.
24. Hatamoto M, Imachi H, Fukayo S, Ohashi A, Harada H. Syntrophomonas palmitatica sp. nov., an anaerobic, syntrophic, long-chain fatty-acid-oxidizing bacterium isolated from methanogenic sludge. Int J Syst Evol Microbiol. 2007;57:2137–42.
25. Neyrinck AM, Delzenne NM. Potential interest of gut microbial changes induced by non-digestible carbohydrates of wheat in the management of obesity and related disorders. Curr Opin Clin Nutr Metab Care. 2010;13:722–8.
26. Jing G, Sun Z, Wang H, Gong Y, Huang S, Ning K, Xu J, Su X. Parallel-META 3: comprehensive taxonomical and functional analysis platform for efficient comparison of microbial communities. Sci Rep. 2017;7:40371.
27. Lima-Mendez G, Faust K, Henry N, Decelle J, Colin S, Carcillo F, Chaffron S, Ignacio-Espinosa JC, Roux S, Vincent F, et al. Ocean plankton. Determinants of community structure in the global plankton interactome. Science. 2015; 348:1262073.
28. Feng XM, Mo YX, Han L, Nogi Y, Zhu YH, Lv J. Qipengyuania sediminis gen. Nov., sp. nov., a member of the family Erythrobacteraceae isolated from subterrestrial sediment. Int J Syst Evol Microbiol. 2015;65:3658–65.
29. Shehzad A, Liu J, Yu M, Qismat S, Liu J, Zhang XH. Diversity, community composition and abundance of Anammox Bacteria in sediments of the north marginal seas of China. Microbes Environ. 2016;31:111–20.

Zhang *et al. BMC Genomics* 2019, **20**(Suppl 2):221
https://doi.org/10.1186/s12864-019-5472-0

BMC Genomics

RESEARCH

Open Access

Genome-wide analysis of epigenetic dynamics across human developmental stages and tissues

Xia Zhang[1†], Yanglan Gan[2*†], Guobing Zou[1], Jihong Guan[3] and Shuigeng Zhou[4]

From The 17th Asia Pacific Bioinformatics Conference (APBC 2019)
Wuhan, China. 14-16 January 2019

Abstract

Background: Epigenome is highly dynamic during the early stages of embryonic development. Epigenetic modifications provide the necessary regulation for lineage specification and enable the maintenance of cellular identity. Given the rapid accumulation of genome-wide epigenomic modification maps across cellular differentiation process, there is an urgent need to characterize epigenetic dynamics and reveal their impacts on differential gene regulation.

Methods: We proposed DiffEM, a computational method for differential analysis of epigenetic modifications and identified highly dynamic modification sites along cellular differentiation process. We applied this approach to investigating 6 epigenetic marks of 20 kinds of human early developmental stages and tissues, including hESCs, 4 hESC-derived lineages and 15 human primary tissues.

Results: We identified highly dynamic modification sites where different cell types exhibit distinctive modification patterns, and found that these highly dynamic sites enriched in the genes related to cellular development and differentiation. Further, to evaluate the effectiveness of our method, we correlated the dynamics scores of epigenetic modifications with the variance of gene expression, and compared the results of our method with those of the existing algorithms. The comparison results demonstrate the power of our method in evaluating the epigenetic dynamics and identifying highly dynamic regions along cell differentiation process.

Keywords: Epigenetic modification, Differential analysis, Hamming distance

Background

Lineage specification and maintenance of cellular identity are complex biological processes [1]. It is now widely accepted that cell phenotypes are significantly regulated by epigenetic states and that chromatin changes during differentiation contribute to the determination of cell fate [2]. Recent evidence further shows that coordinated epigenetic changes influence the maintenance of such cellular memory [3, 4]. DNA methylation and certain epigenetic modifications are essential for chromatin structures and gene expression in proper execution of developmental programs [5, 6]. Therefore, a fundamental question in the field is to exactly answer where and how the epigenetic changes regulate phenotypic changes.

To fully understand the dynamics and regulatory roles of epigenetic modifications, advanced sequencing technologies have generated genome-wide epigenetic maps of diverse developmental stages, lineages and tissues [7, 8]. In previous studies, researchers have differentiated human embryonic stem cells (hESCs) into mesendoderm, neural progenitor cells, trophoblast-like cells, and mesenchymal stem cells and systematically sequenced the transcriptome and epigenetic modifications of these lineages [9, 10]. The first three hESC derivatives reflects critical developmental linages in the embryo [11]. Mesenchymal stem

*Correspondence: ylgan@dhu.edu.cn
†Xia Zhang and Yanglan Gan contribute equally to the paper.
[2]School of Computer Science and Technology, Donghua University, Shanghai, China
Full list of author information is available at the end of the article

cells have the ability of further multi-lineage differentiation to bone, cartilage, adipose, muscle, and connective tissues [12]. Mouse embryonic stem cells were also differentiated into a variety of precursor cell types [13]. The expanding body of epigenomic data permits researchers to study the dynamics of epigenetic marks. This is a key step to reveal regulatory roles of epigenetic modifications, and to understand how global features of epigenetic modifications impact cellular phenotypes across different developmental stages, lineages and tissues.

Most previous works focused on comparing the epigenetic modification profiles between two biological conditions, and further identifying regions that show differential patterns, such as ChIPDiff [14], diffReps [15], dPCA [16], HistoneHMM [17], csaw [18] and HMCan-diff [19]. While some other methods such as dMCA [20] and Yang's method [21], were designed to detect cell-type-specific differential regions. Moreover, there are also some methods that were designed for identifying differential methylated region, such as QDMR [22] and MethylAction [23], whereas QDMR can also be applied to histone modification data analysis. Although several algorithms have been developed to analyze the epigenetic difference between two different conditions, little work devoting to differential analysis of epigenetic modifications among multiple cell types and across different developmental stages.

Here, we presented DiffEM, a computational method to quantify the dynamics of epigenetic marks and identified highly dynamic modification sites (HDMSs) across different human embryonic developmental stages. We applied this method to a public datasets with 6 intensely studied epigenetic marks of 20 different developmental stages and tissues. We identified HDMSs where different cell types exhibit distinctive epigenetic modification patterns, and found that these highly dynamic sites are enriched in genes related to cellular development and differentiation. We further correlated the dynamics scores of these epigenetic marks with those of gene expression levels. The results indicate that the changes of gene expression are closely related to the modification patterns of H3K4me1 and H3K27me3 in promoter regions during cell differentiation process. We compared DiffEM with the existing algorithms for identifying HDMSs. The comparison results show that DiffEM perform better in evaluating the epigenetic dynamics and identifying highly dynamic modification sites. This method is promising for broad applications in evaluating epigenetic dynamics in other complex biological processes.

Materials and methods

Datasets

To analyze the dynamic epigenetic changes during cellular differentiation and lineage specification, we obtained a large panel of epigenetic maps of human embryonic stem cells (hESCs) and the key derivatives, including trophoblast-like cells (TBL), mesendoderm (ME), neural progenitor cells (NPCs), and mesenchymal stem cells (MSCs). The iHMS [24] database has integrated massive genome-wide epigenetic modification maps and RNA expression data spanning different developmental stages and tissues. From iHMS, we downloaded 6 epigenetic modification maps (H3K4me1, H3K4me3, H3K9me3, H3K27ac, H3K27me3 and H3K36me3) of 20 different human developmental stages and tissues, including hESCs, the hESC-derived precursor cell types (TBL, ME, NPCs and MSCs), and 15 human primary tissues (adipose, adrenal gland, adult liver, aorta, esophagus, gastric, left ventricle, lung, ovary, pancreas, psoas muscle, right ventricle, right atrium, sigmoid colon, spleen, thymus, small intestine, breast, brain and bladder). Meanwhile, the RNA expression data and reference gene annotations were also downloaded from iHMS.

Overview of the DiffEM model

To characterize epigenetic dynamics across different development stages, we developed DiffEM, a new method to estimate the dynamics of epigenetic modifications based on hamming distance and identify highly dynamic modification sites. Unlike the previous work [20], we aimed to detect highly dynamic regions of epigenetic modification during cell differentiation process. To evaluate the dynamics across different differential stages and the primary tissues respectively, these 20 cell types were further categorized into three groups, hESC-derived precursor cell types, primary tissues and the whole group. We introduce the following steps to identify HDMSs , which are also shown in Fig. 1.

Data binarization. The raw ChIP-seq data were preprocessed in iHMS database [24]. The whole-genome was first segmented into 200bp bins. For each bin, neighboring read counts were summarized into an integer, indicating the extent of epigenetic modification in this region [25]. To reduce the effect of noise, we transformed these integers into binary values. First, we calculated the binarization threshold, by dividing the total read counts of all bins by the number of bins. If the read count of a bin is higher than the threshold, its binary value is set as 1, otherwise 0, After binarization, we noticed that some bins have no signals in all cell types, which may consist of sequences of low mappability. The consecutive regions with length more than 5 kb were removed from the genome. Finally, for the 6 investigated epigenetic marks, we obtained 6 binary matrices B_K of size T (the number of cell types) by N (the number of 200 bp bins on the whole genome).

Calculation of the dynamics scores for each epigenetic mark. After data binarization, we calculated the dynamics scores for each epigenetic mark among multiple cell types. In particular, we used the hamming distance to measure

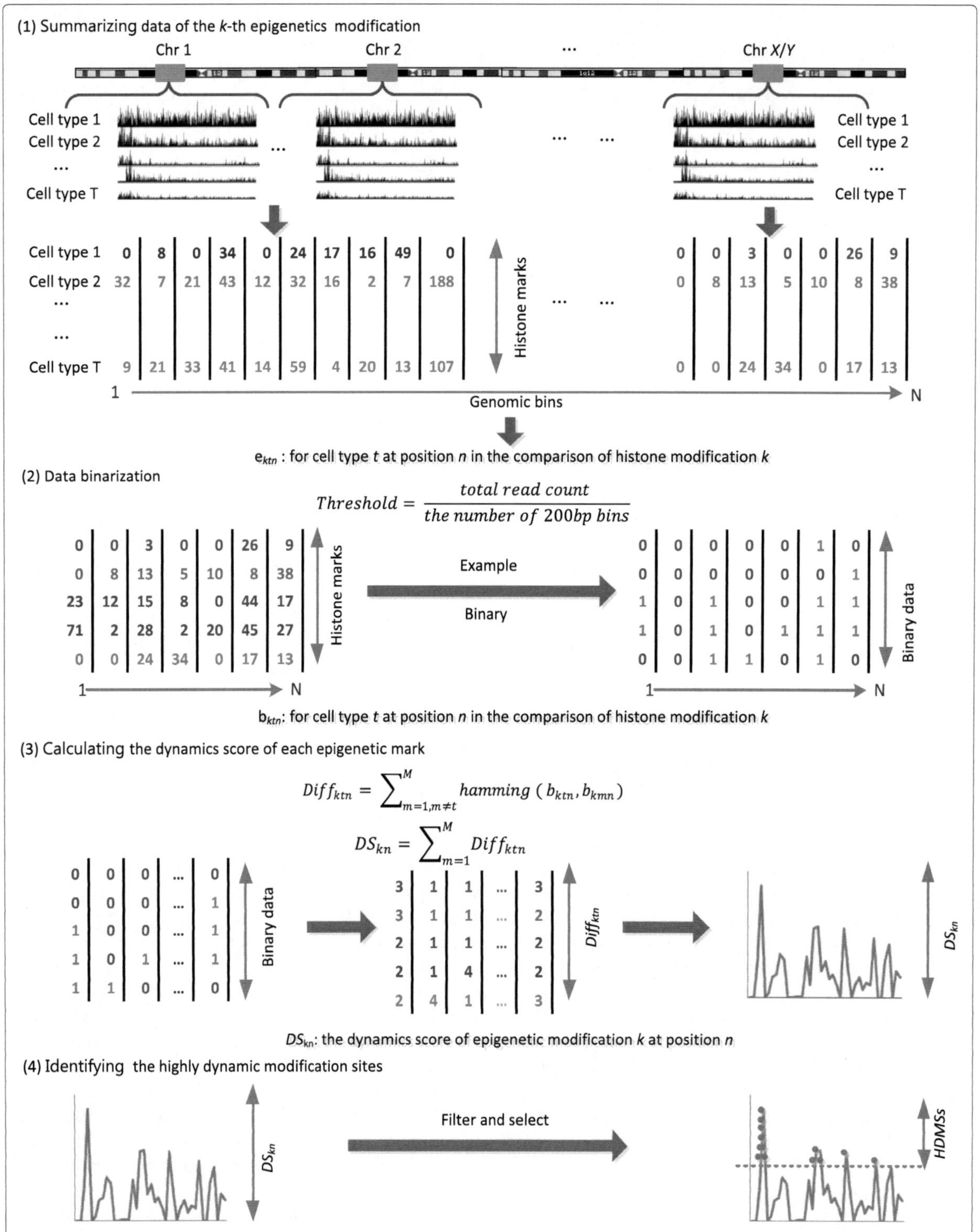

Fig. 1 The flowchart of the DiffEM approach

the dynamics of each epigenetic modification. Here, we respectively calculated the dynamics scores of the 6 investigated epigenetic modifications in three cell type groups. As described above, given M cell types and N bins, we denoted b_{ktn} as the binary profiles of epigenetic modification k for cell type t at position n. Then the difference between cell type t and others are calculated as:

$$Diff_{ktn} = \sum_{m=1, m\neq t}^{M} hamming\,(b_{ktn}, b_{kmn}) \tag{1}$$

Further, the dynamics score of epigenetic modification k at position n was summed as:

$$DS_{kn} = \sum_{t=1}^{M} Diff_{ktn} \tag{2}$$

Identification of the highly dynamic modification sites. For each epigenetic mark, we have obtained the dynamics scores along the genome in each cell type group. The higher the dynamics score is, the greater the difference across these cell types exhibits. The sites with zero score were filtered first. Based on the calculated dynamics scores, we selected those bins whose dynamics scores are significantly higher than the genome background ($p < 0.05$) and merged the adjacent bins into longer regions. These regions are referred to as highly dynamic modification sites (HDMSs).

Functional analysis of the highly dynamic modification sites

To investigate the potential functions of these identified HDMSs, we mapped them to RefSeq genes and some functional regions. According to their relative positions, we related the HDMSs to various genes when the centers of HDMSs are located in gene regions. The number of genes related to HDMSs was counted. Furthermore, we mapped the bins with the highest score to genomic features like promoter, coding region and exon. If a HDMS is not related to any gene, it is labeled as an intergenic sites. For further analysis of the functional relevance of HDMSs, we performed gene ontology (GO) enrichment analysis and pathway enrichment analysis for genes enriched with HDMSs via DAVID bioinformatics resources. The significant enrichment lists are obtained with $p<0.05$.

Comparisons among different epigenetic modifications

Epigenetic modifications play a critical role in cell differentiation process. Different epigenetic modifications may collaborate with each other to execute specific functions. We investigated the relations among different types of epigenetic marks. Based on the identified HDMSs of each epigenetic mark, the common HDMSs between different epigenetic modifications were obtained in the whole genome. Further, we estimated the correlations between the dynamics scores of these epigenetic modifications.

Correlation analysis between the dynamics of epigenetic mark and gene expression

First, we evaluated the dynamic scores of gene expression along the genome in each cell type group, which was calculated as the variance divided by the mean of gene expression. Then, we evaluated the correlation coefficients between the dynamic scores of epigenetic modifications and gene expression levels. For those identified HDMSs, a higher correlation coefficient indicates that gene expression is more easily regulated by the specific epigenetic modification.

Comparison among DiffEM , QDMR and IOD

As there exists no gold standard to benchmark highly dynamic modification sites, we adopted an indirect validation strategy. As previous studies [26], the validation was based on the correlations between the dynamics of epigenetic modifications and gene expression levels. To evaluate the performance in identifying HDMSs, we compared DiffEM with existing methods, QDMR and IOD. Unlike the methods restricted to the differential analysis between two cell types, the above three methods are capable of analyzing three or more cell types. QDMR was proposed for genome-wide differential analysis of epigenetic states based on Shannon entropy [22]. IOD was developed to detect differential regions across multiple cell types [27]. We first normalized the epigenetic data, and used QDMR and IOD to detect highly dynamic modification sites. These methods were compared by the correlations between the dynamics of epigenetic modifications and expression levels of the HDMSs.

Results

To investigate the dynamics of epigenetic modifications during cell differentiation process, we proposed a computational method, DiffEM, to quantify the dynamics score of various epigenetic marks and identify highly dynamic modification sites (HDMSs). We focused on human differentiation-related cell types, consisting of human embryonic stem cell, 4 hESC-derived precursor cell types, and 15 primary tissues. In each cell type, we collected 6 genome-wide epigenetic maps and gene expression datasets. DiffEM was applied to identify HDMSs along cell differentiation process. To evaluate the performance of our proposed method, in this section we analyzed the identified HDMSs to discover their potential biological roles during cell differentiation and development. Furthermore, we compared DiffEM with two previous methods, QDMR and IOD.

Genome-wide characterization of epigenetic modification dynamics

To better explore the dynamic epigenetic changes across different cell differentiation stages, these 20 cell types were further grouped into three groups, hESCs and hESC-derived precursor cell types, primary tissues and the whole group. For each group and each epigenetic modification mark, we quantified the dynamics score for each bin based on hamming distance, and then ranked these bins according to their dynamics scores. We selected those bins whose dynamics scores were significantly higher than the genome background (p <0.05).

After merging the neighboring bins, we obtained the HDMSs for each epigenetic modification in each group. For different epigenetic marks, we found that there exist big overlaps between the HDMSs of different epigenetic modifications. This is consistent with previous finding that the epigenetic modifications collaborated with each other to consummate certain regulatory function. As shown in Fig. 2, we respectively calculated the percentage of overlapping HDMSs among 6 epigenetic modifications in these three groups. On the whole, the overlapping sites make up 20%˜60% of total HDMSs in different groups. In the hESCs and hESC-derived precursor group, the HDMSs of different epigenetic marks overlap more than those of the other two groups. For example, the overlap rates of H3K4me1 with other five epigenetic marks range from 40% to 50% in hESC-derived group, while those overlap rates in the other two groups are not greater than 25% . Specifically, H3K4me3 is highly overlapped with H3K9me3 and H3K27ac. These observations demonstrate that epigenetic modifications collaborate closely to regulate the cell differentiation process [4].

As distinct epigenetic modifications share HDMSs, we further investigated the correlation between the dynamics scores of different epigenetic marks. As shown in Fig. 3, the investigated epigenetic marks demonstrate varied correlation in the three comparison groups. In particular, the epigenetic marks show higher correlation in the hESCs and hESC-derived precursor group. This result indicates that the dynamics of epigenetic modifications are similar during the cell differentiation process, which is conformed to the results of previous overlaps analysis.

Highly dynamic modification sites are related to various genomic features

Further, we mapped the identified HDMSs to RefSeq genes and collected the genes enriched with HDMSs for each epigenetic mark. Here we explored how the dynamic epigenome participates in early embryonic developmental stages and focused on the hESCs and hESC-derived precursor group. To examine the potential functions of those genes, we performed systematic gene ontology enrichment analysis using DAVID tools (https://david.ncifcrf.gov/) and summarized the key biological processes and pathways for each epigenetic mark. Overall, for the aforementioned six epigenetic modification marks, we found that those HDMSs-enriched genes exhibit enrichment for cell differentiation and development functions (Table 1) (p value <0.05). For example, GO terms related to development such as 'nervous system development' are enriched in HDMSs of H3K4me1, H3K4me3, H3K9me3,H3K27ac, H3K27me3, GO terms related to differentiation such as 'neuron differentiation' and 'cerebellar granule cell differentiation' are enriched in HDMSs of H3K4me1, H3K9me3, H3K27me3, H3K36me3. This is consistent with previous finding that regulatory elements essential for cellular identity are often epigenetically modified in parental cells [28, 29]. The results highlight the importance of stage-specific epigenetic modification patterns of transcription factors for defining the developmental potentials.

Also, we noticed that the biological processes of distinct epigenetic marks have overlappings. One possible interpretation for this observation could be that these epigenetic marks may have the same changing trend, collaborating with each other to finish the complex regulatory functions. Taken together, the above results of GO annotation demonstrated the power of our method in identifying the highly dynamic sites of these epigenetic modifications. And, the results strongly suggest that the HDMSs mark critical regulatory regions for cell differentiation and development process. Further characterization

A

100%	50%	43%	57%	49%	40%	H3K4me1
48%	100%	48%	57%	54%	45%	H3K4me3
41%	48%	100%	51%	49%	39%	H3K9me3
48%	49%	43%	100%	48%	39%	H3K27ac
46%	52%	48%	55%	100%	41%	H3K27me3
46%	54%	47%	56%	50%	100%	H3K36me3
H3K4me1	H3K4me3	H3K9me3	H3K27ac	H3K27me3	H3K36me3	

B

100%	25%	22%	25%	20%	25%	H3K4me1
17%	100%	25%	21%	21%	21%	H3K4me3
15%	24%	100%	16%	25%	19%	H3K9me3
23%	29%	24%	100%	18%	25%	H3K27ac
13%	20%	24%	12%	100%	15%	H3K27me3
20%	25%	22%	21%	19%	100%	H3K36me3
H3K4me1	H3K4me3	H3K9me3	H3K27ac	H3K27me3	H3K36me3	

C

100%	21%	19%	22%	21%	24%	H3K4me1
15%	100%	18%	17%	21%	19%	H3K4me3
17%	22%	100%	19%	23%	22%	H3K9me3
22%	23%	20%	100%	21%	24%	H3K27ac
16%	22%	19%	17%	100%	19%	H3K27me3
18%	20%	18%	19%	20%	100%	H3K36me3
H3K4me1	H3K4me3	H3K9me3	H3K27ac	H3K27me3	H3K36me3	

Fig. 2 The overlaps between the top HDMSs of different epigenetic marks. The value in row i column j represents the proportion of HDMSs of epigenetic modification i overlapped by those of epigenetic modification j. **a** The hESCs and hESC-derived precursor group. **b** The primary tissues group. **c** The whole group

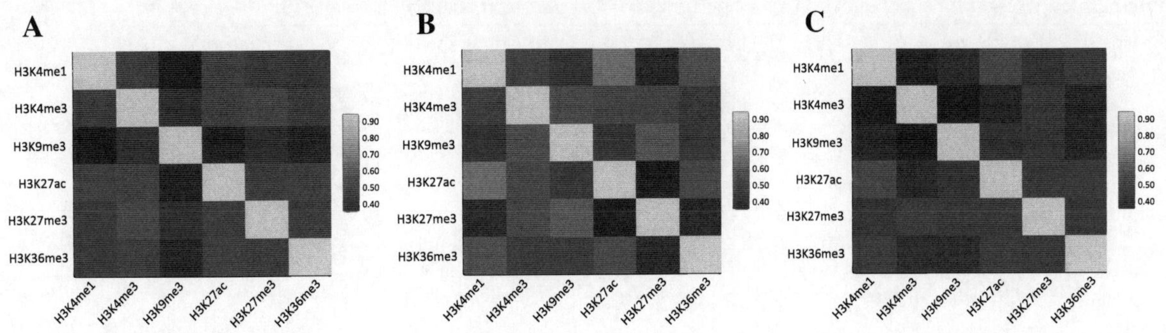

Fig. 3 The heatmaps representing the correlations between each pair of epigenetic modifications. **a** The hESCs and hESC-derived precursor group. **b** The primary tissues group. **c** The whole group

of epigenetic modification patterns and gene expression within HDMSs may provide important insights into the regulatory functions of the specific epigenetic patterns.

Highly dynamic modified sites neighboring genes reveal diverse transcriptional patterns

To analyze the regulatory roles of these dynamic epigenetic patterns, we further explored the epigenetic modification and gene expression patterns within HDMSs. We computed the correlation coefficients between the dynamics of epigenetic modifications and gene expression levels of the HDMSs-enriched genes. We mapped the HDMSs to Ref-Seq genes and obtained gene expression of the associated genes. As these 20 cell types were divided into three groups, the dynamics score of gene expression was assessed using the same method as epigenetic marks (see Methods). For those HDMSs located in promoters, and coding regions, the Pearson correlation coefficients were respectively computed.

As shown in Fig. 4, we noted that there is highly correlation between the dynamics of gene expression level and epigenetic modification in promoter regions. Relatively, the correlation in coding regions is lower. These results indicate that the variance of epigenetic modification patterns in promoter regions has a higher regulatory role than that in coding regions. The three different groups have a similar trend. In detail, the six epigenetic modification marks exhibit different regulatory effect. For the hESCs and hESC-derived precursor group, the dynamics of gene expression levels are highly regulated by the modification patterns of H3K4me1 and H3K27me3 in promoter regions. For the primary tissues, the correlations are much higher for H3K9me3 and H3k27ac.

Comparison with QDMR and IOD in identifying HDMSs

Considering that our method was developed for the differential analysis for multiple cell types, we compared DiffEM with two similar previous methods QDMR and IOD [22, 27], which were also designed for multiple conditions. QDMR is based on Shannon entropy [22], and IOD is defined as the variance divided by the mean value [27]. The performance was measured by the correlation analysis between the epigenetic modification dynamics and gene expression difference.

Firstly, we respectively identified the highly dynamic modification sites using these three methods, and ranked the HDMSs according to the dynamics score. Similarly, we obtained the ranked highly dynamic expression sites. Then, we associated these HDMSs with the highly dynamic expression sites by bitwise matching. To evaluate the performance of these three methods, we define two metrics, *MatchedNum* and *AveDS*. *MatchedNum* is computed as the number of highly dynamic expression sites matching with the top ranked HDMSs, which is similar to recall. *AveDS* represents the average dynamics score of these matched highly dynamic expression sites. Here, for fair comparison among the three methods, we calculated the entropy as the average dynamics score as QMDR.

We compared the performance on the aforemetioned 6 epigenetic modifications, the results are shown in Fig. 5 and Additional file 1. Figure 5 shows the comparison results for the hESCs and hESC-derived precursor group. Figure S1, Figure S2 (see Additional file 1) showed the results of the other two groups. We first compared the matched numbers of all differential gene expression sites output by these methods. Our method could get a higher *MatchedNum* of highly dynamic expression sites than those of QDMR and IOD (Fig. 5a). However, this raises the question that to what extent these matched sites are dynamically expressed. As we noted that changes in epigenetic modifications could cause differential expression of related genes, we further compared the average dynamics of gene expression of these matched sites. Lower ave indicate better performance. As the results showed (Fig. 5b), our method has good performance in *AveDS*. These observations demonstrate that our method always achieves a balance between matched *MatchedNum* and *AveDS*, which means our approach could be applied to

Table 1 Functional enrichment of genes on the whole genome of six histone modifications

Term type	Term name	*P*-value	Term type	Term name	*P*-value
H3K4me1					
BP	Cell adhesion	1.42E-06	CC	Cytoskeleton	2.90E-03
BP	Axon guidance	2.86E-05	CC	Growth cone	2.15E-02
BP	Nervous system development	1.91E-04	KEGG	Arrhythmogenic right ventricular	5.82E-03
BP	Signal transduction	1.92E-04		Cardiomyopathy (ARVC)	
BP	Neuron development	2.68E-02	KEGG	Axon guidance	2.79E-02
BP	Cerebellar granule cell differentiation	4.35E-02	KEGG	Hippo signaling pathway	4.32E-02
H3K4me3					
BP	Intracellular signal transduction	7.16E-04	BP	Adult behavior	6.93E-03
BP	Signal transduction	1.22E-03	MF	Extracellular-glutamate-gated ion channel	2.77E-03
BP	Nervous system development	2.63E-03		Activity	
BP	Chemical synaptic transmission	2.88E-02	KEGG	Neuroactive ligand-receptor interaction	6.24E-03
H3K9me3					
BP	Heterophilic cell-cell adhesion	1.87E-07	BP	Regulation of RNA splicing	9.41E-03
BP	Cell adhesion	1.06E-04	BP	Regulation of alternative mRNA splicing	1.58E-02
BP	Nervous system development	6.27E-04	BP	Chemical synaptic transmission	3.35E-02
BP	Regulation of neuron projection	4.09E-03	BP	Cerebellar granule cell differentiation	3.98E-02
	Development		MF	Calcium ion binding	1.25E-05
BP	Signal transduction	5.41E-03	KEGG	Cell adhesion molecules (CAMs)	2.10E-02
H3K27ac					
BP	Signal transduction	7.03E-05	BP	Regulation of RNA splicing	2.08E-02
BP	Nervous system development	7.47E-05	BP	Cytoskeleton organization	3.57E-02
BP	Neuron cell-cell adhesion	2.31E-04	MF	Actin binding	1.20E-04
BP	Neuron development	5.28E-03	CC	Growth cone	1.83E-04
BP	Glutamate receptor signaling pathway	5.76E-03	MF	Protein kinase activity	1.15E-02
BP	Brain development	1.57E-02	KEGG	Neuroactive ligand-receptor interaction	1.06E-02
H3K27me3					
BP	Social behavior	9.15E-05	BP	Cerebellar granule cell differentiation	3.89E-02
BP	Signal transduction	5.10E-04	MF	Calcium ion binding	7.03E-05
BP	Nervous system development	2.97E-03	MF	Cell adhesion molecule binding	1.65E-04
BP	Regulation of RNA splicing	8.99E-03	CC	Growth cone	1.55E-02
H3K36me3					
BP	Heterophilic cell-cell adhesion	4.04E-07	CC	Neuron projection	4.32E-03
BP	Signal transduction	2.27E-03	MF	Actin binding	7.37E-03
BP	Cell adhesion	1.25E-02	KEGG	Neuroactive ligand-receptor interaction	4.97E-02
BP	Neuron differentiation	3.37E-02			

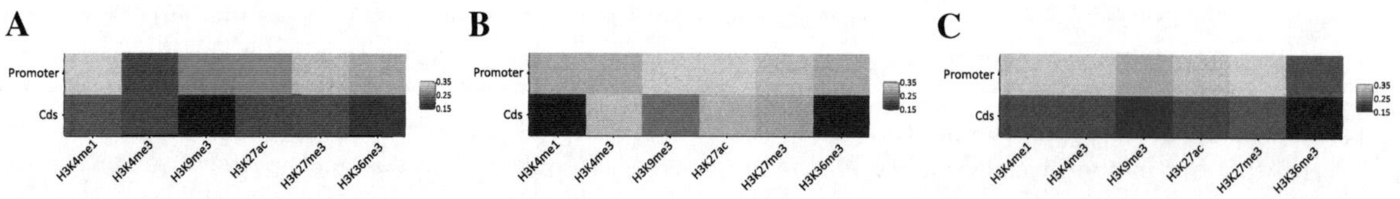

Fig. 4 The correlations between the dynamics of epigenetic modification and that of expression level in HDMSs. **a** The hESCs and hESC-derived precursor group. **b** The primary tissues group. **c** The whole group

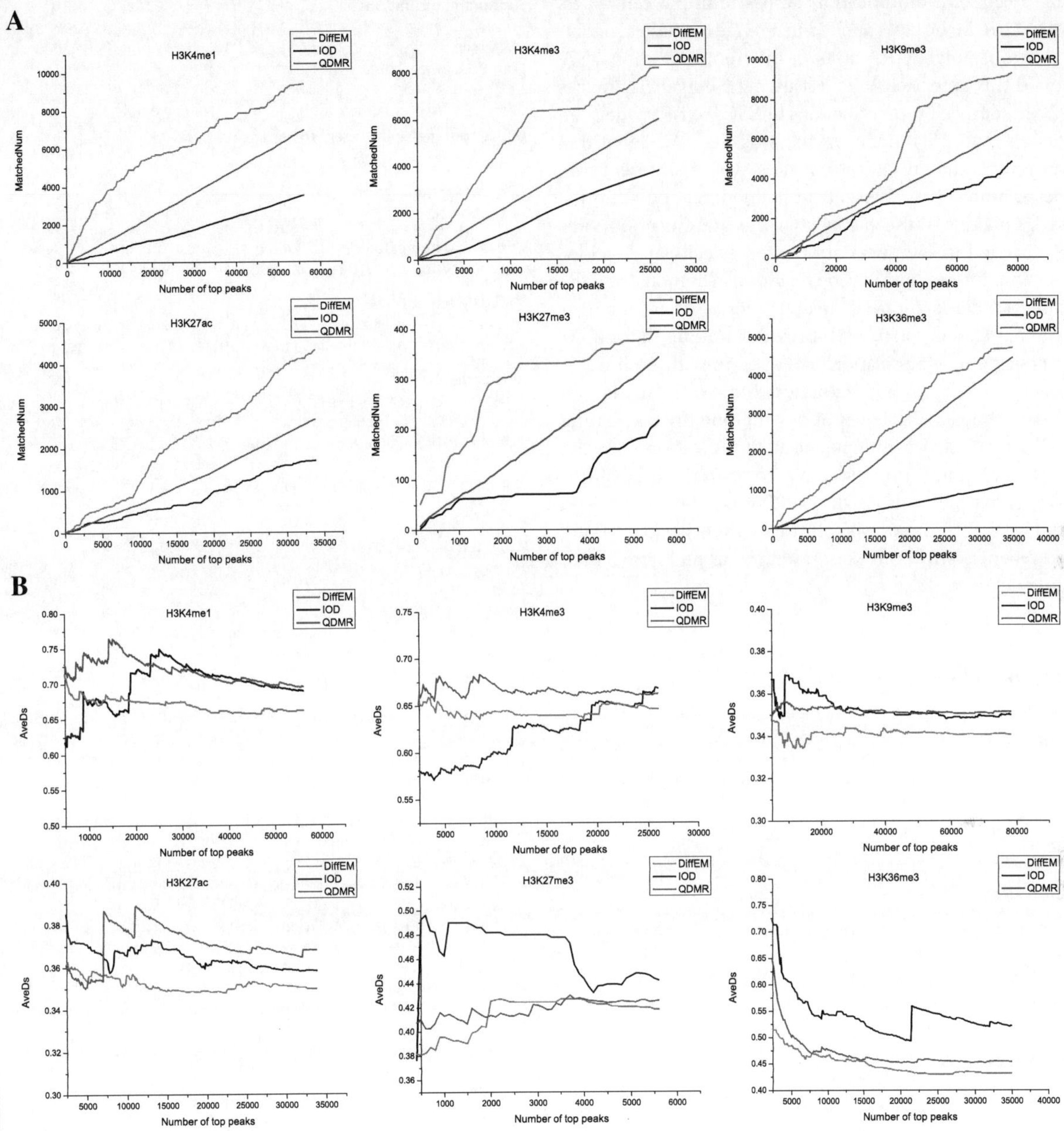

Fig. 5 Performance comparisons among our proposed method, IOD and QDMR for each epigenetic mark. The performance was evaluated by *MatchedNum* and *AveDs*. The higher the Num is, and the lower the Ave is, the better the performance in detecting HDMSs. (**a**) The *MatchedNum*, (**b**) The *AveDss*

find meaningful HDMSs as many as possible. In addition, the overall analysis for *MatchedNum* and *AveDS* shows that IOD may be applicable to detecting the highest HDMSs, because of the commonly small Num but better *AveDS* of related differential gene expression sites. In summary, our method outperforms the two existing methods in identifying the HDMSs across different developmental stages and tissues in the whole genome.

Discussion

In this paper, we proposed a new computational method, DiffEM, based on hamming distance to identify the highly dynamic modification sites that undergo chromatin changes during human cell differentiation process. Different from previous methods that mostly focused on differential analysis between two cell types, our method is designed for differential analysis of genome-wide epigenetic modification across multiple cell types. DiffEM can be broadly applied in a range of studies involving various epigenetic marks in different conditions. We applied this approach to investigating 6 epigenetic marks of 20 human cell types, including hESCs, 4 hESC-derived Lineages and 15 human primary tissues. We identified highly dynamic modification sites where different cell types exhibit distinctive epigenetic modification patterns, and found that these highly dynamic modification sites are enriched in the genes are related to cellular development and differentiation. The results also demonstrate the strong association among the dynamics of different epigenetic marks, consistent with previous finding that different epigenetic modifications collaborate with each other to consummate complex regulatory functions. Further, we evaluated the effectiveness of our method, by correlating the dynamics scores of epigenetic modification with the variance of gene expression. We compared DiffEM with two existing methods, QDMR and IOD. The comparison results indicate the power of our method in quantifying the epigenetic dynamics and identifying highly dynamic regions.

Additional file

Additional file 1: Figure S1. Performance comparisons for primary tissues among our method DiffEM, IOD and QDMR for each epigenetic mark. (A) The MatchedNum, (B) The AveDS. **Figure S2.** Performance comparisons for the whole group among our method DiffEM, IOD and QDMR for each epigenetic mark. (A) The MatchedNum, (B) The AveDS. (PDF 405 kb)

Abbreviations
DAVID: The database for annotation, visualization and integration discovery; GO: Gene ontology; HDMSs: Highly dynamic modification sites; hESCs: Human embryonic stem cells; iHMS: A database integrating human histone modification data across developmental stages and tissues; ME: Mesendoderm; MSCs: Mesenchymal stem cells; NPCs: Neural progenitor cells; TBL: Trophoblast-like cells

Acknowledgements
Authors are grateful to NIH Roadmap Epigenome Project and iHMS website for providing the epigenomic data to carry out this work.

Funding
This work and the publication costs were supported in part by the Fundamental Research Funds for the Central Universities (2232016A3-05), the National Natural Science Foundation of China (61772128, 61772367), National Key Research and Development Program of China (2016YFC0901704) and Shanghai Natural Science Foundation (17ZR1400200,18ZR1414400).

Availability of data and material
Supplementary data are available at Github: https://github.com/xiazhang0513/DiffEM.

About this supplement
This article has been published as part of *BMC Genomics Volume 20 Supplement 2, 2019: Selected articles from the 17th Asia Pacific Bioinformatics Conference (APBC 2019): genomics*. The full contents of the supplement are available online at https://bmcgenomics.biomedcentral.com/articles/supplements/volume-20-supplement-2.

Authors' contributions
YLG and XZ are responsible for the main idea, as well as the completion of the manuscript and experiment. GBZ, SGZ and JHG have coordinated data preprocessing and supervised the effort. All authors have read and approved the final manuscript.

Ethics approval and consent to participate
Not applicable.

Consent for publication
Not applicable.

Competing interests
The authors declare that they have no competing interests.

Author details
[1]School of Computer Engineering and Science, Shanghai University, Shanghai, China. [2]School of Computer Science and Technology, Donghua University, Shanghai, China. [3]Department of Computer Science and Technology,Tongji University, Shanghai, China. [4]Shanghai Key Lab of Intelligent Information Processing, and School of Computer Science, Fudan University, Shanghai, China.

Published: 10 April 2019

References
1. Atlasi Y, Stunnenberg HG. The interplay of epigenetic marks during stem cell differentiation and development. Nat Rev Genet. 2017;18(6990): 643–58.
2. Kouzarides T. Chromatin modifications and their function. Cell. 2007;128: 693–705.
3. Rivera CM, Ren B. Mapping human epigenomes. Cell. 2013;155(1):39–55.
4. Dixon JR, Jung I, Selvaraj S, Shen Y, Antosiewicz-Bourget JE, Lee AY, et al. Chromatin architecture reorganization during stem cell differentiation. Nature. 2015;518(7539):331–6.
5. Mikkelsen TS, Ku M, Jaffe DB, Issac B, Lieberman E, Giannoukos G, Alvarez P, Brockman W, Kim TK, Koche RP. Genome-wide maps of chromatin state in pluripotent and lineage-committed cells. Nature. 2007;448(7153):553.
6. Tsankov AM, et al. Transcription factor binding dynamics during human es cell differentiation. Nature. 2015;518(7539):344–9.
7. Ernst J, Kheradpour P, Mikkelsen TS, Shoresh N, Ward LD, Epstein CB, Zhang X, Wang L, Issner R, Coyne M, et al. Mapping and analysis of chromatin state dynamics in nine human cell types. Nature. 2011;473(7345):43–9.
8. Roadmap EC, Kundaje A, Meuleman W, Ernst J, et al. Integrative analysis of 111 reference human epigenomes. Nature. 2015;518(7539):317–30.
9. Gifford CA, Ziller MJ, Gu H, Trapnell C, Donaghey J, Tsankov A, Shalek AK, Kelley DR, Shishkin AA, Issner R, et al. Transcriptional and epigenetic dynamics during specification of human embryonic stem cells. Cell. 2013;153(5):1149–63.
10. Xie W, Schultz MD, Lister R, et al. Epigenomic analysis of multilineage differentiation of human embryonic stem cells. Cell. 2013;153(5):1134–48.
11. Thomson JA, Itskovitz-Eldor J, Shapiro SS, Waknitz MA, Swiergiel JJ, Marshall VS, Jones JM. Embryonic stem cell lines derived from human blastocysts. Science. 1998;282(5391):1145.
12. Vodyanik MA, Yu J, Zhang X, Tian S, Stewart R, Thomson JA, Slukvin II. A mesoderm-derived precursor for mesenchymal stem and endothelial cells. Cell Stem Cell. 2010;7(6):718.
13. Yu P, Xiao S, Xin X, Song C-X, Huang W, McDee D, Tanaka T, Wang T, He C, Zhong S. Spatiotemporal clustering of the epigenome reveals rules of dynamic gene regulation. Genome Res. 2013;23(2):352–64.
14. Xu H, Wei CL, Lin F, Sung WK. An hmm approach to genome-wide identification of differential histone modification sites from chip-seq data. Bioinformatics. 2008;24(20):2344–9.
15. Shen L, Shao NY, Liu X, Maze I, Feng J, Nestler EJ. diffreps: Detecting differential chromatin modification sites from chip-seq data with biological replicates. PloS ONE. 2013;8(6):65598.

16. Ji H, Li X, Wang QF, Ning Y. Differential principal component analysis of chip-seq. Proc Natl Acad Sci U S A. 2013;110(17):6789.
17. Heinig M, Colom-Tatch M, Taudt A, Rintisch C, Schafer S, Pravenec M, Hubner N, Vingron M, Johannes F. histonehmm: Differential analysis of histone modifications with broad genomic footprints. BMC Bioinformatics. 2015;16(1):1–15.
18. Lun ATL, Smyth GK. csaw: a bioconductor package for differential binding analysis of chip-seq data using sliding windows:. Nucleic Acids Res. 2016;44(5):45.
19. Ashoor H, Louis-Brennetot C, Janoueix-Lerosey I, Bajic VB, Boeva V. Hmcan-diff: a method to detect changes in histone modifications in cells with different genetic characteristics. Nucleic Acids Res. 2017;45(8):58.
20. Chen C, Zhang S, Zhang XS. Discovery of cell-type specific regulatory elements in the human genome using differential chromatin modification analysis. Nucleic Acids Res. 2013;41(20):9230.
21. Yang X, Shao X, Gao L, Zhang S. Systematic dna methylation analysis of multiple cell lines reveals common and specific patterns within and across tissues of origin. Hum Mol Genet. 2015;24(15):4374.
22. Zhang Y, Liu H, Lv J, Xiao X, Zhu J, Liu X, Su J, Li X, Wu Q, Wang F. Qdmr: a quantitative method for identification of differentially methylated regions by entropy. Nucleic Acids Res. 2011;39(9):58.
23. Bhasin JM, Bo H, Ting AH. Methylaction: detecting differentially methylated regions that distinguish biological subtypes. Nucleic Acids Res. 2016;44(1):106–16.
24. Gan Y, Han T, Guan J, Zhou S. ihms: a database integrating human histone modification data across developmental stages and tissues. BMC Bioinformatics. 2017;18(1):103.
25. Gan Y, Tao H, Zou G, Yan C, Guan J. Dynamic epigenetic mode analysis using spatial temporal clustering. BMC bioinformatics. 2016;17(17):537.
26. Zhang Y, An L, Yue F, Hardison RC. Jointly characterizing epigenetic dynamics across multiple human cell types. Nucleic Acids Res. 2016;44(14):6721–31.
27. Pinello L, Xu J, Orkin SH, Yuan GC. Analysis of chromatin-state plasticity identifies cell-type-specific regulators of h3k27me3 patterns. Proc Natl Acad Sci U S A. 2014;111(3):344.
28. Wang H, Liu C, Deng L. Enhanced prediction of hot spots at protein-protein interfaces using extreme gradient boosting. Sci Rep. 2018;8(1):14285.
29. Pan Y, Liu D, Deng L. Accurate prediction of functional effects for variants by combining gradient tree boosting with optimal neighborhood properties. PloS ONE. 2017;12(6):0179314.

Park et al. BMC Genomics 2019, 20(Suppl 2):188
https://doi.org/10.1186/s12864-019-5473-z

BMC Genomics

RESEARCH Open Access

A secure SNP panel scheme using homomorphically encrypted K-mers without SNP calling on the user side

Sungjoon Park[1], Minsu Kim[2], Seokjun Seo[3], Seungwan Hong[4], Kyoohyung Han[4], Keewoo Lee[4], Jung Hee Cheon[4] and Sun Kim[1,2,5*]

From The 17th Asia Pacific Bioinformatics Conference (APBC 2019)
Wuhan, China. 14–16 January 2019

Abstract

Background: Single Nucleotide Polymorphism (SNP) in the genome has become crucial information for clinical use. For example, the targeted cancer therapy is primarily based on the information which clinically important SNPs are detectable from the tumor. Many hospitals have developed their own panels that include clinically important SNPs. The genome information exchange between the patient and the hospital has become more popular. However, the genome sequence information is innate and irreversible and thus its leakage has serious consequences. Therefore, protecting one's genome information is critical. On the other side, hospitals may need to protect their own panels. There is no known secure SNP panel scheme to protect both.

Results: In this paper, we propose a secure SNP panel scheme using homomorphically encrypted K-mers without requiring SNP calling on the user side and without revealing the panel information to the user. Use of the powerful homomorphic encryption technique is desirable, but there is no known algorithm to efficiently align two homomorphically encrypted sequences. Thus, we designed and implemented a novel secure SNP panel scheme utilizing the computationally feasible equality test on two homomorphically encrypted K-mers. To make the scheme work correctly, in addition to SNPs in the panel, sequence variations at the population level should be addressed. We designed a concept of Point Deviation Tolerance (PDT) level to address the false positives and false negatives. Using the TCGA BRCA dataset, we demonstrated that our scheme works at the level of over a hundred thousand somatic mutations. In addition, we provide a computational guideline for the panel design, including the size of K-mer and the number of SNPs.

Conclusions: The proposed method is the first of its kind to protect both the user's sequence and the hospital's panel information using the powerful homomorphic encryption scheme. We demonstrated that the scheme works with a simulated dataset and the TCGA BRCA dataset. In this study, we have shown only the feasibility of the proposed scheme and much more efforts should be done to make the scheme usable for clinical use.

Keywords: SNP panel, Homomorphic encryption, K-mer, Genomic security, Genomic privacy

*Correspondence: sunkim.bioinfo@snu.ac.kr
[1]Department of Computer Science and Engineering, Seoul National University, Seoul, Republic of Korea
[2]Interdisciplinary Program in Bioinformatics, Seoul National University, Seoul, Republic of Korea
Full list of author information is available at the end of the article

Background

Single Nucleotide Polymorphism (SNP) is crucial information in medical sciences than ever before. A single aberrant nucleotide variation can incur dysfunction in a biological process, affecting individual vulnerability to certain diseases. Hence, SNP existence can be utilized to diagnose diseases. Sometimes, SNPs help determine effective treatment, especially in cancer. From The Cancer Genome Atlas (TCGA) project, numerous driver mutations are reported in many cancer types [1] and panels using the curated mutation information have been developed [2]. Furthermore, genetic disorders in Mendelian diseases are very well studied [3] and thus SNP detection can be directly translated into the contribution to actual medical applications. Because of the importance of SNPs to disease, the US NIH has compiled a database called ClinVar [4].

The utility of SNPs goes beyond the medicine domain. People are diverse in the genomic content, thus the difference in genomic content among people can be used to identify a specific person. For such reason, SNPs are often used for legal and forensic purposes. There are Direct-to-Customer SNP kits that are designed for non-medical use such as pedigree search. More people measure their genomes and use the information for various purposes.

As genomic data are widely used and disclosed, a potential threat to genomic privacy is a serious problem. DNA sequence is a blueprint of a human being since the information includes not only medical conditions but to the extent of even reconstructing facial model which is one of the most sensitive information of human being [5]. Our knowledge on the human genome is very limited as of now and much more information encoded in the genome will be disclosed over the years. Thus, genomic information is critical and will cause severe damage to individuals if leaked. This could also lead to a social crisis because genomic data can be used to justify discrimination among people. Unlike other confidential data, genomic data are innate and immutable, making the damage irreversible and permanent throughout one's lifetime. Due to its far-reaching and sensitive nature, the genomic data is prone to be monetized. For these reasons, It is very important to protect the genomic information from hackers, insurance establishments, hospitals, pharmaceuticals, government and all the possible threats yet to come.

Among recent measures to protect data, homomorphic encryption is a technology that gains attention recently. It refers to an encryption scheme that allows the third party to perform computations while not knowing any content of the original source data or the private key. If such computation includes addition and multiplication, then theoretically all computations can be performed and that is called fully homomorphic encryption. The result of the computation is also returned encrypted, thus the third party can provide its inference based on the data and at the same time cannot know the data content or perform inference at all. The first fully homomorphic encryption scheme is proposed by Gentry at 2009 [6], and continually improved its efficiency.

For example, consider that Alice is running a company and Bob provides a cloud storage service. Alice stores her encrypted data on Bob's cloud. When Alice wants to compute aggregate data, an average revenue per month for instance, she has to decrypt the data if it is traditionally encrypted. Decrypting on the cloud storage may reveal both the private key and the source data to Bob. Downloading the encrypted data to local and decrypting it goes against the purpose of utilizing cloud service in the first place. This is when homomorphic encryption comes useful. If the data is homomorphically encrypted, Alice does not have to decrypt the source data on Bob's cloud to do the computation needed. Instead, the encrypted form of aggregate data can be computed as ciphertexts and downloaded to Alice's computer, and then safely decrypted. This way, Alice can exploit Bob's resource to compute the average revenue without giving any information to Bob.

In more formal notation, an encryption scheme is additive homomorphic encryption if and only if

$$\forall p\, \forall q\, \exists \odot : \mathcal{E}(p) \odot \mathcal{E}(q) = \mathcal{E}(p+q)$$

given plaintexts p, q, and $\mathcal{E}(\cdot)$ an encryption procedure. $\mathcal{E}(p+q)$ denotes a ciphertext that can be decrypted to plaintext $p+q$ with the private key. Multiplicative homomorphic encryption works likewise.

Related work

Given the importance of genome information, there has recently been active research on genome security. Naveed et al. [7] presented the history of genomics and related privacy issues including the homomorphic encryption. Dowlin et al. [8] also demonstrated how homomorphic encryption and security could be used in the fields of bioinformatics. In detail, we categorize recent genomic security research works into three major groups: (1) differential privacy, (2) secure system design with traditional encryption scheme and (3) homomorphic encryption scheme.

Differential privacy includes de-identification, which refers to making genomic data unidentifiable by either anonymizing or discarding personally differential information. This aims to perturb the information so that any leak of data itself would not possibly lead to identifying the patient. However, it is shown in many papers that generic de-identification techniques are not powerful enough to prevent reconstructing identity [9–13]. To mitigate the risk, many improvements were made based on the domain knowledge of the genomic data [14–19].

The secure system design with traditional encryption focuses on controlling the flow of sensitive information. It relies on secret sharing of private keys with multiple parties that do not collude to ensure the confidentiality of the data. Canim et al. [20] suggested secure operations based on a cryptographic coprocessor. Kamm et al. [21] adopted multiple third parties to securely perform GWAS analyses in a distributed way. On the other hand, Xie et al. [22] proposed a statistical approach called meta-analysis to recall aggregate features with reduced privacy risks. In the work of Wang et al. [23], secure and efficient computation of genomic edit distance and querying similar sequence based on that is introduced.

The application of homomorphic encryption dates back only to few years since it is a new technology. Troncoso-Pastoriza et al. [24] proposed error resilient private string search algorithm that is specially designed for DNAs using homomorphic encryption. Kantarcioglu et al. [25] also adopted homomorphic encryption to securely share the aggregate data of genome sequence among researchers. Ayday et al. [26, 27] proposed methods to query the disease susceptibility with clinical data encrypted. More recently, Kim et al. [28] showed that homomorphic encryption can be used to obtain minor allele frequencies, χ^2 statistic in GWAS and edit distance of sequences in a secure way. Lu et al. [29] and Zhang et al. [30] encrypted phenotype and genotype homomorphically and then was able to infer typical GWAS statistic. On the other hand, Wang et al. adopted homomorphic encryption on rare variants to perform homomorphic exact logistic regression [31]. Raisaro et al. [32] showed retrieving aggregate data computed under homomorphically encrypted data that is deployed to real world application on i2b2 data warehouse. Jagadeesh et al. [33] also have shown secure SNP data sharing between hospitals to induce meaningful inference of disease. Other most recent works on homomorphic encryption include the works of Jacquez et al. [34], Ghasemi et al. [35], and Cheng et al [36].

As described above, research on genomic privacy has been active and advanced significantly over the years. However, current research has been conducted assuming situations with some compromises. De-identification more or less manipulates the content of data and thus has possibility to contaminate the original information. The secure system design distributes the secret key and the computation to multiple third parties. However, either the trust or the resources for computing may not be available in reality. If the third parties are untrustworthy, they may collude to jeopardize the private system. Likewise, the shortage of resources such as storage, bandwidth and processing power is critical for such system to maintain.

Even the works with homomorphic encryptions have their own limitations, due to its inefficient nature. Most applications [25–27, 30–32] encrypt a clean, annotated SNP existence information on the user side or clinic data as a plaintext. In this case, SNP calling should be done on the user side and then the SNP calling information is sent to the hospital. Some applications send sequences to the hospital [25–29, 37]. In this case, the querying result is limited to aggregate genomic data and clinical data. How to use these techniques for a secure SNP panel has not been explored.

Motivation and contribution

In this paper, we propose a patient-to-hospital SNP panel scheme and its architecture is illustrated in Fig. 1. In our two party model, **the patient** has one's raw sequence and the private key, and **the hospital** has computing power and the SNP panel. SNP panel refers to a tool owned by the hospital that can find the specific combination of SNPs.

A possible scenario is that the patient sends the raw sequence to the hospital to detect SNP existence. In order to protect data from being stolen in the middle, the transferring raw sequence must be encrypted. In traditional patient-to-hospital model, the sequence is encrypted using the hospital's public key. This method reveals the patient's raw sequence. On the other hand, with homomorphic encryption, the patient can send the raw sequence encrypted with his/her own public key. Plus, the homomorphic nature of encryption allows the hospital to match encrypted sequence with its SNP panels without the private key or knowing the raw sequence.

Another thing to note is that, homomorphic encryption is slow and expensive that it cannot be applied to raw sequence for practical usage due to its computational inefficiency. In addition, although additive and multiplicative homomorphic encryption schemes is considered fully homomorphic in theory, its functionality is limited in reality due to their computationally demanding nature. Thus, using domain knowledge of genomic sequence, we devise a patient-to-hospital communication protocol onto which homomorphic encryption is applicable in order to securely detect SNPs without revealing the patient's raw sequence and the hospital's SNP panel assets to each other.

Thus, we propose a secure SNP panel scheme using homomorphically encrypted K-mers without SNP calling on the user side. The major challenge is that there is no known algorithm to align two homomorphically encrypted sequences as whole. The basic idea to overcome this challenge is to utilize the equality testing operation on two homomorphically encrypted K-mers that is computationally feasible under the homomorphic encryption scheme. The major features of our scheme is as below:

1 We introduce a secure scheme that uses homomorphically encrypted K-mer, a short subsequence of the raw sequence, and show that encrypted K-mers can detect SNPs as good as the raw

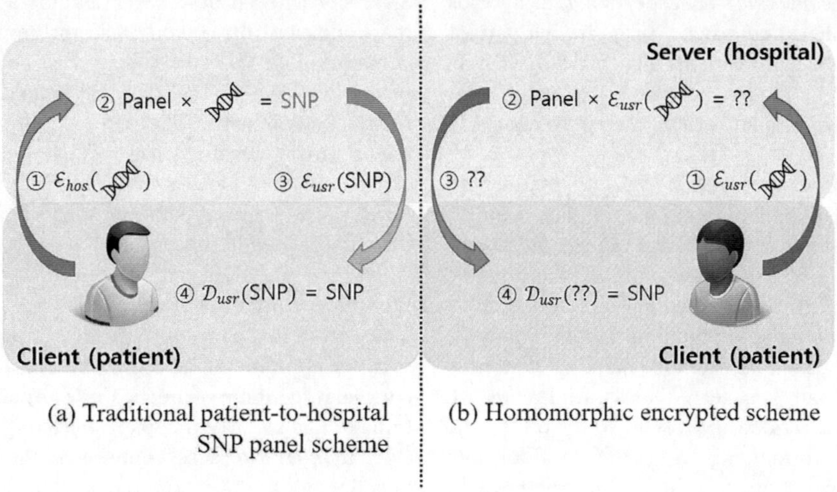

Fig. 1 The traditional and homomorphic SNP detection scheme. In this figure, $\mathcal{E}(\cdot)$ and $\mathcal{D}(\cdot)$ denote encryption and decryption respectively. Encryption needs public key while decryption needs private key. The subscript refers to the owner of the private key. For example, $\mathcal{E}_{hos}(\cdot)$ means the data is encrypted with hospital's public key thus only the hospital can decrypt it. **a** demonstrates traditional way of detecting SNP. ① The patient gives the hospital the raw sequence. In order to protect data from being stolen in the middle, the sequence is encrypted using the hospital's public key. ② The hospital decrypts patient sequence and performs computation to detect SNP. ③ The hospital returns the SNP existence information encrypted with the patient's public key to the patient. ④ The patient can decrypt and get the SNP information. **b** on the other hand, demonstrates the same SNP detection scheme in homomorphic way. ① The patient sends one's encrypted raw sequence but this time with the public key. ② Due to its homomorphic property, the hospital can perform computations on the sequence only with the public key, without decrypting the sequence. ③ The result is acquired in encrypted form, and the hospital returns the result to patient not knowing its content. ④ The patient can decrypt the result in secure environment

sequence does under certain conditions. With properly tuned value of K, exploiting K-mer can achieve small error bounds and practical runtime at the same time.

2 Our method considers genome variations among individuals. We assayed TCGA breast cancer patient data to estimate individual variation ratio. We further define and compute false positive rate and false negative rate of our SNP detection scheme and either suggest a method to control or show that the error is bounded to an ignorable value.
3 Our contribution is also to the extent of providing a SNP panel design guideline. When a hospital selects SNP residues for the diagnosis of a certain disease and wants to use our secure SNP panel scheme, our method suggests guidelines on which SNP residues under consideration can be used.

Methods

Data description and panel preprocessing

In actual clinical case, a SNP *panel* consists of multiple SNP *residues*, where each SNP residue corresponds to a pre-determined disease-associated SNP residue (Fig. 1). A targeted DNA sequencing is then performed to assess SNP status of each residue.

For our study, we generated a synthesized dataset that simulates the aforementioned case. In the dataset, the SNP residues are randomly sampled from coding sequences of the hg19 reference genome [38], where refSeq gene annotation is used to specify coding regions (downloaded from UCSC genome database [39]). First, we chose 1000 SNP residues sampled from the hg19 genome, then randomly combined them into various size of panels. As a result, we generated panels having 10, 20, ..., 100 residues randomly selected from the pool of 1000 SNP residues. Hence, 10 different SNP panels of size ranging from 10 to 100 were generated.

After combining 10 different SNP panels, we simulated a massive parallel sequencing data (or DNA-seq) using WgSim. From the 1300 bp length flanking sequences (650 bp for each side) from each SNP residue, short-read sequences (151 bp × 2) were simulated. Here, the 1300 bp and "151 bp × 2" parameters were set to simulate the actual targeted short-read sequencing condition. The exact parameter for WgSim is "-e 0 -1 151 -2 151 -r 0 -R 0 -X 0 -S 0 -N [VAR]", where all the parameters for random sequencing errors are set to none and the throughput parameter "-N" depends on the size of each panel. The sequencing depth was set to 10, meaning that 10 sequencing reads are expected to cover each SNP residue in average. So, to cover more residues, it requires more sequencing reads to simulate (hence larger "-N"). Lastly, each residue has 50% chance of nucleotide substitution to simulate SNPs by design. For instance, if we check any SNP residue, we can expect to find 5 reads having reference allele and other 5 reads having variant allele as planted. In summary, we generated 10 DNA-seq data corresponding to 10 SNP panels having different combination of SNP residues, each having average 50% substitution rate.

There could be unwanted overlap of flanking sequences between two different SNP residues. The overlap between two SNP residues would require a large value K. Therefore, we either merge the overlapped residues or discard one of them. This problem is described in Fig. 2. If two SNP residues are fully overlapped without mismatch, we merge them into a single entity having extended flanking sequences to prevent unnecessary long search for Ks. And if not, we randomly chose one residue and discarded the other one. To avoid promiscuous merging, we set the cutoff for fully overlap situation to 100 bp. Assuming a uniform random distribution, the probability that two flanking sequences overlap 100 bp by chance is $1 - \left(1 - 2 \cdot 4^{-100}\right)^2$. The computed p-value is negligible. Therefore, it is safe to merge the residues satisfying this condition. However, if an extreme case occurs that flanking sequences of two SNP residues overlap more than 150 bp, which is longer than short-read length, we treat them same as the case of partial overlap situation, hence they are discarded.

The handling of overlapping sequence enables that, even if more than two SNP residues exist near, possibly within the range of K so that there exists a K-mer which has both SNPs, the proposed SNP panel scheme finds both SNPs correctly. It only has to enumerate all possible K-mers and perform equality test on the user K-mers with corresponding SNP labels.

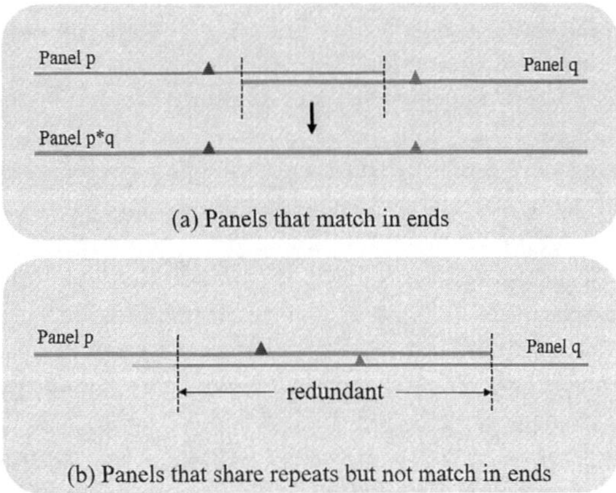

Fig. 2 Two possible cases where different SNP residues have an overlap in their flanking sequences. There can be two different cases where an overlapping occurs between flanking sequences of two different SNP residues. **a** Fully overlapped without mismatch, **b** Partially overlapped with mismatch. In case of (**a**), two SNP residues are merged into a single sequence to prevent unnecessary long search for K's. However, in case of (**b**), we randomly chose one residue and discarded the other one

Now we have a panel that will be used for testing our secure SNP scheme. The panel should be designed by doctors for a specific disease. What we have shown is a computational scheme to test whether a panel with many residues can be used with our secure SNP panel scheme that is proposed in this paper.

Workflow: a bird's eye view

The goal of this paper is to propose a patient-to-hospital secure communication protocol such that a patient would conceal its genome sequence while requesting SNP detections to a hospital with a panel. Specifically, we are interested in separating access between to the SNP panel at the hospital and to the genomic sequence of the patient. The SNP panel information is an asset of the hospital which may not be revealed to the public. On the other hand, the user does not want the hospital to know one's sequence. The goal is to protect both.

In our scheme illustrated in Fig. 3, the patient has one's raw genomic sequence and the hospital has panel sequence. To apply homomorphic encryption to sequences, we chunk both the raw sequence of the patient and the panel sequence of the hospital into K-mers. In other words, we get K-mers from substring of the sequences with a sliding window of stride 1, size K. What we want to do is encrypt K-mers and perform **homomorphic equality tests** on two K-mers. The patient would receive an encrypted result of SNP calling from the hospital. If a K-mer from patient sequence matches a K-mer from the panel sequence annotated with a specific SNP, the patient can tell which SNP one has. Further medical diagnosis can be done based on the test result.

Note that the patient encrypts the K-mers and holds the private key so that the hospital does not gain any information about the patient's sequence. Meanwhile, the computation is done on the hospital's side, securing SNP panel assets from leakage.

K-mer design

Our SNP detection scheme works under specific conditions. To determine SNP existence by performing K-mer equality testing, the corresponding K-mer should be unique throughout the whole SNP panel sequences. Otherwise, K-mer matching cannot guarantee the existence of SNP. Thus the core idea of K-mer design is to set K long enough for all K-mers that contain any panel SNP to be distinguishable. However, while long K ensures unique K-mers, longer K-mers will be computationally expensive.

Therefore, given a panel we computed panel-specific minimum value of K such that all K-mers in the panel

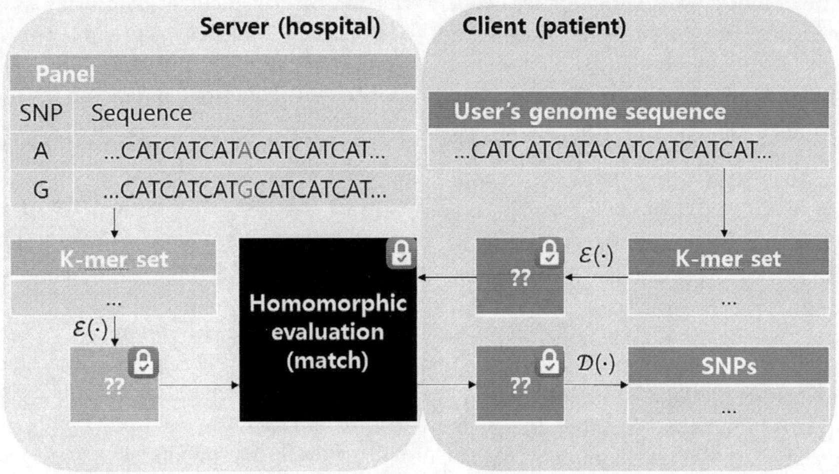

Fig. 3 The overall architecture of our new SNP detection scheme. This figure depicts the general process of proposed SNP detection scheme. The set of panel sequences noted in yellow and the patient genome sequence noted in blue are chunked down into K-mers, respectively. After we encrypt K-mers, the client-side patient's encrypted K-mers are transmitted to the server. The black box depicted in the middle denotes the homomorphic evaluation procedure. Here, we check the matching correspondence between any member in panel K-mer set and in the patient K-mer set. The result noted green is returned to the patient. The patient can decrypt the result with the private key securely on the client-side environment

that have a SNP are distinct. To achieve this goal, first K-merize all panel sequences and group them into two sets: (A) K-mers with SNPs and (B) K-mers without. Then, following conditions must hold for K-mers with SNPs to be distinct.

1 K-mers in set (A) are unique (have no duplicate)
2 K-mers in set (B) do not appear in set (A)

Among the Ks that satisfy the above conditions, the minimum value is chosen.

False positive/negative errors

In our system, K of K-mer is defined to make all K-mers distinguishable and thus no false positive or negative errors. However, we assumed that individual patient sequence derived from SNP panel varies only at the interested SNP residue and the flanking sequences are identical for all patients. This is not the case when *the variations at the population level* are considered. An actual sequence derived by SNP panel may present unannotated deviations from what is known to the panel, other than SNP residues in it. Examples include point mutation, individual genome variation and sequencing errors. Presence of such deviations may result in false positive and false negative errors. False positive error occurs when the scheme identifies a K-mer and declares a SNP existence but actually the patient does not have one. False negative error occurs when the patient indeed has a SNP yet the scheme fails to detect one. We henceforth refer to point deviation as a single nucleotide deviation from intended panel sequence, resulting from either point mutation, individual variation or sequencing error.

False positive occurs when the determined K-mer did not originate from the flanking sequence around the found SNP. Rather, it originated from other irrelevant parts of patient genome sequence that correspond to any other SNP residue in the processing panel. To deal with false positive errors, we devised **Point Deviation Tolerance (PDT) level**. Previously in the process of computing K, we had two conditions on K-mers. Both conditions utilized the equality test to check uniqueness of K-mers in (A) and exclusiveness of K-mers in (A) over those in (B). We generalize the equality test to the hamming distance check with its lower bound being PDT. In other words, we apply strict rules and regard similar sequences ambiguous. Here, being similar is defined by PDT point deviations. The uniqueness condition is the specific case of PDT being 0. Thus we can rewrite the K-mer conditions as below:

1 K-mers in set (A) are distinct (any pairs' hamming distance greater than PDT)
2 K-mers in set (B) are distinct from K-mers in set (A)

Given PDT, one can determine a minimum value K satisfying the updated conditions. PDT works as a safety margin to K-mer ambiguity. The point deviation, namely the sum of point mutations, individual variation and sequencing errors, is allowed up to maximum PDT. Therefore, if the aggregate point deviation occurs less than or equal to PDT, false positive cases do not appear.

On the contrary, we cannot prevent false negative cases. False negative happens when the system cannot determine a SNP when it is truly in the patient's sequence. The major cause of this is also the point deviation in the sequences flanking a SNP. It is infeasible to prepare all the variant K-mers as we did to cope with the false positive errors. In this paper, we assayed TCGA BRCA data to determine the distance distribution of somatic mutations among patients. Based on the data, we estimate the probability of a point mutation lying on K-mers to estimate false negative errors.

RLWE cryptosystem

In this paper, we used HEAAN (Homomorphic Encryption for Arithmetic of Approximate Numbers) library to implement the equality test on two homomorphically encrypted K-mers. HEAAN is based on RLWE (Ring-Learning With Error) encryption scheme. RLWE is a variation of LWE (Learning With Error) problem, which is a lattice-based cryptography. LWE exploits its hardness assumption to ensure security which follows:

$$a \leftarrow \mathbb{Z}_q^n, s \leftarrow \mathbb{Z}_q^n, e \leftarrow \chi^n, r_1, r_2 \leftarrow \mathbb{Z}_q^n : (a, \langle a, s\rangle + e) \approx^c (r_1, r_2)$$

In other words, for some secret key s and some error distribution χ, the relation between a and $\langle a, s\rangle + e$ are computationally indistinguishable from random numbers. RLWE uses polynomial integer rings instead of vectors. Namely, $\mathbb{Z}_q^n$ s are replaced with $\mathbb{Z}_q[X]/\Phi_m(X)$ for $n = \phi(m)$, where $\Phi(\cdot)$ is cyclotomic polynomial and $\phi(\cdot)$ is Euler's phi function. RLWE is estimated to achieve equal or less level of security compared to LWE. Other parameters for the scheme are p for message modulus, q ciphertext modulus and ring $R = \mathbb{Z}/\Phi_M(X)$ for integer M. We further denote by $R_q = R/qR$ and χ for error distribution. The scheme for cryptography used throughout this paper is described in detail below.

SKGen(params) Choose random $s(X) \leftarrow \chi$, and set sk $= \vec{s} = (1, s) \in R_q^2$.

PKGen(params, sk) Choose random $a(X), a'(X) \leftarrow R_q$, $e(X), e'(X) \leftarrow \chi$, and set $b(X) = -a(X)s(X) + pe(X) \in R_q$ and $b'(X) = -a'(X)s'(X) + pe'(X) \in R_q$. The public key is pk $= (b, a) \in R^2$ and the evaluation key is evk $= (b' + ps^2, a') \in R_q$.

Enc(pk, $m \in R_p$**)** Choose $v(X), e_0(X), e_1(X) \leftarrow \chi$ and let $c_1(X) = m(X) + v(X)b(X) + pe_0(X), c_2(X) = v(X)a(X) + pe_1(X)$. Return $\vec{c} = (c_1, c_2) \in R_q^2$.

Dec(sk, $\vec{c}$**)** Return $[\langle \vec{c}, \text{sk}\rangle$.

Add($\vec{c}_1, \vec{c}_2$**)** Return $\vec{c}_{\text{add}} = \vec{c}_1 + \vec{c}_2$.

Mult($\vec{c}_1, \vec{c}_2$**, evk)** For $\vec{c}_1 = (b_1, a_1)$ and $\vec{c}_2 = (b_2, a_2)$. Return $\vec{c}_{\text{mult}} = (b_1 b_2, b_1 a_2 + b_2 a_1) + a_1 a_2 \cdot evk \in R_q$.

RLWE-based homomorphic encryption supports batching (or SIMD) encoding and data array movement. If we call the each element of data array as slot, the scheme has permutation of message slots. This functionality can be used to make our homomorphic evaluation algorithm more efficient and split and merge DNA information.

KeySwitchingMatrixGen(params, sk$_1 \rightarrow$ **sk**$_2$**)**
KeySwitch(c, **KS**$_{\text{sk}_1 \rightarrow \text{sk}_2}$**)**
KeySwitchingMatrixGen(params, sk $(X^k) \rightarrow$ **sk**(X)**)**

Automorphism$\left(c, X \rightarrow X^k, \text{ KS}_{\text{sk}(X^k) \rightarrow \text{sk}(X)}\right)$

Data encoding and encryption

To perform equality tests of K-mers in numerical system, we regard each K-mer as a quaternary number via mapping each nucleobase A, C, G, T to 0, 1, 2, 3, respectively and encode K-mers into integers. In this view, for example, if "GACT" is a K-mer of length 4, then it corresponds to $K = 2013_{(4)} = 2 \times 4^3 + 0 \times 4^2 + 1 \times 4^1 + 3 = 135$.

However, encoding DNA sequence of length L wholly as an integer is inefficient when L is large, requiring a huge set of scheme parameters. To achieve a better performance, we suggest a method of breaking K-mers into smaller blocks of same length and performing equality tests for each block simultaneously. Henceforth for N user-side K-mers and M panel-side K-mers divided into B blocks respectively, we denote the data as following:

- n-th user-side K-mer : $K_{usr}^{(n)}$, or $K^{(n)}$ when obvious. $(n = 0, 1, \cdots, N-1)$
- m-th panel-side K-mer : $K_{ref}^{(m)}$ $(m = 0, 1, \cdots, M-1)$
- b-th block of $K_{usr}^{(n)}$: $K_{usr}^{(n)}[b]$, or $K^{(n)}[b]$ when obvious. $(b = 0, 1, \cdots, B-1)$
- size of a block: L

Once the set of user-side K-mers is ready, we can encode all user K-mers into B vectors. Specifically, b-th blocks of N user-side K-mers $\{K^{(n)}[b]\}_{n \in [N]}$ are encoded into a single vector $\vec{v}_b = (\vec{v}_b[0], \cdots, \vec{v}_b[\texttt{slots}-1])$, where each component of $\vec{v}_b$ is defined as

$$\vec{v}_b[i] = \begin{cases} 1, & \text{if } i = K^{(n)}[b] \cdot N + n \text{ for some } n \in [N] \\ 0, & \text{otherwise} \end{cases}.$$

Here, we choose the dimension of vector `slots` by smallest power of 2 which does not exceed $4^L \cdot N$, or the maximum of $K^{(n)}[b] \cdot N + n$. It is noteworthy that $K^{(n)}[b] \cdot N + n = K^{(m)}[b] \cdot N + m$ if and only if $n = m$, since $n, m < N$.

HEAAN supports a technique to pack k complex numbers in a single polynomial using a variant of the complex canonical embedding map $\phi : \mathbb{C}^k \rightarrow \mathcal{R}$. We make use of the technique and encrypt each of $\vec{v}_b$'s as a single ciphertext $\vec{c}_b$. An example of parallel K-mer encryption is depicted in Fig. 4.

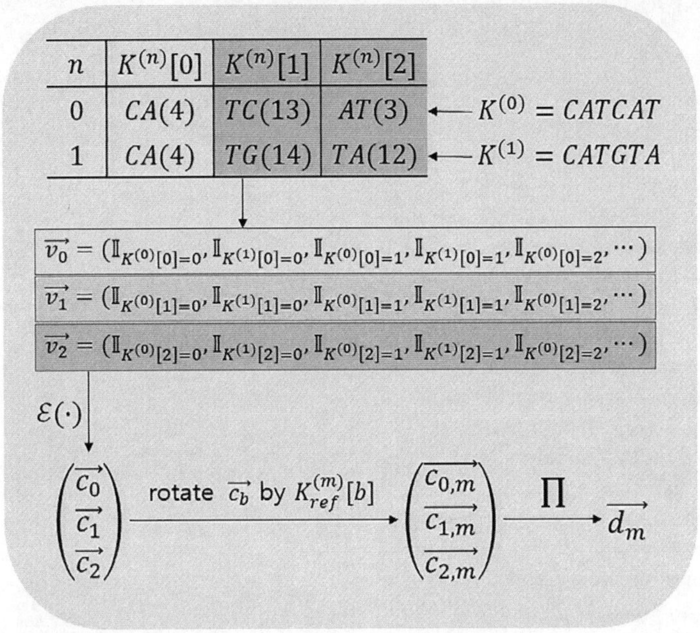

Fig. 4 A detailed example to how multiple K-mers are encoded into vectors in parallel. This figure depicts how K-mers are divided into small blocks and then encrypted to vectors. In this example, two 6-mers $K^{(0)} = CATCAT$ and $K^{(1)} = CATGTA$ are encoded into $B = 3$ blocks each of size $L = 2$ to reduce the size of ciphertext space. Here, the value of `slots` is a power of 2 bounded by $(4^2) \cdot 2$. The subscript b of $\vec{v}_b$ indicates the index of blocks encoded starting from 0. The values of elements in the vectors are indicators to $K^n[b] = \frac{i-n}{2}$. In this sense, $\vec{v}_0[16]$, $\vec{v}_0[17]$, $\vec{v}_1[52]$, $\vec{v}_1[57]$, $\vec{v}_2[12]$, and $\vec{v}_2[49]$ are 1's. The vectors $[\vec{v}_b]$ are encrypted into polynomials $[\vec{c}_b]$ and then rotated by corresponding value of $K_{ref}^{(m)}$. This rotation ensures that the first N blocks indicate the agreement of N b-th block of K-mers and b-th block of $K_{ref}^{(m)}$. Later, these values are multiplied in component-wise manner. Therefore, $\vec{d}[n]$ indicates 1 if n-th K-mer agrees with $K_{ref}^{(m)}$ in all B blocks and 0 if any pair of blocks from both K-mers does not match

Homomorphic equality test of K-mers

In the proposed system, encrypted K-mers are compared in homomorphic way to detect SNPs. The evaluation operation consists of following steps.

Step 1: The HEAAN scheme supports the rotation operation on plaintext slots, i.e., it enables us to securely obtain an encryption of the shifted plaintext vector $(w_r, \ldots, w_{k-1}, w_0, \ldots, w_{r-1})$ from an encryption of $(w_0, \ldots, w_{k-1})$. We denote the rotation operation as $\texttt{Rotate}(\vec{c}; r)$. It outputs a ciphertext $\vec{c}_r$ encrypting the rotated plaintext vector of $\vec{c}$ by r positions. Define and compute

$$\vec{c}_{b,m} = \texttt{Rotate}\left(\vec{c}_b; N \cdot K_{\text{ref}}^{(m)}[b]\right).$$

Note that for $n \in [N]$, $\vec{c}_{b,m}[n]$ is 1 if $K^{(n)}[b] = K_{\text{ref}}^{(m)}[b]$ and 0 otherwise.

Step 2: Define and compute

$$\vec{d}_m = \prod_{b \in [B]} \vec{c}_{b,m},$$

where $\prod$ denotes component-wise multiplication. Note that for $n \in [N]$, $\vec{d}_m[n]$ is 1 if $K^{(n)} = K_{\text{ref}}^{(m)}$ and 0 otherwise.

Step 3: Define and compute

$$\vec{d} = \sum_{m \in [L]} \vec{d}_m.$$

Note that for $n \in [N]$, $\vec{d}[n]$ is the number of m such that $K^{(n)}[b] = K_{\text{ref}}^{(m)}[b]$. However, it is very unlikely K_{ref} to have multiple identical subsequences. We may assume that K_{ref} does not have multiple identical subsequences. If this is the case, $\vec{d}[n]$ has the value 1 if $K^{(n)} = K_{\text{ref}}^{(m)}$ for some m and 0 if not.

Step 4: Define and compute

$$\vec{c_{res}} = \sum_{n \in [N]} \texttt{Rotate}\left(\vec{d}; n\right).$$

Note that the 0th component of $\vec{c_{res}}$ is the encryption of what we wanted: the number of n satisfying $K^{(n)} = K_{\text{ref}}^{(m)}$ for some $m \in [L]$. The fact can be seen by an easy computation below.

$$\vec{c_{res}}[0] = \sum_{n\in[N]} \texttt{Rotate}\left(\vec{d};n\right)[0] = \sum_{n\in[N]} \vec{d}[n]$$

Results

Panel scheme accuracy

As mentioned in the subsection "False positive/negative errors", the false positive risk of our model can be controlled to some extent by setting a PDT parameter. Therefore, we focused on evaluating the false negative risk by performing an additional experiment. In our model, false negative occurs if a patient has individual sequence variations near the panel SNPs. Unlike the conventional unencrypted mutation-calling procedure allowing few mismatches, our model depends on the equality test process, which always needs perfect matching nearby the panel SNPs. Therefore, even a single unexpected variation neighboring a panel SNP can sabotage the calling of the residue. Considering the prevalent somatic mutations observed in cancer patients, we can consider it as a major source of risk. Therefore, we tested the false negative risk caused by somatic mutations by using residues having mutations in actual breast cancer patients, where the data is provided by TCGA BRCA [40]. By this test, we estimated the empirical probability of false negative risk in various experimental conditions.

We collected 116,607 somatic mutations from TCGA BRCA dataset, then computed the distribution of pairwise distances between two mutations in terms of chromosomal positions. The goal of this test is to estimate the probability that any two mutations are located in proximity by chance. We set the threshold of proximity as 32 bp, which requires at least 32 bp as K-mer size to avoid false positive SNP calling. We will discuss about K-mer size later in detail. For now, the total number of all possible SNP pairs is calculated as follows.

$$\binom{116,607}{2} = 6,798,537,921$$

and thus the probability P of a panel with N SNP residues having at least a SNP pair that exists within 32-mer is

$$1 - \left(1 - \frac{1,308}{6,798,537,921}\right)^{\binom{N}{2}}. \quad (1)$$

Table 1 shows the probability of false negative risk calculated by aforementioned formula (Eq. 1). The result indicates that even though the risk gradually increases as the size of panels grows, the false positive risk still remains insignificant even when it reaches to 100 residues. It means that our model can handle panels having large size without significant false negative risk.

As shown in Table 1, the false negative risk of our model caused by somatic mutation is not significant. Even though there can be other source of risks that we did not consider such as germline variations, the chance of false negative caused by germline variation is relatively small compared to the one caused by somatic mutation [41, 42]. Although a SNP residue of interest has a sequence variation in its flanking sequence, there are other adjacent K-mers that might still detect the residue. Only when more than two sequence variations are simultaneously occurred and located in K bp on both sides of the panel residue, the scheme completely fails. Hence the actual hazard of false negative is expected to be small in practice. Moreover, even if the false negative error occurs, the system can still know the occurrence of it because there would be no positive count for that SNP residue at all. Therefore, the model can easily detect false negative errors and can recommend to avoid fatal situations.

Table 1 Probability P of two SNPs residing in a 32-mer given N SNP residues

N	*P* (%)
10	0.00087
20	0.0037
30	0.0084
40	0.015
50	0.024
60	0.034
70	0.046
80	0.060
90	0.070
100	0.095

N indicates the number of residues selected from 116,607 somatic mutations. P indicates the probability that any two residues are located in 32 bp window. Note that P still remains insignificant even when the N reaches to 100. All numbers are rounded down to 2 significant figures

Running time of the proposed panel scheme

In our model, one of the major factors that determine running time is the PDT level. As mentioned, PDT level indicates the tolerance for mismatches. Using a higher PDT guarantees less false positive risk, but it increases the length of K-mer i.e., K, which in turn increases the running time. Especially, the homomorphic equality test procedure is significantly affected by K. As shown in Table 2, the running time of homomorphic equality test has almost linear relationship with K (Pearson correlation coefficient of 0.99).

With 10 SNP residues and PDT level 4, K is found to be 24 (Table 3). In detail, we have 480 24-mers in the user's K-mer set and 960 24-mers in the panel K-mer set. The comparison of the two sets took 603 seconds (Table 2). Considering that the PDT level 4 with finding

Table 2 Length K of product K-mers, time needed to encrypt K-mers and perform equality test to detect SNPs

N	PDT	K	K-mer encryption (ms)	equality test (sec)
10	0	10	45	82
10	1	13	56	141
10	2	17	92	300
10	3	20	101	418
10	4	24	119	603

N indicates the number of SNP residues included in the panel. PDT is the mismatch tolerance, where higher PDT indicates lower false positive risks. The table shows how the changes in PDT and K affect the running time. The result indicates that larger K results in longer running time to encrypt and test equality

Table 4 The number of discarded and remaining panels given the original number of SNP residues randomly selected. N indicates the number of residues in each panel

	$N = 10$	20	30	40	50	60	70	80	90	100
Total	10	20	30	40	50	60	70	80	90	100
Discarded	0	0	2	3	5	6	7	9	9	13
Remaining	10	20	28	37	45	54	63	71	81	87
Survival rate	1.0	1.0	.93	.93	.90	.90	.90	.89	.90	.87

Discarded indicates the number of residues abandoned during preprocessing. During preprocessing, the model checks if there are residues having too much similarity in their neighboring sequences. If two residues have too much similarity in neighboring sequence contexts, they cannot be distinguished by NGS-seq due to the limited read length. Table shows that remaining rate is consistent even when N reaches 100, which indicates that the model can handle panels with large size without losing too much residues

K-mers is overgenerous, despite the wide genome variation our work shows that homomorphic encryption can be computationally feasible to apply on SNP panel scheme.

Panel design guideline for scheme feasibility

The practicality of our scheme relies on shorter K-mers, and minimizing K is one of our best interest. As mentioned in the Method section, K increases as (1) the number of SNP residues included in panels grow and (2) the higher PDT level required. The estimated value of K with the simulated data is shown below (Table 3).

However, the values of K tend to saturate very quickly. There is ignorable difference among the values of K along the number of SNP residues from 20 to 100. The saturating tendency implies that our scheme is likely to be effective even regarding the panel with large number of SNP residues. In the data preprocessing stage, we have discarded SNP residues with redundant flanking sequences to prevent very large number of K. Table 4 shows the number of discarded SNP residues with respect to the initial number of SNP residues of panels. Note that the panels do not drop many SNP residues as their number grows. That is, with carefully chosen SNP residues not to share the redundant flanking sequences, some value roughly around 32 for K would be long enough for any practical number of SNP residues. The value drops to about 21 when PDT is not concerned. Considering the trade-offs these parameters can provide, any user can fully avail oneself to our SNP panel scheme under most circumstances.

Table 3 The estimated values of K given the number of SNP residues and PDT levels

	$N = 10$	20	30	40	50	60	70	80	90	100
PDT = 0	10	21	21	21	21	21	21	21	21	21
1	13	23	23	23	23	23	23	24	24	24
2	17	26	26	26	26	26	26	26	26	26
3	20	27	27	27	27	27	27	28	28	28
4	24	32	32	32	32	32	32	32	32	32

PDT indicates mismatch tolerance and N indicates the panel size. The number inside of table is value of K, which is the minimal length of K-mer for the model. Usually, larger K requires longer running time and more resources. The table shows how the changes in PDT and N affect the value of K. The result indicates that K is positively correlated with both factors respectively

Conclusions

Although homomorphic encryption has a good potential for protecting security of data, the encryption method combined with the current computer systems has not achieved practical performance to fully utilize the power. In the field of genomics, the usage of homomorphic encryption has been mostly limited to querying aggregate or annotated data, requiring the preprocessing of data. However, to preprocess the genomic data, the raw genome sequence should be reveled without adequate protection, thus more reliable scheme for protecting the genome information is much needed. In addition, the SNP panel at the hospital should be protected. In this paper, we propose a secure SNP panel scheme that protects both the user's genome information and the SNP panel information owned by the hospitals. Since the current homomorphic encryption technologies are not computationally efficient, it is not trivial to develop a secure SNP panel scheme that can be used in reality. By chunking the part of genome down into K-mers, we have minimized the size of ciphertext space and overcome the current inefficiency of the homomorphic encryption. The scheme has yet many further possible improvements such as parallel processing and new algorithmic techniques. We expect our method to protect the raw sequence from possible threats and further return the control of genomic data to its owner, and at the same time protect the hospital's SNP panel assets safely. However, we emphasize that our method shows the feasibility of our scheme. Applying our proposed method to hospitals will certainly require extensive evaluation and improvement.

Abbreviations
BRCA: Breast cancer susceptibility gene; DNA: Deoxyribonucleic acid; GWAS: Genome-wide association study; HEAAN: Homomorphic encryption for arithmetic of approximate numbers; LWE: Learning with error; NIH: National institutes of health; PDT: Point deviation tolerance; RLWE: Ring-learning with error; SIMD: Single instruction multiple data; SNP: Single-nucleotide polymorphism; TCGA: The cancer genome atlas; UCSC: University of California, Santa Cruz; US: United states

Acknowledgements
Authors are grateful to anonymous reviewers for their useful comments on the manuscript.

Funding
This research is supported by National Research Foundation of Korea (NRF) funded by the Ministry of Science, ICT (No. NRF-2017M3C4A7065887), The Collaborative Genome Program for Fostering New Post-Genome Industry of the National Research Foundation (NRF) funded by the Ministry of Science and ICT (MSIT) (No. NRF-2014M3C9A3063541), A grant of the Korea Health Technology R&D Project through the Korea Health Industry Development Institute (KHIDI), funded by the Ministry of Health & Welfare, Republic of Korea (grant number: HI15C3224), and Institute for Information & communications Technology Promotion (IITP) grant funded by the Korea government (MSIP) (B0717-16-0098, Development of homomorphic encryption for DNA analysis and biometry authentication). The publication cost will be paid by the Seoul National University Office of Research.

Availability of data and materials
The source data for simulating panels can be downloaded from https://genome.ucsc.edu/. The sequencing reads simulator can be downloaded from https://github.com/lh3/wgsim and its parameters are described in the Method section. HEAAN library can be downloaded from https://github.com/snucrypto/HEAAN. The data used to study somatic mutations can be downloaded from https://portal.gdc.cancer.gov/projects/TCGA-BRCA.

About this supplement
This article has been published as part of *BMC Genomics Volume 20 Supplement 2, 2019: Selected articles from the 17th Asia Pacific Bioinformatics Conference (APBC 2019): genomics*. The full contents of the supplement are available online at https://bmcgenomics.biomedcentral.com/articles/supplements/volume-20-supplement-2.

Authors' contributions
SK conceived of the presented idea. SP designed the algorithm to find K with support from SS. MK processed the experimental data. SH and KH designed the algorithm for K-mer equality test with support from KL. SP conducted the experiment. SP wrote the manuscript with support from MK and SH. SK and JC supervised the project. All authors read and approved the final manuscript.

Ethics approval and consent to participate
Not applicable.

Consent for publication
Not applicable.

Competing interests
The authors declare that they have no competing interests.

Publisher's Note
Springer Nature remains neutral with regard to jurisdictional claims in published maps and institutional affiliations.

Author details
[1]Department of Computer Science and Engineering, Seoul National University, Seoul, Republic of Korea. [2]Interdisciplinary Program in Bioinformatics, Seoul National University, Seoul, Republic of Korea. [3]Hyperconnect Inc, Seoul, Republic of Korea. [4]Department of Mathematical Sciences, Seoul National University, Seoul, Republic of Korea. [5]Bioinformatics Institute, Seoul National University, Seoul, Republic of Korea.

Published: 10 April 2019

References
1. Kandoth C, McLellan MD, Vandin F, Ye K, Niu B, Lu C, et al. Mutational landscape and significance across 12 major cancer types. Nature. 2013;502(7471):333.
2. Easton DF, Pharoah PD, Antoniou AC, Tischkowitz M, Tavtigian SV, Nathanson KL, et al. Gene-panel sequencing and the prediction of breast-cancer risk. N Engl J Med. 2015;372(23):2243–57.
3. Spataro N, Rodríguez JA, Navarro A, Bosch E. Properties of human disease genes and the role of genes linked to Mendelian disorders in complex disease aetiology. Hum Mol Genet. 2017;26(3):489–500.
4. ClinVar Database. Available from: https://www.ncbi.nlm.nih.gov/clinvar/.
5. Lippert C, Sabatini R, Maher MC, Kang EY, Lee S, Arikan O, et al. Identification of individuals by trait prediction using whole-genome sequencing data. Proc Natl Acad Sci. 2017;114(38):10166–71.
6. Gentry C, Boneh D. A fully homomorphic encryption scheme. vol. 20. Stanford University Stanford; 2009.
7. Naveed M, Ayday E, Clayton EW, Fellay J, Gunter CA, Hubaux JP, et al. Privacy in the genomic era. ACM Comput Surv (CSUR). 2015;48(1):6.
8. Dowlin N, Gilad-Bachrach R, Laine K, Lauter K, Naehrig M, Wernsing J. Manual for using homomorphic encryption for bioinformatics. Proc IEEE. 2017;105(3):552–67.
9. Homer N, Szelinger S, Redman M, Duggan D, Tembe W, Muehling J, et al. Resolving individuals contributing trace amounts of DNA to highly complex mixtures using high-density SNP genotyping microarrays. PLoS Genet. 2008;4(8):e1000167.
10. Wang R, Li YF, Wang X, Tang H, Zhou X. Learning your identity and disease from research papers: information leaks in genome wide association study. In: Proceedings of the 16th ACM conference on Computer and communications security. ACM; 2009. p. 534–544.
11. Gymrek M, McGuire AL, Golan D, Halperin E, Erlich Y. Identifying personal genomes by surname inference. Science. 2013;339(6117):321–4.
12. Shringarpure SS, Bustamante CD. Privacy risks from genomic data-sharing beacons. Am J Hum Genet. 2015;97(5):631–46.
13. Humbert M, Huguenin K, Hugonot J, Ayday E, Hubaux JP. De-anonymizing genomic databases using phenotypic traits. Proc Priv Enhancing Technol. 2015;2015(2):99–114.
14. Uhlerop C, Slavković A, Fienberg SE. Privacy-preserving data sharing for genome-wide association studies. J Priv Confidentiality. 2013;5(1):137.
15. Johnson A, Shmatikov V. Privacy-preserving data exploration in genome-wide association studies. In: Proceedings of the 19th ACM SIGKDD international conference on Knowledge discovery and data mining. ACM; 2013. p. 1079–87.
16. Yu F, Fienberg SE, Slavković AB, Uhler C. Scalable privacy-preserving data sharing methodology for genome-wide association studies. J Biomed Inform. 2014;50:133–41.
17. Tramèr F, Huang Z, Hubaux JP, Ayday E. Differential privacy with bounded priors: reconciling utility and privacy in genome-wide association studies. In: Proceedings of the 22nd ACM SIGSAC Conference on Computer and Communications Security. ACM; 2015. p. 1286–97.
18. Simmons S, Berger B. Realizing privacy preserving genome-wide association studies. Bioinformatics. 2016;32(9):1293–300.
19. Simmons S, Sahinalp C, Berger B. Enabling privacy-preserving GWASs in heterogeneous human populations. Cell Syst. 2016;3(1):54–61.
20. Canim M, Kantarcioglu M, Malin B. Secure management of biomedical data with cryptographic hardware. IEEE Trans Inf Technol Biomed. 2012;16(1):166–75.
21. Kamm L, Bogdanov D, Laur S, Vilo J. A new way to protect privacy in large-scale genome-wide association studies. Bioinformatics. 2013;29(7):886–93.
22. Xie W, Kantarcioglu M, Bush WS, Crawford D, Denny JC, Heatherly R, et al. SecureMA: protecting participant privacy in genetic association meta-analysis. Bioinformatics. 2014;30(23):3334–41.
23. Wang XS, Huang Y, Zhao Y, Tang H, Wang X, Bu D. Efficient genome-wide, privacy-preserving similar patient query based on private edit distance. In: Proceedings of the 22nd ACM SIGSAC Conference on Computer and Communications Security. ACM; 2015. p. 492–503.
24. Troncoso-Pastoriza JR, Katzenbeisser S, Celik M. Privacy preserving error resilient DNA searching through oblivious automata. In: Proceedings of the 14th ACM conference on Computer and communications security. ACM; 2007. p. 519–28.

25. Kantarcioglu M, Jiang W, Liu Y, Malin B. A cryptographic approach to securely share and query genomic sequences. IEEE Trans Inf Technol Biomed. 2008;12(5):606–17.
26. Ayday E, Raisaro JL, Hubaux JP, Rougemont J. Protecting and evaluating genomic privacy in medical tests and personalized medicine. In: Proceedings of the 12th ACM workshop on Workshop on privacy in the electronic society. ACM; 2013. p. 95–106.
27. Ayday E, Raisaro JL, Laren M, Jack P, Fellay J, Hubaux JP. Privacy-preserving computation of disease risk by using genomic, clinical, and environmental data. In: Proceedings of USENIX Security Workshop on Health Information Technologies (HealthTech'13). USENIX Security; 2013.
28. Kim M, Lauter K. Private genome analysis through homomorphic encryption. In: BMC medical informatics and decision making. vol. 15. BioMed Central; 2015. p. S3.
29. Lu WJ, Yamada Y, Sakuma J. Privacy-preserving genome-wide association studies on cloud environment using fully homomorphic encryption. In: BMC medical informatics and decision making. vol. 15. BioMed Central; 2015. p. S1.
30. Zhang Y, Dai W, Jiang X, Xiong H, Wang S. Foresee: Fully outsourced secure genome study based on homomorphic encryption. In: BMC medical informatics and decision making, vol. 15. BioMed Central; 2015. p. S5.
31. Wang S, Zhang Y, Dai W, Lauter K, Kim M, Tang Y, et al. HEALER: Homomorphic computation of ExAct Logistic rEgRession for secure rare disease variants analysis in GWAS. Bioinformatics. 2015;32(2):211–8.
32. Raisaro JL, Choi G, Pradervand S, Colsenet R, Jacquemont N, Rosat N, Mooser V, Hubaux J-P. Protecting privacy and security of genomic data in I2B2 with homomorphic encryption and differential privacy. IEEE/ACM Trans Comput Biol Bioinforma. 2018;15(5):1413–26.
33. Jagadeesh KA, Wu DJ, Birgmeier JA, Boneh D, Bejerano G. Deriving genomic diagnoses without revealing patient genomes. Science. 2017;357(6352):692–5.
34. Jacquez GM, Essex A, Curtis A, Kohler B, Sherman R, El Emam K, et al. Geospatial cryptography: enabling researchers to access private, spatially referenced, human subjects data for cancer control and prevention. J Geogr Syst. 2017;19(3):197–220.
35. Ghasemi R, Al Aziz MM, Mohammed N, Dehkordi MH, Jiang X. Private and efficient query processing on outsourced genomic databases. IEEE J Biomed Health Inform. 2017;21(5):1466–72.
36. Cheng K, Hou Y, Wang L. Secure Similar Sequence Query on Outsourced Genomic Data. In: Proceedings of the 2018 on Asia Conference on Computer and Communications Security. ACM; 2018. p. 237–51.
37. Tang H, Jiang X, Wang X, Wang S, Sofia H, Fox D, et al. Protecting genomic data analytics in the cloud: state of the art and opportunities. BMC Med Genom. 2016;9(1):63.
38. Consortium IHGS, et al. Initial sequencing and analysis of the human genome. Nature. 2001;409(6822):860.
39. Casper J, Zweig AS, Villarreal C, Tyner C, Speir ML, Rosenbloom KR, et al. The UCSC genome browser database: 2018 update. Nucleic Acids Res. 2017;46(D1):D762–D769.
40. Network CGA, et al. Comprehensive molecular portraits of human breast tumours. Nature. 2012;490(7418):61.
41. Milholland B, Dong X, Zhang L, Hao X, Suh Y, Vijg J. Differences between germline and somatic mutation rates in humans and mice. Nat Commun. 2017;8:15183.
42. Consortium GP, et al. A global reference for human genetic variation. Nature. 2015;526(7571):68.

Qian *et al. BMC Genomics* 2019, **20**(Suppl 2):189
https://doi.org/10.1186/s12864-019-5474-y

BMC Genomics

RESEARCH Open Access

Microbiota in the apical root canal system of tooth with apical periodontitis

Wenhao Qian[1], Ting Ma[2], Mao Ye[1], Zhiyao Li[1], Yuanhua Liu[2*] and Pei Hao[2*]

From The 17th Asia Pacific Bioinformatics Conference (APBC 2019)
Wuhan, China. 14-16 January 2019

Abstract

Background: Apical periodontitis (AP) is essentially an inflammatory disease of microbial etiology primarily caused by infection of the pulp and root canal system. Variation of the bacterial communities caused by AP, as well as their changes responding to dental therapy, are of utmost importance to understand the pathogensis of the apical periodontitis and establishing effective antimicrobial therapeutic strategies. This study aims to uncover the composition and diversity of microbiota associated to the root apex to identify the relevant bacteria highly involved in AP, with the consideration of root apex samples from the infected teeth (with/without root canal treatment), healthy teeth as well as the healthy oral.

Methods: Four groups of specimens are considered, the apical part of root from diseased teeth with and without root canal treatment, and wisdom teeth extracted to avoid being impacted (tooth healthy control), as well as an additional healthy oral control from biofilm of the buccal mucosa. DNA was extracted from these specimens and the microbiome was examined through focusing on the V3-V4 hypervariable region of the 16S rRNA gene using sequencing on Illumina MiSeq platform. Composition and diversity of the bacterial community were tested for individual samples, and between-group comparisons were done through differential analysis to identify the significant changes.

Results: We observed reduced community richness and diversity in microbiota samples from diseased teeth compared to healthy controls. Through differential analysis between AP teeth and healthy teeth, we identified 49 OTUs significantly down-regulated as well as 40 up-regulated OTUs for AP.

Conclusion: This study provides a global view of the microbial community of the AP associated cohorts, and revealed that AP involved not only bacteria accumulated with a high abundance, but also those significantly reduced ones due to microbial infection.

Keywords: Apical periodontitis, 16S rRNA sequencing, Root canal treatment, Bacterial community, Healthy control

Background

Apical periodontitis (AP) is an endodontic disease, with inflammatory lesion around the apex of a tooth root which is usually caused by microbial invasion of the root canal system [1, 2]. It is mostly the outcome of untreated dental caries when the root canal system is infected by oral microbiota [3]. This root canal infection can become symptomatic and evolve to severe spreading and sometimes life-threatening abscesses [3]. Treatment of AP involves antibiotic therapy and root canal treatment (RCT, also called endodontic therapy) or extraction of the diseased tooth to eliminate the source of infection. Conventional endodontic therapy was reported yielding a highly successful outcome, with the contribution of enhanced diagnostic tool, antibiotics, and advances in biomaterials [4]. Endodontic instrumentation alone can't achieve a sterile condition due to the anatomical complexities of root canals and limitations of medicaments. Initiation of empirical therapy with antibiotics is highly recommended for lesion sterilization and tissue repair. As examples, from the early used EDTA [5], to sodium hypochlorite [6], chlorhexidine and octenidine hydrochloride [7], as well as the triple antibiotic paste (combination of metronidazole, ciprofloxacin, and minocycline) [8, 9], are

* Correspondence: yhliu@ips.ac.cn; phao@ips.ac.cn
[2]Key Laboratory of Molecular Virology and Immunology, Institut Pasteur of Shanghai, Chinese Academy of Sciences, Shanghai 200031, China
Full list of author information is available at the end of the article

widely applied and tested effective for removing microbiome. Even so, quite alarming numbers in the persistence of AP were reported in literature [4], since complete elimination of microorganisms is not always obtained. The spectrum of the antimicrobial agent, based on experience or even microbial susceptibility testing [3], mostly can only cover partial of the pathogenic community (a set of species, usually organized in multispecies biofilm) [10, 11]. This highlights the urgency to uncover the composition of the microbial community involved in AP and hence to elucidate the "unit of pathogenicity" of AP comparing to normal conditions.

Early studies of the microbiota in the root canals of teeth with AP were conducted through broad-range culture/biochemical methods, by which only cultivable and predominant bacteria were reachable, with the risk of missing keystone species [12, 13]. The adoption of 16S rRNA-based microbial diversity analyses allows the detection of a large scale bacterial communities in endodontic infections, including the cultivable/uncultivable and even the uncharacterized bacterial community members, as well as the low abundance taxa [14]. Growing evidence showed that not only one but several species contribute to endodontic infection, which interact one with the other to give rise to disease.

Bacteria located in the apical part of the infected root canals are arguably directly involved in the AP pathogenesis. Researchers were challenged in sampling procedures when evaluating the endodontic microbiota was performed. Paper points are mostly used in early studies, but it is widely accepted that limited region of the root canal can be involved. Gryogenic grinding technique was widely used recently, where the root apex fragment was sectioned and together with the surrounding tissue was cryogenically ground to collect the microbiota samples [15].

Based on the 16S rRNA technique, complex bacterial populations were reported present in AP individuals, mostly classified within phyla of *Proteobacteria, Firmicutes, Bacteriodites, Fusobacteia,* and *Actinobacteria* [12, 16–23], where *Proteobacteria, Firmicutes* and *Bacteriodites* were always found prevalent. With 10 patients (10 extracted teeth) involved, Siqueira's early study identified 187 phylotyes belonging to 84 genera and 10 phyla in apical root canal infections, where *Proteobacteria* and *Firmicutes* together occupied about two thirds of all sequences obtained [16]. Later, in Vengerfeldt's work [18], *Firmicutes* and *Bacteroidetes* were observed prevalent in chromic apical periodontitis. Their work has revealed many of the known root canal pathogens including gram-negtive anaerobes such as *Prevotellasp, Porphyromonassp, Fusobacteriumsp, Tannerellasp,* and *Pyramidobacter piscolens* as well as gram-positive *Dialistersp,* oral spirochete *Treponema socranskii, Solobacterium moorei.* They also noted an increase of *E.faecalis* after treatment failure. Most recently, Rocas [24] found *P.endodontalis, Dialisterinvisus, Olsenellauli* and *F.nucleatum* were most frequently detected in their asymptomatic AP cases.

We note that the profile of the microbial community around the apical part of the root in AP differed at different taxonomic levels with respect to labs (mainly due to geographic differences) [25, 26], and individual diversity was commonly observed even in a single study. Furthermore, researchers aimed to uncover the microbial composition and diversity, and focused on the taxons (OTUs, species or genus) which are predominant in AP to understand the pathogenesis of AP and establish effective antimicrobial therapeutic strategies [15]. However, no microbial profile of the root apex was available for normal condition. This might be due to the fact that it is difficult or even impossible to collect the extracted tooth for normal condition. Thus, questions could arise: is there microbiome located around the root apex in the normal condition? Without the answer of this question, it is ambiguous to consider the microorganisms predominantly present in AP as putative pathogens. In fact, scientists in this field did be challenged for clarifying whether more-virulent communities develop right from the beginning of the infection process or are a result of a shift in the community composition due to environmental changes [3]. Contribution from the side of the normal condition (or close to the normal condition) will aid in unearthing this pathogenic process of AP. Most recently, Coleman and colleagues [21] considered the supragingival samples (supragingival buccal biofilm scraping) from healthy teeth as reference in their work, and revealed an increased abundance of Gram-negative (*Prevotella, Fusobacterium, Treponema, Veillonellaceae, TG5* (*Synergistetes*) bacteria and a decreased abundance of Gram-positive (*Actinomyces, Corynebacterium*) bacteria in the diseased tissue samples. Bacteria decreased in diseased cases were emphasized in this study, somehow indicating the dysbiosis caused by AP. However, the bacterial community located in the supragingival buccal biofilm is too limited to represent the root canal system since distinguished variation of microbial composition and diversity were observed in different regions of human oral cavity [25, 26].

We therefore plan our research in two ways: i) collecting 16S rRNA samples from both healthy wisdom teeth (extracted in case of becoming impacted) and the oral sample (through rubbing a sterile cotton swab across the buccal mucosa) for each contributor, by which to investigate the microbial community close to normal condition and uncover the relationship between oral and the root apex microbiota; ii) collecting root apex specimen from AP to identify the key bacteria relevant to the etiology and pathogenicity of AP. We also involved the diseased teeth failed after conventional root canal therapy (RCT) to explore the effect of RCT on microorganisms in the tooth root canal system.

Materials and methods

Subjects population

We considered the teeth and oral samples from AP patients (31) and healthy controls (9). From the 31 patients, we collected 8 teeth failed after RCT (labeled as RCT) and 23 teeth without RCT (labeled as NRCT), where each patient contributed one. 10 healthy teeth (labeled as HT) were obtained through common surgical extractions of supernumerary or wisdom teeth, where one healthy control contributed two and others each contributed one. We also collected 9 oral samples from healthy people who contribute the wisdom teeth (labeled as HO). The diseased teeth were presented with a 2–8 mm AP lesion as determined radiographically, and clinically classified as LSA (maximum diameter > =5 mm) and SSA (maximum diameter < 5 mm) according to the area of AP lesion. The inclusion criteria for RCT failure cases: i) previous root canal treatment or retreatment performed at least more than 2 year before; ii)the lesion remained the same size or increased as compared with radiographs taken immediately after the initial treatment/retreatment. Exclusion criteria of AP teeth: i) direct connection of the apical lesion to the oral cavity due to lack of gingival tissue, vertical root fracture and fistula; ii) previous surgery on the apical root; iii) treated with antibiotics within one month.

Sampling procedure and clinical data collection

Oral specimen was collected from individual by rubbing a sterile cotton swab across the buccal mucosa and frozen for storage at − 20 °C. Diseased teeth with/without RCT were extracted because of being unrestorable, while wisdom teeth (healthy controls) were extracted in case of being impacted. The apical part of the tooth root was resected at 2 mm from the root end using a sterile Zekrya FG 28 mm bur, as reported in [19]. Before the tooth extraction surgery, the oral cavity was rinsed with 0.12% chlorhexidine for 2 min and the area to be operated was gently scrubbed with the same solution. Care was taken to avoid saliva contamination of the surgical site during the whole surgery. The root apex specimen was soon placed in a sterile flask and immediately frozen at − 20 ° C. The root apexes were finally crushed in a 6750 freezer mill (Spex, Metuchen, NJ, USA), at the liquid nitrogen temperature, as reported in [15]. Apical root powder samples were stored frozen at − 20 °C.

DNA extraction and 16S rDNA amplicon pyrosequencing

Apical root powder samples and oral samples were treated with the Fast DNA SPIN extraction kits (MP Biomedicals, Santa Ana, CA, USA), following the manufacturer's instructions, to extract the total DNA, and stored at − 20 °C prior to further analysis. The extracted DNAs were then quantified and qualified using a NanoDrop ND-2000 spectrophotometer (Thermo Fisher Scientific, Waltham, MA, USA) and agarose gel electrophoresis, respectively. PCR amplifications were performed using bacterial domain-specific primers, targeting variable regions V3-V4 of the 16S rRNA gene, using primers 338F (5'-ACTCCTACG GGAGGCAGCA-3′) and 802R (5'-GGACTACHVGGGT-WTCTAAT-3′). Details of the PCR components and the thermal cycling procedures were described in [27]. Purification and quantification of the PCR amplicons were done by Agencourt AMPure Beads (Beckman Coulter, Indianapolis, IN) and PicoGreen dsDNA Assay Kit (Invitrogen, Carlsbad, CA, USA), respectively. After that, equal amounts of the amplicons were pooled, and 2 × 250 bp paire-end sequencing was performed using the illumina MiSeq platform with MiSeq Reagent Kit v3.

Sequence analysis

Sequences were processed with QIIME following an established illumima sequencing analysis pipeline [28]. Briefly, paired-end reads were firstly joined using SeqPrep method (https://github.com/jstjohn/SeqPrep) where minimum overlap between paired reads is set as 5 bps. Demultiplexing was then performed to assign the assembled reads to individual samples according to the barcodes and filtered through the default strategy for quality filtering of illumina data described in [29]. An open-reference clustering [30] was then used to approach OUT picking, where the Greengenes database (the Greengenes 13_8 97% OTU representative sequences) and denovo clustering of unmapped sequences were combined to build the reference database. Sequences were queried against the reference database and assigned to the reference OTUs if the identity is not less than 97%. The remainder of the sequences that fail to hit the reference database were then be clustered using UCLUST [31] for denovo reference OTUs, and assigned to the new reference OTUs. An OTU table was further generated to record the abundance of each OTU in each sample and the taxonomy of these OTUs. This has singleton OTUs (or OTUs with a total count of 1) removed, as well as OTUs whose representative (i.e., centroid) sequence couldn't be aligned with PyNAST [32].

The OTU table was then used to estimate the community richness (chao1 and OTUs) and diversity indices (Shannon index and simpson index) for individual sample, using R packages. Shannon index and simpson index were calculated using the function *diversity* in package *vegan*; Chao1 was obtained through chao1 in package *fossil*.

Statistical analysis

All statistical analysis starts from the raw counts of the OTU table, and taxa of the same type were summarized at the phylum, class, order, family and genus levels. For comparisons between samples/groups, abundance of each OTUs were normalized by a down-sampling way, where the number of sequences of individual sample was reduced to the minimum of all the 50 samples by random sampling. To identify the essential OTUs relevant to AP, the OTUs were firstly filtered to remove the OTUs present in less than 10% of all samples. Wilcoxon test [33] was then used to perform differential analysis between different groups (e.g. HO, HT, NRCT and RCT). *P*-values were adjusted by Storey's Q value [34]. Significant OTUs were selected using q-value 0.05, 2 fold-change as well as the Q3 quantile from the two groups involved in any comparisons. The Q3 quantile threshold was employed to filter the candidates if the 75th quantile of the high-level group is less than 10 in NRCT versus HT at OTU level and less than 100 at Genus level.

Results

Coverage, richness and diversity

We have 50 16S samples in total including diseased and healthy teeth samples as well as oral samples. The number of the valid reads from 16S RNA sequencing for OTUs detection is ranging from 29,604 to 68,088. 43,839 OTUs in total appear across 50 samples, while the number of OTUs in single sample ranges from 918 to 4436. To study the changes of the richness and evenness of individual samples in terms of locations and healthy status, we calculated the richness (OTUs and Chao1) and diversity indices (Shannon and Simpson), and compare them from different groups (HO, HT, NRCT and RCT). Comparing to the diseased samples, higher richness and diversity were observed in healthy samples, including healthy oral and healthy teeth (Fig. 1). In particular, the numbers of the OTUs in healthy samples (HT and HO) were significantly higher than the diseased samples (RCT and NRCT). This indicates that the balance of the microbial community of the root apex was changed due to the disease as well as RCT treatment. Dynamically saying, i) healthy tooth host plentiful of bacteria (both richness and diversity), ii) AP significantly reduced the OTU richness (*P* values are 0.0011 and 0.0067 for NRCT versus HT and RCT versus HT, respectively) and made the community purer; iii) and after that, RCT treatment further slightly increased the OTU number but reduced the diversity. We therefore speculate that microbial infection in AP caused part of the bacteria increased sharply while part of some other bacteria decreased or even disappear. In other word, there might be some bacteria becoming predominant due to infection. For this, we sorted the OTUs based on their abundance, and calculated the fraction of the abundance accumulation of the top OTUs in the total abundance of

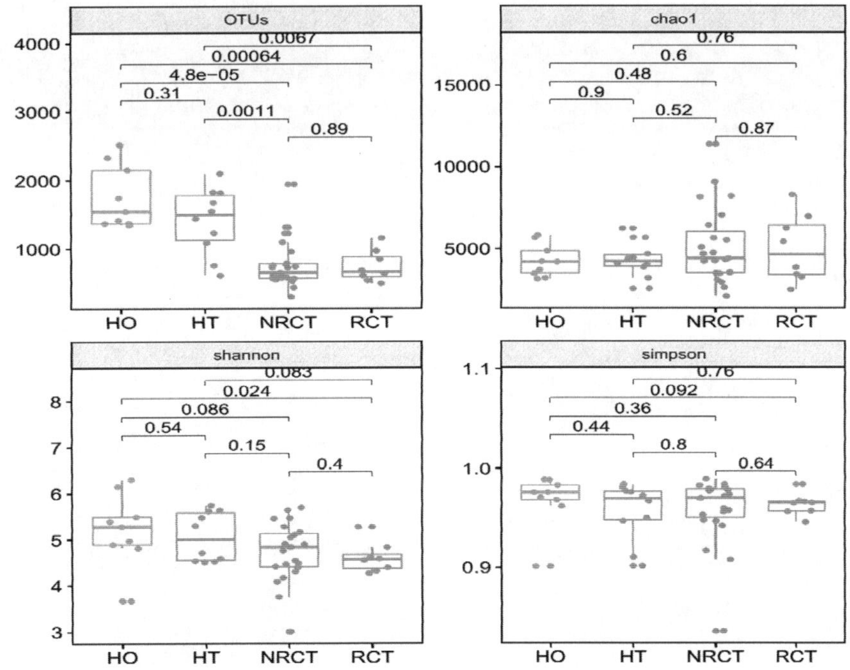

Fig. 1 Bacterial community richness (observed OTUs and chao1) and diversity (Shannon and Simpson index) in the groups of AP (NRCT), AP failed after RCT treatment (RCT), healthy control of root apex (HT), as well as the oral healthy control (HO)

the whole sample. Dynamics of the fraction with the increasing of the number of the top OTUs showed that the fraction in NRCT and RCT groups dramatically increased and close to a steady state much earlier than healthy groups (Additional file 1: Figure S1). This indicates that much fewer bacteria in NRCT and RCT have very high abundance and dominate the bacterial community (the fractions close to 85, 83, 75 and 72% at top 100 OTUs in RCT, NRCT, HT and HO, respectively). Root canal treatment has pulled the richness (OTUs) back slightly (NRCT compared with RCT, from 764 ± 352 to 740 ± 229 in Fig. 1), but it is still far away from that in the healthy status (from 764 ± 352 to 1410 ± 484 in Fig. 1). Furthermore, the lower diversity of RCT as well as the top 100 OTUs occupied 85% of the whole abundance (Additional file 1: Figure S1) somehow told that root canal treatment was not able to change the dominant phenomena.

Global view of the bacterial communities

26 phyla present in the 50 microbiota samples. From a global view, the most abundant phyla are *Firmicutes, Proteroidetes* and *Bacterioidetes*, followed by *Actinobacteria, Fusobacteria, Spirochaetes, Synergistetes, Acidobacteria, TM7, Cyanobacteria, Tenericutes, Chloroflexi, SR1, Elusimicrobia, GN02, Planctomycetes, Gemmatimonadetes, [Thermi]* and *Nitrospirae*. Percentages differed due to individuals as well as location, healthy status or treatment (Fig. 2, details see Additional file 1: Figure S2). Basically, bacterial distributions at phylum level did not show big difference between two healthy groups (HO and HT). *Firmicutes* is the most abundant in healthy samples, averaged at round 40% in both teeth and oral, while it dropped to around 26% in diseased samples (NRCT and RCT). On the contrary, *bacterioidetes* jumps to the top 1 in NRCT (26.6% in NRCT from 11.9% in HT) and a slight decrease in RCT. Although root canal treatment failed in RCT, it has changed the percentage of *Bacterioidetes* (from 26.6 to 17.3%), *Fusobacteria* (4.2 to 8.6%) and *Actinobacteria* (11 to 13.6%).

Clustering results using tSNE (implemented by an R package Rsne [35]) at OTU level showed that oral samples (HO) samples are basically separated from teeth

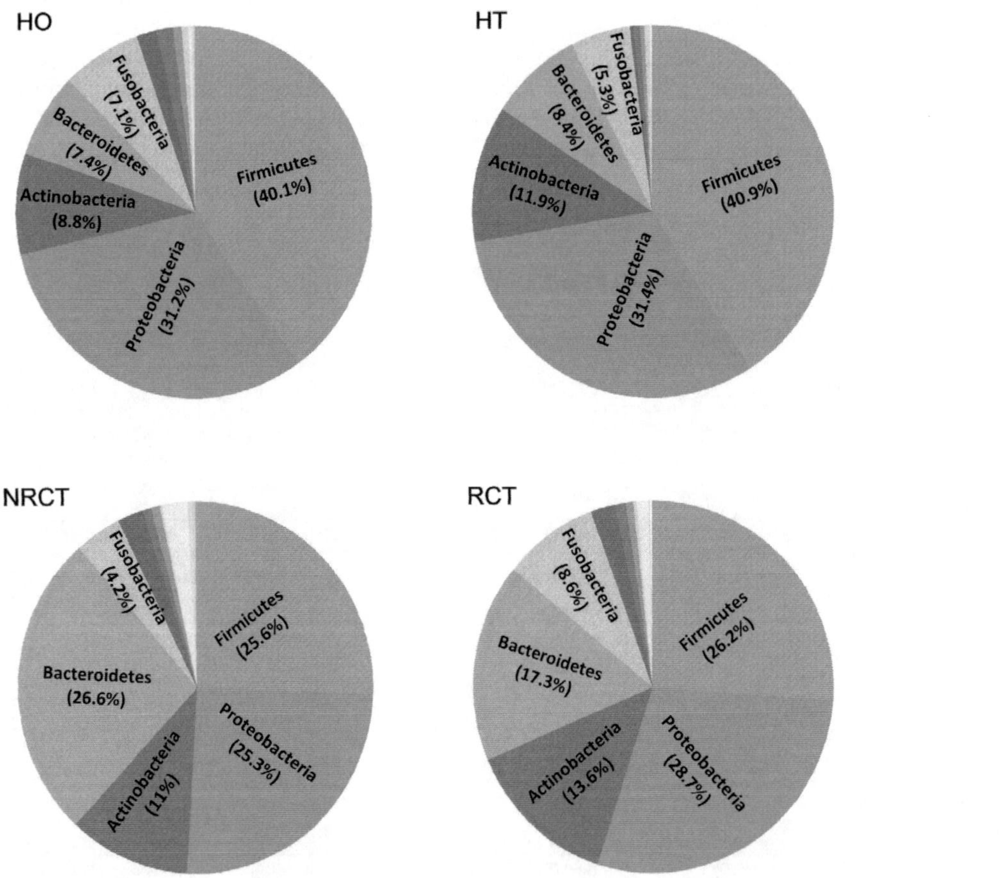

Fig. 2 Piechart of the distributions of the species in individual samples at phylum level. NRCT, AP without RCT treatment; RCT, AP failed after RCT treatment; HT, Healthy control of root apex; HO, the oral healthy control

samples(HT, NRCT and RCT), and healthy teeth (HT) can also be distinguished from teeth without root canal treatment (NRCT), see Additional file 1: Figure S3. However, teeth samples failed after RCT treatments are mixed with other samples.

Bacterial communities around healthy root apex

As stated above, we did identify quite a number of bacterial community round the healthy tooth apex (1410 ± 484 OTUs, 120 ± 49 Genera). This gives some clue that the root canal system itself might not initially clean. On the contrary, there are microorganisms commensal around the tooth apex. That means the bacteria appear in the root apex will not work as a pathogen even if they are predominant. Meanwhile, we observed a very close distribution of taxa at phylum level to the healthy oral case (Fig. 2). We next go deep into the OTU level to investigate the relationship of the microbiota between root apex and oral. 16 paired 16S samples from 8 healthy individuals, each contributed one wisdom teeth (HT) and one oral sample (HO), were used for comparisons in two ways: i) differential analysis based on the OTUs between HT and HO, and ii) similarity based on abundance and richness within and between HT and HO. Paired Wilcox test combined with Stoery's FDR adjustment was used for differential analysis. However, no significant OTUs are identified. For the later way, we calculated the overlaps and uniqueness for every pair of samples from between or within HT and HO. For any two samples from HT and HO, we counted the shared OTUs and those presented uniquely in each samples, and showed it in one piechart in Fig. 3. We then collected the calculated the ratio of the shared OTUs in the piechart for each two samples, which were used to estimate the difference between-group-samples (O2T, overlaps between oral and tooth samples) and within-group-samples similarity (O2O, overlaps between samples from oral; T2T, overlaps between samples from teeth). We did the same way on abundance, but focused on the shared abundance between two samples. Significant difference of the pair-wise overlaps was observed between within oral (O2O), within tooth apex (T2T) and between oral and tooth apex (O2T). Similarity within oral samples is higher than that of within tooth apex samples, and within-group (O2O, T2T) similarity is higher than between-group (O2T) samples. In particular,

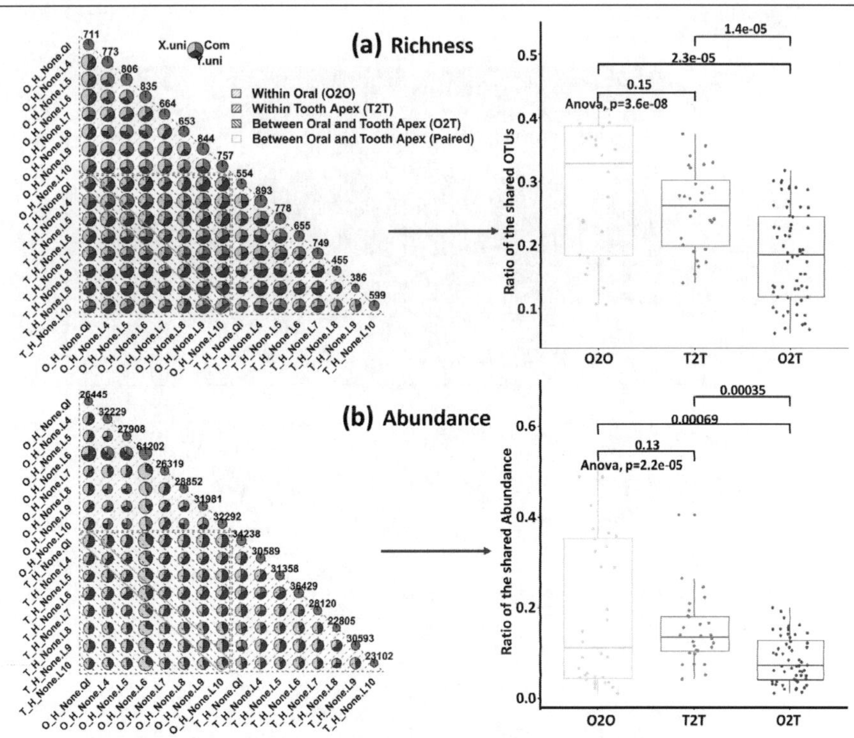

Fig. 3 Overlaps of the richness (**a**) and abundance (**b**) between paired root apex and oral samples from the same patient. Upper panel: piecharts to show the unique and common OTUs (richness) or abundance between two samples (only the patients who contributed both oral and tooth samples were considered). X.uni represents the OTU or abundance unique in the sample from X axis; Y.uni represents the OTU or abundance unique in the sample from Y axis; Com means the OTU or abundance shared in two samples. Bottom panel: boxplots to show the shared OTUs and abundance with respect to location (oral and tooth). O2O represents shared OTUs or abundance within oral samples; T2T represents shared OTUs or abundance within tooth samples; O2T represents shared OTUs or abundance between oral and tooth samples

paired oral and tooth apex samples (oral and tooth samples from the same individual) did not show higher ratio of overlaps compared to non-paired samples. In the abundance case (Fig. 3b), much lower ratio of overlaps was found, and the pair-wise similarity within tooth apex is averagely higher than within oral samples.

Changes of bacterial communities due to apical periodontitis

From above, we found the changes of richness, diversity as well as distribution due to geographical location (oral and teeth), healthy state (diseased versus healthy) and therapy (RCT and NRCT). Next, differential analysis at OTU level and genus level was done for NRCT versus HT, HT versus HO, and RCT versus NRCT, respectively.

We identified 89 significant OTUs in total, details see Additional file 2. 49 of them were significantly decreased in the AP samples compared to healthy teeth. These OTUs are basically from genera of *Streptococcus, Haemophilus* (*parainfuenzae* in particular), *Actinomyces, Granulicatella, Leptotrichia.* The species including *subflava, dispar, nanceiensis, melaninogenica, mucilaginosa were found.* Almost all of these OTUs have a high abundance in the healthy teeth, while very low abundance was observed in AP teeth (close to zero). These OTUs were clustered into two groups using *hclust* in R, C1 and C2 in Fig. 4. C1 showed very high abundance in healthy, but almost empty in NRCT, while OTUs in C2 were found with a low level, low level of these OTUs were also observed in the oral control and RCT group (Fig. 5a). This indicates that there exists high abundance of bacterial community specific in healthy teeth, but killed due to AP, and they are not recovered in most of the RCT samples. In literature, researchers focused on the abundant bacteria appeared in the AP teeth to identify the pathogenic bacteria for AP. No report was found to describe the bacterial profile around the root apex of healthy teeth, due to limitation of sampling. With the healthy controls in our case, these decreased OTUs told that there did exist plentiful of bacteria and they disappear in the AP cases. We therefore speculate that these abundant bacteria specific in health teeth root

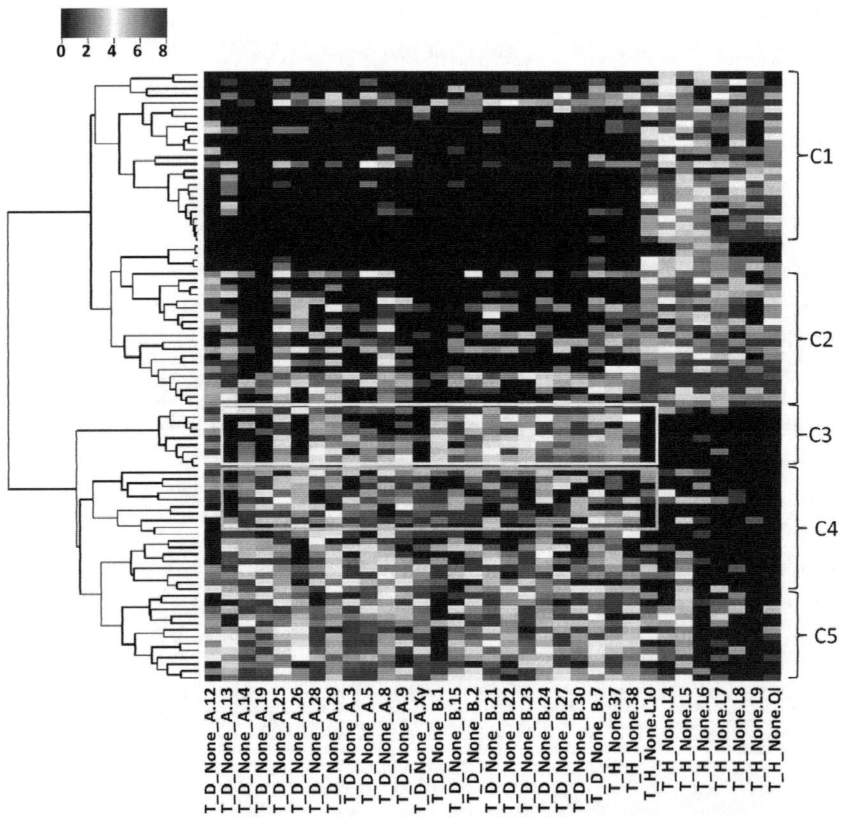

Fig. 4 Heatmap of the key OTUs identified through comparison between AP (NRCT) and health control (HT). Key OTUs are clustered into 5 groups, C1~C5 which differed at their variation patterns due to healthy and disease as well as the subtypes (A and B) in the AP samples, detailed OTUs see Additional file 3

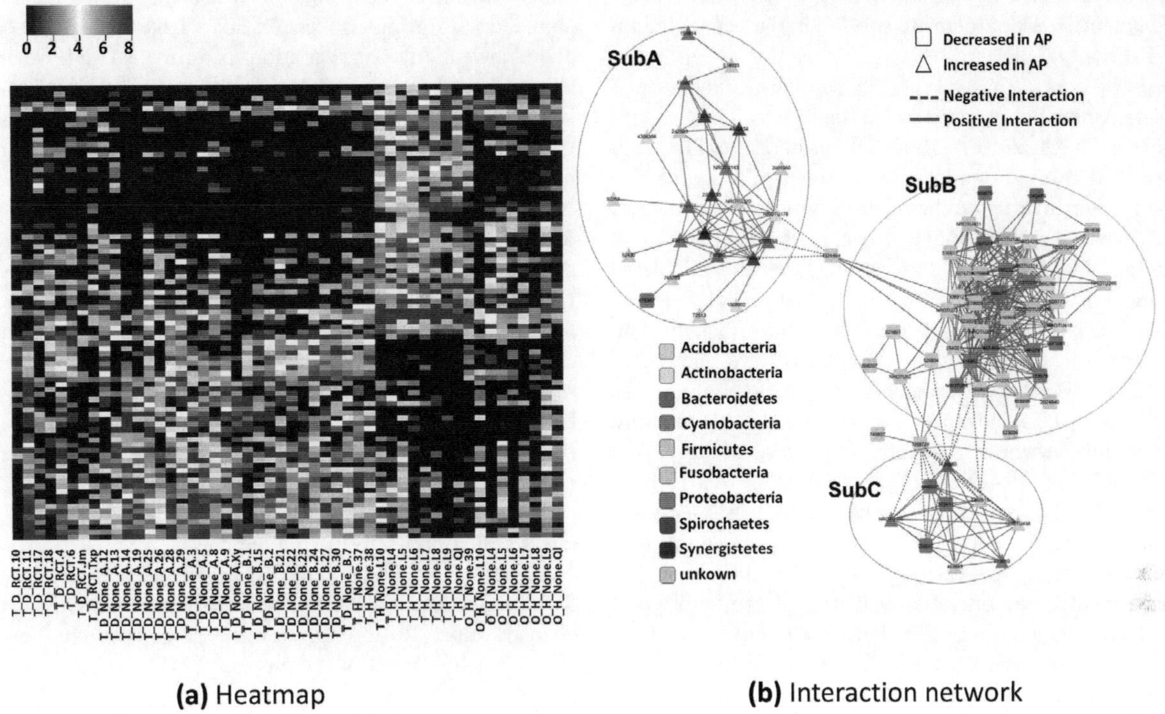

Fig. 5 Heatmap and their interactions of the key OTUs identified through comparison between AP (NRCT) and health control (HT). **a** heatmap of the key OTUs across all 50 samples. **b** Interaction network of the key OTUs is constituted with two parts: SubA, where increased OTUs are inter-connected; SubB, where decreased OTUs are inter-connected; SubC: increased and decreased OTUs interacted

might be essential for keeping a healthy environment. Interestingly, *Streptococcus* and *Granulicatella* were considered as putative pathogens due to the fact that they were detected in acute apical abscesses (belong to AP) [3]. They did be detected in our NRCT samples. But far more abundant of them (*Streptococcus* in particular) was found in our healthy teeth. From our experiment, we say that *Streptococcus* and *Granulicatella* are relevant for AP because of their reduction, other than their presence. *Prevotella* was found in both increased and decreased OTUs we identified, while it was widely reported to be increased in literature. Two species in genera of *Prevotella*, *Prevotella naneiensis* and *Prevotella melaninogenica*, are decreased in NRCT compared to HT.

Forty OTUs are significantly increased in AP. These OTUs were grouped into three clusters, i.e. C3, C4 and C5 in Fig. 4. OTUs in C3 showed higher abundance in A than B (A and B are subtypes in NRCT by samples clustering and match with the large lesion and small lesion, respectively with a high accuracy 0.89). The reverse for OTUs in C4. OTUs in C5 were high in both A and B. All of these 40 OTUs showed very low abundance (close to zero) in healthy teeth but very high level in NRCT, including the OTUs in the genera of *TG5*(*Synergistetes*), *Treponema* (*socranskii*), *Sphaerochaeta*, *Burkholderia*, *Schwartzia*, *Dialister*, *Pseudoramibacter_Eubacterium*, [*Prevotella*], Slackia, *Bifidobacterium* (*longum*) and [*Mogibacteriaceae*] (family). Among them, *treponema*, *prevotella*, *TG5* are wildly reported to be present in periodontitis [21, 36]. In particular, *Prevotella*, *Treponema*, and TG5 (*Synergistetes*) were observed increased in AP root apex compared to supragingival buccal biofilm (HO samples in our case) from healthy teeth [21]. *Sphaerochaeta* was found in subgingival microbiota with a high abundance from chronic periodontitis patients [37] but in our case it is very low in NRCT and empty in healthy teeth. Very high abundance of *Burkholderia* was found in our NRCT while it is not reported in previous researches. Low abundant *Bifidobacterium* (*longum*) was detected in our NRCT, as described in subgingival in [36, 38], but empty in any of the healthy teeth. We found 2 OTUs from *Pseudoramibacter_Eubacterium* genus, both are absent in healthy teeth, while one is with very high abundance and the other is at a low level in NRCT. *Dialister* is frequently recovered from endodontal infections and deep periodontal pockets [3]. *Prevotella* and *Treponema* were also reviewed as factors involving in AP infections [3].

Differential analysis at genus level identified fewer significantly changed genera than that found through OTU level, see Additional file 3 for details, including the decreased *ones Haemophilus, Actinobacillus, Granulicatella, Staphylococcus and Streptococcus, and the increased ones*

Schwartzia, Slackia, Treponema, TG5, Pseudoramibacter_-Eubacterium. All of them are covered in the genera identified at the OTU level.

Relationships between the key OTUs

As described above, significant bacteria (OTUs) are identified through comparison between NRCT and HT. We further investigate the relationships between the key OTUs based on the abundance of each OTU across all the samples. Correlations of these OTUs were calculated using R package *corr.test* with the method of Pearson. The OTUs highly correlated (correlations bigger than 0.6 and adjusted *p*-value less than 0.05) were selected to construct a network using cytoscape (Fig. 5b). In the network, OTUs (nodes in Fig. 5b) are partitioned into three sub-networks, basically consisted of increased OTUs (**SubA**), decreased OTUs (**SubB**) as well as both increased and decreased OTUs (**SubC**), respectively. OUT 43244264 (yellow filled box) works as a connection between SubA and SubB. K-core calculations (the strength to be a core of any cluster) of the whole network showed a bimodal distribution wherein most of the OTUs have a very high value of k-core values and others with very low values. Decreased OTUs from the phyla of *Firmicutes*, *Proteobacteria* as well as some unknown species are tightly connected in **SubB**. We say that these decreased species co-varied with respect to location, AP or RCT therapy with equivalent contribution, and built a robust network. This somehow indicates the difficulty to destroy the balance of the bacterial community through the decreased species. In other word, it is hard to recover once the network is broken. This somehow explained the failure of RCT in our case, where decreased OTUs showed no big changes in RCT compared to NRCT (Fig. 5a). Increased OTUs are interacted in **SubA**, but showed weak relationships (fewer connections and smaller correlation values). **SubB** and **SubC** are negatively connected (blue dashed edges).

Subtypes in diseased AP patients

At the genus level, the AP patients can be partitioned into two groups (labelled as A and B, respectively), using an R package *hclust* (see Additional file 1: Figure S3). This clustering showed significant agreement with the clinical classification based on the lesion area of AP, where A and B match to LSA and SSA with an accuracy of 0.885 ($p < 10^{(-6)}$), respectively (See Table 1, complete clinical information was in Additional file 4). At phylum level, a big difference between A and B was found in the percentages of *Protepbacteria* (19.6% in A and 32.5% in B) and *Actinobacteria* (14.9% in A and 6.1% in B), see Additional file 1: Figure S5. Differential analysis of A versus B identified 53 significant OTUs, basically from phyla *Proteobacteria, Firmicutes, Actinobacteria and Bacteroidetes,* details see Additional file 5. Interestingly, 51 of them showed higher abundance in group B, see Additional file 1: Figure S6, and showed different variation patterns, **P1~P4**. That means these bacteria enriched in the teeth with small lesion area (SSA). In other word, the lesion area will be bigger without these bacteria. Furthermore, a large number of B-abundant OTUs showed similar abundance in the healthy teeth, yellow rectangle in Additional file 1: Figure S6. Among them, *lactococcus* (genus), known as a lactic acid bacteria, was reported to cause teeth decay known as caries. However, previous researches demonstrated that fermentation products produced by lactic acid bacteria contribute to the attenuation of pathogens and improvement of human animal health [39]. We therefore speculate that the abundant lactic acid bacteria in our case (*Lactococcus, Enterococcus* and *Streptococcus*) might act as probiotics to impair the infection.

Effect of RCT on bacterial communities

No significant OTUs was identified between NRCT and RCT. Decreased OTUs due to AP showed no obvious changes after RCT (Fig. 5a). In some patients (eg. T_D_RCT.Jm), most of the increased OTUs (due to AP)

Table 1 Clinical information of the AP patients without RCT treatment

SampleID	Gender	Age	Lesion	Levels	Clusters
T_D_None_B.24	M	74	2×2 mm	SSA	B
T_D_None_B.23	F	62	2×2 mm	SSA	B
T_D_None_A.3	M	72	2×3 mm	SSA	A
T_D_None_A.25	M	65	3×2 mm	SSA	A
T_D_None_B.2	F	41	3×2 mm	SSA	B
T_D_None_B.15	F	52	3×2 mm	SSA	B
T_D_None_B.1	F	74	3×3 mm	SSA	B
T_D_None_B.27	F	72	3×3 mm	SSA	B
T_D_None_B.30	F	62	3×3 mm	SSA	B
T_D_None_A.14	M	45	4x3mm	SSA	A
T_D_None_A.26	M	28	3×4 mm	SSA	A
T_D_None_B.7	F	62	4×3 mm	SSA	B
T_D_None_B.22	M	79	4×4 mm	SSA	B
T_D_None_B.21	F	34	4×4 mm	SSA	B
T_D_None_A.28	M	59	6×3 mm	LSA	A
T_D_None_A.9	F	73	4×5 mm	LSA	A
T_D_None_A.12	F	29	6×4 mm	LSA	A
T_D_None_A.8	F	52	6×5 mm	LSA	A
T_D_None_A.19	F	61	5×6 mm	LSA	A
T_D_None_A.29	M	34	6×5 mm	LSA	A
T_D_None_A.5	M	41	8×5 mm	LSA	A
T_D_None_A.Xy	F	42	8×5 mm	LSA	A
T_D_None_A.13	M	83	6×7 mm	LSA	A

did disappeared, failure of the therapy might be due to the fact that decreased OTUs are not recovered. From this, we say that AP might be a multi-species infection.

Discussion

In this study, we collected microbiota samples from root apex diagnosed as AP with and without RCT treatment, as well as the healthy controls including root apex and supragingival buccal biofilm, and sequenced the 16S ribosomal RNA gene fragments using a next-generation sequencing platform.

Our study focused on the bacterial community around the root apex to investigate the microbial features with respect to apical periodontitis. We start our investigations from the community richness and diversity in AP patients (NRCT) as well as other relevant groups, including failed AP after root canal treatment (RCT), healthy control of root apex (HT), as well as and the oral healthy control (HO). We observed significant decreasing of the richness and diversity from healthy controls to diseased samples (Fig. 1), somehow indicates that AP has broken the microbial balance, where both richness and diversity were reduced. Next, we identified the key bacteria relevant to AP by comparison between AP without root canal treatment (NRCT) and the healthy control (HT). Compared to the healthy root apex, we found a large population of bacteria down-regulated in AP root apex, basically from genera of *Haemophilus* (*parainfuenzae* in particular), *Streptococcus*, *Granulicatella*, *Actinomyces*, and some unknown bacteria, as well as the species including *subflava*, *dispar*, *nanceiensis*, *melaninogenica*.

Our study emphasizes the importance of the healthy control, based on which we revealed plentiful of bacteria around healthy root apex and further identified both up and down regulated bacteria relevant for AP. Furthermore, we involved the supragingival buccal biofilm (HO) as additional healthy control to confirm whether it works instead of the healthy root apex (HT). Global view of the clustering of all the 16S samples using R package *hclust* showed that HO samples are much closer to HT samples than to AP teeth samples (NRCT or RCT), see Additional file 1: Figure S4. Meanwhile, microbial richness and diversity are also similar with higher values in HO and HT samples, and no significant OTUs can be found through differential analysis between HT and HO. However, these are not enough to determine that supragingival buccal biofilm can work as healthy controls. To confirm this we did differential analysis between NRCT and HO, and found lots of bacteria (150 OTUs) significantly up/down regulated. Nevertheless, only 13 out of 150 OTUs overlapped with that of NRCT versus HT. Most of the decreased OTUs and some wildly reported increased OTUs including *Sphaerochaeta*, *Burkholderia*, *Schwartzia*, *Dialister* and [*Prevotella*], missed in NRCT versus HO.

Conclusions

This study provide a global view of the microbial community of the apical periodontitis associated cohorts, and revealed that apical periodontitis involved not only bacteria accumulated with a high abundance, but also those significantly reduced ones due to microbial infection. This provides a more precise spectrum of pathogenic community, aiding for choosing or designing antimicrobial agent used in RCT treatment.

Additional files

Additional file 1: Packages of the additional figures. (PDF 609 kb)

Additional file 2: Differential analysis results between AP samples (NRCT) and healthy teeth (HT), at OTU level. (XLSX 29 kb)

Additional file 3: Differential analysis results between AP samples (NRCT) and healthy teeth (HT), at Genus level. (XLSX 12 kb)

Additional file 4: Clinical information of all the 50 samples. (XLSX 15 kb)

Additional file 5: Differential analysis results between subtypes A and B inNRCT at OTU level. (XLSX 22 kb)

Abbreviations

AP: Apical periodontitis; OTU: Operational taxonomic unit; RCT: Root canal treatment; tSNE: t-distributed stochastic neighbor embedding

Acknowledgements

Not applicable.

Data access

All raw sequences were deposited in the NCBI Sequence Read Archive (BioProject: PRJNA415692).

Funding

This work is supported by Natural Science Foundation of Shanghai (grant no. 17ZR1426700), and Scientific Research Project of Science and Technology Commission of Xuhui Municipality (grant no. SHXH201613). Publication of this article was sponsored by the grant 17ZR1426700.

Availability of data and materials

All raw sequences were deposited in the NCBI Sequence Read Archive (BioProject: PRJNA415692).

About this supplement

This article has been published as part of *BMC Genomics Volume 20 Supplement 2, 2019: Selected articles from the 17th Asia Pacific Bioinformatics Conference (APBC 2019): genomics*. The full contents of the supplement are available online at https://bmcgenomics.biomedcentral.com/articles/supplements/volume-20-supplement-2.

Authors' contributions

WQ and PH contributed conception and design of the study; WQ, MY and ZL recruit patients and collecting metagenome samples; WQ wrote the sections associated with patients recruitments and sample collection. YL performed the data analysis, and interpreted the bioinformatics results with PH; TM contributed preprocessing the sequencing data; YL and PH drafted the manuscript; All authors contributed to manuscript revision, read and approved the submitted version.

Ethics approval and consent to participate

This study was carried out in accordance with the recommendations of the Guide for the Use of Human Samples, Ethics Committee of Shanghai Xuhui Dental Clinic. The protocol was approved by the Ethics Committee of Shanghai Xuhui Dental Clinic. All subjects gave written informed consent in accordance with the Declaration of Helsinki.

The study protocol was approved by the Ethics Committee of Shanghai Xuhui Dental Clinic and written informed consent was obtained from all individuals.

Consent for publication
Not applicable.

Competing interests
The authors declare that they have no competing interests.

Publisher's Note

Author details
[1]Shanghai Xuhui District Dental Center, Shanghai 200032, China. [2]Key Laboratory of Molecular Virology and Immunology, Institut Pasteur of Shanghai, Chinese Academy of Sciences, Shanghai 200031, China.

Published: 10 April 2019

References

1. Cardoso FG, Ferreira NS, Martinho FC, Nascimento GG, Manhaes LRC, Rocco MA, et al. Correlation between volume of apical periodontitis determined by cone-beam computed tomography analysis and endotoxin levels found in primary root canal infection. J Endod. 2015;41:1015–9.
2. Endodontic Microbiology FAF. Pathobiology: current state of knowledge. Dent Clin N Am. 2017;61:1–15.
3. Siqueira JFJ, Rocas IN. Microbiology and treatment of acute apical abscesses. Clin Microbiol Rev. 2013;26:255–73.
4. Mehrazarin S, Alshaikh A, Kang Mo K. Dental Molecular Mechanisms of Apical Periodontitis: Emerging Role of Epigenetic Regulators. Clin N Am. 2017;61(1):17–35.
5. Lee LW, Lan WH. A clinical evaluation of antimicrobial effectiveness of EDTA as an endodontic irrigant. Chin Dent J. 1988;7:106e12.
6. JFJr S, Rôças IN, Favieri A, Lima KC. Chemomechanical reduction of the bacterial population in the root canal after instrumentation and irrigation with 1, 2.5, and 5.25% sodium hypochlorite. J Endod. 2000;26:331–4.
7. Guneser MB, Akbulut MB, Eldeniz AU. Antibacterial effect of chlorhexidine-cetrimide combination, Salvia officinalis plant extract and octenidine in comparison with conventional endodontic irrigants. Dent Mater J. 2016;35: 736–41.
8. Valverde EM, Baca P, Ceballos L, Fuentes MV, Ruiz-Linares M, Ferrer-Luque CM. Antibacterial efficacy of several intracanal medicaments for endodontic therapy. Dent Mater J. 2017;36:319–24.
9. Vijayaraghavan R, Mathian VM, Sundaram AM, Karunakaran R, Vinodh S. Triple antibiotic paste in root canal therapy. J Pharm Bioallied Sci. 2012; 4(Suppl 2):S230–3.
10. Dumani A, Yoldas O, Yilmaz S, Akcimen B, Seydaoglu G, Kipalev A, et al. In vitro susceptibility of e.Faecalis and c.Albicans isolates from apical periodontitis to common antimicrobial agents, antibiotics and antifungal medicaments. J Clin Exp Dent. 2012;4:e1–7.
11. Tziafas D, Alraeesi D, Al Hormoodi R, Ataya M, Fezai H, Aga N. Preparation prerequisites for effective irrigation of apical root canal: a critical review. J Clin Exp Dent. 2017;9:e1256–63.
12. Sakamoto M, Rocas IN, Siqueira JFJ, Benno Y. Molecular analysis of bacteria in asymptomatic and symptomatic endodontic infections. Oral Microbiol Immunol. 2006;21:112–22.
13. JFJr S, Rocas IN. Uncultivated phylotypes and newly named species associated with primary and persistent endodontic infections. J Clin Microbiol. 2005;43:3314–9.
14. Sogin ML, Morrison HG, Huber JA, Mark Welch D, Huse SM, Neal PR, Arrieta JM, Herndl GJ. Microbial diversity in the deep sea and the underexplored "rare biosphere". PNAS. 2006;103:12115–20.
15. Alves FRE, Siquera JFJ, Carmo FL, Santos AL, Peixoto RS, Rocas IN, et al. Bacterial community profiling of cryogenically ground samples from the apical and coronal root segments of teeth with apical periodontitis. J Endod. 2009;35:486–92.
16. JFJr S, Alves FR, Rocas IN. Pyrosequencing analysis of the apical root canal microbiota. JEndod. 2011;37:1499–503.
17. Lee LW, Lee YL, Hsiao SH, Lin HP. Bacteria in the apical root canals of teeth with apical periodontitis. J Formos Med Assoc. 2017;116:448–56.
18. Vengerfeldt V, Spilka K, Saag M, Preem JK, Oopkaup K, Truu J, et al. Highly diverse microbiota in dental root canals in cases of apical periodontitis (data of Illumina sequencing). J Endod. 2014;40:1778–83.
19. JFJr S, Antunes HS, Rocas IN, Rachid CTCC, Alves FRF. Microbiome in the apical root canal system of teeth with post treatment apical periodontitis. PLoS One. 2016;11:e0162887.
20. Hong BY, Lee TK, Lim SM, Chang SW, Park J, Han SH, et al. Microbial analysis in primary and persistent Endodontic infections by using pyrosequencing. J Endod. 2013;39:1136–40.
21. Dingsdag S, Nelson S, Coleman NV. Bacterial communities associated with apical periodontitis and dental implant failure. Microb Ecol Health Dis. 2016; 27:31307.
22. Özok AR, Persoon IF, Huse SM, Keijser BJF, Wesselink PR, Crielaard W, et al. Ecology of the microbiome of the infected root canal system: a comparison between apical and coronal root segments. Int Endod J. 2012;45:530–41.
23. Tatikonda A, Sudheep N, Biswas KP, Gowtham K, Pujari S, Singh P. Evaluation of bacteriological profile in the apical root segment of the patients with primary apical periodontitis. J Contemp Dent Pract. 2017;18:44–8.
24. Rocas IN, JFJr S. Frequency and levels of candidate endodontic pathogens in acute apical abscesses as compared to asymptomatic apical periodontitis. PLoS One. 2018;13:e0190469.
25. Mark Welch JL, Rossetti BJ, Rieken CW, Dewhirst FE, Borisy GG. Biogeography of a human oral microbiome at the micron scale. Proc Natl Acad Sci U S A. 2016;113:E791–800.
26. Wu J, Peters BA, Dominianni C, Zhang Y, Pei Z, Yang L, et al. Cigarette smoking and the oral microbiome in a large study of American adults. The ISME J. 2016;10:2435–46.
27. Song J, Duan C, Sang Y, Wu S, Ru J, Cui X. Effects of graphene on bacterial community diversity and soil environments of haplic Cambisols in Northeast China. Forests. 2018;9:677.
28. Caporaso GJ, Bittinger K, Bushman FD, DeSantis TZ, Andersen GL, Knight R. PyNAST: a flexible tool for aligning sequences to a template alignment. Bioinformatics. 2010;26:266–7.
29. Bokulich NA, Subramanian S, Faith JJ, Gevers D, Gordon JI, Knight R, et al. Quality-filtering vastly improves diversity estimates from Illumina amplicon sequencing. Nat Methods. 2013;10:57–9.
30. Rideout JR, He Y, Navas-Molina JA, Walters WA, Ursell LK, Gibbons SM, et al. Subsampled open-reference clustering creates consistent, comprehensive OTU definitions and scales to billions of sequences. PeerJ. 2014;2:e545.
31. Edgar RC. Search and clustering orders of magnitude faster than BLAST. Bioinformatics. 2010;26:2460–1.
32. Caporaso JG, Kuczynski J, Stombaugh J, Bittinger K, Bushman FD, Costello EK, et al. QIIME allows analysis of high-throughput community sequencing data. Nat Methods. 2010;7:335–6.
33. Conover WJ. Practical nonparametric statistics. 3rd ed. New York: John Wiley & Sons; 1999.
34. Storey JD, Tibshirani R. Statistical significance for genomewide studies. PNAS. 2003;100:9440–5.
35. van der Maaten LJP, Hinton GE. Visualizing high-dimensional data using t-SNE. J Mach Learn Res. 2008;9:2579–605.
36. Schaumann S, Staufenbiel I, Scherer R, Schilhabel M, Winkel A, Stumpp SN, et al. Pyrosequencing of supra- and subgingival biofilms from inflamed peri-implant and periodontal sites. BMC Oral Health. 2014;14:157.
37. Coretti L, Cuomo M, Florio E, Palumbo D, Keller S, Pero R, et al. Subgingival dysbiosis in smoker and non-smoker patients with chronic periodontitis. Mol Med Rep. 2017;15:2007–14.
38. Camelo-Castillo AL, Mira A, Pico A, Nibali L, Henderson B, Donos N, et al. Subgingival microbiota in health compared to periodontitis and the influence of smoking. Front Microbiol. 2015;6:119.
39. Nakamura S, Morimoto Y, Kudo S. A lactose fermentation product produced by Lactococcus lactis subsp. lactis, acetate, inhibits the motility of flagellated pathogenic bacteria. Microbiology. 2015;161:701–7.

Wu et al. BMC Bioinformatics 2019, 20(Suppl 3):134
https://doi.org/10.1186/s12859-019-2644-5

BMC Bioinformatics

RESEARCH Open Access

Prediction of drug-disease associations based on ensemble meta paths and singular value decomposition

Guangsheng Wu[1], Juan Liu[1,2*] and Xiang Yue[1,3]

From The 17th Asia Pacific Bioinformatics Conference (APBC 2019)
Wuhan, China. 14–16 January 2019

Abstract

Background: In the field of drug repositioning, it is assumed that similar drugs may treat similar diseases, therefore many existing computational methods need to compute the similarities of drugs and diseases. However, the calculation of similarity depends on the adopted measure and the available features, which may lead that the similarity scores vary dramatically from one to another, and it will not work when facing the incomplete data. Besides, supervised learning based methods usually need both positive and negative samples to train the prediction models, whereas in drug-disease pairs data there are only some verified interactions (positive samples) and a lot of unlabeled pairs. To train the models, many methods simply treat the unlabeled samples as negative ones, which may introduce artificial noises. Herein, we propose a method to predict drug-disease associations without the need of similarity information, and select more likely negative samples.

Results: In the proposed EMP-SVD (Ensemble Meta Paths and Singular Value Decomposition), we introduce five meta paths corresponding to different kinds of interaction data, and for each meta path we generate a commuting matrix. Every matrix is factorized into two low rank matrices by SVD which are used for the latent features of drugs and diseases respectively. The features are combined to represent drug-disease pairs. We build a base classifier via Random Forest for each meta path and five base classifiers are combined as the final ensemble classifier. In order to train out a more reliable prediction model, we select more likely negative ones from unlabeled samples under the assumption that non-associated drug and disease pair have no common interacted proteins. The experiments have shown that the proposed EMP-SVD method outperforms several state-of-the-art approaches. Case studies by literature investigation have found that the proposed EMP-SVD can mine out many drug-disease associations, which implies the practicality of EMP-SVD.

Conclusions: The proposed EMP-SVD can integrate the interaction data among drugs, proteins and diseases, and predict the drug-disease associations without the need of similarity information. At the same time, the strategy of selecting more reliable negative samples will benefit the prediction.

Keywords: Drug repositioning, Drug development, Meta path, Commuting matrix, Singular value decomposition

*Correspondence: liujuan@whu.edu.cn
[1]School of Computer Science, Wuhan University, Wuhan, 430072, People's Republic of China
[2]Suzhou Institute of Wuhan University, Suzhou, 215123, People's Republic of China
Full list of author information is available at the end of the article

Background

De novo drug discovery is a complex systematic project which is expensive, time-consuming and with high failure risks. As reported, it will take 0.8–1.5 billion dollars and about 10–17 years to bring a small molecule drug into market, and during the development stage, almost 90% of the small molecules can not pass the Phase I clinical trial and finally be eliminated [1, 2]. For the approved drugs in market, their pharmacological and toxicological properties are clear and the drug safeties are often guaranteed, but only some of their indications are found. For example, there are 2589 approved small molecule drugs in DrugBank [3], and more than 25000 diseases in UMLS medical database [4], resulting in over 60 millions of drug-disease pairs. However, only less than 5% of the drug-disease pairs were identified to have therapeutic relationships, and most of the drug-disease relationships are unknown [5]. Therefore, to discover the new indications of approved drugs, known as drug repositioning, can greatly save money and time, especially can improve the success rate, has become a promising alternative for de novo drug development.

Historically, finding a new indiction of a drug is likely to be an accidental event with a bit of luck. For example, Minoxidil, originally for the treatment of hypertension, was found by chance to have the treatment efficacy for hair loss [6]; Sildenafil (trade name: Viagra), originally for the treatment of angina, was occasionally found to have the potential to treat erectile dysfunction [7]. Such occasional findings of the drugs' new indictions suggest a new methodology of drug development. However, the "pot-luck" approach can not promise drug repositioning effectively and efficiently. It is necessary to develop a computational method that helps to redirect approved drugs. Fortunately, with the accumulation of multiple omics data and the development of machine learning methods, it is possible to mine the drugs' potential indications in silico. Up to now, many computational methods have been proposed to find new indictions of drugs by predicting potential treatment relationships of drug-disease pairs.

Based on the hypothesis that the gene expression signature of a particular drug is opposite to the gene expression signature of a disease, some gene expression based methods [8, 9] have been proposed. Noticing that such kind of methods may fail to consider the different roles of genes and their dependencies at the system level, system-level based approach that integrates the gene expressions and related network has recently been proposed [10].

Recently, along with the increase of drugs and diseases related multi-omics data, many methods have been proposed to integrate multiple sources of data to predict the drug-disease interactions based on machine learning techniques. Gottlieb et al. proposed a method (PREDICT) to predict new associations between drugs and diseases by integrating five drug-drug similarities and two disease-disease similarities data [11]. Wang et al. proposed a computational framework based on a three-layer heterogeneous network model (TL-HGBI) by integrating similarities and interactions among diseases, drugs and drug targets [12]. Luo et al. utilized some comprehensive similarities about drugs and diseases, and proposed a Bi-Random walk algorithm (MBiRW) to predict potential drug-disease interactions [13]. Martinez et al. developed a method named DrugNet for drug-disease and disease-drug priorization by integrating heterogeneous data [14]. Wu et al. integrated comprehensive drug-drug and disease-disease similarities from chemical/phenotype layer, gene layer and treatment network layer, and proposed a semi-supervised graph cut method (SSGC) to predict the drug-disease associations [15]. Moghadam et al. adopted the kernel fusion technique to combine different drug features and disease features, and then built SVM models to predict novel drug indications [16]. Liang et al. integrated drug chemical information, target domain information and gene ontology annotation information, and proposed a Laplacian regularized sparse subspace learning method (LRSSL) to predict drug-disease associations [17]. Zhang et al. introduced a linear neighborhood similarity [18] and a network topological similarity [19], then proposed a similarity constrained matrix factorization method (SCMFDD) to predict drug-disease associations by making use of known drug-disease associations, drug features and disease semantic information [20].

However, most of the existed methods are facing two main problems: one is that most of them are based on the hypothesis that similar drugs treat similar diseases, thus they need the similarity information between drugs, proteins, diseases, and so on. However, the similarity data can be not easily obtained. People often need to customize a program to collect data and to calculate the similarities so as to satisfy their own needs. Moreover, the calculation of similarity scores depends on the adopted measures, which may lead that the similarity score of a pair varies dramatically from one method to another. For example, two proteins are similar according to their structures, while they may be dissimilar according to their sequences. Even worse, some features required for calculating the similarities may be unknown or unavailable, resulting that these methods fail to work [21]. The other problem is that supervised learning based methods usually need both positive and negative samples to train the prediction models, whereas the drug-disease pair data, like other biological data, is lack of experimental validated negative samples. To train the models, most of the existing methods randomly select some unlabeled samples as the negative ones. Obviously, such strategy is very rough, for we are not sure whether there are some positive samples uncovered in the unlabeled data.

In this paper, we propose a method, called EMP-SVD (Ensemble Meta Paths and Singular Value Decomposition), to detect drug-disease treatment relations by using drug-disease, drug-protein and disease-protein interaction data. Unlike other methods, EMP-SVD needs no similarity information at all. In order to integrate different kinds of interaction data and consider different dependencies, we introduce five meta paths. For each meta path, we first generate a commuting matrix based on the corresponding interaction data, and then get latent features of drugs and diseases by using SVD (Singular Value Decomposition). All drug-disease pairs can be represented by the features. Finally, we train a base classifier by using the Random Forest algorithm. Five base classifiers are combined as an ensemble model to predict the drug-disease interactions. The framework of our method is shown in Fig. 1. In order to train out a more reliable prediction model, we select more likely negative ones from unlabeled samples under the assumption that non-associated drug and disease pair have no common interacted proteins, which is different from other methods. To evaluate our proposed method, we will compare it with the state-of-the-art methods, and also do case studies by literature investigation.

Materials and methods

Data sets

In this paper, we mainly made use of the interaction data of drug-disease, drug-protein and disease-protein to build the prediction model. We collected such data from DrugBank [3, 22, 23], OMIM [24] and Gottlieb's data set [11]. Concretely, we collected 4642 drug-protein interaction data from DrugBank, involving 1186 drugs and 1147 proteins; 1365 disease-protein interactions from OMIM, involving 449 diseases and 1147 proteins; and 1827 drug-disease interactions from Gottlieb's data set, involving 302 disease, 551 drugs. Obviously, the heterogenous network composed of drugs, proteins, diseases and the known interactions is sparse. The statistic of the data is shown in Table 1.

Although our method does not need the similarity information, most of other machine learning based methods do need. For the convenience of comparison, we still collected the chemical structure of drugs and the sequence data of proteins from DrugBank. We computed the drug-drug chemical similarities according to their SMILES strings [25] via Openbabel tool [26], and the protein-protein similarities according to the sequence data by Smith-Waterman algorithm [27]. Moreover, we directly downloaded the disease-disease similarities from MimMiner [28].

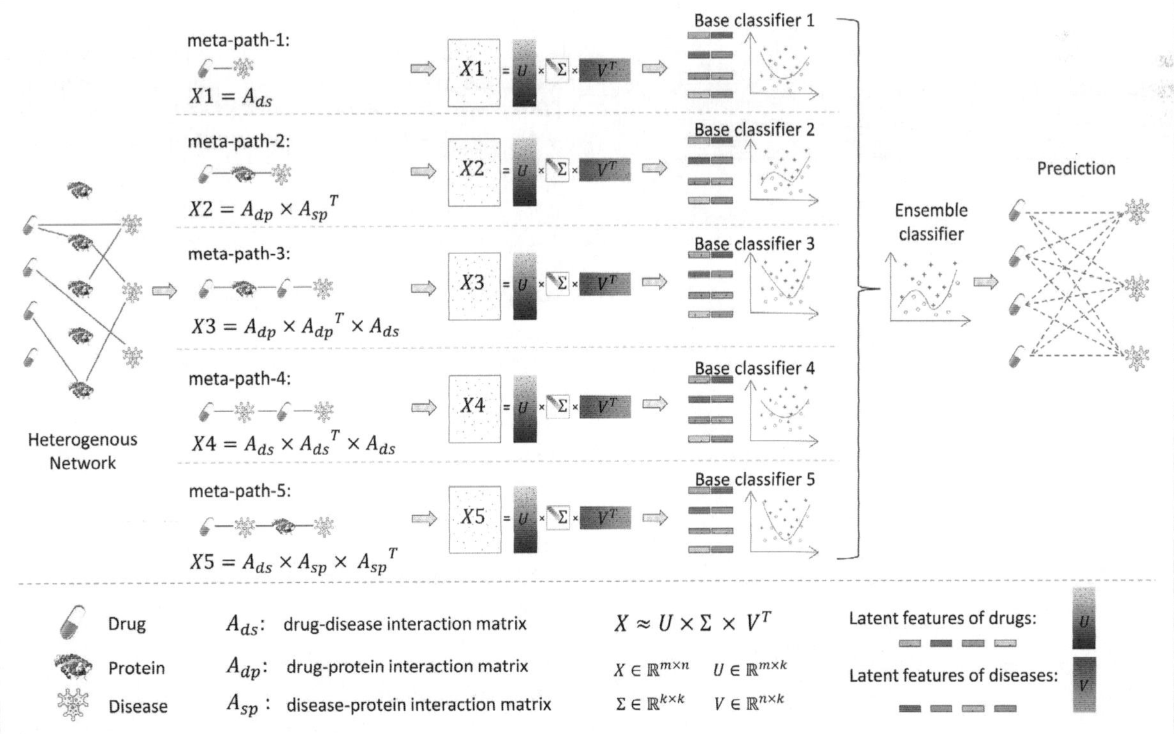

Fig. 1 The framework of our proposed EMP-SVD

Table 1 Statistic information of the drug-protein-disease heterogenous network

Type	Property	Number(Density)
Nodes	Drug	1186
	Protein	1147
	Disease	449
Known interactions	Drug ⟷ Protein	4642 (0.0034)
	Disease ⟷ Protein	1365 (0.0027)
	Drug ⟷ Disease	1827 (0.0034)

Density= #known interactions between node1 and node2 / (#node1 * #node2)

Definitions and notations

In this section, we will give the formal definitions and notations used in this paper.

Definition 1 (Heterogeneous drug-protein-disease network schema) . *For a given heterogenous drug-protein-disease network* $G = (V, E)$*, where* $V = D \cup P \cup S$*,* D*,* P *and* S *are the sets of drug, protein, disease nodes in the network respectively, while* $E = E_{d,p} \cup E_{p,d} \cup E_{p,s} \cup E_{s,p} \cup E_{d,s} \cup E_{s,d}$ *are the sets of heterogeneous links in G, which include the "binds to" link between drugs and proteins, "causes/caused by" link between proteins and diseases, "treats/treated by" link between drugs and diseases. The schema of G can be defined as* $M_G = (\mathcal{T}, \mathcal{R})$*, where* $\mathcal{T} = \{Drug, Protein, Disease\}$ *,* $\mathcal{R} = \{binds\ to, cuases, caused\ by, treats, treated\ by\}$ *,* $\mathcal{T}$ *and* $\mathcal{R}$ *are the sets of node types and link types in G, respectively.*

The network schema M_G severs as a template of a network G. For a drug-protein-disease heterogenous network, the network schema is shown in Fig. 2.

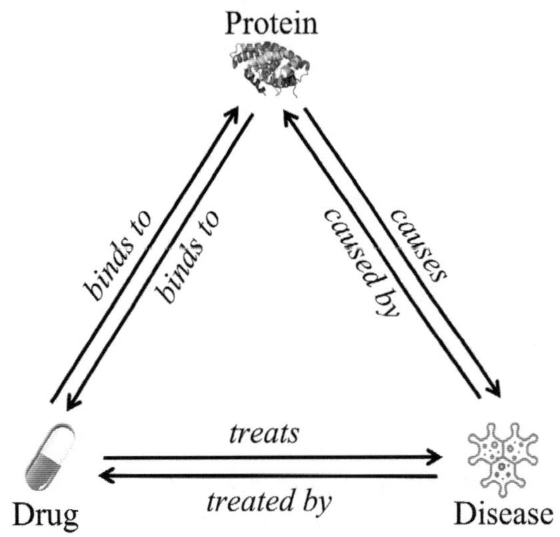

Fig. 2 Schema of drug-protein-disease heterogeneous network

Definition 2 (Heterogenous network meta path) *Based on a given heterogenous network schema* $M_G = (\mathcal{T}, \mathcal{R})$*,* $\mathcal{P} = T_1 \xrightarrow{R_1} T_2 \xrightarrow{R_2} ... \xrightarrow{R_{k-1}} T_k$ *is defined to be a heterogenous network meta path in network G, where* $T_i \in \mathcal{T}$*,* $i \in \{1, 2, ..., k\}$ *and* $R_i \in \mathcal{R}$*,* $i \in \{1, 2, ..., k-1\}$ *and if* $(T_1, T_2, ..., T_k$ *are not all the same)* $\vee$ $(R_1, R_2, ..., R_{k-1}$ *are not all the same).*

For simplicity, we also omit the link types in denoting the meta path if there is no multiple links between the two types, for examples, $\mathcal{P} = T_1 \rightarrow T_2 \rightarrow ... \rightarrow T_k$ denotes the meta path $\mathcal{P} = T_1 \xrightarrow{R_1} T_2 \xrightarrow{R_2} ... \xrightarrow{R_{k-1}} T_k$. The length of $\mathcal{P}$ is the number of links in $\mathcal{P}$.

Definition 3 (Commuting matrix [29]) *Given a network* $G = (V, E)$ *and its network schema* M_G*, a commuting matrix for a meta path* $\mathcal{P} = T_1 \rightarrow T_2 \rightarrow ... \rightarrow T_k$ *is defined as* $X = A_{T_1T_2}A_{T_2T_3}...A_{T_{k-1}T_k}$*, where* $A_{T_iT_j}$ *is the adjacency (interaction) matrix between type* T_i *and type* T_j*.* $X(i,j)$ *represents the number of path instances between object* $u_i \in T_1$ *and object* $v_j \in T_k$ *under meta path* $\mathcal{P}$*.*

Since we want to detect the interactions between the drugs and the diseases, we only consider the cases of $T_1 = Drug$ and $T_k = Disease$.

Now that there are only three kinds of nodes (drug, protein and disease) in the heterogenous network, we think the meta path with length greater than three may be too long to contribute to the prediction. Sun's work also has shown that short meta paths are good enough, and long meta paths may even reduce the quality [29]. Therefore, in this work, we only selected meta paths with length no longer than three. As a result, we select five meta paths described below.

Let A_{ds} be the drug-disease interaction matrix, A_{dp} be the drug-protein interaction matrix, and A_{sp} be the disease-protein interaction matrix, we can get the commuting matrices of the five meta paths as follows:

Meta-path-1: Drug $\xrightarrow{treats}$ Disease. The commuting matrix of it, denoted as $X1$, can be obtained by:

$$X1 = A_{ds} \tag{1}$$

Meta-path-2: Drug $\xrightarrow{binds\ to}$ Protein $\xrightarrow{causes}$ Disease. The commuting matrix of it, denoted as $X2$, can be obtained by :

$$X2 = A_{dp} \times A_{sp}^T \tag{2}$$

By using meta-path-2, we can integrate the drug-protein interaction information and the disease-protein interaction information, that is to say, we easily take the protein related information into account.

Meta-path-3: Drug $\xrightarrow{binds\ to}$ Protein $\xrightarrow{binds\ to}$ Drug $\xrightarrow{treats}$ Disease. The commuting matrix of it, denoted as $X3$, can be obtained by:

$$X3 = A_{dp} \times A_{dp}^T \times A_{ds} \quad (3)$$

By using meta-path-3, we can integrate drug-protein interaction and drug-disease interaction information. What's more, meta-path-3 also indicates that if two drugs share some common proteins, they may have similar indications.

Meta-path-4: Drug $\xrightarrow{treats}$ Disease $\xrightarrow{treated\ by}$ Drug $\xrightarrow{treats}$ Disease. The commuting matrix of it, denoted as $X4$, can be obtained by :

$$X4 = A_{ds} \times A_{ds}^T \times A_{ds} \quad (4)$$

By using meta-path-4, we can integrate the drug-disease interaction information. Besides, meta-path-4 also indicates that if two drugs share some common indications, then the indication of one drug may also be the potential indication of another drug.

Meta-path-5: Drug $\xrightarrow{treats}$ Disease $\xrightarrow{caused\ by}$ Protein $\xrightarrow{causes}$ Disease. The commuting matrix of it, denoted as $X5$, can be obtained by :

$$X5 = A_{ds} \times A_{sp} \times A_{sp}^T \quad (5)$$

By using meta-path-5, we can integrate the drug-disease interaction and the disease-protein interaction information. What's more, meta-path-5 also indicates that if two disease share some common proteins, the drug for treating one disease may also be the potential therapeutical drug for another disease.

As the definition, the element $X(i,j)$ of the commuting matrix X denotes the number of path instances from drug d_i to disease s_j under the corresponding meta path. We show an example in Fig. 3. There are two path instances from drug d_3 to disease s_2 under **Meta-path-2**, $d_3 \rightarrow p_3 \rightarrow s_2$ and $d_3 \rightarrow p_5 \rightarrow s_2$, thus we have $X2(3,2) = 2$ in commuting matrix $X2$.

Feature extraction with singular value decomposition

Now that element $X(i,j)$ in a commuting matrix X denotes the number of path instances from the drug d_i to disease s_j, then row i in the commuting matrix can be used as features of drug d_i, and column j can be used as features of disease s_j. And we can use the concatenation of them to represent the drug-disease pair. Suppose there are m drugs and n diseases, we will have $m+n$ (In this work, $m = 1186, n = 449$) features to represent the drug-disease pair. By contrast, the number of drug-disease pairs is small (We only have 1827 known interactions in this work). Obviously, the feature dimension is relatively high, which is not proper to construct a robust prediction model. Now that the singular value decomposition (SVD) has been successfully used to reduce the dimension in

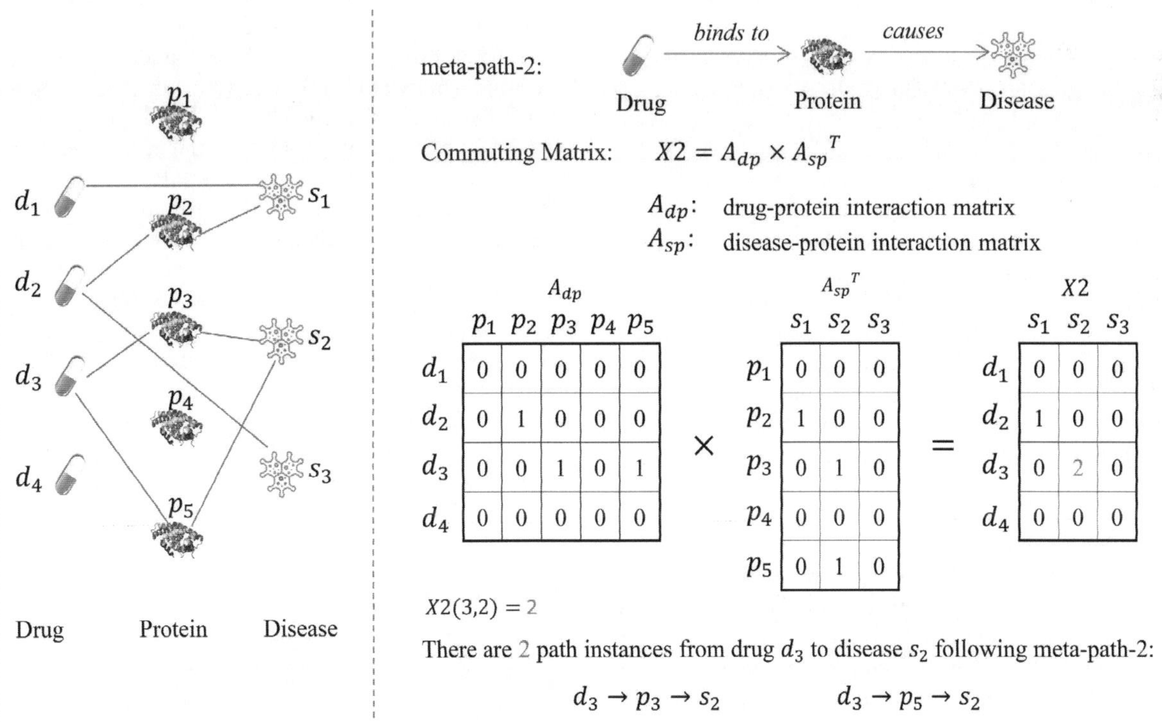

Fig. 3 An example of the meaning of commuting matrix

many researches, we also employed SVD to extract small number of features in our work.

By using SVD, the commuting matrix $X \in \mathbb{R}^{m\times n}$ can be factorized into U, Σ and V such that

$$X = U\Sigma V^T \tag{6}$$

where $U \in \mathbb{R}^{m\times m}$, $\Sigma \in \mathbb{R}^{m\times n}$ and $V \in \mathbb{R}^{n\times n}$. The diagonal entries of Σ are equal to the singular values of X (Other elements in Σ other than diagonal entries are 0). The columns of U and V are, respectively, left- and right-singular vectors for the corresponding singular values.

As is known to all, the magnitude of the singular values represents the importance of the corresponding vectors; and in Σ, the singular values are ordered in descending order. Moreover, in most cases, the sum of the first 10% or even 1% of the singular values is over 99% of the total sum of all singular values. Specifically in this drug-disease associations prediction problem, in the biomedical meaning, the most useful information about drug and disease features will be included in the first 10% even less singular values. In the process of dimensionality reduction, the useful data will not be lost, but the redundant information will be discarded. That is to say, we can use the top r singular values to approximate the matrix X:

$$X \approx U_{m\times r}\Sigma_{r\times r}{V^T}_{r\times n} \tag{7}$$

where $r \ll min(m,n)$.

Row i in U can be used as latent features of drug d_i, and row j in V can be used as latent features of disease s_j. As a result, the dimension of the latent feature vector of each drug-disease pair can be reduced to $2 * r$. In this work, we will introduce a parameter *latent_feature_percent* far less than 1 (say 1%, 2%,...) to control the value of r such that $r = latent\_feature\_percent \times min(m,n)$.

Selection of likely negative samples from unlabeled drug-disease pairs

To build a prediction model by using supervised learning, we need both positive and negative samples. The known drug-disease treatment relations are positive samples. Being lack of validated negative samples, most methods simply select some of unlabeled samples as negative ones by random. However, the unlabeled samples are not necessarily negative, some of them may be positive samples that still remain uncovered by experiments [30]. Different with other methods, we try to find more reliable negative samples from the unlabeled ones in this work.

If a drug shares some proteins with a disease, then the drug may have potential to treat the disease. Intuitively, if a drug and a disease have no common related proteins, we can think the disease is not the indication of the drug, and thus the drug-disease pair is more likely a negative sample. By this means, we can select out more reliable negative samples from the unlabeled pairs based on the drug-protein and disease-protein interactions information. The procedure is listed in Algorithm 1.

Algorithm 1 proteinFilteredNegative

Input:
  Drug-protein interaction matrix $A_{dp} \in \mathbb{R}^{m\times p}$;
  Disease-protein interaction matrix $A_{sp} \in \mathbb{R}^{n\times p}$.
Output:
  Filtered negative samples set *filterNeg* .
$X = A_{dp} \times A_{sp}^T$;
filterNeg $\leftarrow \emptyset$;
for $i = 1$ to m **do**
  for $j = 1$ to n **do**
    if $X(i,j) = 0$ **then**
      filterNeg $\leftarrow$ *filterNeg* $\cup$ (i,j) *pair*;
    end if
  end for
end for
return *filterNeg*;

Construction and ensemble of classifiers

The five meta paths we have selected to integrate heterogeneous data reflect different aspects of the drug-disease treatment relationship, such as two drugs with common proteins having similar indications, two drugs sharing one common indication also sharing another indication, and so on. Thus we can build five base classifiers for the prediction of drug-disease treatment relations from different sides. In our work, the base classifiers are built based on the Random Forest algorithm which was implemented by using the *RandomForestClassifier* function in the scikit-learn package [31], we set the number of trees as 256.

Since ensemble learning can often help to improve the performances [32, 33], after the five base classifiers are constructed, we can obtain an ensemble classifier. For an input of drug-disease pair, each base classifier outputs two probabilities indicating that the pair being negative and positive respectively. Since we want to know whether the pair has treatment relation, we only take the positive probability as considered in the ensemble model.

For a drug-disease pair x with unknown label, suppose the predicted score (probability) of each base classifier be $h_i(x), i = 1, 2, ...5$, we used average strategy to get the final score of the ensemble model:

$$H(x) = \frac{1}{5}\sum_{i=1}^{5} h_i(x) \tag{8}$$

If $H(x)$ is greater than a predetermined threshold, then the sample x is predicted as the positive. Because F_1-measure is a comprehensive metric, in this work, we let the program automatically determine the threshold value when F_1-measure reaches the maximum value, which is the same strategy as the other researchers used.

Experiments and results

We perform 5-fold cross validation to evaluate our method. Since the filtered negative samples are more than the positive ones, we randomly select a subset from them that with size equal to the positives, and use the balanced data to train the models. We first select the appropriate number of features according to the relationship of the model performance and the feature number. Then we did three kinds of evaluation experiments: (1) We investigate whether our negative samples filtering strategy can help to improve the prediction performance; (2) We compare EMP-SVD with other state-of-the-art methods by using the same data; (3) We check the practicality of our method by doing case studies.

Evaluation metrics

Just as most other work, we performed 5-fold cross validation in the experiments. To evaluate performance of a method, there are some common metrics: Precison (*PRE*), Recall (*REC*), Accuracy (*ACC*), Matthews Correlation Coefficient (*MCC*) and F_1-measure (F_1). They can be calculated according to the following equations:

$$PRE = \frac{TP}{TP + FP} \tag{9}$$

$$REC = \frac{TP}{TP + FN} \tag{10}$$

$$ACC = \frac{TP + TN}{TP + FP + TN + FN} \tag{11}$$

$$MCC = \frac{TP \times TN - FP \times FN}{\sqrt{(TP + FP)(TP + FN)(TN + FP)(TN + FN)}} \tag{12}$$

$$F_1 = \frac{2 \times PRE \times REC}{PRE + REC} \tag{13}$$

where *TP*, *FP*, *TN* and *FN* denote the number of true positive samples, false positive samples, true negative samples and false negative samples, respectively.

Since Precision(*PRE*) and Recall(*REC*) have some conflicts, in general, a classifier gets a higher *PRE* will have a lower *REC*, and vise versa. To get a comprehensive performance, Area Under Precison-Recall Curve(*AUPR*) and Area Under Receiver Operating Characteristic Curve(*AUC*) are often used. *AUPR* takes both *PRE* and *REC* into account, *AUC* takes both the true positive rate(*TPR*, the same as *REC*) and the false positive rate (*FPR*) into account, so they are comprehensive metrics. At the same time, with the help of the curves we can intuitively find which classifier is better. Therefore, in this work, we adopted *AUPR* and *AUC* as the main metrics.

Determination of appropriate number of features

Parameters are often used in existing computational methods, which limits the generalization of a model. So, it will be better to use fewer parameters or to get an analytical solution.

In this work, we just need to determine the number of singular values (corresponding to the feature number that is controlled by the parameter *latent_feature_percent*) during the model construction, which is very different with most state-of-the-art methods. Just mentioned above $r \ll min(m, n)$, so we set *latent_feature_percent* as 1%, 2%, 3%, ..., 20% respectively, and the performance curves of five base classifiers and the ensemble one with different *latent_feature_percent* are shown in Fig. 4. The results have shown that the performances of the ensemble classifier are better than other five base classifiers, illustrating that our ensemble rule is effective. Moreover, the performances of the six classifiers are robust across different parameter settings. Anyway, we set *latent_feature_percent* as 3% according to the curves in this work.

We also find that the performances of classifiers based on meta-path-1 and meta-path-4 are the worst. Noticing that both meta-path-1 and meta-path-4 just take drug-disease interactions into consideration, while the other three meta paths contain more information on drug-protein or protein-disease interactions, we think integrating more interaction information into the meta path can help to improve the performance of the classifier.

Investigation of the filtering strategy of negative samples

Being lack of validated negative samples, most of the other methods randomly select unlabeled samples to be negative ones. However, the unlabeled samples are not necessarily negative, some of them may be positive samples still uncovered by experiments. So in this work we selected out more likely negative samples from unlabeled ones according to the common protein information (as described in Algorithm 1). As shown in Table 2, all the classifiers achieve better performances in most metrics when using our negative samples filtering strategy. We also noted that the improvement is little, which may due to the fact that the known drug-protein interactions and disease-protein interactions are too few (with density of 0.0034 and 0.0027, as shown in Table 1), resulting that very few proteins could be used in the filtering process. Anyway, our strategy for selecting more reliable negative samples is useful, feasible and interpretable. We believe that along with the increase of interactions data, we will get more reliable negative

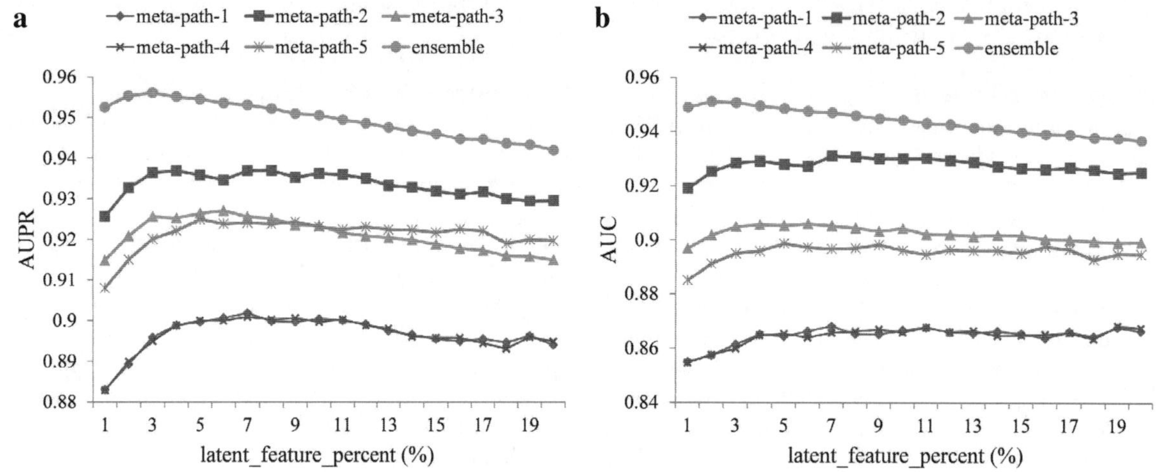

Fig. 4 Influence of different *latent_feature_percent* on the **a** AUPR **b** AUC

samples and thus achieve more great performance improvements.

Comparison with other methods

In this section, we compare EMP-SVD with state-of-the-art methods to demonstrate the superior performance of our method. PREDICT [11] and TL-HGBI method [12] are classical methods used to predict the drug-target and drug-disease interactions. MBiRW [13], LRSSL [17] and SCMFDD [20] are the methods proposed in these two years, and achieved high performance in the prediction of drug-disease interaction. So we choose these state-of-the-art methods to compare.

PREDICT calculates the score of a given drug-disease pair (d_r, d_i) according to all the known drug-disease pairs (d_r', d_i') associated with that given pair by equation $Score(d_r, d_i) = \max_{d_r', d_i' \neq d_r, d_i} \sqrt{S(d_r, d_r') \times S(d_i, d_i')}$, where $S(d_r, d_r')$ is drug-drug similarity and $S(d_i, d_i')$ is disease-disease similarity. TL-HGBI is a three layer heterogenous network model, which makes use of the similarities and interactions of drugs, diseases and targets by iterative update. MBiRW adjusts the similarities of drugs and diseases by correlation analysis and known drug-disease associations, then uses Bi-random walk algorithm to predict the potential drug-disease associations. LRSSL is a Laplacian regularized sparse subspace learning method used to predict the drug-disease associations which integrates drug chemical information, drug target domain information and target annotation information. SCMFDD is a similarity constrained matrix factorization method for the prediction of drug-disease associations by using known drug-disease interactions, drug features and disease semantic information.

We obtained the source code of PREDICT, TL-HGBI and SCMFDD from the authors, the code of MBiRW, LRSSL are publicly available, and the parameters were set according to their papers. The parameter *latent_feature_percent* in EMP-SVD was set 3%. To be fair, the five parts data were kept the same division in all methods when conducting 5-fold cross validation.

As shown in Table 3, compared with other five state-of-the-art methods which make use of several kinds of similarities as well as the interaction data, the proposed

Table 2 Performances comparison with different negative samples selecting strategies (random strategy is denoted "random", our strategy is "reliable")

Methods	AUPR		AUC		PRE		REC		ACC		MCC		F1	
	Random	Reliable	Random	Reliable	Random	Reliable	Random	Reliable	Random	Reliable	Random	Reliable	Random	Reliable
meta-path-1	0.894	0.896	0.859	0.861	0.786	0.771	0.875	0.891	0.835	0.835	0.673	0.677	0.827	0.826
meta-path-2	0.930	0.936	0.925	0.928	0.873	0.850	0.839	0.873	0.850	0.861	0.702	0.722	0.855	0.860
meta-path-3	0.921	0.926	0.902	0.905	0.826	0.832	0.862	0.883	0.843	0.858	0.690	0.719	0.842	0.855
meta-path-4	0.894	0.895	0.858	0.860	0.782	0.790	0.882	0.867	0.836	0.832	0.676	0.667	0.828	0.826
meta-path-5	0.918	0.920	0.892	0.895	0.809	0.800	0.900	0.925	0.859	0.865	0.721	0.737	0.852	0.858
ensemble	0.954	0.956	0.949	0.951	0.924	0.913	0.837	0.854	0.871	0.876	0.745	0.755	0.878	0.882

Table 3 Performances of proposed EMP-SVD and state-of-the-art methods

Methods	AUPR	AUC	PRE	REC	ACC	MCC	F_1
EMP-SVD	0.956	0.951	0.913	0.854	0.876	0.755	0.882
PREDICT	0.908	0.895	0.809	0.850	0.830	0.662	0.828
TL-HGBI	0.852	0.846	0.829	0.750	0.774	0.552	0.787
LRSSL	0.881	0.861	0.864	0.732	0.770	0.553	0.790
SCMFDD	0.836	0.854	0.926	0.713	0.774	0.575	0.805
MBiRW	0.952	0.942	0.867	0.901	0.884	0.769	0.884

classifier EMP-SVD only uses the known interaction data but achieves better performances in most metrics, especially the comprehensive metrics (AUPR and AUC). To make it more intuitively, we plotted the Precison-Recall Curve and ROC curve, which are shown in Fig. 5a and b, respectively. The AUPR and AUC of the proposed EMP-SVD are 0.956 and 0.951, respectively, better than the compared methods. Hence, it shows the simplicity and effectiveness of our method.

Case studies

Here, we test the practicality of EMP-SVD for predicting unknown associations. Except for training set composing of the known 1827 drug-disease associations and randomly selected 1827 negative samples by using our strategy, we used the trained EMP-SVD model to predict the associations for other unknown drug-disease pairs, and validate the results by literature investigation.

The new predicted top 20 drug-disease associations are shown in Table 4. We checked them carefully by literature validation and found that 13 of the top 20 predicted associations have been reported in the literatures. And these predicted associations were not originally in our data set, but we could find it out by our method, thus showing the practicality of our proposed EMP-SVD.

It should be noted that Triamcinolone (DrugBank ID: DB00620) and Betamethasone (DrugBank ID: DB00443), as glucocorticoid, are commonly used in the treatment of various skin diseases such as "Eczema" [34–36], and we find that their predicted associations include the disease "Growth Retardation, Small And Puffy Hands And Feet, And Eczema" (OMIM ID:233810). During the process of literature validation, we also find a case of growth retardation and Cushing's syndrome due to excessive application of betamethasone-17-valerate ointment [37]. In a responsible attitude, we think that whether they can be used to treat the disease "Growth Retardation, Small And Puffy Hands And Feet, And Eczema", or the usage and dosage should be further carefully studied by the chemists and doctors, especially should be with caution when used on children and pregnant women.

In more details, we checked the predicted potential indications of drug "Amitriptyline" (DrugBank ID: DB00321). Amitriptyline is a tricyclic antidepressant which is often used to treat symptoms of depression with the brand name: Vanatrip, Elavil, Endep. As shown in Table 5, we can

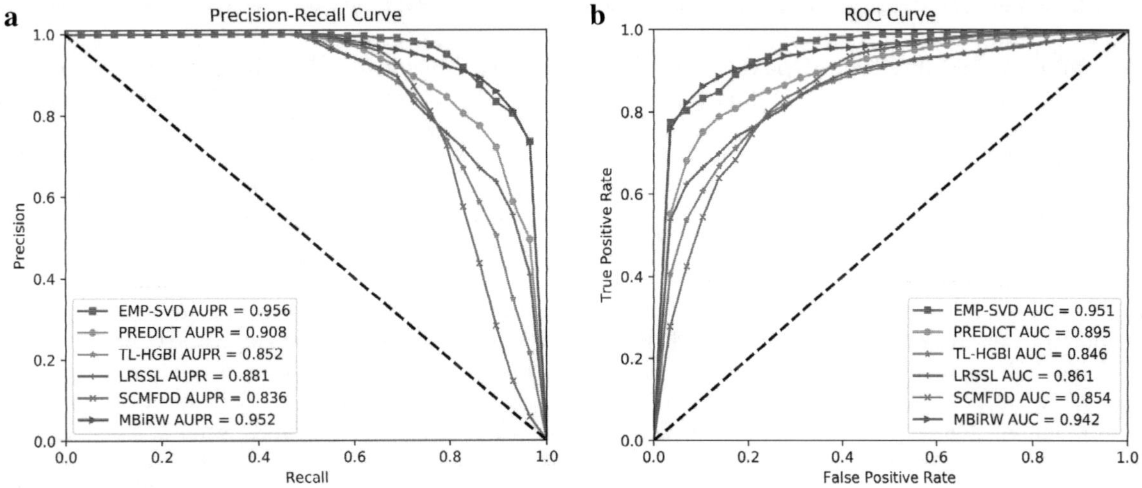

Fig. 5 **a** Precision-Recall Curve **b** ROC Curve of EMP-SVD and compared methods

Table 4 The predicted drug-disease associations (Top 20)

Rank	Score	DrugBank ID	Drug name	OMIM ID	Disease name	Literature validation
1	0.994	DB00776	Oxcarbazepine	239350	Hyperphosphatemia, Polyuria, And Seizures	[38]
2	0.992	DB01234	Dexamethasone	151590	Lichen Sclerosus Et Atrophicus; Lsa	[39]
3	0.991	DB00443	Betamethasone	233810	Growth Retardation, Small And Puffy Hands And Feet, And Eczema	[36]
4	0.991	DB00694	Daunorubicin	236000	Hodgkin Lymphoma	[40, 41]
5	0.987	DB01234	Dexamethasone	146850	Immune Suppression; Is	[42]
6	0.986	DB01013	Clobetasol propionate	233810	Growth Retardation, Small And Puffy Hands And Feet, And Eczema	N.A.
7	0.986	DB00620	Triamcinolone	125600	Dermatosis Papulosa Nigra	N.A.
8	0.986	DB00863	Ranitidine	600263	Helicobacter Pylori Infection, Susceptibility To	[43]
9	0.985	DB00620	Triamcinolone	233810	Growth Retardation, Small And Puffy Hands And Feet, And Eczema	[34, 35]
10	0.984	DB00694	Daunorubicin	267730	Reticulum Cell Sarcoma	[44]
11	0.984	DB00694	Daunorubicin	109543	Leukemia, Chronic Lymphocytic, Susceptibility To, 2	N.A.
12	0.984	DB00773	Etoposide	247640	Lymphoblastic Leukemia, Acute, With Lymphomatous Features; Lall	[45, 46]
13	0.984	DB00214	Torasemide	256370	Nephrotic Syndrome, Early-Onset, With Diffuse Mesangial Sclerosis	N.A.
14	0.983	DB00443	Betamethasone	188030	Thrombocytopenic Purpura, Autoimmune; Aitp	[47]
15	0.981	DB00444	Teniposide	601626	Leukemia, Acute Myeloid; Aml	[48, 49]
16	0.981	DB00481	Raloxifene	215470	Chorioretinal Dystrophy, Spinocerebellar Ataxia, And Hypogonadotropic	N.A.
17	0.980	DB00335	Atenolol	608622	Hypertension, Diastolic, Resistance To	[50]
18	0.980	DB00612	Bisoprolol	608622	Hypertension, Diastolic, Resistance To	[51]
19	0.980	DB00443	Betamethasone	146850	Immune Suppression; Is	N.A.
20	0.980	DB01177	Idarubicin	109543	Leukemia, Chronic Lymphocytic, Susceptibility To, 2	N.A.

N.A.: We haven't found the literature evidence

find literature evidences to support 8 diseases in the top 10 predictions for Amitriptyline.

Breast cancer is a relatively common malignant tumor for female, which seriously endangers women's health and life safety. To discover the potential drugs is of great value. So we also checked the drug list that have been predicted to treat the disease "Breast Cancer" (OMIM ID: 114480). In the top 10 drugs, as shown in Table 6, we found that 8 have been reported to be used in the clinical treatment.

Therefore, the case studies have further shown the practicality of the proposed method EMP-SVD.

Conclusions and discussions

To uncover the potential drug-disease associations is an important step in drug development, but it is time-consuming and costly to uncover them by wet experiments. Along with the accumulation of drug and disease related multi-omics data, as well as the development of machine learning techniques, more and more computational methods have been proposed to predict the potential drug-disease associations. To help the prediction, many methods integrate multiple source of data, including drugs, diseases, targets, side effects, and so on. They achieved good performances and could provide a helpful reference to the drug development. Most of them need the similarities of drug and disease related data. However, the similarity data can not be easily obtained, and people often need to customize a program to crawl data and to compute the similarities to satisfy their own need. Even worse, some features needed to calculate the similarity are unknown or unavailable. These methods will not work facing the incomplete data. Besides, being lack of validated negative samples in the prediction of drug-disease associations, most of the machine learning based methods assume the unlabeled samples to be negative ones in the training of the model. Such strategy may input errors because there may be positive samples uncovered in the unlabeled samples. What's more, most of the existing methods use many parameters in the data integration and the model construction. The parameters are difficult to tune, which limits the generalization ability of the method.

Table 5 Top 10 predictions for the drug "Amitriptyline"

Rank	Score	OMIM ID	Disease name	Literature validation
1	0.880	102300	Restless Legs Syndrome, Susceptibility To, 1; Rls1	[52]
2	0.877	200170	Acanthosis Nigricans With Muscle Cramps And Acral Enlargement	N.A.
3	0.843	143465	Attention Deficit-Hyperactivity Disorder; Adhd	[53]
4	0.837	600631	Enuresis, Nocturnal, 1; Enur1	[54]
5	0.837	600808	Enuresis, Nocturnal, 2; Enur2	[54]
6	0.817	608088	Neuropathy, Hereditary Sensory And Autonomic, Type I, With Cough And Gastroesophageal Reflux	[55]
7	0.803	145590	Hyperthermia, Cutaneous, With Headaches And Nausea	[56]
8	0.774	164230	Obsessive-Compulsive Disorder; Ocd	N.A.
9	0.769	167870	Panic Disorder 1; Pand1	[57]
10	0.745	600082	Prostatic Hyperplasia, Benign; Bph	[58]

N.A.: We haven't found the literature evidence

In this work, we proposed a method named EMP-SVD to predict drug-disease interactions based on ensemble meta paths and singular value decomposition. Five meta paths from source node (drug) to end node (disease) were selected to integrate the interaction information of drugs, proteins and diseases. Then the commuting matrices of these meta paths were calculated out, each element indicates the number of path instances between the corresponding drug and disease pair. By using singular value decomposition on the commuting matrices, we can extract small number of latent features of drugs and diseases. In order to get reliable negative samples, we selected those unlabeled samples as negative under the assumption that if a drug and a disease have no common proteins, then there is smaller probability for them to be treatment relationship. Based on each meta path we first built a base classifier, and then combined them to get an ensemble classifier. The experiments results have shown that our proposed EMP-SVD method outperformed several state-of-the-art methods. Better than other methods, EMP-SVD has few parameters and very easy to set. Further more, case studies have shown the predicted new associations could be useful for further biomedical research, which demonstrate the practicality of our method.

Table 6 Top 10 predictions for the disease "Breast Cancer"

Rank	Score	DrugBank ID	Drug name	Literature validation
1	0.931	DB00541	Vincristine	[59–61]
2	0.924	DB00399	Zoledronate	[62]
3	0.902	DB00987	Cytarabine	[63]
4	0.901	DB00884	Risedronate	[64]
5	0.893	DB01073	Fludarabine	N.A.
6	0.889	DB00755	Tretinoin	[65]
7	0.884	DB00762	Irinotecan	[66]
8	0.884	DB00630	Alendronate	[67]
9	0.880	DB01005	Hydroxyurea	[68]
10	0.878	DB01196	Estramustine	N.A.

N.A.: We haven't found the literature evidence

Although there are meta path based methods in social network and some other networks, to the best of our knowledge, it is the first work in the prediction of drug-disease associations by using ensemble meta paths and singular value decomposition. Different with many existing methods, we do not need the similarity data which are not easily obtained or sometimes unavailable or unknown. Instead, we just use the interaction data which can be easily accessed in many databases to build the prediction model. The other advantage of method is that there is only one parameter that can easily set. Though we use ensemble strategy to improve the performance, each of the five base classifiers can independently act as the model as well to predict the drug-disease interactions. Since there are many computational methods to predict the target proteins for a new drug such as docking methods. For a new drug which has no known interactions with any diseases, we still can predict its interacted diseases by building classifier using meta-path-2 by making use of drug-protein and protein-disease interactions.

Though the results of our methods are promising, there are still some limitations. Firstly, we only use the information of drugs, proteins and diseases, there are many other information could also be integrated in the further work, such as the information of side effects, pathways, tissues, and so on. Secondly, we only make use of common proteins to select out the negative samples, some other information such as gene expression data can also be used for this purpose. Or we can directly build the model by positive and unlabeled samples based learning method. We will address these issues in the future study.

Abbreviations

ACC: Accuracy; AUC: Area under ROC curve; AUPR: Area under Precison-Recall curve; FN: False negative; FP: False positive; FPR: False positive rate; MCC: Matthews correlation coefficient; OMIM: Online Mendelian inheritance in man; PRE: Precision; REC: Recall; ROC: Receiver operating characteristic; SMILES: Simplified molecular input line entry specification; SVD: Singular value decomposition; SVM: Support vector machine; TN: True negative; TP: True positive; TPR: True positive rate; UMLS: Unified medical language system.

Acknowledgements

The authors thank the editors and the anonymous reviewers for their helpful comments and suggestions on the quality improvement of our present paper.

Funding
Publication of this article was sponsored by grant from the National Science Foundation of China [61272274, 60970063]; the program for New Century Excellent Talents in Universities [NCET-10-0644]; and the National Science Foundation of Jiangsu Province [BK20161249].

Availability of data and materials
The data supporting the results of this research paper are included within this article.

About this supplement
This article has been published as part of *BMC Bioinformatics Volume 20 Supplement 3, 2019: Selected articles from the 17th Asia Pacific Bioinformatics Conference (APBC 2019): bioinformatics*. The full contents of the supplement are available online at https://bmcbioinformatics.biomedcentral.com/articles/supplements/volume-20-supplement-3.

Authors' contributions
GW, JL and XY developed the methodology. GW and XY executed the experiments, JL provided guidance and supervision. JL and GW wrote this paper. All authors read and approved the final manuscript.

Ethics approval and consent to participate
Not applicable.

Consent for publication
Not applicable.

Competing interests
The authors declare that they have no competing interests.

Author details
[1]School of Computer Science, Wuhan University, Wuhan, 430072, People's Republic of China. [2]Suzhou Institute of Wuhan University, Suzhou, 215123, People's Republic of China. [3]Department of Computer Science and Engineering, The Ohio State University, Ohio, 43210, USA.

Published: 29 March 2019

References

1. Adams CP, Brantner W. Estimating the cost of new drug development: is it really 802 million? Health Aff. 2006;25(2):420–8.
2. Wu G, Liu J, Wang C. Semi-supervised graph cut algorithm for drug repositioning by integrating drug, disease and genomic associations. In: 2016 IEEE International Conference on Bioinformatics and Biomedicine (BIBM). Institute of Electrical and Electronics Engineers Inc.; 2016. p. 223–228.
3. Wishart DS, Feunang YD, Guo AC, et al. Drugbank 5.0: a major update to the drugbank database for 2018. Nucleic Acids Res. 2017;46(D1):1074–82.
4. Bodenreider O. The unified medical language system (umls): integrating biomedical terminology. Nucleic Acids Res. 2004;32(suppl_1):267–70.
5. Hurle M, Yang L, Xie Q, et al. Computational drug repositioning: from data to therapeutics. Clin Pharmacol Ther. 2013;93(4):335–41.
6. Varothai S, Bergfeld WF. Androgenetic alopecia: an evidence-based treatment update. Am J Clin Dermatol. 2014;15(3):217–30.
7. Novac N. Challenges and opportunities of drug repositioning. Trends Pharmacol Sci. 2013;34(5):267–72.
8. Sirota M, Dudley JT, Kim J, et al. Discovery and preclinical validation of drug indications using compendia of public gene expression data. Sci Transl Med. 2011;3(96):96ra77.
9. Jahchan NS, Dudley JT, Mazur PK, et al. A drug repositioning approach identifies tricyclic antidepressants as inhibitors of small cell lung cancer and other neuroendocrine tumors. Cancer Discov. 2013;3(12):1364–77.
10. Peyvandipour A, Saberian N, Shafi A, et al. A novel computational approach for drug repurposing using systems biology. Bioinformatics. 2018;34(16):2817–2825.
11. Gottlieb A, Stein GY, Ruppin E, et al. Predict: a method for inferring novel drug indications with application to personalized medicine. Mol Syst Biol. 2011;7(1):496.
12. Wang W, Yang S, Zhang X, et al. Drug repositioning by integrating target information through a heterogeneous network model. Bioinformatics. 2014;30(20):2923–30.
13. Luo H, Wang J, Li M, et al. Drug repositioning based on comprehensive similarity measures and bi-random walk algorithm. Bioinformatics. 2016;32(17):2664–2671.
14. Martínez V, Navarro C, Cano C, et al. Drugnet: Network-based drug–disease prioritization by integrating heterogeneous data. Artif Intell Med. 2015;63(1):41–9.
15. Wu G, Liu J, Wang C. Predicting drug-disease interactions by semi-supervised graph cut algorithm and three-layer data integration. BMC Med Genet. 2017;10(5):79.
16. Moghadam H, Rahgozar M, Gharaghani S. Scoring multiple features to predict drug disease associations using information fusion and aggregation. SAR QSAR Environ Res. 2016;27(8):609–28.
17. Liang X, Zhang P, Yan L, et al. Lrssl: predict and interpret drug-disease associations based on data integration using sparse subspace learning. Bioinformatics (Oxford, England). 2017;33(8):1187–1196.
18. Zhang W, Yue X, Liu F, et al. A unified frame of predicting side effects of drugs by using linear neighborhood similarity. BMC Syst Biol. 2017;11(6): 101.
19. Zhang W, Yue X, Huang F, et al. Predicting drug-disease associations and their therapeutic function based on the drug-disease association bipartite network. Methods. 2018;145:51–59.
20. Zhang W, Yue X, Lin W, et al. Predicting drug-disease associations by using similarity constrained matrix factorization. BMC Bioinforma. 2018;19(1):233.
21. Zhang W, Yue X, Chen Y, et al. Predicting drug-disease associations based on the known association bipartite network. In: 2017 IEEE International Conference on Bioinformatics and Biomedicine (BIBM). Institute of Electrical and Electronics Engineers Inc.; 2017. p. 503–9.
22. Knox C, Law V, Jewison T, et al. Drugbank 3.0: a comprehensive resource for 'omics' research on drugs. Nucleic Acids Res. 2011;39(suppl 1):1035–41.
23. Law V, Knox C, Djoumbou Y, et al. Drugbank 4.0: shedding new light on drug metabolism. Nucleic Acids Res. 2013;42(D1):1091–7.
24. Hamosh A, Scott AF, Amberger JS, et al. Online mendelian inheritance in man (omim), a knowledgebase of human genes and genetic disorders. Nucleic Acids Res. 2005;33(suppl 1):514–7.
25. Weininger D. Smiles, a chemical language and information system. 1. introduction to methodology and encoding rules. J Chem Inf Comput Sci. 1988;28(1):31–6.
26. O'Boyle NM, Banck M, James CA, et al. Open babel: An open chemical toolbox. J Cheminformatics. 2011;3(1):33.
27. Smith TF, Waterman MS, Burks C. The statistical distribution of nucleic acid similarities. Nucleic Acids Res. 1985;13(2):645–56.
28. Van Driel MA, Bruggeman J, Vriend G, et al. A text-mining analysis of the human phenome. Eur J Hum Genet. 2006;14(5):535–42.
29. Sun Y, Han J, Yan X, et al. Pathsim: Meta path-based top-k similarity search in heterogeneous information networks. Proc VLDB Endowment. 2011;4(11):992–1003.
30. Wu G, Liu J, Min W. Prediction of drug-disease treatment relations based on positive and unlabeled samples. J Intell Fuzzy Syst. 2018;35(2):1363–73.
31. Pedregosa F, Varoquaux G, Gramfort A, et al. Scikit-learn: Machine learning in python. J Mach Learn Res. 2011;12(Oct):2825–30.
32. Guan D, Yuan W, Lee YK, et al. A review of ensemble learning based feature selection. Iete Tech Rev. 2014;31(3):190–8.
33. Zhang W, Shi J, Tang G, Wu W, Yue X, Li D. Predicting small rnas in bacteria via sequence learning ensemble method. In: 2017 IEEE International Conference on Bioinformatics and Biomedicine (BIBM). Institute of Electrical and Electronics Engineers Inc.; 2017. p. 643–7.
34. Keczkes K, Frain-Bell W, Honeyman A, Sprunt G. The effect on adrenal function of treatment of eczema and psoriasis with triamcinolone acetonide. Br J Dermatol. 1967;79(8–9):475–86.
35. Schmied C, Piletta P-A, Saurat J-H. Treatment of eczema with a mixture of triamcinolone acetonide and retinoic acid a double-blind study. Dermatology. 1993;187(4):263–7.
36. Granlund H, Erkko P, Eriksson E, Reitamo S. Comparison of cyclosporine and topical betamethasone-17, 21-dipropionate in the treatment of severe chronic hand eczema. Acta Derm Venereol. 1996;76(5):371–6.
37. Vermeer B, Heremans G. A case of growth retardation and cushings syndrome due to excessive application of betamethasone-17-valerate ointment. Dermatology. 1974;149(5):299–304.
38. Sillanpää M, Pihlaja T. Oxcarbazepine (gp 47 680) in the treatment of intractable seizures. Acta Paediatr Hung. 1988;29(3-4):359–64.

39. Virdi SK, Kanwar AS. Generalized morphea, lichen sclerosis et atrophicus associated with oral submucosal fibrosis in an adult male. Indian J Dermatol Venereol Leprology. 2009;75(1):56.
40. Willemze R, Peters W, Van Hennik M, et al. Intermediate and high-dose ara-c and m-amsa (or daunorubicin) as remission and consolidation treatment for patients with relapsed acute leukaemia and lymphoblastic non-hodgkin lymphoma. Scand J Haematol. 1985;34(1):83–7.
41. Richardson DS, Kelsey SM, Johnson SA, et al. Early evaluation of liposomal daunorubicin (daunoxome®, nexstar) in the treatment of relapsed and refractory lymphoma. Investig New Drugs. 1997;15(3):247–53.
42. Gustafson MP, Lin Y, New KC, et al. Systemic immune suppression in glioblastoma: the interplay between cd14+ hla-drlo/neg monocytes, tumor factors, and dexamethasone. Neuro-Oncol. 2010;12(7):631–44.
43. Hultén K, Jaup B, Stenquist B, Engstrand L. Combination treatment with ranitidine is highly eff icient against helicobacter pylori despite negative impact of macrolide resistance. Helicobacter. 1997;2(4):188–93.
44. Greenbaum-Lefkoe B, Rosenstock JG, Belasco JB, et al. Syndrome of inappropriate antidiuretic hormone secretion. a complication of high-dose intravenous melphalan. Cancer. 1985;55(1):44–6.
45. Winick NJ, McKenna RW, Shuster JJ, et al. Secondary acute myeloid leukemia in children with acute lymphoblastic leukemia treated with etoposide. J Clin Oncol. 1993;11(2):209–17.
46. Marks DI, Forman SJ, Blume KG, et al. A comparison of cyclophosphamide and total body irradiation with etoposide and total body irradiation as conditioning regimens for patients undergoing sibling allografting for acute lymphoblastic leukemia in first or second complete remission. Biol Blood Marrow Transplant. 2006;12(4):438–53.
47. Christiaens G, Nieuwenhuis H, von Dem Borne AK, et al. Idiopathic thrombocytopenic purpura in pregnancy: a randomized trial on the effect of antenatal low dose corticosteroids on neonatal platelet count. BJOG: Int J Obstet Gynaecol. 1990;97(10):893–8.
48. Ratain M, Rowley J. Therapy-related acute myeloid leukemia secondary to inhibitors of topoisomerase ii: from the bedside to the target genes. Ann Oncol. 1992;3(2):107–11.
49. Rivera GK, Pui C-H, Abromowitch M, et al. Improved outcome in childhood acute lymphoblastic leukaemia with reinforced early treatment and rotational combination chemotherapy. Lancet. 1991;337(8733):61–6.
50. Kamp O, Sieswerda GT, Visser CA. Comparison of effects on systolic and diastolic left ventricular function of nebivolol versus atenolol in patients with uncomplicated essential hypertension. Am J Cardiol. 2003;92(3): 344–8.
51. Williams B, MacDonald TM, Morant S, et al. Spironolactone versus placebo, bisoprolol, and doxazosin to determine the optimal treatment for drug-resistant hypertension (pathway-2): a randomised, double-blind, crossover trial. Lancet. 2015;386(10008):2059–68.
52. Allen RP, Bharmal M, Calloway M. Prevalence and disease burden of primary restless legs syndrome: results of a general population survey in the united states. Mov Disord. 2011;26(1):114–20.
53. Barrickman LL, Perry PJ, Allen A, et al. Bupropion versus methylphenidate in the treatment of attention-deficit hyperactivity disorder. J Am Acad Child Adolesc Psychiatry. 1995;34(5):649–57.
54. Sinha R, Raut S. Management of nocturnal enuresis-myths and facts. World J Nephrol. 2016;5(4):328.
55. Jeyakumar A, Brickman TM, Haben M. Effectiveness of amitriptyline versus cough suppressants in the treatment of chronic cough resulting from postviral vagal neuropathy. Laryngoscope. 2006;116(12):2108–12.
56. Di Lorenzo C, Ambrosini A, Coppola G, Pierelli F. Heat stress disorders and headache: a case of new daily persistent headache secondary to heat stroke. J Neurol Neurosurg Psychiatry. 2008;79(5):610–1.
57. Gawin FH, Markoff RA. Panic anxiety after abrupt discontinuation of amitriptyline. Am J Psychiatry. 1981;138(1):117–118.
58. Beckman TJ, Mynderse LA. Evaluation and medical management of benign prostatic hyperplasia. In: Mayo Clinic Proceedings, vol. 80. Elsevier Inc.; 2005. p. 1356–62.
59. Cassidy J, Merrick MV, Smyth JF, Leonard RC. Cardiotoxicity of mitozantrone assessed by stress and resting nuclear ventriculography. Eur J Cancer. 1988;24(5):935–8.
60. Yap H-Y, Blumenschein GR, Tashima CK, et al. Combination chemotherapy with vincristine and methotrexate for advanced refractory breast cancer. Cancer. 1979;44(1):32–4.
61. Holland JF, Scharlau C, Gailani S, et al. Vincristine treatment of advanced cancer: a cooperative study of 392 cases. Cancer Res. 1973;33(6):1258–64.
62. Gnant M, Mlineritsch B, Schippinger W, et al. Endocrine therapy plus zoledronic acid in premenopausal breast cancer. N Engl J Med. 2009;360(7):679–91.
63. Mego M, Sycova-Mila Z, Obertova J, et al. Intrathecal administration of trastuzumab with cytarabine and methotrexate in breast cancer patients with leptomeningeal carcinomatosis. Breast. 2011;20(5):478–80.
64. Hines SL, Mincey BA, Sloan JA, et al. Risedronate, breast cancer. J Clin Oncol Off J Am Soc Clin Oncol. 2009;27(7):1047.
65. Fenner MH, Elstner E. Peroxisome proliferator-activated receptor-γ ligands for the treatment of breast cancer. Expert Opin Investig Drugs. 2005;14(6):557–68.
66. Ma CX, Ellis MJ, Petroni GR, et al. A phase ii study of ucn-01 in combination with irinotecan in patients with metastatic triple negative breast cancer. Breast Cancer Res Treat. 2013;137(2):483–92.
67. Thamake SI, Raut SL, Gryczynski Z, et al. Alendronate coated poly-lactic-co-glycolic acid (plga) nanoparticles for active targeting of metastatic breast cancer. Biomaterials. 2012;33(29):7164–73.
68. Vaughan W, Reed E, Edwards B, Kessinger A. High-dose cyclophosphamide, thiotepa and hydroxyurea with autologous hematopoietic stem cell rescue: an effective consolidation chemotherapy regimen for early metastatic breast cancer. Bone Marrow Transplant. 1994;13(5):619–24.

Kawaguchi *et al. BMC Bioinformatics* 2019, **20**(Suppl 3):130
https://doi.org/10.1186/s12859-019-2645-4

BMC Bioinformatics

RESEARCH Open Access

reactIDR: evaluation of the statistical reproducibility of high-throughput structural analyses towards a robust RNA structure prediction

Risa Kawaguchi[1,2*], Hisanori Kiryu[2], Junichi Iwakiri[2] and Jun Sese[1,3,4]

From The 17th Asia Pacific Bioinformatics Conference (APBC 2019)
Wuhan, China. 14-16 January 2019

Abstract

Background: Recently, next-generation sequencing techniques have been applied for the detection of RNA secondary structures, which is referred to as high-throughput RNA structural (HTS) analyses, and many different protocols have been used to detect comprehensive RNA structures at single-nucleotide resolution. However, the existing computational analyses heavily depend on the experimental methodology to generate data, which results in difficulties associated with statistically sound comparisons or combining the results obtained using different HTS methods.

Results: Here, we introduced a statistical framework, reactIDR, which can be applied to the experimental data obtained using multiple HTS methodologies. Using this approach, nucleotides are classified into three structural categories, *loop*, *stem/background*, and *unmapped*. reactIDR uses the irreproducible discovery rate (IDR) with a hidden Markov model to discriminate between the true and spurious signals obtained in the replicated HTS experiments accurately, and it is able to incorporate an expectation-maximization algorithm and supervised learning for efficient parameter optimization. The results of our analyses of the real-life HTS data showed that reactIDR had the highest accuracy in the classification of ribosomal RNA stem/loop structures when using both individual and integrated HTS datasets, and its results corresponded the best to the three-dimensional structures.

Conclusions: We have developed a novel software, reactIDR, for the prediction of stem/loop regions from the HTS analysis datasets. For the rRNA structure analyses, reactIDR was shown to have robust accuracy across different datasets by using the reproducibility criterion, suggesting its potential for increasing the value of existing HTS datasets. reactIDR is publicly available at https://github.com/carushi/reactIDR.

Keywords: RNA secondary structure, High-throughput structural analysis, Reproducibility

*Correspondence: kawaguchi-rs@aist.go.jp
[1]Artificial Intelligence Research Center, National Institute of Advanced Industrial Science and Technology, Aomi, Koto-ku, Tokyo, Japan
[2]Department of Computational Biology and Medical Sciences, Graduate School of Frontier Sciences, the University of Tokyo, Kashiwanoha, Kashiwa-shi, Chiba, Japan
Full list of author information is available at the end of the article

Background

RNA secondary structures play diverse roles in the fundamental RNA functions, such as gene expression regulation, translation, and localization, which are mediated by certain structural motifs or regional accessibilities [1, 2]. Due to the difficulties in the experimental determination of the secondary structures of RNAs prior to the development of the next-generation sequencing, a number of computational methods were developed to predict the secondary structures from sequences or sequence alignment data [3], but the time and computer memory requirements remain too high to be applicable at the genome-wide level [4]. To analyze the comprehensive landscape of RNA secondary structures, novel types of high-throughput experimental methods, such as PARS [5] and icSHAPE [6], have been developed using short-read next-generation sequencers, and are referred to as high-throughput RNA structural (HTS) analyses [7–9]. These methods involve the use of certain types of chemical reagents or enzymes that cause probing (e.g., modification or cleavage) at each RNA nucleotide with a different "reactivity" depending on the existence of base pairing. Using these approaches, RNA secondary structures are not directly predicted, instead, some structure-indicating scores which provide information about the molecular structures, such as reactivity scores, are obtained. The reactivity scores provide *in silico* analyses with the information necessary to guide secondary structure prediction [10], as well as to indicate the propensity of structural accessibility directly.

As summarized in Fig. 1a, each experimental method can be used for the detection of different types of structural footprints, in order to determine the single- or double-strandedness of each nucleotide, and finally obtain the structure scores from the distribution of structural footprints. For example, the PARS method involves the activity of RNase V1 and S1, which perform structure-specific cleavage at the stem (base-paired) and loop (unpaired) regions, respectively, of the RNA molecules with a complicated structural context [5]. Conversely, using icSHAPE [6], base modifications are introduced into the loop regions by 2-methylnicotinic acid imidazolide (NAI)-N_3 treatment and then detected according to a frequent drop-off at that location during reverse transcription.

While these HTS methods rely on different approaches and hold fundamental differences with regards to the structure propensity to be detected [11], they suffer from a common problem of experimental noise, such as the spurious signals that are attributed to reasons other than the experimental modifications. To reduce the misdetection of false positives, a naïve, but statistically not justified approach is the computing of the reactivity scores using the (relative) differences or ratios of the read levels between different experimental conditions, such as case/control (i.e., chemically treated samples *vs.* untreated ones) or V1/S1 treated, setting a filtering threshold for each dataset independently.

Another method for precise signal detection from sequencing read distribution is the repeated performance of experiments, since the consistency between independent replicates generally indicates the quality of the experiments beyond stochastic noise, particularly in the sequencing-based experiments [12, 13]. However, most of the existing HTS analytical approaches, including probabilistic models, such as Spats [8], ProbRNA [14], and PROBer [15], do not consider this information in order to improve the robustness of reactivity scoring. Although Mod-seeker [16] and BUMHMM [17] were developed to include the information on the reproducibility of read distribution, these and other existing probabilistic models still have issues with the high dependency on certain target experiments in their statistical or probabilistic framework, such as the empirical distribution fitted to icSHAPE-like analyses in BUMHMM. Due to the diversity of HTS methodologies, the distribution of structural footprints is highly heterogeneous (Additional file 1: Table S2 and Figures S1 and S2), and therefore, the models may not be applicable to other HTS methodologies.

The importance of comprehensive HTS data comparisons has been recently argued, in a study showing that the results obtained using different HTS approaches are less correlated than those obtained by using the same approach, and potentially contain largely non-overlapping conformational information [18]. To overcome HTS dataset heterogeneity for fair multiple experiment comparisons, one promising approach is the combining of a probabilistic model and supervised learning. Reliable structural information has been gathered in previous computational and experimental analyses [2] and many previous studies have applied supervised learning to *in silico* structure prediction to estimate the optimal model parameters [19]. However, to the best of our knowledge, there is no method that computes optimal structure scores by handling HTS raw read counts in a supervised learning manner to improve the reproducibility of HTS analyses (Fig. 1b). Therefore, computational methods for HTS analyses can be further improved, especially by developing approaches for handling multiple HTS experiments simultaneously for comprehensive understanding of the RNA secondary structure.

Here, we developed a novel method, reactIDR, to determine true reactivity signals using the replicated HTS data by calculating a statistically valid reliability score. To evaluate the reliability in a way applicable to a general HTS dataset, we extend a statistical method for chromatin immunoprecipitation (ChIP)-Seq peak detection, named irreproducible discovery rate (IDR) [13]. We applied a

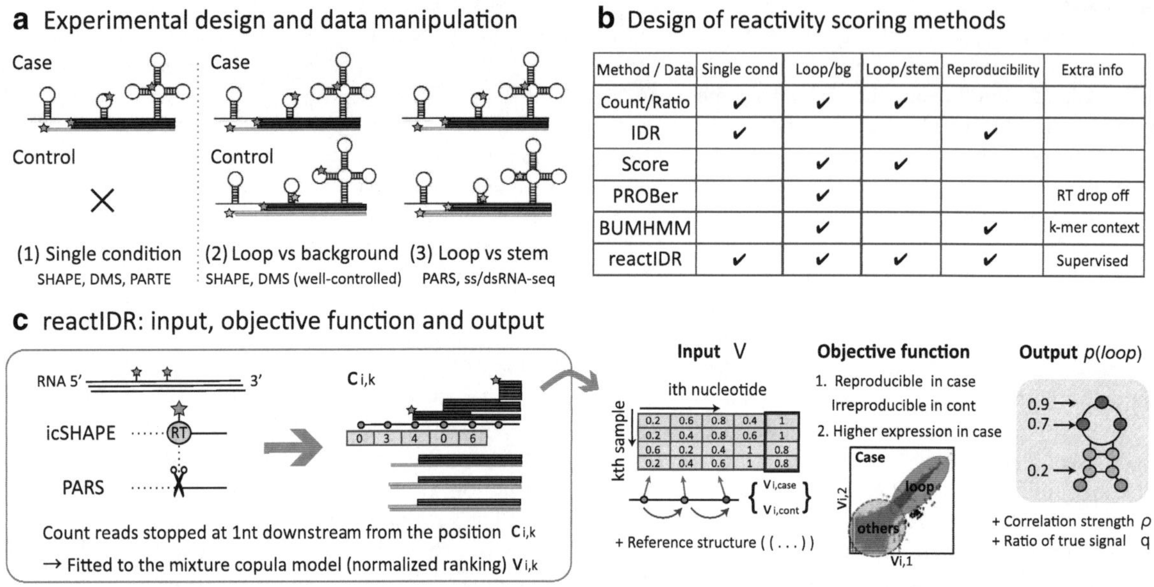

Fig. 1 Schematic illustration of the high-throughput structural (HTS) analytical process and reactIDR. **a** Major types of HTS-associated experimental designs analyzed in this study. All experimental approaches can be classified into the listed groups, according to the number of sample conditions and their detection targets. Red and blue stars represent a probing occurrence at the loop and stem regions, respectively. **b** Applicability of the reactivity-scoring methods to three types of HTS experimental designs (i.e., whether the method is intended to analyze each type of experiment or not), in which *Score* indicates each threshold-based scoring method used for studying the icSHAPE and PARS analyses. **c** Overview of reactIDR procedures. reactIDR receives the input data consisting of 5′-end read-depths repeatedly measured in a single or two conditions. The input data, after trimming of the very ends of the transcripts, is converted into the scaled ranking data to adapt the copula model. Afterward, a posterior loop probability distribution at each site can be computed as an index of reactivity

hidden Markov model (HMM) with the emission probability of IDR, in which the loop and stem regions are automatically segmented by a maximum posterior estimate. reactIDR is the first framework for supervised learning with known secondary structures obtained from replicated HTS sequencing data as well as applying parameter optimization based on the expectation-maximization (EM) algorithm to avoid any arbitrary step of parameter hand-tuning. According to our analyses, reactIDR has statistically one of the best accuracies for classifying loop and stem regions from HTS datasets containing multiple types of experiments. Therefore, reactIDR is suggested to be applicable to various HTS comparisons, including emerging technologies, and has the potential for increasing the value of existing HTS datasets.

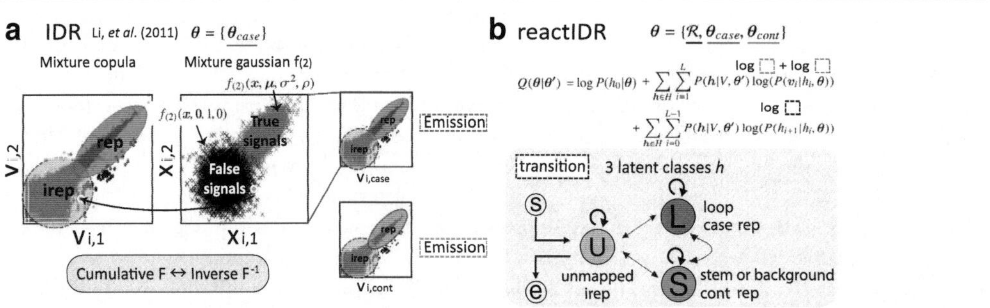

Fig. 2 Comparison between the irreproducible discovery rate (IDR) and reactIDR model. **a** IDR is a method used for the classification of signals into true (high-coverage and reproducible) and false (low-coverage and irreproducible) signals, based on the Gaussian mixture model, by associating the observation with the cumulative joint probability distribution and pseudo-value *x*. **b** In reactIDR, IDR is combined with the hidden Markov model of the three latent classes: *loop*, *stem/bg*, and *unmapped*. The start and end of transcripts should be assigned to the *unmapped* class. These three classes are defined to handle case/control comparisons with the Gaussian mixture copulas for each condition simultaneously, in which each parameter is optimized to the data and reference structure using the expectation-maximization algorithm

Methods

reactIDR overview

A schematic workflow of reactIDR is presented in Fig. 1c and Additional file 1: Figure S3. We aimed to develop reactIDR so that it can be applied for solving HTS problems as follows:

- **Input** – Assuming that K is the number of samples in an HTS experiment, and the sequencing reads were obtained from two $K/2$ samples, under two different conditions (and an optional reference structure), each read can be mapped against a reference sequence, and the indicator of reactivity at each base can be measured by counting the reads that start at the subsequent (3′) base.
- **Output** – The output represents the probability that each nucleotide is the latent class *loop*.
- **Objective function** – To provide reliable structure scores for accurate structure prediction, the outputs should be optimized so that they satisfy the following conditions: (1) the sequential loop regions of the RNAs provide consistently enriched read counts under the case conditions, in contrast to those under the control conditions, while stem motifs should be enriched in the control obtained from the PARS dataset, and (2) the true signals of structural footprints should be consistent among the replicates with the higher expression.

Using the HTS approach, the positions in stem (base paired) or loop (accessible) structures can be detected by the enrichment of chemical or enzymatic footprints, since their reactivity is highly affected by the existence of base pairs. However, to remove many false positives derived from the random occurrences of fragmentation or endogenous modification, reactIDR further considers the ratio of read count enrichment between two-conditional samples. As shown in Fig. 1, *case* (*control*) samples were specifically defined as the first (last) $K/2$ samples with the sequencing reads stochastically truncated at one base downstream from the loop (stem or non-specific background, represented as bg). In reactIDR, a latent class *loop*, *stem/bg*, or *unmapped*, is expected to be assigned to the region corresponding to the case-derived, control-derived, and no sufficient reads obtained, respectively. Here, icSHAPE case and control samples were regent-treated and untreated, respectively, while PARS samples were treated with S1 and V1, respectively, and the control samples were not the untreated samples.

IDR

Our novel algorithm, reactIDR, extends the idea of the IDR [13] criteria to obtain reproducible and irreproducible classification. We first introduce the model for the IDR estimation. Here, as an input of IDR and reactIDR, we considered the read-count data obtained for a single transcript or concatenated multiple transcripts with the total length of L nucleotides. As these data were initially converted into the ranking data across all positions in each experiment, these ranking data in an ascending order were scaled to the range of $[0, 1]$. These normalized data were hypothesized to correspond to the cumulative probability distribution in the copula model, necessarily within that range. After scaling, a vector of the ranks of read counts at the ith position was defined as $\mathbf{v}_i := \{v_{i,1}, \ldots, v_{i,K}\}$ for IDR, and $\mathbf{v}_i := \{\mathbf{v}_{i,case}, \mathbf{v}_{i,cont}\} = \{\{v_{i,1}, \ldots, v_{i,K/2}\}, \{v_{i,K/2+1} \ldots v_{i,K}\}\}$ for reactIDR. The overall input data of the IDR and reactIDR were: $V = \{\mathbf{v}_1, \ldots, \mathbf{v}_L\}$. IDR is based on the use of the mixture copula model to discriminate between one reproducible (represented by *rep*) distribution and the irreproducible (*irep*) distributions, recognizing them as true and false signals, respectively. Each copula can explain the distribution of true, reproducible or false, poorly correlated, signals. This is because true signals obtained from the protein binding sites in ChIP-Seq should exhibit reproducible enrichment with an increase in the read-count depth and a high correlation between the replicates. Based on this, the signals categorized into *rep* are thought to contain many more true signals compared with those in the *irep* group.

For simplicity, we considered an example of the mixture of two Gaussian copulas in two dimensions hereafter, showing that our approach can be used for the analysis of two experimental replicates. When the ith signal is produced by true (false) signals, represented as $r_i = true(false)$, the distribution of a random variable $x_i = (x_{i,1}, x_{i,2})$ corresponding to the ith signal is as follows:

$$\begin{pmatrix} x_{i,1} \\ x_{i,2} \end{pmatrix} \Bigg| h \sim \mathcal{N}\left(\begin{pmatrix} \mu_h \\ \mu_h \end{pmatrix}, \begin{pmatrix} \sigma_h^2 & \rho_h\sigma_h^2 \\ \rho_h\sigma_h^2 & \sigma_h^2, \end{pmatrix}\right) \qquad (1)$$

where μ_h, σ_h, and ρ_h represent mean, variance, and correlation, respectively, with the indicator h for each state of r_i ($h = 1$ when $r_i = true$ and $h = 0$ when $r_i = false$). The distribution of true *rep* signals is assumed to contain a positive mean and correlation ($\mu_1 > 0$ and $0 < \rho_1 \leq 1$), while spurious *irep* signals should be located around the low-coverage regions and show increased variance ($\mu_0 = 0$, $\sigma_0^2 = 1$, and $\rho_0 = 0$). These parameters can be fitted not to the actual read counts, but to the normalized ranks of read coverage for ChIP-Seq data $\mathbf{v}$, such that those ranks are derived from the cumulative marginal probability distribution function of $x_{i,k}$ ($k = 1, 2$), as shown below:

$$F_k(x_k) := v_{i,k} = \int_{-\infty}^{x_k} q\mathcal{N}\left(x'_k|\mu_1, \sigma_1^2\right) + (1-q)\mathcal{N}\left(x'_k|\mu_0, \sigma_0^2\right) dx'_k,$$

where q represents the ratio of true and reproducible signals in the samples and k represents the k-th experiment.

By considering the contribution of the second Gaussian distribution representing irreproducible signals, a probability of each peak derived from the irreproducible signal or IDR, can be estimated.

reactIDR: an algorithm based on the IDR and HMM

While IDR is a powerful method for the evaluation of the reliability of various joint distributions of ChIP-Seq peaks, HTS data used for RNA stem/loop detection must additionally include the consideration of the dependency between the consecutive nucleotides over multiple conditions. The application of IDR to HTS analyses, therefore, requires further specific extensions for the more complicated situations. To this end, we developed reactIDR, a novel method for the extraction of true reactivity signals from the replicated HTS data, based on the determination of statistical reliability score according to the IDR and HMM (Fig. 2).

For L as a total length of transcripts and a vector of scaled rankings at the ith position, $\mathbf{v_i}$ can be given. In reactIDR, a latent variable at the ith position h_i can belong to any element of {*loop, stem/bg,* and *unmapped*}, corresponding to each status associated with the HTS data and RNA secondary structures. The enrichment of $\mathbf{v_i}$ is thought to be most likely observed at the regions that belong to the *loop* class in the case samples and *stem/bg* class in the control samples, while we did not expect any specific enrichment in the case of *unmapped* classes.

In reactIDR, the paths of all latent variables can be represented by $\mathbf{h} = \{h_0, \ldots, h_L\}$, and the likelihoods of V and $\mathbf{h}$ can be obtained as the products of emission and transition probabilities, as formulated below:

$$P(V, \mathbf{h}|\theta) = P(h_0|\theta) \prod_{i=1}^{L} P(\mathbf{v_i}|h_i, \theta) \prod_{i=0}^{L-1} P(h_{i+1}|h_i, \theta) \tag{2}$$

where θ consists of the transition matrix between each latent class pair $\mathcal{R}$, and the set of all of parameters in the copula model for case samples θ_{case} and control samples θ_{cont}, respectively (i.e., μ, σ^2, and ρ, see Additional file 1: Figures S4 and S5). Due to the difficulties in the mapping of reads into the edge of transcripts, h_0 and h_L are always assigned to the *unmapped* class ($P(h_0|\theta) = 1$ and $P(h_L|\theta) = 1$). $P(h_{i+1}|h_i, \theta)$ represents the transition probability between h_i and h_{i+1} in the HMM, represented by $\mathcal{R}$, which is further optimized by the expectation of the transition between h_i and h_{i+1} at each step of the EM algorithm iteration. $P(\mathbf{v}_i|h_i, \theta)$ is an emission probability, which we defined as follows:

$$P(\mathbf{v}_i|h_i, \theta) = P(\mathbf{v}_{i,case}|r_{case}, \theta_{case}) \cdot P(\mathbf{v}_{i,cont}|r_{cont}, \theta_{cont})$$
$$r_{cont} := rep \text{ if } h_i = stem/bg \text{ and } irep \text{ otherwise}$$
$$r_{case} := rep \text{ if } h_i = loop \text{ and } irep \text{ otherwise},$$

where $\mathbf{v}_{case}$ and $\mathbf{v}_{cont}$ are expected to show the enrichment of loop-like and stem-like (or background noise) regions, respectively. These probabilities for *rep* and *irep* signals can be obtained by the mixture Gaussian copula, as when using the IDR.

Joint distribution $f_{(2)}$, cumulative marginal distribution $F_k(= v_{i,j,k})$, and the function of the vectorized inverse of F_k F^{-1} can be defined as follows:

$$f_{(2)}(\mathbf{x}, \mu, \sigma^2, \rho) = \mathcal{N}(\mathbf{x}|\mu, \sigma^2, \rho)$$
$$F_k(x_{j,k}) = \int_{-\infty}^{x_{j,k}} f_k(x'_k)\, dx'_k = \int_{-\infty}^{x_{j,k}} q_j \mathcal{N}(x'_k|\mu_h, \sigma_h^2) + (1 - q_j)\mathcal{N}(x'_k|\mu_0, \sigma_0^2)\, dx'_k$$
$$F^{-1}(\mathbf{v_j}) = \left\{F_1^{-1}(v_{i,j,1}), F_2^{-1}(v_{i,j,2})\right\},$$

where $h = 1$ (2) when $j = case$ (*cont*), and q_j represents the ratio of true signals in j samples. Afterward, the emission probability based on the mixture Gaussian copula consisting of *rep* and *irep* distributions with parameters θ_{case} or θ_{cont} can be formulated as follows:

$$P(\mathbf{v}_{i,j}|\, r_j, \boldsymbol{\theta}_j) = \begin{cases} \dfrac{f_{(2)}(F^{-1}(\mathbf{v}_{i,j}), \boldsymbol{\mu_h}, \sigma_h^2, \rho_h)}{\prod_{k=1}^{2} f_k\left(F_k^{-1}(v_{i,j,k})\right)} & \text{if } r_j = rep \\ \dfrac{f_{(2)}(F^{-1}(\mathbf{v}_{i,j}), \boldsymbol{\mu_0}, \sigma_0^2, \rho_0)}{\prod_{k=1}^{2} f_k\left(F_k^{-1}(v_{i,j,k})\right)} & \text{otherwise,} \end{cases}$$

where $\mu_k > 0$ and $0 < \rho_k \leq 1$ while $\mu_0 = 0$, $\sigma_0^2 = 1$, and $\rho_0 = 0$, the same as those in Eq.1 (further details can be found in Additional file 1). In this way, using reactIDR, we can compute a posterior probability distribution for each latent class at each site as an index of the reactivity and evaluate its reliability. Using the EM algorithm, each parameter can be iteratively optimized to maximize the expectation of likelihood Eq. 2 in reactIDR. Furthermore, reactIDR can incorporate supervised learning at this step, by limiting H so that it is consistent with the reference structure. Specifically, *loop, stem/bg,* and *unmapped* class is associated with loop, stem, and regions, respectively, in which *stem/bg* class is expected to play a role in removing false positives from true loop regions in the icSHAPE datasets. Afterward, trained parameters were used as the initial set of parameters in the re-fitting process for each rRNA sequence of the test set. The details of the optimization process, such as the derivative of Q for each parameter, are described in Additional file 1.

Datasets used and the evaluation of classification based on reactivity indicators

In this study, datasets generated by using two HTS methods, icSHAPE and PARS, were used to validate the accuracy of reactIDR. A whole-transcriptome PARS dataset of the native deproteinized transcriptome of GM12878 cells in vitro was obtained (GEO accession number, GSE50676)

[20]. Here, we used the normalized read counts of two replicates treated with nucleases S1 and V1 (hereafter referred to as S1 and V1, respectively). Analyses were also conducted using the icSHAPE dataset obtained for the HEK293T cells (GEO accession number, GSE74353), which contains sequencing reads obtained for three conditions with two replicates: dimethyl sulfoxide (DMSO)-, in vitro NAI-N3-, and in vivo NAI-N3-treated cells [21]. Computation of the original PARS and icSHAPE scores and the characteristics of these datasets can be found in Additional file 1: Section 2 and Figures S6 and S7.

To construct the reference set of the rRNA sequences, a human ribosomal repeating unit (NT_167214.1) was extracted from the NCBI database, and a 5S rRNA sequence (ENST00000364451) was additionally included. As a rRNA structure reference, cryo-EM-based ribosomal structure (PDB ID: 4v6x) was aligned to our reference sequence [22]. To obtain base-to-base correspondence between the reference sequences and the structure, 5S, 5.8S, 18S, and 28S rRNA reference sequences were aligned to each sequence within the structure dataset, and all bases successfully aligned to the reference were used for the evaluation of classification. Of the four rRNAs in the cryo-EM ribosomal structure, 18S rRNA reference structure was used as the training set for the supervised learning of reactIDR, with the negative alignment data for the minus strand of each rRNA, representing the read counts produced by misalignment. The accessible surface area (ASA) of the ribosome structure was calculated for each nucleotide using NACCESS with default parameter settings [23].

To evaluate the accuracy of structure classification based on reactivity indicators, we constructed a receiver operating characteristic (ROC) curve. In the ROC curve, the y-axis corresponds to the true positive rates, while the x-axis corresponds to the false positive rate. *P*-values were computed in order to compare the area under curve (AUROC) of reactIDR and other scoring methods by using the pROC library in R with a bootstrap method with 100 repetitions. We measured the accuracy of *in silico* structure prediction with or without the structure scores by positive predictive value (PPV) and sensitivity (identical to the true positive rates) for each possible base pair within the transcript except for pseudo-knots. The accuracy of the structure predictions based on multiple HTS datasets was investigated using the support vector machine (SVM) as well, through the scikit-learn library interface. For *in silico* structure prediction assisted by the structure scores, RNAfold from the Vienna RNA package v2.4.0 [3] was applied.

Results

Characteristics of the IDR-based structure classification with or without HMM

We investigated the improvements that can be obtained by combining HMM with IDR as criteria for stem/loop classification in reactIDR. To compare the read coverage distribution of HTS analysis between the case/control conditions and stem/loop positions, in vivo icSHAPE data were aligned to the reference sequence of rRNAs and read count (*Count*) was obtained for duplicated case and control samples. The structure scores based on the *Ratio* (normalized read count ratio versus the number of reads passing through that position), IDR, and reactIDR were also computed from the distribution of *Count* scores. In Fig. 3a, the distributions of the rank orders from the indices of chemical footprint enrichment for case and control samples, and stem and loop locus, individually, are presented. The locations of loop and stem structures were predicted for four human rRNAs using RNAview [24] for the 3D structure of ribosome determined by cryo-EM (details can be found in Additional file 1). The distributions of averaged *Count* and *Ratio* rank scores of the treated samples were shown to contain a larger number of

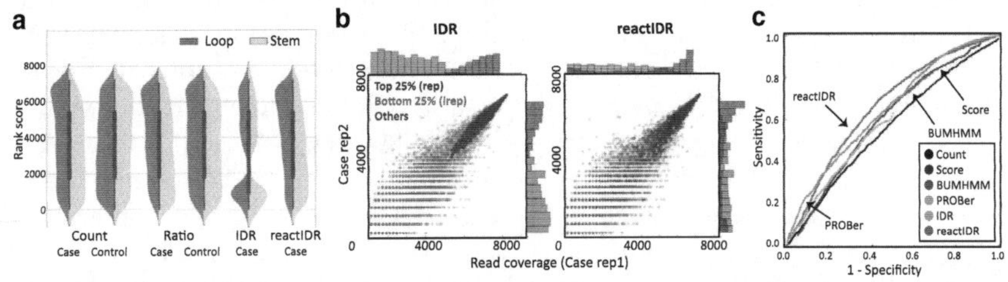

Fig. 3 Distribution of raw read counts and structure scores obtained using reactIDR at stem and loop loci. **a** Violin plots, showing the ascending rank orders of the four indices of chemical footprint enrichment that can be computed for four rRNAs from case and control samples separately: *Count*, raw read count; *Ratio*, read count normalized by the reads passing through the position; irreproducible discovery rate (IDR); and reactIDR. With the increase in the rank of the structure score obtained from the case samples, the likelihood of the nucleotide being in the loop structure increases. **b** Read count distribution of reproducible and irreproducible groups, classified into top 25% and bottom 25% rankings, based on the IDR (left) and reactIDR reactivity indicator(right) criteria. **c** Receiver operating characteristic (ROC) curve, showing stem/loop structure prediction of the test set which is composed of 5S, 5.8S, and 28S rRNAs based on the structure scores obtained using six scoring methods, for the in vivo icSHAPE dataset

high-ranking scores to a similar extent each other, compared with those of the control samples. The distribution of the IDR rank scores demonstrated two separate clusters associated with the reproducibility due to the characteristics of two Gaussian mixture copula models (Fig. 3b). Using reactIDR, more distinguishable distributions of the higher-ranked scores at the loop regions were obtained, compared with those at the stem regions. This may be because the structure scores obtained with reactIDR were estimated based on the case/control ratios and 18S rRNA reference structure, while other three indices were used only for the evaluation of the read count enrichment of the samples belonging to one condition.

Since the top rankings of reactIDR tended to contain the nucleotides with not only high read coverage (Fig. 3b), using reactIDR, we were able to fit more complicated enrichment patterns considering the read coverage of surrounding nucleotides. Moreover, even for the noisy genome-wide analysis of the PARS data, IDR-based classification was successful as well in associating the reproducibility of read enrichment and the strength of stem probability obtained *in silico* (Additional file 1: Figure S8). Taken together, using reactIDR case and control read enrichment criteria and employing the data showing the reproducibility and reference structure, we were able to infer the accessibility of each nucleotide with higher precision than that possible when using the raw read count.

Comparison of reactIDR structure classification accuracy for 2D accessibility

To evaluate the accuracy and robustness of reactIDR for different HTS methodologies, stem/loop classification based on the structure scores using multiple HTS rRNAs datasets was performed. For a fair comparison, only 18S rRNA data were subjected to parameter optimization using reactIDR and other rRNAs were subjected to comparison. In addition to the structure scores used in the previous analyses, three more types of reactivity scoring methods were computed for each dataset: *Score* used in the original HTS studies, BUMHMM [17], and PROBer [15]. Afterward, the AUROCs were computed for rRNA structure classification, while the structure status expected to be enriched (i.e., stem or loop) was set to positive. AUROCs for each classification obtained using icSHAPE and PARS datasets are shown in Table 1. reactIDR was observed to generate the highest AUROC for the icSHAPE datasets and only the prediction accuracy of PROBer was statistically comparable to the reactIDR results obtained using in vivo and in vitro icSHAPE datasets. This consistency was also confirmed for different choices of the training set as well (Additional file 1: Tables S3 and S4). In Fig. 3c, ROC curves of six structure score-based predictions for the in vivo icSHAPE dataset are presented

Table 1 Area under the ROC curve (AUROC), showing stem/loop classification of reactivity scoring methods

	icSHAPE		PARS	
Method	in vivo	in vitro	V1(cont)	S1(case)
Count	*0.555	*0.514	* 0.702	* 0.349
Score	*0.600	*0.564	*0.574	
BUMHMM	*0.592	*0.526	*0.634†	
PROBer	0.645	0.607	*0.639†	
IDR	*0.563	*0.518	* 0.664	* 0.375
reactIDR	**0.665**	**0.626**	**0.715**	0.420

* AUROC is significantly lower than that of reactIDR with *p*-value < 0.05 for the comparison after Bonferroni correction. † AUROCs obtained by methods not originally designed for PARS experiments. Structure scores of *Score*, *BUMHMM*, and *PROBer* are common for the case and control conditions of the PARS dataset. Their AUROC significances were computed for that of reactIDR for the PARS V1 condition

(see also Additional file 1: Figures S9 and S10). Using reactIDR, we achieved the highest accuracy for the medium portions ($0.1 - 0.6$) among the overall range of 1-specificity values, shown along the x-axis. For the stem (V1) prediction from the PARS datasets, reactIDR generated the highest AUROC among all reactivity scoring methods as well. In contrast, the prediction accuracy of all available methods was shown to be lower than that randomly obtained for the PARS S1 dataset, unless the common structure scores were computed for the stem/loop predictions using the S1 and V1 (*Score*, BUMHMM, and PROBer). As the enrichment of the read count was not observed at loop regions in the PARS S1 dataset (Additional file 1: Figure S11), the PARS S1 dataset of rRNAs itself is suggested not to follow the experimental hypothesis .

Moreover, we examined the accuracy of the structure scores from multiple HTS datasets combined with SVM and linear discriminant analysis, a representative machine learning method. Using the same training and test set of rRNAs, reactIDR showed the highest mean prediction accuracy compared with that obtained for the other methods by merging the features of structure scores for all datasets (Fig. 4a and Additional file 1: Figures S12 and S13). These results suggest the robustness of reactIDR when used in combination with different HTS approaches, and the potential of its use for the integrative analyses of various HTS datasets, regardless of the classification method. The reactIDR reliability was further examined for *in silico* structure prediction, particularly for base pair prediction, using RNAfold [3]. We applied one of the previously developed pseudo-free energy methods [10] to perform a hybrid structure prediction of computational and experimental structural analyses, in which the reactivity is accounted for in the log linear form as $\Delta G_{\mathrm{SHAPE}} = m \ln(\text{reactivity} + 1) + b$. In Fig. 4b, PPV and sensitivity of the minimum free energy structure predicted by RNAfold with structure scores of reactIDR are

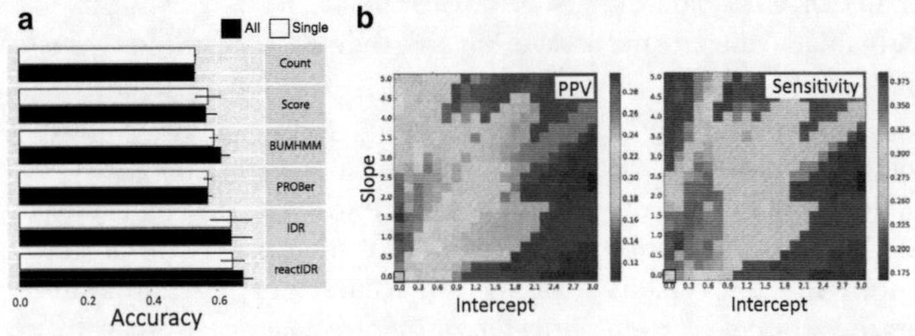

Best PPV and sensitivity

	PPV	Sensitivity
in silico	0.214	0.305
+PROBer	0.249	0.369
+reactIDR	**0.290**	**0.373**

Fig. 4 Application of structure scores for structure prediction using a machine learning approach and *in silico* pseudo-energy model [10] for the test set of three rRNAs. **a** Stem/loop structure prediction accuracy based on the structure scores computed from individual or combined datasets (icSHAPE in vivo and in vitro and PARS datasets) by the support vector machine, with the removal of indeterminable scores. Average and variance of accuracy were computed after 10-fold cross-validation. The predictions based on a single dataset utilized the structure scores from either in vivo icSHAPE dataset, in vitro icSHAPE dataset, or PARS V1 dataset, and the results of the maximum mean accuracy are presented. **b** Positive predictive value (PPV) and sensitivity of *in silico* structure prediction using RNAfold with and without reactivity assisting. The best PPV and sensitivity pair was selected to maximize their sum for each prediction condition

shown, with the progressive changes in the parameters of the slope m and intercept b. The structure scores of reactIDR with RNAfold were shown to perform better than *in silico* prediction and that for PROBer based on the best PPV and sensitivity pair selected to maximize the sum of pair (Fig. 4b and Additional file 1: Figures S14 and S15).

Taken together, the reactIDR scores were able to improve the prediction accuracy of integrative analyses as well as *in silico* structure prediction as conformational constraints. According to the assumptions used for reactIDR, the output structure scores of reactIDR correspond to the probability of the signals being true or false signals, which accounts for the reproducibility. Therefore, it may be theoretically applicable to the cross-comparison of exact reactivity scores between the datasets obtained by different HTS methods to combine them for further structure prediction, such as pseudo-free energy methods [10] and recently developed machine learning methods [19].

Correlation between RNA reactivity and 3D accessibility

The agreement between reactivity and 2D accessibility was demonstrated for rRNAs, but the reactivity of RNA in a chemical probing reaction appears to be affected by the 3D conformation, as observed in some previous studies [25], in particular, the relationship between the presence of the sugar puckers of RNA structures in $C2'$- or $C3'$-end and the modification efficiency of SHAPE reagent [26, 27]. To assess the influence of 3D accessibility on reactivity, we examined the correlation of the structure score with 2D and 3D accessibility, using ASA. In Fig. 5a, the relationship between ASA and structure score is presented, in which ASA is averaged with the 500-nt sliding window estimates for the test set of rRNAs from the in vivo icSHAPE dataset (Additional file 1: Figure S16). All analyzed methods showed a positive correlation between ASAs and structure score ranking (Fig. 5b), with the exception of a partial region of ties that appeared when using the BUMHMM or PROBer. Furthermore, using reactIDR, the highest correlation was obtained across six scoring methods with a statistically significant p-value ($p < 10^{-10}$). This result is consistent with the high accuracy of stem/loop classification because 2D accessibility and 3D accessibility are highly correlated as well (Additional file 1: Figure S17). On the other hand, the correlation between structure scores

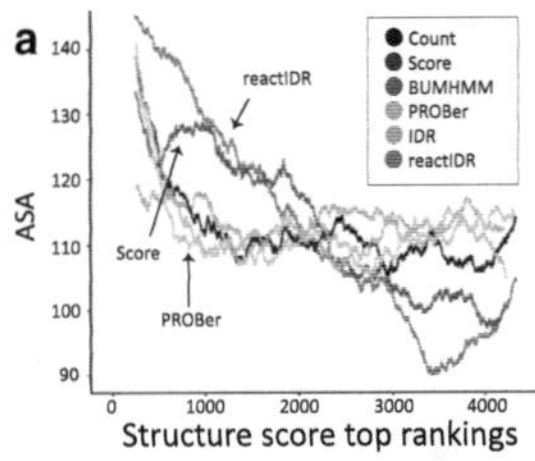

b

	vs 3D		vs 2D	
Method	in vivo	in vitro	in vivo	in vitro
Count	*0.099	*-0.109	*0.184	*0.132
Score	*0.207	-0.033	*0.207	*0.175
BUMHMM	0.021	*-0.155	*0.194	*0.111
PROBer	*0.096	*-0.092	*0.234	***0.205**
IDR	*0.105	*-0.104	*0.181	*0.124
reactIDR	***0.315**	***0.201**	***0.255**	*0.114

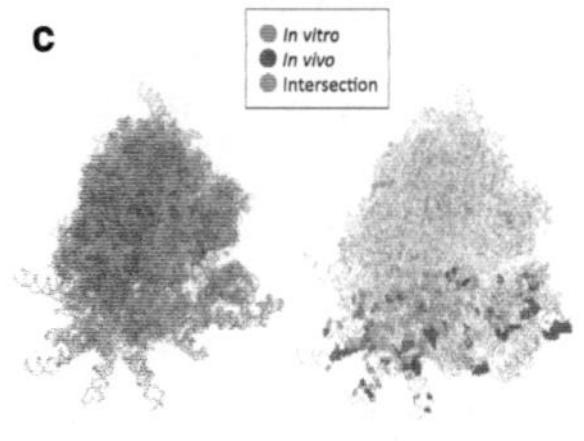

Fig. 5 Correlation between RNA reactivity and 3D accessibility for the test set of three rRNAs. **a** Relation between the accessible surface area (ASA) and reactivity scores computed by six types of scoring methods. y-axis, averaged ASAs within each 500-nucleotide sliding window; x-axis, a ranking of each nucleotide based on the structure scores, in which ties were randomly broken. The structure scores of reactIDR from the in vivo icSHAPE dataset presented the highest correlation with ASAs across the six scoring methods. **b** Correlations between structure scores and 2D or 3D accessibility indices. * p-value $< 10^{-5}$ after Bonferroni correction. **c** Ribosomal 3D structures visualized by PyMOL [29]. Left, ribosome changes its color from red to blue according to the ASAs of each nucleotide (high to low). Right, the highlighted regions indicate the top 7% reactive sites of 18S rRNA from the in vivo dataset (red), in vitro dataset (blue), and their intersection (purple)

and 3D accessibility was lower for the in vitro dataset, except for the moderate decline obtained when using reactIDR, with the subtle differences in the correlation between structure scores and 2D accessibility obtained when setting the loop and stem structures to 1 and 0.

The in vivo reactive sites (represented in red in Fig. 5c) were observed to be located at the outer side of ribosomes, unlike those obtained in vitro, and the regions with the highly reactive nucleotides in vivo and in vitro rarely overlapped, suggesting that the 3D conformation affects the results obtained when using the in vivo dataset. Since the aim of reactIDR is to allow comparisons between different HTS datasets in a fair manner, the reactIDR structure scores can potentially be used for the quantification of conformational changes between different conditions.

Discussion

In this study, reactIDR was shown to achieve an accuracy statistically comparable to that of PROBer (Fig. 3c and Additional file 1: Figure S10). Although reactIDR uses several assumptions for the analysis of HTS datasets, such as rank-based read count evaluation, localization of stem and loop loci, and the arbitrary assignment of latent classes during the supervised learning, our results suggest the robustness and rationality of reactIDR application to the HTS datasets (Additional file 1: Section 1.9).

To examine the robustness of reactIDR, we further examined the structure prediction of reactIDR from the SHAPE and mutational profiling (MaP) datasets obtained in [30], in which the reactivity was measured by capillary electrophoresis. In spite of the severe differences in experimental techniques, reactIDR still showed high and stable AUROCs (> 0.8) for *E. coli* rRNAs with the lower variances compared to the case using the average of the original reactivity scores (Additional file 1: Figure S18).

Importantly, although the increase of AUROCs was not substantial for its computational time (Additional file 1: Table S1), reactIDR is still considered to possess the advantage of helping researchers to set a reasonable threshold based on the probability being irreproducible. For example, when the threshold of reactIDR is set to 0.05 for the SHAPE-MaP dataset, true and false positive rate of the loop prediction is 0.81 and 0.39, respectively. Moreover, reactIDR is expected to be applicable to the heterogeneous HTS datasets which are known to have a specific bias by using post-processed scores after removing the biases. As such, further extensions can be considered to increase the applicability and flexibility of reactIDR for a variety of data, such as accounting for the base type dependency that clearly appeared in DMS-Seq [28] as well as further optimization for the HTS datasets without any replicates or more complicated structural footprints base-pair detection by RNA crosslink [21].

Conclusions

We have developed a novel software, reactIDR, for the prediction of stem/loop regions from the HTS analysis datasets, based on the reproducibility criterion. For the rRNA structure analyses, reactIDR was shown to have robust accuracy across different datasets. Moreover, the structure scores obtained with reactIDR enhanced the prediction accuracy by integrating multiple datasets and *in silico* structure analyses. Considering the 3D accessibility, reactIDR structure scores showed the significant and highest correlation, suggesting the potential of structure scores to reflect both 3D and 2D accessibility. Since reactIDR is the first method for the comparison of the HTS datasets obtained from multiple sources in a single unified model, it may help increase the accuracy of the RNA secondary structure predictions at transcriptome-wide level, allowing further studies.

Additional file

Additional file 1: The detail of mathematical background, experimental design, and results of additional experiments. (PDF 1680 kb)

Abbreviations

IDR: irreproducible discovery rate; HMM: hidden Markov model; HTS: high-throughput structural; EM: expectation-maximization; ROC: receiver operating characteristic; AUROC: area under the receiver operating characteristic curve; SVM: support vector machine; PARS: parallel analysis of RNA structure; SHAPE: selective 2′-hydroxyl acylation analyzed by primer extension; DMS: dimethyl sulfate; MaP: mutational profiling; NAI: 2-methylnicotinic acid imidazolide

Acknowledgements

Computations were performed using the supercomputing facilities at the Human Genome Center, the University of Tokyo (http://supcom.hgc.jp/shirokane.html). We also thank Tony Kuo and Raissa Relator for critically reading the manuscript.

Funding

This work, including software development, use of supercomputing facilities, and writing of the manuscript, was supported by a Grant-in-Aid for JSPS Fellows Grant Number 17J01882 (R.K.), KAKENHI Grant Number 15H01717 and 15H05713 (J.S.), and JST CREST Grant Number JPMJCR1502 and JPMJCR1689 (J.S.). Publication charges for this article have also been covered by the first grant.

Availability of data and materials

Source code of reactIDR is available at https://github.com/carushi/reactIDR. All original data including sequencing data are publicly available (see a subsection "Datasets used and the evaluation of classification based on reactivity indicators" and Additonal file 1).

About this supplement

This article has been published as part of *BMC Bioinformatics Volume 20 Supplement 3, 2019: Selected articles from the 17th Asia Pacific Bioinformatics Conference (APBC 2019): bioinformatics*. The full contents of the supplement are available online at https://bmcbioinformatics.biomedcentral.com/articles/supplements/volume-20-supplement-3.

Authors' contributions

RK and HK developed a method proposed in this study and RK did an implementation of the method. RK, HK, JI, and JS were engaged in the experimental design and wrote the manuscript. RK and JI did an experiment. All of the authors have read and approved the final manuscript.

Ethics approval and consent to participate
Not applicable.

Consent for publication
Not applicable.

Competing interests
The authors declare that they have no competing interests.

Publisher's Note

Springer Nature remains neutral with regard to jurisdictional claims in published maps and institutional affiliations.

Author details
[1]Artificial Intelligence Research Center, National Institute of Advanced Industrial Science and Technology, Aomi, Koto-ku, Tokyo, Japan. [2]Department of Computational Biology and Medical Sciences, Graduate School of Frontier Sciences, the University of Tokyo, Kashiwanoha, Kashiwa-shi, Chiba, Japan. [3]AIST- Tokyo Tech Real World Big-Data Computation Open Innovation Laboratory, Ookayama, Meguro-ku, Tokyo, Japan. [4]Humanome Lab Inc., Shinjuku-ku, Tokyo, Japan.

Published: 29 March 2019

References

1. Borujeni AE, Channarasappa AS, Salis HM. Translation rate is controlled by coupled trade-offs between site accessibility, selective RNA unfolding and sliding at upstream standby sites. Nucleic Acids Res. 2014;42(4): 2646–2659.
2. Nawrocki EP, Burge SW, Bateman A, Daub J, Eberhardt RY, Eddy SR, et al. Rfam 12.0: updates to the RNA families database. Nucleic Acids Res. 2015;43:D130-7.
3. Lorenz R, Bernhart SH, Zu Siederdissen CH, Tafer H, Flamm C, Stadler PF, et al. ViennaRNA Package 2.0. Algoritm Mol Biol. 2011;6(1):26.
4. Kawaguchi R, Kiryu H. Parallel computation of genome-scale RNA secondary structure to detect structural constraints on human genome. BMC Bioinformatics. 2016;17(1):1.
5. Kertesz M, Wan Y, Mazor E, Rinn JL, Nutter RC, Chang HY, et al. Genome-wide measurement of RNA secondary structure in yeast. Nature. 2010;467(7311):103–7.
6. Spitale RC, Flynn RA, Zhang QC, Crisalli P, Lee B, Jung JW, et al. Structural imprints in vivo decode RNA regulatory mechanisms. Nature. 2015;519(7544):486–90.
7. Underwood JG, Uzilov AV, Katzman S, Onodera CS, Mainzer JE, Mathews DH, et al. FragSeq: transcriptome-wide RNA structure probing using high-throughput sequencing. Nat Methods. 2010;7(12):995–1001.
8. Aviran S, Trapnell C, Lucks JB, Mortimer SA, Luo S, Schroth GP, et al. Modeling and automation of sequencing-based characterization of RNA structure. Proc Natl Acad Sci. 2011;108(27):11069–74.
9. Lucks JB, Mortimer SA, Trapnell C, Luo S, Aviran S, Schroth GP, et al. Multiplexed RNA structure characterization with selective 2′-hydroxyl acylation analyzed by primer extension sequencing (SHAPE-Seq). Proc Natl Acad Sci. 2011;108(27):11063–8.
10. Deigan KE, Li TW, Mathews DH, Weeks KM. Accurate SHAPE-directed RNA structure determination. Proc Natl Acad Sci. 2009;106(1):97–102.
11. Strobel EJ, Angela MY, Lucks JB. High-throughput determination of RNA structures. Nat Rev Genet. 2018;19:615–634.
12. Robinson MD, McCarthy DJ, Smyth GK. edgeR: a Bioconductor package for differential expression analysis of digital gene expression data. Bioinformatics. 2010;26(1):139–40.
13. Li Q, Brown JB, Huang H, Bickel PJ. Measuring reproducibility of high-throughput experiments. Ann Appl Stat. 2011;5(3):1752–79.
14. Hu X, Wong TK, Lu ZJ, Chan TF, Lau TCK, Yiu SM, et al. Computational identification of protein binding sites on RNAs using high-throughput RNA structure-probing data. Bioinformatics. 2013;30(8):1049–55.
15. Li B, Tambe A, Aviran S, Pachter L. PROBer Provides a General Toolkit for Analyzing Sequencing-Based Toeprinting Assays. Cell Syst. 2017;4(5): 568–74.
16. Talkish J, May G, Lin Y, Woolford JL, McManus CJ. Mod-seq: high-throughput sequencing for chemical probing of RNA structure. RNA. 2014;20(5):713–20.
17. Selega A, Sirocchi C, Iosub I, Granneman S, Sanguinetti G. Robust statistical modeling improves sensitivity of high-throughput RNA structure probing experiments. Nat Methods. 2017;14(1):83–89.
18. Sexton AN, Wang PY, Rutenberg-Schoenberg M, Simon MD. Interpreting reverse transcriptase termination and mutation events for greater insight into the chemical probing of RNA. Biochemistry. 2017;56(35):4713–21.
19. Ledda M, Aviran S. PATTERNA: transcriptome-wide search for functional RNA elements via structural data signatures. Genome Biol. 2018;19(1):28.
20. Wan Y, Qu K, Zhang QC, Flynn RA, Manor O, Ouyang Z, et al. Landscape and variation of RNA secondary structure across the human transcriptome. Nature. 2014;505(7485):706–9.
21. Lu Z, Zhang QC, Lee B, Flynn RA, Smith MA, Robinson JT, et al. RNA Duplex Map in Living Cells Reveals Higher-Order Transcriptome Structure. Cell. 2016;165(5):1267–79.
22. Anger AM, Armache JP, Berninghausen O, Habeck M, Subklewe M, Wilson DN, et al. Structures of the human and Drosophila 80S ribosome. Nature. 2013;497(7447):80.
23. Hubbard S, Thornton J. NACCESS: Department of Biochemistry and Molecular Biology, University College London. 1993. Software available at http://www.bioinf.manchester.ac.uk/naccess/nacdownload.html.
24. Yang H, Jossinet F, Leontis N, Chen L, Westbrook J, Berman H, et al. Tools for the automatic identification and classification of RNA base pairs. Nucleic Acids Res. 2003;31(13):3450–60.
25. Bindewald E, Wendeler M, Legiewicz M, Bona MK, Wang Y, Pritt MJ, et al. Correlating SHAPE signatures with three-dimensional RNA structures. RNA. 2011;17(9):1688–96.
26. Vicens Q, Gooding AR, Laederach A, Cech TR. Local RNA structural changes induced by crystallization are revealed by SHAPE. Rna. 2007;13(4):536–48.
27. Mlynsky V, Bussi G. Molecular Simulations Reveal an Interplay Between SHAPE Reagent Binding and RNA Flexibilty. J Phys Chem Lett. 2018;9(2): 313–18.
28. Rouskin S, Zubradt M, Washietl S, Kellis M, Weissman JS. Genome-wide probing of RNA structure reveals active unfolding of mRNA structures in vivo. Nature. 2014;505(7485):701–5.
29. Schrödinger LLC. The PyMOL Molecular Graphics System, Version 1.8. 2015. https://pymol.org/2/support.html.
30. Siegfried NA, Busan S, Rice GM, Nelson JAE, Weeks K. RNA motif discovery by SHAPE and mutational profiling (SHAPE-MaP). Nature Methods. 2014;11(9):959-65.

Ryu *et al. BMC Bioinformatics* 2019, **20**(Suppl 3):127
https://doi.org/10.1186/s12859-019-2646-3

BMC Bioinformatics

RESEARCH Open Access

A new class of constitutively active super-enhancers is associated with fast recovery of 3D chromatin loops

Jayoung Ryu[1†], Hyunwoong Kim[1†], Dongchan Yang[2], Andrew J. Lee[1] and Inkyung Jung[1*]

From The 17th Asia Pacific Bioinformatics Conference (APBC 2019)
Wuhan, China. 14-16 January 2019

Abstract

Background: Super-enhancers or stretch enhancers are clusters of active enhancers that often coordinate cell-type specific gene regulation during development and differentiation. In addition, the enrichment of disease-associated single nucleotide polymorphism in super-enhancers indicates their critical function in disease-specific gene regulation. However, little is known about the function of super-enhancers beyond gene regulation.

Results: In this study, through a comprehensive analysis of super-enhancers in 30 human cell/tissue types, we identified a new class of super-enhancers which are constitutively active across most cell/tissue types. These 'common' super-enhancers are associated with universally highly expressed genes in contrast to the canonical definition of super-enhancers that assert cell-type specific gene regulation. In addition, the genome sequence of these super-enhancers is highly conserved by evolution and among humans, advocating their universal function in genome regulation. Integrative analysis of 3D chromatin loops demonstrates that, in comparison to the cell-type specific super-enhancers, the cell-type common super-enhancers present a striking association with rapidly recovering loops.

Conclusions: In this study, we propose that a new class of super-enhancers may play an important role in the early establishment of 3D chromatin structure.

Keywords: Super-enhancer, Gene regulation, 3D chromatin structure, Chromatin loops

Background

Super-enhancers or stretch enhancers are defined by a strong enrichment of mediators and transcription-regulating proteins, appearing to play a deterministic role in cellular identity by controlling the expression of cell-type specific genes [1, 2]. Previous studies have revealed the critical function of super-enhancers during development and differentiation [3]. The enrichment of disease-associated single nucleotide polymorphism (SNP) in super-enhancers compared to that of typical enhancers proposed a substantial link between super-enhancers and many complex human diseases [2]. In addition, recent studies have proposed potential functions of super-enhancers in the extremely long-range chromatin communications and the establishment of 3D chromatin loops [4]. These results suggest a universal role of super-enhancers in genome regulation apart from cell-type specific gene regulation, but little is known about the mechanisms underlying these various functions. To extend the current knowledge of super-enhancers and their biological roles, we conducted a comprehensive analysis of super-enhancer activities across 30 human cell/tissue types. Our analysis suggests that a substantial number of super-enhancers exhibits prevalent activities across cell-types in terms of H3K27ac signals and that these non-canonical super-enhancers are involved in the formation of fast recovering chromatin loops.

* Correspondence: ijung@kaist.ac.kr
†Jayoung Ryu and Hyunwoong Kim contributed equally to this work.
1Department of Biological Sciences, KAIST, Daejeon 34141, South Korea
Full list of author information is available at the end of the article

Results

Genomic landscape of super-enhancer domains

To characterize different modes of super-enhancers across various human cell/tissue types, we first investigated the genomic distribution of super-enhancers across 30 human cell/tissue types, including 5 H1-derived early lineages, 8 immortalized cell lines, and 17 human postmortem tissues. Super-enhancers were defined using ROSE algorithm [2, 5] based on H3K27ac ChIP-seq signal (see Methods), which resulted in 17,916 super-enhancers with an average of 597 super-enhancers per cell/tissue type. Merging the super-enhancers in all of the cell/tissue types resulted in 6039 putative super-enhancer regions, named super-enhancer domains (SE domains) (Fig. 1a, see Methods). The super-enhancer domains presented a mean length of 32Kb, which was slightly longer than the individual super-enhancers with the mean length of 26Kb. Excluding any one of 30 samples did not significantly affect the overall distribution of length of super-enhancer domains (one-way Analysis of Variance, p-value > 0.5), which excludes the possibility of a critical bias in the definition of super-enhancer domains caused by one or more of samples. In total, around 6.32% of the human genome was marked by super-enhancer domains, and 48.3% of the super-enhancer domains consisted of multiple super-enhancers identified in at least two or more cell/tissue types analyzed, suggesting a recurrent formation of super-enhancers in specific genomic regions.

Identification of common super-enhancer domains

In contrast to the previous notion that super-enhancers are highly cell/tissue-type specific, some super-enhancer domains exhibited a strong enrichment of H3K27ac signals in a surprisingly large number of cell/tissue types. To assess cell/tissue-type specificity of the super-enhancer domains, we utilized tau score [6], a measurement of tissue-specificity commonly used in gene expression studies (see Methods). Tau score of 0 indicates ubiquitous super-enhancer domain activity, while 1 represents tissue-specific activation of the domain. The super-enhancer domain activities based on H3K27ac signals showed a wide spectrum of tau scores, reflecting a varying degree of cell/tissue-type specificity for each super-enhancer domain (Fig. 1b). It should be noted that some super-enhancer domains were annotated by only one cell/tissue type, but showed low tau scores. Through manual inspection, we found that these domains presented moderately high H3K27ac signals in other cell/tissue types (See Additional file 1: Figure S1), but the signals were insufficient to pass the threshold for super-enhancer calling by a small margin. For this reason, tau score was used as the main factor to define super-enhancer specificity.

In order to categorize super-enhancer domains with different cell/tissue-type specificity, we defined cell/tissue-type common and cell/tissue-type unique super-enhancer domains. The domains located within 20% or lower percentile of tau score were defined as common super-enhancer domains (1208 domains, 20.0%, Fig. 1c). On the other hand, domains that were called only in one cell/tissue type and located at 80% or higher percentile of tau score were defined as unique super-enhancer domains (1098 domains, 18.2%, Fig. 1e). Remaining domains were defined to be non-unique super-enhancer domains (3733 domains, 61.8%, Fig. 1d). To ensure that cancer cell lines or other immortalized cell lines did not produce critical bias by producing a large portion of common super-enhancers, we conducted hierarchical clustering between samples using the H3K27ac signals in the super-enhancer domains of available cell/tissue types (See Additional file 1: Figure S2). The result demonstrated that, while tissue samples with close relatedness grouped together, cancer cell lines or other immortalized cell lines were not concentrated in the same cluster.

A notable example of unique super-enhancer domains was found in neural progenitor cells derived from human embryonic stem cells (H1-ESC). Two super-enhancer domains were uniquely defined in neural progenitor cells at downstream of *SOX1* (Fig. 1f), a gene known to be highly expressed only in neural progenitor cells and to play an important role in nervous system development. Non-unique super-enhancer domains showed high activity in several cell/tissue types but low activity in the rest. For example, a non-unique super-enhancer domain was called for tissues associated with the digestive system such as stomach, small bowel, and sigmoid colon (Fig. 1g). There was also a high H3K27ac signal in the same domain in esophagus tissue, although it was not called as a super-enhancer. The putative target gene of the domain is *TJP3* (tight junction protein 3), which is involved in junctional integrity of the intestinal cells. This example suggests a possibility that the formation of non-unique super-enhancer domains is a key regulator for shared functions in several tissues. Common super-enhancer domains showed high ubiquitous H3K27ac signals across the cell/tissue types, discordant with the previous notion of cell/tissue-type specificity of super-enhancers. An example of common super-enhancer domains was found near the noncoding RNA genes, *NEAT1* and *MALAT1*, which are highly expressed across all cell/tissue types and previously suggested to function in a general biological process, such as the association with nuclear speckle formation [7] (Fig. 1h).

Distinct biological function of target genes in common super-enhancer domains

Super-enhancers are presumably associated with cell/tissue--type specific gene regulation, but the identification of distinct classes of super-enhancer domains raises the possibility that distinct modes of action exist for super-enhancers. We investigated the expression patterns of putative target genes in each class of super-enhancer domains. As expected, putative target genes of unique super-enhancer domains were highly

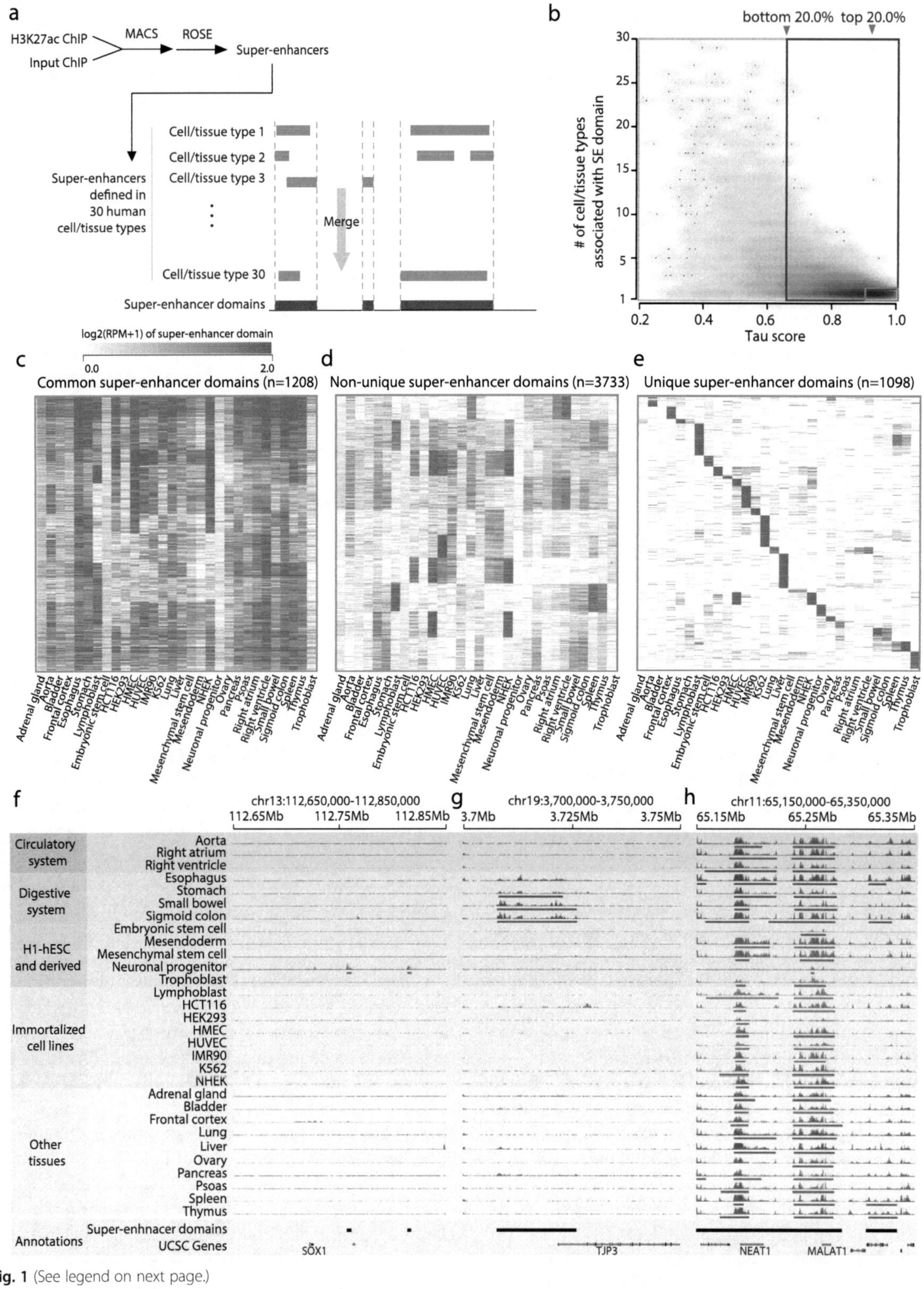

Fig. 1 (See legend on next page.)

(See figure on previous page.)

Fig. 1 Super-enhancer domains can be grouped into distinct categories. **a** A definition of super-enhancer domains. **b** A smooth scatter plot between tau score (x-axis) and the number of cell/tissue types associated with super-enhancer domains. Vertical arrowheads indicate 20 and 80% percentile of tau score. Red, blue, and green colored box indicate unique, non-unique and common super-enhancer domains, respectively. **c-e** Heatmaps of background-subtracted H3K27ac RPM of super-enhancer domains across cell/tissue types. Color intensity indicates log2 transformed H3K27ac RPM of super-enhancer domains. **f-h** Genome browser snapshots showing H3K27ac signals of each class of super-enhancer domain examples. Unique super-enhancer domain nearby *SOX1*, specific to neuronal progenitor cells (**f**). Non-unique super-enhancer domain nearby *TJP3*, specific to digestive system related tissues (**g**). Common super-enhancer domain nearby *MALAT1* and *NEAT1* (**h**). H3K27ac signal is uniformly scaled in the range of 0–20. Cell/tissue types are grouped according to their lineage or functional similarity

expressed in the corresponding cell/tissue type (Fig. 2a and b, see Methods). In contrast, genes associated with common super-enhancer domains showed universally high expression (Fig. 2a and b). Although there was a significant enrichment of housekeeping genes [8] in putative target genes of common super-enhancer domains, the majority (82.4%) were non-housekeeping genes, indicating that housekeeping genes cannot fully explain the function of common super-enhancer domains. Gene set enrichment analysis of genes associated with unique super-enhancer domains showed enrichment in cellular identity functions in the corresponding cell/tissue type (Table 1, see Methods), whereas genes associated with common super-enhancer domains were enriched in basic cellular functions, such as transcriptional regulation, cell motility, and regulation of cell proliferation (Table 2). The overall pattern of expression and the significance of enriched pathways were moderately affected but largely stayed consistent, when an alternative definition of putative target genes was used including the nearest 3 genes from the midpoint of each super-enhancer domain (See Additional file 1: Figure S3a, b and Table S1, S2). Our analysis revealed distinct biological functional enrichment in putative target genes between common and unique super-enhancer domains.

Genomic properties of common super-enhancer domains

To reason that super-enhancers play a key role in gene regulation, one would expect to see a strong correlation between the activity of a super-enhancer domain and the expression of its putative target gene. In general, putative target gene expression was significantly correlated with the activities of super-enhancer domains (one-sample t-test, p-value $< 10^{-15}$ for common, non-unique, and unique super-enhancer domains) (Fig. 3a, see Methods). However, common super-enhancer domains showed the lowest correlation coefficients, when compared to unique and non-unique super-enhancer domains (KS test, *** p-value $< 10^{-15}$). For example, a common super-enhancer domain shows considerable activities with enriched H3K27ac signals, but its putative target gene, *MLL5*, does not show a concordant expression level with the super-enhancer domain activities in all of the cell/tissue types (Fig. 3b). This result raised the possibility that common super-enhancer domains are responsible for an additional biological role, other than the expression regulation of genes within their close proximity.

Further investigation revealed that common super-enhancer domains are often located in evolutionarily conserved genomic regions compared to other super-enhancer domains and random expectation (Fig. 3c, see Methods). However, we did not observe a strong conservation score in the unique super-enhancer domains compared to random expectation. Similarly, genomic areas of common super-enhancer domains also tend to have a low context-dependent tolerance score (CDTS) [9], which represents the frequency of sequence variation across the human population (Fig. 3d, see Methods). The lower CDTS indicates a sequence is more resistant to the expected variation. Thus, low CDTS for common super-enhancer domains suggests, again, a potential function of these domains in universal genome regulation. For example, a common super-enhancer domain nearby *NR2F2* gene presents both a high evolutionary conservation score and a low CDTS compared to the adjacent non-unique super-enhancer domain (Fig. 3e).

A striking association of common super-enhancer domains during early establishment of 3D chromatin loops

A previous study has revealed an association between super-enhancers and 3D chromatin structure [4], where 3D chromatin loops have a wide range of recovery rate following cohesin degradation. Loop domains with a fast recovery rate contained a greater number of super-enhancers compared to slow-recovering loops. We surmised that common super-enhancer domains play a significant role in the loop recovery since rapid recovery of 3D chromatin structure could be more critical in the universal genome regulation compared to the cell/tissue-type specific regulation.

To test our hypothesis, we examined the enrichment of 336 super-enhancer domains called from HCT-116 human colorectal cancer cell lines in both loop anchors (two contact loci) and loop domains (between two anchors). For each time point from 20 min to 180 min after the degradation of cohesin and subsequent start of recovery, we classified chromatin loops into fast, slow, and moderate-recovery loops by the intensity percentile (top 10% defined as fast

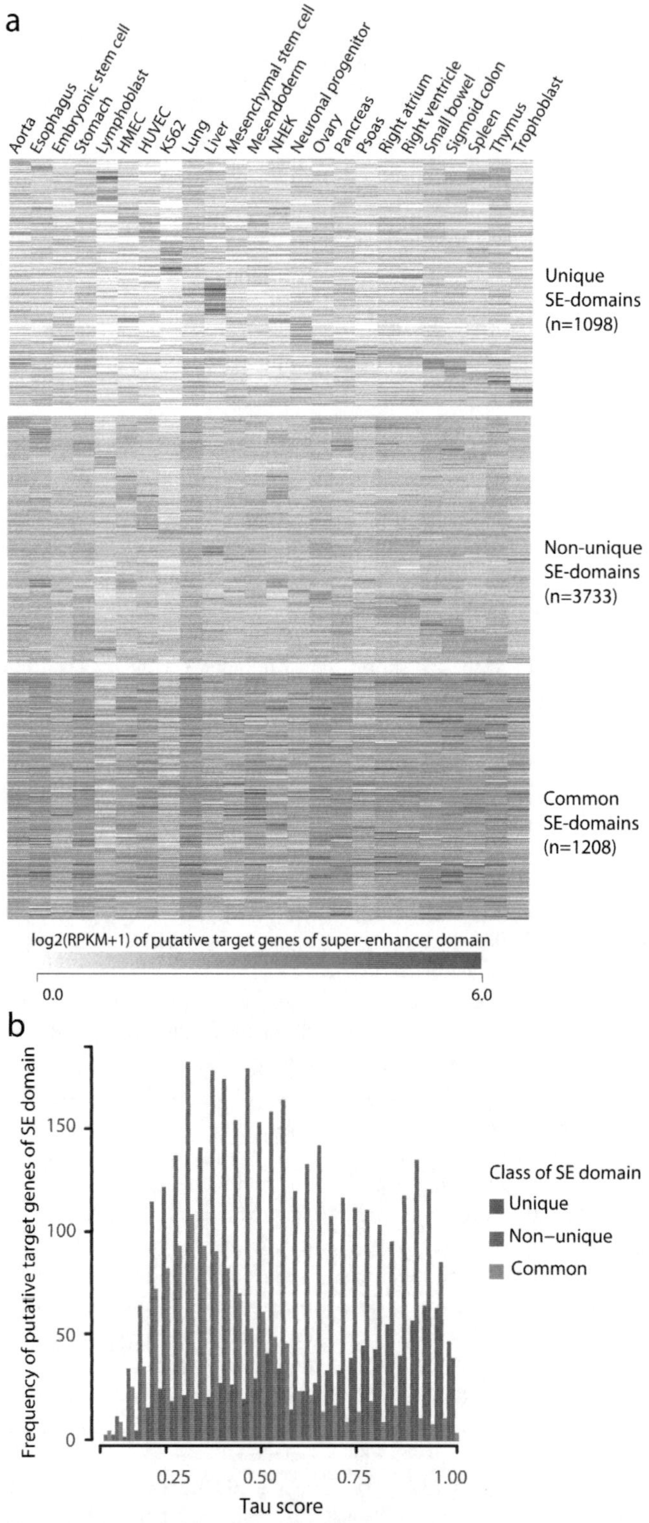

Fig. 2 Functional characterization of super-enhancer domain classes. **a** Heatmaps showing log2(RPKM+ 1) value of putative target gene of unique super-enhancer domains (top), non-unique super-enhancer domains (middle), and common super-enhancer domains (bottom). **b** Histogram of tau score for putative target gene expression in each class of super-enhancer domains. Low tau score indicates a universal expression pattern

Table 1 GO analysis of putative target genes of unique super-enhancer domains in lymphoblast and prefrontal cortex

Cell/tissue type	Term	Fold Enrichment
GM12878 (Lymphoblast)	GO:0006955~immune response	4.50
	GO:0031295~T cell costimulation	12.1
	GO:0071222~cellular response to lipopolysaccharide	8.30
	GO:0000188~inactivation of MAPK activity	22.6
	GO:2001244~positive regulation of intrinsic apoptotic signaling pathway	17.2
	GO:0035556~intracellular signal transduction	3.30
	GO:0001922~B-1 B cell homeostasis	94.3
	GO:0033209~tumor necrosis factor-mediated signaling pathway	6.40
	GO:0032729~positive regulation of interferon-gamma production	12.3
	GO:0031532~actin cytoskeleton reorganization	12.0
Prefrontal Cortex	GO:0051965~positive regulation of synapse assembly	21.4
	GO:0022010~central nervous system myelination	126
	GO:0007409~axonogenesis	13.5
	GO:0007417~central nervous system development	11.0
	GO:0031018~endocrine pancreas development	34.0
	GO:0007155~cell adhesion	3.90
	GO:0032570~response to progesterone	22.7
	GO:0046426~negative regulation of JAK-STAT cascade	22.1

recovery loop and bottom 10% as slow recovery loop) (Fig. 4a upper panel, see Methods). The overlap between super-enhancer domain and loop anchor or loop domain was calculated as the measurement of super-enhancer domain enrichment (Fig. 4a lower panel, see Methods). At an early time point (20 min after cohesin recovery), fast recovering loop anchors and domains were strikingly enriched in common super-enhancer domains but not in unique super-enhancer domains (Fig. 4b upper panels). In contrast, slow-recovering loop anchors were significantly enriched in unique super-enhancer domains but not in common super-enhancer domains. In order to confirm the result not only for super-enhancer domains of HCT-116 but also of the 30 cell/tissue types, we further expanded the analysis and observed concordant results (Fig. 4b, lower panels). This result supports the cell/tissue-type independent role of common super-enhancers in 3D chromatin loop formation.

Table 2 GO analysis of putative target genes of common super-enhancer domains

Term	Fold Enrichment	*p*-value
GO:0098609~cell-cell adhesion	4.14	4.69E-18
GO:0000122~negative regulation of transcription from RNA polymerase II promoter	2.56	2.64E-15
GO:0045944~positive regulation of transcription from RNA polymerase II promoter	2.14	8.07E-12
GO:0045892~negative regulation of transcription, DNA-templated	2.17	2.42E-05
GO:0006351~transcription, DNA-templated	1.52	2.19E-05
GO:0006366~transcription from RNA polymerase II promoter	2.01	5.22E-04
GO:0071364~cellular response to epidermal growth factor stimulus	6.57	4.60E-04
GO:0030036~actin cytoskeleton organization	3.20	1.13E-03
GO:0008360~regulation of cell shape	3.10	1.01E-03
GO:0048870~cell motility	7.53	1.05E-03
GO:0008283~cell proliferation	2.12	1.96E-03
GO:0008285~negative regulation of cell proliferation	2.05	2.48E-03
GO:0045893~positive regulation of transcription, DNA-templated	1.90	2.66E-03
GO:0043547~positive regulation of GTPase activity	1.82	4.17E-03
GO:0043433~negative regulation of sequence-specific DNA binding transcription factor activity	4.22	4.64E-03

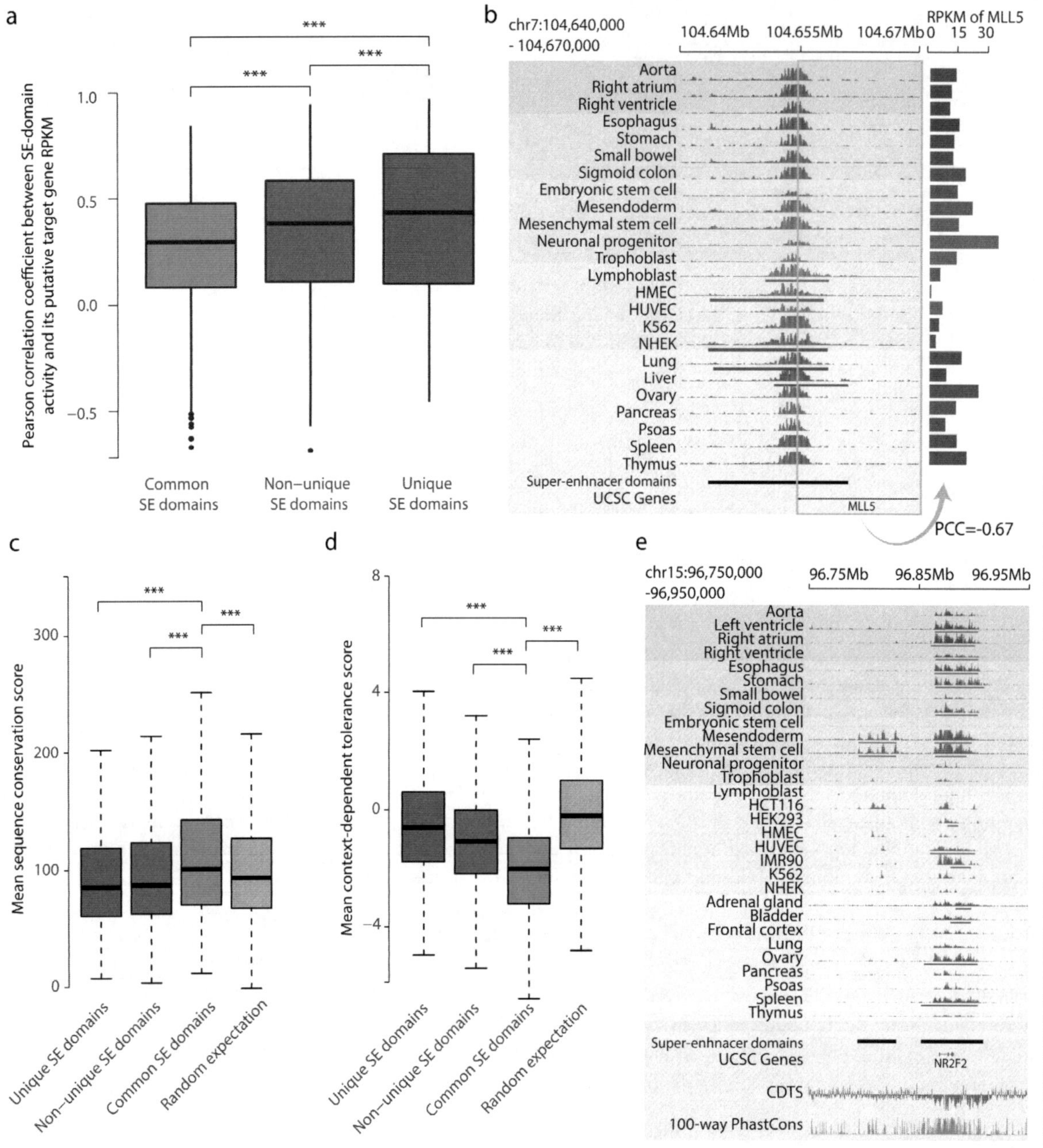

Fig. 3 (See legend on next page.)

(See figure on previous page.)

Fig. 3 Genomic characterization of common super-enhancer domains. **a** Boxplots of correlation between log2(H3K27ac RPM + 1) in super-enhancer domains and RPKM of putative target genes ($n = 1208$ for common SE domains, $n = 3733$ for non-unique SE domains, $n = 1098$ for unique SE domains). Pearson correlation coefficient was calculated for each super-enhancer domain across 24 cell/tissue types. Significance of difference between super-enhancer domain categories was evaluated by KS test (*** p-value $< 5 \times 10^{-12}$). **b** An example of common super-enhancer domain showing high H3K27ac signal across cell/tissue types but showing irregular RNA expression level. Left panel presents genome browser snapshot of H3K27ac signals at super-enhancer domain. Right panel presents RNA-seq RPKM of *MLL5* gene. Pearson correlation coefficient (PCC) between H3K27ac signal at a super enhancer domain and the *MLL5* RPKM is shown below. **c** Boxplots of mean sequence conservation score for each super-enhancer domain ($n = 1098$ for unique SE domains, $n = 3733$ for non-unique SE domains, $n = 1208$ for common SE domains). Significance of difference between super-enhancer domain categories was evaluated with KS test (*** p-value $< 1.1 \times 10^{-5}$ between random and others, p-value $< 5.1 \times 10^{-12}$ for the rest). **d** Boxplots of mean CDTS for each super-enhancer domain ($n = 1098$ for unique SE domains, $n = 3733$ for non-unique SE domains, $n = 1208$ for common SE domains). Significance of difference between super-enhancer domain categories is evaluated with KS test (*** p-value $< 1.7 \times 10^{-9}$ for unique and random, p-value $< 3.0 \times 10^{-14}$ for the rest). **e** An example of adjacent non-unique super-enhancer domain and common super-enhancer domain illustrating evolutionary conservation of common super-enhancer domain. Shown together are 100-way PhastCons and CDTS. 100-way PhastCons track has the scale from 0 to 1, and CDTS track is scaled from −20 to + 20. Common and unique super-enhancer domain are shown as right and left bar, respectively, in 'Super-enhancer domains' annotation

We expanded the analysis to include later time points (see Methods) and revealed that enrichment of common super-enhancer domains at both loop anchors and domains was the highest at the earliest time point (20 min), and gradually decreased in later time points (Fig. 4c). On the other hand, enrichment of unique super-enhancer domains significantly increased in the later time points (two-sample t-test, * p-value < 0.05, ** p-value < 0.01, *** p-value < 0.001). Our result implies that common super-enhancer domains are often associated with the initial establishment of 3D chromatin loops compared to other super-enhancer domains.

Discussion and conclusion

We present a systematic approach to assess and characterize super-enhancers in a wide array of human cell/tissue types. The effective computation of background normalized super-enhancer activity and the use of simple, yet powerful mathematical expression to evaluate the tissue specificity have led us to identify a new class of super-enhancers. These super-enhancer domains exhibit a high degree of universality across many cell/tissue types – a novel aspect of super-enhancers that has never been spotlighted in previous literature. The universally active super-enhancers, which we defined as common super-enhancer domains, display a strong association with fast recovering chromatin loops after a sequential cohesin removal and restoration. Although the enrichment of super-enhancers at fast-recovering loops is previously reported [4], our analysis further reveals that common super-enhancer domains are strikingly enriched in the fast-recovering loops by more than 12-folds, compared to unique super-enhancer domains. The implications that our data bring forth to the function of super-enhancers in shaping chromatin organization may be outlined as follows. First, common super-enhancers may facilitate recruiting a substantial amount of structural proteins, such as cohesin and CTCF, to expedite loop recovery. A sequential model is also an interesting possibility, where the cell-type nonspecific formation of fast recovering loops facilitated by common super-enhancer guide the genome folding in a stepwise manner to lead following cell-type specific loop formation. A hierarchical model where cell-type specific loop formation by unique super-enhancers requires the higher-order loop formation promoted by common super-enhancers is also reasonable. Further experiments will be required to test our hypothesis for the validation.

To conclude, we propose two distinct modes of super-enhancers; one is cell-type specific gene regulation mainly mediated by unique super-enhancer domains and second is 3D genome regulation mediated by common super-enhancer domains as a non-canonical function (Fig. 4d). Although there is a limitation to generalize our conclusion since we investigated only one cell type, we shed light on common super-enhancers as a new potential mechanism underlying the early establishment of 3D chromatin loops.

Methods

Human cell/tissue types

In this study, 30 human cell/tissue types were examined. This includes H1 embryonic stem cell and its derived cell types (H1 embryonic stem cell, mesendoderm, mesenchymal stem cell, neuronal progenitor cell, and trophoblast), 8 cell lines (GM12878, HUVEC, IMR90, K562, NHEK, HMEC, HEK293, and HCT-116), and 17 tissue types (adrenal gland, aorta, bladder, dorsolateral prefrontal cortex, esophagus, gastric, lung, ovary, pancreas, psoas, right atrium, right ventricle, liver, sigmoid colon, small bowel, spleen, and thymus).

Super-enhancer call

H3K27ac and input ChIP-seq reads were downloaded from Roadmap Epigenomics, ENCODE project and GEO database [10, 11]. Sources are specified in Additional file 1: Table S3. Reads were aligned to the human reference genome (hg19 assembly) using BWA-mem [12]. Unmapped and poorly mapped (MAPQ < 10) reads were removed. PCR duplicates were also removed using Picard

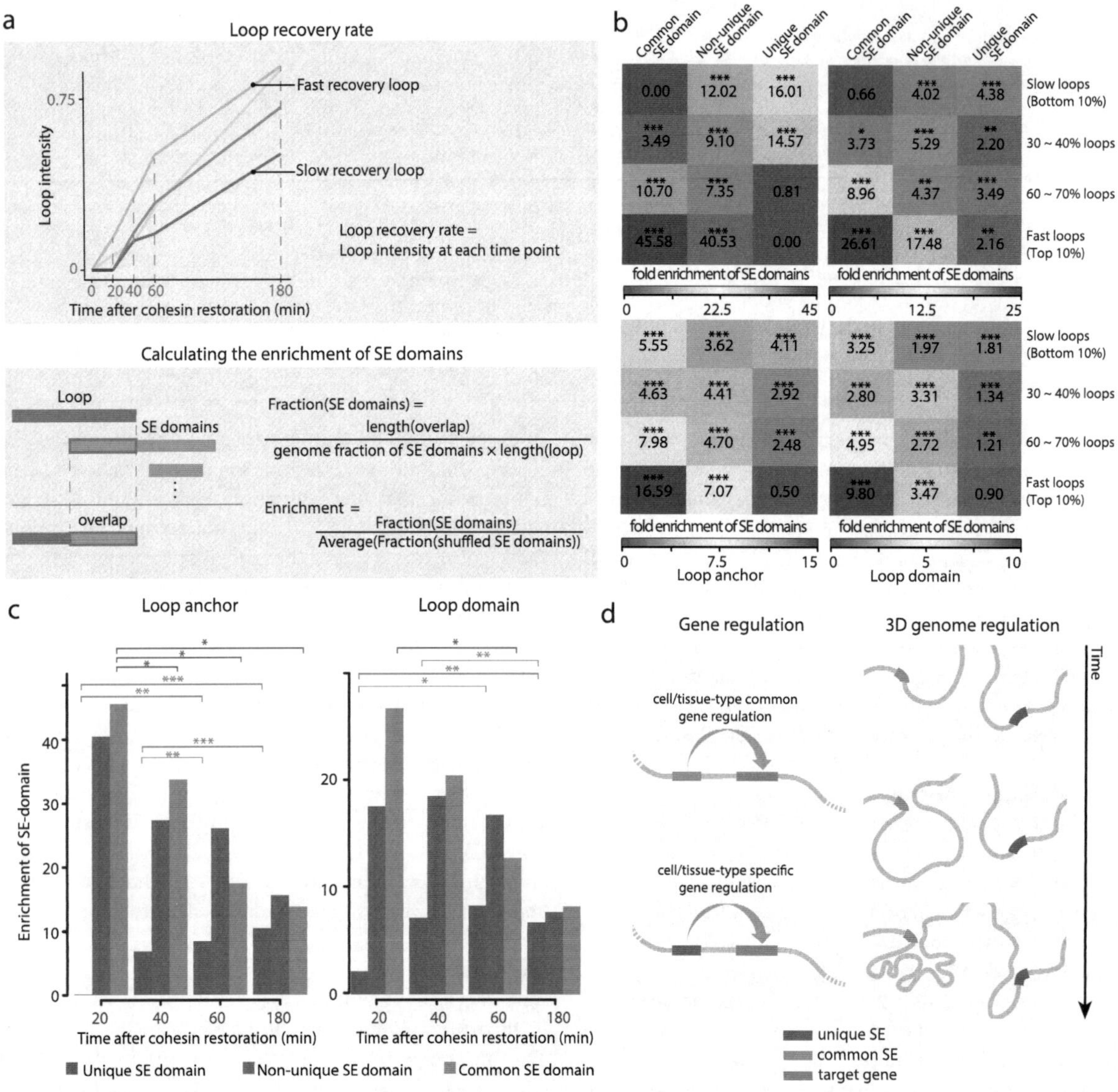

Fig. 4 Common super-enhancer domains are associated with the early establishment of 3D chromatin loops. **a** The schematic overview of calculated loop recovery rate (upper panel) and super-enhancer domain enrichment (lower panel). **b** Heatmaps showing mean enrichment value of super-enhancer domains called from HCT-116 cell line (upper two panels) and all samples (lower two panels) at 20 min after cohesin recovery in loop anchors (left panels) and loop domains (right panels). Empirical *p*-values were calculated through random expectation values based on 10,000 iterations (* *p*-value < 0.05, ** *p*-value < 0.01, *** *p*-value < 0.001). **c** Bar plots of super-enhancer domain enrichment in the top 10% fast recovery loops for each time point in loop anchors (left) and loop domains (right) (two-sample t-test, * *p*-value < 0.05, ** *p*-value < 0.01, *** *p*-value < 0.001). **d** Model diagram of modes of super-enhancer functions. Canonical cell-type specific gene regulation (left) and 3D genome regulation of common super-enhancer domains (right)

Markduplicates. The remaining reads were then used to call peaks using MACS2 [13]. Default parameters were used except q-value 0.01 was replaced with the *p*-value less than 1E-05. Super-enhancers were defined by stitching peaks using ROSE [2, 5] with default parameters, except TSS exclusion zone size was adjusted to 2500 bp. TSS exclusion was used because the H3K27ac signal is enriched for both active enhancers and promoters. Super-enhancers that span ENCODE blacklist regions were discarded.

Calling super-enhancer domains

Super-enhancer domains were defined by merging the super-enhancers called for 30 samples using BEDTools (see Fig. 1a). Super-enhancers with any overlap were merged into one super-enhancer domain.

Cell/tissue-type specificity measurement with tau value

Tau score was utilized to assess the cell/tissue-type specificity of super-enhancer domain activities and gene expression. Tau score is a measure commonly used to quantify tissue-specificity of expression as it enables a threshold-free evaluation of cell/tissue-type specificity. Tau score ranges from 0 to 1, where 0 indicates general, and 1 indicates specific. Tau score is calculated as $\tau = \frac{\sum_{i=1}^{n}(1-\hat{x}_i)}{n-1}$; $\hat{x}_i = \frac{x_i}{\max\limits_{1\le i\le n}(x_i)}$ [6], where x indicates cell/tissue-type specificity and n denotes the number of cell/tissue types.

Gene expression data

Uniformly processed and quantified RPKM of protein-coding genes for 10 cell lines (H1 embryonic stem cell, mesendoderm, trophoblast, mesenchymal stem cell, neuronal progenitor cell, GM12878, HMEC, HUVEC, K562, and NHEK) and 14 tissues (aorta, esophagus, gastric, liver, lung, ovary, pancreas, psoas, right atrium and ventricle, sigmoid colon, small intestine, thymus, and spleen) was obtained from Roadmap epigenomics ("RNA-seq uniform processing and quantification for consolidated epigenomes"). The data had been normalized between cell/tissue types to have the same number of aligned reads on coding exons. Super-enhancer domains called for none of the 24 cell/tissue types are excluded in gene expression analysis.

Target gene assignment of super-enhancer domains

Putative target gene of a super-enhancer domain was defined as the gene with the closest transcriptional start site to the midpoint of the super-enhancer domain. For extended analysis, alternative definition utilized the 3 genes with transcriptional start site closest to the midpoint of super-enhancer domains.

GO enrichment analysis using DAVID

For putative target genes of super-enhancer domains, GO terms were analyzed using DAVID 6.8 [14]. Significant GO_BP annotations were selected using *p*-value corrected by Benjamini-Hochberg procedure with threshold 0.05 and sorted by the *p*-value. For the analysis of the putative target genes of unique super-enhancer domains, we selected lymphoblast and prefrontal cortex as examples. Due to the small number of putative target genes of unique super-enhancer domains in each cell type, *p*-values tend to be insignificant. Thus, we presented up to top 10 enriched GO_BP annotations sorted with *p*-values without cutoff.

Housekeeping genes

The putative target genes of super-enhancer domains were compared to the list of housekeeping genes provided by Eisenberg, E. et al. [8].

Correlation between SE domain activity and expression level of its putative target gene

Pearson correlation coefficients between background-sub tracted log2(RPM + 1) value of H3K27ac ChIP-seq signal on super-enhancer domains and RPKM of their putative target gene were calculated across 24 cell/tissue. Super-enhancer domains called for none of the 24 cell/tissue types were excluded in the correlation analysis.

Conservation and tolerance scores of super-enhancer domains

Evolutionary sequence conservation score calculated from multiple alignments of 100 vertebrates using phastCons method was obtained from the UCSC annotation database (phastCons100way). SumData column of 24 bp resolution conservation score was mapped onto super-enhancer domains, and the mean value of mapped conservation score was used. CDTS referring to the variability of a nucleotide calculated for 11,257 human genomes was obtained from a previous study [9]. CDTS of the 1 bp resolution was mapped onto super-enhancer domains, and the mean of mapped CDTS was used.

Hi-C matrix of HCT-116 following cohesin degradation and subsequent recovery

Hi-C matrix data of HCT-116 with a 25 kb resolution for control, cohesin degraded, and 20, 40, 60, and 180 min since subsequent recovery of cohesion, were obtained from the previous study (GEO accession number GSE104334) [4]. Each matrix was normalized by Knight-Ruiz algorithm. As the original read depth of each matrix was different, we normalized matrices to have the same total interaction frequency within 100 kb distance since short-range interactions are known to be not affected by cohesin degradation. Chromatin loops were identified using HiCCUPS [15] with default parameters for medium resolution matrix. False positive chromatin loops from karyotypic abnormalities of HCT-116 cell line were filtered. The observed/expected enrichment filter of 4.5 was applied, as suggested by the previous study.

Super-enhancer enrichment in loop anchors and domains

The recovery rate of loop anchors (interacting loci) and domains (region between two anchors) was defined as the ratio of loop strength at each time point compared to the strength in untreated cells (see Fig. 4a, upper panel). Loops with monotonically increasing strength with regard to the time were selected for the analysis. Loop anchors were padded by 5 kb in obtaining the super-enhancer domain enrichment. "Fraction" value of super-enhancer domains in loop anchors or domains was calculated by dividing the density of overlapping super-enhancer domains by the expected density (see Fig. 4a, lower panel). The expected density of each class of super-enhancer domain is the proportion of the corresponding domains in the whole hg19 genome. The expected value of fraction for randomly shuffled super-enhancer domains is 1, with an infinite number of trials. The super-enhancer domain enrichment of a loop is obtained by dividing the observed fraction value by the random expectation of the fraction. The random expectation of fraction is the average of fractions calculated for 10,000 shuffled super-enhancer domains. Empirical p-value of a fraction where the distribution is defined as the result of 10,000 trials, is utilized as the significance measure of the enrichment. When comparing two enrichments, the p-value of two-sample t-test was utilized as the significance measure.

Additional file

Additional file 1: Figure S1. Ubiquitous activity of super-enhancers defined in one cell/tissue type. 196 super-enhancer domains are called for only one cell/tissue type but showed constitutive H3K27ac signals across cell/tissue types and also located at less than 20% percentile of tau score. The heatmap of background subtracted log2 transformed H3K27ac signals of the domains across cell/tissue types is shown. **Figure S2.** Hierarchical clustering of samples using H3K27ac signal on super-enhancer domains. Complete-link hierarchical clustering of distance between samples with respect to log2 transformed H3K27ac RPM of SE domains. H1 human embryonic cell line and its derived cell lines are shown in brown and immortalized cell lines are shown in blue. Cancer cell lines are shown in bold text. **Figure S3.** Functional characterization of super-enhancer domain classes with alternative putative target genes. **a**, Heatmaps showing log2(RPKM+ 1) value of putative target gene of unique super-enhancer domains (top), non-unique super-enhancer domains (middle), and common super-enhancer domains (bottom). **b**, Histogram of tau score for putative target gene expression in each class of super-enhancer domains. Low tau score indicates universal expression pattern. **Table S1.** GO analysis of alternative putative target genes of unique super-enhancer domains. **Table S2.** GO analysis of alternative putative target genes of common super-enhancer domains. **Table S3.** Source of H3K27ac and input ChIP-seq reads. (PDF 574 kb)

Abbreviations

CDTS: Context-dependent tolerance score; ChIP-seq: Chromatin immunoprecipitation sequencing; H3K27ac: Histone H3 lysine 27 acetylation; KS test: Kolmogorov-Smirnov test; RPKM: Reads per kilobase million; RPM: Reads per million mapped reads; SE domain: Super-enhancer domain; TSS: Transcription start site

Acknowledgments

We thank all members of Dr. Jung's laboratory for their comments and criticisms on this study.

Funding

Ministry of Science, ICT, and Future Planning through the National Research Foundation in Republic of Korea [2017R1C1B2008838 to I. Jung]; Korean Ministry of Health and Welfare [HI17C0328 to I. Jung]. This work was supported by the KAIST Future Systems Healthcare Project from the Ministry of Science and ICT. Publication charges come from HI17C0328.

Availability of data and materials

All data generated or analyzed during this study are included in this published article.

About this supplement

This article has been published as part of *BMC Bioinformatics Volume 20 Supplement 3, 2019: Selected articles from the 17th Asia Pacific Bioinformatics Conference (APBC 2019): bioinformatics.* The full contents of the supplement are available online at https://bmcbioinformatics.biomedcentral.com/articles/supplements/volume-20-supplement-3.

Author's contributions

JR, HK, and IJ conceived the study. JR led the data analysis with assistance from HK and DY. JR prepared the manuscript with assistance from AJL and IJ. All authors read and commented on the manuscript. All authors read and approved the final manuscript.

Ethics approval and consent to participate

Not applicable.

Consent for publication

Not applicable.

Competing interests

The authors declare that they have no competing interests.

Publisher's Note

Springer Nature remains neutral with regard to jurisdictional claims in published maps and institutional affiliations.

Author details

[1]Department of Biological Sciences, KAIST, Daejeon 34141, South Korea. [2]Department of Bio and Brain Engineering, KAIST, Daejeon 34141, South Korea.

Published: 29 March 2019

References

1. Parker SC, Stitzel ML, Taylor DL, Orozco JM, Erdos MR, Akiyama JA, et al. Chromatin stretch enhancer states drive cell-specific gene regulation and harbor human disease risk variants. Proc Natl Acad Sci U S A. 2013;110(44):17921–6.
2. Whyte WA, Orlando DA, Hnisz D, Abraham BJ, Lin CY, Kagey MH, et al. Master transcription factors and mediator establish super-enhancers at key cell identity genes. Cell. 2013;153(2):307–19.
3. Adam RC, Yang H, Rockowitz S, Larsen SB, Nikolova M, Oristian DS, et al. Pioneer factors govern super-enhancer dynamics in stem cell plasticity and lineage choice. Nature. 2015;521(7552):366–70.
4. Rao SSP, Huang SC, Glenn St Hilaire B, Engreitz JM, Perez EM, Kieffer-Kwon KR, et al. Cohesin Loss Eliminates All Loop Domains. Cell 2017;171(2):305–320 e24.
5. Loven J, Hoke HA, Lin CY, Lau A, Orlando DA, Vakoc CR, et al. Selective inhibition of tumor oncogenes by disruption of super-enhancers. Cell. 2013;153(2):320–34.
6. Kryuchkova-Mostacci N, Robinson-Rechavi M. A benchmark of gene expression tissue-specificity metrics. Brief Bioinform. 2017;18(2):205–14.
7. West JA, Davis CP, Sunwoo H, Simon MD, Sadreyev RI, Wang PI, et al. The long noncoding RNAs NEAT1 and MALAT1 bind active chromatin sites. Mol Cell. 2014;55(5):791–802.
8. Eisenberg E, Levanon EY. Human housekeeping genes, revisited. Trends Genet. 2013;29(10):569–74.
9. di Iulio J, Bartha I, Wong EHM, Yu HC, Lavrenko V, Yang D, et al. The human noncoding genome defined by genetic diversity. Nat Genet. 2018;50(3):333–7.
10. Roadmap Epigenomics C, Kundaje A, Meuleman W, Ernst J, Bilenky M, Yen A, et al. Integrative analysis of 111 reference human epigenomes. Nature. 2015;518(7539):317–30.
11. Leung D, Jung I, Rajagopal N, Schmitt A, Selvaraj S, Lee AY, et al. Integrative analysis of haplotype-resolved epigenomes across human tissues. Nature. 2015;518(7539):350–4.
12. Li H, Durbin R. Fast and accurate short read alignment with burrows-wheeler transform. Bioinformatics. 2009;25(14):1754–60.

13. Zhang Y, Liu T, Meyer CA, Eeckhoute J, Johnson DS, Bernstein BE, et al. Model-based analysis of ChIP-Seq (MACS). Genome Biol. 2008;9(9):R137.
14. Huang d W, Sherman BT, Lempicki RA. Systematic and integrative analysis of large gene lists using DAVID bioinformatics resources. Nat Protoc. 2009; 4(1):44–57.
15. Durand NC, Shamim MS, Machol I, Rao SS, Huntley MH, Lander ES, et al. Juicer provides a one-click system for analyzing loop-resolution hi-C experiments. Cell Syst. 2016;3(1):95–8.

Jammali et al. BMC Bioinformatics 2019, **20**(Suppl 3):133
https://doi.org/10.1186/s12859-019-2647-2

BMC Bioinformatics

RESEARCH Open Access

SplicedFamAlign: CDS-to-gene spliced alignment and identification of transcript orthology groups

Safa Jammali[1,2*], Jean-David Aguilar[1,2], Esaie Kuitche[1] and Aïda Ouangraoua[1]

From The 17th Asia Pacific Bioinformatics Conference (APBC 2019)
Wuhan, China. 14-16 January 2019

Abstract

Background: The inference of splicing orthology relationships between gene transcripts is a basic step for the prediction of transcripts and the annotation of gene structures in genomes. The splicing structure of a sequence refers to the exon extremity information in a CDS or the exon-intron extremity information in a gene sequence. Splicing orthologous CDS are pairs of CDS with similar sequences and conserved splicing structures from orthologous genes. Spliced alignment that consists in aligning a spliced cDNA sequence against an unspliced genomic sequence, constitutes a promising, yet unexplored approach for the identification of splicing orthology relationships. Existing spliced alignment algorithms do not exploit the information on the splicing structure of the input sequences, namely the exon structure of the cDNA sequence and the exon-intron structure of the genomic sequences. Yet, this information is often available for coding DNA sequences (CDS) and gene sequences annotated in databases, and it can help improve the accuracy of the computed spliced alignments. To address this issue, we introduce a new spliced alignment problem and a method called SplicedFamAlign (SFA) for computing the alignment of a spliced CDS against a gene sequence while accounting for the splicing structures of the input sequences, and then the inference of transcript splicing orthology groups in a gene family based on spliced alignments.

Results: The experimental results show that SFA outperforms existing spliced alignment methods in terms of accuracy and execution time for CDS-to-gene alignment. We also show that the performance of SFA remains high for various levels of sequence similarity between input sequences, thanks to accounting for the splicing structure of the input sequences. It is important to notice that unlike all current spliced alignment methods that are meant for cDNA-to-genome alignments and can be used for CDS-to-gene alignments, SFA is the first method specifically designed for CDS-to-gene alignments.

Conclusion: We show the usefulness of SFA for the comparison of genes and transcripts within a gene family for the purpose of analyzing splicing orthologies. It can also be used for gene structure annotation and alternative splicing analyses. SplicedFamAlign was implemented in Python. Source code is freely available at https://github.com/UdeS-CoBIUS/SpliceFamAlign.

Keywords: Spliced alignment; Transcript orthology groups; Splicing structure; Splicing orthology; Gene family

*Correspondence: safa.jammali@usherbrooke.ca
[1]Department of Computer science, Faculty of Science, Université de Sherbrooke, Sherbrooke, Quebec, Canada
[2]Department of Biochemistry, Faculty of medecine and health science, Université de Sherbrooke, Sherbrooke, Quebec, Canada

Background

A spliced alignment is an alignment of a partial or full length spliced transcript sequence against an unspliced genomic sequence [1]. A spliced alignment allows to highlight the boundaries and the alignment of exons of the transcript sequence on the genomic sequence. It is an effective method for gene recognition, gene structure prediction, alternative splicing analyses and the annotation of protein-coding genes in non-model genomes [2–6].

Several methods have been developed to address different versions of the cDNA-to-genome spliced alignment problem, which consists in finding an optimal alignment of a spliced cDNA sequence against an unspliced genomic sequence, given an optimal function [7–12]. A complete overview of these methods is provided in [12] and complemented by [9]. These methods follow a general scheme that consists in first identifying candidate target alignment locations on the genomic sequence for the cDNA sequence, and then computing approximate spliced alignments of the cDNA sequence against each candidate target genomic regions. For computing approximate spliced alignments, the first criterion of optimality used by all spliced alignment methods is the sequence similarity. In addition to sequence similarity, some methods also account for splice signals on the unspliced genomic sequence such as canonical dinucleotide splice signals "GT" and "AG" at extremities of an intron, in their criterion of optimality in order to infer accurate exon boundaries in the alignments. It has been shown that the performance of methods accounting for splice signals is superior to that of only sequence similarity-based methods [9, 12]. Indeed, splice signals constitute strong structural signals that are used by splice signal-based methods for inferring accurate splice sites. Thus, the use of splice signals explains the superiority of splice signal-based methods over only sequence similarity-based methods. However, none of the existing spliced alignment methods takes into account the splicing structure of the input sequences, namely the exon structure of the cDNA sequence and the exon-intron structure of the genomic sequences, in addition to splice signals and sequence similarity. Yet, the information on the splicing structure and known splice sites is often available in annotation of CDS and gene sequences. This information can be used to improve the accuracy of spliced alignments.

In this paper, we re-visit the spliced alignment problem for the purpose of computing accurate CDS-to-gene spliced alignments and identifying transcript orthology groups within a set of transcripts from a coding gene family. Identifying orthologous isoforms at transcript level is a pre-requisite to describing evolutionary relationships between genes in terms of splicing structure and sets of splice variants [13, 14]. Here, we focus on the spliced alignment of full CDS against gene sequences within a gene family, that allows to identify splicing orthologous CDS which are pairs of CDS with similar sequences and splicing structures from genes that have evolved from a common ancestral gene [15, 16]. Splicing orthologs are supposed to have retained the same function in the course of evolution. Identifying splicing orthologs using spliced alignments requires precise alignment of exons and location of their boundaries in the spliced alignments. To achieve this aim, we introduce a new version of the spliced alignment problem that accounts for the splicing structure of the input sequences. For this version of the problem, we propose SplicedFamAlign (SFA), a method for fast and accurate alignment of spliced CDS against unspliced gene sequences (*SFA-align*), and for the identification of splicing orthologs using spliced alignments (*SFA-ortholog*).

The SFA-align algorithm (see overview in Fig. 1) starts by fastly computing local alignments in order to identify highly conserved sequences between the input CDS and gene sequences. These local alignments are used as anchors and each anchor alignment is trimmed at the extremities in order to cover at most one exon on the CDS. Next, a gapped extension algorithm accounting for CDS exon boundaries is used to extend anchor alignments in both direction in order to maximize the coverage of CDS exons. Finally a global spliced alignment algorithm accounting for the exon boundaries of the CDS and the gene sequence can be applied for aligning the remaining segments between the extended anchors. This anchored spliced alignment approach has been used by several other methods such as Splign [9], Spidey [17] and MGAlign [18]. The added value of the SFA method is that it makes use of the splicing structure of the input sequences in its local and global spliced alignment steps, in addition to the splice signals on the genomic sequence, in order to produce more accurate spliced alignments.

The SFA-ortholog algorithm starts by identifying pairwise orthologous CDS. The definition of the orthology relation between two CDS relies on the preservation of their splicing structures. We say that a spliced alignment of a CDS sequence against a gene sequence preserves the exon structure of the CDS, if the spliced alignment induces a sequence conservation of all the exons of the CDS on the gene sequence, and it also induces an intron between any two consecutive exons of the CDS, and no intron within an exon of the CDS (see Fig. 2 for example in which the exon structures of both CDS1 and CDS2 are preserved by their spliced alignments against a gene sequence).

Next, two CDS are *splicing orthologs* if their spliced alignments against the gene of one of the two CDS **a**) preserve their exon structures, and **b**) induce a one-to-one correspondence between the exons of the two CDS (see Fig. 2 for example). Finally, the pairs of orthologous CDS within a gene family are used to define an orthology graph

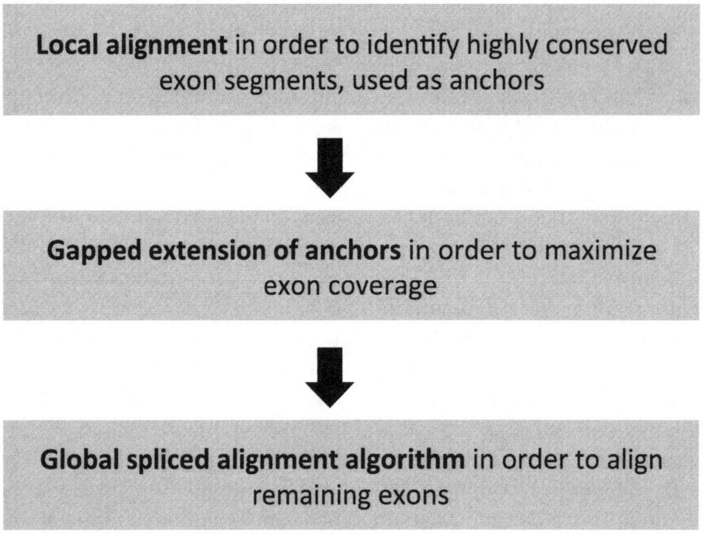

Fig. 1 Overview of the SFA-align algorithm

whose nodes are all the CDS of the family and edges represent pairwise orthology relations between the CDS. Then, assuming perfect pairwise orthology relations, the CDS orthology groups are defined as the connected components of the orthology graph.

The main contributions of SplicedFamAlign can be described as follows:

1 It allows fast computing of accurate CDS-to-gene spliced alignments, and accurate CDS splicing orthology groups within a gene family.
2 Its performance remains high for various levels of sequence similarity, thanks to the use of the splicing structure of input sequences, which allows to detect splicing structure conservation even in the cases of low sequence conservation.
3 In the case where the splicing structure of sequences are not given as input, SplicedFamAlign includes a preliminary step that allows computing the splicing structure of input sequences by aligning each CDS against its own gene.

Results and discussion

Dataset

SplicedFamAlign was evaluated based on a dataset of three real sets of homologous genes and three simulated sets of gene families.

Real data

The dataset contains homologous genes with their CDS sequences from 3 gene families FAM86, MAG and TP53 from the Ensembl-Compara database release 85 [19]. For each family, 8 homologous genes were selected with 14 CDS for FAM86, 26 for MAG and 51 for TP53. For each set of homologous genes, the genes are from 6 different amniote species, *human, chimpanzee, mouse, rat, cow*

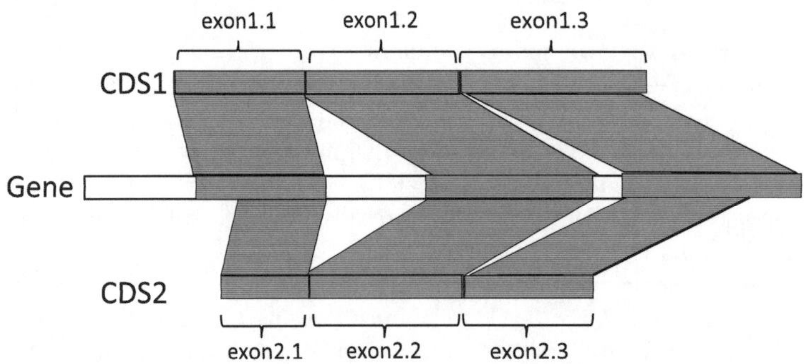

Fig. 2 Two orthologous CDS, according to the definition of splicing orthology: **a**) the exon structures of the two CDS are preserved by their splicing alignments against the gene sequence and **b**) the induced one-to-one correspondence between the exons of CDS1 and CDS2 is (exon1.1,exon2.1),(exon1.2,exon2.2), (exon1.3,exon2.3)

and chicken (for FAM86 and TP53) or *lizard* (for MAG). Table 1 gives more details about the real dataset. FAM86 is a group of genes with sequence similarity and unknown function. MAG is the family of Myelin Associated Glycoprotein genes whose proteins are type I membrane protein and member of the immunoglobulin superfamily. TP53 is a family of genes encoding for tumor suppressor proteins p53.

Simulated data

We generated a dataset of simulated gene families with gene and cDNA sequences accounting for gene splicing structure evolution and alternative splicing events. We designed a simulation method that takes as input a gene tree with branch lengths representing evolutionary rates, generates an ancestral gene at the root of the tree, and makes this gene evolve along branches of the gene tree. The ancestral root gene is generated with an exon-intron structure and an initial set of CDS of alternative transcripts for the gene, based on parameters learned from a dataset of 10.000 vertebrate genes from amniote species from the Ensembl database [19]. The evolution simulated along branches of the tree accounts for two levels of evolution. First, at the level of genes, the following evolutionary events acting on the splicing structure of genes are included, exon duplication, exon gain and exon loss. Evolutionary events acting on the sequence of coding exons, namely nucleotide insertion, deletion and substitution events are also included at the level of genes, using empirical codon evolution models [20]. Second, at the level of transcripts, we account for two events acting on the sets of transcripts and CDS generated by genes, isoform creation, and isoform loss. The isoform creation event corresponds to the acquisition by a gene of the ability to produce a new transcript through a new combination of exons. The isoform loss event corresponds to the loss of the ability to produce a given transcript.

Using the simulation method, we generated three sets of gene family, a first set called *Small* for which the evolutionary rates on the gene tree branches are low, a second set called *Medium* with medium evolutionary rate, and a third set called *Large* with a high evolutionary rates. Each of the three set contains 36 simulated gene families with 5 genes and 5 to 17 CDS in total. For each family in each set, we generated a set of 5 gene sequences with their CDS, all true pairwise spliced alignments between any CDS and any gene of the family, the true multiple sequence alignment of all CDS and gene sequences, and all true splicing orthology relations between CDS. Table 2 gives more details about the simulated datasets used for the evaluation.

Evaluated methods

SplicedFamAlign results were compared with the results of Splign [9], the most recent and current best performing cDNA-to-genome spliced alignment method. In [9], Splign was compared with SIM4 [21], Spidey [17], BLAT [10], GMAP [12], and Spa [11] in terms of the capacity of the method to realize alignments at various levels of similarity between input sequences. Splign was shown to be more performant that other methods in the comparison for all levels of sequence similarity. Three version

Table 1 Detailed description of the real data, from the Ensembl-Comapara database, used for the evaluation

	Family					
	FAM86		MAG		TP53	
Species	Gene_ID	#	Gene_ID	#	Gene_ID	#
Human	ENSG00000158483	3	ENSG00000105492	6	ENSG00000141510	15
	ENSG00000186523	4	ENSG00000142512	7	ENSG00000073282	11
	ENSG00000145002	2	ENSG00000105695	4	ENSG00000078900	9
Chimp.	ENSPTRG00000007738	1	ENSPTRG00000011374	1	ENSPTRG00000008703	1
Mouse	ENSMUSG00000022544	1	ENSMUSG00000051504	4	ENSMUSG00000022510	8
Rat	ENSRNOG00000002876	1	ENSRNOG00000021023	2	ENSRNOG00000010756	4
Cow	ENSBTAG00000008222	1	ENSBTAG00000017044	1	ENSBTAG00000001069	1
Chiken	ENSGALG00000002044	1			ENSGALG00000007324	2
Lizard			ENSACAG00000005408	1		
Total	8 genes	14	8 genes	26	8 genes	51
Avg. CDS length	726		1397.42		1277.41	
Avg. gene length	22782.37		26049.75		109457.75	
Avg. pairwise PID	56.57		41.41		57.21	

For each gene family, the following information are given: the species name, the Ensembl identifier of gene, the number of CDS for each gene (#), the average CDS length, the average gene length, and the average pairwise Percent Sequence Identity (PID). The average pairwise PID were computed based on pairwise alignments of the CDS obtained from the multiple alignments of their proteins families provided by Ensembl-Comapara

Table 2 Detailed description of the simulated data used for the evaluation: For each simulated dataset, the number of families, the number of genes per family, the average number of CDS, the average CDS length, the average gene length, and the average pairwise PID in families are given

	Small	Medium	Large
Number of families	36	36	36
Number of genes per family	5	5	5
Avg. number of CDS	8.86 (2.95)	8.03 (2.78)	9.30 (2.87)
Avg. CDS length	808.24 (175.79)	779.44 (186.66)	799.40 (214.40)
Avg. gene length	2556.83 (537.93)	2358.60 (548.06)	2477.95 (552.02)
Avg. pairwise PID	76.25 (18.92)	39.20 (13.85)	11.42 (3.86)

For the average measures, the standard deviations are also given

of SplicedFamAlign were tested. SFA_L corresponds to the first local alignment step of SplicedFamAlign. SFA_E is SFA_L followed by the gapped extension step of the method. SFA_G is SFA_E followed by the final global alignment step of the method. SFA_E_ws corresponds to SFA_E in the case where the splicing structures of the input sequences are not provided, but predicted in a preliminary step by aligning each CDS against its own gene. A description of the methods used in the evaluation is given in Table 3. A more detailed description of the prediction method of splicing structures is given in Algorithm 1 in Additional file 1.

For SFA_E, the maximum number of gaps in a left or right extension of an anchor alignment is $\alpha = 21$. For SFA_E and SFA_G, the additional aligned segments obtained after the anchor extension step or the global alignment step are kept if and only if they have a Percent Sequence Identity (PID) greater or equal to $\beta = 30\%$ (See "The SFA-align algorithm for the SAP_III problem" section for a more detailed description of the parameters of algorithms used in the anchor extension and the global alignment steps of SFA). For Splign, the parameter min_exon_identity is set to 0 in order to allow Splign achieve the highest coverage of CDS possible. Setting this parameter to higher values makes Splign get worse performance (data not shown).

Discussion

First, we compared the ability of the methods to compute spliced alignments with high coverage of CDS (Fig. 3), relevant Percent Sequence Identity (PID) (Fig. 4), and induced exon extremities corresponding to actual exon extremities in the CDS and the gene sequence (Fig. 5). For the simulated data, for which the true alignment of the sequences in provided, we compared the ability of the methods to recover the true spliced alignments (Fig. 6).

Second, the true splicing orthology relationships between the CDS are also known for the simulated data. We compared the ability of the methods to be used for computing the true CDS orthology groups within a gene family based on spliced alignments (Fig. 7).

Third, we compared the execution time of the methods (Fig. 8).

Evaluation of the quality of spliced alignments

CDS coverage. We compared the CDS coverage of the spliced alignments computed using the methods. The results are shown in Fig. 3. In terms of CDS coverage, SFA methods and especially SFA_G, show the best results and their performances remain high for various levels of similarity between input sequences, covering in average more than 90% of CDS for all simulated datasets (Small, Medium, Large). Splign achieves a high CDS coverage when the sequences have a high level of similarity (Small dataset), but the performance decreases for medium and low similarity levels.

Percent Sequence Identity (PID). The PID of the spliced alignments constitutes also a good criterion to evaluate the quality of the alignments. For all methods and all datasets, the PID computed for their spliced alignment results are shown in Fig. 4. The PID for SFA_L and SFA_E decreases when the input sequence similarity decreases, which is expected. However, for the other methods, especially Splign, the PID remains high for most datasets (more than 80%). For instance, while Splign achieves the lowest CDS coverage for the MAG family (Fig. 3), it

Table 3 Description of the methods used in the evaluation: For each method, the optimization criteria and the specific parameters are given

Method	Optimization criteria	Specific parameters
SFA_L		tblastx e-value= 10^{-7}
SFA_E	sequence similarity splice signal splicing structure	max. number of gaps in extension $\alpha = 21$ min. PID of aligned segments $\beta = 30$
SFA_G		min. PID of aligned segments $\beta = 30$
Splign	sequence similarity splice signal	Default parameters min_exon_identity = 0
SFA_E_ws	same as SFA_E with predicted splicing structure	

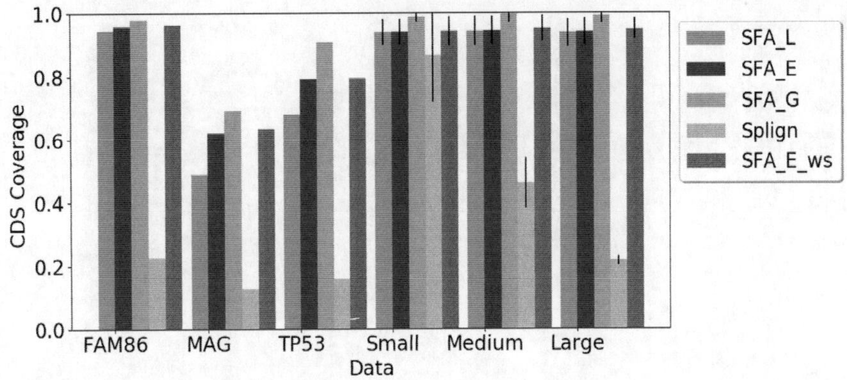

Fig. 3 Average CDS coverage of spliced alignments obtaining using the methods (SFA_L, SFA_E, SFA_G, Splign, SFA_E_ws) on the real and simulated dataset. For the simulated datasets composed of several gene families, the standard deviations are also given

achieves the highest PID for this family. Taken together, the results on the CDS coverage and the PID (Figs. 3 and 4) suggest that Splign is very stringent and aligns only highly sequence-conserved segments, whereas the SFA methods are also able to align less similar segments with conserved splicing structures.

Inference of real exon extremities. We compared the ability of the methods to correctly identify actual exon extremities in the CDS and the gene sequence and canonical splice sites in the gene sequence. We used the following performance metrics. The precision measure represents the fraction of inferred exon extremities that corresponds to real CDS exon extremities (A), real gene exon extremities (B), canonical gene splice sites (C). The recall measure represents the fraction of real CDS exon extremities (A), real gene exon extremities (B), canonical gene splice sites (C) that are correctly inferred by the spliced alignments. The f-score is the harmonic mean of precision and recall.

$$\texttt{f-score} = 2 * \frac{\texttt{precision} * \texttt{recall}}{\texttt{precision} + \texttt{recall}}$$

The results of the comparison are shown in Fig. 5. In terms of precision, we observe that, for all references (A), (B) and (C), all methods achieve a good performance. Globally, they perform better for gene exon extremities (B) and canonical splice sites (C) than for CDS exon extremities (A). In particular, Splign performs almost perfectly for

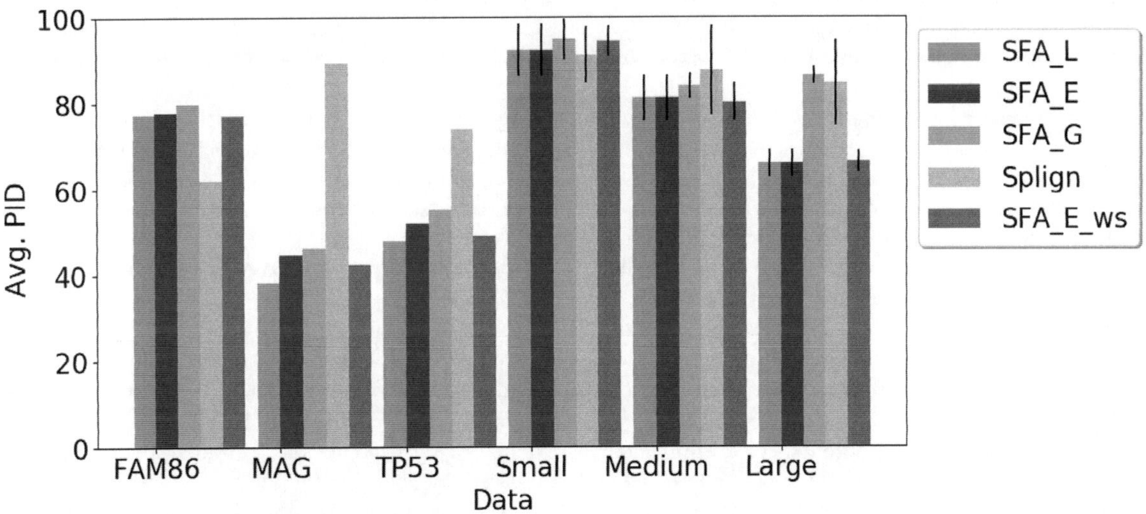

Fig. 4 Average Percent Sequence Identity (PID) for each method and each dataset used in the evaluation. For the simulated datasets composed of several gene families, the standard deviation are also given

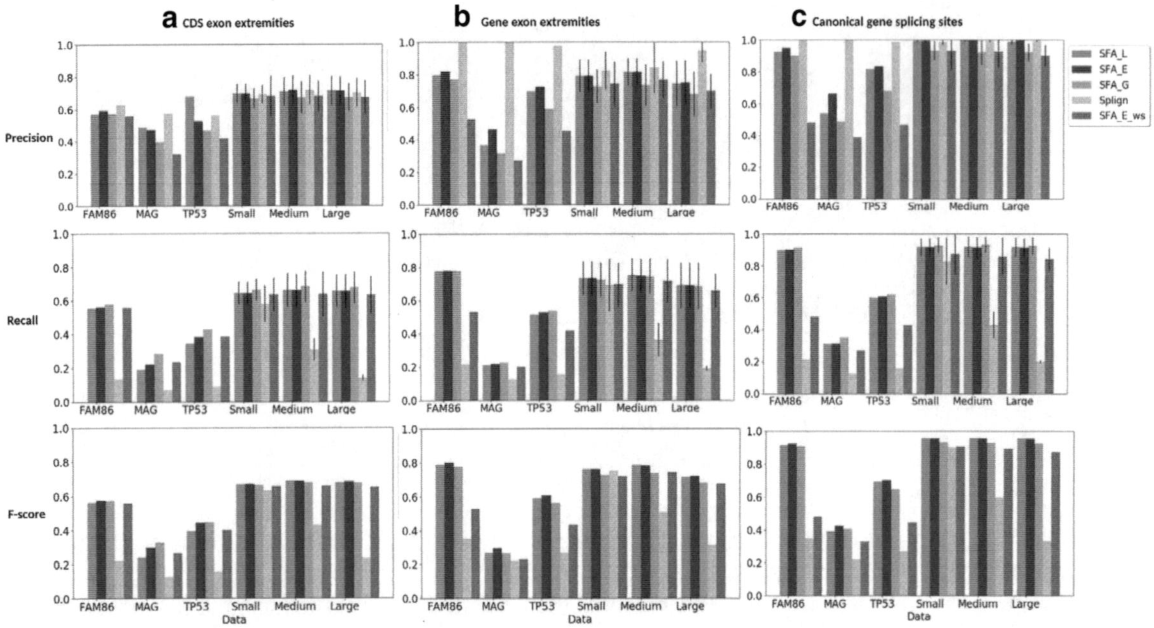

Fig. 5 Precision, recall and f-score measure, for all methods (SFA_L, SFA_E, and SFA_G, Splign, SFA_E_ws), for the comparison of inferred exon extremities with real CDS exon extremities (**a**) , real gene exon extremities (**b**), canonical gene splice sites (**c**). For the simulated datasets composed of several gene families, the standard deviation are also given for precision and recall

(B) and (C). SFA_E is the second one in terms of precision and also performs well.

In terms of recall, SFA_G is the best performing method for all references (A), (B) and (C), closely followed by SFA_E and SFA_L. The performance of SFA methods remains similar for various levels of sequence similarity (Small, Medium and Large datasets). However, the performance of Splign decreases when the level of sequence similarity decreases. This observation is coherent with the results obtained for the evaluation of the CDS coverage achieved by the methods. Indeed, the lower the CDS coverage, the lower the recall measure achieved.

Combining the precision and the recall measures, the f-score measure shows that SFA_E and SFA_L are the best performing methods. Note also that SFA_E_ws, without using prior information on the splicing structure of input sequences, achieves a higher performance than Splign, thanks to its preliminary step of splicing structure prediction. The robustness of the SFA methods to changes in the level of sequence similarity can be explained by the use of prior or predicted splicing structure information that allows to detect structure conservation even when the sequence conservation signal is low.

Comparison with true alignments. For the simulated datasets for which the true spliced alignments of CDS against gene sequences are provided, we evaluated the ability of the methods to correctly recover the true spliced alignments. We evaluated the precision, recall and f-score measure of a computed spliced alignment as follows. The precision measure is the fraction of pairs of aligned nucleotides in the computed alignment that are also aligned together in the true alignment. The recall measure is the fraction of pairs of aligned nucleotides in the true alignment that are also aligned together in the computed alignment. The f-score is a combination of the precision and recall as defined for the evaluation of the ability to infer real exon extremities.

The results of the evaluation are shown in Fig. 6. We can observe that all methods achieve high and comparable precision rates. In terms of recall, the SFA methods also have a high performance for various levels of sequence similarity. Splign also performs very well for high sequence similarities (Small dataset), but the performance decreases with the decrease of sequence similarity.

Evaluation of the quality for CDS orthology groups identification

We applied our algorithm for the identification of CDS orthology groups based on structural similarity, using spliced alignments computed by the methods SFA_L, SFA_E, and SFA_G and Splign. For each method, we then obtained a set of CDS orthology groups. We evaluated and compared the ability of the methods to recover the true CDS orthology groups. Two CDS are true orthologs if the sets of exons composing them are in bijection in such a way that any pair of exons in bijection descend from a

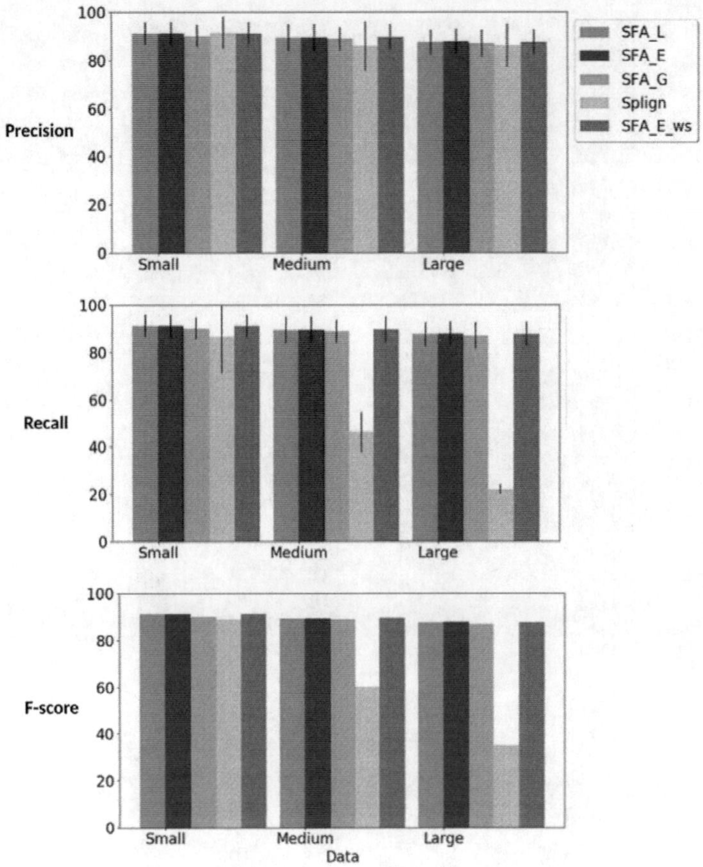

Fig. 6 Precision, recall and f-score measure, for all methods (SFA_L, SFA_E, and SFA_G, Splign, SFA_E_ws), for the comparison of the computed spliced alignments with the true spliced alignments. For each simulated dataset, the standard deviation are also given for precision and recall

same ancestral exon. Thus, we compared the CDS orthology groups obtained using each method with the true CDS orthology groups given in the simulated datasets. The precision, recall and f-score measures of a computed clustering are defined as follows.

The precision represents the fraction of pairs of CDS found as orthologs by the computed clustering that are true orthologs. The recall represents the fraction of true pairs of orthologous CDS that are also found as orthologs by the computed clustering. The f-score is a combination of the precision and recall as defined in "Evaluation of the quality of spliced alignments" section.

Figure 7 shows the results for each method. The precision score for Splign on the dataset Large is N/A because no pair of CDS were found as orthologs by the computed clustering. As for previous comparisons, the precision scores are high for all methods, but smaller for SFA_E_ws. For the SFA methods with prior splicing information, the recall scores are also high and robust to changes in the level of sequence similarity, whereas for Splign, the recall score decreases when the level of sequence similarity decreases. For SFA_E_ws the results are sensitive to the accuracy of the splicing structure prediction, but still less sensitive to sequence similarity than Splign.

Evaluation of the execution time

Finally, we compared the execution time of the methods. The average execution times of the SFA_L, SFA_E, SFA_E_ws and Splign methods to compute spliced alignments for each dataset are shown in Fig. 8. The execution times of SFA_G are not displayed in the same figure as they are more than 500 times higher than the execution times of other methods. The very high execution times of SFA_G are explained by the global alignment step of the method that uses a dynamic programming algorithm of quadratic time complexity with a multiplicative constant $I_{max} - I_{min} = 5000$ (See "The SFA-align algorithm for the SAP_III problem" section, Step 3 for a description of the main recurrence formula used by the dynamic programming algorithm). For instance, in contrast to Splign which stops without achieving its global alignment step when the sequence

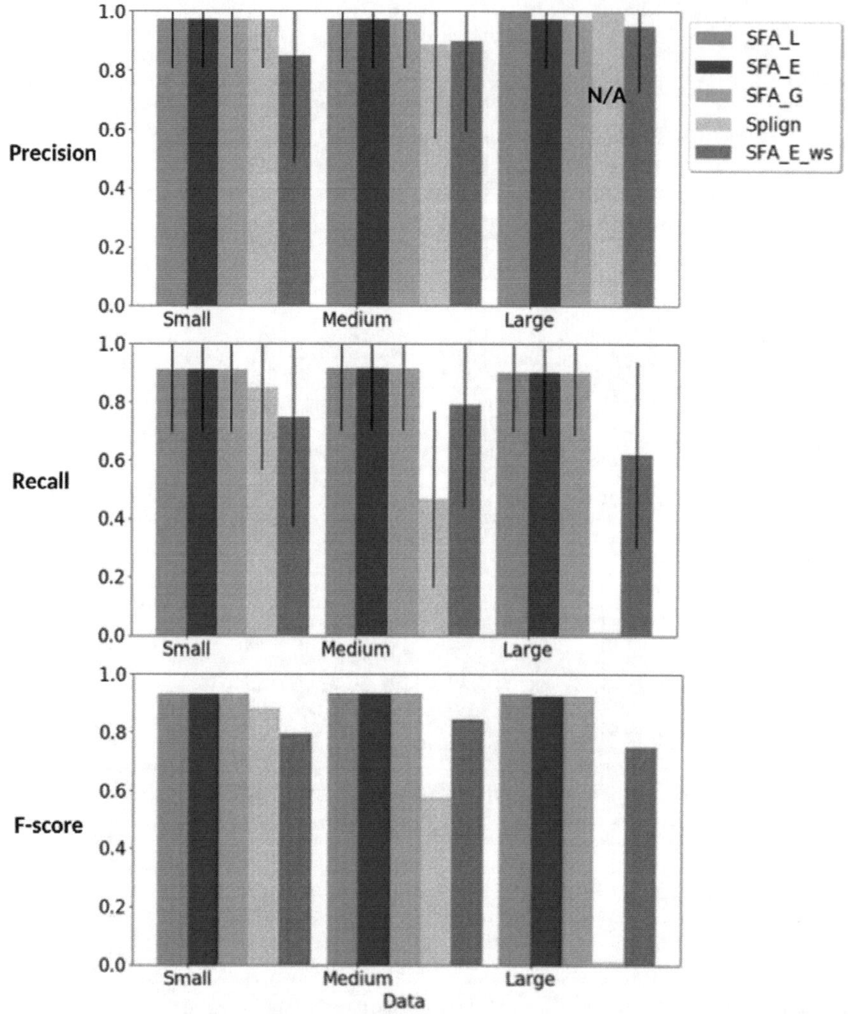

Fig. 7 Precision, recall and f-score measures, for all methods (SFA_L, SFA_E, and SFA_G, Splign, SFA_E_ws), for the comparison of the computed CDS orthology groups with the true CDS orthology groups. For each simulated dataset, the standard deviation are also given for precision and recall

similarity is low, the global alignment step of SFA_G is achieved even if the preceding steps of local alignment and anchor extension end up with a spliced alignment having a low CDS coverage. Thus, we discard SFA_G and consider SFA_E as the default method of the SFA framework, as it achieves similar accuracy as SFA_G, but with fast execution time. Figure 8 shows that the execution times of SFA_E are slightly higher than those of SFA_L. SFA_E_ws has slightly higher execution time than SFA_E due to its preliminary step of splicing structure prediction. Splign has higher execution times than SFA_L, SFA_E and SFA_E_ws for real datasets, but slightly lower execution times for simulated datasets. This can be explained by the smaller difference between CDS lengths and gene lengths in the simulated datasets compared to the real data. (See the average CDS and gene lengths of datasets in Tables 1 and 2). So, when the time-consuming global alignment step of Splign is achieved, it is applied on larger instances in the real dataset than in the simulated dataset.

Methods

In this section, we first give some formal definitions that will be useful for the remaining of the section. In the second subsection, the definitions of three versions of the spliced alignment problem are given under a unified framework allowing to compare theoretically the optimization criteria of the different versions of the problem. All existing cDNA-to-genome spliced alignment methods correspond either to the first or the second version of the problem that account for sequence similarity and splice signals in the input genomic sequence. The last version of the problems introduced in this paper additionally takes

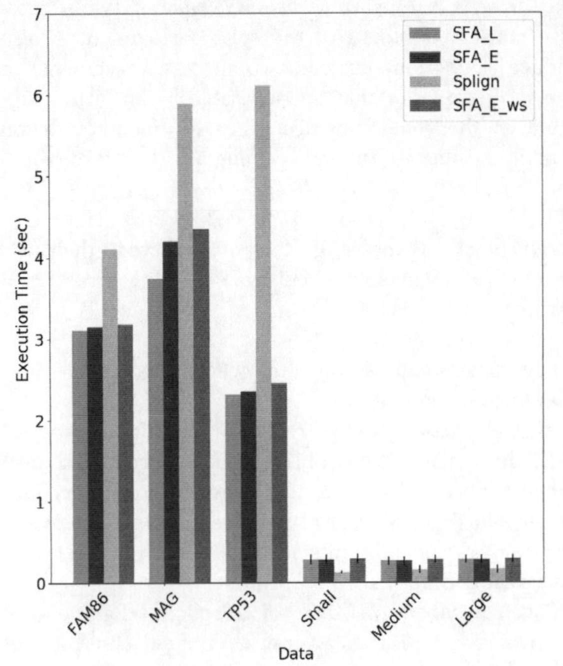

Fig. 8 Average execution times in second (sec) of SFA_L, SFA_E, SFA_E_ws, and Splign to compute spliced alignments for all datasets. For each simulated dataset, the standard deviation are also given

into account the splicing structure of the input sequences. In the third subsection, we describe our SFA-align algorithm for the new version of the spliced alignment problem. In the fourth subsection, we describe SFA-ortholog algorithm for the computation of CDS orthology groups, using spliced alignments.

Preliminary definitions: gene, CDS, spliced alignment

Given a set S, $|S|$ denotes the size of S, and given a sequence T, $\texttt{length}(T)$ denotes the length of T. Note that in the nomenclature used in this section, we call *CDS* the exon structure of a coding sequence and *CDS sequence* the nucleotide sequence of the coding sequence.

Gene and CDS: A gene sequence is a DNA sequence on the alphabet of nucleotides $\Sigma = \{A, C, G, T\}$. Given a gene sequence g, an *exon* of g is represented as a pair of integers (a, b) such that $a \leq b$ and a and b are the start and end locations of the exon on the gene sequence. The sequence of the exon (a, b) is then denoted by $g[a, b]$. A CDS of the gene sequence g is represented as a chain $c = \{(a_1, b_1), \ldots, (a_j, b_j)\}$ of exons of g such that for any two successive exons (a_i, b_i) and (a_{i+1}, b_{i+1}), $b_i < a_{i+1}$. The ith exon of the CDS c is then denoted by $c[i]$. The set of introns induced by the CDS c is denoted by $\texttt{Intron}(c) = \{(b_1, a_2), (b_2, a_3) \ldots, (b_{j-1}, a_j)\}$. The set of known CDS of a gene g is denoted by $\mathcal{C}(g)$ and the set of all exons of all konwn CDS of g is denoted by $\mathcal{E}(g) = \bigcup_{c \in \mathcal{C}(g)} c$ (see Fig. 9 for an illustration).

The sequence of a CDS c of g is denoted by $g[c]$. $g[c]$ is the concatenation of the sequences of the exons composing c. An exon of a CDS sequence $g[c]$ is represented as a pair of integers (k, l) such that $k \leq l$, and k and l are the start and end locations of the exon on the CDS sequence. In this case, the sequence of the exon (k, l) of $g[c]$ is denoted by $g[c]\,[k, l]$. The set of exons composing a CDS sequence $g[c]$ is denoted by $\mathcal{E}(g[c])$ (see Fig. 9 for an illustration).

Spliced alignment: A *spliced alignment* is an alignment of a CDS sequence against a gene sequence that allows to identify conserved exons sequences. Formally, a spliced alignment of a CDS sequence $g[c]$ against a gene sequence h is represented as a chain $A = \{(k_1, l_1, a_1, b_1), \ldots, (k_j, l_j, a_j, b_j)\}$ of quadruplets called blocks such that for any block (k, l, a, b) of A, $k \leq l$ and k and l are the start and end locations of a segment on the CDS sequence, and $a \leq b$ and a and b are the start and end locations of a segment on the gene sequence or $a = b = 0$; The ith block of a spliced alignment A is denoted by $A[i]$, and:

1 $k_1 = 1, l_j = \texttt{length}(g[c])$ and for any two successive blocks $A[i] = (k_i, l_i, a_i, b_i)$ and $A[i+1] = (k_{i+1}, l_{i+1}, a_{i+1}, b_{i+1})$, we have $l_i = k_{i+1} - 1$.

```
pos:000000000111111111122222222223333333333444444444455555555556666
    123456789012345678901234567890123456789012345678901234567890123
g : CGTATGGAATGCGTAAGAAGCAGGTCTGGGTAAGCATACGTGGGTAAGGGGAATGATTGAAAG
c1:         ATGC     AAGCAGGTCTGG                    GGGAATGA
c2:    ATGGAATGC     AAGCAG           CATACGTGG      GGGAATGATTGA
```

Fig. 9 Toy example of a gene sequence g with nucleotides numbered by position from 1 to 63, and a set of two CDS $\mathcal{C}(g) = \{c_1, c_2\}$ such that $c_1 = \{(9, 12), (18, 29), (49, 56)\}$ and $c_2 = \{(4, 12), (18, 23), (35, 43), (49, 60)\}$, inducing a set of exons $\mathcal{E}(g) = \{(4, 12), (9, 12), (18, 23), (18, 29), (35, 43), (49, 56), (49, 60)\}$. The sequence of the CDS c_1 is $g[c_1]$ =ATGCAAGCAGGTCTGGGGGAATGA with a set of exons $\mathcal{E}(g[c_1]) = \{(1, 4), (5, 16), (17, 24)\}$. The first exon of the CDS sequence $g[c_1]$ is $(1, 4)$ with exon sequence $g[c_1]\,[1, 4]$=ATGC

2 for any two blocks $A[i_1]$ and $A[i_2]$ with $i_1 < i_2$, we have $b_{i_1} < a_{i_2}$ if $a_{i_1} \neq 0$ and $a_{i_2} \neq 0$.

A block (k, l, a, b) represents an alignment of a segment $g[c]\,[k, l]$ of the CDS sequence with a segment $h[a, b]$ of the gene sequence. If $a = b = 0$, then the gene segment $h[a, b]$ is empty and the block (k, l, a, b) represents a deletion of the CDS segment $g[c]\,[k, l]$ in the spliced alignment. We call such a block a *deleted block*. Otherwise, the block (k, l, a, b) represents a conservation between a putative CDS exon sequence $g[c]\,[k, l]$ and a putative gene exon sequence $h[a, b]$, and we call such a block a *conserved block* (see Fig. 10 for an illustration).

Condition 1. of the spliced alignment definition implies that the set of conserved and deleted blocks of the spliced alignment covers the entire CDS sequence. Condition 2. implies that the blocks are ordered in the alignment following an increasing order of their location on the CDS sequence, and this order is also preserved on the gene sequence.

For instance, in Fig. 9, the spliced alignment of $g[c_1]$ against g is $A = \{(1, 4, 9, 12), (5, 16, 18, 29), (17, 24, 49, 56)\}$ (see Fig. 10 for more general examples of spliced alignments with conserved and deleted blocks).

A spliced alignment A induces a set of putative gene intron segments. These intron segments are the gene segments that lie between two successive blocks of the spliced alignment that are conserved blocks. Formally, the set of introns induced by a spliced alignment $A = \{(k_1, l_1, a_1, b_1), \ldots, (k_j, l_j, a_j, b_j)\}$ is denoted by $\texttt{Intron}(A) = \{(b_i, a_{i+1})$ *such that* (k_i, l_i, a_i, b_i) *and* $(k_{i+1}, l_{i+1}, a_{i+1}, b_{i+1})$ *are conserved blocks*$\}$. Two successive conserved blocks of the spliced alignment A also induce a junction between two successive segments in the CDS sequence that are separated by an intron segment in the gene sequence. The set of putative exon junctions induced the spliced alignment A is denoted by $\texttt{Junction}(A) = \{l_i$ *such that* (k_i, l_i, a_i, b_i) *and* $(k_{i+1}, l_{i+1}, a_{i+1}, b_{i+1})$ *are conserved blocks*$\}$. Note that if all blocks composing A are conserved, then the number of introns induced by A is $|\texttt{Intron}(A)| = |\texttt{Junction}(A)| = |A| - 1$.

A new constrained version of the spliced alignment problem

In this subsection, we first re-call two existing versions of the spliced alignment problem and we introduce a third more constrained version that takes additionally account of the splicing structure of input sequences. We discuss the motivations and limits of the different versions as we give their definitions.

Given a block (k, l, a, b) of a spliced alignment of a CDS sequence $g[c]$ against a gene sequence h, let $\texttt{sim}(g[c]\,[k, l]\,, h[a, b])$ denote the score of an optimal global alignment between the CDS segment $g[c]\,[k, l]$ and the gene segment $h[a, b]$ in a given scoring scheme. Note that if $a = b = 0$, then $h[a, b]$ is an empty segment. The following is a reformulation of the less constrained version of the spliced alignment problem introduced in [1]. This formulation gives a unified framework to formally define and compare all the versions of the spliced alignment problem formulated thereafter.

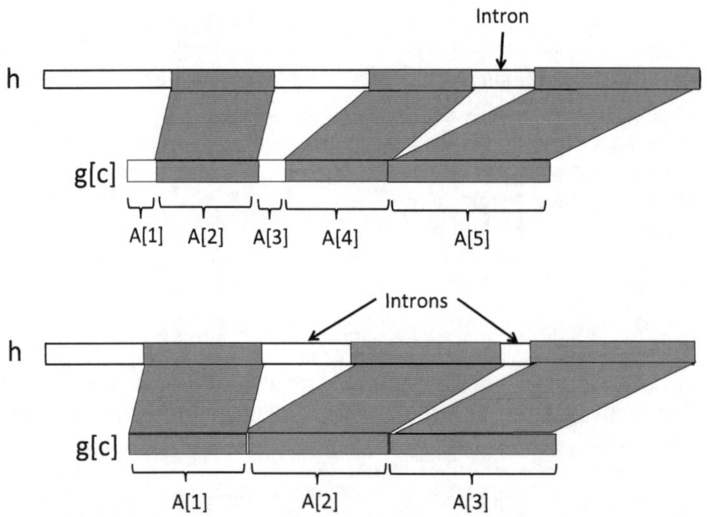

Fig. 10 Top. Illustration of a spliced alignment between a CDS sequence $g[c]$ and a gene sequence h, composed of 5 blocks, 3 conserved blocks ($A[2]$, $A[4]$ and $A[5]$) and 2 deleted blocks ($A[1]$ and $A[3]$). It induces 1 putative intron between the successive conserved blocks $A[4]$ and $A[5]$. The deleted blocks $A[1] = (k_1, l_1, a_1, b_1)$ and $A[3] = (k_3, l_3, a_3, b_3)$ are such that $a_1 = b_1 = 0$ and $a_3 = b_3 = 0$. **Bottom.** A spliced alignment composed of 3 conserved blocks that induce 2 putative introns between $A[1]$ and $A[2]$ and between $A[2]$ and $A[3]$

SPLICED ALIGNMENT PROBLEM I (SAP_I):
Input: A CDS sequence $g[c]$ from a gene sequence g ; a gene sequence h.
Output: A spliced alignment A of $g[c]$ against h that maximizes

$$\sum_{(k,l,a,b)\in A} \texttt{sim}(g[c]\,[k,l]\,,h[a,b]\,)$$

The SAP_I problem accounts only for the sequence similarity between the segments composing the blocks of the spliced alignment. In practice, in more than 99% of real cases of splicing, the spliced intron sequences start with a dinucleotide sequence "GT" and ends with a dinucleotide sequence "AG" corresponding to *canonical splice sites* [22, 23]. They exists also other non-canonical splice site pairs such as GC-AG, AT-AC that occur less frequently. Thus, in order to improve the accuracy of spliced alignments, a more constrained version of the problem allows to account for the extremities of intron segments induced by a spliced alignment. Given an intron (b,a) induced by two successive conserved blocks of a spliced alignment of a CDS sequence $g[c]$ against a gene sequence h, let $\texttt{splicesignals}(h[b,a]\,)$ denote a score of the putative intron segment $h[b,a]$ accounting for the presence or absence of known splice signals at the extremities of $h[b,a]$. A putative intron segment with two canonical splice signals at its extremities has a higher score than a segment with only one which has a higher score than a segment without any canonical splice signal at its extremities. A more constrained version of the spliced alignment problem studied in [9, 12] for instance is the following.

SPLICED ALIGNMENT PROBLEM II (SAP_II):
Input: A CDS sequence $g[c]$ from a gene sequence g ; a gene sequence h.
Output: A spliced alignment A of $g[c]$ against h that maximizes

$$\sum_{(k,l,a,b)\in A} \texttt{sim}(g[c]\,[k,l]\,,h[a,b]\,) + \sum_{(b,a)\in \texttt{Intron}(A)} \texttt{splicesignals}(h[b,a]\,)$$

The SAP_I and SAP_II problems do not account for the splicing structure of the input sequences. In order to further improve the accuracy of spliced alignments, we define a more constrained version of the problem that takes into account the exon structure of the CDS sequence and the exon-intron structure of the gene sequence.

Given a putative intron $(b,a) \in \texttt{Intron}(A)$ induced by two successive conserved blocks of a spliced alignment A of a CDS sequence $g[c]$ against a gene sequence h, let $\texttt{splicesites}_{\mathcal{E}(h)}(b,a)$ denote a score of the putative intron segment $h[b,a]$ accounting for the correspondance of its extremities with known splice sites in the gene sequence h. A putative intron segment whose extremities both correspond to known splice sites in the gene sequence receives a higher score than a segment with only one extremity corresponding to a known splice site which has a higher score than a segment without any correspondance to known splice sites at its extremities.

Similarly, for a putative CDS exon junction $l \in \texttt{Junction}(A)$ induced by two successive conserved blocks of the spliced alignment A, let $\texttt{exonjunction}_{\mathcal{E}(g[c])}(l)$ denote a score of the putative CDS exon junction l accounting for its correspondance with a real exon junction in the CDS sequence $g[c]$. If l corresponds to a real exon junction in $\mathcal{E}(g[c])$ it receives a higher score than if it does not correspond to a junction in $\mathcal{E}(g[c])$. The more constrained version of the problem introduced here is defined as follows.

SPLICED ALIGNMENT PROBLEM III (SAP_III):
Input: A CDS sequence $g[c]$ from a gene sequence g ; the set of exons $\mathcal{E}(g[c])$ of $g[c]$; a gene sequence h ; the set of exons $\mathcal{E}(h)$ of h.
Output: A spliced alignment A of $g[c]$ against h that maximizes

$$\sum_{(k,l,a,b)\in A} \texttt{sim}(g[c]\,[k,l]\,,h[a,b]\,) + \sum_{(b,a)\in \texttt{Intron}(A)} \texttt{splicesignals}(h[b,a]\,) + \sum_{(b,a)\in \texttt{Intron}(A)} \texttt{splicesites}_{\mathcal{E}(h)}(b,a) + \sum_{l\in \texttt{Junction}(A)} \texttt{exonjunction}_{\mathcal{E}(g[c])}(l)$$

Examples of algorithms developed for the SAP_I problem which accounts only for the sequence similarity at the nucleotide or at the amino acid level are BLAT [10], DDS/GAP2 [24], SOAPsplice [8] and HSA [7]. Several algorithms have also been developed for the SAP_II problem, for instance Geneseqer [25], Sipdey [17], Splign [9], SIM4 [21], MGAlign [18], GMAP [12] and Spa [11]. These algorithms account for splice signals in addition to sequence similarity. A comparison of a subset of these tools has demonstrated the superiority of splice signal-based methods compared to only sequence similarity-based methods [12]. Moreover, it was shown in [9] that, among splice signal-based methods, the best performing current spliced alignment method for cDNA-to-gene alignment was Splign. Thus, taking account of more information about the input sequences improves the accuracy of spliced alignments. We then expect that accounting for information on the exon structure of CDS sequences and the exon-intron structure of gene sequences will further improve spliced alignments.

The SFA-align algorithm for the SAP_III problem

In this section, we describe a heuristic algorithm called SFA-align for the SAP_III problem. The algorithm follows a general approach followed by most heuristic methods developed for the SAP_I and SAP_II problems. The algorithm is decomposed into three steps that each accounts for the splicing structure of the input sequences. Note that in the case where the splicing structures of the input sequences are not provided, SFA-align includes a preliminary step in which the splicing structure of input sequences are inferred by computing a spliced alignment of each CDS against its own gene having a Percent Sequence Identity (PID) of 100. An overview of the main steps of SFA-align is depicted in Fig. 1. It starts with a fast local alignment to compute highly conserved local segments used as anchors in the alignment. Next, the anchor alignment are extended in order to maximize the exon coverage. Finally, between the anchors, a global alignment of the remaining segments can be applied to complete the alignment.

Step 1. Local alignments using tblastx. This step is achieved using Translated Blast (tblastx) [26] in order to obtain a preliminary set of local alignments between the input CDS sequence and gene sequence. Tblastx is used in order to account for the translation of the sequences into amino acid sequences. This allows to detect conserved exon segments translated into amino acid sequences even in the presence of translational frameshifts or nucleotide silent mutations.

Given the set of hits obtained using tblastx with a given threshold E-value, the following procedure is applied to obtain the final set of local alignments. i) Each hit is assigned to the exon of the CDS that the hit covers the more, and the hit is minimally trimmed at its extremities to cover only this CDS exon. Thus, the boundaries of a trimmed hit never exceed the boundaries of the CDS exon that it is assigned to ; ii) All hits within the same CDS exon are compared and only a subset of pairwise compatible hits with the lowest E-values is kept ; iii) Finally, the hits within different exons are compared and only compatible hits with the lowest E-values are kept ; iv) For each exon, the remaining compatible hits are gathered into a set of non-overlapping local alignments for the exon. A more detailed description of this procedure is given in Algorithm 2 in Additional file 1. Figure 11 provides an illustration of the result of the procedure on an example. In this example, Step i of the local alignment procedure associates hits 1 and 2 to exon E_1, hits 3 and 4 to exon E_2, hit 5 to exon E_3, and hit 6 to exon E_4. The extremities of the hit 5 are also trimmed so that the hit covers only exon E_3. Step ii removes hit 4 from the list of hits associated to exon E_2 because it is incompatible with hit 3 that has a lower E-value. Next, Step iii removes hit 6 from the list of hits associated to exon E_4 because it is incompatible with hit 5 associated to exon E_3 with a lower E-value. Finally, Step iv ends up with three local alignments covering exons E_1, E_2 and E_3 illustrated by hits 1, 3, 5 in Fig. 12.

Step 2. Gapped extension of anchors. In this step, each local alignment obtained from Step 1 is extended in both directions in order to increase as much as possible the coverage of the CDS exon to which it is associated. The extension procedure allows an extended portion of an alignment to start with a succession of gaps whose number must be a multiple of 3 and shall not exceed a given number α of gaps. For example, a local alignment ("AAUCGGA","AAUCGGA") that partially covers a CDS exon "AAUCGGAUGGGUG" could be extended on th e right until the extremity of the exon following three poss ible configurations. It can be extended as (a) ("AAUCGGAU GGGUG","AAUCGGAUGGGUG") without any gap at the star t of the extension, or (b) ("AAUCGGAUGGGUG","AAUCGGA ---GUG") starting with 3 gaps in the gene sequence, or (c) ("AAUCGGA---UGGGUG", "AAUCGGACCCUGGGUG") starting with 3 gaps in the CDS.

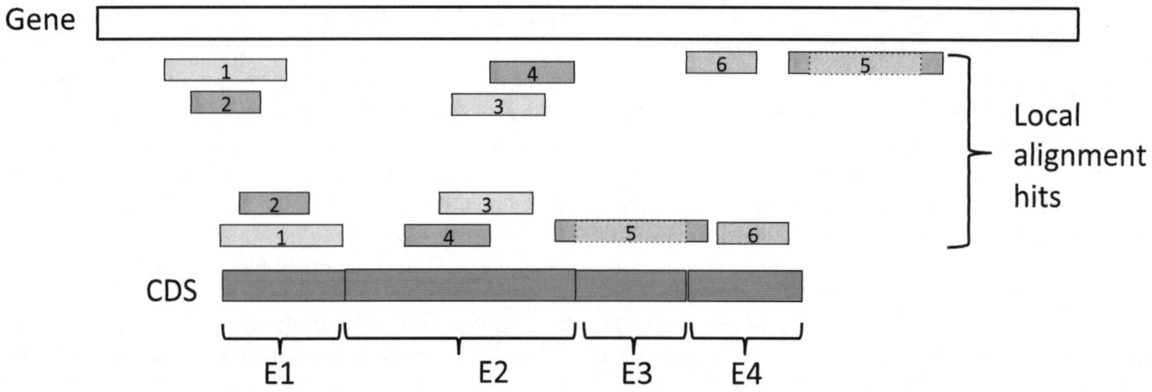

Fig. 11 Example of a set of six hits obtained using tblastx between a CDS and a gene sequence, with hits numbered from 1 to 6 by increasing E-values. The local alignment algorithm ends up with the three local alignments 1, 3, 5 covering exons E_1, E_2 and E_3

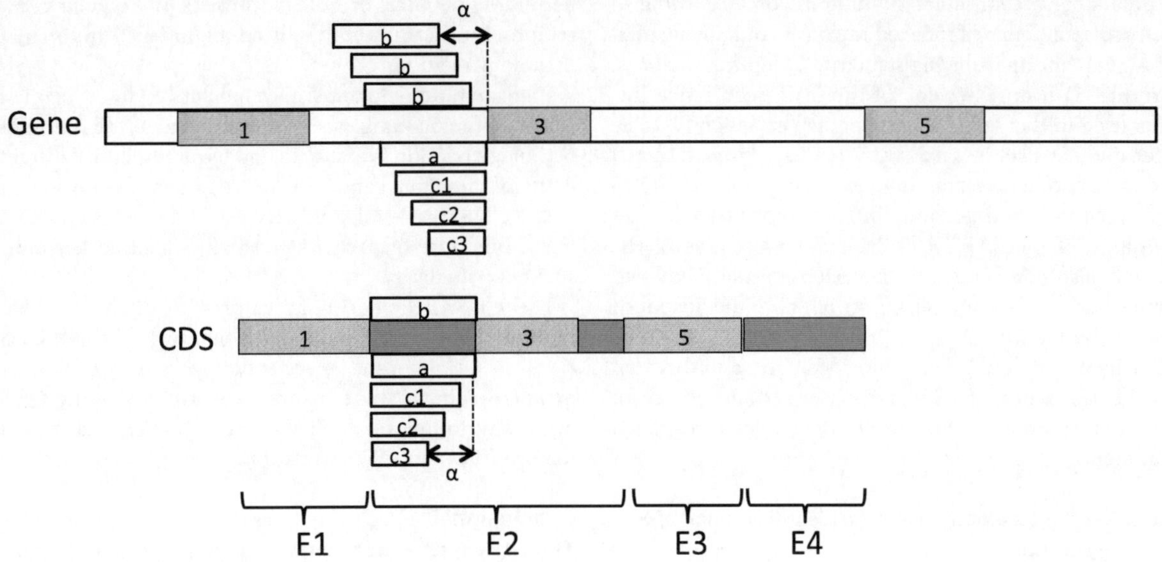

Fig. 12 Example of a set of three local alignments 1, 3, 5 between a CDS and a gene sequence obtained at the end of the local alignment step. All possible configurations (**a**), (**b**) and (**c**) of left extension of the local alignments 3 are illustrated. The maximum number of gaps α for configurations (**b**) and (**c**) is also illustrated

Given a local alignment of a CDS segment and a gene segment represented by a block (k, l, a, b) such that k and l are the start and end positions of the segment in the CDS, and a and b are the start and end positions of the segment in the gene sequence, the gapped extension procedure applied on (k, l, a, b) is as follows. Let (k', l') be the exon of the CDS to which the local alignment (k, l, a, b) is associated. Note that $k' \leq k \leq l \leq l'$. i) If $k' < k$, then the alignment (k, l, a, b) can be extended on the left. The procedure tries all possible configurations of extension, (a) without any gap at the start of the extension, (b) starting with gaps in the gene sequence or (c) starting with gaps in the CDS, such that the extension does not overlap any other local alignment. For configurations (b) and (c), all possible numbers of gaps $3 * i$ with i ranging from 1 and $\alpha/3$ are evaluated, the extension configuration having the highest identity score is returned. If this identity score is above a given identity threshold β, then the alignment is extended on the left, otherwise no extension is applied. i) If $l < l'$, then the alignment (k, l, a, b) can also be extended on the right. A procedure similar to the previous one is applied in order to try all possible configurations of extension on the right. A more detailed description of the procedure is given in Algorithm 3 in Additional file 1. Figure 12 also provides an illustration of the configurations of left extension that are explored by the procedure on an example.

Step 3. Global spliced alignment algorithm. For all CDS exons that were left completely unaligned by the previous steps, a dynamic programming algorithm for global spliced alignment is applied. Restricting the global alignment to remaining unaligned exons of the CDS allows to accelerate the global algorithm step by dividing the dynamic programming space. The following main recurrence formula is used.

$$S(i,j) = max \begin{cases} S(i-1,j-1) + score(i,j) \\ S(i-1,j) + score(i,-); \\ S(i,j-1) + score(-,j); \\ \max_{(j',j)\in\mathcal{E}(h)}\{S(i,j') \\ \quad + \texttt{exonjunction}(i) \\ \quad + \texttt{splicesignals}(h[j',j]) \\ \quad + \texttt{splicesites}(j',j)\}; \\ \max_{l\in[l_{min},l_{max}]}\{S(i,j-l) \\ \quad + \texttt{exonjunction}(i) \\ \quad + \texttt{splicesignals}(h[j-l,j]) \\ \quad + \texttt{splicesites}(j-l,j)\}; \end{cases} \quad (1)$$

In Formula (1), when computing the global spliced alignment between a CDS segment $g[c]\,[k,l]$ and a gene segment $h[a,b]$, $S(i,j)$ is the maximum score of a global spliced alignment of the segment $g[c]\,[k, k+i-1]$ and the segment $h[a, a+i-1]$. For instance, if $g[c]\,[k,l]$ has length m and $h[a,b]$ has length n, then $S(m,n)$ is the maximum score of a global spliced alignment of $g[c]\,[k,l]$ and the segment $h[a,b]$.

In the formula, the three first cases contribute to the computation of sequence alignment scores within blocks of the spliced alignment, given $score(i,j)$, $score(i,-)$, $score(-,j)$ that denote the scores of substitution, insertion and deletion of nucleotides. The two last cases contribute to evaluating the structure alignment score according to the correspondence of induced introns with known splicing signals and the splicing structure of input sequences. Formula (1) is an extension of the main recurrence formula used in the global alignment step of Splign [9]. The extension consists in accounting for `exonjunction`(i) and `splicesites`(j',j) that are the scores for CDS exon junctions and genomic introns introduced for the definition of the SAP_III problem. The last case of the formula also contributes to the extension and allows to further account for the splicing structure of the unspliced genomic sequence.

Finally, any new alignment block computed in this step is kept if and only if its identity score is above a given identity threshold β, otherwise it is not added to spliced alignment.

The SFA-ortholog algorithm for the identification of CDS orthology groups

In this section, we give a definition of CDS orthology groups computed by the SFA-ortholog algorithm based on pairwise spliced alignments. We first start with a definition of orthologous CDS based on spliced alignment.

In [15], an extension of the concept of gene orthology to spliced transcript orthology was introduced. They defined orthologous transcripts as two structurally similar transcripts from two orthologous genes. The orthology relationship between two transcripts relies on the structural similarity between the transcripts. This structural similarity is evaluated using the CDS associated to the transcripts and their spliced alignments against genes.

CDS orthology: Let c_1 and c_2 be two CDS from two homologous genes g and h respectively. Let A_1 be a spliced alignment of the CDS sequence $g[c_1]$ against the gene sequence h. The CDS c_1 and c_2 are orthologs if:

(1) $| c_1 | = | c_2 |$;
(2) $\texttt{Intron}(A_1) = \texttt{Intron}(c_2)$;
(3) for any $i, 1 \leq i \leq | c_1 |$, $[\texttt{length}(c_1[i]) - \texttt{length}(c_2[i])] \% 3 = 0$.

In other terms, c_1 are c_2 as orthologs if (1) they have the same number of exons $| c_1 | = | c_2 |$; (2) the spliced alignment of c_1 against h induces the same introns for c_1 and c_2, $\texttt{Intron}(A_1) = \texttt{Intron}(c_2)$; (3) the lengths of each pair of corresponding exons in c_1 and c_2 are congruent modulo 3. Conditions (1) and (2) ensure that the two CDS have the same splicing structure. Condition (3) ensures that the two CDS are translated in the same codon phase in each pair of corresponding exons in order to generate similar protein sequences.

Note that this definition only requires that one of the spliced alignments of $g[c_1]$ against h or $h[c_2]$ against g supports the orthology relation. An alternative more stringent definition of CDS orthology consists in requiring the reciprocity, i.e. that both spliced alignment support the orthology relation.

CDS orthology groups: Given a set of CDS $\mathcal{C}$ from a set of homologous genes $\mathcal{G}$, the transitivity of the CDS orthology relation is assumed and used to identify distant orthologs in $\mathcal{C}$ that cannot be directlty identified by means of the CDS structural similarity. Such orthologs could be missed because of partial spliced alignments due to low sequence similarity.

The CDS orthology relation on $\mathcal{C}$ is then extended into an equivalence relation such that for any three CDS c_1, c_2, c_3 in $\mathcal{C}$, if c_1 and c_2 are orthologs and c_2 and c_3 are orthologs, then c_1 and c_3 are also orthologs. The CDS orthology groups are defined as the equivalence classes of the resulting equivalence relation.

Conclusion

The article introduces a new version of the spliced alignment problem accounting for the splicing structure of input sequences. It constitutes a new approach to compute accurate CDS-to-gene spliced alignments, by detecting conservation in the splicing structure of input sequences.

We present a heuristic algorithm for the problem, and we show that it is useful to improve the accuracy of spliced alignments. The application of the algorithm to real and simulated datasets shows that the new method outperforms existing spliced alignment methods in terms of accuracy, with comparable execution times for CDS-to-gene spliced alignment. Moreover, its performance is robust to changes in the level of input sequence similarity. The method is particularly useful for identifying CDS splicing orthology groups that are conserved accross a set of homologous genes. It can also be used for the functional annotation of protein-coding genes in non-model genomes. On the algorithmic side, the pairwise spliced alignment problems studied in this paper can be extended to multiple spliced alignment problems that require more in-depth investigation. Future work will include the extension of the SFA method toward multiple spliced alignment. We hypothesize that aligning simultaneously multiple CDS and genes from a gene family can increase the accuracy of spliced alignment and splicing ortholog identification.

Additional file

Additional file 1: Pseudocode of algorithms for SplicedFamAlign steps. A file at the PDF format describing the algorithms used in the SFA method for the preliminary step of splicing structure prediction (Algorithm 1), the first step of local alignment (Algorithm 2), and the second step of gapped extension of anchors (Algorithm 3). (PDF 110 kb)

Abbreviations

CDS: Coding DNA sequences; PID: Percent sequence identity; SFA: SplicedFamAlign

Acknowledgements
We thank the anonymous reviewers and the CoBIUS lab at University of Sherbrooke for their helpful constructive discussion.

Funding
This work has been supported by grants from the Canada Research Chair (CRC Tier2 Grant 950-230577), and the Natural Sciences and Engineering Research Council of Canada (Discovery Grant RGPIN-2017-05552). The CRC Tier2 Grant provided the funding for publication of the article.

Availability of data and materials
Source code, simulated and real data are freely available at https://github.com/UdeS-CoBIUS/SpliceFamAlign.

About this supplement
This article has been published as part of *BMC Bioinformatics Volume 20 Supplement 3, 2019: Selected articles from the 17th Asia Pacific Bioinformatics Conference (APBC 2019): bioinformatics*. The full contents of the supplement are available online at https://bmcbioinformatics.biomedcentral.com/articles/supplements/volume-20-supplement-3.

Authors' contributions
JDA and AO formalized the problem. SJ and AO conceived the methods. SJ and JDA implemented the program and wrote its documentation. EK and AO conceived the simulation method and generated the simulated dataset. SJ ran the experiments. SJ and AO wrote the manuscript. SJ and AO analyzed and interpreted the data and the results. All authors critically revised the manuscript. All authors have read and approved the final manuscript.

Ethics approval and consent to participate
Not applicable.

Consent for publication
Not applicable.

Competing interests
The authors declare that they have no competing interests.

Publisher's Note
Springer Nature remains neutral with regard to jurisdictional claims in published maps and institutional affiliations.

Published: 29 March 2019

References

1. Gelfand MS, Mironov AA, Pevzner PA. In: Hirschberg D, Myers G, editors. Spliced alignment: A new approach to gene recognition. Berlin: Springer; 1996. pp. 141–158. https://doi.org/10.1007/3-540-61258-012.
2. Brendel V, Xing L, Zhu W. Gene structure prediction from consensus spliced alignment of multiple ESTs matching the same genomic locus. Bioinformatics. 2004;20(7):1157–69.
3. Kan Z, Rouchka EC, Gish WR, et al. Gene structure prediction and alternative splicing analysis using genomically aligned ESTs. Genome Res. 2001;11(5):889–900.
4. Kishi RM, dos Santos RF, Adi SS. Gene Prediction by Multiple Spliced Alignment. In: Norberto de Souza O, Telles GP, Palakal M, editors. Advances in Bioinformatics and Computational Biology. BSB 2011. Lecture Notes in Computer Science, vol. 6832. Berlin: Springer; 2011.
5. Rodriguez JM, Maietta P, Ezkurdia I, Pietrelli A, Wesselink JJ, Lopez G, et al. APPRIS: annotation of principal and alternative splice isoforms. Nucleic Acids Res. 2012;41(D1):D110–7.
6. Stanke M, Tzvetkova A, Morgenstern B. AUGUSTUS at EGASP: using EST, protein and genomic alignments for improved gene prediction in the human genome. Genome Biol. 2006;7(1):S11.
7. Bu J, Chi X, Jin Z. HSA: a heuristic splice alignment tool. BMC Systems Biol. 2013;7(2):S10.
8. Huang S, Zhang J, Li R, Zhang W, He Z, Lam TW, et al. SOAPsplice: genome-wide ab initio detection of splice junctions from RNA-Seq data. Front Genet. 2011;2:46.
9. Kapustin Y, Souvorov A, Tatusova T, Lipman D. Splign: algorithms for computing spliced alignments with identification of paralogs. Biol Direct. 2008;3(1):20.
10. Kent WJ. BLAT—the BLAST-like alignment tool. Genome Res. 2002;12(4):656–64.
11. Van Nimwegen E, Paul N, Sheridan R, Zavolan M. SPA: a probabilistic algorithm for spliced alignment. PLoS Genet. 2006;2(4):e24.
12. Wu TD, Watanabe CK. GMAP: a genomic mapping and alignment program for mRNA and EST sequences. Bioinformatics. 2005;21(9):1859–75.
13. Christinat Y, Moret BM. Inferring transcript phylogenies. BMC Bioinformatics. 2012;13(9):S1.
14. Kuitche E, Lafond M, Ouangraoua A. Reconstructing protein and gene phylogenies by extending the framework of reconciliation. In: Proceedings of International Conference on Bioinformatics and Computational Biology (BICOB'17); 2017. p. 79–86.
15. Zambelli F, Pavesi G, Gissi C, Horner DS, Pesole G. Assessment of orthologous splicing isoforms in human and mouse orthologous genes. BMC Genomics. 2010;11(1):1.
16. Blanquart S, Varré JS, Guertin P, Perrin A, Bergeron A, Swenson KM. Assisted transcriptome reconstruction and splicing orthology. BMC Genomics. 2016;17(10):157.
17. Wheelan SJ, Church DM, Ostell JM. Spidey: a tool for mRNA-to-genomic alignments. Genome Res. 2001;11(11):1952–7.
18. Ranganathan S, Lee BT, Tan TW. MGAlign, a reduced search space approach to the alignment of mRNA sequences to genomic sequences. Genome Inform. 2003;14:474–5.
19. Cunningham F, Amode MR, Barrell D, et al. Ensembl 2015. Nucleic Acids Res. 2015;43(D1):D662–9.
20. Kosiol C, Holmes I, Goldman N. An empirical codon model for protein sequence evolution. Mol Biol Evol. 2007;24(7):1464–79.
21. Florea L, Hartzell G, Zhang Z, Rubin GM, Miller W. A computer program for aligning a cDNA sequence with a genomic DNA sequence. Genome Res. 1998;8(9):967–74.
22. Burset M, Seledtsov I, Solovyev V. Analysis of canonical and non-canonical splice sites in mammalian genomes. Nucleic Acids Res. 2000;28(21):4364–75.
23. Sparks ME, Brendel V. Incorporation of splice site probability models for non-canonical introns improves gene structure prediction in plants. Bioinformatics. 2005;21(Suppl_3):iii20–30.
24. Huang X, Adams MD, Zhou H, Kerlavage AR. A tool for analyzing and annotating genomic sequences. Genomics. 1997;46(1):37–45.
25. Usuka J, Zhu W, Brendel V. Optimal spliced alignment of homologous cDNA to a genomic DNA template. Bioinformatics. 2000;16(3):203–11.
26. Kaur H, Singh A, Singh P. Comparison of Variants of BLAST. In: Proceedings of the International MultiConference of Engineers and Computer Scientists. vol. 1; 2008. http://www.iaeng.org/publication/IMECS2008/IMECS2008_pp190-194.pdf.

Legendre *et al. BMC Bioinformatics* 2019, **20**(Suppl 3):128
https://doi.org/10.1186/s12859-019-2648-1

BMC Bioinformatics

RESEARCH

Open Access

RCPred: RNA complex prediction as a constrained maximum weight clique problem

Audrey Legendre, Eric Angel and Fariza Tahi*

From The 17th Asia Pacific Bioinformatics Conference (APBC 2019)
Wuhan, China. 14-16 January 2019

Abstract

Background: RNAs can interact and form complexes, which have various biological roles. The secondary structure prediction of those complexes is a first step towards the identification of their 3D structure. We propose an original approach that takes advantage of the high number of RNA secondary structure and RNA-RNA interaction prediction tools. We formulate the problem of RNA complex prediction as the determination of the best combination (according to the free energy) of predicted RNA secondary structures and RNA-RNA interactions.

Results: We model those predicted structures and interactions as a graph in order to have a combinatorial optimization problem that is a constrained maximum weight clique problem. We propose an heuristic based on Breakout Local Search to solve this problem and a tool, called RCPred, that returns several solutions, including motifs like internal and external pseudoknots. On a large number of complexes, RCPred gives competitive results compared to the methods of the state of the art.

Conclusions: We propose in this paper a method called RCPred for the prediction of several secondary structures of RNA complexes, including internal and external pseudoknots. As further works we will propose an improved computation of the global energy and the insertion of 3D motifs in the RNA complexes.

Keywords: RNA complex, Secondary structure, RNA interaction, Pseudoknot, Maximum weight clique heuristic

Background

RNAs can link to each other and form complexes having catalytic functions. A well known example is the ribosome [1] composed of the 5S, 5.8S, 18S and 28S RNAs (in eucaryotes) and of ribosomal proteins. The RNAs present in the ribosome are involved in the formation of peptid bonds and in the reading of codons in the site A. RNA complexes are formed by canonical interactions (the Watson-Crick base pairs (A-U, G-C) and the Wobble base pair (G-U)) between several RNA strands and by non canonical interactions (tertiary interactions). Canonical interactions are the strongest interactions that stabilize RNA structure and they define the secondary structure. Non-canonical interactions are weaker but more numerous than canonical interactions. They involve chemical H bonds in the Hoogsteen or the sugar edges of the nucleotides and are responsible of the RNA 3D structure. In this paper, we focus only on the strongest interactions involved in RNA complexes, i.e. the canonical interactions.

Many tools exist to predict the joint secondary structure of RNA duplexes (both the base pairs internal to each RNA and the interaction or hybridization base pairs) using either the thermodynamic approach [2–9] or the comparative approach [6, 10–12]. However, to predict the secondary structure of RNA complexes composed of more than two RNAs is difficult and very few dedicated tools exist. The first proposed tool was MultiRNAFold [13]. In this tool, the RNAs are connected as one strand with extra loops between them. The minimum free energy of the complex is computed by a dynamic programming algorithm derived from [14]. The NUPACK package [15], proposed later, includes a software to predict RNA complex secondary structures. It extends the

*Correspondence: fariza.tahi@univ-evry.fr
IBISC, Univ Evry, Université Paris-Saclay, 91025 Evry, France

partition function computation of a single RNA, allowing the computation of the minimum free energy structure, to the multiple RNA case. Later, the RNA complex secondary structure prediction is modeled as a multiple RNA interaction problem [16]. The authors proved that this problem is NP-hard and proposed several approximation algorithms. Then, the RNA complex secondary structure prediction is formulated as a combinatorial optimization problem called *Pegs and Rubber Bands* and an approximation algorithm is proposed [17]. Then the tools NanoFolder [18] HyperFold [19] for RNA complex prediction are proposed. NanoFolder works in two steps, first it computes all the possible helices using a simple energy model and then a greedy algorithm selects the minimum free energy helices and adds them into the RNA complex. HyperFold also generates the possible helices but uses a more sophisticated algorithm than NanoFolder to select them. In MultiRNAFold, NUPACK and the algorithms of [16, 17], it is assumed that the RNAs are linked in a specific order. This order directly impacts on the quality of the predicted structures because the order can forbids some base pairs and so the set of all possible secondary structures is not explored. A solution could be to test all possible RNA linking orders ($n!$ for n RNAs) but it will not guarantee that all the possible structures can be found and the number of orders to test can become high in practice.

In some RNA secondary structures, specific motifs called *pseudoknots* can occur. Pseudoknots are notably involved in the readthrough mechanism of the translation. Therefore, they are important to study the function of RNA complexes. However, pseudoknots are difficult to predict, their prediction often leading to algorithms with high execution times. Then, the prediction of RNA secondary structures with pseudoknots is often restricted to subclasses of pseudoknots [20]. When pseudoknots occur in the interaction of two RNAs, they are called external pseudoknots or crossing interactions. Among the state of the art, only NanoFolder and the algorithms of [16] are able to predict pseudoknots.

All the tools and algorithms presented above are based on different thermodynamic models aiming to minimize the free energy. However, it is now known that the real structure of an RNA is not always the structure of minimum free energy but often a structure close to it. Hence, being able to generate sub-optimal structures is an important feature for the RNA complex prediction problem. Moreover, RNAs can have several structures, as the riboswitches, forming complexes with other RNAs [21] or other molecules to regulate the gene expression. Sub-optimal structures are, especially, used together with SHAPE data to elaborate a conformational ensemble [22] that helps to determine the different states of RNAs or RNA complexes. Sub-optimal structures are also used to identify homologous ncRNAs in bacteria [23]. To our knowledge, only NUPACK provides sub-optimal structures.

Finally, among the state of the art, only NanoFolder, the NUPACK package and MultiRNAFold are available. NanoFolder is available on a web server and the NUPACK package as well as MultiRNAFold are available as sources.

Here, we propose an original approach and a tool for RNA complex prediction including pseudoknots and crossing interactions. Our approach takes advantage of the numerous tools dedicated to RNA secondary structure prediction as well as RNA-RNA interaction prediction. Indeed, an RNA complex can be considered as a set of structured RNAs interacting with each other, where the secondary structure of each RNA can impact the interactions and vice versa. For each RNA and for each pair of RNAs, several possible secondary structures can be predicted. The prediction of an RNA complex can therefore be viewed as the best combination among those different predictions that achieves the minimum free energy. Thus, given a set of RNAs, our method takes as inputs several secondary structures per RNA and several interaction sites per pair of RNAs. Then it returns several possible complexes composed of some of the inputs. The secondary structure of a single RNA, including or not pseudoknots, can be obtained by several tools, which can return sub-optimal solutions [6, 15, 24–26]. There also exist many tools to predict interaction sites between two RNAs. They do not predict crossing interactions but some can return several solutions [6, 27–29].

In this paper, we show that the RNA complex prediction problem can be defined as a combinatorial optimization problem on a graph. The possible secondary structures of each RNA and the possible interactions between each pair of RNAs are the vertices of the graph. Each vertex has a weight equals to the minimum free energy of its corresponding secondary structure or interaction. If some secondary structures and interactions can form a complex, they are said to be *compatible*. This compatibility relation between the inputs is represented with the edges of the graph. The RNA complexes we are looking for can be viewed as the combinations, with the minimum free energies, of the various inputs. Hence, the problem consists in finding the minimum weight subgraph where all the inputs are compatible with each others. This kind of subgraph, called a *clique* or a *complete graph*, is a graph in which all vertices are linked to each other. More precisely, the prediction of RNA complexes corresponds to a constrained version of the well known Maximum Weight Clique Problem (MWCP). Since solving the MWCP is NP-hard [30], several heuristics have been proposed to find good solutions in polynomial time [31]. We propose an heuristic based on Breakout Local Search [32] to find good solutions to our constrained MWCP. This heuristic allowed us to develop a tool, called RCPred, for

RNA Complex Prediction. We show that compared to NanoFolder, NUPACK and MultiRNAFold, RCPred gives better results for a large set of RNA complexes.

The paper is organized as follows: we first present how the RNA complex problem can be modeled as a constrained maximum weight clique problem. Then we present the heuristic we propose to solve this problem and finally present and discuss the results we obtain with RCPred.

Methods

As stated before, the RNA complex prediction problem can be viewed as a graph problem in which we must find a constrained clique. In this section, we first describe the relationship between the RNA complex prediction and the Maximum Weight Clique Problem (MWCP). Then, we propose an heuristic to find good solutions in a polynomial time for the constrained MWCP.

Predicting RNA complexes: a constrained MWCP

RNA complexes

An RNA complex is composed of a set of structured RNAs that interact with each other. It can therefore be considered as a set of RNA secondary structures and of RNA-RNA interactions. A *secondary structure* involves exactly one RNA strand and is composed of a list of base pairs internal to this RNA strand (Fig. 1a). A secondary structure can contain pseudoknots (Fig. 1b). An *interaction site* (Fig. 1c) involves two RNA strands, is composed of a list of base pairs, and does not contain crossing interactions (Fig. 1d).

Constrained MWCP

The RNA complex prediction problem can be formalized using a weighted graph $G(V, E)$ such as:

- V, the vertex set, is composed of two subsets, V^S and V^I, where V^S is the set of vertices representing the secondary structures and V^I is the set of vertices representing the interactions. Each vertex $v \in V$ has a weight equals to the free energy associated to the structure or the interaction.
- E, the edge set, represents the compatibilities between the vertices. An edge exists if and only if two vertices are compatible. We consider that two vertices are not compatible if at least two identical nucleotides are involved in different pairings or if the two vertices are secondary structures involving the same RNA strand. These compatibility rules allow the presence of any motif in RNA complexes: pseudoknots (that can already be present in the secondary structures, which are represented by the vertices V^S) or crossing interactions.

An RNA complex can be viewed as a complete graph, or a *clique*, where each vertex is linked to all the other vertices. This clique is constrained because, for each RNA, there must be exactly one secondary structure vertex. This brings another constraint. In some known complexes, the RNAs do not have internal base pairs, which implies to add for each RNA a vertex corresponding to a secondary structure with no base pairs. However, an RNA complex only composed of secondary structures or interactions with no base pairs at all is not an RNA complex. We call the cliques corresponding to this type of RNA complexes *weak* cliques.

The weight of a clique (constrained or not) is the sum of the weights of the vertices composing the clique. We have therefore a constrained maximum vertex weight clique problem, denoted in the sequel by constrained MWCP, where the clique i) is composed of exactly one secondary structure per RNA, ii) is not weak and iii) has a minimum free energy.

Free energy computation

To each secondary structure and each interaction is associated a free energy represented by the vertex weights in the graph. This energy is computed to unify the different sources of secondary structures and interactions. We use two energy models:

- The first model is the Turner model [33] (with the 2004 parameter release), which can be used for secondary structure without pseudoknots and for interactions.
- The second model is based on the sum of the stacking energies taken from the Turner model (used in [24]). This allows to compute the free energy of pseudoknotted secondary structures.

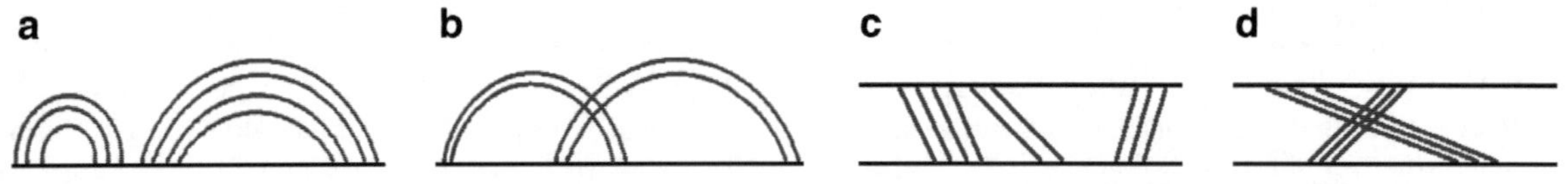

Fig. 1 Secondary structure RNA motifs. **a** Pseudoknot-free secondary structure. **b** Pseudoknotted secondary structure. **c** RNA-RNA Interaction. **d** Crossing interaction (or external pseudoknot)

Once the free energy of each secondary structure and interaction is known, it is used as the weight of each vertex. Then, the free energy of a complex (a constrained clique) is approximated by the sum of the free energies (weights) of the secondary structures and interactions (the vertices in the clique) composing it.

Solving the constrained MWCP

The MWCP is NP-hard. However, this problem is well studied and various methods exist to solve it. Exact methods, which find the optimal solution by optimizing the weight of the clique, are either generalization of methods for the unweight problem [34, 35] or are branch and bound algorithms [36]. Since exact methods are time consuming due to the NP-hardness nature of the problem, a lot of various heuristics have been proposed, either based on local search [37], tabu search [38], both of them [32] or other techniques [39, 40].

In this paper, we propose an adaptation of the heuristic Breakout Local Search (BLS) published in [32] that provides good solutions in a short amount of time. In the following, we first present BLS and then the heuristic we propose for our constrained MWCP that we denote by BLS-CMWCP.

Breakout Local Search heuristic

The BLS heuristic [32] was proposed for the MWCP and is based on both local search and tabu search.

Local search [41] is an heurisitic method to find good solutions for combinatorial optimization problems. The local search is an iterative method. It starts from an initial solution and modifies it at each step by looking in its neighborhood, i.e. a set of neighboring solutions obtained by performing small modifications (movements) on the current solution, for a better solution. When a solution cannot be improved anymore, it is called a local optimum solution.

Tabu search [42] is a metaheuristic based on local search. The main difference with the local search is that at each iteration the best solution in the neighborhood of the current solution is selected, even if it is not better than the current solution. In order to avoid cycling through previously encountered solutions, a tabu list is used.

BLS starts from a random solution and then performs alternatively two phases until the known optimal solution is found or the time limit is reached:

1 Local search: to perform a local search until a local optimum is found.
2 Perturbation: to modify greatly the local optimum solution to escape from it and explore further the search space.

In the local search phase of BLS, all the possible movements are considered and the one optimizing the most the solution weight is chosen. The available movements are either to add a vertex in the clique or to replace a vertex in the clique with one that is not in the clique. To define the movements, some definitions are needed. Let $G(E, V)$ be a graph and C be the current clique.

- PA is the set of vertices that can be directly added in the clique C, because there exist edges between them and all the vertices of the clique C; $PA = \{v : v \notin C, \forall u \in C\ \exists [v, u] \in E\}$.
- OM is the set of pairs of vertices (v, u) where v is not in the clique C but there exist edges between it and all the vertices of the clique C, except u which is in the clique C; $OM = \{(v, u) : v \notin C \text{ and } u \in C, \forall v' \in C \backslash \{u\} \exists [v, v'] \in E \text{ and } [v, u] \notin E\}$. The OM set is used to do the replacement movements.
- OC is the set composed of all the vertices outside C; $OC = \{v : V \backslash C\}$.

The perturbation phase aims to modify the current solution to escape a local optimum. The perturbation can greatly degrade the solution, the strength depending on how many times the solution was not improved in the local search phase. The perturbation strategies are based on four movements that are performed several times: to add, replace or remove a vertex of the clique (weak perturbation) and to restart (strong perturbation). The perturbation phase uses a tabu list in order to avoid to pick up again a vertex for a movement if it was removed from the solution some iterations before. The perturbation phase is a main difference with other local search methods and allows to explore more efficiently and faster the search space.

The authors show that this heuristic provides improved results for a number of MWCP instances and that this heuristic is usable for large graphs in reasonable time. It makes this heuristic a good candidate to develop a tool for the RNA complex prediction since large sets of inputs can be used.

The BLS-CMWCP algorithm

The BLS method is adapted for the constrained MWCP by making some modifications to the initial clique finding phase and to the movements, in order (i) to take into account the different kinds of vertices, (ii) to take into account the constraints related to the secondary structures and (iii) to avoid the weak cliques.

Before describing our BLS-CMWCP algorithm, let us give the new definitions about the sets used (illustrated in Fig. 2). Let C be a clique from the initial graph $G(E, V)$.

- Let PA be the set composed of all the interaction vertices that are outside C and are connected to all the vertices in C; $PA = \{v^I : v^I \notin C, \forall u \in C\ \exists [v^I, u] \in E\}$.

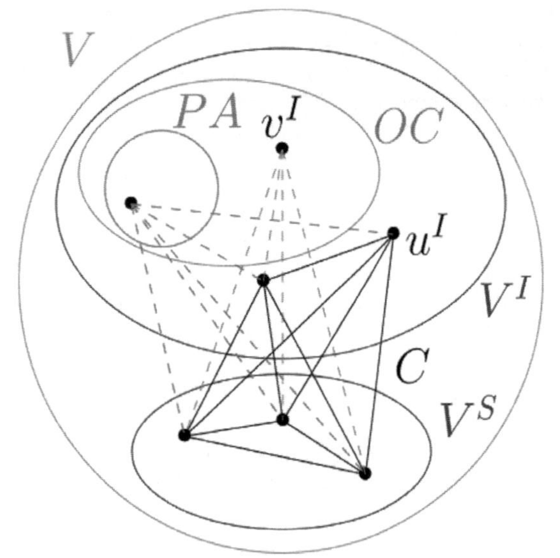

Fig. 2 Sets used in BLS-CMWCP algorithm. *C* is a clique, *PA* is the set composed of interaction vertices that are all linked to all the vertices of the clique, *OC* is the set composed of interaction vertices that are not in the clique

- Let *OM* be the set composed of the interaction vertices pairs (v^I, u^I) (or secondary structure vertices pairs (v^S, u^S)) such that v^I (or v^S) is outside *C* and is connected to all vertices in *C* except to the vertex $u^I \in C$ (or $u^S \in C$).
- Let *OC* be the set composed of all interaction vertices that are outside *C*; $OC = \{v^I : V^I \backslash C\}$.

Having only the interaction vertices v^I in *PA* and in *OC* and having only interaction vertex pairs (v^I, u^I) or secondary structure vertex pairs (v^S, u^S) in *OM* allow only the movements respecting the constraint of having exactly one secondary structure per RNA. In the following, we describe each modification and the differences between our algorithm BLS-CMWCP and BLS:

- **To generate the initial clique:** in BLS, the phase consists in selecting randomly a vertex and then to add iteratively vertices if they form a clique, until no more vertex can be added. In BLS-CMWCP, this phase consists in selecting randomly an interaction vertex and then selecting for each RNA a secondary structure vertex that forms a clique. Forming a clique is always possible thanks to the empty secondary structures which are obviously compatible with any interaction vertex.
- **To add a vertex movement:** an interaction vertex v^I is selected in *PA* and added into *C*.
- **To replace a vertex movement:** a vertex pair (v^I, u^I) (or (v^S, u^S)) is selected in *OM*, v^I (or v^S) is added to *C* and u^I (or u^S) is removed from *C*.
- **To remove a vertex movement:** an interaction vertex v^I is selected in *C* to be removed.
- **To restart the clique movement:** an interaction vertex v^I from *OC* is added to the clique *C*. Then if the structure vertices of the clique, $v^S \in C$, do not form a clique, they are replaced with other structure vertices. Finally, the remaining interaction vertices of the clique, $v^I \in C$, are removed if they do not form a clique anymore.
- **To generate sub-optimal cliques:** contrary to BLS method, we want to return several sub-optimal cliques. In BLS-CMWCP, at each iteration of the local search and of the perturbation phases, any new clique is saved. Then when the solutions are returned, they are sorted according to their free energy.
- **To forbid the weak cliques:** during all the search, if a movement leads to a weak clique, it is not considered.

RCPred: implementation of BLS-CMWCP for RNA complex prediction

The BLS heuristic proposed in [32] has a set of parameters to modulate the strength of the perturbation (L_0 and L_{Max}), the maximum number of non-improving solutions before a strong perturbation is performed (T), the coefficients for accepting non-improving solutions (α_s and α_r), the coefficient for tabu tenure (ϕ) and the probability for applying directed perturbations (P_0). Some parameters were fixed in the BLS heuristic and we used them as such in BLS-CMWCP. We determined the other parameters by performing experiments on a dataset of 30 graphs derived from RNA complexes. We then chose the following parameters: $L_0 = 0.1 * |V|$, $L_{Max} = 0.1 * |V|$, $T = 10$, $\alpha_s = 0.5$, $\alpha_r = 0.5$, $\phi = 7$ and $P_0 = 1$. The stop condition for the local search occurs either when the optimum clique is found or the maximum number of iterations is reached. In RNA complex prediction we do not know the optimum clique, then the stop condition here is a maximum number of iterations (fixed at 500). This parameter can be set by the user.

We implemented in C++ BLS-CMWCP and obtained the tool called RCPred (RNA Complex Prediction). RCPred takes as inputs n sequences of RNAs, several secondary structures per RNA and several interactions per pair of RNAs. First the energies of the secondary structures and the interactions are computed. The compatibilities between the secondary structures and the interactions are determined and the graph is built. BLS-CMWCP returns constrained cliques from which are derived RNA complexes. If some sequences are identical, symmetrical complexes can occur. These lasts are identified and removed to avoid redundancies in the results. Finally, the RNA complexes are sorted according to their free energy. RCPred is available on the EvryRNA platform.

Results

In this section, we present the results we have obtained with RCPred on a large set of RNA complexes. First we detail below the dataset used and how we recovered the secondary structures and interactions. We then give the results of RCPred and compare it with NanoFolder [19], NUPACK [15] and MultiRNAFold [13].

Datasets and RCPred inputs

In the following experiments, we use a dataset composed of 90 non-redundant RNA complexes. The dataset is extracted from the database RNA STRAND [43] that gathers 4,666 secondary structures of single and multi-strand RNAs. All the recovered complexes are experimentally validated by NMR or X-ray and are not composed of modified nucleotides. We are interested here in complexes longer than 20 nucleotides and smaller than 1000 nucleotides. Because of the complexity in time of the tools from the state of the art, we exclude the complexes longer than 1000 nucleotides in order to be able to generate several secondary structures and interactions. The dataset of the benchmark is available on the EvryRNA platform.

For each RNA, we generated the secondary structures using three tools from the literature for RNA secondary structure prediction, namely BiokoP [24], pKiss [26] and RNAsubopt (from the ViennaRNA package [6]). We chose these three tools because they are able to generate sub-optimal solutions. Moreover, they have good performances as shown in [24]. For each of the three tools, we fixed the maximum number of sub-optimal solutions to 30. Note that BiokoP and pKiss can predict pseudoknots while RNAsubopt predicts only pseudoknot-free secondary structures. We merged the results of the three tools to have a diversified set of secondary structures for each RNA. To generate the interactions between each pair of RNAs, we used the tool RNAsubopt [6] which can also predict RNA-RNA interaction sites with numerous sub-optimal solutions. We fixed the maximum number of sub-optimal solutions to 90.

Prediction results

We present in this section the results obtained by our tool RCPred and the comparison with the results of NanoFolder, NUPACK and MultiRNAFold. As RCPred is based on an heuristic, the results presented here are obtained from 5 executions. Among these tools, only RCPred and NUPACK can return sub-optimal solutions corresponding to different predicted complexes.

Statistics used

To evaluate the quality of predicted complexes, we used the sensitivity to measure the ability of finding positive base pairs and the Positive Predictive Value (PPV) to measure the ability of not finding false positive base pairs. We also used the F_1-score which is the harmonic mean between the sensitivity and the PPV and the Mathews Correlation Coefficient (MCC) which is a balanced measure between sensitivity and specificity (that measures the proportion of negatives that are correctly identified). These statistics are computed as follows:

$$Sensitivity = \frac{TP}{TP+FN}, PPV = \frac{TP}{TP+FP}$$

$$\mathrm{F_1\text{-}score} = 2 \times \frac{Sensitivity \times PPV}{Sensitivity + PPV}$$

$$MCC = \frac{TP \times TN - FP \times FN}{\sqrt{(TP+FP) \times (TP+FN) \times (TN+FP) \times (TN+FN)}}$$

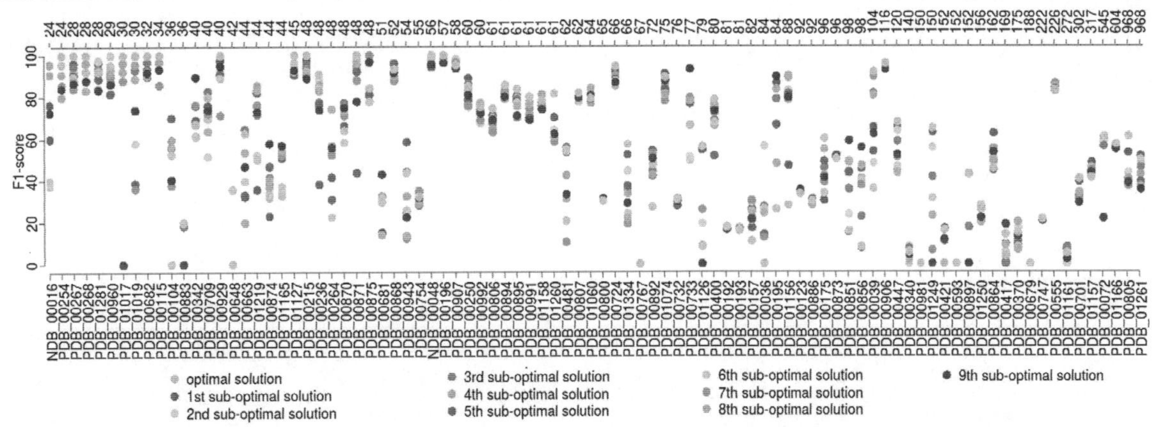

Fig. 3 F_1-score results of RCPred. Results are shown for the 10 first solutions returned by RCPred in average on 5 executions on our dataset. The secondary structure inputs are generated by BiokoP, pKiss and RNAsubopt and the interaction inputs are generated by RNAsubopt. The complex lengths are indicated in nucleotides in the superior axis

where *TP* is the number of true positive base pairs, *FN* is the number of false negative base pairs, *FP* is the number of false positive base pairs and *TN* the number of true negative base pairs.

RCPred evaluation

Sub-optimal solutions We first study for each complex the optimal and sub-optimal predictions of RCPred in order to see the relevance of generating several solutions. We report on Fig. 3 the F_1-score results of the 10 first solutions returned by RCPred in average on 5 . As we can see, the solutions having the highest F_1-scores are in most cases the first returned. However, in many cases, the sub-optimal solutions reach the highest F_1-scores, there are even some complexes for which the best prediction is given by the tenth sub-optimal solution. This confirms that better predictions can be found in the sub-optimal solutions and therefore the need to generate them.

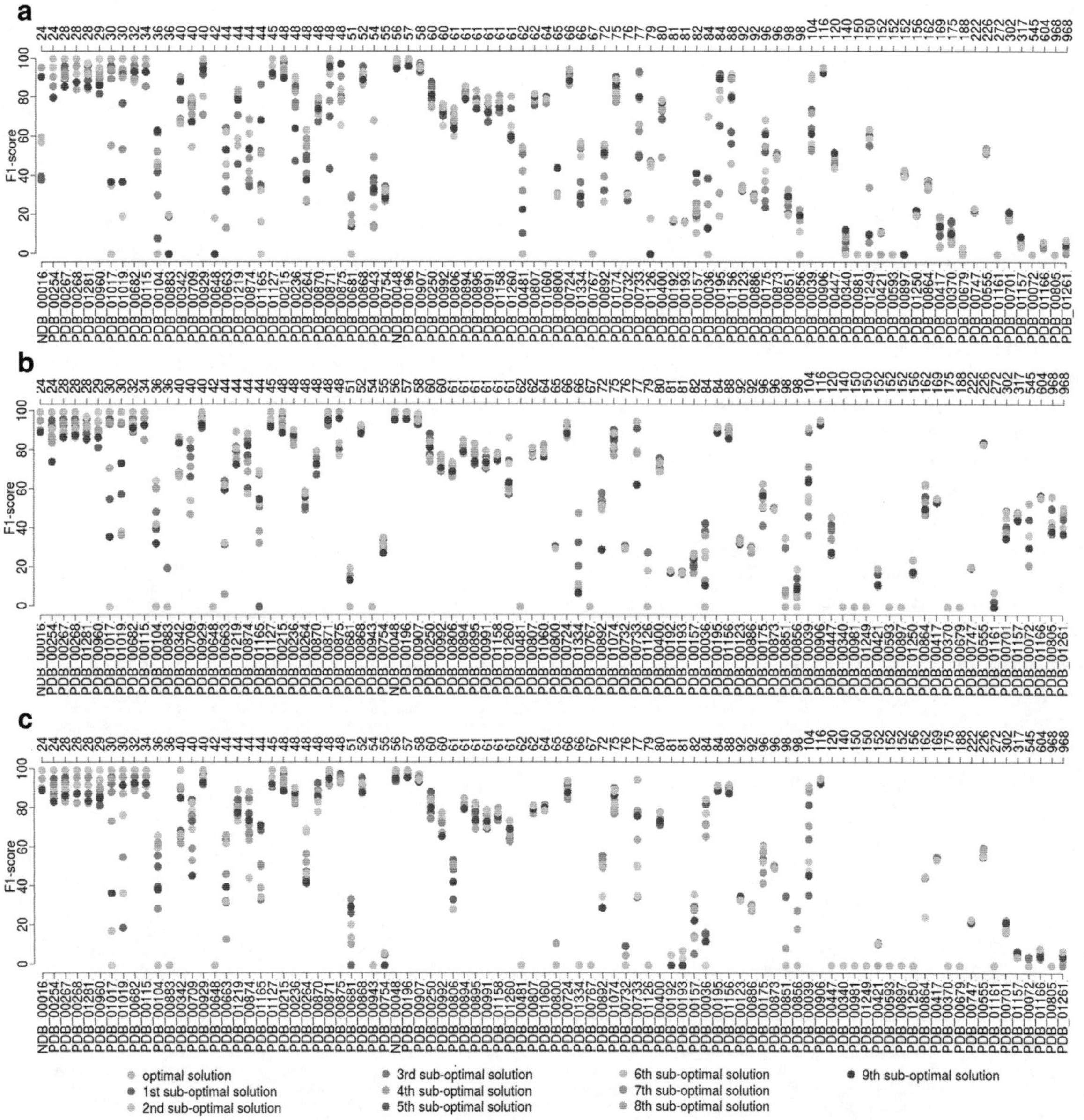

Fig. 4 F_1-score results of RCPred in function of the inputs. Results are shown for the 10 first solutions returned by RCPred in average on 5 executions on our dataset. **a** The secondary structure inputs are generated by BiokoP. **b** The secondary structure inputs are generated by pKiss. **c** The secondary structure inputs are generated by RNAsubopt.The complex lengths are indicated in nucleotides in the superior axis

Influence of the inputs We study here the influence of the inputs by taking either the secondary structures generated by BiokoP, pKiss or RNAsubopt alone. We report on Fig. 4 the F_1-score results of the 10 first solutions returned by RCPred in average on 5 executions, with inputs of BiokoP (A), pKiss (B) and RNAsubopt (C). Knowing that the performances of BiokoP, pKiss and RNAsubopt depends on the size of the RNAs, it is expected that they all predict accurate secondary structures for small RNAs. Hence it is not surprising to observe similar results for the smallest complexes (inferior to 62 nucleotides). For longer complexes the results differ. A comparison of these results with the ones obtained when the input secondary structures generated by the three tools are merged (Fig. 3) shows that we obtain better results when the inputs are merged. This strongly suggests that when the inputs are merged, the best secondary structure inputs are selected by RCPred. This confirms the first interest of RCPred, which is to deal with numerous and varied possible secondary structures and interactions, in order to predict efficiently and in low time computing RNA complexes.

Comparison with the state of the art

We then compare RCPred with NanoFolder, NUPACK and MultiRNAFold. We report the F_1-score results of these tools on our dataset in Fig. 5. We execute RCPred 5 times and, for each execution, we recover the maximum F_1-score solution of the 10 first solutions returned for each complex. We then compute the mean of these recovered solutions. For NUPACK, we report the maximum F_1-scores among the 10 first solutions returned. Note that we could not test NanoFolder (which is usable through a web server) on the longest complexes (of size greater than 550 nucleotides), because of the size limitation of the web server. Also we encounter some difficulties with the tool MultiRNAFold on some complexes.

As we can see on Fig. 5, RCPred obtains better F_1-scores than the other tools in most cases. With the sub-optimal solutions, we can guarantee that at least a predicted structure is close to the referenced one. For almost all complexes smaller than 120 nucleotides, RCPred is able to find accurate predictions with F_1-scores higher than 80%. For longer complexes, RCPred, as well as the other tools, becomes less effective with a maximum F_1-score around 70%. Similar results and findings are obtained with sensitivity, PPV, and MCC statistics (see Additional file 1). We summarize these results by reporting the average on all the dataset in Table 1. As NanoFolder and MultiRNAFold were not able to give results for some complexes, their means are done only on the complexes that were successfully predicted. The results show that in average RCPred is able to predict complexes more accurately than NanoFolder, NUPACK and MultiRNAFold.

Conclusion and perspectives

In this paper, we propose a new method and a tool, called RCPred, to predict secondary structures of RNA complexes composed of several RNAs.

We model the problem of RNA complex prediction with input knowledge as a constrained maximum weight clique problem in a graph and we present an heuristic based on Breakout Local Search to find good solutions, resulting in the tool RCPred. This modeling allows to predict all kinds of RNA complex motifs including pseudoknots or crossing interactions. This is not the case of the tools NUPACK and MultiRNAFold that do not predict at all these motifs.

RCPred is also able to provide sub-optimal solutions. Generating sub-optimal solutions is very important in RNA secondary structure prediction for several reasons. First, because it is known that the real structure is not always the structure of minimum free energy but often a structure close to the structure of minimum free energy.

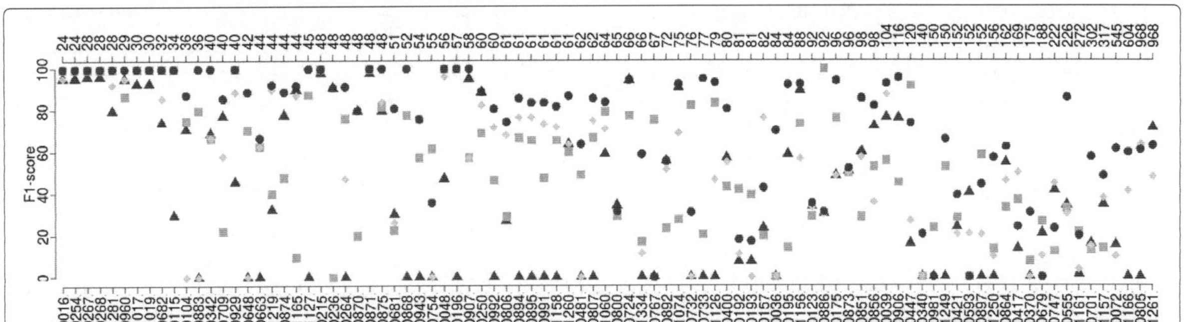

Fig. 5 F_1-score results of RCPred, NanoFolder, NUPACK and MultiRNAFold. For RCPred and NUPACK, the maximum F_1-scores among the 10 first returned solutions are taken for each complex to compute the mean. For RCPred, the mean is computed over 5 executions. The complex lengths are indicated in nucleotides in the superior axis

Table 1 Mean sensitivity, PPV, F_1-score and MCC results of RCPred, NanoFolder, NUPACK and MultiRNAFold on our dataset

	RCPred	NanoFolder	NUPACK	MultiRNAFold
Sensitivity	**65.8**	61.9	37.9	54.6
PPV	**70.5**	50.4	41.6	56.0
F_1 score	**67.3**	54.9	39.2	54.8
MCC	**67.5**	55.0	38.8	54.6

For RCPred and NUPACK, the maximum F_1-scores among the 10 first returned solutions are taken for each complex to compute the mean. The corresponding means of sensitivity, PPV and MCC are given. For RCPred, the mean is computed over 5 executions. Bold text indicates the higher scores

Moreover, a model cannot capture all the subtleties of the minimum free energy computation of an RNA complex of more than two strands. Sub-optimal solutions are then needed to allow to cover the discrepancies of the model. Among the existing tools, only the tool from the NUPACK package can generate sub-optimal solutions.

We test RCPred on a large dataset composed of 90 RNA complexes of various lengths, including or not pseudoknots. We show that RCPred is able to predict accurately RNA complex secondary structures and gives competitive results compared to NanoFolder, NUPACK and MultiRNAFold. Each returned RNA complex has a global free energy resulting from the sum of the free energies of the secondary structures and of the interactions composing it. A perspective could be to improve the global free energy computation of the complexes (by adapting for example the calculation method used in RNAeval from ViennaRNA package [6]) and reorder them accordingly.

The time execution of RCPred varies between 0.05 s in average on 5 executions for the smallest complex and 16.7 seconds for the longest. We are currently working on optimizing the time execution.

In RNA complexes, the tertiary interactions are numerous and have an important role in the stabilization of the global structure. There exist databases gathering 3D motifs appearing in single RNA structures, like the Rna3Dmotif database [44]. Moreover, it has been shown that inserting 3D motifs in RNA secondary structures helps in improving the prediction [45]. A perspective for this project would be to insert 3D motifs of single RNAs and of interacting RNAs in the predicted RNA complex secondary structures.

Additional file

Additional file 1: Supplementary statistical results. Supplementary Figures showing the sensitivity, PPV and MCC results of RCPred compared to the results of NanoFolder, NUPACK and MultiRNAFold. (PDF 61.8 kb)

Abbreviations

BLS: Breakout local search; BLS-CMWCP: Breakout local search-constrained maximum weight clique problem; BLS-MWCP: Breakout local search-maximum weight clique problem; MWCP: Maximum weight clique problem

Acknowledgements

Fariza Tahi would like to thank the Genopole for the financial support given for the platform EvryRNA.

Funding

Publication of this article was sponsored by IBISC laboratory. The platform EvryRNA is partially supported by the Genopole.

Availability of data and materials

RCPred and the benchmark datasets are available on EvryRNA platform (https://EvryRNA.ibisc.univ-evry.fr/).

About this supplement

This article has been published as part of *BMC Bioinformatics Volume 20 Supplement 3, 2019: Selected articles from the 17th Asia Pacific Bioinformatics Conference (APBC 2019): bioinformatics*. The full contents of the supplement are available online at https://bmcbioinformatics.biomedcentral.com/articles/supplements/volume-20-supplement-3.

Authors' contributions

AL developed the tool and tested it. AL recovered the data and analyzed the results. EA contributed to the maximum clique problem. FT was involved in the prediction of RNA complex structure. EA and FT supervised the work. The three authors participated in the writing of the manuscript and approved the final version of the manuscript.

Ethics approval and consent to participate

Not applicable.

Consent for publication

Not applicable.

Competing interests

The authors declare that they have no competing interests.

Published: 29 March 2019

References

1. Correll CC, Freeborn B, Moore PB, Steitz TA. Metals, motifs, and recognition in the crystal structure of a 5S rRNA domain. Cell. 1997;91(5): 705–12.
2. Kato Y, Mori T, Sato K, Maegawa S, Hosokawa H, Akutsu T. An accessibility-incorporated method for accurate prediction of RNA–RNA interactions from sequence data. Bioinformatics. 2017;33(2):202–9.
3. Xu X, Chen SJ. VfoldCPX server: Predicting RNA–RNA complex structure and stability. PloS ONE. 2016;11(9):e0163454.
4. Montaseri S, Zare-Mirakabad F, Moghadam-Charkari N. RNA–RNA interaction prediction using genetic algorithm. Algoritm Mol Biol. 2014;9(1):17.
5. Salari R, Backofen R, Sahinalp SC. Fast prediction of RNA–RNA interaction. Algoritm Mol Biol. 2010;5(1):5.
6. Lorenz R, Bernhart SH, Zu Siederdissen CH, Tafer H, Flamm C, Stadler PF, et al. ViennaRNA Package 2.0. Algoritm Mol Biol. 2011;6(1):26.
7. Poolsap U, Kato Y, Akutsu T. Dynamic programming algorithms for RNA structure prediction with binding sites. Pac Symp Biocomput. 2010;15: 98–107.
8. Aksay C, Salari R, Karakoc E, Alkan C, Sahinalp SC. taveRNA: a web suite for RNA algorithms and applications. Nucleic Acids Res. 2007;35(suppl_2): W325–9.
9. Alkan C, Karakoc E, Nadeau JH, Sahinalp SC, Zhang K. RNA–RNA interaction prediction and antisense RNA target search. J Comput Biol. 2006;13(2):267–82.
10. Li AX, Marz M, Qin J, Reidys CM. RNA–RNA interaction prediction based on multiple sequence alignments. Bioinformatics. 2010;27(4):456–63.
11. Seemann SE, Richter AS, Gesell T, Backofen R, Gorodkin J. PETcofold: predicting conserved interactions and structures of two multiple alignments of RNA sequences. Bioinformatics. 2010;27(2):211–9.
12. Havgaard JH, Lyngsø RB, Gorodkin J. The FOLDALIGN web server for pairwise structural RNA alignment and mutual motif search. Nucleic Acids Res. 2005;33(suppl_2):W650–3.
13. Andronescu M, Zhang ZC, Condon A. Secondary structure prediction of interacting RNA molecules. J Mol Biol. 2005;345(5):987–1001.

14. Zuker M, Stiegler P. Optimal computer folding of large RNA sequences using thermodynamics and auxiliary information. Nucleic Acids Res. 1981;9(1):133–48.
15. Zadeh JN, Steenberg CD, Bois JS, Wolfe BR, Pierce MB, Khan AR, et al. NUPACK: analysis and design of nucleic acid systems. J Comput Chem. 2011;32(1):170–3.
16. Tong W, Goebel R, Liu T, Lin G. Approximating the maximum multiple RNA interaction problem. Theor Comput Sci. 2014;556:63–70.
17. Mneimneh S, Ahmed SA. Gibbs/MCMC Sampling for Multiple RNA Interaction with Sub-optimal Solutions. In: International Conference on Algorithms for Computational Biology. Springer; 2016. p. 78–90.
18. Bindewald E, Afonin K, Jaeger L, Shapiro BA. Multistrand RNA secondary structure prediction and nanostructure design including pseudoknots. ACS Nano. 2011;5(12):9542–51.
19. Bindewald E, Afonin KA, Viard M, Zakrevsky P, Kim T, Shapiro BA. Multistrand structure prediction of nucleic acid assemblies and design of RNA switches. Nano Lett. 2016;16(3):1726–35.
20. Nebel ME, Weinberg F. Algebraic and combinatorial properties of common RNA pseudoknot classes with applications. J Comput Biol. 2012;19(10):1134–50.
21. Green NJ, Grundy FJ, Henkin TM. The T box mechanism: tRNA as a regulatory molecule. FEBS Lett. 2010;584(2):318–24.
22. Woods CT, Lackey L, Williams B, Dokholyan NV, Gotz D, Laederach A. Comparative visualization of the RNA suboptimal conformational ensemble in vivo. Biophys J. 2017;113(2):290–301.
23. Panek J, Krásný L, Bobek J, Ježková E, Korelusova J, Vohradský J. The suboptimal structures find the optimal RNAs: homology search for bacterial non-coding RNAs using suboptimal RNA structures. Nucleic Acids Res. 2010;39(8):3418–26.
24. Legendre A, Angel E, Tahi F. Bi-objective integer programming for RNA secondary structure prediction with pseudoknots. BMC Bioinformatics. 2018;19(1):13.
25. Sato K, Kato Y, Hamada M, Akutsu T, Asai K. IPknot: fast and accurate prediction of RNA secondary structures with pseudoknots using integer programming. Bioinformatics. 2011;27(13):i85–93.
26. Janssen S, Giegerich R. The RNA shapes studio. Bioinformatics. 2014;31(3): 423–5.
27. Wenzel A, Akbaşli E, Gorodkin J. RIsearch: fast RNA–RNA interaction search using a simplified nearest-neighbor energy model. Bioinformatics. 2012;28(21):2738–46.
28. Busch A, Richter AS, Backofen R. IntaRNA: efficient prediction of bacterial sRNA targets incorporating target site accessibility and seed regions. Bioinformatics. 2008;24(24):2849–56.
29. Kato Y, Sato K, Hamada M, Watanabe Y, Asai K, Akutsu T. RactIP: fast and accurate prediction of RNA–RNA interaction using integer programming. Bioinformatics. 2010;26(18):i460–6.
30. Karp RM. Reducibility among combinatorial problems. Boston: Springer; 1972.
31. Wu Q, Hao JK. A review on algorithms for maximum clique problems. Eur J Oper Res. 2015;242(3):693–709.
32. Benlic U, Hao JK. Breakout local search for the quadratic assignment problem. Appl Math Comput. 2013;219(9):4800–15.
33. Turner DH, Mathews DH. NNDB: the nearest neighbor parameter database for predicting stability of nucleic acid secondary structure. Nucleic Acids Res. 2009;38(suppl_1):D280–2.
34. Kumlander D. Problems of optimization: an exact algorithm for finding a maximum clique optimized for dense graphs. In: Proceedings–Estonian Academy of Sciences Physics Mathematics, vol. 54(2). Tallinn Estonia: Estonian Academy Publishers; 2005. p. 79.
35. Östergård PR. A new algorithm for the maximum-weight clique problem. Electron Notes Discrete Math. 1999;3:153–6.
36. Warren JS, Hicks IV. Combinatorial branch-and-bound for the maximum weight independent set problem. Texas A&M University, Citeseer. Pennsylvania State University: Citeseer; 2006.
37. Pullan W. Approximating the maximum vertex/edge weighted clique using local search. J Heuristics. 2008;14(2):117–34.
38. Wu Q, Hao JK, Glover F. Multi-neighborhood tabu search for the maximum weight clique problem. Ann Oper Res. 2012;196(1):611–34.
39. Singh A, Gupta AK. A hybrid heuristic for the maximum clique problem. J Heuristics. 2006;12(1-2):5–22.
40. Massaro A, Pelillo M, Bomze IM. A complementary pivoting approach to the maximum weight clique problem. SIAM J Optim. 2002;12(4):928–48.
41. Aarts Emile LJK. Local Search Algorithms. Hoboken: Wiley; 1997.
42. Glover F. Tabu search Fred Glover, Manuel Laguna. Boston: Kluwer Academic Publishers; 1997.
43. Andronescu M, Bereg V, Hoos HH, Condon A. RNA STRAND: the RNA secondary structure and statistical analysis database. BMC Bioinformatics. 2008;9(1):340.
44. Djelloul M. Algorithmes de graphes pour la recherche de motifs récurrents dans les structures tertiaires d'ARN. Université Paris Sud-Paris XI: HAL; 2009.
45. Reinharz V, Major F, Waldispühl J. Towards 3D structure prediction of large RNA molecules: an integer programming framework to insert local 3D motifs in RNA secondary structure. Bioinformatics. 2012;28(12):i207–14.

Lei et al. BMC Bioinformatics 2019, 20(Suppl 3):131
https://doi.org/10.1186/s12859-019-2649-0

BMC Bioinformatics

RESEARCH Open Access

Protein complex detection based on flower pollination mechanism in multi-relation reconstructed dynamic protein networks

Xiujuan Lei[1*], Ming Fang[1], Ling Guo[2] and Fang-Xiang Wu[3]

From The 17th Asia Pacific Bioinformatics Conference (APBC 2019)
Wuhan, China. 14-16 January 2019

Abstract

Background: Detecting protein complex in protein-protein interaction (PPI) networks plays a significant part in bioinformatics field. It enables us to obtain the better understanding for the structures and characteristics of biological systems.

Methods: In this study, we present a novel algorithm, named Improved Flower Pollination Algorithm (IFPA), to identify protein complexes in multi-relation reconstructed dynamic PPI networks. Specifically, we first introduce a concept called co-essentiality, which considers the protein essentiality to search essential interactions, Then, we devise the multi-relation reconstructed dynamic PPI networks (MRDPNs) and discover the potential cores of protein complexes in MRDPNs. Finally, an IFPA algorithm is put forward based on the flower pollination mechanism to generate protein complexes by simulating the process of pollen find the optimal pollination plants, namely, attach the peripheries to the corresponding cores.

Results: The experimental results on three different datasets (DIP, MIPS and Krogan) show that our IFPA algorithm is more superior to some representative methods in the prediction of protein complexes.

Conclusions: Our proposed IFPA algorithm is powerful in protein complex detection by building multi-relation reconstructed dynamic protein networks and using improved flower pollination algorithm. The experimental results indicate that our IFPA algorithm can obtain better performance than other methods.

Keywords: Protein complex, Dynamic protein-protein interaction (PPI) network, Essential protein, Flower pollination algorithm

Background

Understanding biological processes is an important task in the living organisms. Proteins are vital components in many biological processes, such as metabolism, signaling, transportation and so on. Biological functions are performed by protein complexes composed of proteins interacted with each other, rather than by individual proteins [1, 2]. Detection of protein complexes made great contribution to our knowledge of the molecular mechanisms in cellular life activities. To the best of our knowledge, a large number of works have been done to identify protein complexes from the PPI networks up to now.

As one of the earliest computational methods to predict protein complexes, the Molecular Complex Detection (MCODE) [3] weighted all vertices by using their local neighborhood density and identified the densely connected areas in PPI networks. ClusterONE [4] was utilized to find overlapping protein complexes in the PPI networks. The Clustering-based on Maximal Cliques (CMC) [5] method weighted the interacting protein pairs to identify protein complex. Recent studies TP-WDPIN [6] and NEOComplex [7] were based on the

* Correspondence: xjlei@snnu.edu.cn
[1]School of Computer Science, Shaanxi Normal University, 710119, Xi'an, China
Full list of author information is available at the end of the article

seed-extension idea to mine protein complexes. WG-Cluster [8] considered the edge weights to detect network modules. Markov Clustering (MCL) [9] discovered relatively dense regions based on the random walks. After that, F-MCL [10] used the firefly algorithm into Markov clustering to optimize the parameters and then recognized protein complexes.

It is well known that Gavin et al. [11] introduced the proteins in complexes consist of two types: core and attachment (periphery), namely, core-attachment structure, core represents the proteins that are densely linked and attachment are those proteins that have a few connections to the core. And then, CORE [12] first identified cores and then added proteins that had interactions with the majority of core proteins in the protein complex as attachment proteins. COACH [13] predicted cores in complexes and involved attachments into the cores to obtain protein complexes. Similarly, DCA [14] also used core-attachment feature to identify protein complexes.

In general, the judgment of interactions between two proteins is implemented by using experimental methods. Unfortunately, these methods are not always dependable [1] and it means that this may contain false positive interactions. There are many previous literatures have revealed the fact that the incorporation of additional biological information can improve the accuracy of protein complex prediction to some extent. For example, Zhang et al. [15] proposed CSO method to predict complexes by combining gene ontology (GO) information with PPI networks. InteHC method [16] integrated different types of data sources to predict protein complexes, including PPI data, GO data, gene expression profiles and AP-MS data. Zhao et al. [17] constructed the weighted protein interaction network by using gene expression information for protein complex identification. Zhou et al. [18] utilized GO to measure semantic similarities as the weights.

Since the flower pollination algorithm (FPA) [19] has shown excellent performance in many applications, such as clustering problem [20] and the identification of essential proteins [21], we explore the application of FPA in detecting protein complexes. In this study, we elicit a concept named co-essentiality, whose basic idea is to use the protein essentiality to find essential edges. Then the multi-relation reconstructed dynamic protein networks are built by combining heterogeneous topology and biology information. Next, those closely linked proteins are grouped together as the cores. Finally, based on the core–periphery structure, the modified FPA algorithm is developed to find the optimal pollination plants for pollen, which means that the peripheries attach to the best core to form the predicted protein complex. The experiments between IFPA and several typical algorithms including MCODE, MCL, ClusterONE, CSO, COACH and CORE are performed on three different PPI networks, and the results demonstrate that IFPA algorithm is more robust and powerful than those existing methods in protein complexes recognition.

The remaining part of this paper is organized as follows. Section 2 (Methods) elucidates our proposed new algorithm called IFPA. Section 3 (Results and Discussion) provides the exhaustive analysis and descriptions for the experiments. Finally, Section 4 (Conclusions) is the summary of this study.

Methods

The PPI network can be represented generally as an undirected graph, and the proteins are treated as nodes and the interactions are considered as edges. Here, the static PPI networks are converted into the multi-relation reconstructed dynamic PPI networks. And then we apply IFPA to add attachments to the appropriate cores based on the core-attachment structure.

Building multi-relation reconstructed dynamic PPI networks

The availability of gene expression data enables researchers to reveal the dynamics of molecular networks and improve the identification of protein complexes [22–24]. Hence, based on the study [25], the time course gene expression data is integrated into original static PPI networks to generate dynamic PPI subnetworks so that we can capture the dynamics of protein complexes, that is to say, we split the original static PPI network (OSPN) into twelve dynamic PPI subnetworks (DPSNs), in which all interactions in a DPSN can occur simultaneously, and then perform complex discovery on each DPSN.

First, we use three-sigma method [25] in order to construct dynamic PPI subnetworks with time series gene expression data. The gene expression data involves three metabolic cycles and each cycle contains twelve timestamps. A protein v is considered to be active in DPSN if its gene expression value is not less than the active threshold $Active\_Th(v)$:

$$Active\_Th(v) = \mu(v) + 3\sigma(v)(1-F(v)) \tag{1}$$

$$F(v) = \frac{1}{1+\sigma^2(v)} \tag{2}$$

where $\mu(v)$ is the algorithmic mean of gene expression values of v over times 1 to n and $\sigma(v)$ is the standard deviation of its gene expression values. For each protein, three-sigma method is used to calculate the active threshold $Active\_Th(v)$. A original PPI network can be described as an undirected graph $G(V, E)$, where V denotes node set that are proteins and E presents edge set that are their connections. And the dynamic PPI network can be represented as $G_t (V_t, E_t)$ at timestamp t ($t = 1,2, \ldots, n$). At a certain time point, if two proteins v_i and v_j are active and interact with each other in the original static PPI network, then there is a connection between protein v_i and v_j in a DPSN. After that, twelve dynamic PPI subnetworks are constructed from the original static PPI network.

Moreover, integrating heterogeneous data source into a single network can enhance the reliability of networks, which inspires us that assigning the suitable weights to edges can strengthen the confidence of interactions, and the implementation will be discussed in the following. Figure 1 illustrates an example of multi-relation reconstructed dynamic PPI networks construction.

Definition 1 (Co-essentiality) Essential proteins are indispensable for the survival of an organism. Then we can believe that the interaction between two essential proteins is also necessary. Hence, a concept based on essential protein is extended to measure the essentiality between two proteins, and the essentiality values are considered as their weights.

Before giving the concept of co-essentiality, we first elaborate the definition of an essential edge. Given two proteins v_i and v_j, the edge between them is considered as an essential edge if both of v_i and v_j are the essential proteins, similarly, the edge between them is considered as an uncertain edge if v_i or v_j is the essential protein, and the edge between them is considered as a nonessential edge if neither of v_i and v_j is the essential protein. Only the essential edges are taken into account to reconstruct the networks here. And ee_{ij} is the essential edge between v_i and v_j, the *co-essentiality* between these two proteins can be represented as follows.

$$co\text{-}essentiality\ (i, j) = \frac{ESS_{ij}}{sum\ \left(ESS_j\right)} \tag{3}$$

where ESS_{ij} denotes the weight value of essential edge which equals to one and sum(ESS_j) denotes the sum of the weight values of a column.

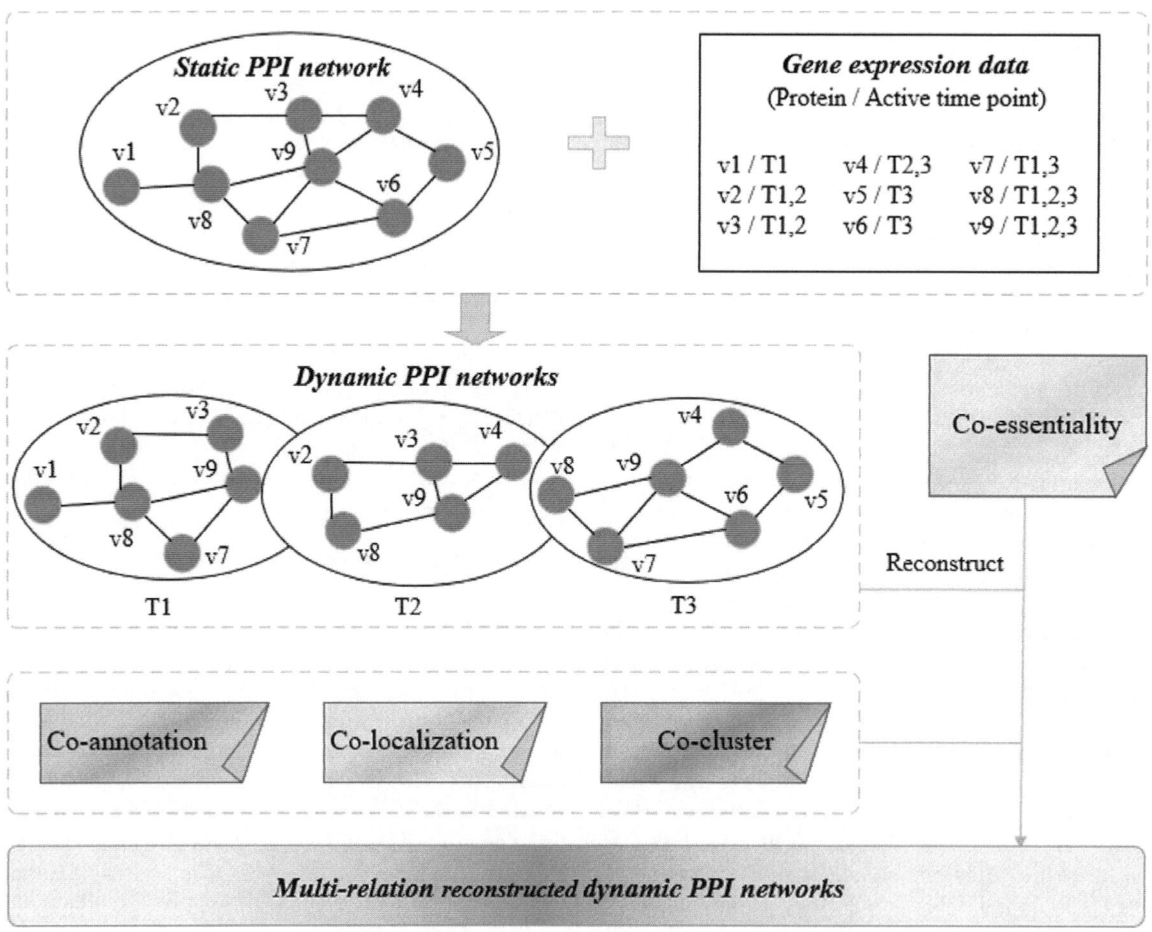

Fig. 1 An example of multi-relation reconstructed dynamic PPI networks construction

Definition 2 (Co-localization) Given two interacting proteins v_i and v_j, the interaction between them will be more reliable if v_i and v_j exist in same subcellular location, its *co-localization* is defined by the following equation.

$$co\text{-}localization\ (i,j) = \frac{|SCL_i \cap SCL_j|^2}{|SCL_i| \cdot |SCL_j|} \tag{4}$$

where $|SCL_i|$ and $|SCL_j|$ are the number of subcellular location of proteins v_i and v_j, respectively.

Definition 3 (Co-annotation) Given two interacting proteins v_i and v_j, they have the similar function if there are some common GO annotations between v_i and v_j, its *co-annotation* is calculated as follows.

$$co\text{-}annotation(i,j) = \frac{|GO_i \cap GO_j|^2}{|GO_i| \cdot |GO_j|} \tag{5}$$

where $|GO_i|$ and $|GO_j|$ are the number of GO annotations of proteins v_i and v_j, respectively.

Definition 4 (Co-cluster) Given two interacting proteins v_i and v_j, its *co-cluster* is measured by using the edge clustering coefficient (ECC) [26] as follows.

$$co\text{-}cluster(i,j) = \frac{Z_{ij}}{min\{|N_i|-1, |N_j|-1\}} \tag{6}$$

where Z_{ij} represents the number of triangles built on edge (v_i, v_j), $|N_i|$ and $|N_j|$ are the degrees of protein v_i and v_j, respectively.

Multiple relation defined above are used to weight the networks. The multi-relation value between v_i and v_j is stands for as follows.

$$\begin{aligned} multi\text{-}relation(i,j) &= co\text{-}essentiality(i,j) \\ &+ co\text{-}localization(i,j) \\ +co\text{-}annotation(i,j) &+ co\text{-}cluster(i,j) \end{aligned} \tag{7}$$

These multi-relation values are regarded as the weights of edges *W(i,j)* to upgrade the credibility of interactions. For an edge, its normalized *W(i,j)* value *NW(i,j)* is expressed by the following formula.

$$NW(i,j) = \frac{multi\text{-}relation(i,j)}{num(multi\text{-}relation)} \tag{8}$$

where *num(multi-relation)* is the total number of the network relations, i.e., the four kinds of relations including coessentiality, colocalization, coannotation, cocluster and the networks are reconstructed by mixing them. Eventually, the dynamic PPI subnetworks (DPSNs) are switched into the multi-relation reconstructed dynamic PPI networks (MRDPNs).

Finding cores

As we all know that protein complex core should be a densely connected subgraph in the PPI network. Thus, we pick the seed proteins in the first stage, and extend seed proteins to the cores in the second stage.

Definition 5 (Weighted Degree) The proteins with weighted degree greater than average weighted degree are sorted in descending order as the candidate core set *CC*. The weighted degree of a protein *i* in the MRDPN is the number of interactions in which this protein is involved, which can be expressed as follows.

$$Weighted\ Degree(i) = \sum_j interactions(i,j) \tag{9}$$

Let first node in the candidate core set *CC* be a seed protein which plays an irreplaceable role in protein complex. The neighbors of the seed protein are inserted into a core set when the condition that the density of core set is greater than a given threshold *DT* is satisfied. The threshold *DT* will be discussed in the next section.

Definition 6 (Density) The density of core set *CS* can measure how close the core is, and its definition is as follows.

$$Density(CS) = \frac{2 \times \sum_{(i,j)} NW(i,j)}{|CS| \cdot (|CS|-1)} \tag{10}$$

where $|CS|$ denotes the number of nodes in core set. Initially, core set *CS* contains one seed protein *i*. A neighbor of seed protein is added to the core set if adding it can make the *Density*(*CS*) greater than the threshold *DT*. This process is repeated until all neighbors of seed protein are sought and the predicted core is generated. Once a complex core is completed, all nodes in it will be labeled with "1" and cannot be extended into any other complex cores. This process will stop when the *CC* is empty.

Finding peripheries

Since the core plays a central role, the periphery plays a supporting role. The key idea behind our presented IFPA algorithm is to utilize the pollination mechanism to mimic the process of pollen falling on suitable flowers, which is completely different from other general methods. In this subsection, we first give a brief introduction to the flower pollination algorithm (FPA) [19], and then we find the optimal cores for peripheries by ameliorating it.

FPA is a nature-inspired optimization algorithm that comprises two main patterns, that is global pollination and local pollination. The global pollination can be represented as:

$$x_i^{t+1} = x_i^t + L\left(x_i^t - G\right) \tag{11}$$

where x_i^t is the pollen i at iteration t, and G is the current best solution. The parameter L is the strength of the pollination, namely a step size, we use a Lévy flight to represent that insects move over a long distance with various distance steps. That is, L is greater than 0 and obeys the Lévy distribution:

$$L \sim \frac{\lambda \Gamma(\lambda) \sin(\pi\lambda/2)}{\pi} \frac{1}{s^{1+\lambda}}, (s \gg s_0 > 0) \tag{12}$$

where $\Gamma(\lambda)$ is the standard gamma function. The local pollination can be defined as:

$$x_i^{t+1} = x_i^t + \Psi\left(x_j^t - x_k^t\right) \tag{13}$$

where x_j^t and x_k^t are pollen from the different flowers of the same plant species. This substantially simulates the flower constancy in a limited neighborhood. Mathematically, if x_j^t and x_k^t come from the same plant species or select from the same population, this can be seen as a local random walk if Ψ obeys the uniform distribution of 0 to 1.

Then, we use IFPA which is an advanced version of FPA algorithm to find the closest cores for peripheries, which is equivalent to finding the most satisfactory flowering plants for pollen. The workflow of IFPA algorithm is shown in Fig. 2. Those proteins not included in the core set are considered as the candidate pollen. In IFPA algorithm, the pollen corresponds to attachments and the pollination plants correspond to cores. The pollen position equals the core sequence numbers. The update of pollen position is expressed as follows.

$$S_{i,j}^{t+1} = \begin{cases} S_{i,j}^t, \mathit{if\ Pollination\ Priority}_{i,j} > \mathit{Thr} \\ \mathit{randperm}(\mathit{Num}, d), \mathit{otherwise} \end{cases} \tag{14}$$

where *Thr* denotes a threshold and it is set as 0.2 here. The function of *randperm* is to return an integer from one to *Num* which means to find new core sequence number, and *Num* is the number of cores and the value of *d* is one.

Definition 7 (Pollination Priority) As a part of an entire protein complex, the attachments maintain relatively close relationship with the core, we call this relationship as *pollination priority*. The "*pollination priority*" of a pollen to its core set *CS* is represented as follows.

$$\mathit{Pollination\ Priority}(\mathit{pollen}, \mathit{CS}) = \sum_{u \in CS} \mathit{co\text{-}cluster}(\mathit{pollen}, u) \tag{15}$$

where u is the protein in core set *CS*. The pollination priority depends on the affinity between the pollen and the flowers. The closer the relationship between pollen and a flower, the higher priority it pollinates on this flower. In the update procedure, if the pollen can find a flower that makes the value of *pollination priority* better, then the pollen falls on this flower, otherwise, the pollen finds a new flower to pollinate.

Algorithm 1: The pseudocode of IFPA algorithm.

Input: The multi-relation reconstructed dynamic PPI subnetworks G_t (W_t, V_t, E_t), t = 1,2,…,12; Density threshold *DT*; The maximal iterations of external loop *maxiter*; Population size *n*.

Output: The set of detected protein complexes *PCSet*

```
Begin:
for each Gt do
  Candidate core set CC ← the proteins with weighted degree greater than average weighted
  degree are sorted in descending order.
  PCSet = ∅
  // Finding cores
  For each protein v ∈ CC
    If protein v has not been visited do
      CS = { v }
      Find the neighbors of v named Nv
      For each protein u ∈ Nv do
        Calculate Density by using Formula (8)
        If Density > TD do
          CS = CS ∪ { u }
    For each protein v ∈ CS do
      Label (v) = 1
    Merge all the CS called CSs
    Output CSs
  End for
  // Finding peripheries
  Initialize population
  Calculate the values of pollination priority and find optimal solution in initial population
  For m = 1 : maxiter
    For i = 1 : n
      Update the position of pollen by using formula (14)
      Measure new solutions by using formula (15)
      If new solutions are better, update them in the population
    End for
  End for
  Find the optimal pollen position, viz., detected candidate protein complexes PCt
End for
Return the final detected protein complexes PCSet = (PC1,…,PCt)
Filtration operation
```

Finally, we further merge all the candidate protein complexes mined in twelve subnetworks and filter highly overlapping complexes, as our final predicted protein complexes. Algorithm 1 outlines the implementation process of our IFPA method.

Results and discussion

Datasets

Three popular datasets, i.e., DIP [27], MIPS [28] and Krogan [29], are used to verify our proposed IFPA algorithm. The DIP dataset contains 5028 proteins and 22,302 interactions, the MIPS dataset composes of 4546 proteins and 12,319 interactions, and the Krogan dataset

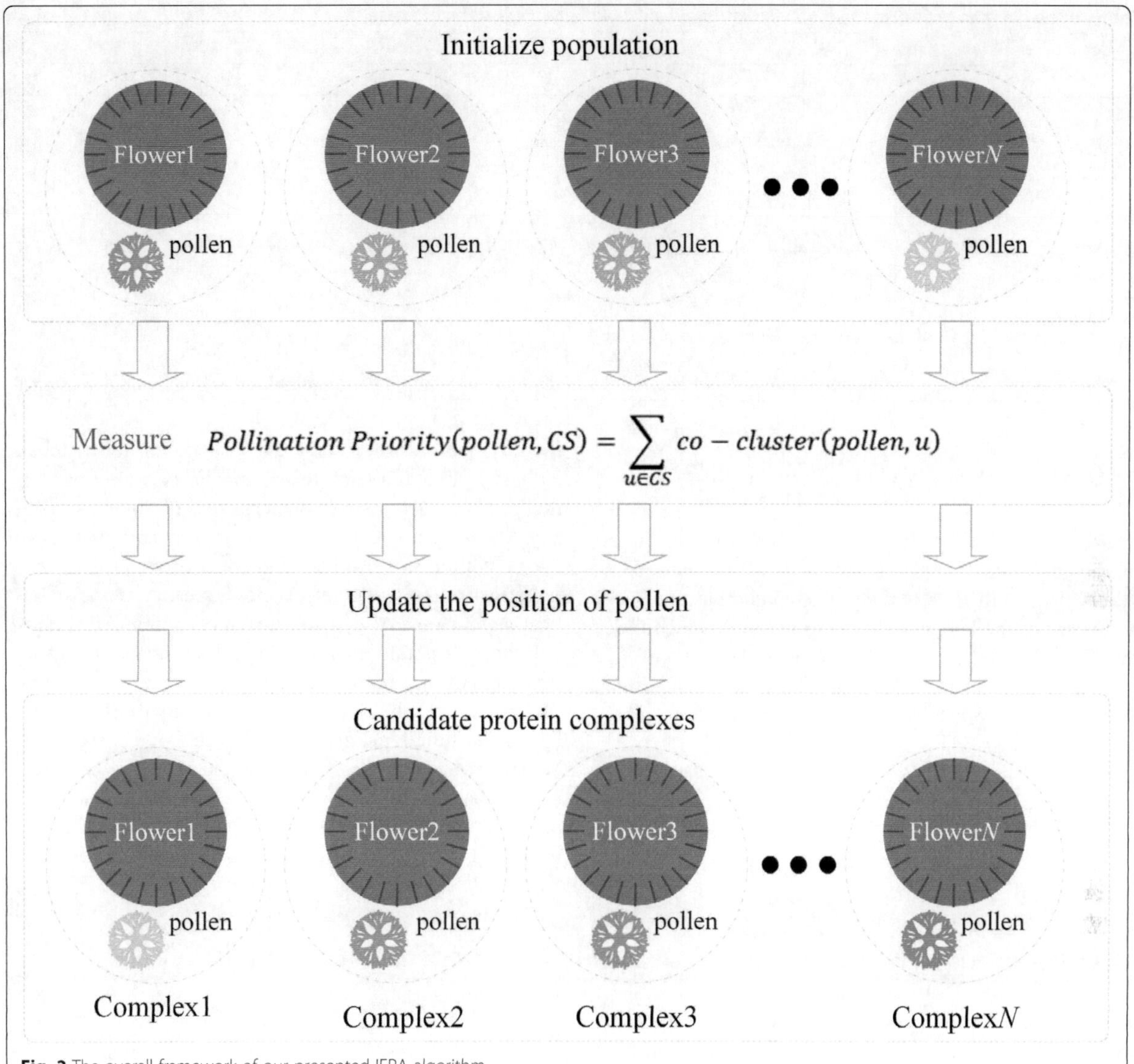

Fig. 2 The overall framework of our presented IFPA algorithm

includes 2674 proteins and 7075 interactions. The gene expression data is obtained from GEO [30] and the dataset contains 9336 genes in three cell life cycles, each cycle having twelve time points. The dynamic PPI networks are built by combining original static PPI networks with gene expression data and the details of dynamic PPI networks on three datasets are presented in Table 1.

The protein subcellular localization dataset is downloaded from the COMPARTMENTS database [31]. There are eleven subcellular localizations as follows: Cytoskeleton, Golgi apparatus, Peroxisome, Nucleus, Extracellular space, Vacuole, Cytosol, Endosome, Mitochondrion, Plasma membrane, Endoplasmic reticulum. After preprocessing, it still includes 6892 subcellular localization records. The GO information was gained from the SGD database [32]. There are 1285 essential proteins are collected from the following databases: MIPS [33], SGD [32], DEG [34], and SGDP (http://sequence.stanford.edu/group/yeast_deletion_project). CYC2008 [35] is used as the benchmark dataset which contains 408 protein complexes.

Evaluation metrics

The most commonly used evaluation metrics are used in our experiments and their specific definitions are described below.

Definition 8 (Overlapping Score) Given a predicted protein complex P and a known protein complex K, the Overlapping Score (OS) between P and K is defined as follows.

Table 1 The number of proteins and interactions in dynamic PPI networks on three datasets

Dataset	Timestamp *t*											
	1	2	3	4	5	6	7	8	9	10	11	12
DIP												
Protein	860	1029	863	671	645	598	530	1000	1194	638	690	489
Interaction	1103	1608	1337	839	835	752	627	1861	2447	950	1026	569
MIPS												
Protein	737	897	781	583	570	531	470	839	1014	523	616	402
Interaction	1097	1443	1183	754	684	642	504	1238	1637	878	1207	700
Krogan												
Protein	336	379	320	256	206	189	202	580	626	304	330	250
Interaction	334	464	331	234	210	184	213	1025	1081	314	373	258

$$OverlappingScore(P,K) = \frac{|V_P \cap V_K|^2}{|V_P| \cdot |V_K|} \quad (16)$$

where $|V_P \cap V_K|$ is the number of common proteins in the predicted protein complex P and the known protein complex K, $|V_P|$ is the size of the predicted complex and $|V_K|$ is the size of the known complex. If $OS \geq 0.2$, we consider that the predicted complex matches with the real one.

Definition 9 (Sensitivity, Specificity and F-measure) Sensitivity (*Sn*), Specificity (*Sp*) and *F-measure* are represented as follows.

$$Sn = \frac{TP}{TP + FN} \quad (17)$$

$$Sp = \frac{TP}{TP + FP} \quad (18)$$

$$F\text{-}measure = \frac{2 \cdot Sn \cdot Sp}{Sn + Sp} \quad (19)$$

where *TP* is the number of the predicted complexes which are matched with the known complexes, *FN* is the number of known complexes which are not matched with any predicted complexes, and *FP* is the number of the predicted complexes which are not matched with any known complexes. *F-measure* is a comprehensive metric combined sensitivity and specificity.

Definition 10 (*p-value*) In order to estimate the statistical significance of detected protein complexes, the researchers annotate their biological functions by using *p-value*. Given a predicted protein complex, the *p-value* [36] is the probability that a protein complex is enriched by a given functional group by chance. Let *k* is the number of proteins of the functional group in the complex, *N* is the size of the whole PPI network, *C* is the size of a protein complex and *F* is the size of a functional group in the network. And the *p-value* is defined as follows.

$$p\text{-}value = 1 - \sum_{i=0}^{k-1} \frac{\binom{F}{i}\binom{N-F}{C-i}}{\binom{N}{C}} \quad (20)$$

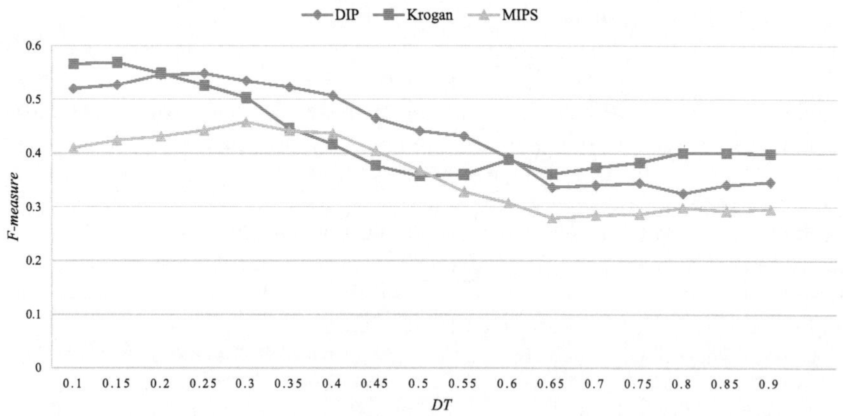

Fig. 3 The impact of *DT* on the performance of IFPA in terms of *F-measure*

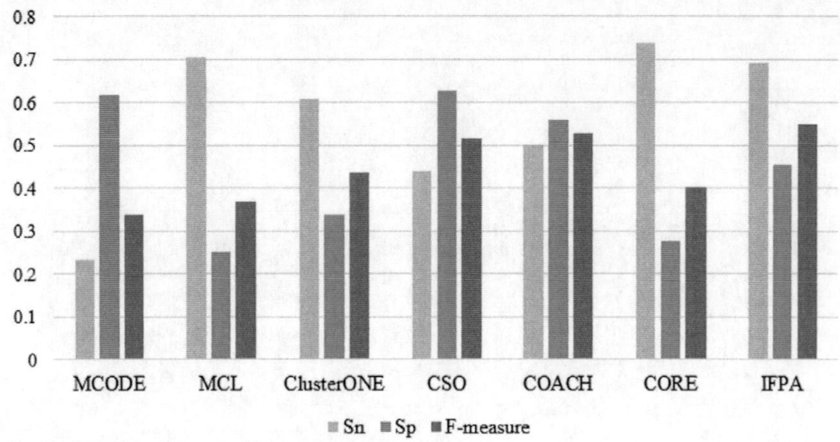

Fig. 4 Comparative performance of IFPA and other methods on DIP dataset

The *p-value* is used to evaluate the biological relevance of the predicted protein complexes. Generally, the protein complex is considered to be meaningless if its *p-value* is greater than 0.01. And for a complex, the smaller its *p-value* is, the more biological significance it has.

Parameter analysis

The density threshold *DT* decides whether a protein could be merged into the current core. How to choose a relatively suitable *DT* should be carefully considered to achieve better performance of our IFPA algorithm. Thus, varying *DT* from 0.1 to 0.9 with the interval 0.05, we calculate *F-measure* to observe the effect of the variation of *DT* on the performance of our IFPA algorithm, so as to choose the relatively appropriate *DT*, as shown in Fig. 3. Obviously, all three datasets show similar trends in most cases from Fig. 3. Especially, the major evaluation metrics *F-measure* obtains the better value when *DT* is set as 0.25 based on the both DIP and MIPS datasets and when *DT* is set as 0.15 based on the Krogan dataset, which means that IFPA is more effective. For this reason, the density threshold *DT* is set as 0.25 in DIP and MIPS datasets and 0.15 in Krogan dataset.

Performance comparison

To prove the validity of our proposed algorithm, we compare IFPA algorithm with MCODE, MCL, ClusterONE, CSO, COACH and CORE on three different PPI networks. Figures 4, 5, 6 show the overall comparison in terms of *Sn*, *Sp* and *F-measure* based on DIP, MIPS and Krogan datasets, respectively. Apparently, IFPA algorithm yields the best *F-measure* in comparison with other

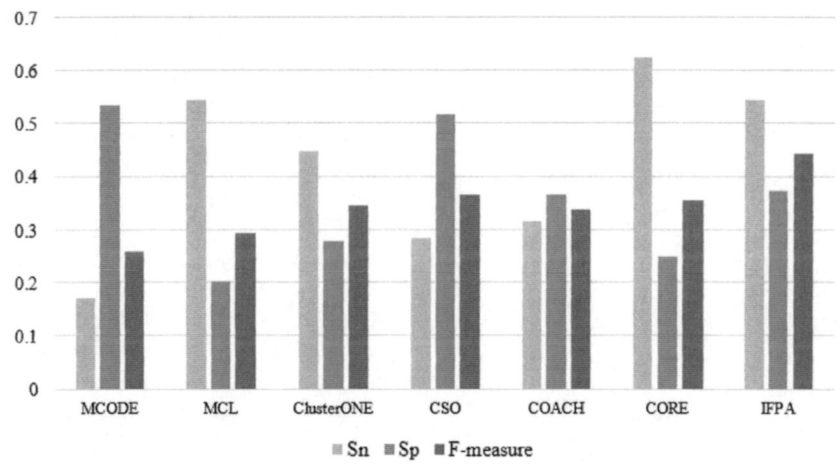

Fig. 5 Comparative performance of IFPA and other methods on MIPS dataset

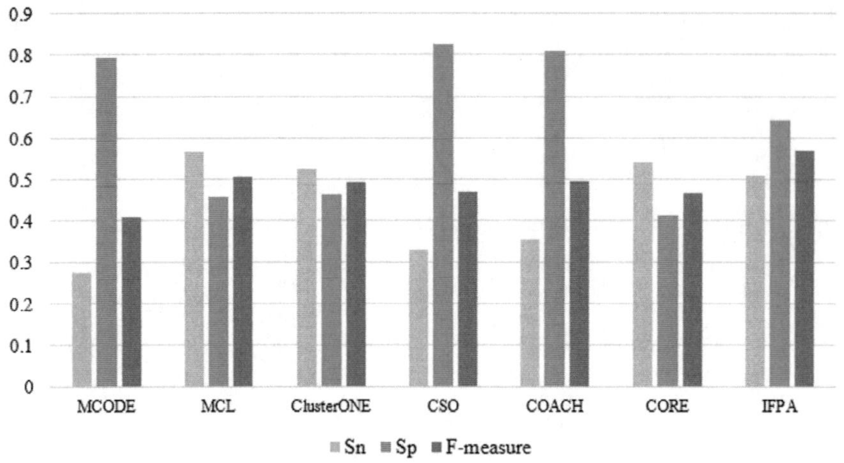

Fig. 6 Comparative performance of IFPA and other methods on Krogan dataset

existing methods in all datasets, which means that our IFPA method remarkably outperforms other methods. Besides, in Table 2, *PC* denotes the total number of detected protein complexes, *MPC* is the number of detected protein complexes which were matched, *MKC* represents the number of matched known protein complexes, *Perfect* denotes $OS = 1$ which means that the predicted protein complexes are perfectly matched with the known protein complexes. As shown in Table 2, the protein complexes detected by our IFPA method dominates other methods in the aspects of *Perfect* both in DIP and MIPS datasets.

Table 2 Comparative performance of IFPA and other methods on three datasets

Dataset	Algorithm	PC	MPC	MKC	Perfect
DIP	MCODE	165	102	70	6
	MCL	1541	386	245	14
	ClusterONE	972	329	197	15
	CORE	1517	420	259	39
	CSO	342	214	136	11
	COACH	474	265	144	13
	IFPA	935	425	219	47
MIPS	MCODE	135	72	60	4
	MCL	1259	254	196	17
	ClusterONE	744	208	152	17
	CORE	1217	303	225	29
	CSO	246	127	87	6
	COACH	396	145	92	5
	IFPA	772	288	167	32
Krogan	MCODE	160	127	73	10
	MCL	658	300	178	40
	ClusterONE	585	271	161	28
	CORE	677	279	172	39
	CSO	189	156	89	10
	COACH	221	179	85	11
	IFPA	447	288	131	21

Many of our detected protein complexes have a good match with the known protein complexes. We consider a detected complex to be biologically significant if its *p-value* is less than 0.01. In order to confirm the biological significance of detected protein complexes, the *p-value* is calculated by using the tool GO::TermFinder (https://www.yeastgenome.org/cgi-bin/GO/goTermFinder.pl). We randomly select some predicted protein complexes to calculate their *p-value* concerning Biological Process ontologies based on Krogan dataset, as shown in Table 3. From Table 3, all of these detected protein complexes obtain smaller *p-value* and it demonstrates that the protein complexes predicted by our IFPA method have strong biological significance. The predicted complexes with strong biological significance can provide help for biology researches to some extent.

Figure 7 visualizes an example of predicted protein complex named golgi transport complex in Krogan dataset so as to display the detection result more obviously. Figure 7 (A) displays a benchmark protein complex, Fig. 7 (B), (C), (D), (E) and (F) illustrate the identified protein complexes by MCODE, MCL, CORE, ClusterONE and IFPA, respectively. The purple nodes are the correctly identified proteins, the blue nodes are proteins that are not recognized, and the pink nodes are the wrongly identified proteins. From Fig. 7, we can see that MCODE and CORE correctly identifies four and two proteins, respectively. And MCL identifies a total of four proteins including

Table 3 Function enrichment analysis of predicted protein complexes detected on Krogan dataset

No.	Gene Ontology term	*p-value*	Cluster frequency	Genes annotated to the term
1	intra-Golgi vesicle-mediated transport	1.91e-16	100.0%	COG3/YER157W, COG7/YGL005C, COG1/YGL223C, COG2/YGR120C, COG8/YML071C, COG6/YNL041C, COG4/YPR105C
2	polyadenylation-dependent snoRNA 3'-end processing	3.32e-18	100.0%	RRP43/YCR035C, RRP45/YDR280W, MTR3/YGR158C, SKI6/YGR195W, LRP1/YHR081W, RRP40/YOL142W, RRP6/YOR001W
3	exonucleolytic trimming involved in rRNA processing	5.02e-20	100.0%	RRP43/YCR035C, RRP42/YDL111C, RRP45/YDR280W, MTR3/YGR158C, SKI6/YGR195W, RRP4/YHR069C, LRP1/YHR081W, CSL4/YNL232W
4	negative regulation of gluconeogenesis	3.00e-17	100.0%	GID7/YCL039W, RMD5/YDR255C, VID30/YGL227W, VID28/YIL017C, FYV10/YIL097W, GID8/YMR135C
5	chromatin disassembly	3.83e-20	100.0%	HTL1/YCR020W-B, RSC6/YCR052W, RTT102/YGR275W, STH1/YIL126W, RSC58/YLR033W, SFH1/YLR321C, RSC2/YLR357W, NPL6/YMR091C
6	positive regulation of transcription from RNA polymerase I promoter	5.96e-16	87.5%	UTP4/YDR324C, UTP5/YDR398W, UTP8/YGR128C, UTP9/YHR196W, UTP10/YJL109C, UTP15/YMR093W, NAN1/YPL126W

one that is misidentified. Albeit ClusterONE recognizes more proteins, it also has misidentifications. IFPA successfully detects the most proteins and all of them are correct, indicating that our predicted complex match very well with benchmark complex and our IFPA method is more accurate than other comparative methods.

Conclusions

Identification of protein complexes from PPI networks is distinctly important in proteomics. In this study, a flower pollination mechanism-based method is proposed to detect protein complexes in multi-relation reconstructed dynamic protein networks. To begin with, we build multi-relation reconstructed dynamic protein networks. Then, according to the core–periphery structure, we group the closely connected proteins as the cores and apply IFPA algorithm to attach peripheries to the optimal cores to form the predicted protein complexes. IFPA algorithm has been carried out on three different multi-relation reconstructed dynamic PPI networks and the experimental results demonstrate that our IFPA algorithm can obtain better clustering performance compared with other methods in most cases. The protein complexes we detected are likely to help the biologists gain some useful biological insights.

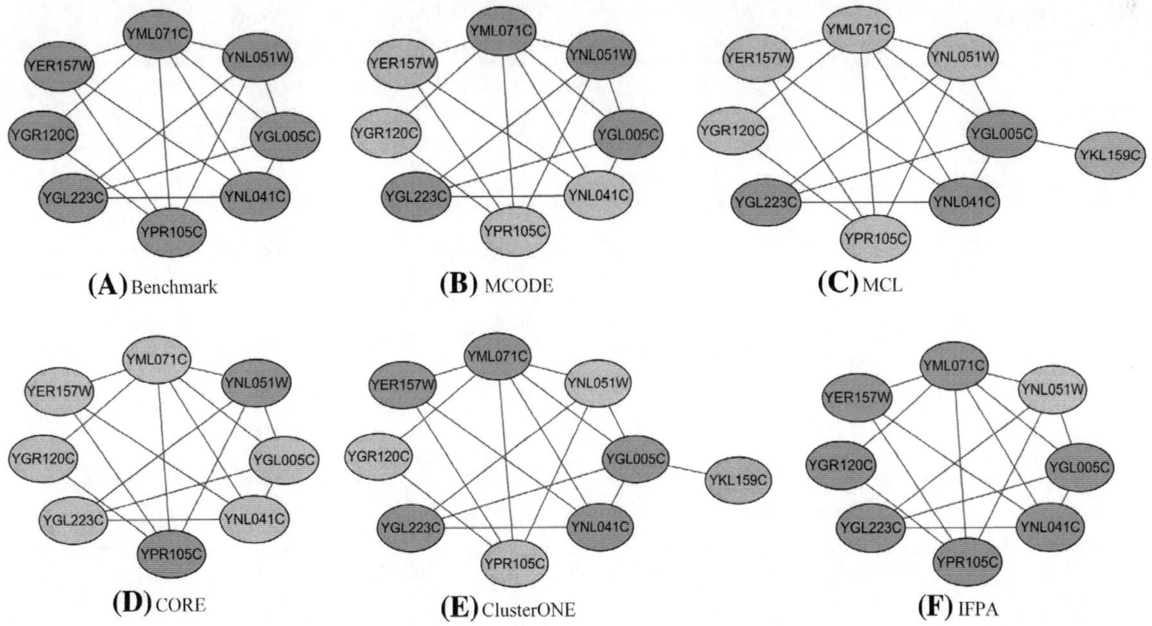

Fig. 7 The golgi transport complex detected by different methods

Abbreviations
ECC: Edge clustering coefficient; FPA: Flower pollination algorithm; GO: Gene ontology; OS: Overlapping Score; PPI: Protein-protein interaction

Acknowledgements
This paper is supported by the National Natural Science Foundation of China (61672334, 61502290, 61401263) and the Fundamental Research Funds for the Central Universities (No. GK201901010).

Funding
Publication of this article was funded by the National Natural Science Foundation of China (61672334).

Availability of data and materials
The datasets used and/or analyzed during the current study are available from the corresponding author on reasonable request.

About this supplement
This article has been published as part of *BMC Bioinformatics Volume 20 Supplement 3, 2019: Selected articles from the 17th Asia Pacific Bioinformatics Conference (APBC 2019): bioinformatics*. The full contents of the supplement are available online at https://bmcbioinformatics.biomedcentral.com/articles/supplements/volume-20-supplement-3.

Authors' contributions
XL conceptualized the algorithm, designed the method and drafted the manuscript, MF designed the method and drafted the manuscript, MF and LG analyzed the data and carried out the experiments, FXW modified the manuscript and polished the English expression. All of the authors have read and approved the final manuscript.

Ethics approval and consent to participate
Not applicable.

Consent for publication
Not applicable.

Competing interests
The authors declare that they have no competing interests.

Publisher's Note
Springer Nature remains neutral with regard to jurisdictional claims in published maps and institutional affiliations.

Author details
[1]School of Computer Science, Shaanxi Normal University, 710119, Xi'an, China. [2]College of Life Sciences, Shaanxi Normal University, 710119, Xi'an, China. [3]Department of Mechanical Engineering and Division of Biomedical Engineering, University of Saskatchewan, Saskatoon, SK S7N 5A9, Canada.

Published: 29 March 2019

References

1. von Mering C, Krause R, Snel B, Cornell M, Oliver SG, Fields S, Bork P. Comparative assessment of large-scale data sets of protein-protein interactions. Nature. 2002;417(6887):399–403.
2. De Las Rivas J, Fontanillo C. Protein-protein interactions essentials: key concepts to building and analyzing interactome networks. PLoS Comput Biol. 2010;6(6):e1000807.
3. Bader GD, Hogue CWV. An automated method for finding molecular complexes in large protein interaction networks. BMC bioinformatics. 2003;4:2-2.
4. Nepusz T, Yu H, Paccanaro A. Detecting overlapping protein complexes in protein-protein interaction networks. Nat Methods. 2012;9(5):471–U481.
5. Liu G, Wong L, Chua HN. Complex discovery from weighted PPI networks. Bioinformatics. 2009;25(15):1891–7.
6. Lei X, Zhang Y, Cheng S, Wu F-X, Pedrycz W. Topology potential based seed-growth method to identify protein complexes on dynamic PPI data. Inf Sci. 2018;425:140–53.
7. Ma C-Y, Y-PP C, Berger B, Liao C-S. Identification of protein complexes by integrating multiple alignment of protein interaction networks. Bioinformatics. 2017;33(11):1681–8.
8. Lecca P, Re A. Detecting modules in biological networks by edge weight clustering and entropy significance. Front Genet. 2015;6:265.
9. Van Dongen S. Graph clustering by flow simulation: Phd thesis University of Utrecht; 2000.
10. Lei X, Wang F, Wu F-X, Zhang A, Pedrycz W. Protein complex identification through Markov clustering with firefly algorithm on dynamic protein-protein interaction networks. Inf Sci. 2016;329:303–16.
11. Gavin A-C, Aloy P, Grandi P, Krause R, Boesche M, Marzioch M, Rau C, Jensen LJ, Bastuck S, Dumpelfeld B, et al. Proteome survey reveals modularity of the yeast cell machinery. Nature. 2006;440(7084):631–6.
12. Leung HCM, Xiang Q, Yiu SM, Chin FYL. Predicting protein complexes from PPI data: a core-attachment approach. J Comput Biol. 2009;16(2):133–44.
13. Wu M, Li X, Kwoh C-K, Ng S-K. A core-attachment based method to detect protein complexes in PPI networks. BMC Bioinformatics. 2009;10:169.
14. Shen X, Yi L, Jiang X, He T, Yang J, Xie W, Hu P, Hu X. Identifying protein complex by integrating characteristic of core-attachment into dynamic PPI network. PLoS One. 2017;12(10):e0186134.
15. Zhang Y, Lin H, Yang Z, Wang J, Li Y, Xu B. Protein complex prediction in large ontology attributed protein-protein interaction networks. IEEE/ACM Transactions on Computational Biology and Bioinformatics. 2013;10(3):729–41.
16. Wu M, Xie Z, Li X, Kwoh C-K, Zheng J. Identifying protein complexes from heterogeneous biological data. Proteins-Structure Function and Bioinformatics. 2013;81(11):2023–33.
17. Zhao JM, Hu XH, He TT, Li P, Zhang M, Shen XJ. An edge-based protein complex identification algorithm with gene co-expression data (PCIA-GeCo). IEEE Transactions on Nanobioscience. 2014;13(2):80–8.
18. Zhou HF, Liu J, Li JH, Duan WC. A density-based approach for detecting complexes in weighted PPI networks by semantic similarity. PLoS One. 2017;12(7):e0180570.
19. Yang X-S. Flower pollination algorithm for global optimization. Berlin: Heidelberg. Springer Berlin Heidelberg; 2012. p. 240–9.
20. Wang R, Zhou Y, Qiao S, Huang K. Flower pollination algorithm with bee pollinator for cluster analysis. Inf Process Lett. 2016;116(1):1–14.
21. Lei X, Fang M, Wu F-X, Chen L. Improved flower pollination algorithm for identifying essential proteins. BMC Syst Biol. 2018;12(4):46.
22. Ou-Yang L, Dai DQ, Li XL, Wu M, Zhang XF, Yang P. Detecting temporal protein complexes from dynamic protein-protein interaction networks. BMC Bioinformatics. 2014;15:335.
23. Zhang YJ, Lin HF, Yang ZH, Wang J, Liu YW, Sang ST. A method for predicting protein complex in dynamic PPI networks. BMC Bioinformatics. 2016;17:229.
24. Li M, Meng X, Zheng R, Wu FX, Li Y, Pan Y, Wang J. Identification of protein complexes by using a spatial and temporal active protein interaction network. IEEE/ACM Transactions on Computational Biology and Bioinformatics. 2017:1–1.
25. Wang JX, Peng XQ, Li M, Pan Y. Construction and application of dynamic protein interaction network based on time course gene expression data. Proteomics. 2013;13(2):301–12.
26. Zhao B, Wang J, Li M, Li X, Li Y, Wu FX, Pan Y. A new method for predicting protein functions from dynamic weighted interactome networks. IEEE Trans Nanobioscience. 2016;15(2):131–9.
27. Xenarios I, Salwinski L, Duan XJ, Higney P, Kim SM, Eisenberg DDIP. The database of interacting proteins: a research tool for studying cellular networks of protein interactions. Nucleic Acids Res. 2002;30(1):303–5.
28. Guldener U, Munsterkotter M, Oesterheld M, Pagel P, Ruepp A, Mewes HW, Stumpflen V. MPact: the MIPS protein interaction resource on yeast. Nucleic Acids Res. 2006;34(Database issue):D436–41.
29. Krogan NJ, Cagney G, Yu H, Zhong G, Guo X, Ignatchenko A, Li J, Pu S, Datta N, Tikuisis AP, et al. Global landscape of protein complexes in the yeast Saccharomyces cerevisiae. Nature. 2006;440(7084):637–43.
30. Tu BP, Kudlicki A, Rowicka M, McKnight SL. Logic of the yeast metabolic cycle: temporal compartmentalization of cellular processes. Science (New York, NY). 2005;310(5751):1152–8.
31. Binder JX, Pletscher-Frankild S, Tsafou K, Stolte C, O'Donoghue SI, Schneider R, Jensen LJ. COMPARTMENTS: unification and visualization of protein subcellular localization evidence. Database-the Journal of Biological Databases and Curation. 2014;2014:bau012.

32. Cherry JM, Adler C, Ball C, Chervitz SA, Dwight SS, Hester ET, Jia Y, Juvik G, Roe T, Schroeder M, et al. SGD: Saccharomyces genome database. Nucleic Acids Res. 1998;26(1):73–9.
33. Mewes HW, Frishman D, Mayer KF, Munsterkotter M, Noubibou O, Pagel P, Rattei T, Oesterheld M, Ruepp A, Stumpflen V. MIPS: analysis and annotation of proteins from whole genomes in 2005. Nucleic Acids Res. 2006; 34(Database issue):D169–72.
34. Zhang R, Ou HY, Zhang CT. DEG: a database of essential genes. Nucleic Acids Res. 2004;32(Database issue):D271–2.
35. Pu SY, Wong J, Turner B, Cho E, Wodak SJ. Up-to-date catalogues of yeast protein complexes. Nucleic Acids Res. 2009;37(3):825–31.
36. Altaf-Ul-Amin M, Shinbo Y, Mihara K, Kurokawa K, Kanaya S. Development and implementation of an algorithm for detection of protein complexes in large interaction networks. BMC Bioinformatics. 2006;7:207.

Li *et al. BMC Bioinformatics* 2019, **20**(Suppl 3):132
https://doi.org/10.1186/s12859-019-2650-7

BMC Bioinformatics

RESEARCH

Open Access

Automatic localization and identification of mitochondria in cellular electron cryo-tomography using faster-RCNN

Ran Li[1†], Xiangrui Zeng[2†], Stephanie E. Sigmund[3], Ruogu Lin[2], Bo Zhou[4], Chang Liu[5], Kaiwen Wang[5], Rui Jiang[1], Zachary Freyberg[6*], Hairong Lv[1*] and Min Xu[2*]

From The 17th Asia Pacific Bioinformatics Conference (APBC 2019)
Wuhan, China. 14-16 January 2019

Abstract

Background: Cryo-electron tomography (cryo-ET) enables the 3D visualization of cellular organization in near-native state which plays important roles in the field of structural cell biology. However, due to the low signal-to-noise ratio (SNR), large volume and high content complexity within cells, it remains difficult and time-consuming to localize and identify different components in cellular cryo-ET. To automatically localize and recognize *in situ* cellular structures of interest captured by cryo-ET, we proposed a simple yet effective automatic image analysis approach based on Faster-RCNN.

Results: Our experimental results were validated using *in situ* cyro-ET-imaged mitochondria data. Our experimental results show that our algorithm can accurately localize and identify important cellular structures on both the 2D tilt images and the reconstructed 2D slices of cryo-ET. When ran on the mitochondria cryo-ET dataset, our algorithm achieved Average Precision > 0.95. Moreover, our study demonstrated that our customized pre-processing steps can further improve the robustness of our model performance.

Conclusions: In this paper, we proposed an automatic Cryo-ET image analysis algorithm for localization and identification of different structure of interest in cells, which is the first Faster-RCNN based method for localizing an cellular organelle in Cryo-ET images and demonstrated the high accuracy and robustness of detection and classification tasks of intracellular mitochondria. Furthermore, our approach can be easily applied to detection tasks of other cellular structures as well.

Keywords: Cryo-ET, Faster-RCNN, Cellular structure detection, Biomedical image analysis

Background

In cells, most biological processes are dominated by intricate molecular assemblies and networks. Analyzing the structural features and spatial organization of those assemblies is essential for understanding cellular functions. Recently, cellular cryo-Electron Tomography (cryo-ET) has been developed as an approach to obtain 3D visualization of cellular structures at submolecular resolution and in a close-to-native state [1]. Cryo-ET has been proven to be a powerful technique for structural biology *in situ* and has been successfully applied to the study of many important structures, including vaults [2], Integrin Linked Kinase (ILK) [3], and the nuclear pore complex (NPC) [4]. However, the systematic structural analysis of cellular components in cryo-ET images remains challenging due to several factors including low signal-to-noise ratio (SNR), limited projection range (leading to the missing wedge effect) and a crowded intracellular environment composed of complex intracellular structures.

*Correspondence: freybergzz@upmc.edu; lihua@ict.ac.cn; mxu1@cs.cmu.edu
[†]Ran Li and Xiangrui Zeng contributed equally to this work.
[6]Departments of Psychiatry and Cell Biology, University of Pittsburgh, Pittsburgh, PA, USA
[1]Department of Automation, Tsinghua University, Beijing, China
Full list of author information is available at the end of the article

Given the critical roles played by mitochondria within mammalian cells, and the distinctive morphology of these organelles, we chose to examine mitochondria imaged by *in situ* cryo-ET [5]. The 3D visualization of mitochondria can provide insights into mitochondrial structure and functionalities. Therefore, methodological improvements in the detection and localization of mitochondria within complex *in situ* cryo-ET datasets may significantly improve accuracy of detection of these organelles and directly impact further structural analyses.

Localization of the subcellular structures of interest can facilitate subsequent study of specific macromolecular components within the selected structures [6]. Such localization can be performed through image segmentation, which are usually performed manually or by specifically designed heuristics. Although some visualization tools have been developed to facilitate these approaches, manual segmentation in Cryo-ET images still requires large amounts of repetitive labor from researchers, and the results of which are subjective. On the other hand, automatic methods are fast and can produce consistent results. Contour-based methods like Watershed yield great results when the image complexity is low, but appear to be sensitive to noise [7]. Threshold-based methods, which usually generate a mask according to the density threshold, can be applied to foreground-background segmentation but still have difficulty in identifying different cellular components [8]. Recently, segmentation methods focusing on specific types of structures including membranes, microtubules and filaments [9–11], have drawn a lot of attention. These methods perform well on specific cellular structures, but lack generality. To date, machine learning approaches to identify intracellular structures appears to be promising. Consequently, we have developed an unsupervised segmentation method based on manually designed heuristic rules [12], and by clustering representative features [13]. Luengo et al. [14] proposed a supervised approach to classify each voxel with a trained classification model. However, both of these methods require manually designed features or rules, which might be time- and effort-consuming while having various limitations. Chen et al. developed another supervised segmentation method, taking advantage of the excellent capability of feature extraction of convolutional neural network (CNN) [15]. But in this way, a separate CNN has to be trained for each type of structural features, and the precise contours need to be manually annotated in the training data, which may not be trivial.

Our goal is to design a simple and generic method of automatic identification and localization of subcellular structures of interest within *in situ* cryo-ET images with weak annotations, which is different from existing segmentation-type methods and can greatly reduce the time and effort cost of detailed manual annotation. We aim to detect all objects of interest in an image and output corresponding bounding box with class prediction simultaneously. Region-based convolutional neural network (RCNN) [16], which generates region proposals using Selective Search, extracts features from all the proposals after normalization with CNNs, and finally feeds the features to a classifier and a regression layer simultaneously to get both classification scores and bounding box coordinates as output, lays the foundation for our goal. And its last incarnation, Faster RCNN [17], has achieved almost real-time detection with a high degree of accuracy. Faster RCNN based localization methods have been applied to biomedical imaging data such as breast mammography [18] and cellular fluorescence imaging [19].

In this work, we proposed an automatic identification and localization method based on Faster-RCNN, which is the first Faster-RCNN based method for localizing an cellular organelle in Cryo-ET images. Our algorithm is trained and validated on 2D projection images of a cryo-ET tomogram for localization and classification tasks of mitochondira. Our experimental results show that our algorithm is able to robustly predict the object's bounding box with classification scores. Moreover, we extended our study to 3D tomogram slices and achieved accurate and robust performance.

Method

Our mitochondria identification and localization method is comprised of two main parts: (1) pre-processing to improve the quality of samples, and (2) object detection using Faster-RCNN. The input of our system is 2D projection images of a tomogram, and the output includes coordinates of the bounding boxes of object of interest, the class of each object and the probability of the classification. A flowchart of our method is shown in Fig. 1. In this section, we willdescribe each part of our system in details.

Preprocessing

Since biological samples are sensitive to radiation damage, only low-dose electrons can be used for electron microscopy imaging [6]. Compared to normal images, electron tomography images are usually noisier and have lower contrast. To make the images suitable for subsequent processing, we first perform noise reduction and contrast enhancement. To reduce noise, considering the edge features are often important for subcellular structures, we chose Bilateral Filtering [20], a nonlinear filtering method that preserves the original edges as much as possible. Bilateral Filtering considers the effects of both spatial distance and gray scale distance, and can be implemented by combining two Gaussian Filters. To improve local contrast and the definition of details, we

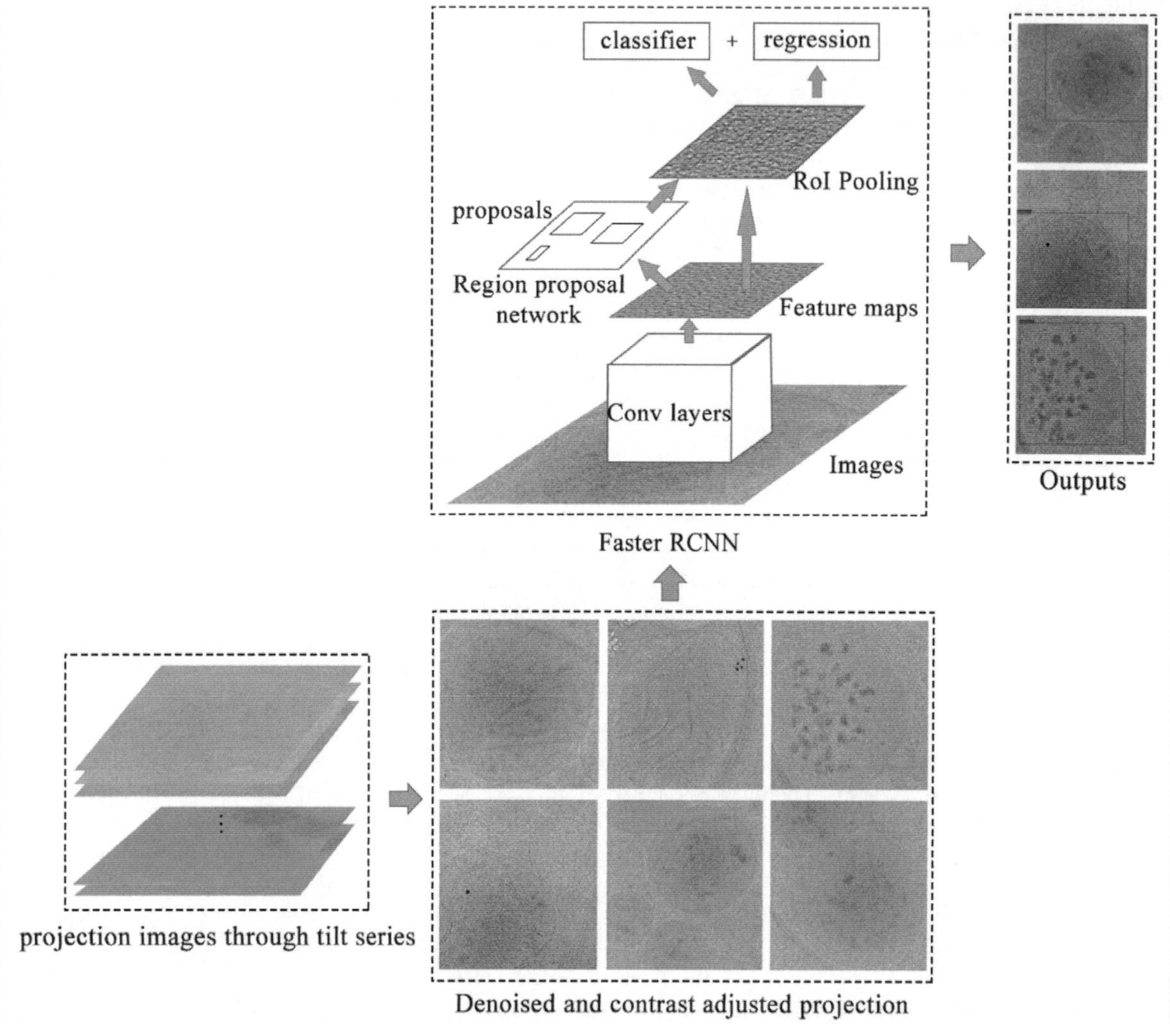

Fig. 1 Flowchart of our Faster-RCNN model. The denoised input image is fed into Conv layers to generate the feature map. Then, region proposal network proposes potential regions that contain object of interest. The proposal regions are passed to 1) classifier for classification, 2) regressor for refine the bounding box location

use Histogram Equalization, which can also balance the brightness of different images.

Object detection in 2D images

The main idea of our method is based on Faster RCNN [17], in which the four modules of feature extraction, proposal generation, RoI Pooling, classification and regression are organically combined to form an end-to-end object detection system.

Feature extraction is the first step of our method. The input of the deep convolutional neural network is the image I, and the output is the extracted feature map. These features will be shared by subsequent modules. The basic feature extraction network in our model, Resnet-50, is based on [21]. He et al. proposed this deep residual learning method in 2015 to make the deeper network train properly. The architecture of our network is shown in Fig. 2. The original Resnet-50 network is split into two parts in our model: part one including layers conv1 to conv4_x is used for extraction of shared features, and part two including layer conv5_x and upper layers further extracts features of proposals for the final classification and regression. The implementation of the model refers to the work of Yann Henon in 2017 [22].

The feature extraction network is followed by a region proposal network (RPN). A window of size $n \times n$ slides onto the feature map, and at each location it stays the features in the window are mapped to a low-dimensional

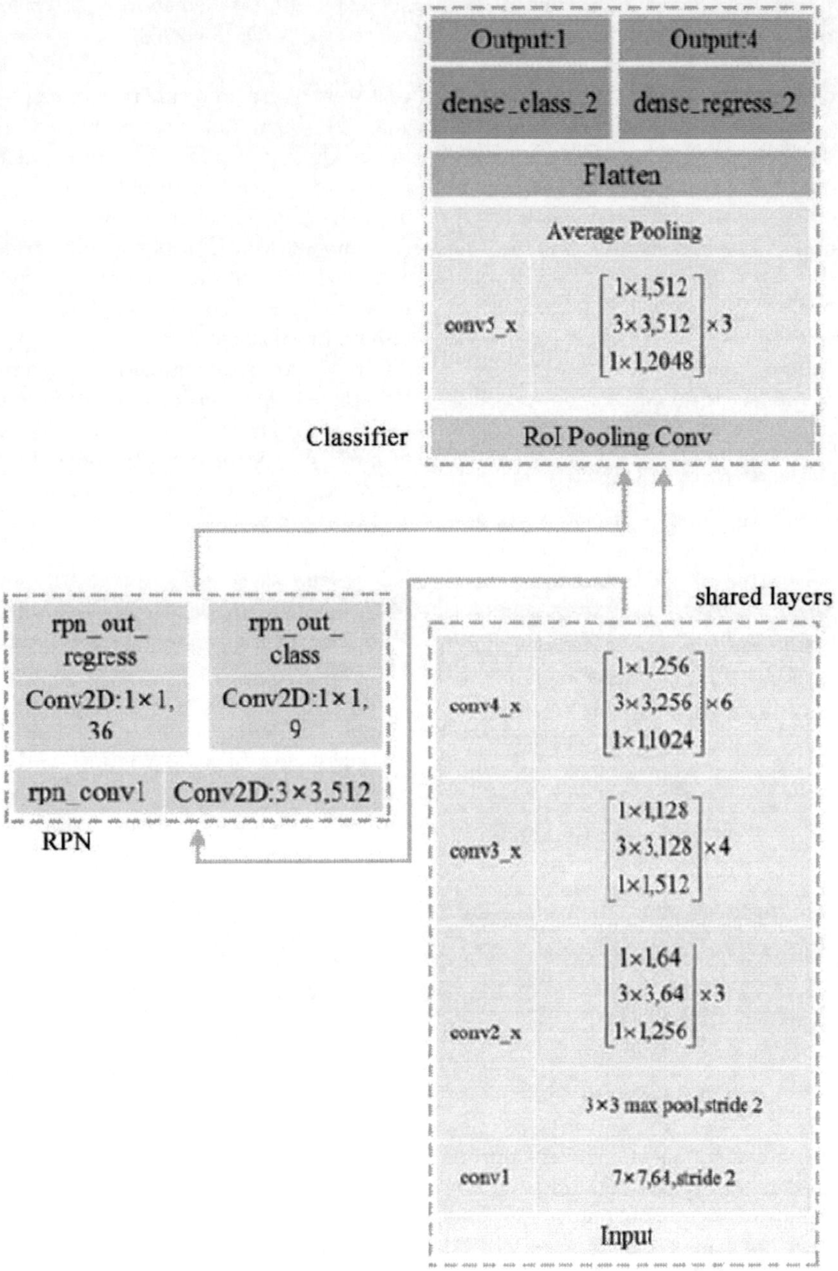

Fig. 2 Detailed Architecture of the Faster-RCNN model. The basic feature extraction network Resnet-50 is split into two parts in our model: 1) layers conv1 to conv4_x is used for extraction of shared features (in the shared layers), 2) layer conv5_x and upper layers further extracts features of proposals for the final classification and regression (in the classifier). And the RPN implemented with three convolutional layers generates proposals from the shared feature map

vector, which will be used for object-background classification and proposal regression. At the same time, k region proposals centered on the sliding window in the original image are extracted according to k anchors, which are rectangular boxes of different shapes and sizes. Moreover, for each proposal, two probabilities for the classification and four parameters for the regression will be achieved, composing the final $6k$ outputs of the classification layer and the regression layer. The sliding window, classification layer and regression layer are all implemented using convolutional neural networks. In practice, we chose $k = 9$ with 3 scales of 128^2, 256^2, and 512^2 pixels and

3 aspect ratios of 1:1, 1:2, and 2:1 as the default in [17]. And non-maximum suppression(NMS) was adopted with the IoU threshold at 0.7, while the maximum number of proposals produced by the RPN was 300.

Features of different scales are then integrated into feature maps of the same size (7×7 in our experiment) via RoI pooling layer, so that the features can be used in final fully connected classification and regression layers. For a region proposal of any size, like $h \times w$, it will be divided into a fixed number, like $H \times W$, of windows of size $h/H \times w/W$. Then max pooling will be performed and a fixed-size ($H \times W$) feature map will be obtained with the maximum of each window.

To train the whole model end-to-end, a multi-task loss function is proposed as follows [17].

$$L\left(p, u, t^u, v\right) = L_{cls}(p, u) + \lambda[u \geq 1] L_{loc}\left(t^u, v\right) \quad (1)$$

Where u is the ground truth label of the proposal, and $v = (v_x, v_y, v_w, v_h)$ represents the regression offset between the proposal and the ground truth.The output of the classification layer, $p = (p_0, p_1, ..., p_K)$, represents the probabilities of the proposal belonging to each one of the $K + 1$ classes and $t^u = \left(t_x^u, t_y^u, t_w^u, t_h^u\right)$ represents the predicted regression offset for a proposal with label u. The loss function of the classification task is defined as:

$$L_{cls}(p, u) = -\log p_u. \quad (2)$$

And the loss function of the regression is a robust L1 loss as follows:

$$L_{loc}\left(t^u, v\right) = \sum_{i \in x,y,w,h} smooth_{L1}\left(t_i^u - v_i\right). \quad (3)$$

Where

$$smooth_L 1\,(x) = \begin{cases} 0.5x^2, & if \; \|x\| < 1 \\ \|x\| - 0.5, & otherwise \end{cases} \quad (4)$$

The hyperparameter λ is used to control the balance between the two losses and is set to $\lambda = 1$ in our experiment. Similarly, the loss function of the RPN during training is also defined in this form. In the training process, the RPN with the shared layers is trained first and then the classifier is trained using proposals generated by the RPN, with the initial weights for both networks given by a pretrained model on ImageNet [17, 23].

Results

Dataset and evaluation metrics

Data Acquisition:

Tissue Culture: Rat INS-1E cells (gift of P. Maechler, Université de Genève) were cultured in RPMI 1640 medium supplemented with 2 mM L-glutamine (Life Technologies, Grand Island, NY), 5% heat-inactivated fetal bovine serum, 10 mM HEPES, 100 units/mL penicillin, 100 μg/mL streptomycin, 1 mM sodium pyruvate, and 50 μM b-Mercaptoethanol as described earlier (insert reference: PMID: 14592952).

EM Grid Preparation: For cryo-ET imaging, INS-1E cells were plated onto either fibronectin-coated 200 mesh gold R2/1 Quantifoil grids or 200 mesh gold R2/2 London finder Quantifoil grids (Quantifoil Micro Tools GmbH, Jena, Germany) at a density of 2×10^5 cells/mL. Following 48 h incubation under conventional culture conditions in complete RPMI 1640 medium, grids were removed directly from culture medium and immediately plunge frozen in liquid ethane using a Vitrobot Mark IV (Thermo Fisher FEI, Hillsboro, OR).

Cryo-Electron Tomography: Tomographic tilt series for INS-1E cells were recorded on a FEI Polara F30 electron microscope (Thermo Fisher FEI) at 300kV with a tilt range of $\pm 60°$ in $1.5°$ increments using the Gatan K2 Summit direct detector (Gatan, Inc.) in super-resolution mode at 2X binned to 2.6Å/pixel; tilt series were acquired via SerialEM.

Datasets: We collected 9 cryo-ET tomograms (786 2D slices) contains mitochondria. 482 out of the 786 slices were selected and annotated manually via LabelImg [24]. Then, the 2D slices were randomly divided into training and testing set with a ratio of 5:1. Details of our dataset are shown in Table 1.

Metrics: To evaluate the performance of our model, we mainly use two metrics from common object detection and segmentation evaluation: AP (average precision) and F_1 *score*. The definitions are as follows:

$$AP = \int_0^1 P(R)\, d(R) \quad (5)$$

$$F_1\ score = \frac{2P \times R}{P + R} \quad (6)$$

where P represents precision, which indicates the ratio of the true positives to all predicted positives; R represents recall, which indicates the ratio of the true positives to all true elements. Neither precision nor recall alone

Table 1 Cryo-ET dataset properties

Tomogram basename	Image size	All slices	Used slices
Unstim_20k_mito1	3708 × 3838	101	75
Unstim_20k_mito2	3708 × 3838	89	44
CTL_Fibro_mito1	3708 × 3838	82	36
M2236_Fibro_mito2	3708 × 3838	90	46
M2236_turemito3	3708 × 3838	86	39
CHX + Glucose Stimulation A2	3708 × 3838	53	51
HighGluc_Mito1	3708 × 3838	101	71
HighGluc_Mito2	3708 × 3838	101	69
INS_21_g3_t10	3708 × 3838	81	51
Total		786	482

is sufficient to fully evaluate the prediction performance. Therefore, the F1 score defined by the weighted harmonic mean of precision and recall is commonly used in the case where both of them need to be high enough. And AP, equivalent to the area under the precision-recall curve, may provide an overall evaluation of the model's performance at different precision/recall rates. As an object detection problem, the correctness of each sample prediction is not only related to classification, but also related to localization. The accuracy of localization is evaluated by (Intersection over Union), which is defined as:

$$IoU = \frac{S_P \cap S_G}{S_P \cup S_G} \tag{7}$$

where S_P is the predicted bounding box and S_G represents the ground truth, and IoU measures the degree of coincidence. In our experiments, different IoU thresholds(0.5, 0.6, 0.7, 0.8, and 0.9) are set, and those samples with mitochondria prediction labels and IoUs higher than the specific threshold are considered. The higher the IoU threshold, the higher the accuracy requirements for localization. Thus we can see the difference in the detection accuracy under different localization accuracy requirements, and judge the localization performance of our model. The precision, recall, F1 score and AP in our experiment are calculated.

Data preprocessing and model training

The 2D projection images we acquired from the original tomograms have low SNR and contrast which interferes with subsequent identification and segmentation of intracellular features. Thus, the images are first denoised via a bilateral filter with $\sigma_r = 1.2$ and $\sigma_d = 100$, suppressing noise and retaining the original edge features as much as possible. This is followed by enhancement of contrast via histogram equalization which improves in the resolution of previously indistinguishable details. Figure 3 shows an example of two images before and after preprocessing. The preprocessing methods and parameters in our method were finally determined based on the single-image SNR estimated according to [25], grayscale distribution histograms and visual effect of the image. Figure 4 shows SNR of the same image with different σ_d and σ_r and the performance of different preprocessing schemes. We found that performing histogram equalization first will increase the noise in the original image, and the contrast will be reduced again after filtering, failing to achieve the desired effect. Furthermore, we found that Gaussian filtering used for noise reduction cannot preserve the edge as well as Bilateral filtering.

All the models in our experiments were trained and tested using Keras [26] with Tensorflow [27] as the backend, using optimizer Adam (Adaptive Moment Estimation) [28] with $\beta_1 = 0.9, \beta_2 = 0.999$ and learning rate of 1×10^{-5} for both RPN and the classifier. The 482 annotated slices were randomly split into a training set of 402 slices and a test set of 80 slices according to a ratio of 5:1. The model would be saved only if the loss after one epoch is less than the best loss before.

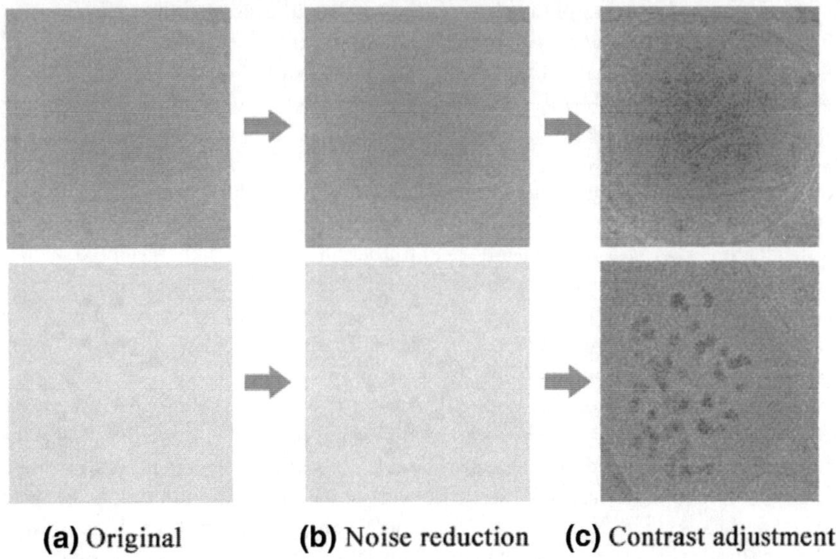

Fig. 3 a Original 2D projection images, **b** Images after noise reduction (Bilateral Filtering with $\sigma_r = 1.2$ and $\sigma_d = 100$), **c** Images after noise reduction and contrast adjustment

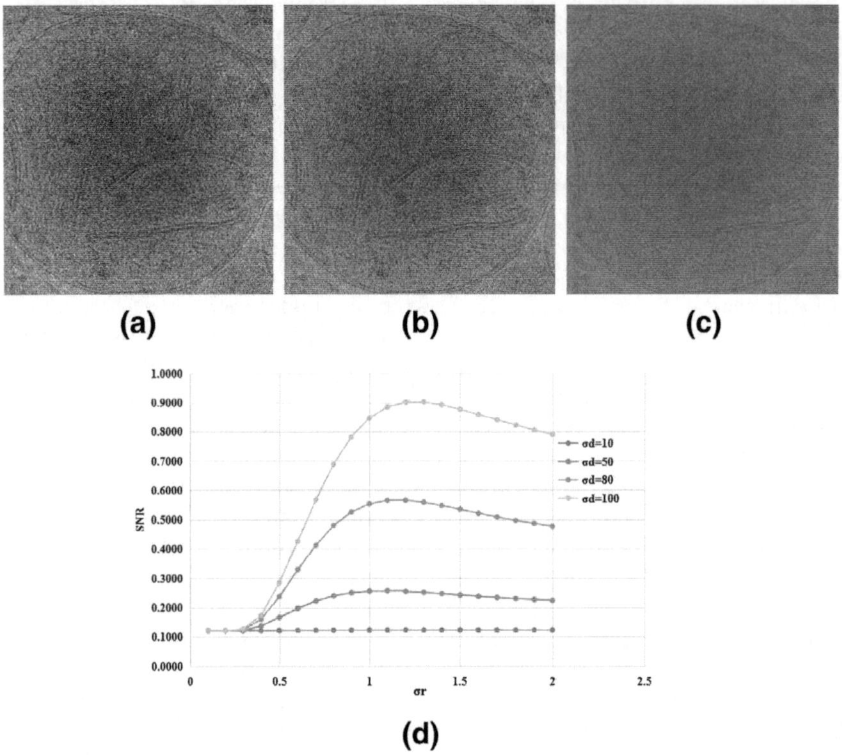

Fig. 4 a Bilateral Filter + Histogram Equalization, **b** Gaussian Filter + Histogram Equalization, **c** Histogram Equalization + Bilateral Filter **d** SNR with different σ_d and σ_r

Prediction performance

We trained the model on the training set and tested it on the test set. Figures 5 and 6 show the test results visually and quantitatively. In addition to the bounding box, our model also gives the most likely category of the object and the probability of it belonging to that category. In Fig. 5, the red bounding box is the manually annotated ground truth, and the blue box is predicted by the model. We notice that the predicted results and the ground truth are highly coincident, and even the regions that cannot be completely overlapped basically contain the entire mitochondria, which means that our system can achieve the goal of automatic identification and localization of mitochondria quite successfully. The area where the mitochondria is located can be separated from the outside by the bounding box, so as to eliminate the influence of the surrounding environment as much as possible, making it possible to analyze the internal structures in more detail.

In Fig. 6, we plotted the precision-recall curve and calculated the APs at different IoU thresholds to measure the detection performance. We noticed that when the IoU threshold is set to 0.7 and below, the AP is close to 1, which means that almost all samples were correctly predicted,indicating that our system can successfully identify the mitochondria in the picture. However, when the IoU threshold is increased to 0.9, the AP drops sharply to around 0.4, which indicates that our system still has some deficiencies in the accuracy of localization. The overlap between the predicted area and the ground truth area can be further improved, which can be an important aspect of our future work. The precision-recall curve for IoU thresholds of 0.7 is also given in Fig. 6. When the IoU threshold is 0.7, all positive samples can be correctly predicted while the precision requirement is not higher than 0.9, that is, all mitochondria can be found in that condition; even with a precision of 1, which means all samples predicted to be positive must be correct, 70% of the mitochondria can still be detected.

In addition, we compared the effect of preprocessing on the prediction results. It is noted that no matter how the IoU threshold is set, the AP value of the model without preprocessing is significantly lower than that of the model containing the preprocessing, which again shows that preprocessing is a necessary step for the overall system. Especially when the IoU threshold is 0.8, the system with or without preprocessing shows a great difference in the average precision of prediction, which indicates that the main contribution of preprocessing to the system is to further improve the accuracy of localization. For the model that does not include preprocessing, the predicted

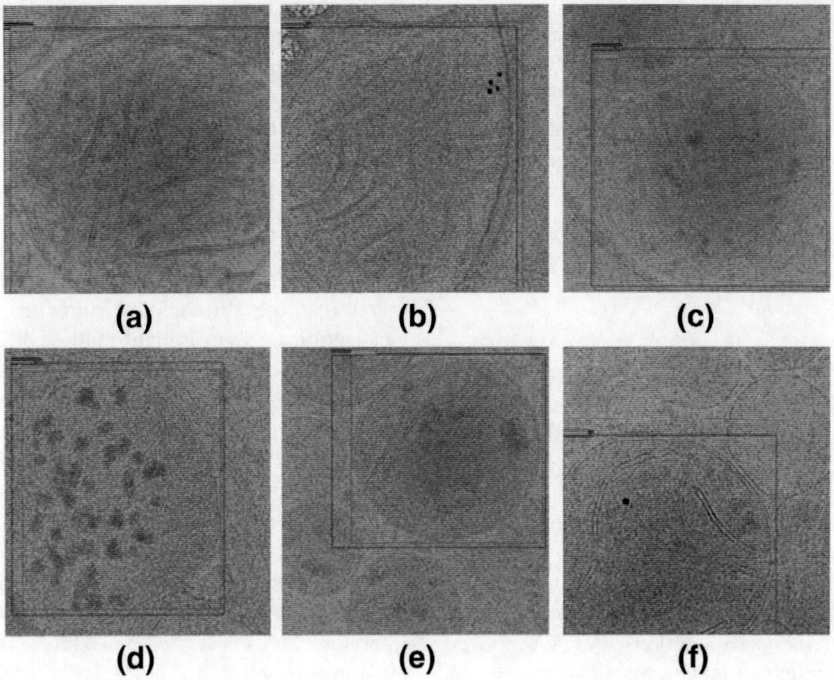

Fig. 5 Examples of detection results: the red boxes are ground truth, and the blue ones are the predicted bounding boxes. Data source: **a** Tomogram: Unstim_20k_mito1 (projection image 63), **b** Tomogram: Unstim_20k_mito2 (projection image 49), **c** Tomogram: HighGluc_Mito2 (projection image 47), **d** Tomogram: CTL_Fibro_mito1 (projection image 44), **e** Tomogram: HighGluc_Mito1 (projection image 48), **f** Tomogram: CHX + Glucose Stimulation A2 (projection image 13)

bounding box that has an IoU no less than 0.8 with ground truth is quite rare, and the average precision calculated in this situation is only 0.3. After the preprocessing step, it becomes common that IoU of the predicted bounding box and the ground truth reaches 0.8, resulting in an increase of the average precision to 0.95 and higher.

Source of error

In order to further analyze the performance of our method, we separately analyzed the prediction results of the system on 9 different *in situ* cryo-ET tomograms (Table 2), and studied the impact of different factors including the quality of the original image, the

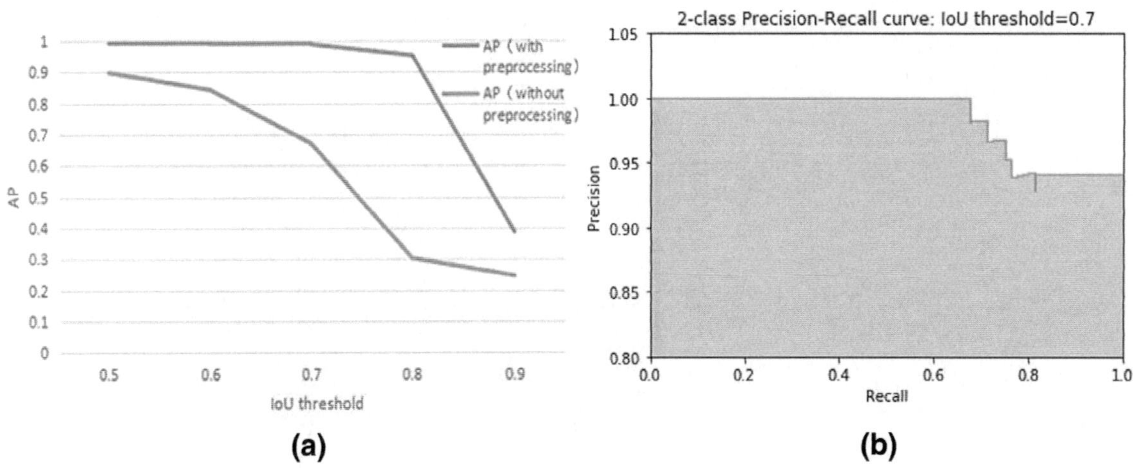

Fig. 6 Prediction performance: **a** AP with different IoU threshold, **b** Precision-Recall curve with IoU threshold=0.7

Table 2 Prediction results on different tomograms

Tomogram basename	F_1 *score*	AP	mIoU	Incomplete mitochondria
Unstim_20k_mito1	0.91	0.98	0.826	YES
Unstim_20k_mito2	1	1	0.864	NO
CTL_Fibro_mito1	0.97	0.99	0.843	NO
M2236_Fibro_mito2	0.96	0.99	0.887	YES
M2236_turemito3	0.91	0.97	0.783	NO
CHX + Glucose Stimulation A2	0.94	1	0.75	YES
HighGluc_Mito1	0.97	0.99	0.843	NO
HighGluc_Mito2	0.97	0.96	0.837	NO
INS_21_g3_t10	0	0	0	YES

The F_1 *score* and AP are calculated at an IoU threshold of 0.7

intactness of the mitochondria etc. The F_1 *score* and AP remain calculated at an IoU threshold of 0.7. In most tomograms, our systems show high accuracy, consistent with the overall results. However, we also found that in INS_21_g3_t10, our system could not accurately detect mitochondria. Therefore, we analyzed the projected image from INS_21_g3_t10 (Fig. 7). We noticed that in all the 2D projection images from that tomogram, the mitochondria included are too small and the structure appeared incomplete, especially the internal structure, which is basically submerged in noise and hard to identify. Even after noise reduction and contrast adjustment, the details of the mitochondria in the image are still too blurred, causing strong interference in the extraction of features. We also calculated the SNR of the two-dimensional projection images in INS_21_g3_t10, which is approximately 0.06 on average. For reference, the SNR of the original projection image from Unstim_20k_mito1 we analyzed in Fig. 4 is 0.12, which is significantly higher than the images in INS_21_g3_t10. It is also worth noting that in Unstim_20k_mito1, the subject of the projection images is the mitochondria we need to detect, while in INS_21_g3_t10, the mitochondria only occupy a very small part of the image. As a result, other components of the image are calculated as signal which may be not that useful for our detection task, making the ratio of effective information to noise even lower than 0.06. This may explain why the detection performance of it is particularly unsatisfactory.

In order to better study the influence of different tomograms on the accuracy of localization, mean Intersection over Union (mIoU) is calculated for each tomogram. It can be noted that, on average, mIoU is higher in the tomograms that contain complete mitochondria, that is, the localization accuracy is higher, although the highest mIoU comes from a tomogram containing incomplete mitochondria. We analyzed the characteristics of this tomogram and found that it is the only one where mitochondria do not appear circular or nearly circular, but instead possess a slanted strip shape (also shown in Fig. 7). Therefore, when the mitochondrion is marked with a rectangular box, the box occupies a larger area and contains more non-mitochondrial regions, which may make the prediction results more easily coincide with the ground truth. Therefore, in general, we can still conclude that complete mitochondria are more easily localized accurately. This is also in consistent with our intuition that the complete mitochondria have a complete outline of a bilayer membrane that approximates a circular shape, which provides a powerful reference for determining its specific boundaries. In fact, the tomogram with best results on the F_1 *score* and AP also contains intact mitochondria. Therefore, the integrity of mitochondria has a certain impact on the detection results of the system.

Prediction on tomogram slices

The ultimate goal is to detect mitonchondria in 3D tomograms. The model trained on 2D projection images can be directly applied to tomogram slices to generate the output. Like projection images, the slices were first preprocessed through Bilateral filtering and histogram

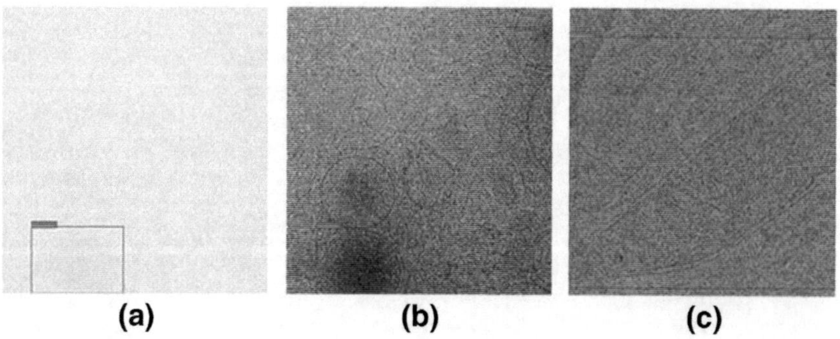

Fig. 7 An example of projection images from tomogram INS_21_g3_t10 (in which the mitochondria is hard to detect): **a** Original image, **b** Image after noise reduction and contrast adjustment, **c** Projection image from M2236_Fibro_mito1

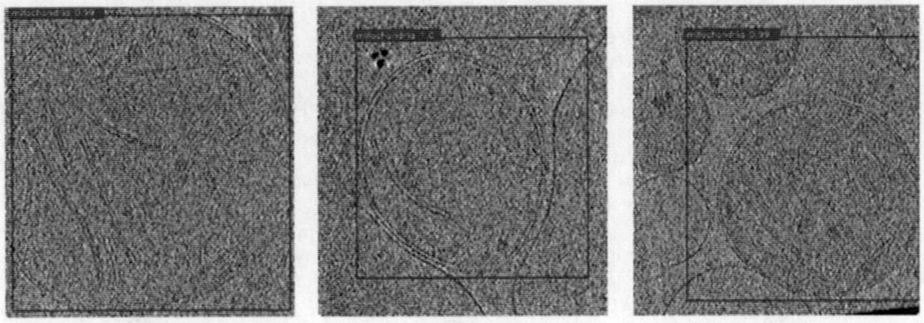
(a) (b) (c)

Fig. 8 Detection results on slices of reconstructed tomograms. Data source: **a** Tomogram: Unstim_20k_mito_1 (slice 26), **b** Tomogram: M2236_truemito3 (slice 97), **c** Tomogram: HighGluc_Mito1 (slice 58)

equalization with the same parameters, and then tested by the Faster-RCNN model. The whole model is applied to the tomogram slice by slice and the output includes all the bounding boxes of mitochondria in the slice with a classification score for each box. And it only takes a few seconds for each slice when tested on CPUs.

As shown in Fig. 8, the mitochondria in tomogram slices can be successfully identified and localized, while the accuracy of localization may be slightly reduced due to higher noise, as compared to 2D projection images. Therefore, it is only necessary to perform annotation and training on the 2D projection images, which can greatly reduce the computational costs, and we can detect mitochondria in 3D tomograms with a tolerable error. And the probability of expanding to different organelles is still retained even in the case of 3D.

Conclusion

In this paper, we proposed an automatic Cryo-ET image analysis algorithm for localization and identification of different structure of interest in cells. To best to our knowledge, this is the first work to applied Faster-RCNN model to Cryo-ET data, which demonstrated the high accuracy ($AP > 0.95$ and $IoU > 0.7$) and robustness of detection and classification tasks of intracellular mitochondria. Furthermore, our algorithm can be generalized to detect multiple cellular components using the same Faster-RCNN model, if annotations of multiple classes of cellular component were provided. For future work, we will further improve the accuracy of localization by collecting more data and we will explore the effects of different network structures to enhance the model.

Abbreviations

Adam: Adaptive moment estimation; AP: Average precision; CNN: Convolutional neural network; cryo-ET: Cryo-electron tomography; ILK: Integrin linked kinase; IoU: Intersection over union; mIoU: Mean intersection over union NMS: Non-maximum suppression; NPC: Nuclear pore complex; SNR: Signal-to-noise ratio; RCNN: Region-based convolutional neural network; RPN: Region proposal network

Acknowledgements

This work was supported in part by U.S. National Institutes of Health (NIH) grant P41 GM103712. MX acknowledges support of the Samuel and Emma Winters Foundation. ZF acknowledges support from the U.S. Department of Defense (PR141292) and the John F. and Nancy A. Emmerling Fund of The Pittsburgh Foundation. This work was partially supported by the National Key Research and Development Program of China (No. 2018YFC0910404), the National Natural Science Foundation of China (Nos. 61873141, 61721003, 61573207, U1736210, 71871019 and 71471016), and the Tsinghua-Fuzhou Institute for Data Technology. RJ is a RONG professor at the Institute for Data Science, Tsinghua University.

Funding

Publication charge for this work has been funded by the National Key Research and Development Program of China (No. 2018YFC0910404), the National Natural Science Foundation of China (Nos. 61873141, 61721003, 61573207, U1736210, 71871019 and 71471016), and the Tsinghua-Fuzhou Institute for Data Technology. RJ is a RONG professor at the Institute for Data Science, Tsinghua University. This work was supported in part by U.S. National Institutes of Health (NIH) grant P41 GM103712. MX acknowledges support of the Samuel and Emma Winters Foundation. ZF acknowledges support from the U.S. Department of Defense (PR141292) and the John F. and Nancy A. Emmerling Fund of The Pittsburgh Foundation.

Availability of data and materials

The datasets used and/or analyzed during the current study are available from the corresponding author on reasonable request.

About this supplement

This article has been published as part of *BMC Bioinformatics Volume 20 Supplement 3, 2019: Selected articles from the 17th Asia Pacific Bioinformatics Conference (APBC 2019): bioinformatics*. The full contents of the supplement are available online at https://bmcbioinformatics.biomedcentral.com/articles/supplements/volume-20-supplement-3.

Authors' contributions

MX, HL and RJ provided guidance and planning for this project. ZF provided the data used in the current study and offered guidance on the data. Ran Li and XZ proposed and implemented the methods, analysed the results and wrote the manuscript. SS, Ruogu Lin, BZ, CL and KW helped with writing and revising the manuscript. All authors read and approved the final manuscript.

Ethics approval and consent to participate

Not applicable.

Consent for publication

Not applicable.

Competing interests

The authors declare that they have no competing interests.

Publisher's Note

Author details

[1]Department of Automation, Tsinghua University, Beijing, China. [2]Computational Biology Department, Carnegie Mellon University, Pittsburgh, PA, USA. [3]Department of Cellular, Molecular and Biophysical Studies, Columbia University Medical Center, New York, NY, USA. [4]Robotics Institute, Carnegie Mellon University, Pittsburgh, PA, USA. [5]Department of Electrical and Computer Engineering, Carnegie Mellon University, Pittsburgh, PA, USA. [6]Departments of Psychiatry and Cell Biology, University of Pittsburgh, Pittsburgh, PA, USA.

Published: 29 March 2019

References

1. Irobalieva RN, Martins B, Medalia O. Cellular structural biology as revealed by cryo-electron tomography. J Cell Sci. 2016;129(3):469–76.
2. Woodward CL, Mendonĉa LM, Jensen GJ. Direct visualization of vaults within intact cells by electron cryo-tomography. Cell Mol Life Sci. 2015;72(17):3401–9.
3. Elad N, Volberg T, Patla I, Hirschfeld-Warneken V, Grashoff C, Spatz JP, et al. The role of integrin-linked kinase in the molecular architecture of focal adhesions. J Cell Sci. 2013;126(18):4099–107.
4. Grossman E, Medalia O, Zwerger M. Functional Architecture of the Nuclear Pore Complex. Annu Rev Biophys. 2012;41(1):557–584. PMID:22577827.
5. Berdanier CD. Mitochondria in health and disease. Boca Raton: CRC Press; 2005.
6. Asano S, Engel BD, Baumeister W. In Situ Cryo-Electron Tomography: A Post-Reductionist Approach to Structural Biology. J Mol Biol. 2016;428(2, Part A):332–343. Study of biomolecules and biological systems: Proteins.
7. Volkmann N. A novel three-dimensional variant of the watershed transform for segmentation of electron density maps. J Struct Biol. 2002;138(1):123–9.
8. Cyrklaff M, Risco C, Fernández JJ, Jiménez MV, Estéban M, Baumeister W, et al. Cryo-electron tomography of vaccinia virus. Proc Natl Acad Sci. 2005;102(8):2772–7.
9. Martinez-Sanchez A, Garcia I, Fernandez JJ. A differential structure approach to membrane segmentation in electron tomography. J Struct Biol. 2011;175(3):372–83.
10. Sandberg K, Brega M. Segmentation of thin structures in electron micrographs using orientation fields. J Struct Biol. 2007;157(2):403–15.
11. Loss LA, Bebis G, Chang H, Auer M, Sarkar P, Parvin B. Automatic Segmentation and Quantification of Filamentous Structures in Electron Tomography. In: Proceedings of the ACM Conference on Bioinformatics, Computational Biology and Biomedicine. BCB '12. New York: ACM; 2012. p. 170–177.
12. Xu M, Alber F. Automated target segmentation and real space fast alignment methods for high-throughput classification and averaging of crowded cryo-electron subtomograms. Bioinformatics. 2013;29(13):i274–82.
13. Zeng X, Leung MR, Zeev-Ben-Mordehai T, Xu M. A convolutional autoencoder approach for mining features in cellular electron cryo-tomograms and weakly supervised coarse segmentation. J Struct Biol. 2018;202(2):150–60.
14. Luengo I, Darrow MC, Spink MC, Sun Y, Dai W, He CY, et al. SuRVoS: Super-Region Volume Segmentation workbench. J Struct Biol. 2017;198(1):43–53.
15. Chen M, Dai W, Sun SY, et al. Convolutional neural Networks for automated annotation of cellular cryo-electron tomograms. Nat Methods. 2017;14(10):983–985.
16. Girshick R, Donahue J, Darrell T, Malik J. Rich feature hierarchies for accurate object detection and semantic segmentation. In: 2014 IEEE Conference on Computer Vision and Pattern Recognition. Columbus: IEEE; 2013. p. 580–587.
17. Ren S, He K, Girshick R, Sun J. Faster R-CNN: Towards Real-Time Object Detection with Region Proposal Networks. In: Cortes C, Lawrence ND, Lee DD, Sugiyama M, Garnett R, editors. Advances in Neural Information Processing Systems 28. Red Hook: Curran Associates, Inc.; 2015. p. 91–99.
18. Xu M, Papageorgiou DP, Abidi SZ, Dao M, Zhao H, Karniadakis GE. A deep convolutional neural network for classification of red blood cells in sickle cell anemia. PLoS Comput Biol. 2017;13(10):e1005746.
19. Wang W, Taft DA, Chen YJ, Zhang J, Wallace CT, Xu M, et al. Learn to segment single cells with deep distance estimator and deep cell detector. arXiv preprint arXiv:180310829. 2018.
20. Tomasi C, Manduchi R. Bilateral filtering for gray and color images. In: Sixth International Conference on Computer Vision (IEEE Cat. No.98CH36271). Bombay: IEEE; 1998. p. 839–846.
21. He K, Zhang X, Ren S, Sun J. Deep Residual Learning for Image Recognition. In: The IEEE Conference on Computer Vision and Pattern Recognition (CVPR). Las Vegas: IEEE; 2016. p. 770–778.
22. Keras-frcnn HY. GitHub . 2017. https://github.com/yhenon/keras-frcnn. Accessed 25 July 2018.
23. Deng J, Dong W, Socher R, Li LJ, Li K, Fei-Fei L. ImageNet: A Large-Scale Hierarchical Image Database. In: CVPR09. Miami: IEEE; 2009.
24. Tzutalin. LabelImg. GitHub. 2015. https://github.com/tzutalin/labelImg. Accessed 05 Apr 2018.
25. Thong JT, Sim KS, Phang JC. Single-image signal-to-noise ratio estimation. Scanning. 23(5):328–336.
26. Chollet F, et al. Keras. GitHub. 2015. https://github.com/fchollet/keras. Accessed 25 July 2018.
27. Abadi M, Barham P, Chen J, Chen Z, Davis A, Dean J, et al. TensorFlow: A system for large-scale machine learning. In: 12th USENIX Symposium on Operating Systems Design and Implementation (OSDI 16). Berkeley: USENIX Association; 2016. p. 265–283.
28. Kingma DP, Ba J. Adam: A Method for Stochastic Optimization. arXiv preprint arXiv:14126980. 2014.

Shan *et al. BMC Bioinformatics* 2019, **20**(Suppl 3):126
https://doi.org/10.1186/s12859-019-2651-6

BMC Bioinformatics

RESEARCH Open Access

Identification of trans-eQTLs using mediation analysis with multiple mediators

Nayang Shan[1,2], Zuoheng Wang[3*] and Lin Hou[1,2,4*]

From The 17th Asia Pacific Bioinformatics Conference (APBC 2019)
Wuhan, China. 14-16 January 2019

Abstract

Background: Mapping expression quantitative trait loci (eQTLs) has provided insight into gene regulation. Compared to cis-eQTLs, the regulatory mechanisms of trans-eQTLs are less known. Previous studies suggest that trans-eQTLs may regulate expression of remote genes by altering the expression of nearby genes. Trans-association has been studied in the mediation analysis with a single mediator. However, prior applications with one mediator are prone to model misspecification due to correlations between genes. Motivated from the observation that trans-eQTLs are more likely to associate with more than one cis-gene than randomly selected SNPs in the GTEx dataset, we developed a computational method to identify trans-eQTLs that are mediated by multiple mediators.

Results: We proposed two hypothesis tests for testing the total mediation effect (TME) and the component-wise mediation effects (CME), respectively. We demonstrated in simulation studies that the type I error rates were controlled in both tests despite model misspecification. The TME test was more powerful than the CME test when the two mediation effects are in the same direction, while the CME test was more powerful than the TME test when the two mediation effects are in opposite direction. Multiple mediator analysis had increased power to detect mediated trans-eQTLs, especially in large samples. In the HapMap3 data, we identified 11 mediated trans-eQTLs that were not detected by the single mediator analysis in the combined samples of African populations. Moreover, the mediated trans-eQTLs in the HapMap3 samples are more likely to be trait-associated SNPs. In terms of computation, although there is no limit in the number of mediators in our model, analysis takes more time when adding additional mediators. In the analysis of the HapMap3 samples, we included at most 5 cis-gene mediators. Majority of the trios we considered have one or two mediators.

Conclusions: Trans-eQTLs are more likely to associate with multiple cis-genes than randomly selected SNPs. Mediation analysis with multiple mediators improves power of identification of mediated trans-eQTLs, especially in large samples.

Keywords: Trans-eQTL, Mediation analysis, Multiple mediators

Background

Expression quantitative trait loci (eQTLs) are genetic variants that influence expression levels of mRNA transcripts. Cis-eQTLs commonly refer to genetic variations that act on local genes (Fig. 1a), and trans-eQTLs are those that act on distant genes and genes residing on different chromosomes (Fig. 1b). Identification of eQTLs can help advance our understanding of genetics and regulatory mechanisms of gene expression in various organisms [1]. Consistent findings suggest that many genes are regulated by nearby single nucleotide polymorphisms (SNPs), and the identified cis-eQTLs are typically close to transcription start sites. In contrast to cis-eQTLs, trans-eQTL identification is much more challenging because a greater number of SNP-gene pairs are tested for trans-association. In order to achieve the same power, analysis of trans-eQTLs requires a much larger sample

* Correspondence: zuoheng.wang@yale.edu; houl@tsinghua.edu.cn
[3]Department of Biostatistics, Yale School of Public Health, New Haven, CT 06510, USA
[1]Center for Statistical Science, Tsinghua University, Beijing 100084, China
Full list of author information is available at the end of the article

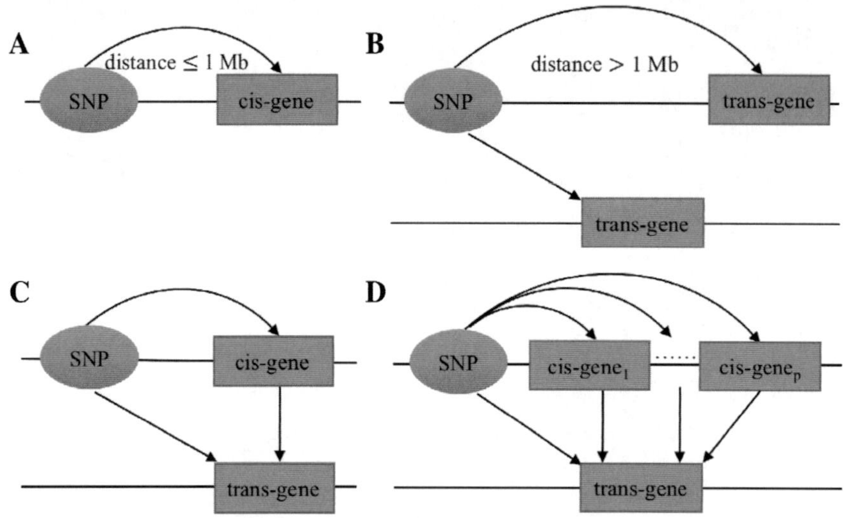

Fig. 1 Graphical representation of eQTLs. **a** cis-eQTL, **b** trans-eQTL, **c** mediated trans-eQTL with a single cis-mediator, and **d** mediated trans-eQTL with multiple cis-mediators

size and/or effect than that in the cis-eQTL analysis. However, trans-eQTLs tend to have weaker effects than cis-eQTLs [2]. Several methods have been developed to improve trans-eQTL detection, such as reducing the multiple-testing burden based on pairwise partial correlations from the gene expression data to increase power [3], and constructing or selecting variables to control for unmeasured confounders that may lead to spurious association [4–6].

Moreover, the biological mechanisms underlying trans-eQTLs are less understood. Previous studies have shown that trans-eQTLs are more likely to be cis-eQTLs than randomly selected SNPs in the human genome [2, 7], suggesting that trans-eQTLs may regulate expression of remote genes by altering the expression of nearby genes. Recently, mediation analysis has become a popular tool to explore trans-association mediated by cis-regulators [2, 6, 8]. These studies used mediation test assuming a single mediator (Fig. 1c). However, gene expression levels are not independent due to the complex regulatory mechanisms. Correlation between genes may violate the assumptions that are required to identify mediation effects if other cis-genes also affect the trans-gene in study. Mediation analysis with multiple mediators has been applied in genomics [9–11], epigenetics [12], and epidemiological studies [13]. Mediation with two mediators was used in [9, 10, 13] and mediation with high dimensional mediators was implemented in [11, 12].

In this paper, we showed that the assumptions in the multivariate extension of mediation analysis are more likely to be satisfied than that in the single-mediator model (Additional file 1). We also found that trans-eQTLs are more likely to associate with more than one cis-gene than randomly selected SNPs in various tissues from the GTEx database. Then, we developed a computational method to identify trans-eQTLs that are mediated by multiple mediators (Fig. 1d). In simulation studies, we demonstrated that the multiple mediator approach increases the statistical power of identification of mediated trans-eQTLs. The improvement is more pronounced in large sample size. We applied the method to the HapMap3 dataset and identified 11 mediated trans-eQTLs that were not detected by the single mediator analysis in the combined samples of African populations. Lastly, we illustrated that mediated trans-eQTLs are more likely to be trait-associated SNPs in genome-wide association studies (GWAS). These findings advance our knowledge of gene regulation.

Methods

Dataset description

Genotype and gene expression data were retrieved from six HapMap3 populations, LWK (Luhya in Webuye, Kenya), MKK (Maasai in Kinyawa, Kenya), YRI (Yoruba in Ibadan, Nigeria), CEU (Utah residents with Northern and Western European ancestry from the CEPH collection), CHB (Han Chinese in Beijing, China), and JPT (Japanese in Tokyo, Japan) [14]. There are 83, 135, 107, 107, 79, and 81 individuals in each population, respectively. The greater genetic diversity in African populations (LWK, MKK, YRI) tends to increase the power of eQTL detection [15]. Therefore, we performed analyses in the three populations separately and in the combined samples. Due to the sample size below 100 in CHB and JPT, we combined the two populations into a sample of Asian populations. Processed expression data profiling on the Illumina Human-6 v2 Expression BeadChip array for the HapMap3 samples were downloaded from ArrayExpress (accession numbers E-MTAB-264 and E-MTAB-198). We also downloaded the V6p release from the GTEx database, which provides a complete list of cis- and trans-eQTLs identified in the GTEx study [16].

Genotype data processing

In the quality control step, a series of filters were applied to remove samples and SNPs with poor quality in each population. We removed samples with a call rate less than 0.97; next we retained autosomal SNPs with a missing rate less than 0.08 and minor allele frequency (MAF) greater than 0.10; finally, SNPs that failed the Hardy-Weinberg test (p-value $<10^{-5}$) were removed. We then converted the SNP coordinates according to the human reference genome hg38. In addition, we found that some SNPs were in complete Linkage Disequilibrium (LD) with each other or mapped to identical genome positions. For such cases, we randomly selected one SNP to be included in the analysis. The number of individuals and SNPs before and after quality control were listed in Table 1. In total, there are 740,158 SNPs retained in the combined samples of African populations and 540,684 SNPs in the combined samples of Asian populations.

Gene expression data processing

There are 21,800 probes in the microarray gene expression in the HapMap3 samples. Among them, 20,439 probes were mapped to the reference genome. We then removed probes that were mapped to multiple genes or non-autosomes, resulting in 19,832 probes corresponding to 19,643 unique genes. We further removed probes with low variance or low intensity and performed quantile normalization to reduce inter-individual variation [17]. Mediation analysis was applied to the probe level data, mainly because multiple probes in a gene represent different isoforms of this gene and merging them may lose information. More importantly, probes mapped to the same gene were weakly correlated in the HapMap3 data.

Population stratification and confounders in gene expression data

In the single population analysis (LWK, MKK, YRI, CEU), we adopted the strategy of [18] to correct for population admixture in LWK and MKK. We used the EIGENSTRAT program [19] to select the top 10 principal components (PCs) generated from the SNP genotype data as covariates. In the combined samples of African populations and Asian populations, 20 PCs from the genotype data were included in the analysis. To adjust for batch effects and unmeasured confounders in the gene expression datasets, we used the probabilistic estimation of expression residuals (PEER) method [20]. Following the GTEx analysis [21], the number of factors for PEER was determined by the sample size. We included 15 factors for datasets with less than 150 samples, 30 factors for datasets with sample size between 150 and 250, and 35 factors for datasets with more than 250 samples. Gender was also included as a covariate in all analyses.

eQTL analysis

We conducted genome-wide eQTL analysis using the R package, Matrix eQTL [22]. SNPs and probes within 1 Mb were tested for cis-association. All inter-chromosomal SNP-probe pairs as well as intra-chromosomal SNP-probe pairs that are more than 1 Mb apart were tested for trans-association.

Enrichment analysis

The motivation of our work was based on the observation that many trans-eQTLs are also identified as cis-eQTLs and they are often associated with more than one cis-gene in the GTEx database. In order to test whether the association with multiple cis-genes is over-represented in trans-eQTLs, we compared the proportion of trans-eQTLs that are associated with more than one cis-gene with that in the human genome. We considered the trans-eQTLs reported in the GTEx V6p dataset and those identified in the HapMap3 dataset. Permutation tests were used to assess significance. To

Table 1 The number of individuals and SNPs before and after quality control in the HapMap3 data

	Sample Size	# SNPs before quality control	# SNPs after quality control
LWK	83	1533540	953834
MKK	135	1541375	989807
YRI	107	1505108	943161
LWK+MKK+YRI	325	NA[a]	740158
CEU	107	1416121	787357
CHB	79	1332120	675811
JPT	81	1300764	643419
CHB+JPT	160	NA	540684

[a]*NA*: data is not available

elaborate, for the trans-eQTLs reported in the GTEx V6p dataset, we randomly sampled the same number of SNPs with matched MAF from the 1000 Genomes Project [23] and calculated the proportion of SNPs that are associated with multiple cis-genes. The empirical *p*-value was obtained by resampling 1000 times. The same test procedure was applied to the trans-eQTLs identified in the HapMap3 dataset.

To understand the role of mediated trans-eQTLs in disease association, we performed Fisher's exact test to assess the enrichment of trait-associated SNPs in the trans-eQTLs identified by our method. The trait-associated SNPs were obtained from the NHGRI GWAS catalog [24].

Mediation analysis

To identify trans-eQTLs that are mediated by one or more cis-genes, we first selected candidate trios, composed of SNP, one or multiple cis-genes, and trans-gene. The trios were selected based on the following criteria. First, trans SNP-gene pairs were selected if their *p*-value is less than 10^{-6}. The *p*-value cutoff was chosen to reduce the multiple-testing burden [6]. Second, cis-genes that are associated with the SNPs identified from the first step at a genome-wide false discovery rate (FDR) less than 0.05 were selected as candidate cis-mediators.

In all the analyses described below, we assume gene expression data have been normalized and transformed so that the expression values approximately follow a normal distribution. In mediation analysis with a single mediator, we followed the test procedure in [25]. The bootstrap *p*-value was used to assess significance when testing the single mediation effect (SME). In mediation analysis with multiple mediators, we considered the following model. For the ith subject, let Y_i be the expression level of a trans-gene, X_i be the SNP genotype coded by the number of minor alleles, $\boldsymbol{M}_i = (M_{i1}, \cdots\cdots, M_{ip})^T$be the expression levels of the p cis-genes, $\boldsymbol{C}_i = (C_{i1}, \cdots\cdots, C_{iq})^T$be the q covariates. The mediation model is stated below:

$$Y_i = \beta_0 + X_i\beta_X + \boldsymbol{M}_i^T\boldsymbol{\beta}_M + \boldsymbol{C}_i^T\boldsymbol{\beta}_C + \varepsilon_{Y_i}$$

$$M_{ij} = \alpha_{0j} + X_i\alpha_{Xj} + \boldsymbol{C}_i^T\boldsymbol{\alpha}_{Cj} + \varepsilon_{M_{ij}} \tag{1}$$

where $\boldsymbol{\beta}_M = (\beta_{M_1}, \cdots\cdots, \beta_{M_p})^T$ is the effect of the p cis-genes on the trans-gene adjusting for the SNP and covariates, $\boldsymbol{\alpha}_X = (\alpha_{X1}, \cdots\cdots, \alpha_{Xp})^T$ is the effect of the SNP on the p cis-genes adjusting for covariates. ε_{Y_i} and $\varepsilon_{M_{ij}}$ are measurement errors on gene expression. Here we assume $\varepsilon_{Y_i} \sim N(0, \sigma^2)$, $\boldsymbol{\varepsilon}_{M_i} = (\varepsilon_{M_{i1}}, \cdots\cdots, \varepsilon_{M_{ip}})^T \sim N_p(\boldsymbol{0}, \boldsymbol{\Sigma})$, and ε_{Y_i} and $\varepsilon_{M_{ij}}$ are independent, but we allow dependence among cis-genes, i.e., the off-diagonal elements in the covariance matrix $\boldsymbol{\Sigma}$ can be non-zero [11].

Denote the total mediation effect (TME) as $\Delta = \boldsymbol{\alpha}_X^T\boldsymbol{\beta}_M$ and the component-wise mediation effects (CME) as $\boldsymbol{\delta} = (\delta_1, \cdots\cdots, \delta_p)^T$, where $\delta_j = \alpha_{Xj}\beta_{M_j}$ [11]. In the following, we focus on the hypothesis tests of TME and CME:

$$H_0 : \Delta = 0 \tag{2}$$

$$H_0 : \boldsymbol{\delta} = \boldsymbol{0} \tag{3}$$

where (2) consists of a broader class of null than (3). For example, when δ_j 's are nonzero in different directions and sum to 0, the TME is zero while the CME is not. Thus, the CME test is of particular interest in the presence of the cancellation effect, which is evident in the HapMap3 dataset. That is, if a SNP has a positive mediation effect through one cis-gene and a negative mediation effect through another cis-gene, the CME test can be more powerful than the TME test, as demonstrated in the simulation studies.

Conventional multivariate tests for CME, such as likelihood ratio test, have limited power when there are a large number of mediators [26]. In our problem, we were less concerned because there are 1 or 2 cis-mediators in majority of the trios (see results in Additional file 2, Additional file 3, Additional file 4, Additional file 5, Additional file 6 and Additional file 7). We used the bootstrap method to assess significance. For comparison, we also tested the SME for each mediator in the trios that have multiple mediators, and the mediation effect was considered to be significant if at least one of the SME tests is significant.

Simulation setup

We conducted simulation studies to evaluate the impact of model misspecification on type I error and statistical power. In detail, we considered three types of model misspecification: Scenario I, the true model has only one mediator while the analysis includes the true mediator and another irrelevant variable as the mediators; Scenario II, the true model has two mediators and the mediation effects are in the same direction; Scenario III, the true model has two mediators and the mediation effects are in opposite direction. In all three scenarios, the performance of the TME, CME, and SME tests are evaluated and compared. We considered sample size of 100 and 300 to mimic the sample size in the HapMap3 single population analysis and combined analysis.

Scenario I: The MAF of the SNP is set to 0.3. For the cis-regulatory effect in model (1), α_{X1} varies from 0.2 to 1, α_{X2} is fixed at 0.6, and $\alpha_{01} = \alpha_{02} = \beta_0 = 0.5, \beta_{M_2} = 0$, $\beta_X = 0.3$. We assume an exchangeable covariance structure for $\boldsymbol{\varepsilon}_M$ with the variance being 1 and the correlation coefficient being 0.2, and ε_Y follows the standard normal distribution. We set $\beta_{M_1} = 0$ in the type I error experiments and $\beta_{M_1} = 0.1$ in the power evaluation. The parameters are chosen to mimic the effects estimated in the HapMap3 dataset.

Scenario II: We set $\beta_{M_2} = 0$ and 0.1 to evaluate the type I error and the power respectively. The other parameters are set the same as in Scenario I.

Scenario III: We set $\beta_{M_2} = 0$ and –0.1 to evaluate the type I error and the power respectively. The other parameters are set the same as in Scenario I.

Results

Trans-eQTLs are more likely to associate with multiple cis-genes

Previous studies showed that trans-eQTLs are more likely to associate with cis-genes [2, 7], which lays the foundation for the employment of mediation analysis in trans-eQTL studies. To justify multiple mediators, we hypothesized that trans-eQTLs tend to associate with more than one cis-gene, and validated this hypothesis in the GTEx dataset. In 14 out of the 22 tissues available in the GTEx database, trans-eQTLs were found to be significantly associated with two or more cis-genes, and the sample sizes are all greater than 100 (Table 2). In the remaining 8 tissues, sample size is less than 100 in 4 tissues, and no more than 3 trans-associations were observed in 5 tissues. Consistent with the GTEx dataset, we also observed an enrichment of multiple cis-genes in trans-eQTLs in MKK, YRI, CEU, and the combined samples of African populations and Asian populations in the HapMap3 dataset (Table 3). The only exception is the LWK population, possibly because the power of identifying cis- and trans-eQTLs is limited at the sample size of 83. Thus, multiple mediators are prevalent among trans-eQTLs. In the upcoming sections, we developed and evaluated statistical tests to identify trans-eQTLs in a multiple-mediator setup, and then applied the method in the HapMap3 dataset.

Simulation studies

In simulations, we studied the effect of model misspecification in three scenarios of trans-eQTL identification (see Simulation setup in Methods) from two perspectives, the type I error and the statistical power, and we compared three tests, TME, CME, and SME.

In Scenario I, the type I error rates did not differ significantly from the nominal level of 0.05 even though a

Table 2 Enrichment results in different tissues in the GTEx database

Tissue	Sample size	Cis-association (FDR<0.05)	Trans-association (FDR<0.1)	Trans-association[a]	Trans-association[b]	Empirical *p*-value
Adipose subcutaneous	298	1282841	45	10	10	0.009
Adrenal gland	126	396098	1	0	0	1
Artery aorta	197	853794	288	2	243	<0.001
Artery tibial	256	1210709	12	3	1	0.156
Brain hypothalamus	81	150415	2	0	0	1
Brain nucleus accumbens (basal ganglia)	93	244929	2	0	0	1
Brain putamen (basal ganglia)	82	183240	11	0	0	1
Cells transformed fibroblasts	272	1283340	658	19	376	<0.001
Colon transverse	169	581854	18	0	6	0.006
Esophagus mucosa	241	1089061	980	122	145	<0.001
Esophagus muscularis	218	997653	15	5	10	0.001
Heart left ventricle	190	605253	3	3	0	1
Lung	278	1068860	98	0	8	0.041
Muscle skeletal	361	1100532	59	0	35	<0.001
Nerve tibial	256	1454889	30	10	1	0.288
Pancreas	149	515665	283	0	244	<0.001
Prostate	87	177994	1	0	0	1
Skin not sun-exposed (suprapubic)	196	722868	11	0	6	0.004
Skin sun-exposed (lower leg)	302	1306762	64	13	25	<0.001
Testis	157	1121727	203	34	68	0.013
Thyroid	278	1551668	2120	230	1390	<0.001
Whole blood	338	1036239	35	2	24	<0.001

[a]Trans-association in which the trans-eQTL is also associated with 1 cis-gene
[b]Trans-association in which the trans-eQTL is also associated with 2 or more cis-genes

Table 3 Enrichment results in the HapMap3 data

Population	Sample size	Cis-association (FDR<0.05)	Trans-association (FDR<0.1)	Trans-association[a]	Trans-association[b]	Empirical *p*-value
LWK	83	6838	7	1	0	1
MKK	135	17889	46	6	3	<0.001
YRI	107	18239	51	18	10	<0.001
LWK+MKK+YRI	325	56437	192	35	64	<0.001
CEU	107	26506	210	18	60	<0.001
CHB+JPT	160	42953	135	56	29	<0.001

[a]Trans-association in which the trans-eQTL is also associated with 1 cis-gene
[b]Trans-association in which the trans-eQTL is also associated with 2 or more cis-genes

second mediator was falsely included in the analysis (Fig. 2a), and the results were consistent in all three tests we considered. In terms of power, as we expected, the SME test (true model) achieved the highest power, while the TME and CME tests had reduced power due to falsely including a cis-gene that does not mediate the trans-association (Fig. 3a). However, the power difference between SME and TME quickly diminishes as the mediated effect of the true mediator increases. In the mediated trans-eQTL problem, we pre-select trios for mediation test, and there is no guarantee that false mediators are excluded at this step. However, as shown in simulations, the type I error was under control, at the expense of power loss.

In Scenarios II and III, the null hypotheses are identical for each test respectively, thus we presented the type I error results in one graph (Fig. 2b). We can see that the type I error was under control when either one or both mediators are included in the model. In terms of power, the TME test was more powerful than the CME test in Scenario II when the two mediation effects are in the same direction. In contrast, the TME test was less powerful than the CME test in Scenario III when the two mediation effects are in opposite direction. The SME test lost power due to leaving out one of the two cis-mediators, and its power fell between that of TME and CME (Fig. 3b, c). It is noteworthy that for the SME test, we actually performed two separate tests for each of the mediators, and the rejection of either one leads to the final rejection. The results were consistent with [11], where similar directionality effect was reported. The power difference of the three tests increases as the sample size increases. When the mediation effects were in different direction, the power of the TME test declined until α_{X1} reached 0.6 when the two mediation effects were cancelled. After that, the power of the TME test

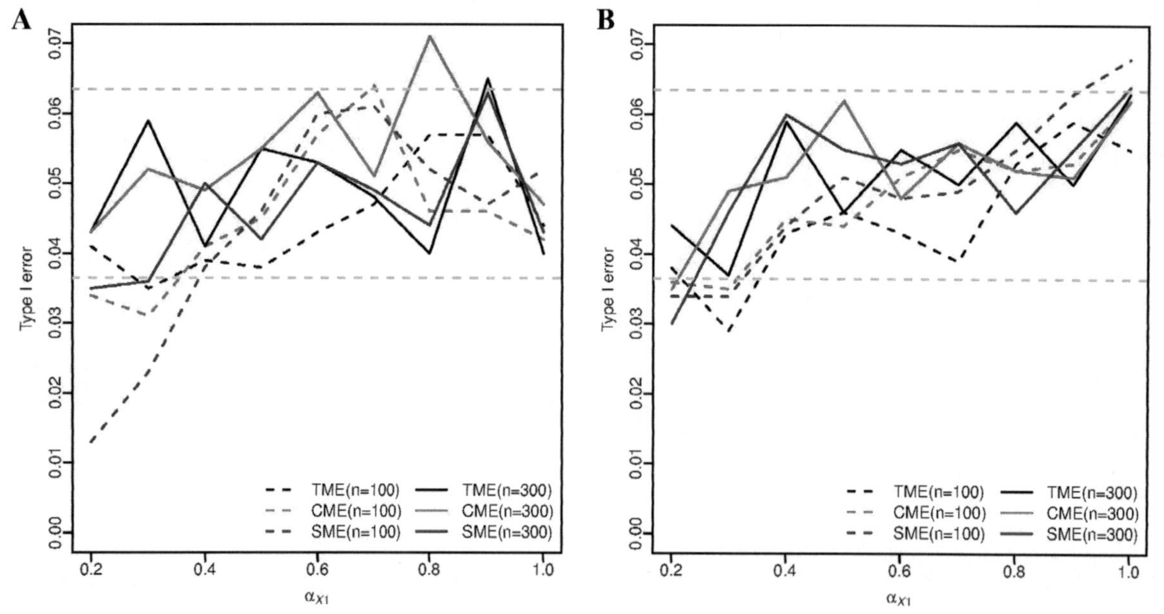

Fig. 2 Empirical type I error of the TME, CME, and SME tests, based on 1,000 simulation replicates, α=0.05. **a** type I error in Scenario I, and **b** type I error in Scenario II/III. The two horizontal grey dashed lines are the 95% confidence interval (0.0365-0.0635)

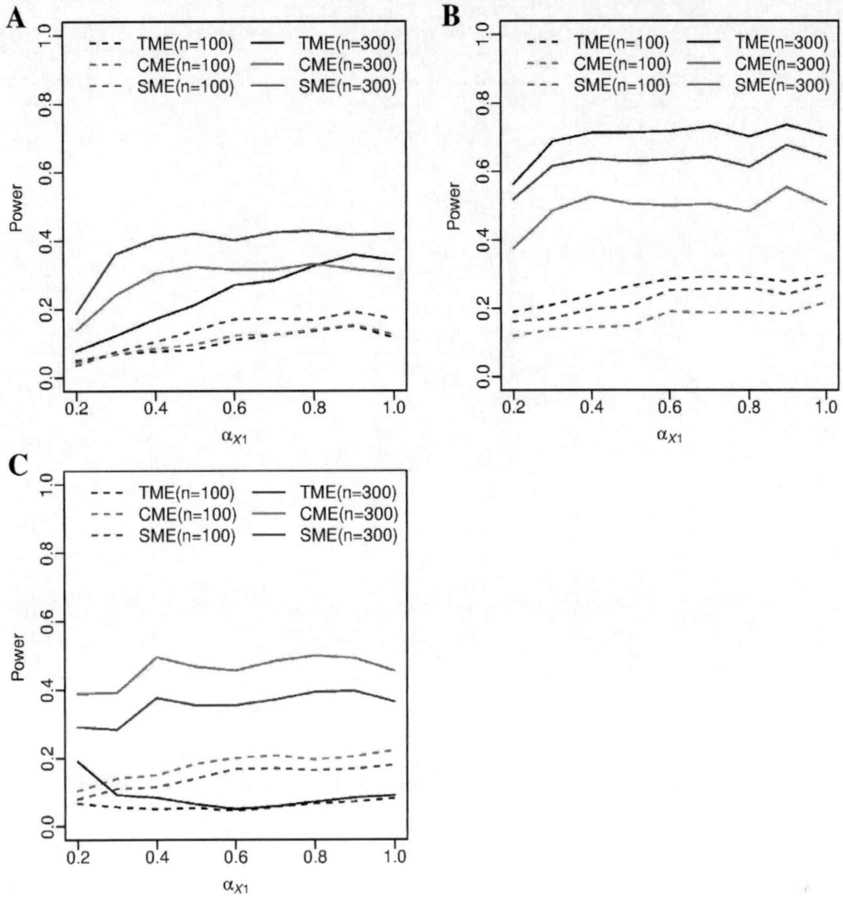

Fig. 3 Empirical power of the TME, CME, and SME tests, based on 1,000 simulation replicates, α=0.05. **a** power in Scenario I, **b** power in Scenario II, and **c** power in Scenario III

rose with the value of α_{X1}, but was still inferior to that of CME and SME (Fig. 3c). In summary, the single mediator model loses power when multiple mediators are present, and the optimal choice of the hypothesis test depends on the unknown directionality of the mediation pathways.

Identification of mediated trans-eQTLs: application to the HapMap3 dataset

We applied the mediation tests to LWK, MKK, YRI, CEU, and the combined samples of African populations and Asian populations in the HapMap3 dataset. In each population, mediated trans-eQTLs with p-values less than 0.05 are shown in Table 4 (more details in Additional file 2, Additional file 3, Additional file 4, Additional file 5, Additional file 6 and Additional file 7). The three tests gave similar results in the single population analysis, perhaps due to the small sample size. In the combined samples of African populations, 291 (24.3%) trans-eQTLs were associated with two or more cis-genes. Among the 248 trans-eQTLs associated with two cis-genes, 70 trios were identified by both the TME and CME tests, 13 trios in which the estimated mediation effects were in the same direction were identified by the TME test but not the CME test, and 17 trios in which the estimated mediation effects were in opposite direction were identified by the CME test but not the TME test. All the 89 trios detected by the SME test were also identified by either the TME or CME test. In total, we identified 11 mediated trans-eQTLs that were not detected by the single mediator analysis (Table 5). In the Asian populations, 254 (26.1%) trans-eQTLs were associated with two or more cis-genes. Among the 195 trans-eQTLs associated with two cis-genes, 33 trios were identified by both the TME and CME tests, 12 trios in which the estimated mediation effects were in the same direction were identified by the TME test but not the CME test, 2 trios were identified by the CME test but not the TME test, and 4 trios in which the estimated

Table 4 Mediated trans-eQTLs with *p*-value <0.05 in the HapMap3 data

Population	Sample Size	# cis-mediators	# trios tested	# trios with *p*-value<0.05			
				TME	CME	TME+CME	SME
LWK	83	1	139	NA[a]	NA	NA	8
		2	26	5	3	5	4
MKK	135	1	426	NA	NA	NA	54
		2	32	4	6	6	7
		4	2	0	0	0	0
YRI	107	1	463	NA	NA	NA	43
		2	47	10	11	11	11
		3	8	5	5	5	5
LWK+MKK+YRI	325	1	905	NA	NA	NA	77
		2	248	83	87	100	89
		3	35	6	11	11	11
		4	7	1	2	2	2
		5	1	1	1	1	1
CEU	107	1	527	NA	NA	NA	60
		2	215	57	61	75	73
		3	14	1	3	3	2
CHB+JPT	160	1	721	NA	NA	NA	64
		2	195	45	39	51	45
		3	41	4	3	4	3
		4	13	1	1	1	1
		5	5	0	0	0	0

[a]*NA*: the tests are not applicable

mediation effects were in opposite direction were identified by the CME test but not the TME test. All the 45 trios detected by the SME test were also identified by either the TME or CME test. There are 6 mediated trans-eQTLs that were not detected by the single mediator analysis. Similar results were obtained when the trans-eQTL *p*-value threshold was set at 10^{-7} (data not shown).

Replication of trans-eQTLs and mediated trans-eQTLs

We demonstrated the replication of trans-eQTLs from LWK, MKK, YRI, and the combined samples of African

Table 5 Mediated trans-eQTLs that were detected by the multiple mediator analysis but not detected by the single mediator analysis in the combined samples of African populations

SNP	Chr (SNP)	Position	Cis-gene$_1$	Cis-gene$_2$	Chr (trans-gene)	Trans-gene
rs2024679	6	29259340	*ZKSCAN3*	*PGBD1*	17	*NCOR1*
rs3117327	6	29271373	*ZKSCAN3*	*PGBD1*	17	*NCOR1*
rs3135392	6	32441465	*HLA-DRB5*	*HLA-DRB1*	4	*RPL34*
rs2239804	6	32443746	*HLA-DRB5*	*HLA-DRB1*	4	*RPL34*
rs9270623	6	32597554	*HLA-DRB5*	*HLA-DRB1*	4	*RPL34*
rs642093	6	32614298	*HLA-DRB5*	*HLA-DRB1*	4	*RPL34*
rs2097431	6	32623056	*HLA-DRB5*	*HLA-DRB1*	12	*ATP5MFP5*
rs9272105	6	32632222	*HLA-DRB5*	*HLA-DRB1*	4	*RPL34*
rs10987642	9	127411687	*SLC2A8*	*ZNF79*	17	*RPL12P38*
rs10511793	9	26924623	*CAAP1*	*IFT74*	7	*BRI3*
rs2835187	21	35967194	*SETD4*	*CBR1*	3	*PCOLCE2*

populations. When the FDR was controlled at 0.1, the trans-eQTLs identified in LWK, MKK, YRI, and the combined samples have a large overlap (Additional file 8). Among the 7 trans-eQTLs identified in LWK, all of them were also identified in another population or the combined samples. Among the 46 trans-eQTLs identified in MKK, 23 of them were also identified in another population or the combined samples. Among the 51 trans-eQTLs identified in YRI, 30 of them were also identified in another population or the combined samples. Additionally, we compared the results with that from a previous study in which the FDR of trans-eQTLs was set at 0.05 [2]. There were 2, 5, 5, and 20 trans-eQTLs identified by our method in LWK, MKK, YRI, and the combined samples respectively that were previously reported (more details in Additional file 9). The relatively low rates of replication with the previous study may be explained by genetic and environmental differences between populations [2]. Next, we evaluated the replication of mediated trans-eQTLs across populations (Additional file 10). 7 of 13 mediated trans-eQTLs identified in LWK were also identified in another population or the combined samples. 13 of 61 mediated trans-eQTLs identified in MKK were also identified in another population or the combined samples. 10 of 59 mediated trans-eQTLs identified in YRI were also identified in another population or the combined samples. For those trans-eQTLs that have inconsistent mediation across populations, it may be due to different gene regulatory mechanisms between populations [27]. Lastly, we observed that trait-associated SNPs are enriched in the mediated trans-eQTLs identified in the combined samples of African populations and Asian populations (Table 6).

Table 6 Enrichment results of trait-associated SNPs in the mediated trans-eQTLs identified in the HapMap3 data

Population	Mediated trans-eQTL	Trait association		Enrichment *p*-value
		Yes	No	
LWK	Yes	2	11	0.013
	No	13180	940641	
MKK	Yes	3	56	0.055
	No	14365	975383	
YRI	Yes	2	53	0.172
	No	12836	930270	
LWK+MKK+YRI	Yes	18	155	1.484×10^{-10}
	No	10901	729084	
CEU	Yes	24	106	$<2.2 \times 10^{-16}$
	No	13403	773824	
CHB+JPT	Yes	15	98	1.019×10^{-9}
	No	9144	531427	

Examples of mediated trans-eQTLs

The identified trans-eQTLs that are mediated by multiple mediators may bring new biological insight in gene regulation. For example, the *RPL34* gene on chromosome 4 was found to be trans-associated with 5 SNPs on chromosome 6 through the mediation of *HLA-DRB5* and *HLA-DRB1* (Table 5). *RPL34* was previously reported to be trans-associated with the SNP rs2395185 in human monocytes [28], and the association was found unique to TLR4 activation, which plays a key role in innate immunity [29]. However, the biological mechanism underlying this trans-association is unknown. Our study identified the mediated trans-association of *RPL34* and the SNP rs2239804 which is in LD with rs2395185 (r^2 = 0.53), suggesting a mediating pathway of the previously reported trans-regulation of *RPL34* (Fig. 4). The SNP rs2395185 and the two cis-mediators, *HLA-DRB5* and *HLA-DRB1*, were reported to be susceptible to ulcerative colitis [30], and the two HLA genes were also identified in the rheumatoid arthritis GWAS [31]. The dysfunction

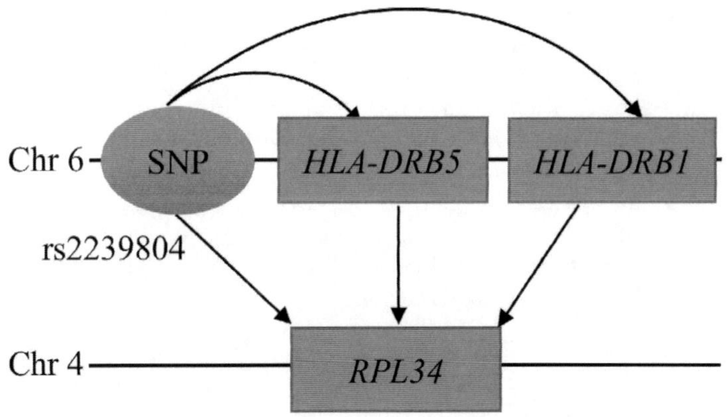

Fig. 4 Mediation diagram of the trans-association between rs2239804 and *RPL34*

of innate immunity is critically important in the pathogenesis of ulcerative colitis [32] and rheumatoid arthritis [33]. Thus, the identified mediated trans-eQTLs not only suggest a biological mechanism for the trans-association of rs2239804 and *RPL34*, but also suggest a role of the mediated pathway in the disease etiology of ulcerative colitis and rheumatoid arthritis.

Discussion

eQTL studies have shed enormous light on gene regulatory mechanisms. Significant progress has been made to integrate eQTL information with genome-wide association signals to explain SNP-phenotype associations and prioritize genes and variants for functional studies [34–36]. The ongoing efforts such as GTEx and the HapMap Project have greatly expanded current knowledge of eQTLs. However, the identification and interpretation of trans-eQTLs remain a challenging yet important topic. In this work, we developed a computational method to identify trans-eQTLs that are mediated by multiple mediators, and demonstrated its superiority to the single mediator test in mediation analysis.

Previous studies considered the identification of cis-transcripts that mediate the effects of trans-eQTLs on distant genes in a single-mediator setting [2, 6, 8], and may be subject to potential model misspecification. One innovative aspect of our work is to employ the multiple-mediator analysis to identify mediated trans-eQTLs. We observed that the associations of trans-eQTLs with more than one cis-gene are prevalent in the GTEx and HapMap3 datasets. Thus, mediation analysis allowing for multiple mediators would be less sensitive to model misspecification, and as a result improve the statistical power of the tests. Applied to the HapMap3 data, our approach allowing for multiple mediators identified 11 mediated trans-eQTLs that were not detected in the single mediator analysis.

There are several caveats in our work. First, unmeasured confounders may not be fully accounted for in the mediation analysis due to the biological complexity in gene regulatory networks. The influence of potential confounders was further evaluated in the single-mediator setting [37]. Sensitivity analysis in the mediation with multiple mediators will be investigated in future studies. Second, we cannot make causal claims based on the detected mediation effects because the observed mediations simply explain trans-associations but not establish causal relationships. Third, the selection of cis-gene mediators is completely data-driven in the current study. It would be of great interest to integrate the knowledge of gene networks into the mediation framework.

Conclusions

We implemented a multiple-mediator analysis approach to identify mediated trans-eQTLs. In simulation studies, we illustrated that our method improves the statistical power of identification of mediated trans-eQTLs compared to the single mediator analysis. Furthermore, we identified 11 mediated trans-eQTLs that were not detected by the single mediator analysis in the HapMap3 data.

Additional files

Additional file 1: Assumptions in mediation analysis. This document explaining that the assumptions in the multivariate extension of mediation analysis are more likely to be satisfied than that in the single-mediator model. (DOCX 14 kb)

Additional file 2: Mediated trans-eQTLs in LWK. (XLS 64 kb)

Additional file 3: Mediated trans-eQTLs in MKK. (XLS 73 kb)

Additional file 4: Mediated trans-eQTLs in YRI. (XLS 71 kb)

Additional file 5: Mediated trans-eQTLs in the combined samples of African populations. (XLS 106 kb)

Additional file 6: Mediated trans-eQTLs in CEU. (XLS 94 kb)

Additional file 7: Mediated trans-eQTLs in the combined samples of Asian populations. (XLS 91 kb)

Additional file 8: Venn diagram of trans-eQTLs in African populations. (PNG 81 kb)

Additional file 9: Overlap of trans-eQTLs with a previous study [2]. (XLS 52 kb)

Additional file 10: Venn diagram of mediated trans-eQTLs in African populations. (PNG 82 kb)

Abbreviations

CEU: Utah residents with Northern and Western European ancestry from the CEPH collection; CHB: Han Chinese in Beijing, China; CME: Component-wise mediation effect; eQTL: Expression quantitative trait locus; FDR: False discovery rate; GWAS: Genome-wide association study; JPT: Japanese in Tokyo, Japan; LD: Linkage disequilibrium; LWK: Luhya in Webuye, Kenya; MAF: Minor allele frequency; MKK: Maasai in Kinyawa, Kenya; PC: Principal component; PEER: Probabilistic estimation of expression residuals; SME: Single mediation effect; SNP: Single nucleotide polymorphism; TME: Total mediation effect; YRI: Yoruba in Ibadan, Nigeria

Acknowledgements

We thank two reviewers for providing thoughtful and constructive comments to improve the manuscript.

Funding

This study was supported by the National Natural Science Foundation of China grant No. 11601259 (LH), and the National Institutes of Health grant K01AA023321 (ZW). The publication of this article was sponsored by National Natural Science Foundation of China (Grant No. 11601259).

Availability of data and materials

The gene expression data analyzed in this study are available at ArrayExpress E-MTAB-264, https://www.ebi.ac.uk/arrayexpress/experiments/E-MTAB-264/ and E-MTAB-198, https://www.ebi.ac.uk/arrayexpress/experiments/E-MTAB-198/. The genotype data analyzed in this study are available at HapMap phase III, ftp://ftp.ncbi.nlm.nih.gov/hapmap/genotypes/2009-01_phaseIII/plink_format/.

About this supplement

This article has been published as part of *BMC Bioinformatics Volume 20 Supplement 3, 2019: Selected articles from the 17th Asia Pacific Bioinformatics Conference (APBC 2019): bioinformatics*. The full contents of the supplement are available online at https://bmcbioinformatics.biomedcentral.com/articles/supplements/volume-20-supplement-3.

Authors' contributions
LH conceived the project, LH and ZW designed the project, NS performed the analysis. All authors interpreted the data and wrote the manuscript. All authors have read and approved the final version of the manuscript.

Ethics approval and consent to participate
Not applicable.

Consent for publication
Not applicable.

Competing interests
The authors declare that they have no competing interests.

Publisher's Note

Springer Nature remains neutral with regard to jurisdictional claims in published maps and institutional affiliations.

Author details
[1]Center for Statistical Science, Tsinghua University, Beijing 100084, China. [2]Department of Industrial Engineering, Tsinghua University, Beijing 100084, China. [3]Department of Biostatistics, Yale School of Public Health, New Haven, CT 06510, USA. [4]MOE Key Laboratory of Bioinformatics, School of Life Sciences, Tsinghua University, Beijing 100084, China.

Published: 29 March 2019

References

1. Veyrieras JB, Kudaravalli S, Kim SY, Dermitzakis ET, Gilad Y, Stephens M, et al. High-resolution mapping of expression-QTLs yields insight into human gene regulation. PLoS Genet. 2008;4:e1000214.
2. Pierce BL, Tong L, Chen LS, Rahaman R, Argos M, Jasmine F, et al. Mediation analysis demonstrates that trans-eQTLs are often explained by cis-mediation: a genome-wide analysis among 1,800 South Asians. PLoS Genet. 2014;10:e1004818.
3. Weiser M, Mukherjee S, Furey TS. Novel distal eQTL analysis demonstrates effect of population genetic architecture on detecting and interpreting associations. Genetics. 2014;198:879–93.
4. Stegle O, Parts L, Durbin R, Winn J. A Bayesian framework to account for complex non-genetic factors in gene expression levels greatly increases power in eQTL studies. PLoS Comput Biol. 2010;6:e1000770.
5. Rakitsch B, Stegle O. Modelling local gene networks increases power to detect trans-acting genetic effects on gene expression. Genome Biol. 2016;17:33.
6. Yang F, Wang J, Consortium G, Pierce BL, Chen LS. Identifying cis-mediators for trans-eQTLs across many human tissues using genomic mediation analysis. Genome Res. 2017;27:1859–71.
7. Westra HJ, Peters MJ, Esko T, Yaghootkar H, Schurmann C, Kettunen J, et al. Systematic identification of trans eQTLs as putative drivers of known disease associations. Nat Genet. 2013;45:1238–43.
8. Yao C, Joehanes R, Johnson AD, Huan T, Liu C, Freedman JE, et al. Dynamic role of trans regulation of gene expression in relation to complex traits. Am J Hum Genet. 2017;100:571–80.
9. Zhao SD, Cai TT, Li H. More powerful genetic association testing via a new statistical framework for integrative genomics. Biometrics. 2014;70:881–90.
10. Huang YT. Integrative modeling of multi-platform genomic data under the framework of mediation analysis. Stat Med. 2015;34:162–78.
11. Huang YT, Pan WC. Hypothesis test of mediation effect in causal mediation model with high-dimensional continuous mediators. Biometrics. 2016;72:402–13.
12. Zhang H, Zheng Y, Zhang Z, Gao T, Joyce B, Yoon G, et al. Estimating and testing high-dimensional mediation effects in epigenetic studies. Bioinformatics. 2016;32:3150–4.
13. Huang YT, Yang HI. Causal mediation analysis of survival outcome with multiple mediators. Epidemiology. 2017;28:370–8.
14. The International HapMap3 Consortium. Integrating common and rare genetic variation in diverse human populations. Nature. 2010;467:52–8.
15. Brynedal B, Choi J, Raj T, Bjornson R, Stranger BE, Neale BM, et al. Large-scale trans-eQTLs affect hundreds of transcripts and mediate patterns of transcriptional co-regulation. Am J Hum Genet. 2017;100:581–91.
16. The GTEx Consortium. Genetic effects on gene expression across human tissues. Nature. 2017;550:204–13.
17. Gentleman RC, Carey VJ, Bates DM, Bolstad B, Dettling M, Dudoit S, et al. Bioconductor: open software development for computational biology and bioinformatics. Genome Biol. 2004;5:R80.
18. Stranger BE, Montgomery SB, Dimas AS, Parts L, Stegle O, Ingle CE, et al. Patterns of cis regulatory variation in diverse human populations. PLoS Genet. 2012;8:e1002639.
19. Price AL, Patterson NJ, Plenge RM, Weinblatt ME, Shadick NA, Reich D. Principal components analysis corrects for stratification in genome-wide association studies. Nat Genet. 2006;38:904–9.
20. Stegle O, Parts L, Piipari M, Winn J, Durbin R. Using probabilistic estimation of expression residuals (PEER) to obtain increased power and interpretability of gene expression analyses. Nat Protoc. 2012;7:500–7.
21. Consortium G, et al. The Genotype-Tissue Expression (GTEx) pilot analysis: Multitissue gene regulation in humans. Science. 2015;348:648–60.
22. Shabalin AA. Matrix eQTL: ultra fast eQTL analysis via large matrix operations. Bioinformatics. 2012;28:1353–8.
23. The 1000 Genomes Project Consortium. An integrated map of genetic variation from 1,092 human genomes. Nature. 2012;491:56–65.
24. Welter D, MacArthur J, Morales J, Burdett T, Hall P, Junkins H, et al. The NHGRI GWAS Catalog, a curated resource of SNP-trait associations. Nucleic Acids Res. 2014;42:D1001–6.
25. Mackinnon DP, Lockwood CM, Williams J. Confidence limits for the indirect effect: Distribution of the product and resampling methods. Multivariate Behav Res. 2004;39:99.
26. Huang YT, Vanderweele TJ, Lin X. Joint analysis of SNP and gene expression data in genetic association studies of complex diseases. Ann Appl Stat. 2014;8:352–76.
27. Pai AA, Pritchard JK, Gilad Y. The genetic and mechanistic basis for variation in gene regulation. PLoS Genet. 2015;11:e1004857.
28. Kim S, Becker J, Bechheim M, Kaiser V, Noursadeghi M, Fricker N, et al. Characterizing the genetic basis of innate immune response in TLR4-activated human monocytes. Nat Commun. 2014;5:5236.
29. Beutler BATLR. innate immunity. Blood. 2009;113:1399–407.
30. Silverberg MS, Cho JH, Rioux JD, McGovern DP, Wu J, Annese V, et al. Ulcerative colitis loci on chromosomes 1p36 and 12q15 identified by genome-wide association study. Nat Genet. 2009;41:216–20.
31. Eyre S, Bowes J, Diogo D, Lee A, Barton A, Martin P, et al. High-density genetic mapping identifies new susceptibility loci for rheumatoid arthritis. Nat Genet. 2012;44:1336–40.
32. Geremia A, Biancheri P, Allan P, Corazza GR, Di Sabatino A. Innate and adaptive immunity in inflammatory bowel disease. Autoimmun Rev. 2014;13:3–10.
33. Gierut A, Perlman H, Pope RM. Innate immunity and rheumatoid arthritis. Rheum Dis Clin North Am. 2010;36:271–96.
34. Montgomery SB, Dermitzakis ET. From expression QTLs to personalized transcriptomics. Nat Rev Genet. 2011;12:277–82.
35. Gusev A, Ko A, Shi H, Bhatia G, Chung W, Penninx BW, et al. Integrative approaches for large-scale transcriptome-wide association studies. Nat Genet. 2016;48:245–52.
36. Wu M, Lin Z, Ma S, Chen T, Jiang R, Wong WH. Simultaneous inference of phenotype-associated genes and relevant tissues from GWAS data via Bayesian integration of multiple tissue-specific gene networks. J Mol Cell Biol. 2017;9:436–52.
37. Imai K, Keele L, Yamamoto T. Identification, inference and sensitivity analysis for causal mediation effects. Stat Sci. 2010;25:51–71.

Wang et al. BMC Bioinformatics 2019, 20(Suppl 3):135
https://doi.org/10.1186/s12859-019-2652-5

BMC Bioinformatics

RESEARCH Open Access

Constructing effective energy functions for protein structure prediction through broadening attraction-basin and reverse Monte Carlo sampling

Chao Wang[1,2†], Yi Wei[1,2†], Haicang Zhang[1,2], Lupeng Kong[1,2], Shiwei Sun[1,2], Wei-Mou Zheng[3,2*] and Dongbo Bu[1,2*]

From The 17th Asia Pacific Bioinformatics Conference (APBC 2019)
Wuhan, China. 14-16 January 2019

Abstract

Background: The *ab initio* approaches to protein structure prediction usually employ the Monte Carlo technique to search the structural conformation that has the lowest energy. However, the widely-used energy functions are usually ineffective for conformation search. How to construct an effective energy function remains a challenging task.

Results: Here, we present a framework to construct effective energy functions for protein structure prediction. Unlike existing energy functions only requiring the native structure to be the lowest one, we attempt to maximize the attraction-basin where the native structure lies in the energy landscape. The underlying rationale is that each energy function determines a specific energy landscape together with a native attraction-basin, and the larger the attraction-basin is, the more likely for the Monte Carlo search procedure to find the native structure. Following this rationale, we constructed effective energy functions as follows: *i*) To explore the native attraction-basin determined by a certain energy function, we performed reverse Monte Carlo sampling starting from the native structure, identifying the structural conformations on the edge of attraction-basin. *ii*) To broaden the native attraction-basin, we smoothened the edge points of attraction-basin through tuning weights of energy terms, thus acquiring an improved energy function. Our framework alternates the *broadening attraction-basin* and *reverse sampling* steps (thus called BARS) until the native attraction-basin is sufficiently large. We present extensive experimental results to show that using the BARS framework, the constructed energy functions could greatly facilitate protein structure prediction in improving the quality of predicted structures and speeding up conformation search.

Conclusion: Using the BARS framework, we constructed effective energy functions for protein structure prediction, which could improve the quality of predicted structures and speed up conformation search as well.

Keywords: Protein structure prediction, Energy function, Attraction-basin, Reverse Monte Carlo sampling, Monte Carlo search, Linear program

*Correspondence: zheng@itp.ac.cn; dbu@ict.ac.cn
†Chao Wang and Yi Wei contributed equally to this work.
[3]Institute of Theoretical Physics, Chinese Academy of Sciences, 55, Zhongguancun East Road, 100190 Beijing, China
[2]University of Chinese Academy of Sciences, 19-1, Yuquan Road, Shijingshan, 100049 Beijing, China
Full list of author information is available at the end of the article

Background

Determination of protein structure is important for understanding protein functions [1]. The classical techniques for protein structure determination include X-ray crystallography, nuclear magnetic resonance, and electron microscopy. These determination techniques, however, often suffer from the limitations in both expensive costs and long determination period, leading to the ever-increasing gap between the number of known protein sequences and that of solved protein structures [2]. Computational approaches to protein structure prediction from sequences are becoming increasingly important to narrow down the gap [3].

The protein structure prediction approaches can be categorized into two families, namely, template-based modeling [4–10] and *ab initio* approaches [11–15]. Recently the predicted contacts have also been shown to be invaluable to protein structure prediction [16–21]. Unlike the template-based modeling approaches, the *ab initio* prediction approaches work without requirements of known similar protein structures. Briefly speaking, most *ab initio* prediction approaches are based on the hypothesis that the native structure of a protein should be the highly-populated one with sufficiently low energy; thus, *ab initio* approaches usually perform conformation search to find a structural conformation with sufficiently low energy. For example, Rosetta employs the Monte Carlo technique to search conformations assembled from fragments of known structures, and finally reports the centroid of a large cluster of low-energy conformations [11].

For the *ab initio* prediction approaches, one of the key issues is designing an effective energy function [11, 14, 15]. Typically, an energy function is a weighted-sum of multiple energy terms. The energy terms characterize specific structural features, especially the interplay between local and global interactions of residues. For example, the hydrophobic interaction term was designed to capture the observed tendency of non-polar residues to aggregate in aqueous solution and exclude water molecules. Van der Waals force term is the sum of the attractive or repulsive forces among residues. Hydrogen bonding term describes the electromagnetic attractive interaction between polar molecules in which hydrogen is bound to highly electronegative atom oxygen in the carboxyl [1]. In Rosetta, a total of 13 energy terms were used at the residue level, and over 140 terms were used at the full-atom level; therefore, it is important to find the optimal weighting of so many energy terms [11]. This study focuses on designing an optimal weighting of the 13 energy terms used in Rosetta.

Ideally, an effective energy function is expected to be able to distinguish the native structure from non-native conformations (called *decoys*), and could drive as much as possible initial conformations to the native-like one during the conformation search process. To achieve these two objectives, a widely-used strategy for designing energy function is to maximize the correlation between energy and quality of decoys [22]. Here the quality of a decoy refers to the structural similarity between the decoy with the native structure, which is measured using root mean square deviation (RMSD) of backbone atoms in this study. Inspired by the idea of "funnel-shaped free energy surface", Levitt et al. proposed a funnel sculpting technique to construct energy functions that allow the conformation search procedure to easily "roll" into the native structure from a random starting conformation [23]. In another study, Shell et al. attempted to smooth energy function to make the energy landscape a funnel [24].

In this study, we present a framework that constructs effective energy function for protein structure prediction. Our framework, called BARS, consists of two procedures, i.e., *broadening attraction-basin* where the native structure lies in the energy landscape (hereafter denoted as *native attraction-basin*), and *reverse sampling*. The underlying rationale is that the larger the attraction-basin is, the more likely for the Monte Carlo procedure to find the native structure. To explore the attraction basin, we performed reverse sampling starting from the native structure. Subsequently, we tuned the weights of energy terms to broaden the native attraction-basin and thus acquired an improved energy function. We showed that both the possibility of successful search and the quality of predicted conformation increase when using the improved energy function.

The manuscript is organized as follows: "Methods" section describes the whole framework of our method, and the linear program model to optimize protein energy weights as well. "Results and discussion" section lists experimental results of the optimized energy function. In "Conclusion" section, we will discuss some limitations of our method and possible future works.

Methods

To construct an effective energy function, our BARS framework alternates two procedures, i.e., for an energy function, we first explored the native attraction-basin in the corresponding energy landscape using reverse sampling, and then improved the energy function through broadening the attraction-basin. These two procedures were alternated until the energy function changes sufficiently small between successive iterations as below.

The details of the two procedures are described as follows.

Exploring the native attraction-basin using *reverse sampling*

When applying Monte Carlo technique to search the structural conformation with the lowest energy, the *ab initio* approaches might finally end with success if starting

Algorithm 1 BARS framework

Set weights of energy terms W' with initial values;
repeat
    $W = W'$;
    Run *reverse sampling* procedure to identify edge point conformations of the energy landscape corresponding to W;
    Run *broadening attraction-basin* procedure to smoothen the identified edge point conformations and generate a new weight scheme W';
until the difference between W and W' is sufficiently small
return W;

from some initial structural conformations, and might end with failure if starting from other initial structural conformations. An initial structural conformation is said to lie in the native attraction-basin if the conformation can finally evolve into the native structure during the conformation search process.

To explore the native attraction-basin under a specific energy function, we propose a technique called *reverse Monte Carlo sampling.* Here, the term "reverse" comes from the fact that the sampling process works essentially reverse to the general Monte Carlo technique used for conformation search. Specifically, the general Monte Carlo search procedure starts from a random initial conformation and moves towards the native structure, during which the energy of conformation is reducing. For each inter-mediate structural conformation, a perturbation is made to generate a new conformation. Some popular perturbation techniques include fragment replacing used by Rosetta [11] and torsion angle sampling used by FALCON [13]. The newly-generated conformation is accepted if it has lower energy relative to the original conformation; otherwise the new conformation will be rejected with a probability according to the Metropolis-Hasting rule [25] (Fig. 1). To emphasize the difference between the general Monte Carlo search technique and the sampling technique used in this study, we denoted the former one as *forward Monte Carlo* technique.

On the contrary to the forward Monte Carlo search technique, our reverse Monte Carlo sampling procedure starts from the native structure and moves towards the edge of the native attraction-basin, during which the energy of conformations is increasing (Fig. 2). The reverse sampling process ends at a conformation if any perturbation of this conformation cannot leads to increase of energy. Intuitively, this conformation lies at the edge of the native attraction-basin and thus is denoted as *edge point conformation* in this study.

Formally, each execution of reverse sampling will generate a path of conformations $P = S_0 \rightarrow S_1 \rightarrow S_2 \rightarrow \ldots \rightarrow S_n$, where S_0 denotes the native structure, S_i $(1 \leq i \leq n)$ denotes the ith inter-mediate conformations along the reverse sampling path, and S_n denotes the final edge point conformation. The energy of the inter-mediation conformations increases along the path, suggesting the monotonicity of energy within the native attraction-basin. The RMSD between S_0 and S_n is calculated as a rough measure of the radius of native attraction-basin. For the edge point conformation S_n, we perform m times of perturbation, thus acquiring m perturbation neighbors of S_n,

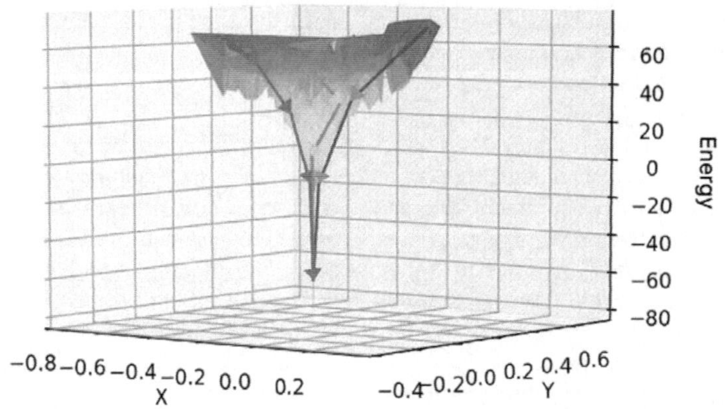

Fig. 1 Conformation searching process using the Monte Carlo technique. The protein structure prediction approaches usually employ the Monte Carlo technique to search the conformation with the lowest energy. An execution of conformation search will generate a path of conformations, e.g., the lines in blue and yellow. Here the energy landscape was drawn using 1000 decoys of protein `1ctfA`: we calculated the RMSD among all possible pairs of decoys, and then performed principal component analysis of the generated RMSD matrix [23]. The *x* and *y* axises represent the first and second principal components, respectively. Decoy energy was calculated using *score3* of Rosetta

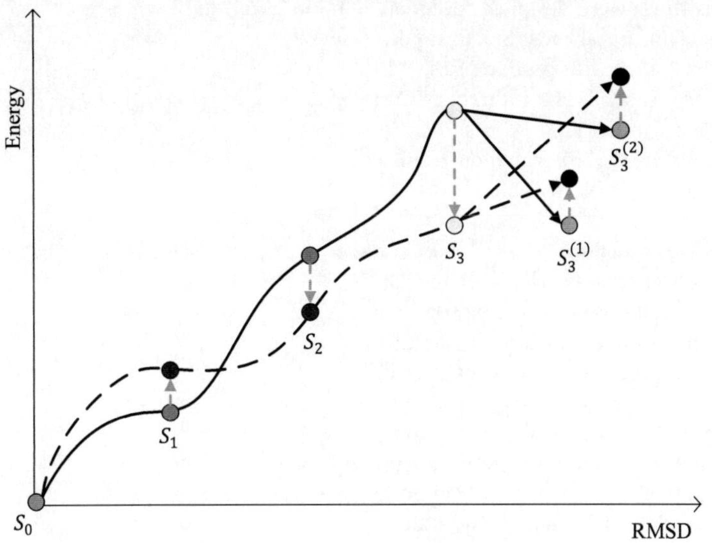

Fig. 2 Reverse Monte Carlo sampling and tuning weights of energy terms. Reverse Monte Carlo sampling starts from the native structure and moves towards the edge of the native attraction-basin, during which the energy of conformations is increasing. Here the solid line shows a path of conformations generated by reverse Monte Carlo sampling $S_0 \rightarrow S_1 \rightarrow S_2 \rightarrow S_3$. S_3 represents an edge point conformation as its perturbation neighbors, e.g., S_3^1 and S_3^2, have lower energy than S_3. The dash line shows these conformations after tuning weights of energy terms, where S_3 is no longer the edge point conformation as one of its perturbation neighbors, S_3^2 has larger energy than S_3. Note that during weight-tuning, the monotonicity of energy is maintained within the native attraction-basin

denoted as $N(S_n) = \left\{S_n^{(1)}, S_n^{(2)}, ..., S_n^{(m)}\right\}$ (Fig. 2). In our study, m is set as 1000.

Broadening the native attraction-basin by smoothening the edge point conformations

Intuitively, if we can "smoothen" the edge point conformations, the native attraction-basin will be broadened since the reverse sampling process will not be stuck at these edge point conformations. We accomplished the "smoothening" operation through tuning weights of energy terms such that the energy of S_n is decreased to be less than at least one of its perturbation neighbors $S_n^{(i)}$ $(1 \leq i \leq m)$. To make the native structure still the lowest one under the new energy function, we imposed a constraint on weight-tuning such that after tuning weights, the new energy of S_i should be lower than that of the conformation S_{i+1}. In other words, the monotonicity of energy are maintained within the attraction-basin, and thus the shape of the native attraction-basin will also be maintained. Figure 2 shows how the conformation path changes after improving energy function.

We designed a linear program to calculate the optimal weights that satisfy this constraint.

$$\min \quad ||W - W_0||$$
$$s.t. \quad W \cdot E_i \leq W \cdot E_{i+1}, \quad 0 \leq i \leq n-1 \tag{1}$$

$$W \cdot E_n \leq \frac{1}{m}\sum_{j=1}^{m} W \cdot E_n^{(j)} \tag{2}$$

$$W \geq 0 \tag{3}$$

$$|W| = |W_0| \tag{4}$$

Here the vector W denotes the weights of energy terms, and W_0 denotes the original weights before tuning. For an inter-mediate conformation S_i in the reverse sampling path, E_i denotes the vector of its energy terms, i.e. $E_i =< e_i^{(1)}, e_i^{(2)}, \ldots, e_i^{(13)} >$, where $e_i^{(k)}$ represents the k-th energy term. The objective of the linear program is to find a new weighting scheme with change as small as possible. Formula (1) describes a constraint that the original relative order of S_i and S_{i+1}, i.e., the monotonicity of energy within the native attraction-basin, should be kept after tuning weights. Formula (2) was designed to "smoothen" the edge point conformation, i.e. at least one of the m perturbation neighbors of the edge point conformation has a higher energy; thus, S_n is no longer an edge point conformation under the new energy function.

Results and discussion

Data set

We evaluated the BARS framework on `Test101` dataset that contains a total of 101 benchmark proteins. The criteria for selecting these proteins are: *i)* The length of these proteins are less than 120 amino acids as the energy terms used in this study were designed for small proteins [11]. *ii)* These proteins cover most SCOP superfamilies as energy functions differ with protein class. We used three proteins as representatives of three SCOP classes (Table 1) to explain the working process of the BARS framework and summarized experimental results on Test101 dataset in Additional file 1: Table S1.

Analysis of evolution of energy functions

As mentioned above, the BARS framework alternates the *broadening attraction-basin* and *reverse sampling* steps to gradually improve energy function. In our experiment, the initial weights of energy terms were set as the weights used by Rosetta in the scoring function *score*3.

Table 2 shows how the energy functions evolve as iteration proceeds for protein 1ctfA. From this table, it can be observed that the weights of energy terms are almost fixed after 6 iterations. In addition, although the difference between consecutive iterations are not very large as expected, the final weights are quite different from the initial ones (Manhattan distance: 15.16).

We also compared the final weights for proteins in different classes. As shown in Table 3, the final weights exhibited considerable difference for proteins from different classes. For example, the weight of Env term in 1iieA (class A) is 5.69, about 4 times larger than that of 1ctfA (class D), and over 2.5 times larger than that of 1iloA (class C). This is consistent with the fact that the environment local geometrical term is more important for all-α proteins, since local residue-residue interactions dominate the helix formation process. In addition, 1ctfA (class D) can be distinguished from the other two proteins at the sheet term: the final weight of this term is 3.48 for 1ctfA, much larger than that of 1iloA (1.90) and 1iieA (1.14). This is also reasonable as α+β proteins usually contain anti-parallel β-sheets, whereas α/β proteins contains β-α-β motifs. Taken together, the table supports the view point that different energy terms are emphasized for proteins in different classes.

Broadening attraction-basin as iteration proceeds

We further investigated whether the native attraction-basin could be broadened after smoothening the edge point conformations. To measure the size of the native attraction-basin, we performed reverse sampling for 50 times, thus acquiring 50 edge point conformations. The mean RMSD between these edge point conformations and the native structure is calculated and used to measure the

Table 1 Three benchmark proteins used in the study

PDB ID	Chain	Class	#Residues	#α helices	#β strands
1ctf	A	α+β	68	4(38)	3(18)
1ilo	A	α/β	77	3(27)	4(18)
1iie	A	all α	75	3(42)	-

The 3 proteins come from 3 different SCOP classes: all α (Class A), α/β (Class C), α+β (class D). Residue numbers are 68, 77, 75, respectively. Columns 5 and 6 shows the number and total length of α helices and β strands

Table 2 Weights of energy terms during the iteration process of BARS

		Weights at iteration					
Energy terms	Initial weights	#1	#2	#3	#4	#5	#6
Env	1.00	1.00	0.97	0.48	0.49	1.26	1.30
Pair	1.00	1.23	1.22	0.64	2.71	2.62	2.59
Vdw	1.00	1.06	0.88	0.61	0.55	0.55	0.55
Hs	1.00	1.70	1.70	1.06	0.09	0.09	0.09
Ss	1.00	1.00	1.00	0.48	0.48	0.48	0.48
Sheet	1.00	1.00	1.00	1.00	3.48	3.48	3.48
R-sigma	1.00	1.00	1.00	0.53	0.53	0.53	0.53
Cb	1.00	0.02	0.01	0.41	0.00	0.00	0.05
Rg	3.00	3.00	3.14	6.05	0.49	0.49	0.49
Contact order	1.00	1.00	1.05	0.74	0.00	0.00	0.00
Ramachandran	0.00	0.00	0.00	0.01	0.91	1.01	1.02
Hb-srbb	0.00	0.00	0.02	0.00	3.27	2.49	2.41
Hb-lrbb	0.00	0.00	0.00	0.00	0.00	0.00	0.00

Here, the initial weights were set as the weights used by Rosetta in *score*3. A total of 6 iterations are shown here. The Manhattan distances of each adjacent weighting are 1.97, 0.44, 6.62, 17.47, 1.74, 0.21, respectively. The iteration process stopped when the Manhattan distance is less than a threshold of 0.3. The cosine of angle of weight vectors at iterations #4 and #5 is 0.98, while that of iterations #5 and #6 reaches 0.99

Table 3 The optimized weights of energy terms calculated by BARS for protein 1ctfA (after 6 iterations), 1iloA (after 5 iterations), 1iieA (after 5 iterations)

Energy terms	Initial weights	Final weights for 1ctfA	Final weights for 1iloA	Final weights for 1iieA
Env	1.00	1.30	1.98	5.69
Pair	1.00	2.59	2.02	1.09
Vdw	1.00	0.55	0.72	0.20
Hs	1.00	0.09	0.82	1.28
Ss	1.00	0.48	0.37	0.20
Sheet	1.00	3.48	1.90	1.14
R-sigma	1.00	0.53	1.11	0.19
Cb	1.00	0.05	0.24	1.62
Rg	3.00	0.49	2.63	0.31
Co	1.00	0.00	0.00	0.62
Ramachandran	0.00	1.02	0.43	0.27
Hb-srbb	0.00	2.41	0.80	0.36
Hb-lrbb	0.00	0.00	0.00	0.00

The Manhattan distance of the weights of 1ctfA and 1iieA is 15.56, and that between 1ctfA and 1iloA is 12.39, while that between 1iloA and 1iieA is only 8.95. The cosine values of weight vectors are 0.63, 0.47, 0.77, respectively size of the native attraction-basin. Intuitively the calculated mean RMSD can be treated as radius of the native attraction-basin.

As shown in Fig. 3, the mean RMSD was 6 Å initially, and increased to nearly 14 Å at the final iteration. This clearly suggested that the attraction basin was really significantly enlarged as iteration proceeds.

Protein structure prediction using the improved energy function

We further investigated whether the improved energy function could facilitate protein structure prediction or not. For this aim, we compared the predicted structures by running Rosetta with different energy functions. Specifically, for each iteration step of BARS, we used the corresponding weighting scheme of energy terms to construct an energy function, and then run Rosetta using this energy function to generate 1000 decoys. We investigated two aspects of these predicted decoys: *i) Quality of final prediction results:* Among these decoys, a clustering procedure was executed and the centroid of the largest cluster was reported as the final prediction. Then we analyzed RMSD of the final prediction with the native structure. *ii) Good decoy ratio:* Among these generated decoys, we calculated the ratio of "good" decoys. Here, we adopted the widely-used criterion that for small proteins, a decoy is called *good decoy* if its RMSD to the native structure is less than 6 Å [13].

As shown in Fig. 4, the quality of the final prediction results improved as iteration proceeded. Taking protein `1iloA` as an example, the final prediction had a RMSD of 2.7 Å when using the original weighting scheme (Fig. 5, left panel). In contrast, when using the optimized weighting scheme, the quality of the final prediction structure improved (RMSD: 1.3 Å, Fig. 5, right panel). We repeated this experiment on the 101 benchmark proteins in `Test101` set and observed that for 82 proteins, the quality of final prediction structure improved (Fig. 6 and Additional file 1: Table S1). For example, the quality of predicted structure for protein `1pchA` was low (RMSD: 11.721 Å) when using the original energy weight; in contrast, when using the optimized weighting scheme, the predicted structure significantly improved (RMSD: 3.417 Å).

Besides the quality of the final prediction, the good decoy ratio also increased considerably (Fig. 7). For example, if using the initial weights for protein `1iloA`, only 11 decoys among the 1000 decoys were good decoys. In contrast, over 200 good decoys were generated when using the optimized weighting scheme. Taken together, these results clearly suggest that using the energy function constructed with BARS, the general Monte Carlo search

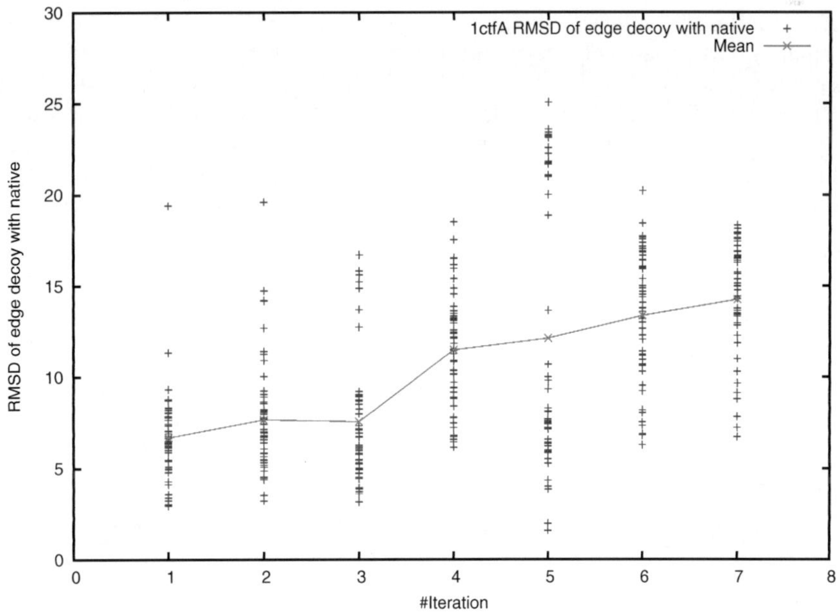

Fig. 3 Broadening the native attraction-basin of protein `1ctfA` as iteration proceeds. Here, *x*-axis denotes the number of iterations, and *y*-axis denotes the RMSD between the native structure and the edge point conformations. For the weighting scheme at each iteration, a total of 50 paths were generated, thus yielding 50 edge point conformations. The mean RMSD increases as iteration proceeds, suggesting that the native attraction-basin was broadened

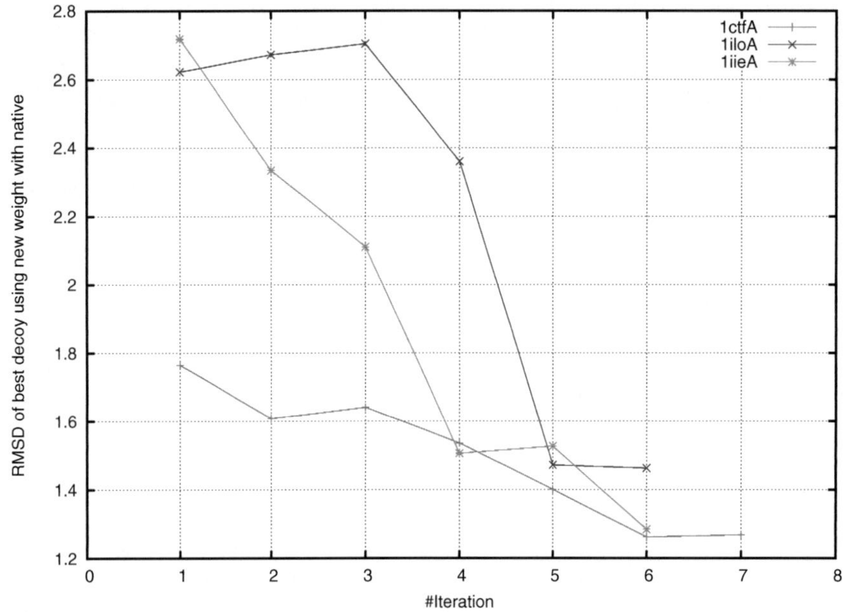

Fig. 4 Quality of the final prediction for protein 1ctfA, 1iloA, and 1iieA. At each iteration step, a total of 1000 decoys were generated by Rosetta with corresponding weights of energy terms. We run clustering procedure for the 1000 decoys and finally selected the centroid of the largest cluster as the best decoy. The RMSD of the best decoys reduces as iteration proceeds, suggesting that the quality of prediction results increases

procedure significantly improved in both success possibility and the quality of prediction results. This improvement should also be attributed to the broadened native attraction-basin.

Application range of the constructed energy functions

Application range is one of the key issues of energy functions. An ideal energy function is expected to be applicable on a large amount of proteins rather than a single protein. To examine the application range of the energy functions acquired using BARS, we run Rosetta with the energy function acquired from protein 1ctfA on other seven benchmark proteins in the same class to 1ctfA. As shown in Table 4, on all of the seven proteins, the predicted structures have high quality (RMSD less than 7 Å). More importantly, on five out of the seven proteins, the prediction results using the energy function acquired from 1ctfA are much better than those predicted using the original weights.

Similarly, we run Rosetta with the energy function acquired from protein 1iieA on other 20 benchmark proteins. As shown in Table 5, on 19 out of the 20 benchmark proteins, the predicted structures have RMSD less than 5 Å. On 18 out of these benchmark proteins, the

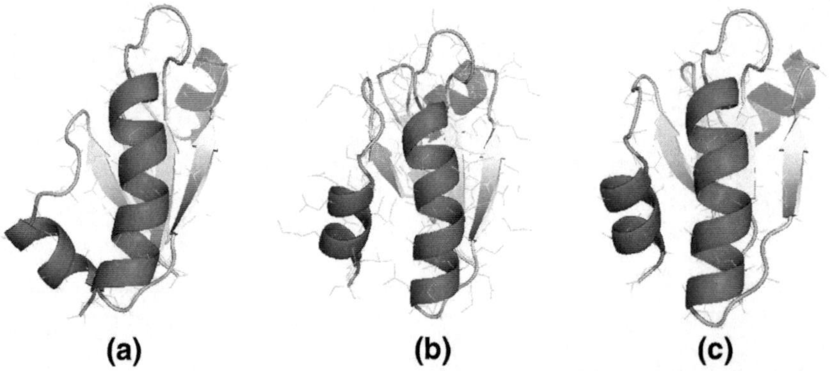

Fig. 5 Predicted structures for protein 1iloA. Left panel: the predicted structures using the initial weights (RMSD: 2.7 Å). Middle panel: native structure. Right panel: the predicted structures using the optimized weights (RMSD: 1.3 Å). Thus, the optimized weights help improve the quality of predicted structures

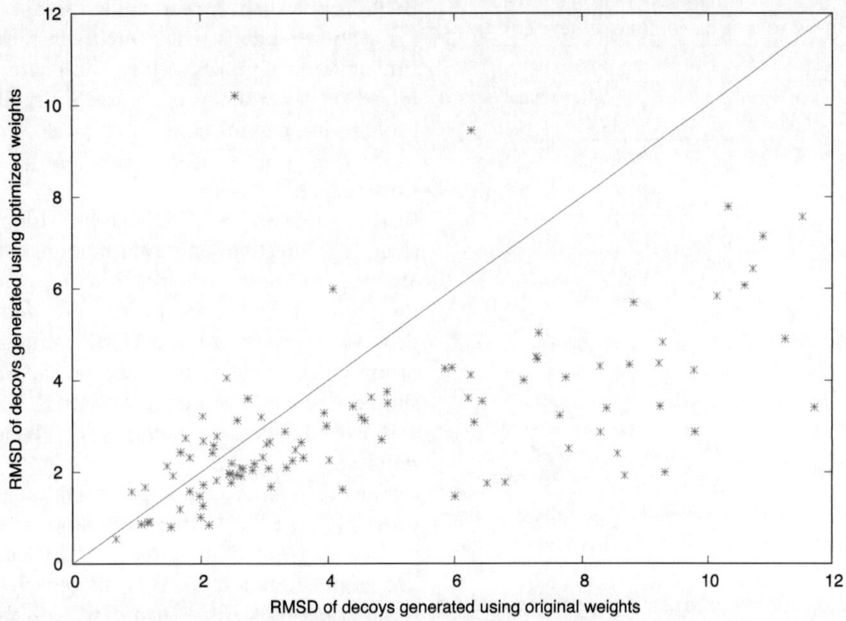

Fig. 6 Quality of predicted structures using the original weights and the optimized weights of energy terms on 101 benchmark proteins of the Test101 dataset. For each benchmark protein, we predicted structures using both original weights and optimized weights of energy terms, and showed RMSD of the predicted structures as *x*-axis and *y*-axis, respectively. Most proteins fall below the diagonal line, suggesting that when using the optimized weights of energy terms, the predicted structures usually have smaller RMSD

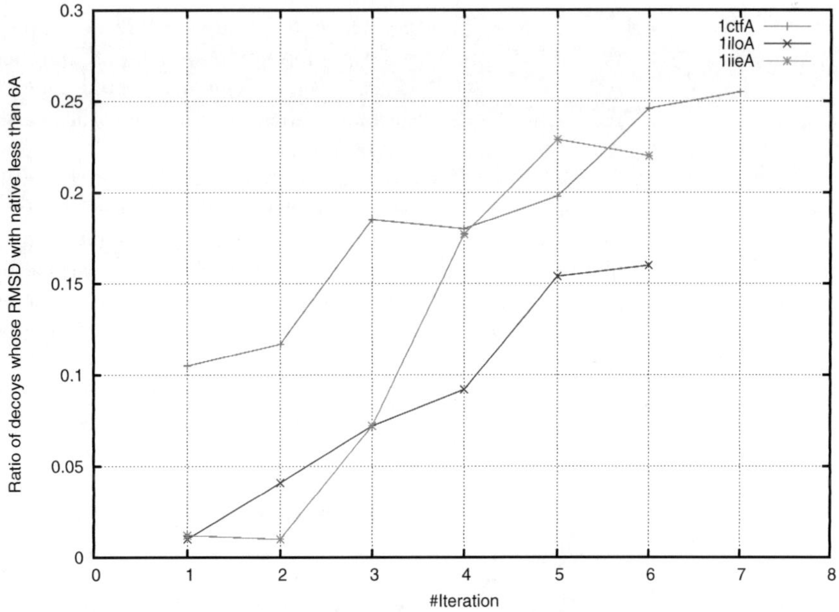

Fig. 7 Good decoy ratio increases as iteration proceeds. At each iteration step, a total of 1000 decoys were generated by Rosetta using corresponding weights of energy terms. Here a decoy is called "good decoy" if it has a RMSD less than 6 Å to the native structure. The figure suggests that the "good decoy ratio" significantly increases, e.g. the ratio increases from 0.01 to over 0.2 for protein `1iloA`. Thus, Rosetta can generate high-quality decoys more efficiently

Table 4 Quality of predicted structures using the original weights of energy terms and the weights acquired from protein 1ctfA

Protein	SCOP family	Quality of the predicted structure	
		Using original weights	Using weights acquired from 1ctfA
1osdA	d.58.17.1	3.221Å	2.053Å
1cpzA	d.58.17.1	2.810Å	2.054Å
1eigA	d.9.1.1	1.786Å	2.071Å
4ubpA	d.8.1.1	7.723Å	3.286Å
1ekzA	d.50.1.1	4.411Å	3.448Å
1dtjA	d.51.1.1	3.726Å	6.249Å
1ulrA	d.58.10.1	10.322Å	6.948Å

On all of the seven benchmark proteins, the predicted structures have RMSD less than 7 Å. On five out of the seven benchmark proteins, the predicted structures using the weights acquired from 1ctfA are better than those predicted using the original weights

predicted structures using the weights acquired from 1iieA are better than those predicted using the original weights.

In summary, these results demonstrate the wide application range of the energy functions acquired using the BARS framework. A reasonable explanation of this wide application range is that proteins in a class might share similar folding process; thus, the energy function optimized on a certain protein is also applicable for other proteins in the same class.

Table 5 Quality of predicted structures using the original weights of energy terms and the weights acquired from protein 1iieA

Protein	SCOP family	Quality of the predicted structure	
		Using original weights	Using weights acquired from 1iieA
1p7iA	a.4.1.1	0.688Å	0.549Å
1am9A	a.38.1.1	2.129Å	1.074Å
1cktA	a.21.1.1	1.675Å	1.147Å
1oqpA	a.39.1.5	1.697Å	1.273Å
1n1jB	a.22.1.3	2.029Å	1.567Å
1nkpA	a.38.1.1	1.991Å	1.645Å
1nkpB	a.38.1.1	2.444Å	1.685Å
1bw5A	a.4.1.1	2.273Å	1.891Å
1of9A	a.64.1.4	1.945Å	2.053Å
1q08A	a.6.1.3	2.584Å	2.316Å
1dgnA	a.77.1.3	10.376Å	2.577Å
1ow5A	a.60.1.2	3.082Å	2.659Å
1aoyA	a.4.5.3	3.035Å	2.756Å
1hstA	a.4.5.13	3.987Å	3.031Å
3ygsP	a.77.1.3	8.804Å	3.14Å
1psrA	a.39.1.2	4.700Å	3.451Å
1pueE	a.4.5.21	9.321Å	3.545Å
1ngrA	a.77.1.2	8.480Å	3.942Å
1hb6A	a.11.1.1	4.138Å	4.542Å
1ctjA	a.3.1.1	9.693Å	6.29Å

On 19 out of the 20 benchmark proteins, the predicted structures have RMSD less than 5 Å. On 18 out of these benchmark proteins, the predicted structures using the weights acquired from 1iieA are better than those predicted using the original weights

Conclusion

In this study we report the BARS framework for constructing effective energy functions. The framework attempts to improve energy function gradually such that the native attraction-basin was broadened. During this process, a reverse Monte Carlo sampling strategy was proposed to explore the native attraction-basin. Extensive experimental results demonstrate both effectiveness and wide application range of the constructed energy functions.

It has been reported that protein folding is a hierarchical process. According to this observation, Rosetta employs a multi-step prediction strategy. In particular, Rosetta first uses score function $score_0$ with only hydrophobic core terms, then uses $score_2/score_5$ with secondary structure terms, and finally uses $score_3$ to incorporate a total of 13 energy terms [11]. This study focuses on the optimization of weights for the third step. How to design effective energy functions for the first and second steps remains as one of our future works. We also noticed that on 19 out of the 101 benchmark proteins, the quality of prediction results using the optimized weighting scheme were low. How to design better energy functions for these proteins is another future work. It should be pointed out that the RMSD deviation of the 50 edge point conformations is large at iteration 5 for protein 1ctfA (Fig. 3). This might imply the irregular shape of the native attraction-basin. We will investigate this issue in future work.

The application of the BARS framework is not limited to protein structure prediction. Constructing an effective scoring function is usually the first important step for optimization problems in various domains such as RNA structure prediction, natural language processing, etc. How to optimally combine multiple terms into a scoring function is a challenging task. Our BARS framework should greatly facilitate designing effective scoring functions for a large variety of problems.

Additional file

Additional file 1: Quality of predicted structures using the original weights and the optimized weights of energy terms on 101 benchmark proteins. (PDF 82 kb)

Abbreviations

BARS: Broadening attraction-basin and reverse sampling; RMSD: Root mean square deviation

Acknowledgements
We would like to thank Torsten Julich for discussion.

Funding
We would like to thank the National Key Research and Development Program of China (2018YFC0910405), and the National Natural Science Foundation of China (31671369 and 31770775) for providing financial supports for this study and publication charges.

Availability of data and materials
The datasets used in this study is available via http://protein.ict.ac.cn/BARS/.

About this supplement
This article has been published as part of *BMC Bioinformatics Volume 20 Supplement 3, 2019: Selected articles from the 17th Asia Pacific Bioinformatics Conference (APBC 2019): bioinformatics*. The full contents of the supplement are available online at https://bmcbioinformatics.biomedcentral.com/articles/supplements/volume-20-supplement-3.

Authors' contributions
DB and WZ conceived the study. CW and YW performed the computation and analysis. HZ and SS analyzed the new framework. CW, LK, WZ, and DB wrote the manuscript. All authors read and approved the final manuscript.

Ethics approval and consent to participate
Not applicable.

Consent for publications
Not applicable.

Competing interests
The authors declare that they have no competing interests.

Publisher's Note

Author details
[1]Key Lab of Intelligent Information Processing, Institute of Computing Technology, Chinese Academy of Sciences, 6, Kexueyuan South Road, Zhongguancun, 100190 Beijing, China. [2]University of Chinese Academy of Sciences, 19-1, Yuquan Road, Shijingshan, 100049 Beijing, China. [3]Institute of Theoretical Physics, Chinese Academy of Sciences, 55, Zhongguancun East Road, 100190 Beijing, China.

Published: 29 March 2019

References

1. Branden C, Tooze J. Introduction to protein structure. New York: Garland Publishing; 1999.
2. Li X, Hu C, Liang J. Simplicial Edge Representation of Protein Structures and Alpha Contact Potential with Confidence Measure. Protein Struct Funct Bioinform. 2003;53(4):792–805.
3. Dill KA. Additivity principles in biochemistry. J Biol Chem. 1997;272:701–4.
4. Schwede T, Guex N, Peitsch MC. SWISSMODEL: an automated protein homology-modeling server. Nucleic Acids Res. 2003;31(13):3381–5.
5. Altschul SF, Madden TL, Schäffer AA, Zhang J, Zhang Z, Miller W, et al. Gapped Blast and PsiBlast: a new generation of protein database search programs. Nucleic Acids Res. 1997;25(17):3389–402.
6. Zhou H, Zhou Y. Fold recognition by combining sequence profiles derived from evolution and from depth-dependent structural alignment of fragments. Proteins. 2004;58(2):321–8.
7. Söding J. Protein homology detection by HMM-HMM comparison. Bioinformatics. 2005;21:951–60.
8. Xu J, Li M, Lin G, Kim D, Xu Y. Protein Threading by Linear Programming. In: Biocomputing: Proceedings of the 2003 Paci Symposium. vol. 8. Pacific Symposium on Biocomputing; 2003. p. 264–75.
9. Wu S, Szilagyi A, Zhang Y. Improving protein structure prediction using multiple sequence-based contact predictions. Structure. 2011;19(8):1182–91.
10. Ma J, Wang S, Wang Z, Xu J. MRFalign: protein homology detection through alignment of Markov random fields. PLoS Comput Biol. 2014;10(3):e1003500.
11. Simons KT, Kooperberg C, Huang E, Baker D. Assembly of protein tertiary structures from fragments with similar local sequences using simulated annealing and Bayesian scoring functions. J Mol Biol. 1997;268(1):209–25.
12. Pauling L, Corey RB. The Pleated Sheet, A New Layer Configuration of Polypeptide Chains. Proc Natl Acad Sci. 1951;37(5):251–6.
13. Li SC, Bu D, Xu J, Li M. Fragment-HMM: A new approach to protein structure prediction. Protein Sci. 2008;17(11):1925–34.
14. Zhang Y, Arakaki AK, Skolnick J. TASSER: An automated method for the prediction of protein tertiary structures in CASP6†. Proteins. 2005;61:91–98.
15. Xu D, Zhang Y. Ab initio protein structure assembly using continuous structure fragments and optimized knowledge based force field. Proteins. 2012;80(7):1715–35.
16. Andreatta M, Laplagne S, Li SC, Smale S. Prediction of residue-residue contacts from protein families using similarity kernels and least squares regularization. 2013. arXiv preprint arXiv:13111301.
17. Di Lena P, Nagata K, Baldi P. Deep architectures for protein contact map prediction. Bioinformatics. 2012;28(19):2449–57.
18. Wang Z, Xu J. Predicting protein contact map using evolutionary and physical constraints by integer programming. Bioinformatics. 2013;29(13):i266–73.
19. Michel M, Hayat S, Skwark MJ, Sander C, Marks DS, Elofsson A. PconsFold: improved contact predictions improve protein models. Bioinformatics. 2014;30(17):i482–8.
20. Zhang H, Gao Y, Deng M, Wang C, Zhu J, Li SC, et al. Improving residue-residue contact prediction via low-rank and sparse decomposition of residue correlation matrix. Biochem Biophys Res Commun. 2016;472(1):217–22.
21. Zhu J, Wang S, Bu D, Xu J. Protein threading using residue co-variation and deep learning. Bioinformatics. 2018;34(13):i263–73.
22. Zhang Y, Kolinski A, Skolnick J. TOUCHSTONE II: A new approach to ab initio protein structure prediction. Biophys J. 2003;85:1145.
23. Fain B, Levitt M. Funnel sculpting for in silico assembly of secondary structure elements of proteins. Proc Natl Acad Sci. 2003;100(19):10700–5.
24. Dill KA, Ozkan SB, Shell MS, Weikl TR. The Protein Folding Problem. Annu Rev Biophys. 2008;37(1):289–316.
25. Anderson HL. Metropolis, Monte Carlo, and the MANIAC. Fall. 1986;14:96–108.

Chen and Zhu *BMC Bioinformatics* 2019, **20**(Suppl 3):129
https://doi.org/10.1186/s12859-019-2653-4

BMC Bioinformatics

RESEARCH Open Access

Detecting virus-specific effects on post-infection temporal gene expression

Quan Chen[1,2] and Jun Zhu[1,2,3*]

From The 17th Asia Pacific Bioinformatics Conference (APBC 2019)
Wuhan, China. 14-16 January 2019

Abstract

Background: Different types of viruses have different envelope proteins, and may have their shared or distinctive host-virus interactions which result in various post-infection effects in humans and animals. These effects often do not appear at once but take time to unfold. To characterize the virus-specific effects, we applied a Multivariate Polynomial Time-dependent Genetic Association (MPTGA) method, previously proposed for detecting differences in temporal gene expression traits, to test for the differences in mouse lung transcriptome response to infection of different subtypes of influenza A viruses.

Results: We compared two methods: the Multivariate Polynomial Time-dependent Genetic Association (MPTGA) method, and the conventional modified t-test, to study the virus-specific effects on mouse lung gene expression. Both methods found H3N2 to be the most different virus among the three viruses tested, with the largest number of genes with H3N2-specific effects. However, the MPTGA method demonstrated much higher power of detection, and the detected genes with virus-specific effects showed better biological relevance.

Conclusions: Transcriptome response to virus infection is dynamic. MPTGA which leverages temporal gene expression traits showed increased power in detecting biologically relevant virus-specific effects comparing with conventional t-test method.

Keywords: Temporal association, Dynamic response, Flu virus

Background

Influenza viruses have various types and subtypes that differ in their morbidity, virulence and pathogenicity. Influenza A virus is the most virulent type and can be further divided into several subtypes [1]. These subtypes also include different strains which may evolve over time and vary in their manifestations: zoonotic, pandemic and seasonal [2]. The highly pathogenic avian influenza H5N1 virus, for example, is a zoonotic influenza virus that is transmitted from poultry to humans. The H5N1 virus strain A/Vietnam/1203/04 was the most pathogenic strain from the 2004 outbreaks of H5N1 influenza viruses [3]. The pandemic H1N1/09 virus strain, on the other hand, has led to the swine-origin H1N1 flu pandemic in 2009. The seasonal H3N2 vaccine seed strain A/Wyoming/03/2003 represents H3N2 viruses isolated during the last three seasons [4].

Previous studies [5] have shown in time course experiments that seasonal H3N2, pandemic H1N1, and zoonotic H5N1 viruses differ strongly in the spatial and temporal dynamics of infection and associated lesions in the ferret respiratory tract. Therefore, to characterize and to differentiate the infection processes with the diverse categories and subtypes of influenza viruses, it is important to acknowledge its highly dynamic nature.

In this paper, we investigated temporal mouse lung gene expression traits in response to infection with influenza A/California/04/09 (H1N1), A/Wyoming/03/03 (H3N2), and A/Vietnam/1203/04 (H5N1) HALo virus (GSE98527) and aimed to identify the virus-specific effects. To

*Correspondence: jun.zhu@mssm.edu
[1]Department of Genetics and Genomic Sciences, Icahn School of Medicine at Mount Sinai, 10029 New York, NY, USA
[2]Icahn Institute for Genomics and Multiscale Biology, Icahn School of Medicine at Mount Sinai, 10029 New York, NY, USA
Full list of author information is available at the end of the article

compare temporal gene expression traits, we proposed a multivariate polynomial temporal genetic association (MPTGA) method [6] previously in the context of genetic association with temporal gene expression. We further extended this method to a more general setting that compares the temporal gene expression trait under different conditions, which applies to the specific aim in this study—to detect the differences in the way each virus affects temporal gene expression traits over time. We also compared the MPTGA method with the conventional modified t-test to detect differential gene expression post infection of different viruses, and demonstrated significant increase in the power and specificity of our temporal method.

Results

Mouse lung dataset

To characterize the cellular transcriptome response of mouse lung to infection of H1N1, H3N2, and H5N1 influenza virus, six to eight week-old female C57BL/6 mice were infected with influenza A/California/04/09 (H1N1), A/Wyoming/03/03 (H3N2), and A/Vietnam/1203/04 (H5N1) HALo virus in 3 experimental groups respectively. The Influenza A/Vietnam/ 1203/04 (H5N1) HALo mutant virus is an attenuated H5N1 virus generated from wild-type Influenza A/Vietnam/1203/04 (H5N1) virus [7]. Lungs were collected from each experimental group with 3 mice infected with the same virus at 12h, 1d, 2d, 3d and 4d post infection. The cellular transcriptome at each time point was profiled by mRNA-Seq technology (GSE98527).

Detecting virus-specific effects on post-infection temporal expression

We applied MPTGA to all 24,421 genes profiled in the mouse lung dataset, testing for differential temporal post-infection gene expression between infection with each virus and the other two. In addition, we applied the moderated t-test using an empirical Bayes method implemented in the R package *Limma* [8] to test for differential expression based on the last time point.

Genes have different temporal trajectories in response to different subtypes of viruses. For example, Fig. 1 shows expression of Mill2 (MHC I like leukocyte 2), an mouse ortholog for the MICB gene in human. Mill2 is highly expressed in mouse lung according to Mouse ENCODE transcriptome data [9]. It involves in immunoregulatory interactions between a lymphoid and a non-lymphoid cell and innate immune system. Its related GO processes include immune response and immune response-activating cell surface receptor signaling pathway. Figures 1 and 2 illustrate the performance of different methods in detecting virus-specific effects on post-infection temporal expression of Mill2. Previous

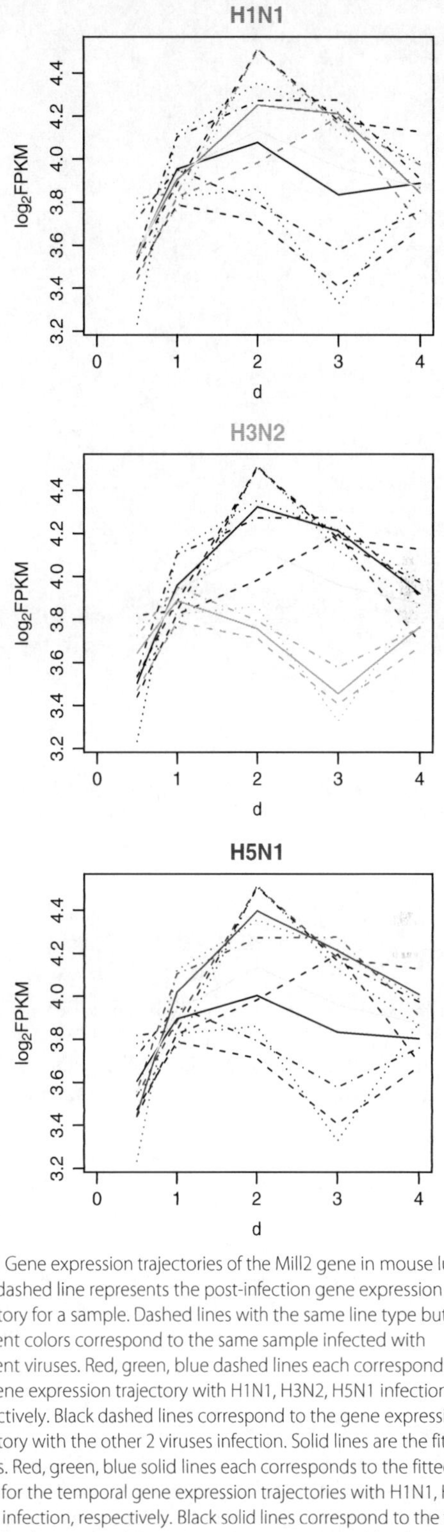

Fig. 1 Gene expression trajectories of the Mill2 gene in mouse lung. Each dashed line represents the post-infection gene expression trajectory for a sample. Dashed lines with the same line type but different colors correspond to the same sample infected with different viruses. Red, green, blue dashed lines each corresponds to the gene expression trajectory with H1N1, H3N2, H5N1 infection, respectively. Black dashed lines correspond to the gene expression trajectory with the other 2 viruses infection. Solid lines are the fitted curves. Red, green, blue solid lines each corresponds to the fitted curve for the temporal gene expression trajectories with H1N1, H3N2, H5N1 infection, respectively. Black solid lines correspond to the fitted curve for the temporal gene expression trajectories with other 2 viruses infection. Yellow solid lines correspond to the fitted curves using the reduced model taking all samples together

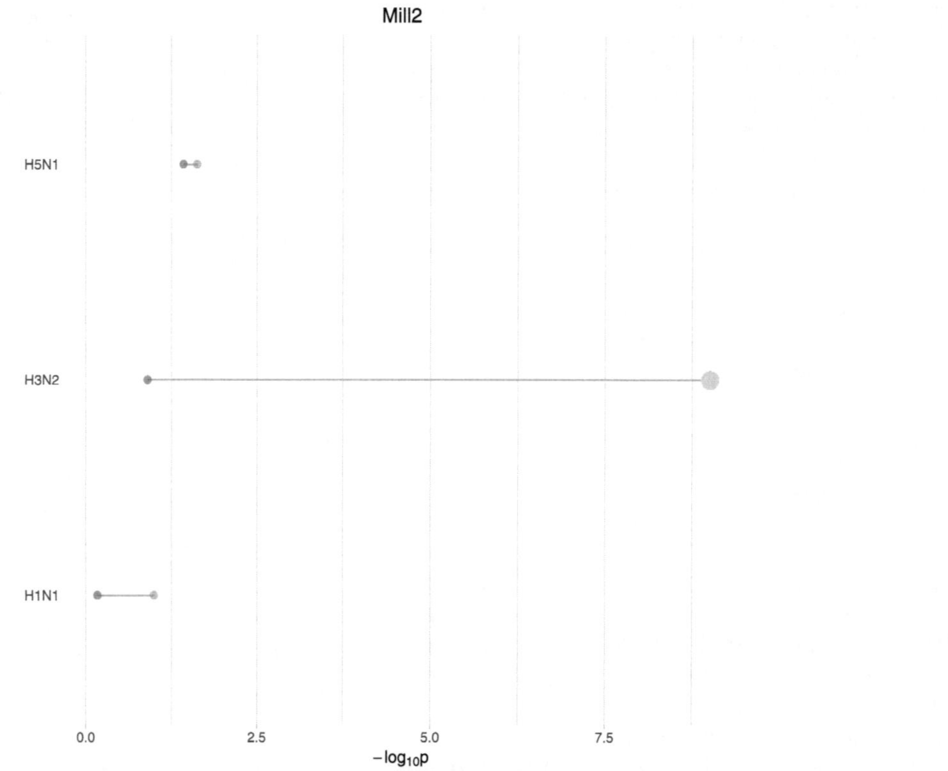

Fig. 2 MPTGA v.s. standard differential expression test on Mill2 gene. The *p*-values of differential temporal gene expression tests on Mill2 post-infection trajectories using MPTGA and the *p*-values of standard differential gene expression tests on Mill2 based on the last time point. Green represents MTPGA, red represents standard differential expression test. Significant *p*-values are highlighted

studies [10–14] have investigated the immune-related functions of Mill2 (or its orthologs) and have characterized the underlying mechanisms. MICB activates the cytolytic response of natural killer (NK) cells and CD8 T-cells. As is shown in Fig. 1 with H3N2 infection, Mill2 expression trajectories demonstrate an evidently distinct pattern from the trajectories post-infection with the other two viruses, consistent with previous observations that H3N2 is less effective in induction of immune response in lung [15]. However, at the last time point, all trajectories return to similar levels of gene expression values, which suggests that the standard differential expression test based on the last time point was not able to detect the difference, whereas the MPTGA method was able to detect the subtype specific effect of post-infection temporal expression of Mill2, as shown in Fig. 2.

Figure 3 provides a summary of the genes identified by MPTGA with differential post-infection temporal gene expression specific to each virus. Figure 4 provides a summary of the genes identified by the conventional differential gene expression test implemented by the R package *Limma* [8] as a moderated t-test based on the

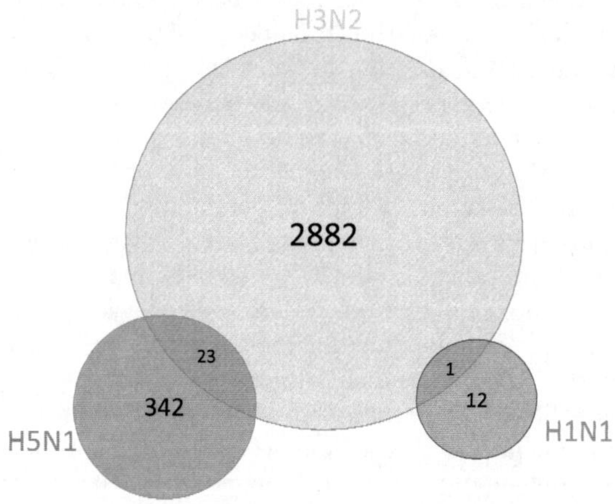

Fig. 3 Venn diagram of MPTGA detected genes with virus-specific effects. Venn diagram of the identified genes with significant differential post-infection temporal gene expression specific to each virus, detected by MPTGA. Bonferroni correction was applied to correct for multiple testing

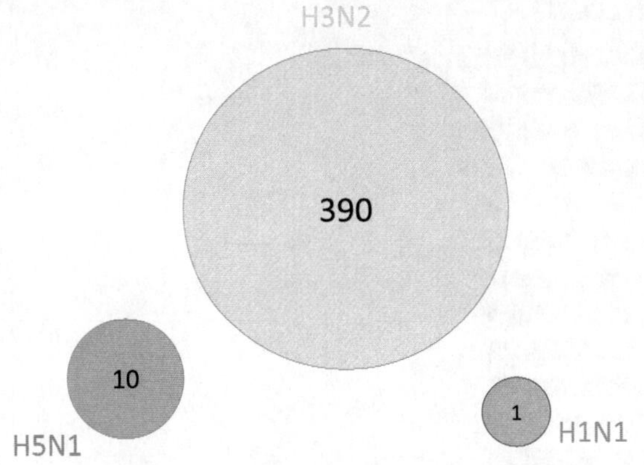

Fig. 4 Venn diagram of *Limma* detected genes with virus-specific effects. Venn diagram of the identified genes with significant differential post-infection gene expression specific to each virus, detected by *Limma* based on the last time point. Bonferroni correction was applied to correct for multiple testing

last time point. MPTGA identified more temporal differential expression genes than the conventional test, with 3260 genes showing differential temporal gene expression specific to at least one virus, while only 401 genes were identified by the conventional method. In addition, the conventional test results in three disjoint sets of genes with differential gene expression specific to each virus, suggesting that the test is only able to distinguish the post-infection effect on the gene expression between at most one of the viruses and the others. On the contrary, MPTGA identified 24 genes with differential temporal expression specific to multiple viruses. These results demonstrate significant superiority of the MPTGA method over the conventional method in both the power of detection and the specificity of virus-specific effects.

In spite of the differences, both MPTGA and the conventional test resulted in the largest number of differentially expressed genes specific to H3N2. The same observation also holds for the genes uniquely specific to each virus, with H3N2 having the largest number of genes that are only found to take on a post-infection effect specific to H3N2 but not to H5N1 nor H1N1. On the contrary, H1N1 appears to be the least specific virus, with the smallest number of a temporal effect specific to H1N1.

Functional enrichment

We looked into the functional enrichment in the identified differentially expressed genes specific to H3N2 only (without being specific to H1N1 or H5N1), as H3N2 was shown to be the virus with the most specific post-infection effects by both the MPTGA and the conventional method. The MPTGA method identified 2882 genes of only H3N2-specific effects, whereas the conventional method identified 390 genes of only H3N2-specific effects. Among them, 381 (98%) genes were in the common between the two. We compare the significance level of the top enriched GO categories for the 2882 genes identified by MPTGA with the significance level of these GO categories for the 390 genes identified by the conventional method in Fig. 5. The majority of the top enriched GO terms from the MPTGA identified 2882 genes of only H3N2-specific effects were related to immune or defense mechanisms. Comparing with *Limma*, the MPTGA method shows a consistently stronger enrichment in the immune related categories. We also look into the 9 genes identified by *Limma* but not by MPTGA and find no function enrichment in any immune or defense related GO categories. Next, we show the GO enrichment of the 2501 genes among the set of 2882 genes identified by MPTGA but not by *Limma* in Fig. 6. Most of the top enriched GO terms are related to immune/defense mechanisms or cell cycle regulation.

Previous studies [16–22] have compared the pathogenesis of different viruses in human cases and animal models. The H5N1 and H1N1 subtype viruses are more pathogenic and have a larger effect on lower respiratory tract comparing the H3N2 subtype virus. The pandemic 2009 H1N1 virus is shown to have similar pathogenesis to other influenza A viruses with high virulence such as the H5N1 viruses, with extension of its inflammation process of the larger airways into the alveoli causing diffuse alveolar damage, while the less virulent seasonal H3N2 virus is found to have less extension of its inflammatory process to alveoli. The difference in the morbidity and pathogenesis observed in these viruses is consistent with what we find in comparing the effect of each of the viruses on the temporal gene expression after infection, and finding the most number of genes showing specific effects unique to H3N2. Our results showed that H3N2 specific effects in lung were related to immune response and cell cycle regulation, indicating that the MPTGA method did not only detect more genes with virus-specific effects, but also revealed significant biological pathways underlying virus-specific effects.

It has been shown that innate lymphoid cells (ILCs) accumulated in lung after virus infection to promote lung tissue homeostasis [23]. Monticelli et al. [23] reports a ILC enriched signature consisting of 121 genes. We compared the signature with our virus specific genes and identified 41, 1, and 0 genes in the overlap with H3N2, H1N1, and H5N1 specific genes, respectively. The significant overlap with H3N2 specific genes (Fisher's Exact test p-value = 5.0e-10) suggests a potential mechanism explaining why H3N2 is least pathogenic to lung tissues.

Discussion

Virus infection in humans and animals is a dynamic process, with wildly different virulence and pathology specific

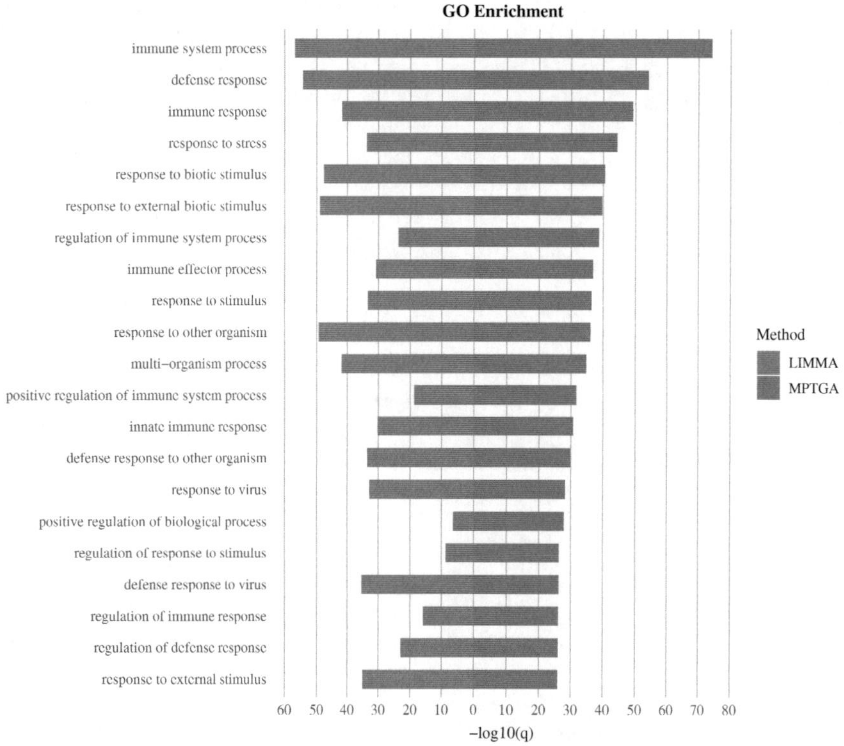

Fig. 5 GO enrichment of H3N2-specific genes. Comparison of GO enrichment from the 2882 genes showing H3N2-specific effects only (without being specific to H1N1 or H5N1) identified by MPTGA and from the 390 genes showing H3N2-specific effects only identified by *Limma*. GO categories are ranked by the significance level obtained from the MPTGA identified 2882 genes

to types or subtypes of viruses. For health-care professionals, it is essential to understand the virus-specific effects over a temporal dimension in order to accurately identify the virus and to proceed with appropriate treatment strategies.

In this paper, we studied the temporal gene expression response in mouse lung to infection with a seasonal H3N2 virus strain, a pandemic H1N1 virus strain, and a zoonotic H5N1 virus strain. We aimed to identify the virus-specific effects on the post-infection gene expression. For this purpose, we compared two methods, a Multivariate Polynomial Time-dependent Genetic Association (MPTGA) method proposed for temporal gene expression traits, and the conventional modified t-test for differential gene expression detection. Both methods showed that the seasonal H3N2 virus was the most different among the three tested, showing distinct response in the expression of the largest number of genes. The MPTGA method identified significantly more genes with virus-specific effects for each of the three viruses comparing with the modified t-test. We looked into specific examples and observed that the effects of the viruses infection varied over time, which explained why the modified t-test failed to capture the post-infection effects for many of the genes based on a single static time point. Moreover, by comparing the genes with virus-specific effects identified by the two methods, we showed that the MPTGA method identified more genes involved in immune response and cell cycle regulation pathways, consistent with previous studies comparing patho-physiological effects of these virus subtypes. Nevertheless, what virus-host interactions drive the virus specific effects can not be derived from this study. Further studies are needed to elucidate causal regulations leading to the virus specific effects.

We also generalized the MPTGA model based on two genotypes/groups to three or more genotypes/groups (the MPTGA models based on 2 or 3 genotypes/groups are noted as MPTGA2 and MPTGA3, respectively). We applied MPTGA3 to the mouse lung data set and compared results based on MPTGA2 and MPTGA3 (Fig. 7). MPTGA3 identified more genes differentially regulated among 3 subtypes of viruses. There were also genes identified by MPTGA2 but not by MPTGA3, suggesting the model selection should be based on underlying data distributions. For results based on MPTGA2, it is easier to interpret common of viruses and unique to

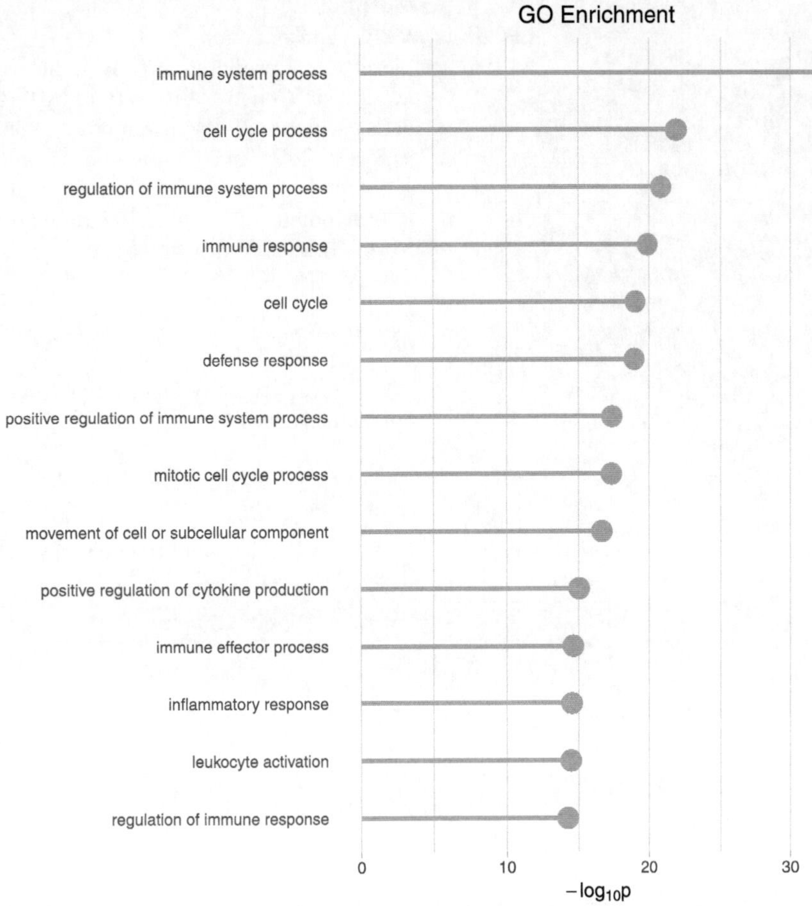

Fig. 6 GO enrichment of H3N2-specific genes detected by MPTGA only. GO enrichment from the 2501 genes showing H3N2-specific effects only (without being specific to H1N1 or H5N1) identified by MPTGA but not identified by *Limma*. GO categories are ranked by the significance level. General GO terms with more than 1500 genes associated are not displayed here

each specific virus. However, it is not easy to biologically interpret the results based on MPTGA3. Thus, we focused on the results based on MPTGA2 in the "Results" section.

In the MPTGA model, we assumed expression levels of a gene followed a group mean trajectory with variance at each time following a normal distribution. To assess the performance of MPTGA in the cases where variances deviate from a normal distribution, we simulated data sets based on the empirical patterns identified in the mouse lung data and added residuals sampling from different distributions (normal, uniform, exponential distributions, and Student's t-distribution with 3, 5, and 7 degrees). Then, we applied MPTGA to the simulated data sets. The accuracy were greater than 99% for T3, T5, and exponential distributions and 100% for other distributions tested. The results are summarized in Fig. 8. In general, MPTGA was robust to residuals deviated from a normal distribution.

It is worth to note that there were lots of assumptions made in the MPTGA model, such as cubic polynomial function for mean trajectories and covariances related in first order auto correlation. The choices were made mainly due to limited number of time points in a time series data set. If a long time series is available, high degree polynomial functions and higher order correlation structure should be explored.

Conclusions

Transcriptome response to virus infection may vary between viruses and over time. The Multivariate Polynomial Time-dependent Genetic Association (MPTGA) method can be applied to detect virus-specific effects on temporal gene expression traits. It is shown to enhance the power and specificity of detection subtype specific effects, which could result a more accurate target gene set for understanding the virus pathology.

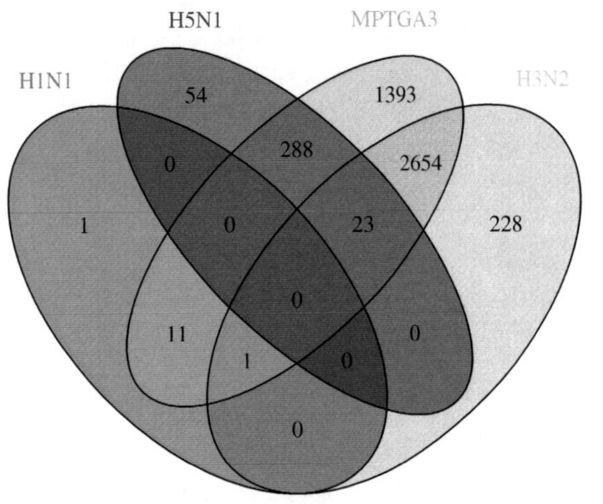

Fig. 7 Venn diagram of MPTGA2 and MPTGA3 detected genes with virus-specific effects. Venn diagram of the identified genes with significant differential post-infection gene expression, detected by MPTGA2 and MPTGA3. Bonferroni correction was applied to correct for multiple testing

Methods

MPTGA and extension

Lin et al. [6] proposed a Multivariate Polynomial Time-dependent Genetic Association (MPTGA) method to detect genetic effects in temporal gene expression trajectories. The MPTGA method assumes that for each genotype *j*, the temporal gene expression trait $\mathbf{y}$ across *m* time points follows a multivariate normal distribution $\mathcal{N}(\mathbf{g}_j, \Sigma)$, with density function

$$f_j(\mathbf{y}) = \frac{1}{(2\pi)^{m/2}|\Sigma|^{1/2}} \exp\left[-\left(\mathbf{y}-\mathbf{g}_j\right)\Sigma^{-1}\left(\mathbf{y}-\mathbf{g}_j\right)^T/2\right].$$

The mean vector $\mathbf{g}_j$ for genotype *j* is modeled with a polynomial curve

$$\mathbf{g}_j = \left[\mathbf{g}_j(t)\right]_{1\times m} = \left[\Sigma_{k=0}^{K}\beta_{kj}t^k\right]_{1\times m},$$

where K is the degree of the polynomial function. In this study, K was set to be 3.

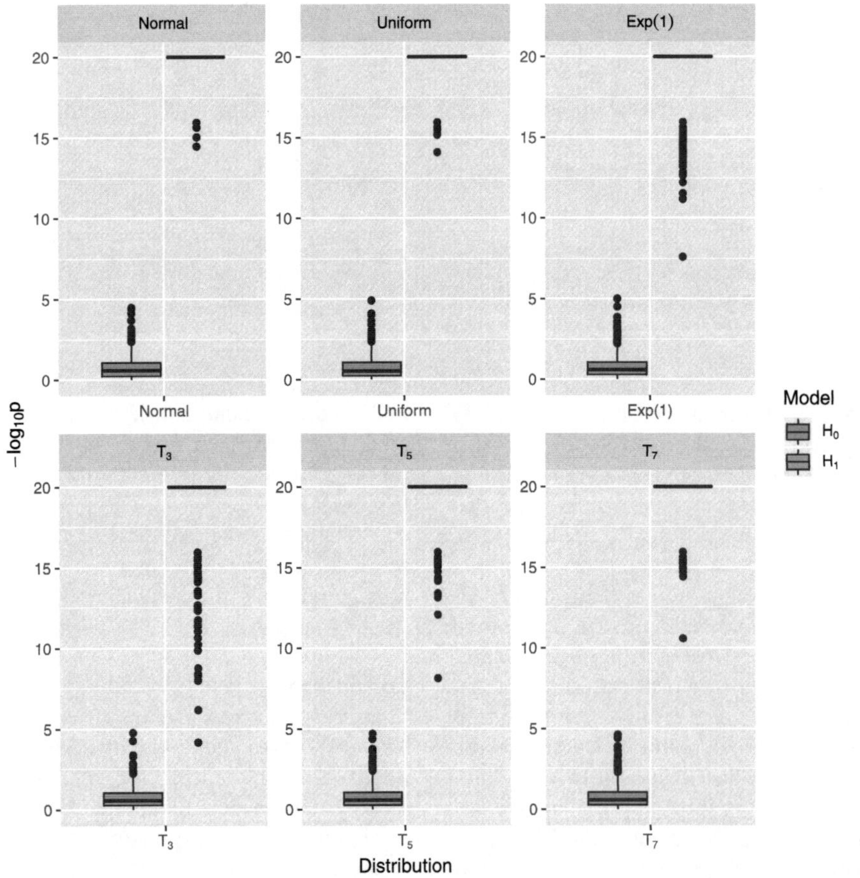

Fig. 8 Evaluation of MPTGA performance robustness to deviations from a normal distrubtion. Summary of the MPTGA test performance when the underlying assumption of normal distribution is met or when the expression traits follow different distributions (normal, uniform, exponential, Student's t with 3,5, and 7 degrees)

In particular, MPTGA assumes a first order auto-regression model to take into account the auto-correlation between different time points, with the covariance matrix specified as

$$\Sigma = \sigma_e^2 \begin{bmatrix} 1 & \rho & \dots & \rho^{m-1} \\ \rho & 1 & \dots & \rho^{m-2} \\ \vdots & \vdots & \ddots & \vdots \\ \rho^{m-1} & \rho^{m-2} & \dots & 1 \end{bmatrix}.$$

For each sample i, its gene expression trajectory $\mathbf{y}_i(t)$ across m time points can be written as

$$\mathbf{y}_i(t) = \delta_{i0} \sum_{k=0}^{K} \beta_{k0} t^k + \delta_{i1} \sum_{k=0}^{K} \beta_{k1} t^k + \epsilon_i,$$

where δ_{i0} and δ_{i1} are the indicator variables of sample i taking genotype 0 or 1.

Given the joint likelihood

$$\mathcal{L}(\Theta) = \prod_{i=1}^{N} [\delta_{i0} f_0(\mathbf{y}_i) + \delta_{i1} f_1(\mathbf{y}_i)]$$

The maximum likelihood estimates of the parameter set $\Theta = \left(\beta_{kj}, \rho, \sigma_e^2\right)$ can be derived as described in the Lin et al. [6] paper.

Lin et al. [6] compared longitudinal trajectories between two different conditions (e.g. two possible genotypes in a haploid system). The method can be extended to accommodate three or more conditions (e.g. effects of three possible genotypes in a diploid system; e.g. effects of different viruses infection). This comparison can be done using done using a pairwise test between conditions. Alternatively, we can construct a full model to detect the (genetic) effects:

$$y_i(t) = \delta_{i0} \sum_{k=0}^{K} \beta_{k0} t^k + \delta_{i1} \sum_{k=0}^{K} \beta_{k1} t^k + \delta_{i2} \sum_{k=0}^{K} \beta_{k2} t^k + \epsilon_i$$

for a given trait Y, then the reduced model H_0 (single gene expression trait curve)

$$H_0 : \beta_{k0} = \beta_{k1} = \beta_{k2} \qquad \text{for all} k$$

can be compared against the full model H_1 (different gene expression trait curve for different conditions (genotypes/viruses infection)):

H_1 : at least one of the equalities does not hold

to test the hypothesis of the existence of eQTL at a locus or difference between effects of viruses by estimating these parameters with a MLE procedure and performing a likelihood ratio test. The full model approach differs from the pairwise test in that it imposes an extra condition of a shared covariance structure among all three conditions, which is estimated in the full model taking all three conditions into account; whereas the pairwise test only requires the covariance structure to be shared between each pair, which is estimated separately for each pair.

Similar to the MPTGA model with two genotypes, we assume that for each genotype $j = 0, 1, 2$, the mean vector $\mathbf{g}_j$ for genotype j is modeled with a polynomial curve

$$\mathbf{g}_j = \left[\mathbf{g}_j(t)\right]_{1\times m} = \left[\Sigma_{k=0}^{K} \beta_{kj} t^k\right]_{1\times m} \qquad j = 0, 1, 2.$$

Given the joint log-likelihood

$$\log \mathcal{L}(\Theta) = \sum_{i=1}^{N} \delta_{i0} \log f_0(\mathbf{y}_i) + \sum_{i=1}^{N} \delta_{i1} \log f_1(\mathbf{y}_i) + \sum_{i=1}^{N} \delta_{i2} \log f_2(\mathbf{y}_i),$$

the maximum likelihood estimates of the parameter set $\Theta = \left(\beta_{0j}, \beta_{1j}, \beta_{2j}, \beta_{3j}, \rho, \sigma_e^2\right), j = 0, 1, 2$ can be derived by first looking for the critical point of $\log \mathcal{L}(\Theta)$ by taking its derivative with respect to β's and σ_e^2, finding that both $\beta's, \sigma_e^2$ can be expressed as functions of ρ at the critical point, therefore $\log \mathcal{L}(\Theta)$ can be expressed as functions of ρ at the critical point as well; then the MLE of ρ can be derived by taking derivative of $\log \mathcal{L}(\Theta)$ with respect to ρ, and thus β's and σ_e^2 can be obtained accordingly.

The detailed derivation is as follows. Define $\mathbf{T}_j = \sum_{i=1}^{N} \delta_{ij} \mathbf{y}_i, j = 0, 1, 2$; $\mathbf{I}_0 = [1 \cdots 1]_{1\times m}, \mathbf{I}_1 = [1 \cdots m]$, $\mathbf{I}_2 = \mathbf{I}_{1.}^2 = [1 \cdots m^2]$, $\mathbf{I}_3 = \mathbf{I}_{1.}^3 = [1 \cdots m^3]$; $Q(\rho, \mathbf{U}, \mathbf{V}) = \frac{1}{1-\rho^2}(U_1 V_1 + U_m V_m) - \frac{\rho}{1-\rho^2}\left[\sum_{i=1}^{m-1} U_i V_{i+1} + U_{i+1} V_i\right] + \frac{1+\rho^2}{1-\rho^2} \sum_{i=2}^{m-1} U_i V_i$, where $\mathbf{U} = [U_1, \cdots, U_m]$ and $\mathbf{V} = [V_1, \cdots, V_m]$.

Taking derivative of $\log \mathcal{L}(\Theta)$ with respect to $\beta_{.0}$'s gives the following linear system:

$$\begin{bmatrix} \alpha_{11}\beta_{00} + \alpha_{12}\beta_{10} + \alpha_{13}\beta_{20} + \alpha_{14}\beta_{30} = b_1 \\ \alpha_{21}\beta_{00} + \alpha_{22}\beta_{10} + \alpha_{23}\beta_{20} + \alpha_{24}\beta_{30} = b_2 \\ \alpha_{31}\beta_{00} + \alpha_{32}\beta_{10} + \alpha_{33}\beta_{20} + \alpha_{34}\beta_{30} = b_3 \\ \alpha_{41}\beta_{00} + \alpha_{42}\beta_{10} + \alpha_{43}\beta_{20} + \alpha_{44}\beta_{30} = b_4 \end{bmatrix},$$

where $\alpha_{ij} = n_0 Q(\rho, \mathbf{I}_{i-1}, \mathbf{I}_{j-1})$, $b_i = Q(\rho, \mathbf{T}_0, \mathbf{I}_{i-1})$, $n_0 = \sum_{i=1}^{N} \delta_{i0}$. Then $\beta_{.0}$'s can be estimated by

$$\begin{bmatrix} \beta_{00} \\ \beta_{10} \\ \beta_{20} \\ \beta_{30} \end{bmatrix} = \begin{bmatrix} \alpha_{11} & \alpha_{12} & \alpha_{13} & \alpha_{14} \\ \alpha_{21} & \alpha_{22} & \alpha_{23} & \alpha_{24} \\ \alpha_{31} & \alpha_{32} & \alpha_{33} & \alpha_{34} \\ \alpha_{41} & \alpha_{42} & \alpha_{43} & \alpha_{44} \end{bmatrix}^{-1} * \begin{bmatrix} b_1 \\ b_2 \\ b_3 \\ b_4 \end{bmatrix}.$$

Similarly, define $n_1 = \sum_{i=1}^{N} \delta_{i1}$, $\alpha_{ij}^{(1)} = n_1 Q(\rho, \mathbf{I}_{i-1}, \mathbf{I}_{j-1})$, $b_i^{(1)} = Q(\rho, \mathbf{T}_1, \mathbf{I}_{i-1})$; $n_2 = \sum_{i=1}^{N} \delta_{i2}$, $\alpha_{ij}^{(2)} = n_2 Q(\rho, \mathbf{I}_{i-1}, \mathbf{I}_{j-1})$, $b_i^{(2)} = Q(\rho, \mathbf{T}_2, \mathbf{I}_{i-1})$; then $\beta_{.}$'s can be derived accordingly.

Next, taking derivative with respect to σ_e^2 gives

$$\sigma_e^2 = \left\{ \sum_{i=1}^{N} Q(\rho, \mathbf{y}_i, \mathbf{y}_i) + \sum_{i=0}^{2} [Q(\rho, \mathbf{T}_i, \mathbf{T}_i) - \sum_{j=0}^{3} 2\beta_{ji} Q\left(\rho, \mathbf{T}_i, \mathbf{I}_j\right) + \sum_{j=0}^{3}\sum_{k=0}^{3} n_i \beta_{ji}\beta_{ki} Q(\rho, \mathbf{I}_j, \mathbf{I}_k)] \right\} /(mN)$$

With the previously derived estimators of β's and σ_e^2 also as functions of ρ, the log-likelihood can now be written as

$$\log \mathcal{L}(\Theta) = -\frac{mN}{2}\log 2\pi - \frac{N}{2}\left[(m-1)\log\left(1-\rho^2\right) + m\log\sigma_e^2\right] - \frac{mN}{2}$$

Based on the MLE of the unknown parameters, a LRT $\left(-2\log\Lambda \sim \chi_8^2\right)$ can be conducted comparing the aforementioned full model H_1 with the null model H_0 to test for virus-specific effects.

Detecting virus-specific effects on post-infection temporal expression

The MPTGA method can be applied to a broader setting that compares the temporal gene expression traits between two different conditions, whereas the original setting proposed by Lin et al. [6] is a special case that compares two possible genotypes in a haploid system. In this paper, we are interested in detecting the virus-specific effects on the post-infection temporal gene expression trajectories. For a given virus v, we define $j = 0, 1$ as the indicator variable of a sample infected by virus v or other viruses. The null hypothesis H_0 assumes that the temporal gene expression trait $\mathbf{y}_j$ share the same trajectory pattern after infected by virus v and other viruses:

$$H_0 : \beta_{k0} = \beta_{k1} \quad \text{for all } k$$

which is tested against the full model H_1, assuming that the temporal post-infection gene expression trait $\mathbf{y}_j$ with virus v has a unique trajectory pattern distinct from the post-infection trajectories with other viruses:

$$H_1 : \text{at least one of the equalities does not hold}$$

The hypothesis can be tested using a likelihood ratio test.

Similarly, MPTGA can be applied to cases with three or more groups as outlined above.

Abbreviations

ENCODE: Encyclopedia of DNA elements; GEO: Gene expression omnibus; GO: Gene ontology; ILC: Innate lymphoid cell; LRT: Likelihood ratio test; MLE: Maximum likelihood estimation; MPTGA: Multivariate polynomial time-dependent genetic association; NK: Natural killer

Acknowledgements

We thank members of Zhu's lab and Dr. García-Sastre for productive discussion.

Funding

This work was partially supported by the U19 AI118610 and U01 HG008451 grants. Publication of this article was supported by NIH-U19 AI118610.

Availability of data and materials

The data used in study is from the public database GEO with accession number GSE98527.

About this supplement

This article has been published as part of *BMC Bioinformatics Volume 20 Supplement 3, 2019: Selected articles from the 17th Asia Pacific Bioinformatics Conference (APBC 2019): bioinformatics*. The full contents of the supplement are available online at https://bmcbioinformatics.biomedcentral.com/articles/supplements/volume-20-supplement-3.

Authors' contributions

QC and JZ designed the study, QC performed data analysis. QC and JZ wrote the manuscript. Both authors have read and approved the final manuscript.

Ethics approval and consent to participate

Not applicable.

Consent for publication

Not applicable.

Competing interests

The authors declare that they have no competing interests.

Publisher's Note

Author details

[1]Department of Genetics and Genomic Sciences, Icahn School of Medicine at Mount Sinai, 10029 New York, NY, USA. [2]Icahn Institute for Genomics and Multiscale Biology, Icahn School of Medicine at Mount Sinai, 10029 New York, NY, USA. [3]Sema4, a Mount Sinai venture, 06902 Stamford, CT, USA.

Published: 29 March 2019

References

1. Hay AJ, Gregory V, Douglas AR, Lin YP. The evolution of human influenza viruses. Philos Trans R Soc Lond Ser B. 2001;356(1416):1861–70. https://doi.org/10.1098/rstb.2001.0999.
2. Reperant LA, Moesker FM, Osterhaus ADME. Influenza: from zoonosis to pandemic. ERJ Open Research. 2016;2(1). https://doi.org/10.1183/23120541.00013-2016.
3. Govorkova EA, Rehg JE, Krauss S, Yen H-L, Guan Y, Peiris M, Nguyen TD, Hanh TH, Puthavathana P, Long HT, Buranathai C, Lim W, Webster RG, Hoffmann E. Lethality to Ferrets of H5n1 Influenza Viruses Isolated from Humans and Poultry in 2004. J Virol. 2005;79(4):2191–8. https://doi.org/10.1128/JVI.79.4.2191-2198.2005.
4. Schweiger B, Bruns L, Meixenberger K. Reassortment between human A(H3n2) viruses is an important evolutionary mechanism. Vaccine. 2006;24(44-46):6683–90. https://doi.org/10.1016/j.vaccine.2006.05.105.
5. Brand JMAvd, Stittelaar KJ, Amerongen Gv, Reperant L, Waal Ld, Osterhaus ADME, Kuiken T. Comparison of Temporal and Spatial Dynamics of Seasonal H3n2, Pandemic H1n1 and Highly Pathogenic Avian Influenza H5n1 Virus Infections in Ferrets. PLoS ONE. 2012;7(8): 42343. https://doi.org/10.1371/journal.pone.0042343.
6. Lin L, Chen Q, Hirsch JP, Yoo S, Yeung K, Bumgarner RE, Tu Z, Schadt EE, Zhu J. Temporal genetic association and temporal genetic causality methods for dissecting complex networks. Nat Commun. 2018;9(1):3980. https://doi.org/10.1038/s41467-018-06203-3.
7. Steel J, Lowen AC, Pena L, Angel M, Solórzano A, Albrecht R, Perez DR, García-Sastre A, Palese P. Live attenuated influenza viruses containing NS1 truncations as vaccine candidates against H5n1 highly pathogenic avian influenza. J Virol. 2009;83(4):1742–53. https://doi.org/10.1128/JVI.01920-08.
8. Ritchie ME, Phipson B, Wu D, Hu Y, Law CW, Shi W, Smyth GK. limma powers differential expression analyses for RNA-sequencing and

microarray studies. Nucleic Acids Res. 2015;43(7):47. https://doi.org/10.1093/nar/gkv007.

9. Yue F, Cheng Y, Breschi A, Vierstra J, Wu W, Ryba T, Sandstrom R, Ma Z, Davis C, Pope BD, Shen Y, Pervouchine DD, Djebali S, Thurman RE, Kaul R, Rynes E, Kirilusha A, Marinov GK, Williams BA, Trout D, Amrhein H, Fisher-Aylor K, Antoshechkin I, DeSalvo G, See L-H, Fastuca M, Drenkow J, Zaleski C, Dobin A, Prieto P, Lagarde J, Bussotti G, Tanzer A, Denas O, Li K, Bender MA, Zhang M, Byron R, Groudine MT, McCleary D, Pham L, Ye Z, Kuan S, Edsall L, Wu Y-C, Rasmussen MD, Bansal MS, Kellis M, Keller CA, Morrissey CS, Mishra T, Jain D, Dogan N, Harris RS, Cayting P, Kawli T, Boyle AP, Euskirchen G, Kundaje A, Lin S, Lin Y, Jansen C, Malladi VS, Cline MS, Erickson DT, Kirkup VM, Learned K, Sloan CA, Rosenbloom KR, Lacerda de Sousa B, Beal K, Pignatelli M, Flicek P, Lian J, Kahveci T, Lee D, Kent WJ, Ramalho Santos M, Herrero J, Notredame C, Johnson A, Vong S, Lee K, Bates D, Neri F, Diegel M, Canfield T, Sabo PJ, Wilken MS, Reh TA, Giste E, Shafer A, Kutyavin T, Haugen E, Dunn D, Reynolds AP, Neph S, Humbert R, Hansen RS, De Bruijn M, Selleri L, Rudensky A, Josefowicz S, Samstein R, Eichler EE, Orkin SH, Levasseur D, Papayannopoulou T, Chang K-H, Skoultchi A, Gosh S, Disteche C, Treuting P, Wang Y, Weiss MJ, Blobel GA, Cao X, Zhong S, Wang T, Good PJ, Lowdon RF, Adams LB, Zhou X-Q, Pazin MJ, Feingold EA, Wold B, Taylor J, Mortazavi A, Weissman SM, Stamatoyannopoulos JA, Snyder MP, Guigo R, Gingeras TR, Gilbert DM, Hardison RC, Beer MA, Ren B, Mouse ENCODE Consortium. A comparative encyclopedia of DNA elements in the mouse genome. Nature. 2014;515(7527):355–64. https://doi.org/10.1038/nature13992.
10. Martínez-Naves E, Lafuente EM, Reche PA. Recognition of the ligand-type specificity of classical and non-classical MHC I proteins. FEBS Lett. 2011;585(21):3478–84. https://doi.org/10.1016/j.febslet.2011.10.007.
11. Kajikawa M, Baba T, Tomaru U, Watanabe Y, Koganei S, Tsuji-Kawahara S, Matsumoto N, Yamamoto K, Miyazawa M, Maenaka K, Ishizu A, Kasahara M. MHC class I-like MILL molecules are beta2-microglobulin-associated, GPI-anchored glycoproteins that do not require TAP for cell surface expression. J Immunol (Baltimore, Md.: 1950). 2006;177(5):3108–15.
12. Zhang B, Wei H, Zheng X, Zhang J, Sun R, Tian Z. The inhibitory effects of synthetic short peptides, mimicking MICA and targeting at NKG2d receptors, on function of NK cells. Peptides. 2005;26(3):405–12. https://doi.org/10.1016/j.peptides.2004.10.008.
13. Ritter C, Fan K, Paulson KG, Nghiem P, Schrama D, Becker JC. Reversal of epigenetic silencing of MHC class I chain-related protein A and B improves immune recognition of Merkel cell carcinoma. Scientific Reports. 2016;6:21678. https://doi.org/10.1038/srep21678.
14. Rabinovich BA, Ketchem RR, Wolfson M, Goldstein L, Skelly M, Cosman D. A role for the MHC class I-like Mill molecules in nutrient metabolism and wound healing. Immunol Cell Biol. 2008;86(6):489–96. https://doi.org/10.1038/icb.2008.41.
15. Wu W, Booth JL, Duggan ES, Wu S, Patel KB, Coggeshall KM, Metcalf JP. Innate immune response to H3n2 and H1n1 influenza virus infection in a human lung organ culture model. Virology. 2010;396(2):178–88. https://doi.org/10.1016/j.virol.2009.10.016.
16. Guarner J, Falcón-Escobedo R. Comparison of the Pathology Caused by H1n1, H5n1, and H3n2 Influenza Viruses. Arch Med Res. 2009;40(8):655–61. https://doi.org/10.1016/j.arcmed.2009.10.001.
17. Gu J, Xie Z, Gao Z, Liu J, Korteweg C, Ye J, Lau LT, Lu J, Gao Z, Zhang B, McNutt MA, Lu M, Anderson VM, Gong E, Yu ACH, Lipkin WI. H5n1 infection of the respiratory tract and beyond: a molecular pathology study. Lancet. 2007;370(9593):1137–45. https://doi.org/10.1016/S0140-6736(07)61515-3.
18. Sreta D, Kedkovid R, Tuamsang S, Kitikoon P, Thanawongnuwech R. Pathogenesis of swine influenza virus (Thai isolates) in weanling pigs: an experimental trial. Virol J. 2009;6(1):34. https://doi.org/10.1186/1743-422X-6-34.
19. Fislová T, Gocník M, Sládková T, Ďurmanová V, Rajčáni J, Varečková E, Mucha V, Kostolanský F. Multiorgan distribution of human influenza A virus strains observed in a mouse model. Arch Virol. 2009;154(3):409–19. https://doi.org/10.1007/s00705-009-0318-8.
20. Tang X, Chong KT. Histopathology and growth kinetics of influenza viruses (H1n1 and H3n2) in the upper and lower airways of guinea pigs. J Gen Virol. 2009;90(2):386–91. https://doi.org/10.1099/vir.0.007054-0.
21. Munster VJ, de Wit E, van den Brand JMA, Herfst S, Schrauwen EJA, Bestebroer TM, van de Vijver D, Boucher CA, Koopmans M, Rimmelzwaan GF, Kuiken T, Osterhaus ADME, Fouchier RAM. Pathogenesis and transmission of swine-origin 2009 A(H1n1) influenza virus in ferrets. Science. 2009;325(5939):481–3. https://doi.org/10.1126/science.1177127.
22. van den Brand JMA, Stittelaar KJ, van Amerongen G, Rimmelzwaan GF, Simon J, de Wit E, Munster V, Bestebroer T, Fouchier RAM, Kuiken T, Osterhaus ADME. Severity of pneumonia due to new H1n1 influenza virus in ferrets is intermediate between that due to seasonal H1n1 virus and highly pathogenic avian influenza H5n1 virus. J Infect Dis. 2010;201(7):993–9. https://doi.org/10.1086/651132.
23. Monticelli LA, Sonnenberg GF, Abt MC, Alenghat T, Ziegler CGK, Doering TA, Angelosanto JM, Laidlaw BJ, Yang CY, Sathaliyawala T, Kubota M, Turner D, Diamond JM, Goldrath AW, Farber DL, Collman RG, Wherry EJ, Artis D. Innate lymphoid cells promote lung-tissue homeostasis after infection with influenza virus. Nat Immunol. 2011;12(11):1045. https://doi.org/10.1038/ni.2131.

AUTHOR INDEX

AUTHOR INDEX